Advanced Sciences and Technology for Security Applications

Series Editor

Anthony J. Masys, Associate Professor, Director of Global Disaster Management, Humanitarian Assistance and Homeland Security, University of South Florida, Tampa, USA

Advisory Editors

Gisela Bichler, California State University, San Bernardino, CA, USA

Thirimachos Bourlai, Lane Department of Computer Science and Electrical Engineering, Multispectral Imagery Lab (MILab), West Virginia University, Morgantown, WV, USA

Chris Johnson, University of Glasgow, Glasgow, UK

Panagiotis Karampelas, Hellenic Air Force Academy, Attica, Greece

Christian Leuprecht, Royal Military College of Canada, Kingston, ON, Canada

Edward C. Morse, University of California, Berkeley, CA, USA

David Skillicorn, Queen's University, Kingston, ON, Canada

Yoshiki Yamagata, National Institute for Environmental Studies, Tsukuba, Ibaraki, Japan

Indexed by SCOPUS

The series Advanced Sciences and Technologies for Security Applications comprises interdisciplinary research covering the theory, foundations and domain-specific topics pertaining to security. Publications within the series are peer-reviewed monographs and edited works in the areas of:

- biological and chemical threat recognition and detection (e.g., biosensors, aerosols, forensics)
- crisis and disaster management
- terrorism
- cyber security and secure information systems (e.g., encryption, optical and photonic systems)
- traditional and non-traditional security
- energy, food and resource security
- economic security and securitization (including associated infrastructures)
- transnational crime
- human security and health security
- social, political and psychological aspects of security
- recognition and identification (e.g., optical imaging, biometrics, authentication and verification)
- smart surveillance systems
- applications of theoretical frameworks and methodologies (e.g., grounded theory, complexity, network sciences, modelling and simulation)

Together, the high-quality contributions to this series provide a cross-disciplinary overview of forefront research endeavours aiming to make the world a safer place.

The editors encourage prospective authors to correspond with them in advance of submitting a manuscript. Submission of manuscripts should be made to the Editor-in-Chief or one of the Editors.

More information about this series at http://www.springer.com/series/5540

Ashok Vaseashta · Carmen Maftei
Editors

Water Safety, Security and Sustainability

Threat Detection and Mitigation

Editors
Ashok Vaseashta ⓘ
International Clean Water Institute
Manassas, VA, USA

Institute of Biomedical
and Nanotechnologies
Riga Technical University
Riga, Latvia

Institute of Electronic Engineering
and Nanotechnologies "D. Ghitu"
ASM, Chisinau, Moldova

Carmen Maftei
Faculty of Civil Engineering
Transylvania University of Braşov
Brasov, Romania

ISSN 1613-5113   ISSN 2363-9466 (electronic)
Advanced Sciences and Technologies for Security Applications
ISBN 978-3-030-76010-6   ISBN 978-3-030-76008-3 (eBook)
https://doi.org/10.1007/978-3-030-76008-3

© Springer Nature Switzerland AG 2021
This work is subject to copyright. All rights are reserved by the Publisher, whether the whole or part of the material is concerned, specifically the rights of translation, reprinting, reuse of illustrations, recitation, broadcasting, reproduction on microfilms or in any other physical way, and transmission or information storage and retrieval, electronic adaptation, computer software, or by similar or dissimilar methodology now known or hereafter developed.
The use of general descriptive names, registered names, trademarks, service marks, etc. in this publication does not imply, even in the absence of a specific statement, that such names are exempt from the relevant protective laws and regulations and therefore free for general use.
The publisher, the authors and the editors are safe to assume that the advice and information in this book are believed to be true and accurate at the date of publication. Neither the publisher nor the authors or the editors give a warranty, expressed or implied, with respect to the material contained herein or for any errors or omissions that may have been made. The publisher remains neutral with regard to jurisdictional claims in published maps and institutional affiliations.

This Springer imprint is published by the registered company Springer Nature Switzerland AG
The registered company address is: Gewerbestrasse 11, 6330 Cham, Switzerland

Foreword by Prof. Juhna

With dynamic global urban development and evidence of climate change, it is quintessential to address the daunting task of water safety and security for sustainable development. Water contamination produced by anthropogenic activities along with limited freshwater resources has become one of the most challenging problems of human society on our planet. This problem is compounded by the fact that developed nations have aging water infrastructure, while developing nations add tremendous water stress, due to rapid industrialization and urbanization. Listed as one of the top 15 global challenges, safety, security, and long-term sustainability of water is a challenge for scientists to find ways to provide clean water for drinking, agriculture, and industrial processing.

The book *Water Safety, Security and Sustainability* is a collection of chapters with up-to-date information, original research, reviews, and discussions by several authors from different parts of the world, presenting a global picture of what is presently being done in the field of water safety, security and sustainability. The book presents various topics ranging from sources of water to technological solutions to capture and remove contaminants from water. Due to the multidisciplinary nature of this subject, the topics also include nanotechnology, water management, geoinformatics, and advanced sensors, not only to assess contamination distribution, but also for security of water resources. The book also reflects on the current trends in the specific areas of research, aimed at proposing solutions to emerging challenges by applying a collection of methods and approaches from chemistry, biology, physics, mathematics, and geology. The more specific problems considered in this book include water quality, biological monitoring, and data analytics methods of assessment of water quality.

Undoubtedly, this book addresses topics which are not only timely but will need to be revisited soon due to fast changing global dynamics. In this respect, I foresee that the editors will be busy again in near future to prepare updated versions in the coming years. This book is important to a wide, interdisciplinary variety of students and researchers involved in water science, ecological chemistry, and related subjects, as well as to people in health care, industry, and agriculture, engaged in the pollution and human safety issues. I highly commend Prof. Ashok Vaseashta, currently serving Riga Technical University as a Visiting Professor in the Institute of Biomedical Engineering and Nanotechnology, for his admirable efforts to prepare this edited book.

Prof. Dr. Talis Juhna
Vice Rector for Research
Professor of Water Technologies
Riga Technical University
Riga, Latvia

Foreword by Prof. Burton

Water is a life giver—even a life creator. Water provides many useful functions—both for the earth and for humans—that help produce as well as maintain the abundance of life around us every day. Without even one of these functions, our lives would be unimaginably different. Over the past few decades, technological developments have advanced immensely, even to the extent that they are often overwhelming, particularly in terms of their impact on water resources. The problem of water stress is increasing globally due to the development of mega-cities, increased industrialization and urbanization, aging water infrastructure, and with new and emerging contaminants, for which no known remediation methodologies exist. The quantity, management, interpretation, and understanding of water quality in the context of water treatment is becoming an ever increasing challenge, even for the most experienced water experts. With increased dependence on computer networks managing our critical infrastructures, securing water resources and distribution is a major challenge, since the systems extend over vast areas, and ownership and operation responsibility are both public and private and are non-federal to a large extent. Hence, it is important to discuss current trends, challenges, and case studies of topics related to water safety, security, and sustainability.

I am pleased to provide this foreword for the book *Water Safety, Security and Sustainability*, edited by Prof. Ashok Vaseashta, whom I have known initially as my Ph.D. student, and afterward as a professor and researcher. He has reached out to many contributors, as extended expertise was needed to cover the expanded scope of the book, and eventually the end result is a book prepared by senior, globally recognized experts with specialties including field investigation, analysis and modeling of groundwater flow, geochemistry, security, and forensics analysis of pollution. The book was subjected to comprehensive peer review and consequently, the material in this book represents the type of interdisciplinary knowledge that is urgently needed to transfer knowledge and raise awareness on this critical topic. Due to the vast nature of this subject, it may be necessary to continue to update topics in this book frequently.

Many books have been published in the past decade drawing attention to the looming global water crisis. Although many are focused on the immense importance of freshwater for humanity and the multitude of water problems people face, this edited book is a collection of articles from researchers, practitioners in the field, hydrologists, water science experts, and security professionals, to cover many topics comprehensively for a broad readership. Such an approach in this field is refreshing, as it combines field knowledge with mathematics, modeling, process engineering, microbiology, physics, and biochemistry in a balanced way, providing theoretical and fundamental information to the extent required for the solution of practical problems, regularly demonstrated by one or more examples. This book will make a major contribution toward better addressing these issues and to bridging the gap between science and technology and their practical applications. This book is a breath of fresh air in the field of water, since the approach is surprisingly direct and transparent, and this knowledge is genuinely shared. The usefulness of this book to all stakeholders in the field is undoubted, and it will be used by its intended audience and, even, may become a compulsory, 'must have', item for the collection of water scientists and professionals. I am delighted that the editor and authors have made such a tremendous effort to create this book.

Larry C. Burton
Professor Emeritus
College of Engineering
Penn State University
University Park
PA, USA

Preface

While growing up, I always observed worship routine of my mother, which concluded by turning toward the sun and pouring water slowly over plants with her hands folded overhead, while reciting prayers. It was her spiritual way of demonstrating nexus of life and nature—sun, earth, water, and plants—nourishment for the body. Water represents heavenly purity in most rituals, languages, and symbolisms in almost all religions. The Holy Verses of several Holy Books contain many references to water as the "fountains of living water", since water is fundamentally essential to our survival. From the prehistoric times, human settlements have developed or connected near or around the sources of water. The Euphrates, the Indus, the Ganges, the Tiber, the Yan-Tse are the main rivers, along which the first human civilizations appeared. According to archeological evidence about the existence of centralized water supply systems in Samaria about 5000 years ago, during the Nippur civilization, an arched drain was discovered with stones fixed by descending feathers for collecting water through fountains and cisterns. The oasis of Kashkadarya, in the Ancient East and currently a major historical and cultural centers of Sogd, Tajikstan (bordering with Jizzakh, Namangan, Samarkand, and Fergana regions of Uzbekistan, and the Osh and Batken regions of Kyrgyzstan) was the basis of formation of cities during first quarter of the I millennium BC, undoubtedly due to irrigation facilities. Scholars explain Mexico's ancient city of Teotihuacán by stating that it was built as an aquatic sanctuary dedicated to the worship of water after finding canals and pool-like cavities located beneath the square, alongside statues of water gods. One of the scholars concluded that *Water is the true protagonist of Teotihuacán*. In the 16–17th centuries, with the appearance of capitalized civilization, water supply techniques were developed. Commercially, most large cities evolved around large bodies of water. In fact, the Romans were renowned for water engineering marvels, among which is the aqueduct that carried water for many miles to supply relatively safe water for potable, agriculture, and other aquatic uses for a crowded urban population. In the 18–19th centuries, the first centralized water supply systems were built in cities in England, Germany, France, and Russia.

Due to rapid growth of industrialization and urbanization, our natural surroundings and their components, including the atmosphere, surface and surrounding ocean waters, lithosphere, soil, flora, and fauna, are continuously subjected to an unprecedented level of usage and anthropogenic impact. Polluting chemicals and industrial wastes entering the environment disrupt their steady-state nature, homeostatic cycles, characteristic times, and other chemical-physical parameters due to the circulation of foreign substances in the biosphere and render negative impacts on habitats. With emergence of new and emergent chemicals, dozens of biological species are disappearing, and many of the remaining species are changing their usual habitats. These discharges modify the chemical composition of the environment and may render long lasting negative impacts on the biosphere. The pollutants penetrate water and intrude in the food supply-chain and ambient air, impacting living organisms and human health. Booming global economic growth may be humanity's greatest challenge to balance the standard of living and associated infrastructures with enormous burden on environmental ecology. For long-term safety and sustainability of the habitat, it is critical to sustain living standards, while preserving the ecological integrity of the environment by using strategic approaches. Additionally, due to increased dependence on computer networks and Internet for managing our critical infrastructures, securing water resources and distribution has become a major challenge, since the systems extend over vast areas, and ownership and operation responsibility are both public and private. Hence, it is quintessential to manage these challenges, which appear seemingly different at a first glance, nevertheless point to one and the same problem, viz., how to ensure safe, secure, and sustainable water resources for the existence and further development of human civilization on our planet.

The thematic objective of this edited book, Water Safety, Security and Sustainability is to present several ongoing research viewpoints, with the objective of formulating new concepts, approaches, and pathways to mitigate water-stress, enhance overall safety and security and make water resources more sustainable. While similar efforts are in progress, unfortunately due to complex dynamics and interplay of many factors—the issue of water safety, security, and sustainability still remains a major global challenge, which requires governments, businesses, universities, and citizens around the world to strategically address such challenges. Each year brings more solutions—such as using nanotechnologies to filter out new and emergent contaminants, water forensics to trace source of contamination, tracing waterborne viruses and pathogens, viz., COVID-19, *E-coli, protozoans*, etc., in wastewater streams, use of wastewater for energy, restoration to bring water back to dry topographies, and monitoring groundwater levels extra closely. This area of scientific study seems to be among the most significant ones, playing a key role in modern sciences and revealing the relations between the living world and the surrounding aquatic environment. As compared to other fields of science, this domain requires a comprehensive investigation. However, even the best solutions will not implement themselves. In addition to much needed research on topics related to water, it requires political will, nexus of technological innovations, data analytics for data-driven decisions, foresight, and public support to ensure clean water as a critical resource for a sustainable future for all.

This book consists of 30 interlinked chapters that cover topics ranging from overview of the challenges, contamination remediation, security issues, sustainable, modeling, and a few case studies. The book begins with a brief overview of challenges related to water (Vaseashta), such as nature of contaminants (Man et al.) and its impact on water quality and human health challenges. Due to limited resources, sensor placement is of critical importance (Adedoja et al.) and a chapter on applied Geology and Geomatic tools (Papatheodorou) provides remote sensing capability to chemical analysis and exposure research. A successful implementation of such tools will provide pathways to moving toward a healthy and sustainable future.

Several chapters provide the theoretical basis and practical means to mitigate contaminants from water. Since society is exposed to hundreds of pollutants in water in the form of particulate matter (nanoparticles), multiple drugs and chemicals that have been either reported to be toxic or just suspected of being toxic. These confirmed or alleged toxicants can enter the human body via ingestion of contaminated water or consumption of foods or crops which are contaminated. The importance of the toxicological study of mixtures cannot be understated and hence contamination remediation if of critical importance. Use of nanomaterials (Ionete et al.) plays a significant role in contamination remediation by adsorption, nanophotocatalysis and also as membranes (Mohanty et al.). Use of electrospinning (Bolgen et al.) provides yet another way to capture contaminants, including returned pharmaceuticals from water. Use of green chemistry by way of coagulation (Karnena et al.) and bio-absorbents (Demir et al.) describes sustainable ways to remove contaminants in water. A system-of-systems approach in water safety plan development (Vasovic et al.) is crucial to its implementation.

It is worth noting that all chapters are connected via a main theme of water-related issues, which is especially relevant in terms of safety and security, considering that humanity faces ever-growing water pollution contamination every single day, which can be accidental, unintentional, nefarious, or even intentional. Since the quality of life and health is directly dependent on water's availability and quality, it is critical to understand security challenges and means to identify the source of contaminants. Environmental forensics (Breceski et al.) is an important and upcoming subject to trace the contaminants to the source. Since controls are interconnected by Internet-of-things, it is critical to discuss security standards (Glevitzky et al.) and water management solutions (Man et al.).

A holistic approach to sustainable development is introduced through a chapter on scarcity (Belhassan). Several chapters on sustainable development complement ecological processes by treating wastewater using metal-organic frameworks (Konni et al.), hydrocarbons removal by expanded graphite (Pokropivny et al.) and by reverse-osmosis desalination (Roy et al.). In other chapters, basic methods such as household water treatment and safe storage (Spiridon et al.) and riverbank filtration systems (Satsangi et al.) are described. As there is great interest among both researchers, readers, and practitioners, several case studies from Romania and India are presented by several authors, including some studies that use web-based applications and data analytic tools. The well-known fact is that without proven analytical and assessment methods, research results may be inappropriate and may trigger unreliable conclusions.

It is vital to note that changes in the environment are occurring faster than our collective ability to develop tools to counter them. Hence, as also mentioned above that policymakers, scientists, and concerned citizens play a collective role to ensure that decision-makers implement policies to slow the progression of these changes. The second important role is by way of education of Environmental Science and Ecological Chemistry which includes the study of mass transfer processes in the environment; chemical kinetics, such as pollutant transformation mechanisms; analytical chemistry, viz., physico-chemical methods of analysis; and biochemistry for illustrating molecular mechanisms regarding the influence of pollutants on vital processes.

It is hoped that this book will be helpful in enhancing awareness, education, and understanding—as one of the most important steps to mitigate such ecological impacts. In addition to testing physico-chemical methods to assess pollution levels, it is crucial to implement principles of green chemistry along with practices to reduce water-stress and green-house gas emissions in the environment. This multidisciplinary subject further includes the study of anthropogenic and natural contaminants, that is, the extent to which pollutants are harmful to human beings, innocent animals and plants; the rates of degradation, migration pathways of hazardous materials and ways to neutralize them; and the scientific risk assessment and uncertainties—all focused on what we can do to protect all life on our unique planet. To investigate the reasons for stressed water quality, it is critical to study how subjects as, Environmental Science, Ecological chemistry, Green Chemistry, Physical Science can help the human race.

The book is intended for professional scientists, practitioners, policymakers, futurists, stakeholders, and students; however, a much wider audience may find contents interesting and useful. The editor in no way claims that contents completely represent the challenge. In fact, subsequent editions will address additional ongoing and upcoming issues. The editor acknowledges the contributions of several authors and reviewers who assisted the editorial efforts by providing their independent assessment of chapters. Such feedback was critical to the quality of this book. The editor immensely acknowledges the publisher Springer Nature, and in particular, Ms. Annelies Kersbergen for her guidance, patience, and help with the book. The editor acknowledges kind forewords written by Prof. Juhna, Riga, Latvia, and Prof. Burton, Pennsylvania, USA. The editor certainly hopes that many researchers, students, and scientists will benefit from these chapters and look forward to the possibility of them becoming contributors to my subsequent editions.

Disclaimers: The opinions, recommendations, and propositions presented in various chapters represent the findings, views, and opinions of the authors of those chapters and may not represent the position(s), views(s), and/or opinion(s) of the editor. The editor does not promote, endorse, or express his dissent against any product(s) that may have been mentioned in chapters. Although due precautions are exercised before publication, the editor, however, does not take any responsibility for infringement of any copyrighted material of either their own or copied from elsewhere.

Poem written for this book by Dr. Kaltoum Belhassan

Life is Water
When there was enough water
Life was easy
When water was for all
Life was peaceful
When water was potable
Life was healthy
Life is water
But because of those factors:
Megacities
Climate change
Agriculture
Pollution
Today there is not enough water
Today life is hard
Today water is not for all
Today life is war
Today water is not potable
Today life is sick
Life is water
What shall humans do to protect water?
What shall humans do to protect life?
Humans need to
Manage Water Scarcity through:
Preventing water pollution
Water harvesting
Seawater desalination
Recharging aquifers
Water reuse
Humans need to
Make a water resource policy through:
Many organizations

Strong awareness
Life is water

Manassas, VA, USA/Riga, Latvia Ashok Vaseashta, Prof. Dr. Acad.

Contents

Introduction

Introduction to Water Safety, Security and Sustainability 3
Ashok Vaseashta

Applied Geology and Geoinformatics for Ground Water Exploration, Protection and Management 23
Konstantinos Papatheodorou

The Roles of Sensor Placement in Water Quality Monitoring in a Water Distribution System 47
Oluwaseye Samson Adedoja, Yskandar Hamam, Baset Khalaf, and Rotimi Sadiku

Exposome, Biomonitoring, Assessment and Data Analytics to Quantify Universal Water Quality 67
Ashok Vaseashta, Gor Gevorgyan, Doga Kavaz, Ognyan Ivanov, Mohammad Jawaid, and Dejan Vasović

Water Resources, Nature of Contaminants, Impact on Health and Water Quality 115
Adrian Lucian Cococeanu and Teodor Eugen Man

Contamination Detection and Mitigation

Nanomaterials and Their Role in Removing Contaminants from Wastewater—A Critical Review 135
Violeta-Carolina Niculescu, Marius Gheorghe Miricioiu, and Roxana-Elena Ionete

Polymeric Nanocomposite Membranes for Water Filtration 161
Jnyana Ranjan Mishra, Sukanya Pradhan, Smita Mohanty, and Sanjay K. Nayak

Electrospun Nanomaterials: Applications in Water Contamination Remediation 197
Nimet Bölgen and Ashok Vaseashta

Water Treatment by Green Coagulants—Nature at Rescue 215
Manoj Kumar Karnena and Vara Saritha

Application of the Systems Approach and System Standards in Water Safety Plan Development and Implementation 243
Dejan Vasović, Goran Janaćković, and Ashok Vaseashta

Chitosan–Gelatin Cryogels as Bio-Sorbents for Removal of Dyes from Aqueous Solutions 263
Didem Demir, Nimet Bölgen, and Ashok Vaseashta

Macroporous Cryogels for Water Purification 275
Didem Demir, Ashok Vaseashta, and Nimet Bölgen

Materials and Processes for Treatment of Microbiological Pollution in Water 291
Marwa Alazzawi and Hilal Turkoglu Sasmazel

Methods and Characteristics of Conventional Water Treatment Technologies 305
Adrian Lucian Cococeanu and Teodor Eugen Man

Water Security

Environmental Forensic Tools for Water Resources 333
Ilija Brčeski and Ashok Vaseashta

Security Standards Applied to Drinking Water 371
Maria Popa and Ioana Glevitzky

Water Security Safeguarded by Safe, Secure and Smart Water Management Solutions 395
Adrian Lucian Cococeanu and Teodor Eugen Man

Groundwater Safety by Monitoring Quality Parameters in Transylvania, Romania 421
Maria Popa and Ioana Glevitzky

Sustainable Development, Management

Water Scarcity Management 443
Kaltoum Belhassan

Sustainable Approaches for the Treatment of Industrial Wastewater Using Metal-Organic Frame Works 463
Madhavi Konni, Saratchandra Babu Mukkamala, R. S. S. Srikanth Vemuri, and Manoj Kumar Karnena

Household Water Treatment and Safe Storage 495
Stefan-Ionut Spiridon, Eusebiu Ilarian Ionete, and Roxana Elena Ionete

**Hydrocarbons Removal from Contaminated Water by Using
Expanded Graphite Sorbents** 523
Anatoly Kodryk, Alexander Nikulin, Alexander Titenko,
Fedor Kirchu, Yurii Sementsov, Kateryna Ivanenko, Yuliia Grebel'na,
Alex Pokropivny, and Ashok Vaseashta

**Sustainable Approach to Water-Energy Nexus: Sea-Water Reverse
Osmosis Desalination and its Future Directions** 547
Anirban Roy and Asim K. Ghosh

**River Bank Filtration System: Cost Effective Water Supply
Alternative** 565
Sachin Saxena, Aparna Satsangi, and Vuppulury Soamidas

Case Studies

**Drought Land Degradation and Desertification—Case Study
of Nuntasi-Tuzla Lake in Romania** 583
Carmen Maftei, Gabriel Dobrica, Constantin Cerneaga, and Nicusor Buzgaru

**Statistical Assessment of the Water Quality Using Water Quality
Indicators—Case Study from India** 599
Alina Bărbulescu, Lucica Barbeş, and Cristian-Ştefan Dumitriu

**Developing Village-Level Water Management Plans Against
Extreme Climatic Events in Maharashtra (India)—A Case Study
Approach** 615
Aman Srivastava and Pennan Chinnasamy

**Web Application Tool for Assessing Groundwater
Sustainability—A Case Study in Rural-Maharashtra,
India** 637
Aman Srivastava, Leena Khadke, and Pennan Chinnasamy

Modeling

**On the Semiconductor Spectroscopy for Identification of Emergent
Contaminants in Transparent Mediums** 663
Surik Khudaverdyan, Ashok Vaseashta, Gagik Ayvazyan,
Mane Khachatryan, Aigars Atvars, Mihail Lapkis, and Sergey Rudenko

Modeling the Evolution of Surface and Groundwater Quality 691
Erika Beilicci, Robert Beilicci, and Mircea Visescu

Author Index 717
Subject Index 719

Editors and Contributors

About the Editors

Ashok Vaseashta (M'79–SM'90) received Ph.D. in Materials Science and Engineering (minor in Electrical Engineering) from Virginia Polytechnic Institute and State University, Blacksburg, Virginia, USA. He is Executive Director of Research for International Clean Water Institute in VA, USA, Chaired Professor of Nanotechnology at the Academy of Sciences of Moldova and Professor, Nanotechnology and Biomedical Engineering at the Faculty of Mechanical Engineering, Transport and Aeronautics at the Riga Technical University. Prior to his current position, he served as Vice Provost for Research at the Molecular Science Research Center in Orangeburg, South Carolina. He served as visiting professor at the 3 Nano-SAE Research Centre, University of Bucharest, Romania and visiting scientist at the Helen and Martin Kimmel Center of Nanoscale Science at the Weizmann Institute of Science, Israel. He served the U.S. Department of State in two rotations, as strategic S&T advisor and U.S. diplomat. His research interests span nanotechnology, environmental/ecological science, and safety and security. His research on nanotechnology has been on improving the understanding, design, and performance of nanofibers and sensors/detectors, mainly for applications such as wearable electronics, target drug delivery, detection of biomarkers and toxicity of nano and xenobiotic materials. In the security arena, he has worked on counterterrorism, countering unconventional warfare and hybrid threats, critical-Infrastructure protection, biosecurity, dual-use research concerns, and mitigating hybrid threats. He has authored over 250 research publications, edited/authored eight books on nanotechnology, and presented many keynotes and invited lectures worldwide. He serves on the editorial board of several highly reputed international journals. He is an active member of several national and international professional organizations. He is a fellow of the American Physical Society, Institute of Nanotechnology, and the New York Academy of Sciences. He has earned several other fellowships and awards for his meritorious service including 2004/2005 Distinguished Artist and Scholar award.

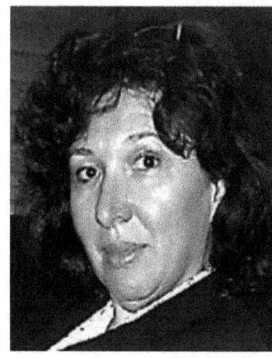

Carmen Maftei is Professor at the Building Services Department of Transylvania University of Brasov, Romania. She attended Polytechnic Institute of Iasi where she majored in Land Reclamation in 1988 and from 1996–1998 she attends a specialization in Water resources and protection of water resources, equivalent of Master of Science at Technical University of Construction Bucharest. In 2002 she finished the Ph.D. studies in Water Science in Continental Environment in joint supervision at Montpellier University and Ovidius University. She's research activity is focused on: (i) hydrology and hydraulic modeling; (ii) Geographic Information Systems applications in environmental sciences; (iii) remote Sensing applications in environmental sciences. As a result of research, she holds 11 books and books chapter, over 100 scientific papers published in different journals or conference proceedings, 1 patent, 3 international projects, 18 national grants, 10 research contracts with economic partners. She is a member in many editorial boards of scientific journals. Her 27 years of teaching experience includes hydrology, hydraulics, Geographic Information Systems & Applications at "Ovidius" University of Constanta and Transylvania University of Brasov.

Contributors

Oluwaseye Samson Adedoja Department of Electrical Engineering/French South African Institute of Technology (F'SATI), Tshwane University of Technology, Pretoria, South Africa

Marwa Alazzawi Department of Biomedical Engineering, Al Nahrain University, Baghdad, Iraq

Aigars Atvars Photonics Laboratory, Institute of Astronomy, University of Latvia, Riga, Latvia

Gagik Ayvazyan National Polytechnic University of Armenia, Yerevan, Armenia

Lucica Barbeş Department of Chemistry and Chemical Engineering, Ovidius University of Constanta, Constanţa, Romania

Erika Beilicci Faculty of Civil Engineering, Department of Hydrotechnical Engineering, Polytechnic University Timisoara, Timisoara, Romania

Robert Beilicci Faculty of Civil Engineering, Department of Hydrotechnical Engineering, Polytechnic University Timisoara, Timisoara, Romania

Kaltoum Belhassan Dewsbury, West Yorkshire, UK

Ilija Brčeski Faculty of Chemistry, University of Belgrade, Belgrade, Serbia; European Academy of Science and Arts, Salzburg, Austria

Nicusor Buzgaru Doctoral School of Applied Sciences, Ovidius University of Constanta, Constanta, Romania

Nimet Bölgen Engineering Faculty, Chemical Engineering Department, Mersin University, Mersin, Turkey

Alina Bărbulescu Department of Civil Engineering, Transilvania University of Braşov, Braşov, Romania

Constantin Cerneaga Doctoral School of Applied Sciences, Ovidius University of Constanta, Constanta, Romania

Pennan Chinnasamy Centre for Technology Alternatives for Rural Areas, Rural Data Research and Analysis (RuDRA) Lab, Indian Institute of Technology Bombay, Mumbai, India

Adrian Lucian Cococeanu Civil Engineering Faculty-Department of Hydrotechnical Engineering, Politehnica University of Timişoara, Timişoara, Romania

Didem Demir Chemical Engineering Department, Engineering Faculty, Mersin University, Mersin, Turkey

Gabriel Dobrica Doctoral School of Applied Sciences, Ovidius University of Constanta, Constanta, Romania

Cristian-Ştefan Dumitriu S.C. Utilnavorep S.A., Constanţa, Romania

Gor Gevorgyan Scientific Center of Zoology and Hydro-Ecology, National Academy of Sciences of Republic of Armenia, Yerevan, Armenia

Asim K. Ghosh Desalination & Membrane Technology Division, Bhabha Atomic Research Centre, Trombay, Mumbai, India

Ioana Glevitzky Doctoral School of the "Lucian Blaga", University of Sibiu, Sibiu, Romania

Yuliia Grebel'na Chuiko Institute of Surface Chemistry of NAS Ukraine, Kiev, Ukraine

Yskandar Hamam Department of Electrical Engineering/French South African Institute of Technology (F'SATI), Tshwane University of Technology, Pretoria, South Africa;
ESSIE, Noisy-le-grand, France

Eusebiu Ilarian Ionete National Research and Development Institute for Cryogenics and Isotopic Technologies, Ramnicu Valcea, Romania

Roxana-Elena Ionete National Research and Development Institute for Cryogenic and Isotopic Technologies – ICSI Ramnicu Valcea, Ramnicu Valcea, Romania

Roxana Elena Ionete National Research and Development Institute for Cryogenics and Isotopic Technologies, Ramnicu Valcea, Romania

Kateryna Ivanenko Institute of High-Molecules Compounds Chemistry, National Academy of Science of Ukraine, Kyiv, Ukraine

Ognyan Ivanov GIS Transfer Center, Electro-magnetic sensors, Bulgarian Academy of Sciences, Sofia, Bulgaria

Goran Janaćković Faculty of Occupational Safety in Niš, University of Niš, Niš, Serbia

Mohammad Jawaid Universiti Putra Malaysia, Seri Kembangan, Selangor, Malaysia

Manoj Kumar Karnena Department of Environmental Science, GITAM Institute of Science, GITAM (Deemed to be) University, Visakhapatnam, India

Doga Kavaz Environmental Research Centre, Cyprus International University, Nicosia, Nicosia, Turkey

Mane Khachatryan National Polytechnic University of Armenia, Yerevan, Armenia

Leena Khadke Department of Civil Engineering, Indian Institute of Technology Bombay, Mumbai, India

Baset Khalaf Department of Electrical Engineering/French South African Institute of Technology (F'SATI), Tshwane University of Technology, Pretoria, South Africa

Surik Khudaverdyan National Polytechnic University of Armenia, Yerevan, Armenia

Fedor Kirchu Department of Aviation Engines, National Aviation University, Kiev, Ukraine

Anatoly Kodryk The Ukrainian Civil Protection Research Institute, Kiev, Ukraine

Madhavi Konni Department of Basic Science and Humanities, Dadi Institute of Engineering, Visakhapatnam, India

Mihail Lapkis RD Alfa Microelectronics, Riga, Latvia

Carmen Maftei Transilvania University of Brasov, Brasov, Romania

Teodor Eugen Man Civil Engineering Faculty, Department of Hydrotechnical Engineering, Politehnica University of Timişoara, Timişoara, Romania

Marius Gheorghe Miricioiu National Research and Development Institute for Cryogenic and Isotopic Technologies – ICSI Ramnicu Valcea, Ramnicu Valcea, Romania

Jnyana Ranjan Mishra Laboratory for Advanced Research in Polymeric Materials (LARPM), School for Advanced Research in Polymers (SARP), Central Institute of Petrochemicals Engineering & Technology (CIPET), Bhubaneswar, Odisha, India

Smita Mohanty Laboratory for Advanced Research in Polymeric Materials (LARPM), School for Advanced Research in Polymers (SARP), Central Institute of Petrochemicals Engineering & Technology (CIPET), Bhubaneswar, Odisha, India

Saratchandra Babu Mukkamala Department of Chemistry, GITAM Institute of Science, GITAM (Deemed to be) University, Visakhapatnam, India

Sanjay K. Nayak Laboratory for Advanced Research in Polymeric Materials (LARPM), School for Advanced Research in Polymers (SARP), Central Institute of Petrochemicals Engineering & Technology (CIPET), Bhubaneswar, Odisha, India

Violeta-Carolina Niculescu National Research and Development Institute for Cryogenic and Isotopic Technologies – ICSI Ramnicu Valcea, Ramnicu Valcea, Romania

Alexander Nikulin The Ukrainian Civil Protection Research Institute, Kiev, Ukraine

Konstantinos Papatheodorou Applied Geology and Geoinformatics, Surveying and Geoinformatics Engineering Department, International Hellenic University, Serres University Campus, Serres, Greece

Alex Pokropivny Frantsevich Institute for Problems of Materials Science of the NAS of Ukraine, Kiev, Ukraine

Maria Popa Faculty of Economic Sciences, "1 Decembrie 1918" University of Alba Iulia, Alba Iulia, România

Sukanya Pradhan Laboratory for Advanced Research in Polymeric Materials (LARPM), School for Advanced Research in Polymers (SARP), Central Institute of Petrochemicals Engineering & Technology (CIPET), Bhubaneswar, Odisha, India

Anirban Roy Department of Chemical Engineering Goa, BITS Pilani, Goa, India

Sergey Rudenko RD Alfa Microelectronics, Riga, Latvia

Rotimi Sadiku Department of Chemical, Metallurgy and Material Engineering/Institute for Nano-Engineering Research (INER), Tshwane University of Technology, Pretoria, South Africa

Vara Saritha Department of Environmental Science, GITAM Institute of Science, (Deemed to Be) University, Visakhapatnam, India

Aparna Satsangi Department of Chemistry, Faculty of Science, Dayalbagh Educational Institute, Dayalbagh, Agra, India

Sachin Saxena Department of Chemistry, Faculty of Science, Dayalbagh Educational Institute, Dayalbagh, Agra, India

Yurii Sementsov Chuiko Institute of Surface Chemistry of NAS Ukraine, Kiev, Ukraine

Vuppulury Soamidas Department of Chemistry, Faculty of Science, Dayalbagh Educational Institute, Dayalbagh, Agra, India

Stefan-Ionut Spiridon National Research and Development Institute for Cryogenics and Isotopic Technologies, Ramnicu Valcea, Romania

R. S. S. Srikanth Vemuri Department of Basic Science and Humanities, Vignan's Institute of Engineering for Women, Visakhapatnam, India

Aman Srivastava Centre for Technology Alternatives for Rural Areas, Indian Institute of Technology Bombay, Mumbai, India

Alexander Titenko The Ukrainian Civil Protection Research Institute, Kiev, Ukraine

Hilal Turkoglu Sasmazel Department of Metallurgical and Materials Engineering, Atilim University, Golbasi, Ankara, Turkey

Ashok Vaseashta International Clean Water Institute, Manassas, VA, USA; Institute of Biomedical and Nanotechnologies, Riga Technical University, Riga, Latvia; Institute of Electronic Engineering and Nanotechnologies "D. Ghitu", ASM, Chisinau, Moldova

Dejan Vasović Faculty of Occupational Safety in Niš, University of Niš, Niš, Serbia

Mircea Visescu Faculty of Civil Engineering, Department of Hydrotechnical Engineering, Polytechnic University Timisoara, Timisoara, Romania

Introduction

Introduction to Water Safety, Security and Sustainability

Ashok Vaseashta

Abstract Based on review of data and information concerning water stress, bio-physico-chemical interactions with human body and nexus of water with food, energy, safety, sustainability and energy, this chapter makes some recommendations for the future. The recommendations are consistent with the United Nations Sustainable Development Goals. A Delphi based survey identified global challenges and a correlation is drawn as how the top twenty priorities correspond to the Sustainable Development Goals. Using risk assessment modalities, the chapter presents a sustainable landscape of water going forward and how to make drinking water systems safe, secure, and sustainable to meet current and future needs. This chapter also serves as an introductory chapter to this edited book on Water Safety, Security and Sustainability with interesting chapters ranging from fundamental concepts, novel materials and their applications, regional cases studies and device modeling to enhance understanding of the subject matter.

Keywords Water · Safety · Security · Sustainability · Sustainable development goals

1 Water—Current Status and Global Challenges

At the first glance, water appears to be one of the most simple, abundant and pure substance, which in its pure state is practically colourless, odourless, and tasteless. However, from scientific standpoint, water possesses quite mysterious and complex physical and chemical characteristics, with many yet to be explored. Furthermore, water is vital for sustaining life on Earth, and in fact, where there is water, there is life and where water is scarce, sustaining life is complex. Water is considered a purifier

---

A. Vaseashta (✉)
International Clean Water Institute, Manassas, VA, USA
e-mail: prof.vaseashta@ieee.org

Institute of Electronic Engineering and Nanotechnologies "D. Ghitu", ASM, Chisinau, Moldova

Institute of Biomedical and Nanotechnologies, Riga Technical University, Riga, Latvia

© Springer Nature Switzerland AG 2021
A. Vaseashta and C. Maftei (eds.), *Water Safety, Security and Sustainability*,
Advanced Sciences and Technologies for Security Applications,
https://doi.org/10.1007/978-3-030-76008-3_1

in most faiths and religions, and particular sources or bodies of water are considered to be sacred or at least auspicious. Water is critical to sustaining basic functions of human body such as homeostasis, maintaining equilibrium between exogenous and endogenous water, regulating metabolism and temperature, filtering out toxins, just to name a few. Water (Hydrogen-oxide), mostly, is a chemical compound with two atoms of hydrogen and one atom of oxygen in a molecule (H_2O). The isotope ratio of deuterium and protium (D/H 150 ppm), that once formed water—a "matrix" of life, facilitated the large variety of biological species. Life on Earth became possible, due to exactly this isotope composition. A particular ratio of hydrogen and oxygen isotopes, determines physical and chemical properties of water, viz., boiling and freezing points, viscosity, density, refractive index, density, surface tension, and several other characteristics, including chemical reaction rate constants and biological processes taking place in living matter. These functions are due to some of the unique characteristics of water. However, the huge amount of water on our planet is not just simple H_2O but water complex that has anomalous properties. Among many other interesting characteristics, water has anomalous thermal expansion property which is responsible for protecting our delicate aquatic ecosystem. Water has strong solvent properties since it mixes with and dissolves a wide range of factors and is therefore easily contaminated. Hence, chemicals and industrial wastes entering the environment pollute water, thus impacting food supply-chain and forages and ambient air that impacts living organisms and human health. On the other hand, due to dilution of contaminants, unless quantities are significantly large, the solvent characteristics of water tend to protect nature. Research on eco- and aqua-toxicity and its impact on human health has accomplished significant progress and the subject is still unfolding and needs attention from a holistic viewpoint by scientific community for our survival and path forward. Due to these and several other unique characteristics, water remains an intriguing substance.

1.1 Current Status and Global Challenges

Water covers ~71% of the Earth's surface and the remaining 29% consists of continents and islands. Furthermore, 96.5% of all the Earth's water is contained in the oceans as salt water, while the remaining 3.5% is freshwater in lakes and frozen, inaccessible being in glaciers and the polar ice caps—about 69% of the total freshwater, and is large enough to raise the sea levels to an altitude of approximately 2.7 m, if all of it were to melt and of course, if the surface of the Earth was perfectly smooth. Unfortunately, even with such a vast reserve of water body on Earth, there is a global shortage of clean water for public consumption. According to the WHO, over 2.1 billion people, which is 3 in 10 worldwide, do not have access to on-premise sources of water, approx. 845 million people do not have access to a water sources which are within about 30 min or less round trip and may not necessarily be always free from contamination or accessible when needed, 265 million people have to travel over 30 min just to access water that isn't even clean, and 159 million drink from untreated surface water sources [1–3]. Some additional alarming statistics show that humans play a large role in the disruption of the hydrologic cycle, irrespective of the reasons, may it be self-serving, intentional or accidental. It is further estimated that by 2025, more than half of the world population be living in water stressed areas due to formation of mega-cities and increasing world population, which is expected to rise to 9.7 billion by 2050, causing further stress on water globally [4–7]. One of the reasons of water stress due to lack of sanitation services and it is estimated that 2.3 billion people, that is 1 in 3 persons worldwide, lack access to even basic sanitation services. Those without safe systems run the risk of having their water supply become contaminated with human and animal waste.

According to WHO, 1 in 3 persons consume water from a source that is contaminated with feces, which is a major cause of deadly waterborne diseases such as *Hepatitis A, Norovirus*, and *E. Coli*. Hence 1.6 million people die every year from waterborne diseases [8]. Nearly 1 out of 5 deaths under the age of 5 worldwide is due to a water-related disease [9]. In developing countries, as much of 80% of illnesses are linked to poor water and sanitation conditions. Also, 443 million school days are lost each year due to water-related diseases and nearly half of the world's hospital beds are filled with people suffering from a water-related disease [10]. It is further estimated that 80% of diseases are carried by water, resulting in loss of life of about 1 child every 8 sec., i.e. approximating 5–7 million people annually and with $125 billion in economic loss due to workday losses/yr.

There are socio-economic aspects, such as, women end up taking responsibility of managing household and hence in some instances walk several miles to fetch water for family, likely contaminated, for cooking and washing. In schools, girls are more likely to drop out of schools than boys, as disproportionate consequences of lack of sanitation services [11, 12]. It is further estimated that ~75% of world population live in water stressed areas, as world continues to lose irrigated land by as much as 30% by 2025 to 50% by 2050. Interestingly, in Asia, over 75% of population live in areas where 50% of rainfall occurs for 20% times of the year.

1.2 Water-Energy-Food Nexus

Water, food and energy are closely coupled, and water is a critical common component to making food and energy systems work. Due to rapid growth in population and associated economic growth in conjunction with accelerated urbanization, the dynamics of demand for water, food and energy is quite complex and constantly changing. Since, the natural resources are limited and water scarcity affects almost one third of world population, it requires a careful planning for the three, known as the Water-Energy-Food Nexus, also known as Food-Energy-Water (FEW) nexus. To highlight its relevance, it is critical to understand their interdependencies and how they collectively impact food, water and energy security.

In order to highlight some interdependences, it should be noted that agricultural processes in the U.S. account for 80% of freshwater consumption. The Water

Footprint Network (WFN) has calculated how much water it takes—called a water footprint—for a large number of food items [13]. It is noteworthy that meats (such as pork and beef), require the highest amount of water to produce and consume [14]. Over and above this consumption, an average person requires one-gallon water per person, as a daily water requirement for proper hydration, which accounts for over 7 billion gallons of water per day for everyone globally. Just in the US alone, ~140,000 million gallons of water per day is used for irrigation and livestock, out of which ~85% is not returned to the local water source. Furthermore, 25% of all freshwater consumed in the US is associated with discarded food, which is about as much as the volume, as that of Lake Erie, New York. New technological advances, such as on-demand irrigation [15], genetically modifying seed to consume less water [16], and similar scientific breakthroughs can dramatically enhance agriculture water conservation and enhance recovery and reuse of irrigation runoff and livestock wastewater and provide much needed impact on future water consumption.

Water-related energy use accounts for 20% of all electricity use in the state of California and 15% of it goes toward moving water large distances for irrigation. Nearly half of all water withdrawals—both freshwater and ocean water—in the U.S. are used for cooling at thermoelectric power plants [17]. By 2030, generating capacity is expected to increase by over 20%, resulting in an expected increase of freshwater consumption by as much as 49%. As per Nuclear Energy Institute (NEI) estimates, a nuclear power reactor consumes between 1,514 and 2,725 L of water per megawatt-hour [18]. In comparison, the consumption is 1,220–2,270 L per MWh for coal and 700–1,200 L per MWh for gas [19]. As of 2016, the U.S. had 99 operating nuclear reactors and 61 plants across the country, with a capacity-weighted average age of 37 years, with additional four reactors are currently under construction. About 25% of these reactors are in water stressed areas and may face potential shutdown due to shortage of water and concerns over environmental issues. The NEI has recommended that a majority of future nuclear power plants should be built at coastal locations that can utilize ocean saltwater to minimize the probability of drought-related shutdowns and to provide the opportunity for the construction of desalination plants to offset their freshwater consumption. This trend would further improve perception of nuclear power plant water withdrawal and consumption and may render it the most viable thermal electricity generation method from a sustainability viewpoint. Yet another example is in producing corn, which is the main ethanol feedstock in the country. Using potential food supplies for energy generation can put us in direct competition and both will depend on large reserves of water.

1.2.1 Interdependence on Environment, Health, Safety

The impact of water reaches far beyond human consumption, as water supports basic pillars of our life and survival, viz. health, environment and safety. In the context of human health, it is estimated that water contents constitute between 60 and 75% of human body weight. Being a polar molecule, water interacts with other polar molecules, such as itself, leading to the cohesion of water molecules, which helps regulate body temperature. Water's capability to dissolve a variety of other molecules makes it an invaluable life-sustaining substance. On a biological level, water's role as a solvent helps cells transport substances such as oxygen or nutrients to the necessary locations within the body that has a major influence on the ability of drugs to reach their targeted location. Additionally, water buffers cells from the dangerous effects of acids and bases. Water contributes to the formation of membranes surrounding cells and these bilayers selectively allow salts and nutrients to enter and exit the cell and perform filtration functions. Furthermore, water molecules encapsulate DNA in an ordered manner to support its characteristic of double-helix conformation to follow instructions encoded in the DNA or to pass the instructions onto future cells, making human growth, reproduction, and, ultimately, survival viable. Although, we have yet to determine the physiological impacts of these properties, but it is interesting to note that such a simple molecule, is vital and universally critical for organisms with vastly diverse needs.

As per environmental considerations, aquatic ecosystems include aspects, such as biodiversity, habitat for fisheries, recreation and tourism. Unfortunately, many freshwater ecosystems and the services they provide are vulnerable to changes due to land use, environmental pollution, water diversion, overfishing and the introduction of foreign species that harm freshwater resources. Changes in temperature and precipitation patterns significantly influence the movement of water through the atmosphere and landscape. Effects associated with climate change continue to disrupt lakes, rivers and wetlands, such as increased evaporation, decreased summer precipitation, and warmer water temperatures. A combined effect of these influences can destructively impact both water quality and quantity.

Contaminants negatively impact the water quality and have direct environmental consequences. New and emerging contaminants, such as microplastics, household cleaning agents, volatile organic compounds (VOCs), returned pharmaceuticals and engineered nanomaterials have adverse health and environmental consequences which are hereto completely unknown and are subject of ongoing investigations [20]. In fact, U.S. Environmental Protection Agency (US EPA) has now listed over 90 of such contaminants on the Candidate Contaminants List (CCL), which water utilities will be required to detect and mitigate. Unfortunately, most of the water filtration facilities use conventional methods for removing sediments and salinity, followed by activated carbon technology to remove only selected contaminants. Aquifers provide reserve for clean water, however only few acquirers have been surveyed for available water reserves. Moreover, impact of water quality in aquifers due to fracking is yet unknown. Detection of new and emerging contaminants by state-of-the-art scientific methods, contamination remediation and mitigation strategies, and impact of contaminants on human health and environment, is an area that will require much more aggressive approach to increase utilization of currently available resources [21]. In addition, facilities developments of new resources of water is necessary to maintain a delicate balance of nexus of clean water with environment, human health and safety.

1.2.2 Transdisciplinarity in Water Research

> Water is a basic human requirement and a matter of sheer survival, it is a natural raw material and an economic commodity, it is a matter of course and a lifestyle product, it is an instrument of power and a scarce good—too much or too little of it may lead to disaster. Anne Dombrowski.

Etymology of transdisciplinarity is the understanding of the present world and goes beyond the compartmentalization among disciplines. According to Nicolescu [22], the prefix trans-indicates that, transdisciplinarity concerns both, between disciplines (transition or transformation) and across different disciplines, and beyond. According to Barbier [23], transdisciplinarity is not accumulation of disciplinary mandates and, in fact, it is an epistemological posture, for which the purpose is the understanding of the present world, and the imperatives being the harmony of knowledge. It also indicates its characteristics of transversality and transcendence, believing that the synergistic encounter among disciplines is an activity, that is both transformative and formative of a new field of research. In a progression from monodisciplinary to transdisciplinary, as defined by Piaget [24], the latter represents an extreme stage of what a discipline can represent, which in the context of water, using the predominant notion of transdisciplinarity, interdisciplinary and participatory research, can be developed further by disciplinary experts, stakeholders and regular citizens.

The term 'transdisciplinarity' was originally coined by Piaget [24], during his lecture: 'The epistemology of interdisciplinary relations at the colloquium L'interdisciplinarité - Problèmes d'enseignement et de recherche'. In his section, he states that "… *we can hope to see a higher stage which would be 'transdisciplinary', which would not be satisfied with achieving interactions or reciprocities between specialized research, but would situate these links within a total system, without stable borders between disciplines*". He wanted to indicate the need to go beyond the interdisciplinary logic towards a more encompassing and integrated vision of knowledge, as transdisciplinarity attempts to respond to a new vision of human and nature, by going beyond the current paradigm in an integrative way. It is not usual to use the term transdisciplinarity in the context of water, however, transdisciplinary research is drawn from a variety of concepts that parallels those found in water research. The motivation for transdisciplinary water research is commonly socially relevant issues where facts are uncertain, values are in dispute, stakes are high, and decisions are urgent. In response to these types of problems, participatory methods have been developed to foster coproduction of knowledge and social learning among different players. Although a detailed discussion of transdisciplinarity and it's societal impact is beyond the scope of this chapter, it is worth noting that using transdisciplinarity, it is possible to cover a broad range of water research from different perspectives for a balanced conceptualization and study of human–water relations in this and other scientific contexts.

1.3 Sustainable Development Goals

Over past couple of decades, water has transitioned from "taken for granted" to "fundamental human right" status. Water also appears among the top ten challenges of the twenty-first century [25]. The United Nations has proposed a series of sustainable development goals for a brighter, self-sustaining future plan for the earth and water is listed in the Sustainable Development Goals (SDG) 6, which focuses on ensuring a clean and stable water supply and effective water sanitation for all people by the year 2030. The goal is a reaction to the fact that many people throughout the world lack these basic services. Some of the poorest countries in the world are affected by drought, resulting in famine and malnutrition. Throughout the world, approx. 2 billion people live in a watershed where water is used faster than the replenishment rate. According to some estimates, if such trends continue, one in four people, or more, might experience water shortages on a regular basis by the year 2050. SDG6 also calls for adequate sanitation and water services available to all people by the year 2030. Almost a billion people, or more, would require the construction of such facilities to provide routine clean water and waste removal facilities. For such missions to succeed, the U.N. has developed a series of targets that include restoring and protecting river ecosystems throughout the world, eliminating sources of water pollution, and increasing international cooperation to bring services throughout the world. In fact, Millennium Development Goals (MDGs), preceded the SDGs and had 8 goals which ranged from reducing extreme poverty rates to half to halting the spread of HIV/AIDS and providing universal primary education, all by the target date of 2015. Water was a part of the development goals, however, was integrated in development goals 6 and 7, to limit waterborne diseases and to ensure environmental sustainability. Increased water scarcity has galvanized unprecedented efforts to meet the needs of the world's poorest and the U.N. is also working with governments, civil organizations and other partners to build on this momentum generated by the MDGs to carry on with an ambitious post-2015 development agenda.

Despite of such ambitious goals, the U.N. SDGs have drawn sharp criticism since their adoption, primarily due to the fact that the goals are not legally binding and with countries left to set up their own frameworks for implementation. Furthermore, for lack of well-defined metrics, the goals may appear aspirational and perhaps even, contradictory. This leaves room for debate among the environmental and societal dimensions, but without really solving the root causes in the system responsible for such problems. In political space, there are discussion about access to water as fundament human rights and democracy. As shown in Fig. 1 and stated above, water forms the nexus of food, energy, safety, security, sustainability and economic well-being as one of the most fundamental necessity of sustaining life on earth, yet the progress on this front has not been at the proportionate scale. It should also be pointed out that in current crisis of COVID-19, the priority is compounded due to need for handwashing, increasing further demand on water for hygiene and also to mitigate contaminants from water supply is equally critical since trace amounts of COVID-19 strain has been detected in water, in several locations.

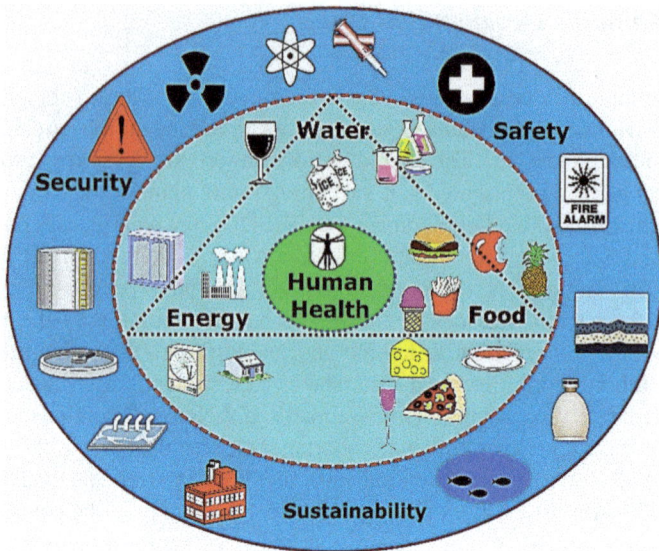

Fig. 1 Water-food-energy nexus and association with safety, Security and Sustainability

While implementing reasonable plans for providing clean water and rebuilding infrastructure are necessary for addressing the world's concerns, it certainly won't be inexpensive for SDG6 to address the global water crisis. Although these goals are comprehensive and well-informed, attaining these goals will be time-consuming, expensive, and may face further political division. It is estimated that in order to reach these goals, it will cost ~$114 billion dollars every year until 2030. Undoubtedly, maintaining clean water sources is an expensive proposition and even for large corporations, whilst motives are mostly internally focused, they already spend millions of dollars a year maintaining clean water sources for their production. However, in an era of perseverance based unmanned Mars exploration, artificial intelligence (AI), virtual and augmented reality (VR/AR), the efforts are incomparable, especially when the progress of development goals must depend on cross sector policies and local actions. Figure 2 provides a pictorial overview of global challenges, based on a Delphi survey, and Table 1 provides SDGs corresponding to the global challenges.

It is imperative that governments need to overcome issues that span individual sector silos and go beyond policies that are aimed at strengthening institutions, mobilizing financing, and delivering sustainable solutions for addressing water scarcity and building resilient water infrastructures for populations. This objective also helps protection of the Earth's biosphere and natural biodiversity. Closing the water access gap and solving intermittent water supply must be a high priority for universal and equitable access to safe drinking water for all. Implementing the principles of sustainable development, social environmental (ecological) responsibility, pro-ecological reforms of implementing eco-innovations in economic processes should become one of the important factors of globalization processes in the twenty-first century. The

Fig. 2 Global challenges with weight % as per Delphi survey

development of the implementation of eco-innovation and the implementation of the sustainable development goals does not necessarily mean reducing the scale of globalization. Hence, international cooperation, exchange of pro-ecological technologies, implementation of global eco-innovation, organizations coordinating and integrating on a global scale for the development of green circular economy, pro-ecological processes of the energy sector transformation, and development of renewable energy sources will only progress towards a more positive aspects of pro-environmental green globalization. If history is our guide, *homo sapiens* could have also been *homo economicus* and *homo ecologicus*.

2 Emerging Trends and Future Pathways for Global Water Safety, Security and Sustainable Development

Access to safe, secure and sustainable drinking water resources is of paramount importance and an existential challenge worldwide, for governments and the scientific community, particularly in a time of climate change and dynamic urban and economic development, globally. The issue is shared among both developed and developing nations, since all nations require adequate levels of good quality water

Table 1 Top global challenges and corresponding actions of SDGs

Global Challenges	Corresponding Sustainable Development Goals	SDGs
Climate change	1, 2, 6, 7, 8, 9, 11, 12, 13, 14	1
Pollution	3, 6, 11, 12, 13	2
Violence	3, 8, 10, 12, 16	3
Security	3, 8, 10, 12, 16	4
Education	4, 5, 8, 10	5
Unemployment	1, 2, 4, 8, 9	6
Corruption	8, 9, 10, 16, 17	7
Malnourishment and Hunger	1, 2, 3, 8, 10	8
Substance Abuse	1, 2, 3, 4	9
Terrorism	1, 2, 4, 16	10
Clean water	6, 12, 14, 15	11
Population	2, 7, 11, 12, 15	12
Democracy	16, 17	13
Equality	1, 4, 8, 17	14
Global Health	2, 3, 15	15
Disease	2, 3, 6, 15	16
Energy	7, 9, 12	17
War	1, 4, 8, 10	
Misinformation	8, 9, 10	
IT Dominance	8, 9, 10	

UN SDGs: https://www.un.org/sustainabledevelopment/

at a reasonable cost. A fundamental question is as to how to make drinking water systems safe, secure, and sustainable to meet current and future needs using emerging trends in contamination monitoring and remediation. Since the issue of water supply safety, security, and sustainability is highly diverse, the chapter presents the myriad of water supply challenges from a holistic viewpoint.

Water safety spans a wide spectrum of safe usage of water to meet the soaring demand at present and in the future. Water safety also refers to the procedures, precautions and policies associated with safety in, on, and around bodies of water, however, for the present discussion, the focus is on human health, agriculture and industries that require high volumes of safe sources of water for their daily operations, such as, environmental remediation, hydrocarbon processing, catalysis and chemical processing, food and beverage, mining and hydrometallurgy, pharmaceuticals, power generation and semi-conductors, just to name a few. The high volume of fresh or processed water required for manufacturing processes can make businesses unsustainable in areas with limited water availability. At the same time, these industries also have stringent requirements to control wastewater discharge in the environment. As industries grow incorporating recently developed more advanced technologies, clean water will become ever more critical to advance these operations and increasing the demand for the limited water resources. Table 2 provides several contaminants that are commonly present in water and need to be isolated from water supply lines for human safety. From policy standpoint, it is critical to research and develop effective water treatment and better distribution network management systems, as the pillars of sustainable water supplies. Creating better disinfection and purification technologies could significantly reduce many human health issues that much of the world currently faces and, equally importantly, certain economically disadvantaged regions of the world.

From a security standpoint, drinking water distribution systems are largely exposed and vulnerable to intentional and/or inadvertent contamination. Such contaminants can be classical and chemical, including but not limited to non-traditional agents (NTAs), toxic industrial chemicals (TICs), and/or toxic industrial materials (TIMs). It is, thus, extremely important to monitor, control, and mitigate contaminants using state-of-the-art materials and technologies to maintain "water quality"—typically defined as physical, chemical, and biological characteristics of water in relationship to a set of standards [26]. Water quality is a rather complex subject and is intrinsically tied to the regional ecology, application, and point-of-use. Detection methodologies based on nanomaterials and other advanced strategies are capable of ppb/ppt and atto-molecule resolution with high selectivity, sensitivity and specificity [27]. For mixed compositions, such methodologies are particularly important in identifying constituent chemical composition. Additional complication arises due to presence of unused pharmaceuticals present in water supply since most filtration systems are not designed to filter such contaminants. Due to large number of contaminants that greatly differ in chemical structure and chromatographic/spectroscopic behavior, it is extremely difficult, if not impossible, to apply a common method to interrogate in real-time.

Securing safe and sustainable drinking water supplies is a prominent challenge, especially when the water supply lines are mostly above ground. In fact, transport of supply of water has been a great challenge since the ancient Roman times, as water was transported by pack-animals to a great distance. Disruption of water supply can serve as a significant and strategic deterrent. In recent military rotations, water supply trucks were provided extra security protection. Also, extra protection was also provided for aquifers, especially for the ones that are located at or near transboundary regions. It was reported by Russian Times that aquifer near Libya was contaminated causing shortage of clean water and drinking water was transported in Libya from

Table 2 List of contaminants

Categories	Sources	MRL/MCL	Health effect
Pesticides			
Atrazine	Runoff from herbicides used on crops	3 μg/L	Cardiovascular diseases; reproductive problems
Carbofuran	Insecticide widely used on field crops	40 μg/L	Blood, nervous, and reproductive system diseases
Methoxychlor	Runoff/leaching from insecticide use	40 μg/L	Reproductive difficulties
Glyphosate	Runoff from herbicide use	700 μg/L	Kidney problems, reproductive difficulties, increased risk of cancer
2,4-Dichlorophenoxyacetic acid	Runoff from herbicide used on crops	70 μg/L	Kidney, liver and adrenal gland pathologies
Pharmaceuticals			
Antibiotics (metronidazole, tinidazole, etc.)	semisynthetic modifications of various natural compounds	Physiological conditions vary	Long term impairment of immunity, loss of probiotics
Steroids	Semisynthetic modifications of various natural compounds	Physiological conditions vary	Weaken the body's immune system, risk of insomnia, infertility, excess fluid retention
Testosterone	Semisynthetic modifications of various natural compounds	Physiological conditions vary	Cancer, especially prostate cancer, aggressive/criminal behavior
Acetaminophen	Semisynthetic modifications of various natural compounds	Physiological conditions vary	Harmful to liver (base on dose level)
Ibuprofen	Semisynthetic modifications of various natural compounds	Physiological conditions vary	Nausea, heartburn, and stomach pain including heart attack
Microorganisms			
Cryptosporidium	Human and animal fecal waste	–	Gastrointestinal illness (e.g., diarrhea, vomiting, cra mps)
Legionella	Found naturally in water; multiplies in heating systems	–	Legionnaire's Disease, a type of pneumonia

(continued)

Table 2 (continued)

Categories	Sources	MRL/MCL	Health effect
Enteric viruses	Human and animal fecal waste	–	Gastrointestinal illness (e.g., diarrhea, vomiting, cramps)
Total coliform (including fecal coliform and *E. coli*)	Indicator if other potentially harmful bacteria may be present	–	Fecal coliforms and *E. coli* come from human and animal fecal waste
Metals, organics, VOCs			
Arsenic	Erosion of natural deposits; runoff from orchards,	0.010 mg/L	Skin damage or problems with circulatory systems, and may increase risk of cancer
Styrene	Discharge from rubber and plastic factories; leaching from landfills	0.1 mg/L	Liver, kidney, or circulatory system problems
Trichloroethylene (1,1,1; 1,1,2)	Chemical factories/degreasing agents	0.2 and 0.005 mg/L	Liver, nervous system, circulatory problems, kidneys and immune systems
Vinyl Chloride	Leaching from PVC pipes; discharge from plastic factories	0.002 mg/L	Increased risk of cancer
Xylenes	Discharge from petroleum factories; discharge from chemical factories	10 mg/L	Nervous system damage
Uranium	Erosion of natural deposits	30 μg/L	Increased risk of cancer, kidney toxicity
Emerging contaminants			
Micro/nano plastics	Discarding plastics bottles, bags and several products that use plastic microbeads		Lungs, dermal, genetic, carcinogen
COVID-19 virus	Untreated discharge from COVID-19 patients		Effects still unfolding but flu is a companied by serious cardiovascular and respiratory consequences
Petroleum products, VOCs, long chain hydrocarbons	Discarded oil, oil wells	–	Carcinogen, irritant

(continued)

Table 2 (continued)

Categories	Sources	MRL/MCL	Health effect
Nanomaterials	Broad classification of ultrafine particulate matter used in more than 1,800 consumer products and biomedical applications		Translocate into the circulatory system through the lungs, accumulation of compounds in the liver, spleen, kidney, and brain
Trichloro propane (TCP)	Chemical intermediate, solvent, and cleaning product		Carcinogen
Perfluoro-octane sulfonate (PFOS) and Perfluorooctanoic	Used in additives and coatings, non-stick cookware, waterproof clothing, cardboard		Carcinogen, may cause high cholesterol, increased liver enzymes, and adverse
Polybrominated diphenyl ethers (PBDEs)	Flame retardant and used in plastics, furniture, and other household products		Endocrine disruptor as well as caecinogenic, also, may cause neural, liver, pancreatic, and thyroid toxicity
Dioxane	Stabilizer of chlorinated solvents, manufacturing of PET, manufacturing by-product		Disruption of lung, liver, kidney, spleen, colon, and muscle tissue, may be toxic to developing fetuses and is a potential carcinogen
Hexa hydro-trinitro-triazane (RDX)	Explosive	–	Kidney and liver damage, possible carcinoma, insomnia, nausea, and tremor

Lampedusa, during the second Libyan civil war that was fought between different armed groups.

Yet another security challenge, is Internet of Things (IoT) operation of water plants for high-level process supervisory management via graphical user interfaces (GUI). As recent as January 2021, water services in the state of Florida were hacked and chemicals were remotely discharged in disproportionate amounts. Similar incidents were previously reported in Michigan by breaching Supervisory control and data acquisition (SCADA) control system, an architecture comprising of computers, programmable logic controllers (PLC) and other peripheral devices networked for data communications. Disabling and/or tampering with urban water distribution systems can impact millions of customers, while intermixing with waste water systems can render them as tools for weaponization by a variety of industrial and commercial contaminants.

Funded through a NATO Science for Peace activity, a GPS/GIS based Contamination Identification and Level Monitoring Electronic Display Systems (CILM-EDS) prototype was developed [27] to spatially monitor contaminants and water levels using advanced (nano) technology based platforms for inadvertent and intentional contaminants. The prototype is also capable of sending signals to provide situational awareness for water distribution system using a smart and connected communities project and advance warning for sudden rising water levels for areas prone to flash floods and Tsunamis. The U.S. EPA initiated a WaterSentinel program for the design, development and deployment of an integrated water contamination monitoring incorporating real time system wide water quality monitoring contaminant sampling and analysis.

Obviously, the basic need for clean drinking water and water resources is a national security and policy concern. There are numerous national security threats emanating from the growing crisis of global freshwater scarcity. Many of the earth's freshwater ecosystems are being critically depleted and used unsustainably to support growing residential and industrial demands; thereby increasing ecological destabilization, and creating a greater regional divide thus directly impacting current political, economic and social landscapes. Finally, in striking contrast to water scarcity are threats to clean water systems posed by high water, as experienced in some flesh flood prone regions and tsunamis in the Asian-Pacific Rim, South-East Asia, and many regions in Europe and even in America when contaminated water infiltrates clean water supplies.

3 Proposed Solutions, Policy Implications and Legal Framework

It is evident that, the society has ignored the problem of use, reuse and recycle water for a long period of time and at the time of realization, it seems that it is already too late. Due to safety, security and sustainability aspects, the scientists and policymakers must work together to develop a paradigm shift plan with built in resiliency, redundancy and incremental improvement. The challenge to overcome the cost of late realization requires a long term vision of critical problems that need to be addressed in plans that cover the next 1, 5, 10 and 25 years.

Conventional methods employed to sense/detect such contaminants use commercial-off the shelf (COTS) systems and broad-spectrum analytical instruments with interpretive algorithms to detect and characterize toxic contaminants. Due to a large number of contaminants that greatly differ in chemical structure and chromatographic/spectroscopic behavior, it is extremely difficult, if not impossible, to apply a common method to interrogate in real-time. Use of nanomaterials based sensors/detectors have been demonstrated to be able to interrogate contaminants in small quantities in mixed environments. Nanomaterials based sensors, in conjunction with plasmonics have demonstrated remote detection capability—which is desirable for biological samples, especially those exhibiting toxic characteristics. It has already been demonstrated that output from these sensors/detectors can remotely be displayed using GIS/GPS based technology.

In addition, a comprehensive effort aimed at improving water management systems with aim towards "capacity building" as a holistic approach, which requires an assessment of current water management systems and creating a managerial and engineering environment that will support improvement in water management systems. As such, we must first analyze the overall environment and working culture of an organization tasked with Water Quality Management and propose systematic measures aimed at making the organization more effective, resilient, responsive and efficient. Other pathways being the source water protection, such as in aquifers. The plans should include sustainable withdrawal capacity. We should identify new sources of water, such as atmospheric water sources, reuse of wastewater by smart plumbing at each and every site, leak reduction (which accounts for almost 30% wastage of water) through smart metering, potable water supplies through reuse of existing wastewater and development of brackish and saline sources and use of cisterns for agriculture. Conservation via improved efficiencies and reduction in waste can significantly reduce overall cost and water savings in conjunction with minimizing the withdrawals for direct draw applications. Furthermore, efforts of scalability, ramp-up and advancing technologies from laboratory to production should be encouraged through academia-industry partnerships. Far too often, the innovations do not advance to protype due to patent considerations. By rapid attribution mechanism, one should facilitate the testing and advancing of new materials and technologies development to marketplace.

Lastly, the concept of water as a human right has been introduced in various contexts. Water has not always explicitly been referred to the worldwide bill of Human Rights, global human rights regulation composed via the popular assertion of Human Rights (1948), the International Covenant on Civil and Political Rights (1966), and the worldwide Covenant on financial, Social, and Cultural Rights (1966). However, many activists have argued its validity based on the UN SDG6, which is to "ensure access to water and sanitation for all,". Water, while not explicitly recognized as an independent human right in international treaties, international human rights law contains specific requirements regarding individuals' right to access safe drinking water [28]. These requirements call on governments to ensure that citizens have access to adequate safe drinking water for personal and domestic use, drinking water, personal hygiene, laundry, cooking, and water necessary for personal and home hygiene. These requirements also call on governments to gradually ensure access to adequate sanitation facilities, a fundamental element of privacy and human dignity, and maintain the quality of drinking water supplies and resources [29].

4 Conclusions and Future Pathways

A vision of Water Sector's Security is a safe, secure, resilient and sustainable drinking water and wastewater infrastructure that provides clean and safe water as an integral part of life. This vision assures public safety and the economic vitality of and public confidence through a layered defense of effective preparedness and security practices in the sector. The vision through research and communication plan will enable better understanding of water borne contaminants, their remediation strategies, improved information-sharing and exchange mechanisms, and a platform to inform certain owners and operators of critical Water Sector infrastructure to be able to prevent, detect, respond to, and recover from attacks, such as intentional acts, natural disasters, and other hazards. Also, the broad problem of water can be solved through technologies, policies and improved laws.

The following directives are aimed at research, development and engineering processes, and also to support policymakers with water quality management:

- Conduct an analysis of state-of-the art water management systems to identify technologies that provide the most efficient and economical means to manage water systems.
- Adopt new engineering and managerial processes for water management systems through process reengineering.
- Conduct an analysis of the current water management system to identify shortcomings and performance standards.
- Use state-of-the-art contamination monitoring/detection systems.
- Design smart water management systems.
- Mandatory installation of smart water leak detection systems with appliances.
- Invest in atmospheric water harvesting systems.
- Invest in state of the art membranes capable of desalination, removing pharmaceuticals and metals from water.
- Reduction on use of single use plastics bottles. Alternatives materials from single use plastic bottles to multi-use containers.
- Monitor new and emerging contaminants and fund research to study impact of such contaminants on human health.
- Enhance cyber security to protect SCADA system.
- Define a Universal Water Quality standard.
- Conduct a resource analysis to determine what resources are needed to perform a certain job effectively.
- Develop a plan to address deficiencies in physical infrastructure, equipment, and machinery.
- Develop a short and long-term (1, 5, 10, 25 years) strategic plan for addressing human resources, infrastructure, equipment, and machinery needed for value efficient water management systems.
- Introduce a performance monitoring system which collects information against predetermined targets and performance standards for the water management system.

References

1. United Nations World Health Organization. https://www.who.int/en/news-room/fact-sheets/detail/sanitation. Accessed 28 Feb 2021
2. United Nations World Health Organization. https://www.who.int/news/item/18-06-2019-1-in-3-people-globally-do-not-have-access-to-safe-drinking-water-unicef-who. Accessed 28 Feb 2021
3. Erickson MC, Ortega YR (2006) Inactivation of protozoan parasites in food, water, and environmental systems. J Food Prot 69(11):2786–2808. https://doi.org/10.4315/0362-028x-69.11.2786. PMID: 17133829
4. Esen Ö, Yıldırım DÇ, Yıldırım S (2020) Threshold effects of economic growth on water stress in the Eurozone. Environ Sci Pollut Res Int 27(25):31427–31438. https://doi.org/10.1007/s11356-020-09383-y. Epub 2 June 2020. PMID: 32488700
5. Varis O (2006) Megacities, development and water. Int J Water Resour Dev 22(2):199–225. https://doi.org/10.1080/07900620600648399
6. Niemczynowicz J (1996) Megacities from a Water Perspective. Water Int 21(4):198–205. https://doi.org/10.1080/02508069608686515
7. Deren L, Ma J, Cheng T, van Genderen JL, Shao Z (2019) Challenges and opportunities for the development of megacities. Int J Digit Earth 12(12):1382–1395. https://doi.org/10.1080/17538947.2018.1512662
8. Moreira NA, Bondelind M (2017) Safe drinking water and waterborne outbreaks. J Water Health 15(1):83–96. https://doi.org/10.2166/wh.2016.103. PMID: 28151442
9. Liu L, Johnson HL, Cousens S, Perin J, Scott S, Lawn JE, Rudan I, Campbell H, Cibulskis R, Li M, Mathers C, Black RE (2012) Child health epidemiology reference group of WHO and UNICEF. Global, regional, and national causes of child mortality: an updated systematic analysis for 2010 with time trends since 2000. Lancet 379(9832):2151–2161. https://doi.org/10.1016/S0140-6736(12)60560-1. Epub 11 May 2012. Erratum in: Lancet. 13 Oct 2012. 380(9850):1308. PMID: 22579125
10. Prüss-Üstün A, Bos R, Gore F, Bartram J (2008) Safer water, better health: costs, benefits and sustainability of interventions to protect and promote health. World Health Organization, Geneva
11. Balazs CI, Ray I (2014) The drinking water disparities framework: on the origins and persistence of inequities in exposure. Am J Pub Health 104(4):603–611
12. Yu W, Bain RE, Mansour S et al (2014) A cross-sectional ecological study of spatial scale and geographic inequality in access to drinking-water and sanitation. Int J Equity Health 13:113. https://doi.org/10.1186/s12939-014-0113-3
13. Water Footprint Network (2017) Product gallery. Water Footprint Network. http://waterfootprint.org/en/resources/interactive-tools/product-gallery/. Accessed 1 Mar 2021
14. Mekonnen MM, Hoekstra AY (2021) The green, blue and grey water footprint of farm animals and animal products, vol. 1: main report. Table 6, No. 48. UNESCO-IHE, p 29. Accessed 1 Mar 2021. http://waterfootprint.org/media/downloads/Report-48-WaterFootprint-AnimalProducts-Vol1.pdf
15. Payero J, Nafchi A, Davis R, Khalilian A (2017) An Arduino-based wireless sensor network for soil moisture monitoring using decagon EC-5 sensors. Open J Soil Sci 7:288–300. https://doi.org/10.4236/ojss.2017.710021
16. Jannat F et al (2019) A proposed unmanned and secured nursery system for photoperiodic plants with automatic irrigation facility. In: Corrales J, Angelov P, Iglesias J (eds) Advances in information and communication technologies for adapting agriculture to climate change II. AACC 2018. Advances in intelligent systems and computing, vol 893. Springer, Cham. https://doi.org/10.1007/978-3-030-04447-3_9
17. Averyt K, Fisher J, Huber-Lee A, Lewis A, Macknick J, Madden N, Rogers J, Tellinghuisen S (2011) Freshwater use by U.S. power plants: electricity's thirst for a precious resource. (2011). Report of the energy water in a warming world initiative, union of concerned scientists

18. Diehl TH, Harris MA (2014) Withdrawal and consumption of water by thermoelectric power plants in the United States, 2010. U.S. geological survey scientific investigations report 2014–5184, 28 pp. https://doi.org/10.3133/sir20145184
19. Spang ES, Moomaw WR, Gallagher KS, Kirshen PH, Marks DH (2014) Multiple metrics for quantifying the intensity of water consumption of energy production. Environ Res Lett 9:105003. https://doi.org/10.1088/1748-9326/9/10/105003
20. Vaseashta A (2013) Emerging sensor technologies for monitoring water quality. In: Van der Bruggen (ed) Applications of nanomaterials for water quality, pp 66–84. https://doi.org/10.4155/ebo.13.208
21. Vaseashta A (2009) Nanomaterials for chemical—biological—physical integrity of potable water. In: Václavíková M, Vitale K, Gallios GP, Ivaničová L (eds) Water treatment technologies for the removal of high-toxicity pollutants. NATO science for peace and security series C: environmental security. Springer, Dordrecht. https://doi.org/10.1007/978-90-481-3497-7_1.
22. Nicolescu B (2021) Methodology of transdisciplinarity. Accessed 20 Feb 2021. www.academia.edu/14441459/methodology_of_transdisciplinarity
23. Barbier R (1977) La Recherche-Action Dans L'institution Educative. Paris. Gauthier-Villars; Da. (1977), (262p.). Abs. Angl.; Bibl. Dissem. (Hommes Organ.). http://pascal-francis.inist.fr/vibad/index.php?action=getRecordDetail&idt=PASCAL7850273475
24. Piaget J (1972) The epistemology of interdisciplinary relationships. In: Centre for Educational Research and Innovation (CERI). Interdisciplinarity: problems of teaching and research in universities, pp 127–139. Organisation for Economic Co-operation and Development, Paris, France
25. Sawka MN, Cheuvront SN, Carter R (2005) Human water needs. Nutr Rev 63(6.2):S30–9. https://doi.org/10.1111/j.1753-4887.2005.tb00152.x. PMID: 16028570
26. Vaseashta A (2013) Ecosystem of innovations in nanomaterials based CBRNE sensors and threat mitigation. In: Vaseashta A, Khudaverdyan S (eds) Advanced sensors for safety and security. NATO science for peace and security series B: physics and biophysics. Springer, Dordrecht. https://doi.org/10.1007/978-94-007-7003-4_1
27. Vaseashta A, Khudaverdyan S, Tsaturyan S, Bölgen N (2020) Cyber-physical systems to counter CBRN threats—sensing payload capabilities in aerial platforms for real-time monitoring and analysis. In: Petkov P, Achour M, Popov C (eds) Nanoscience and nanotechnology in security and protection against CBRN threats. NATO science for peace and security series B: physics and biophysics. Springer, Dordrecht. https://doi.org/10.1007/978-94-024-2018-0_1.
28. Lefers R, Maliva RG, Missimer TM (2015) Seeking a consensus: water management principles from the monotheistic scriptures. Water Policy 17(5):984–1002
29. Luh J, Baum R, Bartram J (2013) Equity in water and sanitation: developing an index to measure progressive realization of the human right. Int J Hyg Environ Health 216(6):662–671

Ashok Vaseashta (M'79-SM'90) received Ph.D. in Materials Science and Engineering (minor in Electrical Engineering) from Virginia Polytechnic Institute and State University, Blacksburg, Virginia, USA. He is Executive Director of Research for International Clean Water Institute in VA, USA, Chaired Professor of Nanotechnology at the Academy of Sciences of Moldova and Professor, Nanotechnology and Biomedical Engineering at the Faculty of Mechanical Engineering, Transport and Aeronautics at the Riga Technical University. Prior to his current position, he served as Vice Provost for Research at the Molecular Science Research Center in Orangeburg, South Carolina. He served as visiting professor at the 3 Nano-SAE Research Centre, University of Bucharest, Romania and visiting scientist at the Helen and Martin Kimmel Center of Nanoscale Science at the Weizmann Institute of Science, Israel. He served the U.S. Department of State in two rotations, as strategic S&T advisor and U.S. diplomat. His research interests span nanotechnology, environmental/ecological science, and safety and security. His research on nanotechnology has been on improving the understanding, design, and performance of nanofibers and sensors/detectors, mainly for applications such as wearable electronics, target drug delivery, detection of biomarkers and toxicity of nano and xenobiotic materials. In the security arena, he has worked on counterterrorism, countering unconventional warfare and hybrid threats, critical-Infrastructure protection, biosecurity, dual-use research concerns, and mitigating hybrid threats. He has authored over 250 research publications, edited/authored eight books on nanotechnology, and presented many keynotes and invited lectures worldwide. He serves on the editorial board of several highly reputed international journals. He is an active member of several national and international professional organizations. He is a fellow of the American Physical Society, Institute of Nanotechnology, and the New York Academy of Sciences. He has earned several other fellowships and awards for his meritorious service including 2004/2005 Distinguished Artist and Scholar award.

Applied Geology and Geoinformatics for Ground Water Exploration, Protection and Management

Konstantinos Papatheodorou

Abstract Groundwater protection leading to a sustainable management of the resource is greatly supported by the use of groundwater monitoring systems. The respective conceptual models representing the groundwater cycle as well as the quality of data input to those systems, defines their efficiency. The chapter is based on previous successful case studies regarding groundwater exploration for resource allocation, vulnerability and risk assessment, management and protection using groundwater information systems. Key outputs show that it is feasible to: (i) identify and accurately map water flow paths and plan efficient exploration while at the same time, identifying and delineating recharge areas for protection; (ii) successfully assess vulnerability and risk using existing models and plan protection from surface pollution/contamination; (iii) use geo-informatics science and technologies to acquire data and information at the required level in order develop reliable conceptual groundwater models, to support groundwater protection and management from the aquifer to consumption and reuse; (iv) protect groundwater throughout its cycle by making informed decisions based on developing groundwater monitoring systems for early warning and management.

Keywords Groundwater exploration · Groundwater management · Remote sensing · Lineament mapping · Groundwater conceptual model

1 Introduction

Water resources are under a gradually growing pressure due to climate change and in support of economic growth, which build continuously increasing demands and use, leading to overexploitation, contamination and pollution [68]. Groundwater has always been a key factor for sustainable development since it offers a number of advantages as compared to surface water; quality and extraction near the point of

---

K. Papatheodorou (✉)
Applied Geology and Geoinformatics, Surveying and Geoinformatics Engineering Department, International Hellenic University, Serres University Campus, 62124 Serres, Greece
e-mail: conpap@ihu.gr

use, being among the most important ones. Groundwater is usually connected to surface water so its quantity and/or quality can be affected and at the same time, it has an impact on related surface water and terrestrial ecosystems [18, 31]. For those reasons, the preservation of groundwater as a natural resource of great importance is absolutely critical.

In the European Union (EU), the legal framework towards the sustainable use of water resources has already been defined over a number of European Commission Directives based on a set of principles [17, 47] which imply for solid scientific knowledge, well informed decisions, focusing on prevention, remediation at the source and transferring the cost of remediation to polluters. It also implies for water monitoring of parameters related to water safety in terms of quantity and quality as identified by the European Environmental Agency [54].

Groundwater sustainable use requires interventions on three major targets: (i) exploration for identifying available resources, (ii) protection and (iii) management with the same essential requirement; the physical model describing the groundwater regime in the specific area [18] from rainfall, through the aquifer system to the abstraction points and its use. Actions towards achieving those targets include a grid of interlinked activities (Fig. 1) which demand a multi-disciplinary approach to develop the groundwater physical (conceptual) model and then, based on it, to develop the monitoring targets and requirements, define the operational and technological challenges and then design and build a groundwater monitoring system. To that end, there has to be an integration of information regarding numerous parameters related to groundwater interaction with other environmental factors affecting the groundwater regime over a specific study area and this is the main target of the "exploration" stage.

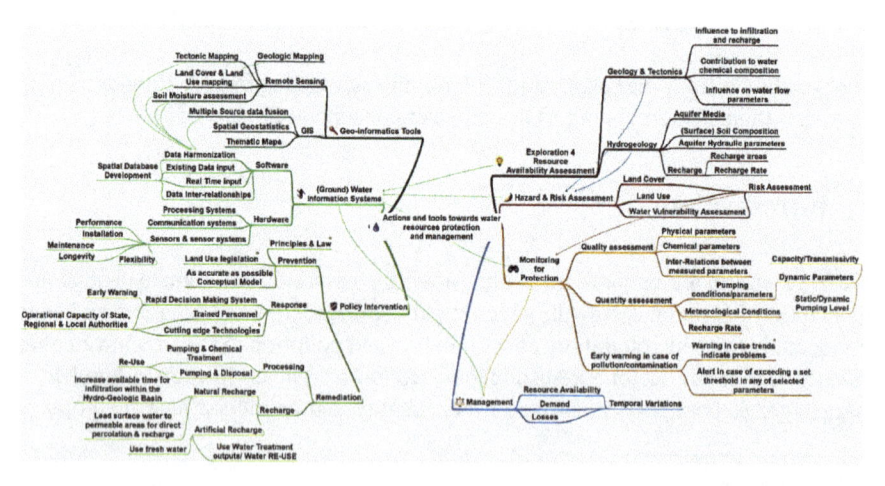

Fig. 1 A Schematic of interlinked groundwater protection and management related tasks and requirements

1.1 Groundwater Conceptual Models

Understanding all the involved parameters related to the groundwater flow, to its dynamic, physical and chemical parameters as well as to their potential temporal variations and their interaction with aquifer media (chemical transfer, and biological functioning of the water body), with surface water bodies and terrestrial eco-systems, provides the ability to build the respective to the specific area groundwater conceptual model (GCM), which is essential for developing an efficient monitoring system [34, 44] valid only for the specific area it has been developed for. Such a system must also include groundwater pressure assessment taking into account all possible pressures, hazards and risks (Fig. 2). In essence, the GCM includes the characteristics that define the nature of the aquifer system including its physical, dynamic, hydraulic and chemical characteristics and provides capability for assessing the potential consequences of pressures, taking always into consideration the existing legislation regarding any of its components [20, 32, 34, 60, 76].

Since there is an indirect access to the aquifer system where physical interactions related to groundwater take place, justifiable assumptions and/or generalizations have to be made. These impose a level of uncertainty, which can be reduced when actual measurements are used to calibrate the conceptual model in an iterative way and adapt it to local hydrogeologic conditions.

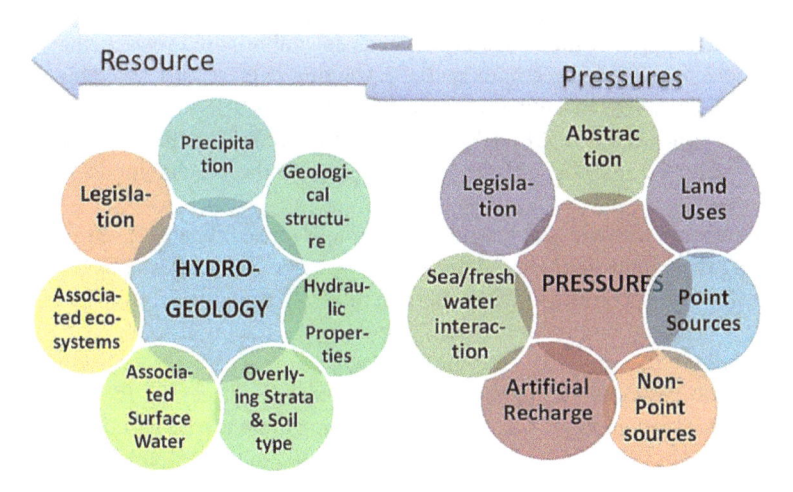

Fig. 2 Schematic representation of information required for building a groundwater conceptual model taking into consideration various parameters affecting resource sustainability and the applied pressures (after Papatheodorou 2018)

1.2 Remote Sensing for Groundwater Exploration

The volume and quality of data and information needed to build an efficient GCM can be a limiting or even blocking factor when considering costs in terms of funds, effort and especially time. The adverse effects of these factors can be greatly reduced with the use of geo-informatics science and technologies [44, 55]. Remote sensing provides the means to acquiring reliable land cover data and information with minimal costs in very little time as compared to classic, geologic mapping techniques. Geology and tectonics play a critical role in the groundwater regime, especially in areas where theoretically impermeable geologic formations outcrop. These are usually mountainous areas, where water scarcity is an important issue that blocks development. Nevertheless, there are specific locations where rain water percolates into the ground through rock fractures and especially fractured high-permeability zones, which allow water to percolate into the ground and accumulate thus forming exploitable aquifers. Considering the importance of those aquifers for such an area and their high vulnerability from surface pollution due to their recharge scheme, the identification and delineation of their recharge areas is essential for their protection.

Impermeable rock formations are easy to identify both using maps and in situ; the problem that exists is related to mapping fractured zones in those rocks and then at a second stage, to identifying water actual flow paths through them. Field mapping of large rock fractures is an extremely demanding, costly and difficult task but it can be greatly supported by using remote sensing science and technologies. Since the length of those outcropping features usually exceeds several hundred meters, they can be traced and mapped using medium spatial resolution data, readily available for free [15, 57, 67]. Open source software and freeware (QGIS [49], SAGA GIS [9], GRASS GIS, the Sentinel Toolbox [58], ESA SNAP, Multispec© [7], ORFEO, OSSIM) can also be used for the analysis [23], thus making the use of these technologies, extremely accessible.

Geologic and tectonic mapping using remotely sensed data is based on various processing techniques including the use of false color composites (FCC) for visual interpretation of spectral and geometric characteristics and the use of band ratios. A number of band combinations to develop FCC for tracing lineaments have been proposed [29, 53].

Faults and large fractured zones can also be traced by seeking for their impact on the outcropping geologic formations so mapping them can be based on detecting and mapping alteration minerals which are related to faulting [6, 28]. Mineralogical content variations of rocks and soils are reported to be best seen in the SWIR part of the electromagnetic spectrum [6, 12, 72].

Based on the fact that moisture absorbs electromagnetic energy more drastically towards the infra-red part of the electromagnetic (EM) spectrum (Fig. 3), the combination of SWIR I and SWIR II bands with respective bands towards its visible part, can provide information regarding soil moisture content provided that the recording system has the required spectral and the radiometric resolution measurement capabilities.

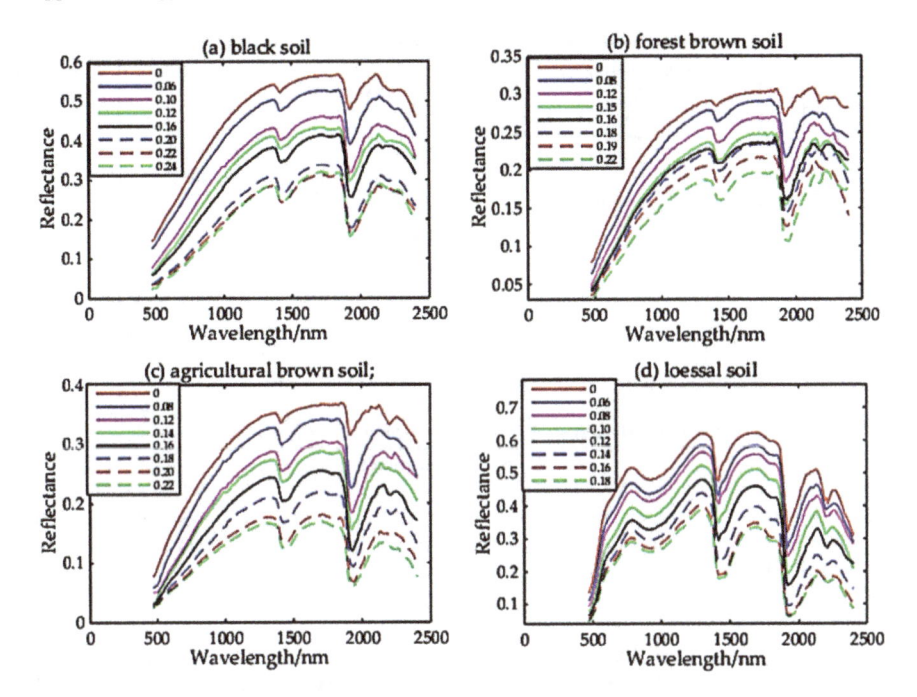

Fig. 3 The higher impact of soil moisture towards the NIR and SWIR part of the E/M spectrum (wavelengths greater than 700 μm) can be seen [30]. Reflectance spectra for four different soil types: **a** black soil; **b** forest brown soil; **c** agricultural brown soil; and **d** loessial soil (after Jing Yuan et al. 2019)

Band ratios are the means of such a correlation, and they provide outputs where topography and sun angle effects are weakened or completely removed thus producing a much greater contrast between geologic formations. Various band ratios have been proposed for tracing lineaments in mafic and ultramafic rocks which can also be related to large fractures [36, 62]. A similar approach can be used to indicate fractured zones using soil/rock moisture assessment [8, 14, 35]. False color composites using spectral band ratios have also been used (R-G-B) including [SWIR I]/[SWIR II]-[SWIR I/NIR]-[Red/Blue] as the mineral composite [36] and [Red/Blue]-[NIR/Red]-[SWIR I]/[SWIR II] to identify altered zones [38, 65].

Since fractures in rocks permit water infiltration, rock weathering in those areas is much faster than in any other part of the same rock. Soil is developed faster, forming a thicker layer and the presence of water helps vegetation to grow faster and remain more vigorous. For that reason, the presence of geologic structures affecting surface and groundwater flow may be inferred from observations related to vegetation cover changes. These become more evident towards the near infra-red (NIR) and short wave infra-red (SWIR) part of the electromagnetic spectrum, due to the respective vegetation spectral response changes [10, 11, 74].

Near infra-red reflections from leaves result from chlorophyll at the leaf surface whereas SWIR reflections originate from internal parts of the leaves [61], thus being

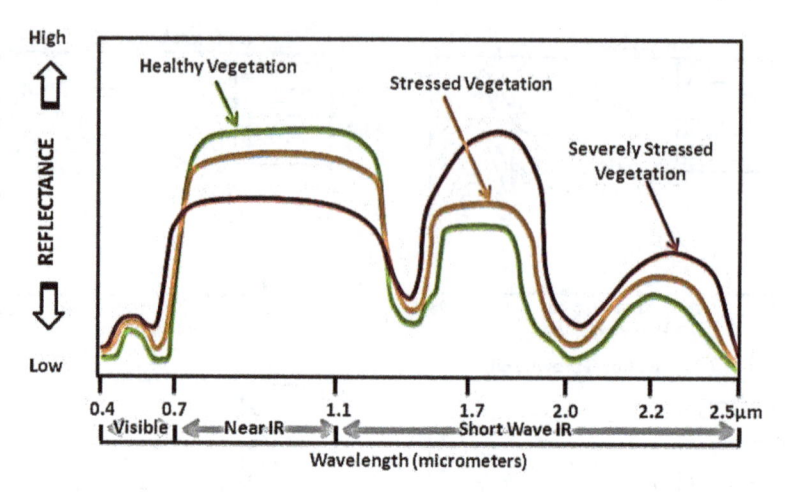

Fig. 4 A schematic representation of the vegetation health impact on its respective Spectral Response. Greatest differences of healthy versus stressed vegetation are located in the NIR and SWIR part of the Spectrum [39] (*Source* http://rst.gsfc.nasa.gov/)

influenced by leaf moisture content. For the same vegetation species, stressed vegetation shows a lower NIR and a comparatively higher SWIR reflectance as compared to healthier and more vigorous vegetation (Fig. 4).

Numerous vegetation band combinations based on the spectral response of most vegetation species have been suggested as indicators of plant physiology changes including the normalized difference vegetation index (NDVI); the soil adjusted vegetation index (SAVI), the ratio vegetation index (RVI), the photochemical reflectance index [22, 25, 75]. Since vegetation is the subject of investigation, spring or summer scenes are the best to provide this kind of information.

Recorded reflectance data (digital numbers; DN) do not provide a direct measure of the actual vegetation health status but instead, they have a meaning as comparative values so, band ratios represent in a more meaningful way the differences in vegetation health status. Moreover, in order to enhance the effect of vegetation physiology changes on their respective response spectra, spectral band ratios between NIR and SWIR to visible light spectral bands (blue, green and red visible light) can be used [35, 69] since vegetation physiology changes, cause smaller spectral response changes in the visible part of the electromagnetic spectrum (Fig. 4). In addition, the effect of morphology shading is also reduced thus providing a "cleaner" output which is easier to interpret. For those reasons, band ratios which can be used to trace lineaments corresponding to rock fractures include the: [NIR]/[SWIR], [NIR]/[Red] and [Green]/[Red]. In all these relations, a higher result indicates a comparatively better vegetation health status. These spectral band ratio outputs can be combined in order to enhance the respective indications so, they can be multiplied taking into consideration their respective meaning. Greater [NIR]/[SWIR], [NIR/[Red], [Green]/[Red] ratio values indicate healthier vegetation and in order to combine those outputs and enhance the discrimination potential of this approach, the ratio multiplication product can be used ([NIR]/[SWIR] x [NIR/[Red] x [Green]/[Red]. Case studies show that it is feasible to locate and map fractured zones which permit groundwater accumulation and flow, even in impermeable rocks [45, 64, 73] using the aforementioned band ratios and their combinations.

Taking into consideration the size of fractured zones and especially their width which can reach tenths of meters (given that their length extends for several hundreds of meters or even ilometres), consistent differences can be shown even on medium spatial and radiometric resolution data. In general, remote sensing data with spatial resolutions of about half the width of the investigated features would be a good choice. Towards that end, pan sharpening techniques to improve the spatial resolution of multispectral data can greatly help.

As there are many landforms associated with faults, fault scarps, intersections of bedding and topography as well as other man-made linear features (roads, railroads, channels etc.), additional information needs to be considered in order to remove non-tectonic linear features and consider only information related to outcrop geology as for instance, displacements of geologic formations, to finally interpret lineaments traced as rock fractures.

Another means of indicating the reliability of interpreting traced lineaments as fractures and/or faults is their comparison to the known tectonic regime of the area under investigation. Comparison of rose diagrams that relate the lineament analysis results to the tectonic history of the area may add to evaluating the credibility of results [48, 69] as can lineament classification according to their estimated influence on the groundwater regime. Lineament influence can be deducted from their geometry (length and width) so their hydrogeological function can even be inferred [12, 72].

Areas where lineaments intersect are most significant to groundwater exploration as they are associated with intensively fractured zones with considerable storage capacity. They may also correspond to fault displacements where one fault acts as groundwater conduit which is interrupted by being displaced due to the second fault. In such a case, groundwater may be accumulated in this zone.

Lineament density maps have also been suggested as indicators of the possible groundwater potential of an area [2, 6, 27, 66]. This has been shown [45] to be an insufficient approach when the target is to define exact locations for boring. In those cases, the role of the tectonic network of the area as potential water flow network has to be assessed [43, 45, 73] considering both its geometry and its physical characteristics including the type of bedrock, the type of filling material and extend.

1.3 Verification of Remote Sensing Outputs

Remote Sensing outputs based assumptions regarding the groundwater regime, can be interpreted and verified by ancillary information related to the groundwater dynamic, physical and chemical properties. This kind of information can be retrieved from sampling points including borings, springs and water wells.

Environmental tracers have the potential to provide flow related information in semi-confined, faulted aquifer systems, provided that there's a high density of sampling points available [39, 42] and the same stands for chemical analyses data which can also provide valuable information regarding the groundwater physical and chemical parameters as well as its characteristics which result from its interaction with the aquifer media and thus they can help define its origin and residence time in the aquifer [5, 26, 33]. Ion ratio [Mg/Ca], can be used to indicate groundwater flow through limestones when ([Mg/Ca] < 0.7), dolomites ([Mg/Ca] 0.7–0.9) or ophiolitic formations ([Mg/Ca] > 1.0) ([40]). Ion ratio [(Ca + Mg)/(Na + K)] provides information about groundwater's residence time in the aquifer and indicates if a part of the aquifer is under continuous groundwater flow [26].

1.4 Groundwater Vulnerability and Risk Assessment

Groundwater protection and preservation as a resource, strongly depends on the balance between availability, consumption and potential threats which may lead to the deterioration of its quality. Groundwater risks are related to both natural and anthropogenic reasons. Seawater intrusion or the establishment of hydraulic conductivity of fresh water aquifers with confined old waters with poor quality and dissolution of hazardous, naturally occurring substances due to groundwater level fluctuations or flow direction changes. Even when water quality deterioration is due to natural reasons, the "triggering" factor in many cases, is anthropogenic: seawater/fresh water transition zone enters land due to over exploitation/abstraction of fresh water (a common case in coastal areas especially during summer) causing sea water intrusion; drilling through confined lens-shaped brackish water aquifers can establish a hydraulic conductivity between them and fresh water aquifers, thus affecting water quality [37].

Numerous groundwater vulnerability from surface pollution assessment approaches have been proposed and any of them could fit the existing in a specific area conditions, constrains and problems and provide reliable and accurate enough outputs to support decisions. Methods proposed include PI [24], ERIK [13], aquifer vulnerability index [69], relating Nitrogen sources and aquifer susceptibility to nitrate approach [41] and DRASTIC [3, 4, 52, 65] and its modifications [50, 51, 56, 75, 76]. Local geological, tectonic and hydrogeological conditions and potential hazards define at large the suitability of any model for an area.

Important parameters considered include the permeability and composition of the surface layers to the aquifer, the aquifer recharge zone and recharge rates and the aquifer hydraulic parameters. The more accurately the spatial distribution of those parameters is defined for the area of interest, the less the uncertainty of the final output. Uncertainty in groundwater vulnerability assessment is closely related to the quality of input information provided by the respective hydrogeologic conceptual model of the area and can be greatly reduced during the operation stage of a monitoring system as new information is acquired.

Groundwater pollution risk assessment by both point and non-point pollution sources is the next step towards groundwater protection. Potential groundwater non-point pollution sources and pressures, which present a high temporal variation (agriculture for instance) should be assessed on a temporal, year-round basis, taking into consideration the respective groundwater recharge parameters which also vary with time. Potential contaminants/pollutants and the groundwater abstraction scheme over a hydrologic year, should also be linked to specific cultivation practices.

On the other hand, potential point-pollution sources must be identified and spatially registered. Their operational requirements in terms of water demands, type and composition of produced wastes and their disposal procedures should be registered for each source.

Even, non-polluting pressures which may have a significant impact on pollutant dispersion within the aquifer system must be considered including various discharges like water pumping sites, springs or water wells which play a critical role on groundwater flow direction and velocity, always in relation to the aquifer hydraulic parameters, recharge rate and capacity.

2 Methodologies Implemented for Groundwater Protection

2.1 Remote Sensing Applications to Groundwater Protection and Management

Groundwater exploration can be supported by remote sensing especially when investigating rock formations as shown by the results of a research carried out in the wider area of Vertiskos mountain, Greece (Fig. 6). Landsat TM and ETM+ (Landsat 7) scenes 184/31 and 184/32 recorded during May (5.05.2000) and June (26.06.2001) were used to trace lineaments and map fractures of the theoretically impermeable formations of the area [70]. Data processing included band ratios ([NIR/Red], [SWIR-I/Blue], [SWIR I/Red] and [SWIR II/Red]; false color composites [SWIR II—Red -Blue], [SWIR II—NIR—SWIR I], [NIR—SWIR II -Blue] and principal component analysis (PCA) which was used to detect low frequency entities in the area. [PCA.4] and [PCA.5] were found to provide particularly useful information.

Lineaments mapped were filtered using the road and the channel networks in order to remove manmade structures irrelevant to rock fractures. Moreover, tectonic and non-tectonic lineaments corresponding to bedding outcrops were also filtered out.

Lineaments mapped were classified according to their length in three classes; 200–1000 m, 1001–3000 m, 3001–6000 m and lineaments with a length greater than 6000 m (Fig. 5). They were statistically processed and compared to the tectonic regime of the area [42, 59, 71].

Comparison with available tectonic investigation data showed that lineament orientations are in full compliance with the tectonic regime of the area. The tectonic

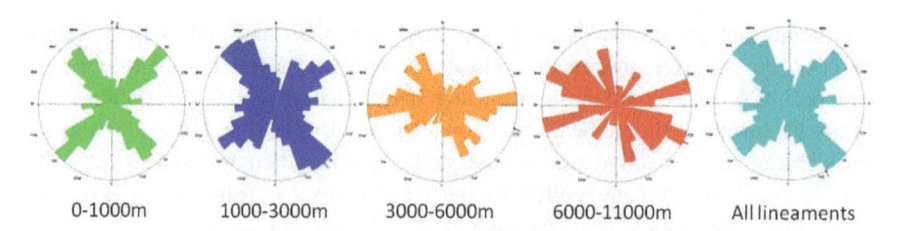

Fig. 5 Rose diagrams of lineament orientations classified according to their length. Vertiskos mount area, Northern Greece (after Papatheodorou et.al, 2018). Longer lineaments (6–11 km) correspond to fractures with an almost E-W and to NW_SE strike. Lineament data were found to fit to the tectonic regime of the area

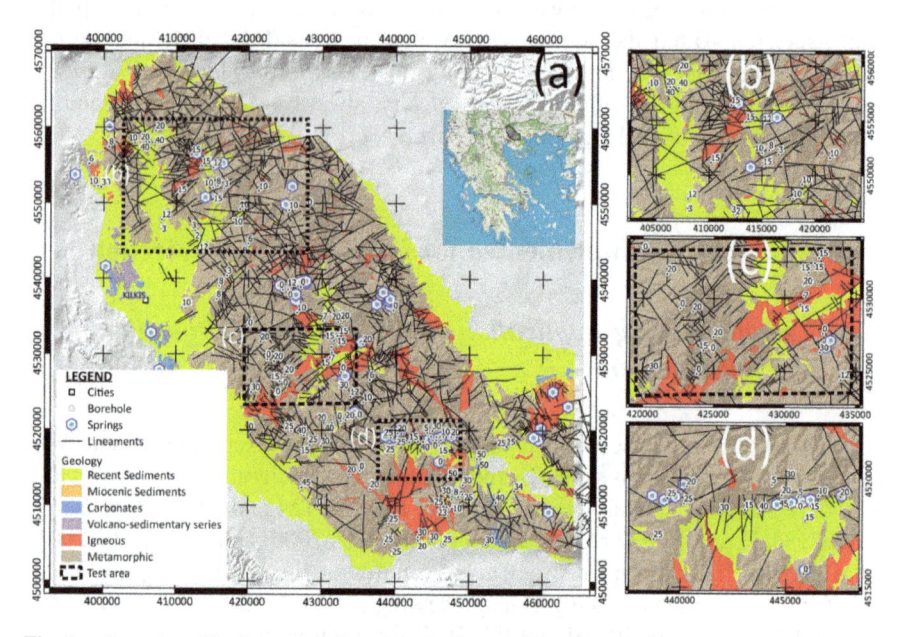

Fig. 6 **a** Location of borings plotted against the lineament map (numbers indicate water yields in m^3/h). Vertiskos mount area, Northern Greece ([45]). Zoomed parts **b**, **c** and **d** indicate the presence of tectonic fractures running through the impermeable rocks as high yield boreholes and springs are located at or near the intersections of traced lineaments

density map of the area was also developed, to provide indications regarding potentially higher groundwater capacity zones.

To investigate the role lineaments play on the groundwater regime, groundwater data from 120 borings and 42 springs were additionally considered (Fig. 6) to develop the respective hydrogeologic conceptual model.

As it finally appeared, higher yield borings are geographically related to tectonic intersections rather than be located in high tectonic density areas [45], a fact suggesting that this approach can lead to more efficiently defining the best locations for constructing water wells. On the other hand, tectonic density maps can be used to indicate areas with a relatively higher permeability where rain water can percolate and reach groundwater aquifers so in those areas, there's a high vulnerability of groundwater and an increased risk for contamination/pollution from surface [21, 76].

2.2 Remote Sensing and Chemical Analyses Data to Detect Groundwater Flow

The western parts of Thessaly plain in Greece (Fig. 8) has always been an intensively cultivated area. Due to the higher demand during the last decades which led to groundwater over-exploitation serious problems emerged including a lowering of water level and settlement due to aquifer consolidation. Two main aquifer systems are developed in the area; one is developed in the carbonate formations of Koziakas mountain to the east and another, within the alluvial sediments that fill the basin of Trikala plain. Previous investigations on the groundwater regime of the area [80] suggested that there is very small or not at all hydraulic connection between the two aquifer systems due to the presence of an ophiolitic complex that forms an impermeable barrier, separating those formations (Fig. 8a and b).

Groundwater recharge in the aquifers developed in Trikala basin is therefore suggested to take place directly from surface water percolation through the permeable surface formations [30]; a fact which does not fully agree with the groundwater balance in the area. Another important consideration is that, based on the assumption that the ophiolitic complex is practically impermeable, a number of potentially hazardous to groundwater land uses could be installed in the area where these formations outcrop.

To investigate the potential role of the ophiolitic complex in the groundwater regime of the area, a remote sensing investigation was applied. Landsat TM scene data were processed using band ratios [NIR/Red]; [SWIR II/NIR/4)] and [SWIRR II/Red] [48], principal component analysis (PCA), and the development of false color composites for visual interpretation (R-G-B): [PCA.2]-[NIR/SWIR II]-[NIR/Red], [NIR/SWIR II]-[NIR]/Red]-[NIR] [43]. Lineament statistical analysis and lineament density maps created were also considered in order to indicate the tectonic regime of the area.

Additional information related to outcrop geology as for instance, displacements of geologic formations (Fig. 7) were considered in order to finally interpret lineaments traced as rock fractures and/or faults.

To verify groundwater flow through the ophiolitic complex, groundwater chemical analyses data from 60 sampling points in the wider area, collected during the same hydrologic year, were used as indicators of water origin and residence time in the sediments of the plain [26]. Ion ratio [Mg/Ca], was used to indicate groundwater flow

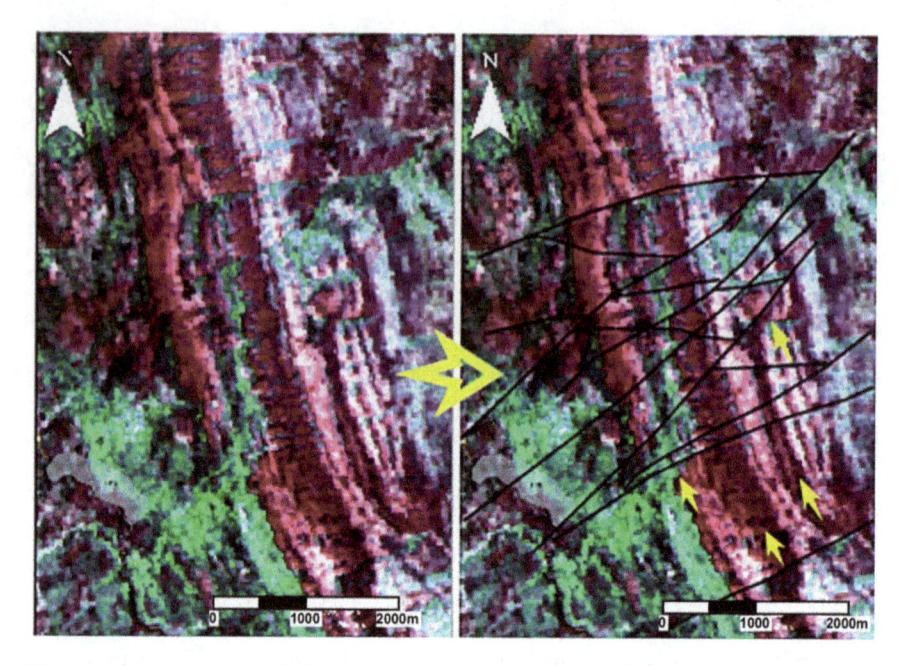

Fig. 7 A false color composite made by using Landsat TM scene band ratios (R-G-B): [SWIR I]/[Red]—[NIR/Red]—[SWIR II/Blue] and the result of the interpretation with a number of lineaments already mapped (right). The outcropping geologic formations are limestones and chert (after [43])

through limestones ([Mg/Ca] < 0.7), dolomites (0.7–0.9) or ophiolitic formations ([Mg/Ca] > 1.0). Ratio values in the area range from 0.1 to 5.39 with the greater values measured towards the western edge of the plain, a fact indicating groundwater inflow from the ophiolitic formations towards the plain sediments (Fig. 8c).

Ion ratio [(Ca + Mg)/(Na + K)] values range from 0.5 towards the center of the plain, a value indicative of high residence time, to 5.0 at specific locations at the western edge of the plain, indicating continuous groundwater flow (Fig. 8d). As is evident (Fig. 8c and d) ion ratio value spatial distribution combined with the lineament and the lineament density maps, provide strong indications of groundwater inflow towards the basin sediments and main recharge areas can be defined both within the ophiolitic complex and inside the plain sediments.

Based on these indications, the recharge area, which contributes to the recharge of the exploitable aquifers inside the plain can be delineated and protected. Additionally, the same area could also be considered for artificial aquifer recharge if needed.

These new findings helped develop the conceptual model of the area which formed the basis for the development of a prototype, groundwater information system [44, 47] to monitor various physical and chemical parameters, capable of providing forecasts and early warnings in case set thresholds are exceeded. The system is designed in a way that selected water physical and chemical parameters can be constantly or

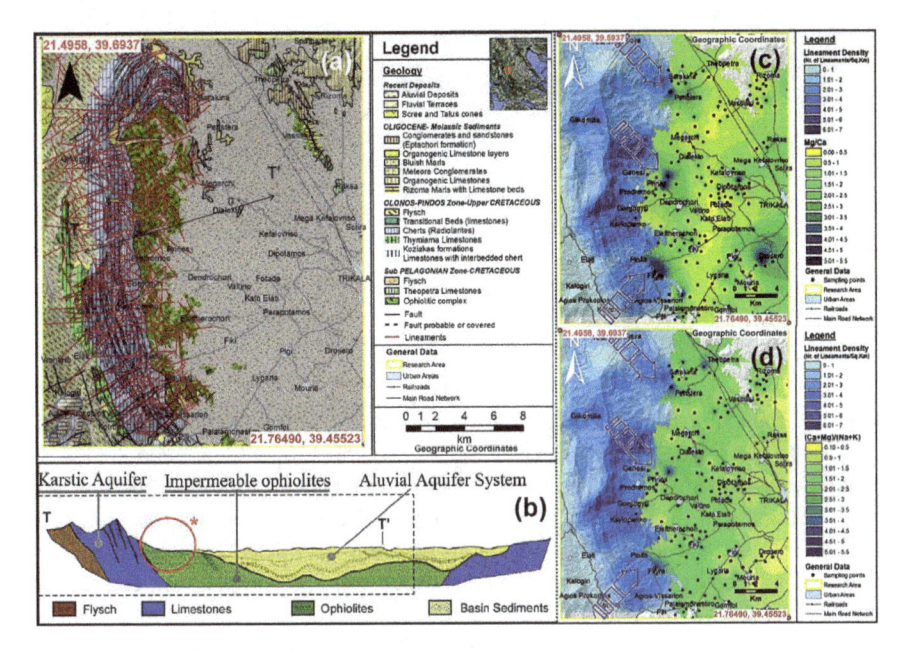

Fig. 8 a Geological map and schematic geological cross section **b** showing the impermeable ophiolites (green formations) separating and hydraulically isolating the limestones from the alluvial deposits. Red lines correspond to lineaments mapped using Remote Sensing data. **b** Geologic cross section TT'. **c** [Mg/Ca] ion spatial distribution. **d** [(Ca + Mg)/(Na + K)] values spatial distribution. Lineament density shown as shades of blue. Arrows indicate the groundwater flow direction through the rocks towards the alluvial deposits

selectively measured and data can be instantly transferred by using telecommunication technology. The system has a multi-tier architecture consisting of three main tiers (layers): (i) data collection, which ensures all possible data inputs including data over GRPS/LTE networks; (ii) services/application, which includes the database and assisting services, WebGIS applications, groundwater simulation models, content management and user access control and; (iii) the presentation tier which provides custom outputs tailored to user demands (Fig. 9). The developed prototype was populated with data from more than 300 sampling points (mostly borings) in the area.

2.3 Groundwater Vulnerability Assessment

Groundwater is subjected to pressures from various natural and anthropogenic sources which may have as result, the deterioration of its quality. For that reason, vulnerability against pollution from ground surface must be carefully considered when planning for groundwater protection.

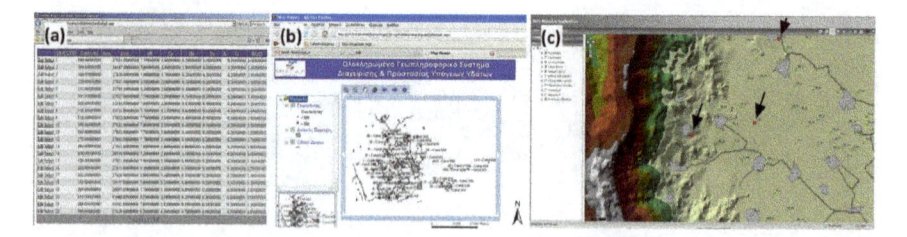

Fig. 9 Screenshots from a prototype groundwater management system: **a** Data input (tabular data are shown); **b** sampling points/potential monitoring stations; **c** alarm (red points) appearing on exceedance of user defined thresholds. (after Papatheodorou et al. 2008)

From the numerous groundwater vulnerability assessment methodologies suggested, the DRASTIC method [1, 4, 16, 46, 50, 56] was applied to assess groundwater vulnerability in an intensively cultivated area in Northern Greece where there are numerous installations producing potentially hazardous to the environment wastes. The methodology takes into consideration parameters related to aquifer hydro-geology, all of which are also considered when developing the conceptual model of an area.

DRASTIC is in fact, the acronym of D(epth to groundwater), R(echarge rate), A(quifer media), S(oil media), T(opography), I(mpact of vadose zone) and C(onductivity of the aquifer). The method involves the calculation of the DRASTIC Index (DI) as:

$$\text{DRASTIC Index} = Dr \times Dw + Rr \times Rw + Ar \times Aw + Sr \times Sw + Tr \times Tw + Ir \times Iw + Cr \times Cw,$$

where "r" denotes the weight rating per parameter ($Dr = 5$; $Rr = 4$; $Ar = 3$; $Sr = 2$; $Tr = 1$; $Ir = 5$; $Cr = 3$).

DRASTIC was modified to Pesticide DRASTIC [16, 52] to "focus" on nitrate pollution (weight ratings per parameter changed to: $Dr = 5$; $Rr = 4$; $Ar = 3$; $Sr = 5$; $Tr = 3$; $Ir = 4$; $Cr = 2$) and further modified (Modified DRASTIC) to incorporate land use [56, 63], using the equation:

$$MDI = DI + Lr \times Lw,$$

where DI is the DRASTIC index, Lr the Land use parameter and Lw the land use parameter weight.

Data from a total number of 350 boreholes were used to assess respective DRASTIC parameters whereas a high resolution digital elevation model with 5×5 m pixel size provided the "topography" related parameter. Groundwater chemical analyses data from a number of borings in the area, provided information regarding the nitrate concentration in groundwater. The area of implementation covers a systematically cultivated area which extends to Emathia, Pieria and Pella (Northern Greece).

Additionally, a large number of potentially hazardous to the environment activities are also located in the same area and especially in its northern part around the towns of Veria and Naoussa.

The modified DRASTIC index values display the potential hazard from surface pollution and the nitrate values at specific sampling points indicate the reliability of the assessment. It must be pointed out that sampling data were collected during a number of years and moreover, water was sampled from the entire borehole column and not from the freatic aquifer so, analyses data should only be considered as indicative of nitrate pollution. Given that under normal conditions, the deeper, confined aquifers have not been infected to nitrate pollution but still, they contribute to the water column of a boring that is sampled, actual nitrate concentration values in the phreatic aquifers may be much higher. Since a high vulnerability value is also related to a high effective rainwater infiltration, there may be significant variations of nitrate concentration in the phreatic aquifers during the year, depending on temporal variations of rainfall and cultivation practices followed.

Having said that, since nitrate concentration values used were based on data collected throughout a period of several years, a detailed sampling and analysis campaign is necessary to help assess the level of pollution at this point. To indicate the potential role of pollution point sources, a list of anthropogenic activities which could potentially pollute groundwater was compiled (Fig. 10). Listed data include location, type of activity, type of wastes, waste physical and chemical characteristics and additional details (ownership, operational status etc.).

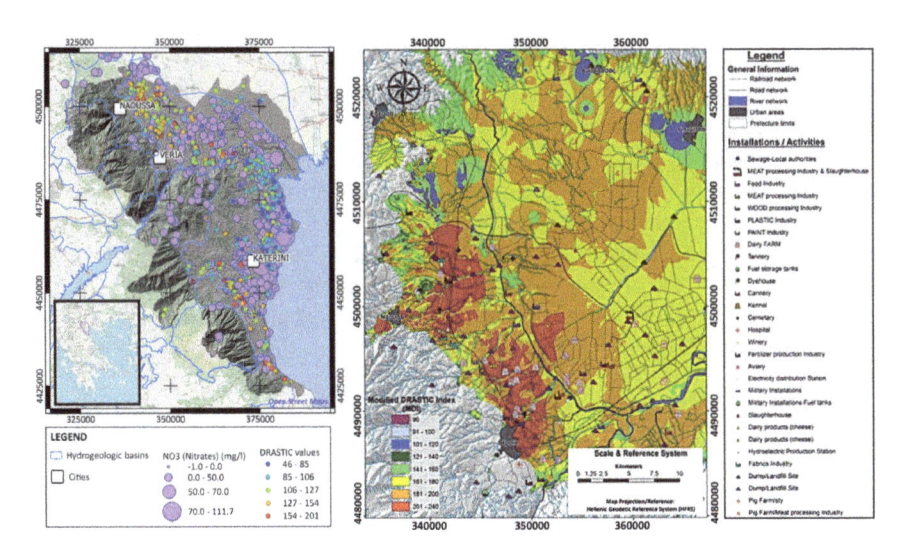

Fig. 10 Groundwater modified DRASTIC values at the sampling locations used, plotted against nitrate (NO^3) concentration (left). Zoom-in to the northern part of the area (right): the location of potentially hazardous to the environment installations plotted against the modified DRASTIC index spatial distribution [46]

The superposition of the location of potential pollution point sources on the vulnerability map reveals the way land uses can affect groundwater quality and indicates the existing potential risks. Using the type of products per installation, the type and composition of wastes and the operational practices and constrains, the groundwater pollution Risk can be assessed and considered when compiling the hydrogeologic conceptual model of the area.

3 Discussion

Planning for groundwater protection and management requires the development of an efficient groundwater monitoring and early warning system [19, 76] which can help achieve the entire set of targets set by the EU Commission according to existing Member States Law. A groundwater monitoring system follows a conceptual model which combines the characteristics of the aquifer system with the potential consequences of pressures, taking always into account the existing legislation regarding any of its components. Such a system must therefore be based on a "tailored-to-case" approach where the respective to location and potential pollution source parameters are continuously monitored and at the same time, it must be dynamic, continuously improving with time as new data and information are obtained.

The efficiency of a groundwater monitoring system strongly depends on the completeness and accuracy of the conceptual model developed to model the aquifer system operation, which can provide well documented information leading to justified decisions regarding "why", "what", "how", "where" and "when" to monitor and in turn, to support planning "how to" design, develop, deploy and maintain an efficient monitoring system with all of its components.

Tracing lineaments to identify which of those correspond to rock fractures permitting groundwater to flow through them, is a demanding task incorporating a high level of uncertainty accumulated from every stage of the respective analysis. Remote sensing outputs can provide information about potentially important locations regarding the groundwater regime thus leading to efficiently selecting the optimal water sampling locations. Groundwater chemical analyses data can then be used to verify the outputs of the remote sensing analysis thus reducing the existing uncertainty.

Groundwater vulnerability assessment methods have been successfully applied but still, the incorporated uncertainty reduces their potential to support decisions regarding land use planning and water protection measures. To increase the reliability and accuracy of these methods, denser, more complete, current and more accurate data input is necessary.

Data availability is a serious concern when planning for groundwater protection and management. Data collection can be a costly, time consuming process, which imposes serious constrains and even blocks attempts towards groundwater systematic monitoring. Geo-informatics technologies and especially remote sensing can be used to acquire important information in little time, with minimal costs.

Data sharing is another important factor which can greatly help overcome the problem but there's always the question of data suitability (reliability, accuracy, completeness etc.) for any specific case. For that reason, the capacity to develop efficient groundwater monitoring systems would be greatly supported by a strategy which includes:

- Development of a common understanding of the entire spectrum of actions towards groundwater protection and management. This common understanding should apply to the monitoring program development team and to the stakeholders including the administration, industry, water companies, farming companies and farmers and of course the public;
- Harmonization of approaches/methods used to collect and manage data.
- Development of open standards for ground water monitoring.
- Development of technical guidance based on best practice cases, to disseminate and enlarge the number of collaborators.
- Sharing competencies to build capacity of researchers. Exchanging experiences brings new ideas, offers more potential solutions and saves time and effort.
- Development of more affordable yet efficient monitoring systems. High maintenance costs (including sensor replacement) have been in many cases, the reason for aborting the use of installed systems. Low cost sensors with minimal maintenance requirements should be designed and produced.
- Systematically assess groundwater vulnerability and pollution risk. Potential pollution sources must be identified and registered together with the physical and chemical characteristics of their respective products and wastes.
- Raising awareness is always a very effective preventive measure. Web based technologies (portals with near real-time information about the status of the resource, the problems, solutions applied etc.) can greatly help towards this scope.

Moreover, the implications of operating such a system have to be considered taking into consideration the impact on social life and on any anthropogenic activity an alarm for water contamination has. Any decision for triggering an alarm must be based on absolutely reliable, cross checked information and should not rely solely on measured values of the monitored parameters since monitoring only upper and lower thresholds of specific parameters is not enough to prove pollution or contamination. Instead, information leading to such a decision must take into consideration the interaction of more parameters related to a potential contamination or pollution. As for any information and communication technologies based system, groundwater monitoring systems are also exposed to internet risks so, security issues also arise and need to be considered.

4 Conclusions

Groundwater exploration can be efficiently supported by geoinformatics science and technologies and groundwater flow paths can be identified and mapped by considering ancillary information. Groundwater recharge zones can be identified and mapped in order to be protected. Furthermore, these zones can be considered as potential locations for artificial groundwater recharge. Uncertainties accumulated during the remote sensing processing and the interpretation stages can be reduced by using ancillary hydrogeologic and hydrochemical information.

In theoretically impermeable hard rock formations, the relation between the location of mapped fractures and higher yield borings shows that exploitable aquifers are geographically related to tectonic intersections. Groundwater pollution vulnerability and risk can be assessed and used to make decisions regarding land use and groundwater protection policies. Vulnerability and risk assessment must take into consideration the information related to potential pollution sources (activities) including the disposal practices they apply on their wastes as well as the physical and chemical characteristics of their products and wastes.

Groundwater protection and management, according adopted EU principles and issued Law, can be achieved if being supported by continuous monitoring of groundwater and aquifer related dynamic, physical and chemical parameters. To that end, reliable hydrogeologic conceptual models, tailored to the specific characteristics of the aquifer system are necessary. Building a sound conceptual model is a critical, multi-disciplinary, complex, time consuming and costly process. The reliability and accuracy of such a model is related to the respective quality of data and information considered. The cost of acquiring data at the required quality level poses a serious problem to efficient groundwater planning.

To improve data availability and thus to support groundwater sustainable management, a strategy to promote data and information sharing is necessary to reduce the operational cost of groundwater monitoring (including system design, development and maintenance). The strategy should include the development of a common understanding of the entire spectrum of actions towards groundwater protection and management, which will lead to the harmonization of approaches/methods, used to collect and manage data and will promote (conditional) data sharing. To that end, the development of open standards for ground water monitoring and technical guidance based on best practice cases can greatly help.

References

1. Al-Adamat RAN, Foster IDL, Baban SMJ (2003) Groundwater vulnerability and risk mapping for the Basaltic aquifer of the Azraq basin of Jordan using GIS, remote sensing and DRASTIC. Appl Geogrv 23, 303–324
2. Al-Djazouli MO, Elmorabiti K, Rahimi A (2020) Delineating of groundwater potential zones based on remote sensing, GIS and analytical hierarchical process: a case of Waddai, eastern Chad. GeoJournal https://doi.org/10.1007/s10708-020-10160-0
3. Al-Zabet T (2002) Evaluation of aquifer vulnerability of contamination potential using the DRASTIC method. Environ Geol 43, 203–208. American Public Health Association
4. Aller L, Bennett T, Lehr J, Hackett G (1987) DRASTIC: a standardized system for evaluating ground water pollution potential using hydrogeologic settings. Robert Kerr Environmental Research Lab., United States Environmental Protection Agency, EPA, p 622

5. Batlle-Aguilar J, Banks EW, Batelaan O, Kipfer R, Brennwald MS, Cook PG (2018)Groundwater residence time and aquifer recharge in multilayered, semi-confined and faulted aquifer systems using environmental tracers. J Hydrol 546, 150–165
6. Benjmel K, Amraoui F, Boutaleb S, Ouchchen M, Tahiri A, Touab A (2019) Mapping of groundwater potential zones in crystalline terrain using remote sensing, GIS techniques, and multicriteria data analysis (Case of the Ighrem Region, Western Anti-Atlas, Morocco). Water 2020, 12(2):471. https://doi.org/10.3390/w12020471
7. Biehl L, Landgrebe D (2021) Multispec: a freeware Multispectral image data analysis system. Purdue University. 27 Feb 2021. https://engineering.purdue.edu/~biehl/MultiSpec/index.html
8. Bowers S, Hanks RJ (1965) Reflectance of radiant energy from soils. Soil Sci 100:130–138
9. Conrad O, Bechtel B, Bock M, Dietrich H, Fischer E, Gerlitz L, Wehberg J, Wichmann V, Bohner J (2015) System for automated geoscientific analyses (SAGA) v. 2.1.4. Geosci Model Dev 8:1991–2007. https://doi.org/10.5194/gmd-8-1991-2015
10. Curran PJ (1989) Remote sensing of foliar chemistry. Remote Sens Environ 30(3):271–278
11. D' Odorico P, Besik A, Wong CY, Isabel N, Ensminger I (2020) High-throughput drone based remote sensing reliably tracks phenology in thousands of conifer seedlings. New Phytol 226(6). https://doi.org/10.1111/nph.16488
12. Djeuda Tcapnga HB, Ekodek GE (1990) Relations entre la fracturation des Roches et les Systemes d'ecoulement. In: Parriaux A (ed) Water resources in mountainous regions. Memoires International Association of XXII, part 2, pp 821–829
13. Doerfliger N, Jeannin PY, Zwahlen F (1999) Water vulnerability assessment in karst environments: a new method of defining protection areas using a multi-attribute approach and GIS tools (EPIK method). Environ Geol 39:165–176. https://doi.org/10.1007/s002540050446
14. Dwivedi, R.S. 2017. "Spectral Reflectance of Soils." In: Remote Sensing of Soils. Springer, Berlin, Heidelberg. https://doi.org/10.1007/978-3-662-53740-4_6
15. Earth Explorer US, Geological Survey (2000) Earth explorer: US geological survey fact sheet 083-00, 1 p https://pubs.usgs.gov/fs/2000/0083/
16. Engel B, Kumar N, Cooper B (1996) Estimating groundwater vulnerability to nonpoint source pollution from nitrates and pesticides, on a regional scale. In: HydroGIS 96, Application of geographic information systems in hydrology and water resources management (Proceedings of the Vienna Conference, April 1996). AHSPublication vol 235
17. EU Commission (2000) Directive 2000/60/EC of the European parliament and of the council of 23 October 2000 establishing a framework for community action in the field of water policy. Off J L 327, 0001 0073. 22 Dec 2000
18. European Commission, Directorate-General for the Environment (2008) Groundwater protection in Europe. In: The new groundwater directive, consolidating the EU regulatory framework. Information centre (BU9 0/11) B-1049 Brussels. ISBN 978-92-79-09817-8, https://doi.org/10.2779/84304
19. European Commission (2004) Groundwater risk assessment. Technical report on groundwater risk assessment issues as discussed at the workshop of 28th Jan 2004–12 Oct 2004
20. Ferronato N, Torretta V (2019) Waste mismanagement in developing countries: a review of global issues. 2017, Int J Environ Res Public Health 16, 1060. https://doi.org/10.3390/ijerph16061060, www.mdpi.com/journal/ijerph
21. Gamon JA, Huemmrich K, Fred W, Christopher YS, Ensminger I, Garrity S (2016) A remotely sensed pigment index reveals photosynthetic phenology in evergreen conifers. In: Proceedings of the national academy of sciences of the USA.
22. GISGeography (2021) Thirteen open source remote sensing software packages. 27 Feb 2021. https://gisgeography.com/open-source-remote-sensing-software-packages/
23. Goldscheider N, Klute M, Sturm S, Hotzl H (2000) The PI method"a GIS-based approach to mapping ground water vulnerability with special consideration of karst aquifer. Zeitschnft Angewandte Geologie 46(3):153–166

24. Hollinger DY, Noormets A, Peñuelas J (2016) A remotely sensed pigment index reveals photosynthetic phenology in evergreen conifers. PNAS 113(46):13087–13092. 15 Nov 2016, first published 1 Nov 2016. https://doi.org/10.1073/pnas.1606162113
25. Hounslow A (1995) Water quality data analysis and interpretation. Lewis Publishers, p 397
26. Huajie D, Zhengdong D, Feifan D, Daqing W (2016) Assessment of groundwater potential based on multicriteria decision making model and decision tree algorithms. Math Probl Eng 2064575. https://doi.org/10.1155/2016/2064575
27. Hunt G, Salisbury J (1971) Visible and near-infrared spectra of minerals and rocks: II, carbonates. Mod Geol 2:23–30
28. Inzana J, Kusky T, Higgs G, Tucker R (2003) Supervised classifications of Landsat TM band ratio images and Landsat TM band ratio image with radar for geological interpretations of central Madagascar. J Afr Earth Sci 37(1–2):59–72
29. Jing Y, Xin W, Chang-xiang Y, Shu-Rong W, Xue-Ping, J, Yi L (2019) Soil moisture retrieval model for remote sensing using reflected hyperspectral information. Remote Sens 11, 366. https://doi.org/10.3390/rs11030366
30. Kallergis Y (1970) Hydrogeological study of the Kalabaka sub-basin (W. Thessaly) [In Greek], PhD dissertation, National Technical University of Athens, Athens
31. Kupfersberger H, Pulido-Velazquez M, Wachniew P (2011) Conceptual models and first simulations, Deliverable D5.2, Project: groundwater and dependent ecosystems: new scientific and technological basis for assessing climate change and land-use impacts on groundwater-GENESIS. Project funded under FP7-ENVIRONMENT
32. Lalbat F, Blavoux B, Banton O (2007) Description of a simple hydrochemical indicator to estimate groundwater residence time in carbonate aquifers. Geophys Res Lett 34(19)
33. Lukjana A, Swasdib S, Chalermyanonta T (2016) Importance of alternative conceptual model for sustainable groundwater management of the Hat Yai Basin, Thailand. In: 12th international conference on hydroinformatics, HIC 2016. Elsevier Sciencedirect, Procedia Engineering 154, pp 308–316
34. Maimaitiyiming M, Ghulam A, Bozzolo A, Wilkins JL, Kwasniewski MT (2017) Early detection of plant physiological responses to different levels of water stress using reflectance spectroscopy. Remote Sens 9(7):745. https://doi.org/10.3390/rs9070745
35. Meijerink AMJ (2007) Remote sensing applications to groundwater. IHP-VI, series on groundwater no. 16. UNESCO
36. Meladiotis I, Papatheodorou K, Popa I (1994) Distribution in time and space of the Chloride ion concentration in the exploitable aquifers in the eastern part of the Plain of Thessaloniki (Greece). In: Proceedings of the international hydrogeologic symposium "Impact of industrial activities on groundwater", Constanta, Romania, pp 617–632
37. Nalbant S, Alptekin O (1995) The use of landsat thematic mapper imagery for analysing lithology and structure of Korucu-Dugla area in western Turkey. Int J Remote Sens 16:2357–2374
38. National Aeronautics and Space Administration, Science Mission Directorate (2010) Reflected near-infrared waves. 10 Sept 2020, NASA Science website: http://science.nasa.gov/ems/08_nearinfraredwaves
39. Neshat A, Pradhan B, Pirasteh S (2014) Estimating groundwater vulnerability to pollution using a modified DRASTIC model in the Kerman agricultural area, Iran. Environ Earth Sci 71, 3119–3131. https://doi.org/10.1007/s12665-013-2690-7
40. Nolan BT (2005) Relating nitrogen sources and aquifer susceptibility to nitrate in shallow ground waters of the United States. Groundwater 39(2):290–299. March 2001, https://doi.org/10.1111/j.1745-6584.2001.tb02311.x
41. Papanikolaou D (2015) Geology of greece. Patakis editions
42. Papatheodorou C (2010) Ground water flow paths delineation using Remote Sensing techniques and GIS. In: 30th EARSel symposium, remote sensing for science, education, natural and cultural heritage. Paris, France, June 2010

43. Papatheodorou K, Evangelidis K (2008) Ground water information system: a digital tool for groundwater resources protection and management. In: 4th international environmental conference of the Balkan environmental association (BENA), life quality and capacity building in the frame of a safe environment. Katerini, Greece
44. Papatheodorou K, Veranis N (2018) Geomatics technologies for groundwater prospecting in hard rocks. J Environ Prot Ecol 19(3):1421–1430
45. Papatheodorou K, Veranis N, Patsiaros N (2010) Groundwater vulnerability in the plain of Emathia prefecture: an implementation of the DRASTIC methodology. Choro-Graphies 1(1),17–24. Series, ISSN: 1792-3913
46. Papatheodorou K, Evangelidis K, Ntouros K (2017) Geomatics technologies for environmental protection and resource management. J Environ Prot Ecol 18(1):168–180
47. Papatheodorou K, Theocharis D, Fountoulis I (2012) A remotely sensed contribution to the Western Attica (Greece) tectonic Geology. In: 32th EARSEL symposium "Advances in geosciences", 4th workshop on remote sensing and geology. Mykonos, Greece
48. QGIS Development Team (2020) QGIS Geographic information system. Open source geospatial foundation project. http://qgis.osgeo.org
49. Rajput H, Goyal R, Brighu U (2020) Modification and optimization of DRASTIC model for groundwater vulnerability and contamination risk assessment for Bhiwadi region of Rajasthan, India. Environ Earth Sci 79, 136. https://doi.org/10.1007/s12665-020-8874-z
50. Rashid U, Izrar A, Fakhre A (2009) Mapping groundwater vulnerable zones using modified DRASTIC approach of an alluvial aquifer in parts of central Ganga plain, Western Uttar Pradesh. J Geol Soc India, Springer, 73.
51. Rupert MG (2001) Calibration of the DRASTIC vulnerability method. Ground Water 625–630
52. Sabins FF (1997) Remote sensing: principles and interpretation, 3rd ed xiii 494 p New York: W. H. Freeman and Co
53. Scheidleder A, Grath J, Winkler G, Stark U, Koreimann C, Gmeiner C, Gravesen P, Leonard J, Elvira M, Nixon S, Casillas J, Lack TJ (1999) Groundwater quality and quantity in Europe: data and basic information. Technical report No 22, European Environment Agency
54. Scheidleder A, Grath J, Winkler G, Stark U, Koreimann C, Gmeier C, Nixon S, Casillas J, Gravesen P, Leonard J, Elvira M (1999) Groundwater quality and quantity in Europe. Report prepared under the supervision of the European Environment Agency
55. Secunda S, Collin ML, Melloul A (1998) Groundwater vulnerability assessment using a composite model combining DRASTIC with extensive agricultural land use in Israel Sharon region. J Environ Manage 54:3957
56. Sentinel-2 Digital Object Identifier. https://doi.org/10.5066/F76W992G
57. Sentinel-2, ESA's Optical High-Resolution Mission for GMES Operational Services (ESA), SP-1322/2 (2012)
58. Sidiropoulos N (1991) Lithology, Geochemistry, Structure and Metamorphism of the Northwestern part of Mountain Vertiskos. The Kroussia Mount area. PhD thesis, Aristotle University of Thessaloniki
59. Spijker J, Lieste R, Zijp M, de Nijs T (2010) Conceptual models for the water framework directive and the groundwater directive. Report 607300015/2010. RIVM, Postbus 1, 3720 BA Bilthoven
60. Suganthi S, Elango L, Subramanian SK (2013) Groundwater potential zonation by remote sensing and GIS techniques and its relation to the groundwater level in the coastal part of the arani and koratalai river basin, Southern India. Earth Sci Res J 17(2):87–95
61. Sultan M, Arvidson RE, Sturchio N (1987) Lithologic mapping in arid regions with Landsat thematic mapper data: meatiq dome, Egypt. Geol Soc Am Bull 99:748–762
62. Thitumalaivasan D (2001) Aquifer vulnerability assessment using analytical hierarchy process and GIS for upper palar watershed. In: 22nd Asian conference on remote sensing. Singapore. 5–9 Nov 2001
63. Travaglia C, Dainelli N (2003) Groundwater research by remote sensing: a methodological approach. In: Environment and natural resources service, sustainable development department. food and agriculture organization of the united nations, Rome

64. Tsiamis A, Papatheodorou K, Perakis K (2015) Geologic indices in locating mineral resources. In: Fourth hellenic conference of planning and regional development. University of Thessaly, Volos, Greece. 24–27 Sept
65. Twana OA, Salahalddin SA, Nadhir AA-A, Sven K (2015) Groundwater vulnerability mapping using lineament density on standard DRASTIC model: case study in Halabja Saidsadiq Basin, Kurdistan Region, Iraq. Engineering, vol 7, pp 644–667. http://www.scirp.org/journal/eng
66. U.S. Geological Survey (2021) Landsat collection 2 (ver. 1.1, 15 Jan 2021): U.S. Geological survey fact sheet 4 p 30 Feb 2021. https://doi.org/10.3133/fs20213002
67. United Nations World Water Assessment Programme Special Report (2009) Climate change and water: an overview from the world water development report 3: water in a changing world. In: Published by the united nations world water assessment programme. programme office for global water assessment, division of water sciences, UNESCO, 06134 Colombella, Perugia, Italy
68. Ustin SL, Gitelson AA, Jacquemoud S, Schaepman M, Asner GP, Gamon JA, Zarco-Tejada P. (2009) Retrieval of foliar information about plant pigment systems from high resolution spectroscopy. Remote Sens Environ 113(1):67–77
69. Stempvoort V, Dale; Ewert, Lee & Wassenaar, Leonard. (1993) Aquifer vulnerability index: a GIS compatible method for groundwater vulnerability mapping. Can Water Resour J/Rev Can RessourS Hydr 18(1):25–37. https://doi.org/10.4296/cwrj1801025
70. Veranis N, Kalousi E, Lazaridou M, Pratanopoulos A, Chatzikyrkou A (2010) Hydro-geological study of the aquifer systems of central macedonia region, CSF Project 2003–2009. In: Inventory and assessment of hydrogeological character of the aquifer systems of the country. RUCM-IGME, 15 issues, maps
71. Waters P (1988) Methodology of lineament analysis for hydrogeological investigations. In: Satellite remote sensing for hydrology and water management, the mediterranean coasts and Islands. Gordon and Breach Science Publishers, London, UK
72. Waters P, Greenbaum D, Smart PL, Osmanston H (1990) Applications of remote sensing to groundwater hydrology. Remote Sens Rev 4(2):223–264
73. Wessman CA, Aber JD, Peterson DL (1989) An evaluation of imaging spectrometry for estimating forest canopy chemistry. Int J Remote Sens 10(8)
74. Wong CYS, D'Odorico P, Bhathena Y, Arain MA, Ensminger I (2019) Carotenoid based vegetation indices for accurate monitoring of the phenology of photosynthesis at the leaf-scale in deciduous and evergreen trees. Remote Sens Environ 233, 111407
75. World Meteorological Organization (2013) Planning of water quality monitoring systems. Publications Board World Meteorological Organization (WMO), Chair
76. Zaporozec A, Conrad JE, Hirata R, Johansson PO, Nonner JC, Romijn E, Weaver JMC (2002) Groundwater contamination inventory: a methodological guide. IHP-VI, series on groundwater No. 2, UNESCO

Konstantinos Papatheodorou received his Ph.D. in Engineering Geology (1991) from the Civil Engineering Dept. (Faculty of Engineering), Aristotle University of Thessaloniki. He's Professor of Applied Geology & Geoinformatics at the Surveying & Geoinformatics Dept, International Hellenic University (Greece). His research interests include engineering Geology, groundwater management & protection, Geo-informatics Science & Technologies (Remote Sensing, Geographic Information Systems) and contemporary surface & subsurface mapping methods (DinSAR, LiDAR, Ground Penetrating radar). With an over 20 years teaching experience in higher education he's the author of four books and more than 70 scientific articles. As a freelance Engineering Geologist (1986-2004) he was awarded by the Hellenic Ministry of Public Works (1997) with a C' class Designer's Degree (highest degree awarded to a physical person/professional geologist). As a Coordinator and/or Senior Researcher (1990-2021) he was involved in more than 50 competitive Research Projects (Grants) covering the entire spectrum of his research interests (engineering geology; disaster prevention and management; groundwater exploration, protection and management; Remote Sensing, GIS, Web-based systems etc.), with a total budget of coordinated Grants during the last 5 years of more than 2 mil euros. Prof. Konstantinos Papatheodorou has Chaired International Conferences, has been a member of the Scientific Committee of numerous others, presented keynotes and invited lectures and is a reviewer in numerous scientific journals.

The Roles of Sensor Placement in Water Quality Monitoring in a Water Distribution System

Oluwaseye Samson Adedoja, Yskandar Hamam, Baset Khalaf, and Rotimi Sadiku

Abstract Globally, the quest to meet water demand and the continuous depletion of quality water have increased. Human existence and its sustenance are subject to the availability of water, particularly potable water. However, the possibility of attacks on the water systems, which are some of the critical infrastructures, has attracted interest of researchers. In recent times, adequate monitoring and safeguarding of the water distribution systems, are of utmost importance, in order to suitably distribute potable water to the society. However, the complexity of water distribution networks permits human and renders them vulnerable to both accidental and intentional attacks. These attacks often compromise the water quality and have often resulted in contaminated water, which may eventually be consumed by the society. The consumption of contaminated water has gross public health implication as well as socio-economic consequences. While it is scientifically possible to monitor the quality of water and detect contaminants, with the deployment of water quality monitoring sensors, this has not come without challenges, which often are related to limited available resources; this is besides the technical issues, sometimes encountered. Consequently, this chapter presents the roles of sensor placement for water quality monitoring purposes, in water network/distribution systems. The prospects and associated challenges are presented as well as future trends.

Keywords Sensor placement · Water quality monitoring · Contaminant detection · Safeguarding · Critical infrastructure · Water distribution system

---

O. S. Adedoja (✉) · Y. Hamam · B. Khalaf
Department of Electrical Engineering/French South African Institute of Technology (F'SATI), Tshwane University of Technology, Pretoria 0001, South Africa

Y. Hamam
e-mail: hamama@tut.ac.za

Y. Hamam
ESSIE, 2 Boulevard Blaise Pascal, 93160 Noisy-le-grand, France

R. Sadiku
Department of Chemical, Metallurgy and Material Engineering/Institute for Nano-Engineering Research (INER), Tshwane University of Technology, Pretoria 0001, South Africa

© Springer Nature Switzerland AG 2021
A. Vaseashta and C. Maftei (eds.), *Water Safety, Security and Sustainability*, Advanced Sciences and Technologies for Security Applications, https://doi.org/10.1007/978-3-030-76008-3_3

1 Introduction

Water is an essential commodity for every human being and there is no substitute for water, particularly, potable water, without which, our existence is limited [1]. Undeniably, the demand for water globally, has continued to rise due to population growth and urbanization, owing to the increase in demand from the industries agriculture and the household consumptions. Additionally, there is a projected report that this demand will increase in the near future [2]. The combination of these with the dwindling depletion of the delivery of potable water, make this resource socio-political attractive that has significantly gained the attention of the research community across the globe. In spite of these hurdles, water utility managers remain resolute to the provision of quality water and distribute the proportionate quantities to the consumers when needed.

Conventionally, water is, transported through a water distribution network, from the treatment plants, where it is usually treated and its quality examined, to the end users. However, water quality can still be, technically contaminated, during transportation from the treatment plant to the consumer's taps, through pipe leakage, cross-connection and even during maintenance processes. This may be attributed to the complex infrastructural nature of the water distribution network, which comprises of hundreds of pipes that are interconnected together at the nodes (junctions), including other components, such as: pumps, valves and reservoirs. A schematic illustration of a typical water distribution system, busted at a junction where contamination intrusion is possible and it is, depicted in Fig. 1. This simple water distribution network consists of a reservoir at node 1 and contaminant intrusion at node 5. The other nodes and pipes, linked to residential and industrial sites, are indicated accordingly.

Some of these components are either laid on the surface or buried at a relatively low depth in the ground, where a human intervention is possible and it is therefore, vulnerable to accidental and/or intentional attacks [3]. These attacks sometimes compromise the water quality, which has already, been examined at the treatment plant and can lead to illness and sometimes loss of life, if such contaminated water is consumed. The attacks on the water distribution system have been conceivably possible and reported. However, these attacks can in turn, result in the contamination of water, which can be, transported to the consumers' taps and eventually be, consumed by the society.

The consumption of contaminated water has serious health implications as well as gross socio-economic consequences. For instance, in 2001, an intrusion occurrence of a water distribution network in the United States, was reported by Bush [4]. This intrusion occurrence and others have posited this problem, in a front-line research endeavour that has significantly gained global attention. While the consumption of contaminated water, which occurred in Walkerton, Ontario, Canada, left about 2,300 hospitalized [5], the accidental intrusion of industrial chemical in West Virginia, USA affected more than 300,000 users [6]. The public health and socio-economic consequences of consuming contaminated water are completely unnecessary and preventable, given the available state-of-the-art technology [7]. These contamination

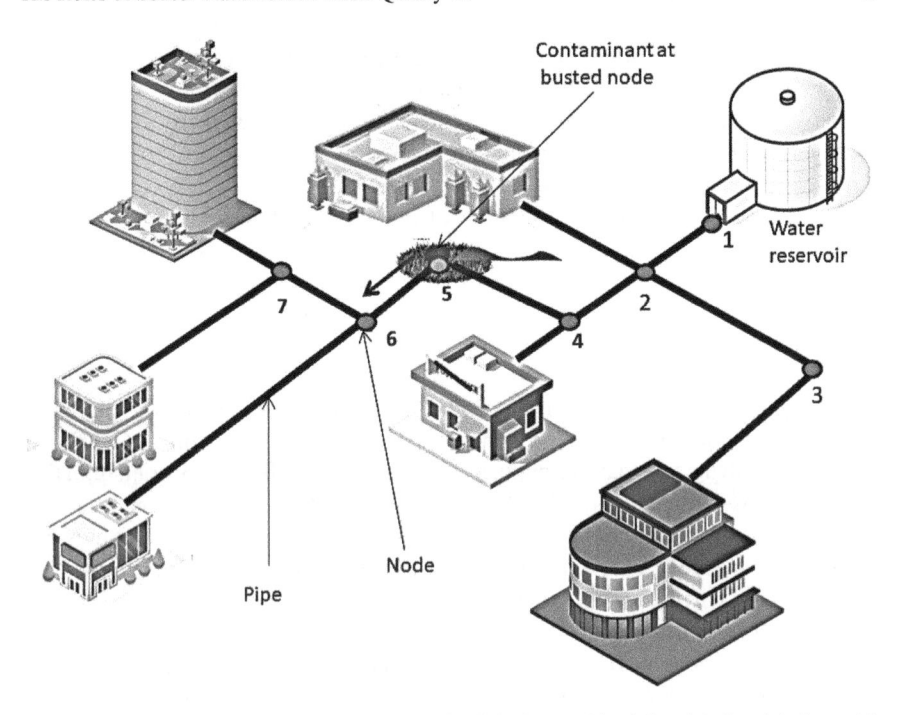

Fig. 1 Example of a water distribution network, linked to residential and industrial sites with contaminant intrusion

occurrences, which are subject to accidental or intentional contamination of the water resources, remain unhealthy to the society and has constituted a serious menace that must be addressed [3, 6, 7].

As a result of this, safeguarding of the water distribution system has become a subject of intense interest and its continuous monitoring is an utmost priority of both governmental and non-governmental organizations, especially, for the utility managers and the environmentalists. Even by enhancing the physical security of a water distribution plant, such as the perimeter fencing is necessary, which can be accessed only by the authorised personnel, worthy as it may be, it can be compromised by saboteurs [8]. However, another promising way of protecting and monitoring water quality, is the deployment of the water quality monitoring sensors across the water network. With this approach, reasonable results can be attained if the sensors are optimally deployed. From an economic perspective, it is practically impossible to deploy sensors across thousands of nodes in the water network due to limited available resources, upon which a modest degree of safety can be achieved, assuming all nodes can be thoroughly monitored [9]. The technique of water sensor placement can be compared to the placement of pressure control valves and water leakage detection techniques in the water distribution networks [10–14]. Interestingly, the use of sensor placement is not limited to water resources system since it has also been employed in other areas of research such as, structural health monitoring, estimation and prediction, localisation, and intelligent buildings, [15–19].

However, the multifaceted nature of a water distribution network, its susceptibility to attacks and the unavoidable demand by the public, among other features, make it more distinctive when compared to other areas of application. The combination of these, led to the appreciable number of research activities, which have aimed at minimizing the number of sensors needed for the contaminant detection due to budget constraints [20–24]. Since its discovery and the use of water quality monitoring sensors for contaminant detection, various techniques have been developed and deplored [25–27].

Technically speaking, sensor placement is designed to perform two important purposes, which are: to reduce the number of sensors that is required in the water network and to minimize the associated negative consequences that may arise. This is also in line with the fact that sensor placement and the identification of contamination source, are inseparable entities [3, 28]. More importantly, identifying the source of contamination helps the utility managers to initiate immediate control measures that can possibly avert the consequences of consuming contaminated water in the case of any such occurrence. Consequently, the sensor placement should be optimally deployed for an effective performance and is highly consequential.

From the foregoing and based on the notable consequences associated with to the consumption of contaminated water on the society and as a contribution to knowledge that is grounded on the systematic investigations, this chapter attempts to present an up-to-date evaluation of the roles of the sensor placement for water quality monitoring in the water distribution system. This concise chapter the existing approaches of sensor placement problem. In addition, it examines the technical challenges associated with the approaches and provides plausible solutions to some of these challenges. This chapter is intended to serve as a source of information to support the need and advocacy for the provision of quality water, which is one of the endorsed goals of the Sustainable Developments Goals (SGDs) of the United Nations and a call for an adequate protection of the water resources and its sustainability, which will be highly beneficial to human kind.

2 Threat to Water Distribution Systems and Components of the Contaminant

Historically, threats to the water distribution system may be traced to as far back as the warfare days. The predominant threat and contamination intrusion to the sources of drinking water are not without records of conflicts and warfare history [29–31]. However, the rate of its occurrence, has drastically reduced, especially, in the developed countries. In spite of its socio-economic landscape due to the recent technological advancements, incessant rural to urban migrations, recent global warming, population growth and projected water scarcity have in turn, posited challenges to the water security structure and its management appears to be on a sharp escalation. This is in addition to the influence of the contributing factor of natural disasters, such as: floods and hurricanes. These, among other factors, constitute the possibility of deliberate and accidental attacks on the water distribution system, which is a growing concern for the utility managers and the scientists, when considering the safety of the society.

These threats may be, in the forms of physical attack, cyber attack, influence of natural misadventures and may even be unintentional or intentional attacks, which sometimes, are caused by trespassers and terrorists [30]. Often, the nature of contaminants can be biological, chemical and even radiological and the consumption of these contaminants, may either result in sickness or loss of life. Therefore, monitoring, detection and possible identification of contaminants will be helpful for early proactive and protective actions. In addition, it would be useful in the design of immediate control measures and warning systems, in order to avert the possible consequences of consuming contaminated water.

3 Why Sensor Placement?

The safeguarding of the water distribution network and the continuous monitoring of water quality, have become the subject of interest among environmentalists. Relative to the roles of sensor placement, various performance objectives have been designed and developed according to a swift detection of contamination, i.e. time to detection (TD), reducing the impact of volume consumed (VC), minimising the size of the population exposed (PE) to the contaminated water and the extent of contamination (EC). The ultimate goal of these objectives is to abate the negative effects that may emanate from the use of contaminated water on the public. A thorough description of each of these performance objectives and other related issues, can be accessed in the work of Adedoja et al. [32].

4 Classification of Role of Sensor's Placement Objective

Subsequent to the significance of constant water quality monitoring for the safety of the public and bearing in mind the role of sensor placement, the performance objective is categorized as: single and multi-objective. However, these classifications are established according to types of the hydrology and hydraulic water quality models used, and the quantities and types of the sensors employed, which are discussed in subsequent subsections.

4.1 Single Objective Strategies

A proper distribution of the adequate volume and quality of the water to consumers remains an utmost priority of the water service providers, which is a global standard for safety purposes. In fact, an early detection of a contamination incident in a water distribution network helps providing a speedy and measured decision that may forestall the possibility of negative implications of consuming contaminated water. Therefore, the time to detection (TD) is an important parameter and could be described as the time passed between the start and end-time of contamination detection in the network [33]. In the quest to determine the best location for sensor placement in a water distribution network, an approach, based on the use of water quality model in a place of the conventional optimisation method was described, in the work of Chastain [34]. The study reflected on the time to detection, which utilized a small water network and has a reasonable prospect for improvements.

Rathi and Gupta [35] employed a heuristic technique that included the minimum travel time in order to determine an appropriate time to detection. Owing to the limited number of sensors, it became a challenging situation to obtain a reasonable level of service because of the direct relationship to the quantity of the monitoring stations deployed. However, a water distribution network from Nagpur City, India was used to test the performance of the technique [36] A co-evolutionary optimisation method that concentrated on minimizing the average time to detection is discussed, in the study of Hu et al. [37]. The performance of this method is examined with a water distribution network and satisfactory results are obtained. The results of the comparison of this method with swarm optimization and generic algorithm also show that the former, performs better than the latter. Nonetheless, improvement on the method that will combine and account for multiple objectives was suggested.

The demand coverage (DC) may be described as the sum of demands of those node locations where the quality of water at the monitoring junction is presumed guaranteed [38]. Therefore, the desire to provide quality water may be increased if a large portion of supply nodes is secured. As part of the effort to increase DC, Lee and Deininger [38] used an integer programming approach with focus on a selected area. The result of the two cases considered, showed the feasibility of the approach, but requires an extension. The negligence of the impact of a temporal distribution, while collecting demand coverage of each pattern during the calculation of demand coverage, can lead to inaccurate result during analysis, which is a weakness of the approach. In order to tackle this drawback, Liu et al. [39] developed a demand coverage index (DCI). In their approach, they combined a generic algorithm with network analysis tools, developed by Rossman [40]. The authors compared the results obtained from their applications with the work of Liu et al. [41].

Al-Zahrani and Moied [42] formulated a genetic algorithm (GA) to address the same problem. They argued that more unique characteristics of the water network, such as network conditions and constituent concentrations, be included. In addition, a mixed integer approach under an assumption of steady condition of water network, was discussed in the work of Kumar et al. [43]. The concepts of DC in relation to time-dependence of demand, has been reported by some authors [44–47]. Ghimire and Barkdoll [48] formulated a demand-based approach and demonstrated the technique on two water networks. The principle of locating the sensors with regards to DC indicates a more useful and better applicability for unintentional attack scenarios. Furthermore, restraining the volume of contaminated water consumed is yet another concern that must be, addressed. Accordingly, Kessler et al. [49] defined the summation of volume of contaminated water consumed (VC) before the detection of contaminant, as a way to estimate the impact of the consumption of contaminated water. Ostfeld and Salomons [50] used a random multiple approach with a focus on stochastic demands in an effort to determine the sensor location in a water distribution network. Recently, Giudicianni et al. [51], described a topological approach without the use of hydraulic analysis to position sensors in a water distribution network. They relied on the installation of the sensors in a central position of a cluster section of the network. The approach is tested on a water network that supplies a community near Naples, Italy. Based on the results obtained on the topological principle, the method is, reported to be effective in the possibility of detection, redundancy and the population exposed.

4.2 Multi-objective Strategies

Unlike the single objective strategy that considers just one performance objective, in the formulation at a time, in here, just as the name implies, more than one performance objectives are combined in the multi-objective strategy, for the sensor placement in a water distribution network. For instance, a mixed integer programming (MIP) strategy, based on a multi-objective is discussed in the work of Berry et al. and Propato et al. [52, 53]. Thereafter, Berry et al. [54] integrated a case of an undetected occurrence into the MIP. In a multi-objectives strategy, two circumstances are often considered.

The first case keeps the objective function separately, while the output is presented in a Pareto form [46, 48, 55–57]. Dorini et al. [48] combined TD, VC, EC and PE in their multi-objective formulation. They used a Noisy Cross-Entropy Sensor Locator (n-CESL)-based algorithm. The authors examined their algorithm with two water networks and a computational burden was encountered. In an effort to maximise the probability of detection, Aral et al. [58] combined TD and VC. They employed the principle of sub-domain as a method to minimize the computational burden in the case of a large network application. This approach is, established on the Non-dominated Sorting Generic Algorithm (NSGA-II). The results of the two water networks used, show an improvement of the method. An interactive deepening Pareto approach that combined five objective is presented [48]. The authors demonstrated the feasibility of the method on two water distribution networks. The use of a generic algorithm (GA) with the data mining approach was described and three performance objectives were incorporated in the investigation carried out by Huang et al. [48]. In this method, the data mining strategy is used to compute the network information, which are ranked based on the sensitivity analysis. This then formed the procedure for the generic algorithm. In order to test the proposed method, two water networks are used and the results obtained are presented.

As the search for solutions to this problem persists, Wu and Walski [59] describe an optimisation technique that combines four performance objectives. The authors use a GA and Montel Carlo principle as a way to deal with the technical problem. They demonstrated this technique on two water distribution networks. However, the inability of the method to cater for a randomness circumstance, e.g., a case of intentional occurrence, is a weakness of the method. A combination of reducing the consumption of contaminated water and minimizing the length of time-taken to eject the contamination from the system after detection with the use of NSGA-II is considered by Preis and Ostfeld [60]. Excessive computational time and disintegration of sensor database were, reported as issues of concern to this method. Weickgenannt et al. [61] adopted the NSGA-II to minimise the quantity of sensors and the risk of contamination in their approach. They employed the Almelo water distribution network, in the Netherland, to test the feasibility of their approach. The authors also compare the use of the NSGA-II with the Monte Carlo method-based sampling, in the work.

There are other multi-objective strategies that rely on the combination of different performance objectives together and manage them as a single entity with the use of the optimisation solver [46, 62–64]. For example, Krause et al. [62] combines TD and an exposure to contamination in their work. The study was, established on the sub-modular principle as a way to navigate from the computational stress. They implemented their approach on two water networks and their results were, presented. In addition, Aral et al. [65], integrates a four-performance objective, based on simulation-optimisation principle. The authors employed a progressive generic algorithm in order to limit the processing time. The work of Margarida and Antunes [63] delved on the combination of different objectives as a single function with the use of the NSGA-II. They employed different water networks to test the viability of their approach. However, they pointed out the need for further examination, based on application perspective. Table 1 below, shows the single, multiple objective methods and the different sets of objectives envisaged/employed by different authors.

5 Comparison of Single and Multi-objectives Strategies of the Sensor Placement

Following the global recognition of the roles of sensor placement for contaminant detection in a water distribution network, the associated research activities, have spanned across the globe. Researchers have continued to strive in the development of different approaches, viz: single and multi-objectives. The global comparison of research activities, based on how the continents of the world, performed in August 2020, is presented in Fig. 2.

Table 1 Single and multiple objective methods

Category	Method	Objective	References
Single Objective	A simple heuristic	To maximize the demand coverage	[66]
	Heuristic principle	To minimise the expected time to detection	[67]
	A co-evolutionary optimisation	To minimize the time to detection	[37]
	An integer programming	To maximise demand coverage	[38]
	A mixed integer programming	To maximise demand coverage	[43]
	A random multiple approach	To minimise the volume consumed	[68, 69]
Multiple Objective	Noisy Cross-Entropy Sensor Locator (nCESL)	To minimise the: time to detection, volume consumed, extent of contamination and population exposed	[48]
	Non-dominated Sorting Generic Algorithm (NSGA-II)	To minimise the time to detection and the volume consumed	[58]
	NSGA-II	To minimise the amount of contaminated consumed and flush out contaminant from the system	[70]
	NSGA-II	To reduce the quantity of sensors and impact of contamination	[55]
	Sub-modularity approach	To reduce the time to detection and population exposed	[62]
	SLOTS	To increase the probability of detection and reduce population exposure to contamination	[71]

From the comparison results in Fig. 2, may be observed that America, Europe and Asia have reasonable works done in this field, more than the other continents. This can be attributed to their desire to provide potable water for their citizens and the level of standardization. In addition, this may be attributed to their readiness to attain the goal number six (6) of the Sustainable Development Goals of the United Nations, which advocates for the provision of potable water for the populace by the year, 2030. On the contrary and at the moment, Africa have little such data are yet to be published in research articles in this domain, which shows their lack of readiness to this growing global distress. Therefore, more efforts in this field of study, is encouraged for a global sustainability.

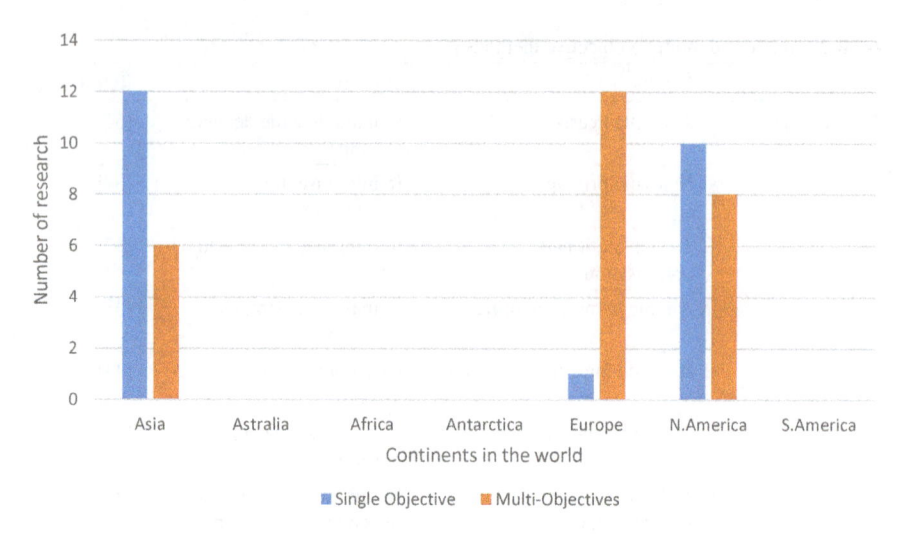

Fig. 2 Comparison of single and multi-objectives strategies, based on the accessible published articles as of August 2020

6 Present Solution Approaches

Researchers have employed various approaches to address the issue of sensor placement in a water distribution network, however, there still exists, challenges in the field. However, these approaches may be classified according to opinion-based, theoretical-based, rule-based and optimisation-based.

In an opinion-based scenario, personal experience takes the wheel of decision-making on sensor placement [48]. The report of the comparison study between an optimisation-based approach and opinion-based showed the latter was ineffective [47]. This renders the relatively less considered when compared to others. Consequently, a theory-based technique has also been employed [72, 73]. An approach aimed to ease the decision-making consequence to optimisation, was presented, based on a controllability theory [72]. The work of Shen and McBean [73] used a diminishing return theory to evaluate the difference in the level of sensor's time delay to detect contaminant. This was done in order to magnify the sensor detection redundancy in the case of inaccurate information. The insignificant recognition of the hydraulic and quality models, makes the rule-based technique relevant for a large network [74]. Chang et al. [75] described the use of the rule-based decision support system (RBDSS). They infer on its suitability for a large network because of the associated diminished computational burden. However, the exclusion of nodal demands renders this approach to be unrefined and the consideration of optimisation-based approach is therefore necessary.

Optimisation-based has been broadly used in sensor placement approaches and it may be classified as deterministic, stochastic and robust optimisations. The lack of random consideration, while a future state of a system is, being developed and can be, presumed as a deterministic system. Krause et al. [48] adopted a deterministic approach in order to locate the best positions for sensors placements. They reported the observation of computational stress in the application of a large water network. The use of a mixed integer stochastic approach was reported by Rico-Ramirez et al. [76]. They inferred that different uncertainties added to the stress in the optimal positioning of sensors. Cozzoling et al. [77] employed the stochastic approach and focused on the random selection of water demand at different hydraulic cases in order to improve the possibility of detections.

An assessment of the different levels of uncertainties and stochastic approach using Monte Carlo is presented in [78]. While the Monte Carlos approach is often employed to generate the initial contamination incident in a stochastic method, it is however, not suitable for critical circumstances. This is because, conventionally, Monte Carlo approach leads to random incidences with uniform distribution, which makes it unfit for critical cases. In order to avoid these associated issues, an importance-based sampling technique, based on a cross-entropy, was formulated in the work of Perelman et al. [79]. On the basis of entropy principle, Christodoulou et al. [80] presented an approach to solve the issues of sensor placement. In spite of the use of stochastic approach to deal with this problem, the inability to discretely pinpoint contaminant occurrence and dealing with estimation values of objective function, remains the drawback of this approach.

Another means to handling these technical drawbacks is the use of a robust optimisation approach. Consequently, an extensive review of the robust approach was presented [81]. In order to properly situate this approach, a robust formulation that considered different uncertainties, related to the determination of the sensor location, was described [82]. Xu et al. [64] adopted a robust optimisation approach in their formulation. They maintained that the output of their investigations was effective. A mixed-integer programming technique, established on the robust approach, was described in the study by Ma et al. [83]. They employed the Tabu search technique in order to solve the problem. The results of the comparison of their investigation with that of the conventional stochastic approach, showed that the former was effective than the latter. Langowski et al. [84] considered the different water demands for robust sensor positioning strategies. The use of simulation-optimisation approach is necessary so that hydraulic analysis model, such as the solver, developed by Rossman [85], can be integrated. With this, the hydraulic flow and the pressure drop, can be estimated and this information will assist in formulating where the water quality sensors can be optimally, positioned in a water distribution network.

7 Challenges and Way Forward

The recommended method of water quality monitoring and contaminant detection in a water distribution network is the application and deployment of sensor placement and globally, its role is highly consequential. In a bid to proffer, solutions to optimally, position these sensors in a water distribution network, different technical drawbacks were observed from the current approaches. Therefore, this section highlights the challenges and presents the probable and promising way forward.

Based on the complexity of the water distribution networks, where hundreds of pipes are, connected at junctions (nodes), the number of computing nodes increases and hence, the computational stress becomes tedious and more cumbersome. Scientific evidence has shown that this has been one of the technical issues that are, sometimes encountered with the current approaches. In order to get rid of this issue and without trading the accuracy of the sensor placement, different approaches have been described and presented [83, 86, 87].

In spite of these efforts and regardless of the approach engaged, it is not certain that an optimal solution will be, attained. Accordingly, the use of node cluster technique has a tendency to minimize the stress and may be explored [88]. In addition, Khorshidi et al. [89] described a theoretical approach and employed the Larmerd water distribution to test their approach. The output of their investigations, when compared to previous method, showed a lesser computational stress with the former. In fact, computational complexity is a critical technical challenge in this domain and therefore, more research efforts should be, channelled in order to resolve the challenge.

Furthermore, the use of imperfect sensors may lead to inaccurate information of contamination detection of water quality sensors. It is possible for sensors to generate false positive signal due to factors, such as: water reaction to pipe walls and even corrosion. Although, some studies have employed imperfect sensors, while formulating a sensor placement approach, however, further examination on its usage is necessary. Another technical hurdle is the selection of the performance objective. The choosing of a single objective, may not be sufficient, however, by considering more than one objective, may result to an excessive computational issue. Therefore, a trade-off between the complimentary and competitive objectives, should be rigorously studied and possibly, be sustained [65].

The types of techniques adopted, can also pose a challenge. This can be related to, amongst other issues: fluctuated water demand, erroneous sensor usage and sensor's reactions, which have the tendency of contributing to the challenges sometimes encountered during computational analysis. As such, by incorporating some of these objectives in a manner as to reduce the probable sensor locations, as well as the network size with the use of heuristic approach, may be explored [48, 90, 91]. Another alternative to curbing it, is the use of the graph theory technique [92]. The use of MIP theory for large network has also been described and embraced, particularly, the multi-objectives approach. Notwithstanding, the description of the heuristic and evolutionary approaches, a technique that will deal with the complex networks, is mostly preferred. In addition, there are prospects with the use and application of TEVA-SPOT and S-PLACE tools. These tools are reported to be suitable to handle complex water networks [93, 94]. Investigation into the use of artificial neural network (ANN), should also be examined in the clipping of the complex nature of water network [95, 96]. The partitioning of a large water network may be a promising idea in order to achieve the deployment of water quality sensor deployment [97].

Nonetheless, further technical investigations are essential so as to overcome these challenges.

8 Future Research Works on Sensor Placement

To an extent, appreciable efforts have been put in place by many scholars as a way to address this pertinent area of research. Still, the field is yet to be fused as efforts to address it can in turn, lead to different technical challenges. Therefore, continuous investigation of sensor placement is germane and this section highlights some of the areas that could be explored for future research works. It is possible to integrate some of the existing approaches in a hybridized form, where the strength of one can complement the limitation of another. For instance, deterministic approach was reported to be well suited for a small network case, while heuristic approach can give a near optimal solution for the application of a large water network. Therefore, research should be conducted on integrating these approaches in the form of hybrid solution. In addition, comparison of these techniques should be useful. There are certain associated limitations to the use of static sensors, such as: procurement cost and difficulties to access some underground components. The discovery of newly developed technologies, e.g. software defined network, has a tendency to overcome some of these difficulties [93, 97–99]. Therefore, the use of these developing technologies should be maximally explored. In addition, the integration of fixed and mobile sensors can improve the possibility of detecting contamination occurrences and hence, minimize the time to detection [100].

With the development of cloud computing, it is possible to navigate, away from the computational stress that is commonly, encountered, especially with large network cases. This is because of its evidence of a wide pool of resources and the tendency to negate the challenges [101]. Nonetheless, more efforts should be devoted to the use of these techniques and hence, its sustainability, is highly advised.

9 Conclusions

The problem of water quality sensor placement in a water distribution network has received escalated attention in the last few years and it, remains an open research problem and a technical concern for the utility managers and researchers, especially, the environmentalists. In addition to this, both governmental and non-governmental organisations remain restless and are determined to find solutions to the problem. Excitingly, the provision and the distribution of potable water is part of the interconnected and recommended goals of the United Nation Sustainable Development Goals (UN-SDGs). These goals are set to be achieved by the year, 2030, for the benefit of humankind. The goals aim to exterminate poverty and enhance the socio-economic situations of the society and an instant call by all nations-developed and developing countries. The combination of these formed the basis for the remarkable effort that researchers have put in place in order to ensure potable water is delivered to the consumers.

However, the unavoidable and highly possible technical intrusion of contaminants in a WDN-a means through which water is, transported from the treatment plant to the consumers' taps, is a central distress. This has, in turn complicated the concern for safe delivery of the drinking water to the end-users. Over time, researchers have taken this as a worthy responsibility and water quality sensor, has been recognised and recommended as a means of detecting on time contaminant in a water distribution network. In addition, there is the possibility of protecting the society from consuming contaminated water, if these sensors are, diligently and well positioned. Unfortunately, the associated procurement and maintenance costs of these sensors come with huge investments, which are rarely available, particularly, to the developing countries. Hence, it has resulted to the optimal sensor placement, which is needed in order, in the efforts to achieve the targeted goals and it has received the keen interest of researchers. Consequently, various techniques have been proposed to address this problem. Even at that, the problem is yet to be fully resolved, since each of the proposed techniques, comes with its associated technical hitches. Therefore, there is a need for continuous investigation.

In the end, if this tempo is, sustained by scholars and the relevant practitioners, viable technical solutions are possible in the near future. These are a part of the driven forces for the presentation of the roles of the sensor placement for water quality monitoring purposes in a water distribution network, which is the focus of this chapter. In conclusion, this chapter has unveiled two major approaches that have been used in the course of seeking solution to this problem. In addition to this, the associated technical hiccups as well as the probable solutions in order to, possibly overcome these challenges are, also presented. Based on the literature review and the evidence available in the public space, the chapter also compared the research efforts by scholars by considering the different continents of the world. The output showed appreciable efforts by some of these scientists, based on their resident continents, while little or yet-to-be publicly recorded works, are envisaged in a few continents, including the Africa continent. Therefore, more research efforts in this domain are required and it is, highly recommended for sustainability purposes.

References

1. Hickman DC (1999) A chemical and biological warfare threat: USAF water systems at risk
2. Biswas AK, Tortajada C (2019) Water crisis and water wars: myths and realities
3. Adedoja O, Hamam Y, Khalaf B, Sadiku R (2018) Towards development of an optimization model to identify contamination source in a water distribution network. Water 10:579
4. Bush GW (2002) The national security strategy of the United States of America
5. Hrudey SE, Payment P, Huck PM, Gillham RW, Hrudey EJ (2003) A fatal waterborne disease epidemic in Walkerton, Ontario: comparison with other waterborne outbreaks in the developed world. J Water Sci Technol 47:7–14

6. Cooper WJ (2014) Responding to crisis: the West Virginia chemical spill
7. Corso PS, Kramer MH, Blair KA, Addiss DG, Davis JP, Haddix AC (2003) Costs of illness in the 1993 waterborne Cryptosporidium outbreak, Milwaukee, Wisconsin. Emerg Infect Dis 9:426
8. Haimes YY, Matalas NC, Lambert JH, Jackson BA, Fellows JFR (1998) Reducing vulnerability of water supply systems to attack. J Infrastruct Syst 4:164–177
9. Yazdi J (2018) Water quality monitoring network design for urban drainage systems, an entropy method. Urban Water J 15:227–233
10. Adedeji KB (2018) Development of a leakage detection and localisation technique for real-time applications in water distribution networks. Tshawane University of Technology, Pretoria
11. Gomes R, Sá Marques A, Sousa J (2012) Identification of the optimal entry points at District Metered Areas and implementation of pressure management. Urban Water J 9:365–384
12. Rathi S, Gupta R (2017) Optimal sensor locations for contamination detection in pressure-deficient water distribution networks using genetic algorithm. Urban Water J 14:160–172
13. Hindi KS, Hamam YM (1991) Pressure control for leakage minimization in water supply networks: Part 2 Multi-period models. Int J Syst Sci 22:1587–1598
14. Farley B, Mounce SR, Boxall JB (2010) Field testing of an optimal sensor placement methodology for event detection in an urban water distribution network. Urban Water J 7:345–356
15. Yi T-H, Zhou G-D, Li H-N, Zhang X-D (2015) Optimal sensor placement for health monitoring of high-rise structure based on collaborative-climb monkey algorithm. Struct Eng Mech 54:305–317
16. Vlasenko I, Nikolaidis I, Stroulia E (2015) The smart-condo: optimizing sensor placement for indoor localization. IEEE Trans Syst Man, Cybern Syst 45:436–453
17. Lee J-H, Kim K, Lee S-C, Shin B-S (2014) An efficient localization method based on adaptive optimal sensor placement. Int J Distrib Sens Netw 10:
18. Welter G, Lechevallier M, Spangler S (2009) Guidance for decontamination of water system infrastructure. Water Research Foundation
19. Younis M, Akkaya K (2008) Strategies and techniques for node placement in wireless sensor networks: a survey. Ad Hoc Netw 6:621–655
20. Hart WE, Murray R (2010) Review of sensor placement strategies for contamination warning systems in drinking water distribution systems. J Water Resour Plan Manag 136:611–619
21. Rathi S, Gupta R (2014) Sensor placement methods for contamination detection in water distribution networks: a review. Procedia Eng 89:181–188. https://doi.org/10.1016/j.proeng.2014.11.175
22. Ung H, Piller O, Gilbert D, Mortazavi I (2017) Accurate and optimal sensor placement for source identification of water distribution networks. J Water Resour Plan Manag 143:4017032
23. Zechman EM, Ranjithan SR (2009) Evolutionary computation-based methods for characterizing contaminant sources in a water distribution system. J Water Resour Plan Manag 135:334–343
24. Berry J, Carr RD, Hart WE, Leung VJ, Phillips CA, Watson J-P (2009) Designing contamination warning systems for municipal water networks using imperfect sensors. J Water Resour Plan Manag 135:253–263
25. Murray R, Haxton T, Janke R, Hart WE, Berry J, Phillips C (2010) Sensor network design for drinking water contamination warning systems: a compendium of research results and case studies using the TEVA-SPOT software. Natl Homel Secur Res Center, US Environ Prot Agency, Cincinnati, OH, USA
26. Rathi S, Gupta R, Kamble S, Sargaonkar A (2016) Risk based analysis for contamination event selection and optimal sensor placement for intermittent water distribution network security. Water Resour Manag 30:2671–2685
27. Ehsani N, Afshar A (2010) Optimization of contaminant sensor placement in water distribution networks: multi-objective approach. Water Distrib Syst Anal 2010:338–346
28. Shahra EQ, Wu W (2020) Water contaminants detection using sensor placement approach in smart water networks. J Ambient Intell Humaniz Comput 1–16

29. Foran JA, Brosnan TM (2000) Early warning systems for hazardous biological agents in potable water. Environ Health Perspect 108:993–995
30. Kornfeld IE (2002) Terror in the water: threats to drinking water and infrastructure. Widener L Symp J 9:439
31. Meinhardt PL (2005) Water and bioterrorism: preparing for the potential threat to US water supplies and public health. Annu Rev Public Heal 26:213–237
32. Adedoja OS, Hamam Y, Khalaf B, Sadiku R (2018) A state-of-the-art review of an optimal sensor placement for contaminant warning system in a water distribution network. Urban Water J 15:985–1000
33. Kumar A, Kansal ML, Arora G, Ostfeld A, Kessler A (1999) Detecting accidental contaminations in municipal water networks. J Water Resour Plan Manag 125:308–310
34. Chastain JR Jr (2006) Methodology for locating monitoring stations to detect contamination in potable water distribution systems. J Infrastruct Syst 12:252–259
35. Rathi S, Gupta R (2014) Locations of sampling stations for water quality monitoring in water distribution networks. J Environ Sci Eng 56:169–178
36. Gupta AD, Kulat K (2018) Leakage reduction in water distribution system using efficient pressure management techniques. Case study: Nagpur, India. Water Supply 18:2015–2027
37. Hu C, Tian D, Liu C, Yan X (2015) Sensors placement in water distribution systems based on co-evolutionary optimization algorithm. Ind Networks Intell Sys (INISCom), 2015 1st International Conference, 2015, pp 7–11
38. Lee BH, Deininger RA, Clark RM. Locating monitoring stations in water distribution systems. J Am Water Works Assoc 1991:60–66
39. Liu S, Liu W, Chen J, Wang Q (2012) Optimal locations of monitoring stations in water distribution systems under multiple demand patterns: a flaw of demand coverage method and modification. Front Environ Sci Eng 6:204–212
40. Rossman LA et al. (2000) EPANET 2: users manual
41. Liu S, Butler D, Brazier R, Heathwaite L, Khu S-T (2007) Using genetic algorithms to calibrate a water quality model. Sci Total Environ 374:260–272
42. Al-Zahrani MA, Moied K (2003) Optimizing water quality monitoring stations using genetic algorithms. Arab J Sci Eng 28:57–75
43. Kumar A, Kansal ML, Arora G (1997) Identification of monitoring stations in water distribution system. J Environ Eng 123:746–752
44. Harmant P, Nace A, Kiene L, Fotoohi F (1999) Optimal supervision of drinking water distribution network. WRPMD'99 Prep. 21st Century, pp 1–9
45. Woo H-M, Yoon J-H, Choi D-Y (2001) Optimal monitoring sites based on water quality and quantity in water distribution systems. Bridg. Gap Meet. World's Water Environment Resources Challenges, pp 1–9
46. Propato M (2006) Contamination warning in water networks: general mixed-integer linear models for sensor location design. J Water Resour Plan Manag 132:225–233
47. Berry JW, Hart WE, Phillips CA, Uber JG, Walski TM (2005) Water quality sensor placement in water networks with budget constraints. Impacts Glob Clim Chang 1–11
48. Ghimire SR, Barkdoll BD (2006) A heuristic method for water quality sensor location in a municipal water distribution system: mass-released based approach. Water Distrib Syst Anal Symp 2008:1–11
49. Kessler A, Ostfeld A, Sinai G (1998) Detecting accidental contaminations in municipal water networks. J Water Resour Plan Manag 124:192–198
50. Ostfeld A, Salomons E (2003) An early warning detection system (EWDS) for drinking water distribution systems security. World Water Environ Resour Congr 2003:1–7
51. Giudicianni C, Herrera M, Di Nardo A, Greco R, Creaco E, Scala A (2020) Topological placement of quality sensors in water-distribution networks without the recourse to hydraulic modeling. J Water Resour Plan Manag 146:4020030
52. Berry JW, Fleischer L, Hart WE, Phillips CA, Watson JP (2005) Sensor placement in municipal water networks. J Water Resour Plan Manag 131:237–243

53. Propato M, Piller O, Uber JG (2005) A sensor location model to detect contaminations in water distribution networks. Impacts Glob Clim Chang1–12
54. Murray R, Haxton T, Janke R, Hart WE, Berry JW, Phillips C (2010) Sensor network design for drinking water contamination warning systems 82
55. Weickgenannt M, Kapelan Z, Blokker M, Savic DA (2010) Risk-based sensor placement for contaminant detection in water distribution systems. J Water Resour Plan Manag 136:629–636
56. Naserizade SS, Nikoo MR, Montaseri H (2018) A risk-based multi-objective model for optimal placement of sensors in water distribution system. J Hydrol 557:147–159
57. de Winter C, Palleti VR, Worm D, Kooij R (2019) Optimal placement of imperfect water quality sensors in water distribution networks. Comput Chem Eng 121:200–211
58. Guan J, Aral MM, Maslia ML, Grayman WM (2008) Optimization model and algorithms for design of water sensor placement in water distribution systems. Water Distrib Syst Anal Symp 1–16
59. Wu ZY, Walski T (2008) Multi objective optimization of sensor placement in water distribution systems. Water Distrib Syst Anal Symp
60. Preis A, Ostfeld A (2008) Multiobjective contaminant response modeling for water distribution systems security. J Hydroinformatics 10:267–274
61. Weickgenannt M, Kapelan Z, Blokker M, Savic DA, Sawodny O (2010) Optimization of sensor locations for contaminant detection in water distribution networks. In: Control Appl (CCA), 2010 IEEE international conference, pp 2160–2165
62. Krause A, Leskovec J, Guestrin C, VanBriesen J, Faloutsos C (2008) Efficient sensor placement optimization for securing large water distribution networks. J Water Resour Plan Manag 134:516–526
63. Margarida D, Antunes CH (2015) Multi-objective optimization of sensor placement to detect contamination in water distribution networks. In: Proceedings of the Companion Publication 2015 Annual Conference on Genetic and Evolutionary Computation, pp 1423–1424
64. Xu J, Johnson MP, Fischbeck PS, Small MJ, VanBriesen JM (2010) Robust placement of sensors in dynamic water distribution systems. Eur J Oper Res 202:707–716
65. Aral MM, Guan J, Maslia ML (2008) A multi-objective optimization algorithm for sensor placement in water distribution systems. World environment water resource congress 2008 Ahupua'A, pp 1–11
66. Kansal ML, Dorji T, Chandniha SK, Tyagi A (2012) Identification of optimal monitoring locations to detect accidental contaminations. World Environment Water Resources Congress 2012 Crossing Boundaries, pp 758–776
67. Rathi S, Gupta R (2014) Monitoring stations in water distribution systems to detect contamination events. ISH J Hydraul Eng 20:142–150
68. Ostfeld A, Salomons E (2004) Optimal layout for early warning detection stations for water distribution systems security. Water Resour Manag 130:377–385
69. Ostfeld A, Salomons E (2005) Optimal early warning monitoring system layout for water networks security: inclusion of sensors sensitivities and response delays. Civ Eng Environ Syst 22:151–169
70. Romero-Gomez P, Choi CY, Lansey KE, Preis A, Ostfeld A (2008) Sensor network design with improved water quality models at cross junctions. Water Distrib Syst Anal 2008:1–9
71. Dorini G, Jonkergouw P, Kapelan Z, Savic D (2010) SLOTS: Effective algorithm for sensor placement in water distribution systems. J Water Resour Plan Manag 136:620–628
72. Diao K, Rauch W (2013) Controllability analysis as a pre-selection method for sensor placement in water distribution systems. Water Res 47:6097–6108
73. Shen H, McBean E (2011) Diminishing marginal returns for sensor networks in a water distribution system. J Water Supply Res Technol 60:286–293
74. Schal S, Lothes A, Bryson LS, Ormsbee L (2013) Water quality sensor placement guidance using teva-spot. World environmental and water resources congress 2013 showcasing future, pp 1022–1032
75. Chang N-B, Pongsanone NP, Ernest A (2012) A rule-based decision support system for sensor deployment in small drinking water networks. J Clean Prod 29:28–37

76. Rico-Ramirez V, Frausto-Hernandez S, Diwekar UM, Hernandez-Castro S (2007) Water networks security: a two-stage mixed-integer stochastic program for sensor placement under uncertainty. Comput Chem Eng 31:565–573
77. Cozzolino L, Della Morte R, Palumbo A, Pianese D (2011) Stochastic approaches for sensors placement against intentional contaminations in water distribution systems. Civ Eng Environ Syst 28:75–98
78. Comboul M, Ghanem R (2012) Value of information in the design of resilient water distribution sensor networks. J Water Resour Plan Manag 139:449–455
79. Perelman L, Maslia ML, Ostfeld A, Sautner JB (2008) Using aggregation/skeletonization network models for water quality simulations in epidemiologic studies. J Am Water Works Assoc 100:122–133
80. Christodoulou SE, Gagatsis A, Xanthos S, Kranioti S, Agathokleous A, Fragiadakis M (2013) Entropy-based sensor placement optimization for waterloss detection in water distribution networks. Water Resour Manag 27:4443–4468
81. Carr RD, Greenberg HJ, Hart WE, Konjevod G, Lauer E, Lin H et al (2006) Robust optimization of contaminant sensor placement for community water systems. Math Program 107:337–356
82. Watson J-P, Murray R, Hart WE (2009) Formulation and optimization of robust sensor placement problems for drinking water contamination warning systems. J Infrastruct Syst 15:330–339
83. Ma X, Song Y, Huang J, Wu J (2010) Robust sensor placement problem in municipal water networks. Comput Sci Optim (CSO), 2010 third international Jt. conference, vol. 1, pp 291–294
84. Łangowski Rafałand Brdys MA, Qi R (2012) Optimised robust placement of hard quality sensors for robust monitoring of quality in drinking water distribution systems. In: 10th World Congress on Intelligent Control and Automation (WCICA 2012), pp 1109–1114
85. Rossman LA (2000) EPANET 2 users manual, Cincinnati. US Environmental Protection Agency, OH
86. Berry J, Hart WE, Phillips CA, Uber JG, Watson J-P (2006) Sensor placement in municipal water networks with temporal integer programming models. J Water Resour Plan Manag 132:218–224
87. Afshar A, Mariño MA (2012) Multi-objective coverage-based ACO model for quality monitoring in large water networks. Water Resour Manag 26:2159–2176
88. Perelman L, Ostfeld A (2011) Water-distribution systems simplifications through clustering. J Water Resour Plan Manag 138:218–229
89. Khorshidi MS, Nikoo MR, Sadegh M (2018) Optimal and objective placement of sensors in water distribution systems using information theory. Water Res 143:218–228
90. Davis MJ, Janke R, Phillips CA (2013) Robustness of designs for drinking water contamination warning systems under uncertain conditions. J Water Resour Plan Manag 140:4014028
91. Schwartz R, Lahav O, Ostfeld A (2014) Optimal sensor placement in water distribution systems for injection of chlorpyrifos. In: World environment water resources congress on 2014, pp 485–494
92. Xu J, Fischbeck PS, Small MJ, VanBriesen JM, Casman E (2008) Identifying sets of key nodes for placing sensors in dynamic water distribution networks. J Water Resour Plan Manag 134:378–385
93. Hart WE, Berry JW, Boman EG, Murray R, Phillips CA, Riesen LA, et al (2008) The TEVA-SPOT toolkit for drinking water contaminant warning system design. World Environment Water Resource Congress on 2008 Ahupua'A, pp 1–12
94. Eliades DG, Lambrou TP, Panayiotou CG, Polycarpou MM (2014) Contamination event detection in water distribution systems using a model-based approach. Procedia Eng 89:1089–1096
95. Hamam YM, Hindi KS (1992) Optimised on-line leakage minimisation in water piping networks using neural nets. Proc IFIP Work Conf Dagsch Ger 28:57–64

96. Hamam YM, Brameller A (1971) Hybrid method for the solution of piping networks. Proc Inst Electr Eng 118:1607–1612
97. Ciaponi C, Creaco E, Di Nardo A, Di Natale M, Giudicianni C, Musmarra D et al (2018) Optimal sensor placement in a partitioned water distribution network for the water protection from contamination. Water 2:670
98. Zhuiykov S (2012) Solid-state sensors monitoring parameters of water quality for the next generation of wireless sensor networks. Sens Act B Chem 161:1–20
99. Shen J, Tan H-W, Wang J, Wang J-W, Lee S-Y (2015) A novel routing protocol providing good transmission reliability in underwater sensor networks. 網際網路技術學刊 16:171–178
100. Perelman L, Ostfeld A (2013) Application of graph theory to sensor placement in water distribution systems. World Environment Water Resource Congress on 2013 Showcasing Future, pp 617–625
101. Tabaa Y, Medouri A, Tetouan M (2012) Towards a next generation of scientific computing in the cloud. Int J Comput Sci 9:177–183

Oluwaseye Samson Adedoja received the Bachelor of Engineering degree from the University of Ilorin in 2007, and the Master of Engineering degree from the University of Nigeria, Nsukka in 2015. He is a registered professional engineer with the Council for the Regulation of Engineering in Nigeria (COREN). He is currently a Doctorate student at Tshwane University of Technology, Pretoria, South Africa.

Yskandar Hamam received the bachelor's degree from the American University of Beirut in 1966, the M.Sc. and Ph.D. degrees from the University of Manchester Institute of Science and Technology in 1970 and 1972, respectively, and the "Diplôme d'Habilitation à Diriger des Recherches" degree from the Université des Sciences et Technologies de Lille in 1998. He conducted research activities and lectured in England, Brazil, Lebanon, Belgium, and France. He was the Head of the Control Department and the Dean of the Faculty with ESIEE Paris, France. He was an Active Member in modelling and simulation societies. He was the President of EUROSIM. He was the Scientific Director of the French South African Institute of Technology, Tshwane University of Technology (TUT), South Africa, from 2007 to 2012. He is currently a Emeritus Professor with the Department of Electrical Engineering, TUT and Emeritus Professor at ESIEE-Paris, France. He has co-authored four books, and 40 chapters in edited books. He has also authored or co-authored more than 400 papers in peer reviewed archived journals and in peer-reviewed conference proceedings.

Baset Khalaf completed his B.Sc. degree in Biomedical Engineering in 1984. In 1995 he completed two years of postgraduate studies in medical technology in Germany and in 2004 he was awarded his master's degree in Clinical Engineering at Tshwane University of Technology. In 2012 he was awarded his Doctorate degree from Tshwane University of Technology and Ph.D. from Universite De Versailles Saint-Quentin-En-Yvelines, France in 2013. Prof. Khalaf is the author and co-author of numerous research papers in medical equipment maintenance models and strategies published in accredited international journals.

Emmanuel R. Sadiku studied (B.Sc. Hons, 1980 and Ph.D. 1986) at the University of Strathclyde, Glasgow, Scotland, UK. He lectured at the Federal University of Technology, Owerri, Imo State, Nigeria from June 1986 to Sept. 1989. He then, took-up a 1½ years Postdoctoral Research Fellowship (1989/90) position at the University of Genova/CNR, Genova, Italy (1989-1990). He returned to Nigeria and joined the services of the Federal University of Technology, Minna, Niger State, Nigeria as a Senior Lecturer, in January 1991. In 1997, he spent a year, as a Research Fellow at the KTH Royal Institute of Technology, Stockholm, Sweden. He also had a 3-month research exposure at the Faculty of Physics, University of Freiburg, Germany. He returned to the Federal University of Technology, Minna. In July 1999, he went to the University of Stellenbosch, as a Research Fellow, where he spent 2 years, following which he returned to the Federal University of Technology, Minna, in June 2001. After a year (in June 2002), he took-up a position as a Principal Scientist with Sasol, in Sasolburg, RSA. He spent 1½ years with Sasol, after which, took-up his present position, at the Tshwane University of Technology (TUT), Pretoria, RSA in Jan. 2004, as a Research Professor in Polymer Physics & Engineering. His fields of expertise include Polymer Physics, Polymer Rheology and Polymer Composites/Nanocomposites and Polymers & Polymer Hydrogels as substrates in Drug delivery for BioMedical Applications for Cancer and Diabetics Therapies.

Exposome, Biomonitoring, Assessment and Data Analytics to Quantify Universal Water Quality

Ashok Vaseashta, Gor Gevorgyan, Doga Kavaz, Ognyan Ivanov, Mohammad Jawaid, and Dejan Vasović

Abstract The objective of this chapter is to evaluate assessment methodology of one of the most basic parameters of water, termed as water quality, also defined as Water Quality Index, in view of recent technological advancements. The chapter focusses on drinking water only, however various other applications, such as in agriculture, transportation and industries, viz. pharmaceuticals, power generation, hydrocarbon, catalysis, chemical processing, mining, hydrometallurgy, and semiconductor processing can benefit from a similar approach. The quality of water varies from region to region and is highly dependent on regional water stressors

---

A. Vaseashta (✉)
International Clean Water Institute, Manassas, VA, USA
e-mail: prof.vaseashta@ieee.org

Institute of Electronic Engineering and Nanotechnologies "D. Ghitu", ASM, Chisinau, Moldova

Institute of Biomedical and Nanotechnologies, Riga Technical University, Riga, Latvia

G. Gevorgyan
Scientific Center of Zoology and Hydro-Ecology, National Academy of Sciences of Republic of Armenia, Yerevan, Armenia
e-mail: gev_gor@mail.ru

D. Kavaz
Environmental Research Centre, Cyprus International University, Nicosia, Mersin 10, Nicosia, Turkey
e-mail: dkavaz@ciu.edu.tr

O. Ivanov
GIS Transfer Center, Bulgarian Academy of Sciences, Sofia, Bulgaria
e-mail: ognyan.ivanov@gis-tc.org

M. Jawaid
Universiti Putra Malaysia, 43400 UPM Serdang Seri Kembangan, Selangor, Malaysia
e-mail: jawaid@upm.edu.my

D. Vasović
Faculty of Occupational Safety, University of Niš, Čarnojevića 10a, Niš, Serbia
e-mail: dejan.vasovic@znrfak.ni.ac.rs

© Springer Nature Switzerland AG 2021
A. Vaseashta and C. Maftei (eds.), *Water Safety, Security and Sustainability*, Advanced Sciences and Technologies for Security Applications,
https://doi.org/10.1007/978-3-030-76008-3_4

and available water purification methods. Despite of recent advances in water filtration mechanisms, new and emerging contaminants and local water stressors, are not filtered effectively. Conventional methods use commercial-off the shelf systems and broad-spectrum analytical instruments with interpretive algorithms to sense/detect and characterize contaminants. Since contaminants differ greatly in chemical structure and chromatographic/spectroscopic behavior, it is extremely difficult, if not impossible, to apply generic methods to interrogate most contaminants in real-time. Several regulations have been implemented to control the use of several chemicals that are toxic to human health. We provide an assessment strategy and Internet of Things based sensor platform, combined with geolocation capability, as Smart and Connected Systems approach for interconnectivity. We introduce a Universal Water Quality index, which is based on an algorithmic determination based on a series of contaminants that may be present in quantities significantly less than the acceptable quantities. The methodology presented here is likely to provide Universal Water Quality Index, in compliance with policy frameworks for chemicals that remain in processed water after a series of efficient contamination mitigation strategies, leading to enhanced water quality for better quality of life and better living standards. We provide an overview of new materials and technologies with unforeseen capabilities to sense/detect multiple exposome in label-free and highly multiplexed format, with ppb/ppt sensitivity, high selectivity and specificity, all with streaming data analytics in real-time. Additionally, there are multiple approaches to analyze solutions for regional sustainable development and the one most frequently employed approach is factor analysis for the regional industrial sector as well as approaches based on the monitoring of core determinants of economic growth rates for sustainable development. These approaches are influenced by numerous parameters, which need to be analyzed, systematized, and processed before they are used to create plans of development. Factor analysis approach, consistent with millennium development goals of the United Nations Development Program, was conducted and methodology and conclusion are described.

Keywords Universal water quality index · Contamination · Monitoring · Remediation · Policy · Management

1 Introduction

Water is vitally essential for life on Earth - where there is water, there is life and where water is scarce, sustaining life is quite complex. Clean water is fundamentally essential to sustaining quality of life (QoL) times higher than and supporting basic functions of human body [1], such as homeostasis, maintaining equilibrium between exogenous and endogenous water, regulating metabolism and temperature and filtering out toxins. These functions are due to some of the unique characteristics of water. It is estimated that worldwide, approximately 1.2 billion people lack access to clean drinking water [2]. For many years, most developed countries have had easy access to high-quality fresh water that was relatively easy to obtain, treat and transport, however, global water shortage problems are becoming more widespread due to population growth, drought, and industrial expansion. Due, in part, to these reasons and associated burden on water supply, there is a significant challenge in maintaining adequate yet acceptable water quality for human consumption and in various sectors. In general, safe, secure, and functional water supplies that support industries, irrigation and agriculture, aquatic systems for fisheries and recreation, thermal management, processing, cleaning and similar functions that are indispensable to sustaining modern urban living. Hence, there is increasing concern about the availability of usable water for many communities and industries facing dramatic reductions in traditional freshwater resources due to deteriorating water quality in contaminated wells and contamination from runoff discharges in surface water resources. Global water scarcities have increased the urgency for preservation of freshwater resources and increased water reclamation efforts since high volumes of wastewater are generated daily from residential and industrial discharge. Management of global water resources is, therefore, critical to creating sustainable water supplies for residential, agricultural, and industrial applications. For geographic locations with water scarcity, the primary use of reclaimed water has been primarily for agricultural and landscape irrigation, however for sustainable water sources to meet the ever increasing demand for clean water, the push for alternatives has propelled more innovative, advanced water treatment technologies to efficiently and safely convert wastewater into high quality water for reuse for such applications as, industrial processes and groundwater replenishment [3–5]. Furthermore, water quality index (WQI), as defined using several protocols, is continuously declining, thereby leading to potential adverse health consequences. The term water quality, however, remains elusive, since many countries, regions and even local municipalities use different protocols to assess WQI, i.e., if processed water meets or exceeds their interpretation of water quality. Most experts working on water safety, security, and policy implications believe that to meet the soaring demand for clean water, humankind must find smarter ways of using our water resources and must use a global water standard for human consumption. Implicitly, instead of a regional interpretation and water quality, it is critical to introduce a universal water quality index (UWQI), determined by a set of standards, using state-of-the-art measurements apparatus and cloud-based real-time verification. Although the scope of this study encompasses only for drinking water, the protocol and standards are equally well valid for commercial and agricultural applications.

1.1 Water Quality Index

Water quality typically refers to the physical, chemical and biological state of water, with respect to its suitability for a designated use. The term WQI is vital to the hydro ecosystem and has been in use for quite some time, yet remains an ambiguous term, due to different protocols and methods to calculate the index, since water is used for social and economic purposes as well as for vital human needs [6]. From a historical standpoint, the development of civilizations has led to a shift in a pattern of water use ranging from rural-agricultural to urban-industrial, which naturally mimics a sequence, viz. drinking and personal hygiene, fisheries, navigation and transport, livestock watering and agricultural irrigation, hydroelectric power generation, industrial production, industrial cooling water, recreational activities and wildlife conservation. In the late 1800s and early 1900s, acute waterborne diseases, such as cholera and typhoid fever, prompted proliferation of filtration and chlorination of water. With the realization that various epidemics resulted from and/or spread by waterborne contamination, people learned that the quality of drinking water could not be accurately judged by the natural perception senses, viz. appearance, taste, and smell. As a result, in 1852, a law was passed in London stating that all waters should be filtered. In 1855, epidemiologist Dr. John Snow was able to prove empirically that cholera indeed was a waterborne disease. In the late 1880s, Pasteur demonstrated the germ theory of disease, which was based upon the science of bacteriology. Only after a century of generalized public health observations of deaths due to waterborne disease, the cause-and-effect association was firmly established. In the United States, federal authority to establish drinking water regulations, was originated in 1893 with the enactment by Congress called the Interstate Quarantine Act, under which, the surgeon general of the U.S. Public Health Service (USPHS) was empowered "…. *to make and enforce such regulations as in his judgment are necessary to prevent the introduction, transmission, or spread of communicable disease from foreign countries into the states or possessions, or from one state or possession into any other state or possession.*" This provision of the act resulted in promulgation of the interstate quarantine regulations in 1894 [7].

The first formal and comprehensive review of drinking water concerns was launched in the U.S. in 1913. The first federal drinking water standards were adopted in 1914 [8]. The U.S. Public Health Service (USPHS) was then part of the U.S. Treasury Department and was charged with the task of administering a health care program for sailors in the Merchant Marine. The surgeon general recommended, and the U.S. Treasury Department adopted standards that applied to water supplied to the public by interstate carriers. These standards were commonly referred to as the "Treasury Standards." They included a 100 organisms/mL limit for total bacterial plate count [9]. Further, they stipulated that no more than one of five 10/cc portions of each sample examined could contain *B. coli* (now called *E. coli*). Since the commission that drafted the standards was unable to agree on specific physical and chemical requirements, the provisions of the 1914 standards were limited to the bacteriological quality of water. The standards were revised several times and in 1962 standards, covering 28 constituents, were the most comprehensive pre-Safe Drinking Water Act (SDWA) federal drinking water standards at that time [10]. In 1969, the USPHS's Bureau of Water Hygiene undertook a comprehensive survey of water supplies in the United States, known as the Community Water Supply Study (CWSS) with objective to determine whether the U.S. consumer's drinking water met the 1962 standards [11]. The water quality was determined to be inadequate and that people were being supplied potentially dangerous drinking water. The results of the CWSS generated congressional interest in federal safe drinking water legislation. The first series of bills to give the federal government power to set enforceable standards for drinking water were introduced in 1970 [12]. New legislative proposals for a safe drinking water law were introduced and debated in Congress and after over four years of effort by Congress, federal legislation was enacted to develop a national program to protect the quality of the nation's public drinking water systems [13]. The U.S. House of Representatives and the U.S. Senate passed a safe drinking water bill in November 1974. The SDWA was signed into law on December 16, 1974, as Public Law 93–523, which mandated a major change in the surveillance of drinking water systems by establishing specific roles for the federal and state governments and for public water suppliers [14]. Many other organizations around the globe had similar scientific and engineering work in progress for developing standards for water.

The regulations discussed above provided a mandate to water authorities to assess and compare the detected values of variables with a desirable limit. In fact, many variables are needed to be determined to assess the quality of water. Hence, water quality index (WQI) is defined as the transformation of the measured values of different parameters into a dimensionless number (scale from 0 to 100), using an aggregation method. Horton [15] developed the first WQI based on 10 parameters. Various other indices have been suggested by different institutions and authors. Some of the these indices are the National Sanitation Foundation Water Quality Index [16], Prati's Index [17], Walskie-Parker index [18], Bascaron WQI [19], Bhargava's Index [20], Second Dinius' Index [21], Smith index [22], Aquatic Toxicity Index [23], the Modified Oregon Water Quality Index [24], Overall Index of Pollution [25], Recreational Water Quality Index (RWQI) [26], River Ganga Water Index [27], and Universal Water Quality Index [28]. Each of these indices describes the water quality for general or specific uses. A detailed description of most of these indices is described in a greater detail in Abbasi and Abbasi [29]. In addition, many international water quality standards were developed by several different organizations such as World Health Organization (WHO), the Canadian Council of Ministers of the Environment Water Quality Index (CCME WQI) [30], British Columbia Water quality index (BCWQI) [31] and many similar studies [32–34].

As stated above that there are many procedures and algorithms for calculating the WQI, however, one of the simplest procedures is to analyze water samples in terms of physical, chemical and biological characteristics with different units and weights and to transform the data using sub-indices through statistical analysis into values with similar units [35]. Weights are assigned to each variable according to their importance and an aggregation of sub-indices generates a Cumulative Index.

$$WQI = \Sigma\, q_n w_n / \Sigma w_n \qquad (1)$$

where q_n and S_n are quality rating for the nth water quality and standard permissible value of nth parameter and W_n is the assigned weight value.

Another most commonly used method for WQI is the modified Oregon Water Quality Index (OWQI) [24]. Using enhanced computational and visualization tools, coupled with a better understanding of water quality, the OWQI was updated in 1995 by integrating values of eight water quality variables, including the use of the Delphi method to develop recreational water quality index [26]. To deal with a complex water quality problem, this technique can be used for structuring information based on a group of experts for hierarchical concurrence [36]. For OWQI, the WQI is represented as:

$$\text{WQI} = \sqrt{n / \sum_{i=1}^{n} \frac{1}{SIi}^2} \quad (2)$$

where SI_i = sub-index of each parameters for W_i = weighting factor and n = number of sub-indices. The weighted arithmetic mean formula can be improved by the unweighted harmonic square mean formula to aggregate sub-index results. The advantage of this WQI is that it allows the most impaired variable to impart the greatest influence on the WQI and it gives the significance of different variables on overall water quality at different times and locations.

To simplify complex water quality data, a WQI was developed by the Canadian Council of Ministers of the Environment (CCME) [30], which is a science-based communication tool that tests multi-variable water quality data against specified water quality benchmarks. The CCME WQI mathematically combines three measures of variance (scope, frequency and magnitude) to generate a single unitless number that represents overall water quality at a site relative to the chosen benchmark. In brief, the Canadian Water Quality Index (CWQI) equation is calculated using three factors as follows:

$$\text{WQI} = 100 - \sqrt{((F_1^2 + F_2^2 + F_3^2)/1.732)} \quad (3)$$

F_1 represents scope, i.e. the percentage of variables above the guideline, F_2 represents frequency by which the objectives are not met, and F_3 represents amplitude - the range to which the failed tests are above the guidelines. The constant, 1.732 is a scaling factor to conform within an index span.

There are several other WQIs, including British Columbia WQI (BCWQI) [31], where parameters are measured, and their deviation is determined by comparison with a predefined limit. The River Ganga Index, which is calculated using a weighted sum aggregation function for water quality parameters, such as dissolved oxygen, biochemical oxygen demand (BOD), pH and fecal coliform—all assessed using Delphi or other decision support tools. Since, extensive aquatic toxicity databases are available, aquatic toxicology [37] provides information about toxicology as it relates to environmental aquatic toxicity index (ATI) [23]. The physical water quality parameters employed were pH, dissolved oxygen and turbidity, while the chemical determinant included ammonium, total dissolved salts, fluoride, potassium and orthophosphates and hazardous metals selected were zinc, manganese, chromium, copper, lead and nickel concentrations. Dinius WQI (DWQI) [21] uses a multiplicative water quality index using Delphi for decision making and includes 12 parameters: viz., dissolved oxygen, 5-day BOD, coliform count, *E-coli* count, pH, alkalinity, hardness, chloride, specific conductivity, temperature, color and nitrate.

Achieving universal access to clean water and sanitation has been overarching goal of the Millennium Development Goals (MDGs) [38], Sustainable Development Goals (SDGs) [39] and for many national and international bodies. Protecting America's drinking water is a top priority for Environmental Protection Agency (EPA) in the United States, which established protective drinking water standards for over 90 contaminants, including drinking water regulations issued since the 1996 amendments to the Safe Drinking Water Act (SWDA), that Congress passed in 1974 to protect and strengthen public health protection. A Safe Drinking Water Information System (SDWIS) offers the capability to query Federal Data Warehouse via report filters and various reporting options. Considerable progress has been achieved to meet this goal, however based on anecdotal evidence and the actual number of people in South Asia, Africa and even in Americas, yet without access to clean and good quality water and sanitation facilities on a sustained basis is significantly large. One of the objectives that needs to be integrated in such development goals is a unit of measure, such as Universal Water Quality Index (UWQI). Unami and Kawachi [40] developed a universal optimization scheme to determine a management strategy for controlling water quality for a generic body of water, which can be refined to accomplish the UWQI. Using ubiquitous access to cloud-based technology coupled with a wide spectrum computational algorithm and streaming data analytics to reduce complexity and enhance uniformity for setting standards of the UWQI methodology, UWQI standard is feasible by selecting a finite and relevant number of parameters and contaminants, commonly found in water and processed at treatment plants for drinking water. The steps for assessing UWQI are (a): Selection of water quality parameters to be analyzed; (b): streaming data from measurement of each selected water quality parameter and real-time streaming data analytics; (d): Logical flow of water based on measured parameters; (e): Transform the measurement of water quality parameters into a dimensionless number, and (f): Assign a weight to each water quality parameter and aggregation of sub-indices to obtain the final UWQI value. The flow diagram of the proposed methodology, as depicted in Fig. 1a, will be more efficient since the rating is based on a maximum allowed limit for each contaminant that can be present in the water. Also, since contaminants vary from region to region, the algorithm will only be scanning for those contaminants that reach the maximum allowed limit for a desired water process function (WPF) and bypass scanning for the ones that were not detected in the initial scan. Furthermore, since currently wastewater contains returned pharmaceuticals, volatile organic compounds, radionuclides and Toxic industrial chemicals (TICs)/Toxic industrial materials (TIMs), the algorithmic determination must be based on a series of sensing, computational algorithm and artificial intelligence platforms with decision making capability, to assess UWQI, before water is allowed for consumption. With growing urbanization, yet another strategy to increase efficiency of water filtration system is to use a dual-filtration mechanism separating water for human consumption from the potable use. Additionally, specialized processes, such as granular activated carbon (GAC) adsorption and ion exchange units are occasionally applied for water treatment to control taste- and odor-causing compounds to remove contaminants such as nitrates. A table showing region specific contamination identification remediation

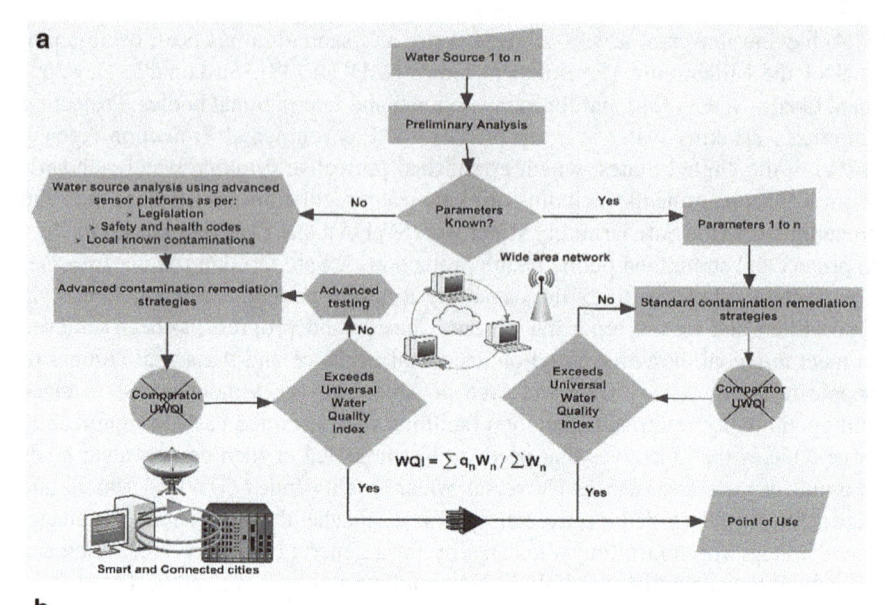

Fig. 1 a Heuristics of proposed universal water quality index methodology. **b**: Heuristic of region and application specific contamination remediation

schemes is shown in Fig. 1b. To get a meaningful information from multivariate data, statistical approaches like Cluster Analysis (CA), Factor Analysis (FA), Principal Component Analysis (PCA), in addition to Discriminate Analysis (DA) are used. A statistical approach lessens the number of assumptions and improves the accuracy of the index.

2 Contaminants in Water

2.1 Exposome and Water Stress

The term exposome was used in early 2005 to highlight the importance of the impact of environmental factors to human health. With increasing population, urbanization, industrialization and emergence of megacity projects worldwide, the contaminants in water and resulting water-stress level have become a major global challenge. Consequently, it has necessitated an integrated understanding of the totality of exposome factors and its overall impact on biota and human genome. The ability to characterize environmental exposures through advanced sensors platforms, Internet of Things (IoT), advanced and streaming data analytics, predictive intelligence, bioinformatics and biomonitoring modality is the key to the framework of exposome research [41–44]. The exposome, generally defined as the environmental exposures from conception onward, will advance our understanding of environmentally induced disease surveillance, by fully assessing the multitude of human exposures over the span of lifetime [45]. A review by Wild et. al. [46] assessed the evidence on people exposed to environmental pollution by a combination of recent technique such as genomics, transcriptomics, epigenomics and metabolomics for several specific profiles of omics. Through exposure assessments and using an interdisciplinary approach, the underlying mechanisms of carcinogenesis was proposed, leading to early detection, identification and an advanced molecular classification through broad spectrum disease surveillance. Vrijheid et al. [49] termed a cooperative project, funded by European Union framework project (E.U. FP7)—called Human Early-Life Exposome (HELIX) [47], which was aimed at implementing novel exposure assessment and biomarker methodology to characterize early-life exposure to multiple environmental factors and relate these with omics biomarkers, as the "early-life exposome". A subsequent study [48] further concluded that a holistic approach is needed to include knowledge of the exposome through a new framework which assimilates disease pathways. An etiological lifecycle analysis can eventually lead to better prevention strategies. It, thereby necessitates, a need to develop technologies for longitudinal, internal, kinetic and external exposure monitoring, and to use wide-spectrum bioinformatics tool [49] to generate, collect, integrate and analyze relevant datasets. The capability and technology to support these types of studies has significantly expanded, yet many challenges to exposome research still remain primarily due to lack of coordinated support from the new technologies [45]. Several recommendations [45–49] reflect on the collective efforts to promote the environmental health and well-being of entire populations and the need for greater data sharing and coordination, development of methodology and minimization of assessment faults, thus encouraging researchers to conduct thorough exposome research by reviewing theoretical and practical challenges, analytical methods, differentiating genetic variation from environmentally induced changes, and addressing the challenge of distinction between susceptibility and resilience.

With increased environmental contamination and limited organization to process and mitigate such contaminants, several water assessments tools and stewardship initiatives continue to emerge with their underlying approaches and methodologies. To put exposome research in perspective, there is a proliferation of sometimes conflicting interpretations of key terms associated with water exposome, such as "Water Scarcity", "Water Stress", and "Water Risk", as they relate to geographic locations, where water-related challenges may be more pronounced. As per the CEOs of Water Mandate, Alliance for Water Stewardship, Carbon Disclosure Project (CDP) Global, Ceres, The Nature Conservancy, Pacific Institute, Water Footprint Network, World Resources Institute, and Water World Federation, the term "Water Stress" refers to the ability, or deficiency thereof, to meet human and ecological demand for fresh water. The term consists of several physical aspects related to water resources, including water availability, water quality, and the accessibility of water, which is often dependent on the adequacy of infrastructure and affordability, among other things. Both water consumption and water withdrawals provide useful information that offer insight into relative water stress, having subjective elements and are assessed differently depending on societal values. The term "Water Scarcity" refers to the volumetric abundance, or deficiency thereof, of freshwater resources. "Scarcity" is human-driven and hence is a function of the volume of human water consumption relative to the volume of available water resources, in a given area, while "Water Risk" refers to the likelihood of an entity experiencing a water-related challenge, and is related to the probability of a specific challenge occurring and the severity of its impact.

Signs of deteriorating water quality and increased stress can be observed all over the world, in both developed and developing countries. The types, magnitudes and extent of water quality problems differ from one country to another and vary even from one part of a country to another. At present almost all water bodies within and near the urban centers of developing countries are seriously contaminated with known and unknown pollutants, thus rendering exposome research a higher priority. However, the situation is getting worse, since all indications point to the fact that it is likely to become increasingly more complex in the future. One of the main reasons, as pointed earlier, is that the number of new and emergent chemicals and returned pharmaceuticals that are being introduced in water supply globally each year. The quantities are significantly large and mostly unknown. It is impossible to reliably assess the health and environmental consequences of all the thousands of new chemicals that are being introduced recently and the new ones that are likely to be introduced in the coming years. It is thus difficult to make evidence-based decisions as to how many water quality parameters should be measured regularly so that a cost-effective management system can be formulated and strictly implemented.

By persistent monitoring the exposome data using the latest state-of-the-art technological advancements, cyber-physical systems, bio-informatic platforms, data analytics and artificial intelligence, it is possible to understand the current limitations and future needs. Advancements in fields of big data analytics for standardization for exposure and data analyses, data sharing, metabolomics and bioinformatics are the driving force in biomonitoring and hence current emphasis is placed on monitoring the entirety of exposures over the lifetime of individuals. More recently, reports by Sarigiannis et al. [50, 51] provided a comprehensive overview of the modelling and data analytics needed for advanced internal and external exposure assessments, by examining the human personal exposome, starting from more accurate methods of measurement, such as use of wearable sensors, to extrapolation of the data to larger population groups. Exposure reconstruction algorithms, coupled with bio-toxicokinetic, support biomonitoring data interpretation and assimilation. A recent report [52] provides an outline of the field of computational exposure biology and shows how the bioinformatics techniques and algorithms are coupled with big data analytics to perform studies of exposome-wide associations with putative adverse health outcomes to decipher the molecular and metabolic pathways of induced toxicity related to environmental chemical stressors through multiple exposure pathways.

Despite of these technological advances, water scarcity remains a major challenge. It is estimated that nearly one-fifth of all countries worldwide will face water scarcity problems by 2040. According to an analysis conducted by the World Resources Institute (WRI) [53], a non-profit organization that investigates the planet's natural resources, most countries in the Middle East, South Asia and Africa are among those with the most severe water stress. The most vulnerable countries are in the Middle East with the high temperatures and minimum rainfall. In South America, currently Chile has deteriorated from average water stress in 2010 to now being considered as one of the locations, likely to experience significant water stress by 2040, mainly resulting from rising temperatures and changing rainfall patterns in the region. Spain and Greece top the ranks in European countries largely affected by water stress, joining their neighboring North African countries.

2.2 Common Contaminants in Drinking Water

Drinking water can be received from two basic sources, viz. surface waters (rivers, reservoirs, etc.) and groundwater. The body of water contains both incidental and anthropogenic contaminants. In this regard, groundwater is safer than surface waters since it is less vulnerable to pollution. Natural contaminants enter water from the geological strata through which the water flows. Anthropogenic contaminants pollute the water from different sources such as industrial waste, agriculture nutrients runoffs, sewage treatment, etc. [54]. It is estimated that there are ~1407 species of microbial and viral pathogens that can infect humans through drinking water. These pathogens include bacteria (~538 species), viruses (~208 types), fungi (~317 species), parasitic protozoa (~57 species) and helminths (~287 species). *Vibrio cholerae, Salmonella, Shigella, Escherichia coli, Yersinia, Campylobacter* are several known examples of common bacterial pathogens in drinking water, while *Hepatitis A virus, Hepatitis E virus, Rotaviruses, Human caliciviruses, Enteric adenoviruses, Astroviruses* are common viral pathogens and *Giardia, Cryptosporidium, Cyclospora, Microsporidia, Entamoeba histolytica, Naegleria, Toxoplasma gondii* are common protozoan pathogens. Human and animal wastes are the main sources of microbial and viral contamination in water [55]. There are many chemicals in drinking water that are regulated by the WHO, the U.S. EPA and the European Union Council [56]. Some chemical contaminants such as nitrate, nitrite, fluoride, undesirable organic compounds (benzene and polycyclic aromatic hydrocarbons (PAHs), pesticides and pesticide residues, bromate, Trihalomethanes (THMs), heavy metals in drinking water are of particular concern [54, 57]. Among them, nitrate, due to its potential strong health effects, which is one of the most common chemical contaminants detected in drinking water throughout the world.

The sources of fluoride are minerals like topaz, fluorite, fluorapatite, theorapatite, cryolite, phosphorite, etc. Pesticides and their degradation products enter water mainly via agriculture. The water sources can be polluted by PAHs through anthropogenic and several natural inputs or biological conversion of fossil fuel products. Water treatment by chloride is the main source of THMs formation due to the reaction of available chlorine with natural organic matters and inorganic substances in water [57]. Groundwater and surface water are being polluted with heavy metals induced by anthropogenic sources (e.g. industrial, agricultural, mining and traffic activities) and natural processes [58]. Although many water quality parameters, e.g. acidity, turbidity, major ions, total coliform bacteria, can be easily and routinely controlled, some parameters such as naturally occurring and man-made radionuclides require more specialized analysis [59]. Drinking water contains naturally occurring radionuclides (i.e. ^{238}U, ^{226}Ra, ^{40}K, etc.) induced by minerals such as pitch blend, monazites, as well as artificial radionuclides (e.g. ^{137}Cs, ^{134}Cs, ^{90}Sr, etc.) coming from fallout from nuclear weapons testing and the accidents at nuclear reactors [60, 61]. Table 1 provides a list of chemicals that are commonly present in surface and ground water and are monitored using sensor platforms. The list is quite exhaustive and not complete, since new chemicals are routinely added via various discharges, especially with rapid pace of innovations in materials design and technologies. These new and emerging chemicals need to be detected, monitored and added to the table.

2.2.1 Microplastics in Aquatic Environment

Growing scientific and public awareness of microplastic debris in the marine and freshwater environment is of paramount concern due to its impact on aquatic ecosystems and public health by transport through drinking water. The source of marine plastics is mainly from rivers, industrial and urban effluents and land runoff [62], and can be found in the rivers and lakes, in municipal and industrial wastewater [63] and in marine environment [62–64]. The total quantity of waste plastics in the marine environment is estimated to be ~250 KTons [64]. Microplastics are produced due to physical disintegration and chemical or biological degradation of plastic materials, arising mainly of plastic tableware, single-use beverage bottles, cosmetics, plastic composites, which disintegrate into microfibers and microspheres [65]. Microplastics

Table 1 List of chemicals that are commonly present in surface and ground water

Category	Contaminants	MCLG[1] (mg/L)[2]	MCL(mg/L)[2] MCLG, MRDLG	Potential health effects from long-term exposure > MCL (unless otherwise specified as short term)	Sources of contaminant in drinking water
Initial water quality parameters	Visual, smell, taste				
Microorganisms	*Cryptosporidium*[a]	Zero		Gastrointestinal (diarrhea, vomiting, cramps)	Human and animal fecal waste
	Giardia lamblia[b]	Zero		Gastrointestinal (diarrhea, vomiting, cramps)	Human and animal fecal waste
	Heterotrophic plate count (HPC)[c]	n/a		HPC is an analytic method used to measure the variety of bacteria that are common in water. The lower the concentration of bacteria, the better is the drinking water	HPC measures a range of bacteria that are naturally present in the environment
	Legionella[d]	Zero		Legionnaire's Disease - a type of pneumonia	Found naturally in water; multiplies in heating systems
	Total Coliforms (including fecal coliform and *E. Coli*)[e]	Zero	5.00%	Not a health threat in itself; it is used to indicate whether other potentially harmful bacteria may be present5	Coliforms are naturally present in the environment and feces. Fecal coliforms and *E. coli* come from human and animal fecal waste
	Turbidity[f]	n/a		Higher turbidity levels ~ higher levels of disease-causing microorganisms, viz. viruses, parasites and some bacteria. Causes nausea, cramps, diarrhea and headaches	Soil runoff
	Viruses (enteric)[g], COVID-19[h]	Zero		Gastrointestinal illness (viz. diarrhea, vomiting, and cramps), fever chills, COV-SARS2	Human and animal fecal waste, Respiratory

(continued)

Table 1 (continued)

Category	Contaminants	MCLG[1] (mg/L)[2]	MCL(mg/L)[2] MCLG, MRDLG	Potential health effects from long-term exposure > MCL (unless otherwise specified as short term)	Sources of contaminant in drinking water
Disinfection Byproducts	Bromate[i]	Zero	0.01	Increased risk of cancer	Byproduct of drinking water disinfection
	Chlorite[i]	0.8	1	Anemia in young children: nervous system	Byproduct of drinking water disinfection
	Haloacetic acids (HAA5)[i]	n/a6	0.06	Increased risk of cancer	Byproduct of drinking water disinfection
	Total Trihalomethanes (TTHMs)[i]	– > n/a6	> 0.080	Liver, kidney or central nervous system problems; increased risk of cancer	Byproduct of drinking water disinfection
Disinfectants	Chloramines (as Cl_2)[i,j]	MRDLG = 41	MRDL = 4.01	Eye/nose irritation; stomach discomfort, anemia	Water additive used to control microbes
	Chlorine (as Cl_2)[i,j]	MRDLG = 41	MRDL = 4.01	Eye/nose irritation; stomach discomfort	Water additive used to control microbes
	Chlorine dioxide (as ClO_2)[i,j]	MRDLG = 0.81	MRDL = 0.81	Anemia; infants and young children: nervous system effects	Water additive used to control microbes
Inorganic chemicals	Antimony[k]	0.006	0.006	Increases cholesterol; decreases blood sugar	Discharge from petroleum refineries; fire retardants; ceramics; electronics; solder
	Arsenic[l]	0	0.010 as of 01/23/06	Skin damage or problems with circulatory systems, increased risk of getting cancer	Erosion of natural deposits; runoff from orchards, glass and electronics wastes
	Asbestos[m] (fiber > 10 μm)	7 million fibers/liter	7 million fibers/liter	Increased risk of developing benign intestinal polyps	Decay of asbestos cement in water mains; erosion of natural deposits
	Barium[n]	2	2	Increase in blood pressure	Discharge from drilling wastes and metal refineries; erosion of natural deposits

(continued)

Table 1 (continued)

Category	Contaminants	MCLG[1] (mg/L)[2]	MCL (mg/L)[2] MCLG, MRDLG	Potential health effects from long-term exposure > MCL (unless otherwise specified as short term)	Sources of contaminant in drinking water
	Beryllium[o]	0.004	0.004	Intestinal lesions	Discharge from metal refineries and coal-burning factories; and from electrical, aerospace, and defense industries
	Cadmium[p]	0.005	0.005	Kidney damage	Corrosion of galvanized pipes; erosion of natural deposits; discharge from metal refineries and waste batteries and paints
	Chromium[q] (total)	0.1	0.1	Allergic dermatitis	Discharge from steel and pulp mills; erosion of natural deposits
	Copper[r]	1.3	TT7; Action Level = 1.3	Short term exposure: Gastrointestinal distress, Long term exposure: Liver or kidney damage, Aggravates Wilson's Disease	Corrosion of household plumbing systems; erosion of natural deposits
	Cyanide[s] (as free cyanide)	0.2	0.2	Nerve damage or thyroid problems	Discharge from steel/metal factories and from plastic and fertilizer factories
	Fluoride[s]	4	4	Bone disease (pain and tenderness of the bones); Children may get mottled teeth	Water additive which promotes strong teeth; erosion of natural deposits; discharge from fertilizer and aluminum factories
	Lead[t]	Zero	Action Level = 0.015	Delays in physical or mental development; children could show slight deficits in attention span and learning abilities; Kidney problems; high blood pressure	Corrosion of household plumbing systems; erosion of natural deposits

(continued)

Table 1 (continued)

Category	Contaminants	MCLG[1] (mg/L)[2]	MCL (mg/L)[2] MCLG, MRDLG	Potential health effects from long-term exposure > MCL (unless otherwise specified as short term)	Sources of contaminant in drinking water
	Mercury[u] (inorganic)	0.002	0.002	Kidney damage	Erosion of natural deposits; discharge from refineries and factories; runoff from landfills and croplands
	Nitrate[v] (measured as Nitrogen)	10	10	Infants < 6 mo. could become seriously ill and, if untreated, may die. Symptoms include shortness of breath and blue-baby syndrome	Runoff from fertilizer use; leaking from septic tanks, sewage; erosion of natural deposits
	Nitrite[v] (measured as Nitrogen)	1	1	Infants < 6 mo. could become seriously ill and, if untreated, may die. Symptoms include shortness of breath and blue-baby syndrome	Runoff from fertilizer use; leaking from septic tanks, sewage; erosion of natural deposits
	Selenium[w]	0.05	0.05	Hair or fingernail loss; numbness in fingers or toes; circulatory problems	Discharge from petroleum refineries and mines; erosion of natural deposits
	Thallium[x]	0.0005	0.002	Hair loss; changes in blood; kidney, intestine, or liver problems	Leaching from ore-processing; discharge from electronics, glass, and drug process
Organic contaminants	Acrylamide[y,z,a1]	Zero		Nervous system, blood problems; risk of cancer	Added during wastewater treatment
	Alachlor[y,z,a1]	Zero	0.002	Eye, liver, kidney, spleen; anemia; risk of cancer	Runoff from herbicide used on row crops
	Atrazine[y,z,a1]	0.003	0.003	Cardiovascular and reproductive problems	Runoff from herbicide used on row crops
	Benzene[y,z,a1]	Zero	0.005	Anemia, decreased blood platelets, cancer	Discharge from factories; leaching gas storage tanks and landfills
	Benzo(a)pyrene (PAHs)[y,z,a1]	Zero	0.0002	Reproductive difficulties; risk of cancer	Leaching from of water tanks and lines

(continued)

Table 1 (continued)

Category	Contaminants	MCLG[1] (mg/L)[2]	MCL (mg/L)[2] MCLG, MRDLG	Potential health effects from long-term exposure > MCL (unless otherwise specified as short term)	Sources of contaminant in drinking water
	Carbofuran [y,z,a1]	0.04	0.04	Problems with blood, nervous system, or reproductive system	Insecticide, leaching of soil fumigant used on rice and alfalfa
	Carbon tetrachloride [y,z,a1]	Zero	0.005	Liver problems; increased risk of cancer	Discharge from chemical plants and other industrial activities
	Chlordane [y,z,a1]	Zero	0.002	Liver and nervous system; risk of cancer	Residue of banned termiticide
	Chlorobenzene	0.1	0.1	Liver or kidney problems	Discharge from chemical and agricultural chemical factories
	2,4-Dichlorophenoxyacetic acid[y]	0.07	0.07	Kidney, liver, or adrenal gland problems	Runoff from herbicide used on row crops
	Dalapon [y,z,a1]	0.2	0.2	Minor kidney changes	Runoff from herbicides on rights of way
	1,2-Dibromo-3-chloropropane (DBCP) [y,z,a1]	Zero	0.0002	Reproductive difficulties; increased risk of cancer	Runoff/leaching from soil fumigant on soybeans, cotton, pineapples in orchards
	o-Dichlorobenzene [y,z,a1]	0.6	0.6	Liver, kidney, or circulatory system problems	Release from chemical factories
	p-Dichlorobenzene [y,z,a1]	0.075	0.075	Anemia; liver, kidney or spleen damage; changes in blood	Discharge from industrial chemical sites
	1,2-Dichloroethane [y,z,a1]	Zero	0.005	Increased risk of cancer	Discharge from chemical factories
	1,1-Dichloroethylene [y,z,a1]	0.007	0.007	Liver problems	Discharge from chemical factories
	cis-1,2-Dichloroethylene [y,z,a1]	0.07	0.07	Liver problems	Discharge from chemical factories
	trans-1,2-Dichloroethylene [y,z,a1]	0.1	0.1	Liver problems	Discharge from chemical factories
	Dichloromethane [y,z,a1]	zero	0.005	Liver problems; increased risk of cancer	Discharge from drug/chemical factories

(continued)

Table 1 (continued)

Category	Contaminants	MCLG[1] (mg/L)[2]	MCL(mg/L)[2] MRDLG	MCLG, MRDLG Potential health effects from long-term exposure > MCL (unless otherwise specified as short term)	Sources of contaminant in drinking water
	1,2-Dichloropropane [y,z,a1]	Zero	0.005	Increased risk of cancer	Discharge from chemical factories
	Di(2-ethylhexyl) adipate [y,z,a1]	0.4	0.4	Weight loss, liver, reproductive difficulties	Discharge from chemical factories
	Di(2-ethylhexyl) phthalate [y,z,a1]	Zero	0.006	Reproductive; liver problems; risk of cancer	Release from rubber, chemical factories
	Dinoseb [y,z,a1]	0.007	0.007	Reproductive difficulties	Herbicide used on soybeans, vegetables
	Dioxin (2,3,7,8-TCDD) [y,z,a1]	Zero	0.00000003	Reproductive difficulties; increased risk of cancer	Emissions from waste incineration and combustion; Runoff from chem. factories
	Diquat [y,z,a1]	0.02	0.02	Cataracts	Runoff from herbicide use
	Endothall [y,z,a1]	0.1	0.1	Stomach and intestinal problems	Runoff from herbicide use
	Endrin [y,z,a1]	0.002	0.002	Liver problems	Residue of banned insecticide
	Epichlorohydrin [y,z,a1]	Zero		Increased cancer risk, and over a long period of time, stomach problems	Discharge from chemical factories; from certain water treatment chemicals
	Ethylbenzene [y,z,a1]	0.7	0.7	Liver or kidneys problems	Discharge from petroleum refineries
	Ethylene dibromide [y,z,a1]	Zero	0.00005	Problems with liver, stomach, reproductive system, or kidneys; increased risk of cancer	Discharge from petroleum refineries
	Glyphosate [y,z,a1]	0.7	0.7	Kidney problems; reproductive difficulties	Runoff from herbicide use
	Heptachlor [y,z,a1]	Zero	0.0004	Liver damage; increased risk of cancer	Residue of banned termiticide

(continued)

Table 1 (continued)

Category	Contaminants	MCLG[1] (mg/L)[2]	MCL (mg/L)[2] MCLG, MRDLG	Potential health effects from long-term exposure > MCL (unless otherwise specified as short term)	Sources of contaminant in drinking water
	Heptachlor epoxide [y,z,a1]	Zero	0.0002	Liver damage; increased risk of cancer	Breakdown of heptachlor
	Hexachlorobenzene [y,z,a1]	Zero	0.001	Liver or kidney problems; reproductive difficulties; increased risk of cancer	Discharge from metal refineries and agricultural chemical factories
	Hexachlorocyclopentadiene [y,z]	0.05	0.05	Kidney or stomach problems	Discharge from chemical factories
	Lindane [y,z,a1]	0.0002	0.0002	Liver or kidney problems	Runoff/leaching from insecticide used on cattle, lumber, gardens
	Methoxychlor [y,z,a1]	0.04	0.04	Reproductive difficulties, risk of cancer	Runoff/leaching from insecticide used on fruits, vegetables, alfalfa, livestock
	Methylene blue, 3,7-bis (Dimethyl amino) phenazathionium chloride, tetramethyl thionine chloride [y]	Based on physiological conditions	Based on physiological conditions	Dye, endoscopic polypectomy biological stains, electroplating	Hypertension, dizziness, CNS, narcosis, fecal discoloration, bladder inflammation, anemia
	MTBE (methyl-t-butyl ether) [y,z]	Research, unregulated contaminant	Research, unregulated contaminant	Fuel oxygenates added to fuel to increase its oxygen content to reduce carbon monoxide and ozone levels caused by auto emissions	Increased amounts detected in surface water, health effects unknown at this time
	Oxamyl (Vydate) [y,z,a1]	0.2	0.2	Slight nervous system effects	Runoff/leaching from insecticide used on apples, potatoes, and tomatoes
	Polychlorinated biphenyls (PCBs) [y,z,a1]	Zero	0.0005	Skin, thymus gland, immune deficiencies, reproductive or nervous, risk of cancer	Runoff from landfills; discharge of waste chemicals
	Pentachlorophenol [y,z,a1]	Zero	0.001	Liver or kidney problems; increased cancer risk	Discharge from wood preserving factories

(continued)

Table 1 (continued)

Category	Contaminants	MCLG[1] (mg/L)[2]	MCL(mg/L)[2] MCLG, MRDLG	Potential health effects from long-term exposure > MCL (unless otherwise specified as short term)	Sources of contaminant in drinking water
	Picloram [y,z,a1]	0.5	0.5	Liver problems	Herbicide runoff
	Simazine [y,z,a1]	0.004	0.004	Problems with blood	Herbicide runoff
	Styrene [y,z,a1]	0.1	0.1	Liver, kidney, or circulatory system problems	Rubber and plastic sites; landfills
	Tetrachloroethylene [y,z,a1]	Zero	0.005	Liver problems; increased risk of cancer	Discharge from factories and cleaners
	Toluene [y,z,a1]	1	1	Nervous system, kidney, or liver problems	Discharge from petroleum factories
	Toxaphene [y,z,a1]	Zero	0.003	Kidney, liver, or thyroid problems; cancer	Runoff/leaching from insecticides
	2,4,5-TP (Silvex) [y,z,a1]	0.05	0.05	Liver problems	Residue of banned herbicide
	1,2,4-Trichlorobenzene [y,z,a1]	0.07	0.07	Changes in adrenal glands	Discharge from textile finishing factories
	1,1,1-Trichloroethane [y,z,a1]	0.2	0.2	Liver, nervous system, or circulatory problems	From metal degreasing factories
	1,1,2-Trichloroethane [y,z,a1]	0.003	0.005	Liver, kidney, or immune system problems	Discharge from chemical factories
	Trichloroethylene [y,z,a1]	Zero	0.005	Liver problems; increased risk of cancer	Metal degreasing sites and factories
	Vinyl chloride [y,z,a1]	Zero	0.002	Increased risk of cancer	Leaching from PVC pipes, plastic sites
	Xylenes (total) [y,z,a1]	10	10	Nervous system damage	Discharge from petroleum, chemical site

(continued)

Table 1 (continued)

Category	Contaminants	MCLG[1] (mg/L)[2]	MCL (mg/L)[2] MCLG, MRDLG	Potential health effects from long-term exposure > MCL (unless otherwise specified as short term)	Sources of contaminant in drinking water
Radionuclides	Alpha particles[b1,c1]	None >> zero	15 pCi/L	Increased risk of cancer	Erosion of natural deposits of certain minerals that are radioactive and Photons and beta radiation emission
	Beta particles and photon emitters[b1,c1]	None >> zero	4 mrems/year	Increased risk of cancer	Decay of natural and man-made deposits of radioactive minerals, Photons and beta radiation emission
	Radium 226/228 (combined)[c1]	None >> zero	5 pCi/L	Increased risk of cancer	Erosion of natural deposits
	Uranium[b1,c1]	Zero	30 ug/L	Increased risk of cancer, kidney toxicity	Erosion of natural deposits
Pharmaceutical	Antibiotics (metronidazole, tinidazole, and trimethoprim-sulfamethoxazole)[d1]	Physiological conditions	Physiological conditions	Semisynthetic modifications of various natural compounds. Presence in water from recycling	Long term impairment of immunity, loss of probiotics
	Steroids[e1]	Based on physiological conditions	Based on physiological conditions	Semisynthetic modifications of various natural compounds. Presence in water mostly from recycled water	Weaken the body's immune system, risk of insomnia, infertility, and excess fluid retention
	Testosterone[e1]	Based on physiological conditions	Based on physiological conditions	Semisynthetic modifications of various natural compounds. Presence in water mostly from recycled water	Weaken the body's immune system, risk of insomnia, infertility, and excess fluid retention
	Acetaminophen[f1]	Physiological conditions	Physiological conditions	Semisynthetic modifications of various natural compounds. Presence in water from recycling	Harmful to liver (base on dose level)
	Ibuprofen[g1]	physiological conditions	physiological conditions	Semisynthetic modifications of various natural compounds. Presence in water from recycling	Nausea, heartburn, and stomach pain including heart attack
	Oxandrolone[h1]	Physiological conditions	Physiological conditions	Semisynthetic modifications of various natural compounds. Presence in water from recycling	allergic reaction, edema, acne, headache,

(continued)

Table 1 (continued)

Category	Contaminants	MCLG[1] (mg/L)[2]	MCL(mg/L)[2] MRDLG	Potential health effects from long-term exposure > MCL (unless otherwise specified as short term)	Sources of contaminant in drinking water
	Oxymetholone[ii]	Physiological conditions	Physiological conditions	Semisynthetic modifications of various natural compounds. Presence in water recycling	Abnormal skin sensations, acne, anxiety, baldness, lethargy, body and headache

a): https://www.epa.gov/sites/production/files/2015-10/documents/cryptosporidium-report.pdf s): https://www.atsdr.cdc.gov/phs/phs.asp?id=70&tid=19
b): https://www.haguewaterofmd.com/effects-drinking-dirty-andor-contaminated-water/ t): https://www.cdc.gov/niosh/topics/lead/health.html
c): https://www.who.int/water_sanitation_health/dwq/WSH02.10.pdf u): https://www.epa.gov/mercury/health-effects-exposures-mercury
d): https://www.epa.gov/ground-water-and-drinking-water/legionella v): https://www.healthline.com/nutrition/are-nitrates-and-nitrites-harmful
e): https://www.healthline.com/health/e-coli-infection w): https://pubmed.ncbi.nlm.nih.gov/28339083/
f): https://www.who.int/water_sanitation_health/publications/turbidity-information-200217.pdf x): https://www.medicinenet.com/thallium/article.htm
g): https://www.ncbi.nlm.nih.gov/pmc/articles/PMC1197419/ y): https://www.epa.gov/ccl/types-drinking-water-contaminants
h): https://www.rti.org/insights/covid-19-impact-on-us-systems z): https://standard.wellcertified.com/water/organic-contaminants
i): https://www.cdc.gov/safewater/chlorination-byproducts.html#usepa a1): https://pubmed.ncbi.nlm.nih.gov/8279720/
j): https://www.who.int/water_sanitation_health/dwq/chemicals/antimony.pdf b1: https://www.epa.gov/dwreginfo/radionuclides-rule
k): https://www.who.int/water_sanitation_health/dwq/chemicals/chloramine-disinfection.html c1): https://www.ncbi.nlm.nih.gov/pmc/articles/PMC3261972/
l): https://www.ncbi.nlm.nih.gov/books/NBK230891/ d1): https://thegrownetwork.com/antibiotics-water-supply/
m): https://www.ecomena.org/environmental-health-impacts-of-asbestos/ e1: https://www.healthline.com/health/steroids-and-vision
n): https://www.who.int/water_sanitation_health/water-quality/guidelines/chemicals/barium-background-jan17.pdf
o): https://www.who.int/water_sanitation_health/dwq/chemicals/beryllium_in_drinking2.pdff1: https://www.webmd.com/drugs/2/drug-362/acetaminophen-oral/details
p): https://www.epa.gov/sites/production/files/2016-09/documents/cadmium-compounds.pdf g1): https://www.drugs.com/sfx/ibuprofen-side-effects.html
q): https://www.epa.gov/sdwa/chromium-drinking-water h1): https://www.drugs.com/mtm/oxandrolone.html
r): https://www.cdc.gov/healthywater/drinking/private/wells/disease/copper.html i1): https://pubchem.ncbi.nlm.nih.gov/compound/oxymetholone

are now considered as "contaminants of emerging concern" in freshwater environments [66], thus garnering interest of many recent investigators. Most common polymers found in aquatic environment are polyurethane, polycarbonate, polystyrene, polyvinylchloride, polyethylene and polypropylene. Due to floatability of polyethylene and polypropylene, the floating microparticles of these plastics are not isolated during water and wastewater treatment processes, because during such processes, filtration occurs through rapid sand filters with typical size of sand ~500–1200 μm or using slow sand filter with the size of sand ~150–300 μm. These filters, even in conjunction with activated carbon, do not retain the particles that are less 100 μm from drinking water treatment facilities [67] and hence, do not settle in the sedimentation tanks. Therefore, such particles are returned to the aquatic environment after wastewater treatment or transported through the drinking water treatment plant via the distribution system.

Disintegration and dispersion of plastic in aquatic environment and formation of fine plastic particulates releases certain additives to plastics during production process, such as phthalates, polybrominated diphenyl ethers and alkylphenols, all of which are carcinogenic, (neuro) toxic, mutagenic and have endocrine disrupting characteristics. Additionally, plastics microbeads are abundantly used in commercial facial and body scrubs, toothpastes and cosmetic products. Floating microplastics can also be considered as vectors for hydrophobic organic chemicals [68]. Hydrophobic pollutants of water are concentrated near water–air interphase and adsorb on hydrophobic surface of floating microplastics. The concentration of hydrophobic substances associated with microplastics, typically range on the orders of magnitude greater than surrounding seawater [69], since plastic concentration in whole ocean is $\sim 2*10^{-9}$ g/L^2 and as compared to $8*10^{-2}$ g/L^2 in highly contaminated rivers [70].

Microplastics are normally ingested by aquatic species as they enter through their tissues and cells with potentially adverse effects causing physical damage and intoxication by chemicals released by microplastics [71]. It was discovered recently that the sea turtles are attracted to the smell emanating from the bio-fouled plastic and are not attracted by the smell from pure microplastic [72]. Microplastics are readily ingested by zooplankton, that is a food source for the secondary consumers [73]. Microplastic particles that are coated with microbial biofilm appear to be attractive feed for zooplankton, which is normally consumed by fish, and finally by humans through the food supply chain. Although much of the marine debris, as described before, focuses on floating plastic debris, it is important to recognize that only approximately half of all plastic is positively buoyant, which is dependent on the density of the material and the presence of entrapped air. After floating, plastic debris may become sufficiently fouled with biological growth that the density becomes greater than seawater, and it sinks. Therefore, more datasets from similar investigations are necessary to comprehend ecological significance of adsorbed hydrophobic water pollutants and their potential impacts on aquatic and marine wildlife, and for humans from secondary exposure via the food chain.

Presently, there is limited information on human and non-human models. Many factors likely affect the absorption of microplastics, viz. size, shape, polymer, charge, hydrophobicity and physiological factors. Major toxicokinetic mechanisms are identified as endocytosis and persorption, while limited toxicodynamic data using animal studies show inflammation during liver histology and oxidative stress, energy and lipid metabolism [74]. It is further known that microplastics are present in food and those with size <20 μm would be able to penetrate organs, while the microplastics with the sizes from <0.1 to 10 μm would be able to penetrate through the cell membranes, the blood–brain barrier and the placenta and can cause cytotoxic effects [75]. However, the fate of microplastics in the human body after ingestion of the particles and their toxicokinetic are also not well understood [76, 77]. The problem with toxicity of microplastic particles may be considered as acute as is the case with nanoparticles [78] since both are considered as emerging environmental contaminants of global concern and their characteristics change with their reduction in dimensions [78]. Our earlier investigations of nanoparticles [78–80] in aquatic media along with this investigation potentially may serve as a Segway for toxicokinetic investigations for mixed pollutants in environment, including Persistent Bioaccumulative and Toxic (PBT) chemicals. Even at significantly low concentrations, PBTs can be illusive in the environment due to their ability to bio-magnify, leading to toxic effects at higher trophic levels even though their ambient concentrations are well below toxic thresholds. A subgroup of PBTs are known as persistent organic pollutants (POPs), such as Dichlorodiphenyltrichloroethane (DDT), dioxins, and Polychlorinated biphenyl (PCBs), which are subject of our ongoing investigations. Microplastics can leach in a set of toxic compounds, such as bisphenol A, phthalates with endocrine disruption and carcinogenic properties [74, 76]. It is for this reason, some plastic additives mimic estrogen, with Estradiol Equivalency Factor (EEF) $\sim 2*10^{-4}$ [81]. Additionally, due to the large specific surface area, the polyethylene and polypropylene microparticles exhibit high and reversible sorption of hydrophobic pollutants from aquatic environment, thus making the microplastics even more toxic with significant adverse impact on water quality.

2.3 Influence of Environmental Factors on Water Quality

Due to omnipresence of water in the environment, its survival requirement for all living organisms, and ubiquitous applications requiring reliable and resilient resources of water, it is imperative that good quality of water is a quintessential requirement. However, the quality of water is constantly impacted by several direct, incidental and anthropogenic activities. These activities increase the concentrations of dissolved or suspended contaminants and change the acidity of water, along with several environmental and anthropogenic factors. In fact, anthropogenic activities increasingly provoke deleterious impacts in aquatic ecosystems, due to its profound effect on organisms. As stated earlier that the most common anthropogenic pollution that impacts aquatic ecosystems are sewage, nutrients and terrigenous materials, crude oil, heavy metals and microplastics. Several emerging contaminants include volatile organic compounds, toxic industrial chemicals (TIC), toxic industrial metals

(TIMs), returned pharmaceuticals, pesticides and Styrofoam. Impact of water on the environment is typically the most severe during extreme events, such as floods and droughts, effecting human health, pets, wildlife, aquatic life, agriculture and vegetation, infrastructure, industry and transportation. A lot of progress has been made over the last 30 years in reducing the human impacts to our water resources, yet a widespread and serious problem still exists from non-point sources.

Recent disruption caused by the nCOV-SARS2 (COVID-19) pandemic experienced by over 187 countries has resulted in unprecedented impact on the environment. There is a considerable drop in travel resulting in a 25–30% reduction in carbon emissions and 50% reduction in nitrogen oxides emissions. From governance standpoint, the European Union's seven-year plan calls for €1 trillion budget proposal and €750 billion recovery plan "Next Generation EU" which seeks to reserve 25% of EU spending for climate-friendly expenditure. Notwithstanding, the outbreak has resulted in additional waterborne disease propagation due to sewage water, which remains largely uncontrolled, deforestation of the Amazon rainforest and reports of poaching in Africa. The excessive amount of personal protective equipment, masks, cleaning agents, and extra packaging material for food, shipping and other services, is causing landfill overrun at several locations. Among other anthropogenic activities, microplastics contamination in water is an imminent and emerging challenge followed by returned pharmaceutics, genetically modified organisms (GMOs) and nanoparticles. While the focus here remains on quality of drinking water, the topic of water quality is of paramount importance due to its significance in society and entire ecosystem, especially as contaminants in the environment continue to increase stressing surface and ground water resources.

2.4 Health Impacts of Common Contaminants in Water

A good quality of water has positive impact of human health while a poor quality of water can have negative health effects, since it may contain microbiological and chemical contaminants [55] and other emerging contaminants. It is suggested that 80% of all human illness in developing countries is associated with the inadequate quality of water and consumption of polluted water is a major worldwide cause of deaths, causing approximately over 14,000 deaths/day. According to the WHO World Health Report, over 1 Billion people did not have an adequate and safe water supply in early 2000, of which 800 million were in rural areas [54] and currently the number without access to clean water is over 2 Billion people, as demographics of people in rural area continues to change. Although, access to safe drinking water is considered essential for all, the political pressure argues, such access as a basic human right [82]. At the beginning of the 20th Century, cholera and typhoid fever were the major waterborne disease of concern to human health. Although, the seventh Cholera pandemic started in 1961 and caused 4700 deaths in South America [54], it should be noted that microbiological contamination of drinking water, continues to significantly cause diarrheal diseases and still remains a major cause of deaths in developing countries [82]. Hepatitis is yet another waterborne disease that is caused by viruses [55]. The list of waterborne microbial and viral diseases is quite extensive and is described by Bitton [55]. Although microbiological contamination of drinking water still remains one of the biggest health threats globally, several other chem.-bio agents in water supplies cause adverse health effects. Many chemicals are regulated by their regulatory standards, however, there are uncertainties about the safety of current regulatory limits for some chemicals and is an integral part of UWQI. Regulatory guidelines require periodic review to be updated according a recent article by Villanueva et al. [56]. It is further noted, that recently discovered presence of COVID-19 SARS virus in water poses further challenge, since there are no regulations and three new vaccines have recently been invented to mitigate the virus and are made available on allocated priority.

Excess concentration of arsenic in drinking water is yet another major concern in many parts of the world [54], as it is well known that arsenic in drinking water causes cancer in human urinary bladder, lung and skin, however risk due to low level exposure remains open for ongoing discussion. Studies conducted in different countries have reported confirmed associations of bladder, colon and rectal cancers with exposure to Di-n-butyl phthalate (DBP) - one of the environmental chemicals. There are studies classifying nitrates, as probably carcinogenic to humans [56]. Nitrate and nitrite are leading cause of methaemoglobinaemia, or blue-baby syndrome. Presence of excessive Fluoride can cause dental and skeletal fluorosis and bone fractures. High intakes of selenium can give rise to loss of hair, weakened nails and skin lesions, changes in peripheral nerves and decreased prothrombin, leading to inadequate generation of thromboplastin. Uranium has toxic effects on kidney and cause an increase in fractional calcium excretion and increased microglobulinurea [54], as listed in Table 1, chemicals with potential health consequences. However, it should be noted, that the potential health effects of unregulated or emerging chemical contaminants are largely unknown [56], e.g. cyanobacterial xenobiotic toxins, such as microcystin variants (MCs) and anatoxin-a (ANA), nodularins and cylindrospermopsins (CYN) can cause liver and kidney damage, cytotoxicity, neurotoxicity, skin toxicity, gastrointestinal disturbances, and many other health problems [83], but the potential health effects of sucralose, benzotriazole, olytriazole, metal oxanes, titanium dioxide, perchlorate and many other emerging chemical contaminants are unknown [56]. Radiation exposure through the ingestion of drinking water results from naturally occurring radionuclides in drinking water sources. There are studies reporting positive associations between uranium concentrations in drinking water and indicators for cytotoxic damage to the proximal tubule of the kidney nephron, the alteration of the renal absorption function, serum carboxyterminal telopeptide [84]. Many other non-carcinogenic possible health effects as well as possible cancer and death incidences of naturally occurring radionuclides in drinking water are described by Canu et al. [84].

2.5 Impact of Contaminants in Water on Various Industries

The demand for clean water extends beyond residential and municipal needs. High volumes of high purity water are critical for most industries and laboratories, e.g., hydrocarbon processing, catalysis and chemical processing, food and beverage, mining and hydrometallurgy, pharmaceuticals, power generation, aviation and transportation and semiconductor processing. For example, microelectronic facilities require approximately 2000 gallons of water to manufacture one 200-mm wafer [79]. Ultrapure water is generally considered to be of >18.2 MΩ-cm resistivity at 25 °C, less than 50 ppt in inorganic anions and ammonia, less than 0.2 ppb in organic anions, and below 1 ppb total organic carbon and silica, and is essential for semiconductor devices processing plants and microelectronic manufacturers and is the primary solvent used to remove contaminants from semiconductor wafers. The high volume of water required for such manufacturing processes can make businesses unsustainable in areas with limited water availability. At the same time, these industries also have stringent requirements to control wastewater discharge in the environment, for which a similar innovative solution, as proposed here, can be employed to isolate contaminants from discharge and wastewater reclamation for manufacturing processes and water quality monitoring prior to re-use. With an exponential growth in industries that employ advanced technologies, the need for pure water will become ever more critical to advance these operations, thus increasing the demand for already limited water resources. Furthermore, by a survey conducted by the International Council for Science, environmental experts ranked freshwater scarcity as a twenty-first century challenge as second only to global warming.

3 Water Quality Exposure Assessment and Biomonitoring

Since contaminants in water pose serious challenges to human health, marine life and critical infrastructure, water quality assessment and monitoring using state-of-the-art equipment for label free, sampled and stand-off detection are critically important. The assessment studies for water quality includes, (a) categories of exposures that can be assessed externally, (b) the current state of the science in external exposure assessment, (c) current tools available for external exposure assessment, and (d) priority research needs.

Biomonitoring measures chemicals in human tissues and fluids and is a critical assessment tool since contaminated water can cause severe and sometimes irreversible health challenges. Due to low concentrations of the chemicals detected in human samples, typically in the parts per billion (ppb) or even parts per trillion (ppt) range, biomonitoring is critically essential to measure low concentrations of chemicals in a human tissue or fluid which otherwise are often difficult to relate in animal toxicity studies [85]. Analysis and understanding of biomonitoring data along with physiologically based pharmaco-kinetic (PBPK) modeling is described as reverse dosimetry having the potential to measure concentration of various biomarkers. This is done to understand human biomonitoring data from the perspective of exposure scenario reconstruction and risk characterization. While biomarkers have been a cornerstone of medical practice, recently identified new biomarkers have the potential to measure adverse health effects that may arise due to exposure to environmental pollutants. In fact, samples in biobanks, are also normally analyzed for meaningful omics study, so long as they satisfy collection and storage criteria. In fact, biomonitoring and epidemiology are reliable tools for exposure metrics necessary for risk assessment and regulatory decision-making [86] and are currently used as exposure scenarios reconstruction and risk assessments tools to reduce uncertainties in the source-to-outcome continuum [87]. Biomonitoring framework can also be used as a guide for understanding biomarker data and developing new risk analysis methodology. An integrative cross-omics platform has been investigated using a combination of transcriptomics, metabolomics and cell biology tools to evaluate physiological variability of different peripheral measurable, as a feasibility study [88]. It must be noted that both traditional and on-traditional (exposomic) biomonitoring strategies are critical studies that are aimed at exposome monitoring, including use of hybrid approaches and emerging techniques that conjoin biomonitoring including exposomic approaches, which differ from traditional biomonitoring methods in that they include all endogenous or exogenous exposures.

Several investigations in literature use the latest available hand-held and mobile technologies for real-time monitoring and assessment, such as use of geographic information systems (GIS); remote sensing; global positioning system (GPS) and geolocation technologies. Usually, portable sensing mechanisms include smart phone or wearable sensors with self-reporting assessments, as some of the platforms for data analysis and interpretation, data sharing, and other practical considerations, including improved assessment of exposure variability. A review of emerging technologies such as pollution sensors, smartphones based platforms, and pollution models for exposure assessment and health research are outlined by Larkin et al. [89]. In another study Steckling et al. [90], examined and summarized the availability of specific biomarkers of exposure (BoE) for a broad range of environmental stressors and exposure determinants, as well as the corresponding reference and exposure limit values and biomonitoring equivalents as part of the Human Biomonitoring (HBM) tool which greatly contributes to the understanding of human exposures. Several investigations [91–93] track human movement and physical activity using tactile and smart based phone technology. In their studies, Loh [94] and Vaseashta [95], lay out important criteria for using mobile technology, nanomaterials based sensors, cyber-physical systems, the internet of Everything (IoET) in conjunction with smartphone apps and wireless devices with external exposome features for exposure studies. They also address various limitations and challenges and discuss ways of combining "traditional" measurement data with gathering personal exposure-factor data, such as location and activities, including a multidisciplinary approach for epidemiological research and prevention measures, while considering ethical, privacy, logistical, and certain data science challenges. Advanced environmental pathogen contamination and detection of infection by novel or emerging disease and bio-threats may become an essential requirement, especially in a current COVID-19 pandemic environment. Particularly for personnel operating in field environments outside of their home base, where the risk of unanticipated detection is greater. Testing for unexpected or bioengineered agents represents a significant logistics challenge. DNA/RNA sequencing can detect/identify/classify a wider range of agents of interest. However, since manual analysis is time consuming, a reliable automated approach would be superior. Vaseashta et al. [49, 95], have designed automated sequencing analytics suited to the detection/identification and classification of mixtures of agents of concern. The method is usable regardless of the sequencer vendor and require no special user skill or foreknowledge to operate. A prototype will automate data mining and provide brief, simple Go/No-Go coupled with decision support analytics for pathogen detections. This tool will be very useful for water quality assessment and monitoring.

4 Discussion, Scenarios and Path Forward

4.1 Cyber Physical Systems for Internet of Things (IoT) Based Monitoring Strategies

Cyber-Physical Systems (CPS) is multi-disciplinary engineering discipline which integrates the dynamics of the physical processes with the software-based data collection and networking of physical systems, conjoining abstractions, modeling, design, and analysis. One of the key tasks is to integrate constructs of modeling physical processes with abstractions in computer science, which significantly contrast over timescales. The Cyber-Physical domain consists of integrated and networked system-of-systems where physical processes are monitored or controlled by algorithms. Computers and networks monitor and even control the physical processes, with feedback loops. The physical and software components are operating on different spatial and temporal scales, exhibiting multiple, yet distinct behavioral modalities, and interacting with each other in ways that change with context, within a domain of interest (DOI). Based on our earlier capabilities, to monitor, sense, detect and identify several air-borne and water-borne contaminants, we developed and deployed an integrated real-time IoT based prototype to monitor contaminants in water. The system uses GPS/Geographical Information System (GIS) based Contamination Identification and Level Monitoring Electronic Display Systems (CILM-EDS) to spatially monitor contaminants and water levels [96], as shown in Fig. 2a. In addition, multi-sensors as payload on an unmanned aerial vehicle (UAV) based surveillance platform was deployed, as an emerging and disruptive technology.

The construct of a platform in conjunction with a ground based mobile operations center presents tremendous opportunities and impact of technology for safety, security, emergency response, monitoring management etc., notwithstanding having

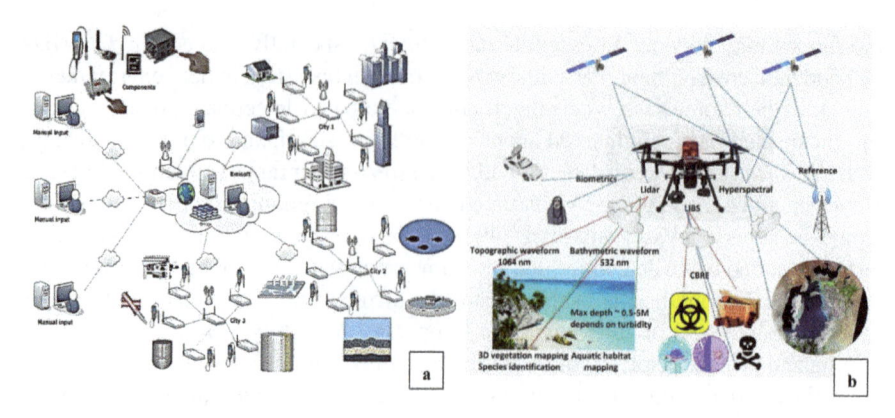

Fig. 2 a: IoT based contamination identification and Level Monitoring electronic display systems. **b**: UAV based sensor platform for detecting and monitoring water quality

broader social and political context. A UAV based sensor platform in conjunction with hyperspectral imaging, light imaging, detection and ranging (LIDAR) and Laser Induced Breakdown spectroscopy (LIBS) is being designed for full operational deployment and is shown in Fig. 2b. The UAVSP in conjunction with CILEM-EDS provides an unprecedented capability using several commercial-off the shelf (COTS) sensors along with several laboratory-based sensors, for detecting and monitoring water quality in a stand-off detection mode. Several other applications for this integrated platform include Safe Drinking Water Information System Federal Reporting Services, Annual Water Quality Reports, National Occurrence Database, Drinking Water Mapping Application, Aircraft Drinking Water Rule Compliance Reports, and Government Performance and Results Act verification tool.

4.2 Big Data Analytics and Factor Analysis—Exposome, Bioinformatics and Health-Informatics

Big data analytics refers to the strategy of analyzing large volumes of data, which is gathered from a wide variety of sources, including text from social networks, videos, digital images, sensors, and transactional records. The aim in analyzing all this data is to uncover patterns, weak signal and connections that might otherwise be invisible, and might provide valuable insights into trends and patterns. Big data requires sophisticated software programs to store and process unstructured data, thus data analytics environments and technologies for identifying patterns in the data is drawing a lot of attention. There are four types of analytics—prescriptive, predictive, diagnostics and descriptive. Depending upon its use, data analytics is used for contamination detection and monitoring, disease propagation and health informatics, water quality monitoring and emergency management. Having high performance computational platforms, it is imperative to use Big Data Analytics in Bio-

and health informatics in academics, commercial applications, and health industry for overall well-being, psychological and social needs [96]. Prototype applications include the design process of an effective and efficient dashboard which displays management information for an Electronic Health Record (EHR) [97]. Currently, novel analytic methods associate the exposome with critical health outcomes and provide recommendations on how to inform the community and scientists who can bridge health data with genomics, and biomedicine in informatics and statistics [98]. The data analytics will play a valuable role in exposome monitoring and automated UWQI monitoring for water supply.

The factor approach, however, is one of the basic, yet reliable, approaches in the analysis and for solution proposal for a given problem. It is methodologically close to the systems approach, which analyses the impact of each factor of a specific problem or target quantity [99]. The first step in factor analysis is to determine the availability of individual factors that impact the target quantity. To analyze exposome, the approach involves factors, viz., population with all of its demographic parameters; degree of urbanization of a settlement; natural resources and their distribution, utilization, and sustainability; water management, energy, and transport infrastructure; economic branches and their technical and technological development; and use of up-to-date management methods and knowledge [100, 101]. Each target quantity, according to its particularities and environment, has more prominent contributing factors, which need to be analysed separately. As can be readily concluded, these factors have a different measurement system, which also makes their impact on the target quantity difficult to measure. Target quantity impact analysis of a single factor is also an unrealistic endeavour, as the factor's impact depends on its constellation with other factors from the impact group of factors. The problem of expressing multiple factors as the target quantity has been successfully tackled over the recent years, with the introduction of standard mathematical, numerical, and statistical methods as well as with the development of numerous simulation and optimisation models, the most prominent of which are the models for multi-criteria optimisation [102]. With these models, factor analysis is now able to reduce all dimensional incongruences and the impact variations of multiple contributing factors into a unifying numerical model and to offer an integral solution to a given problem [103]. Multi-criteria analysis and optimisation is used when there is a larger number of criteria and alternatives for the solution of a target task [104, 105].

The procedure of multi-criteria optimization involves multiple stages of solution and decision making is shown in Fig. 3 and comprises of the following:

1. A definition of system goals and purpose and the identification of the ways to achieve the desired goals (alternatives), accompanied by the selection of criteria and their weights, which are crucial for any alternative solution and decision making.
2. A mathematical description of the system and a definition of the way to evaluate criterion functions.
3. The use of existing optimisation methods supported by software applications for evaluation and ranking.
4. The adoption of a final solution and the making of a decision.

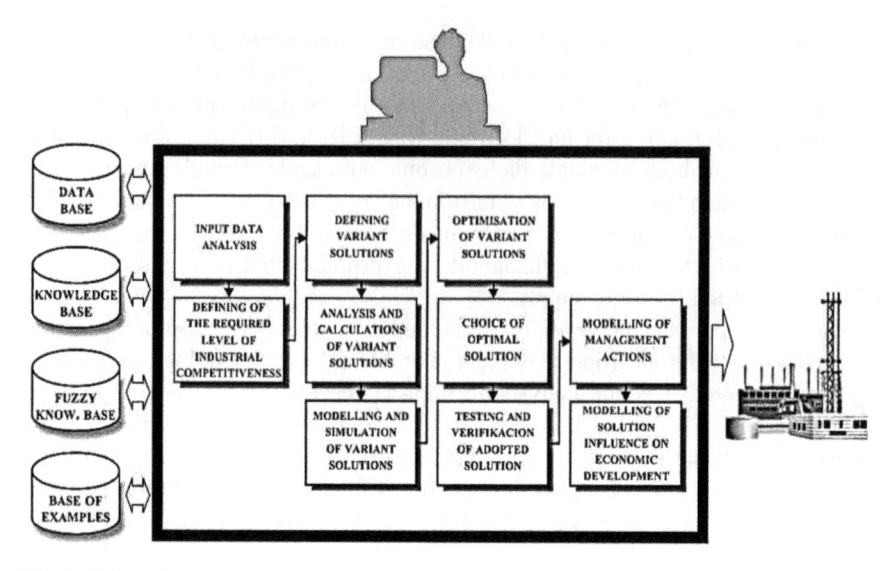

Fig. 3 Schematic representation of the factor analysis optimisation process

Out of several multi-criteria decision analysis (MCDA) methods (PROMETHEE, TOPSIS, MAX–MIN, AHP, ELECTRE, PROTRADE, VICOR), it is better to use at least two different methods, which would provide a higher degree of certainty when reaching the final solution to a given problem. In practice, the use of MCDA and optimization methods is commonly found in the planning of complex systems, e.g. regional development, water management and electric power systems development, urban planning, and environmental protection [106]. Due to its complexity, multi-disciplinarity and numerous contributing factors, it is preferable to solve the integral management of water resources, especially for protection, risk and resiliency against pollution, by using factor analysis and mathematical methods based on MCDA and optimization using advanced computation resources. Wastewater treatment plants play a vital role in the system for water quality protection, both in terms of selecting the optimal technological solution and their efficiency and economy.

The initial factors for the implementation of factor analysis when selecting the optimal solution for a wastewater treatment plant commonly included the following: receiver of processed water; site conditions; mass characteristics of wastewater; desired water quality by key parameters (specific to application); and sludge disposal. Alternative solutions for such treatment plants include the possible technological schemes for wastewater treatment, while the most prominent criteria for their ranking include the following: economic (investments, work and maintenance, amortization); environmental (pollution reduction percentage by key physical, chemical, and biological parameters; environmental impact), and social criteria (viz. introduction of new tariffs, readiness to cover water treatment costs). Due to the multitude, variety, and weights of contributing factors, it can be concluded that the selection of the optimal alternative solution for a wastewater treatment plant should be considered

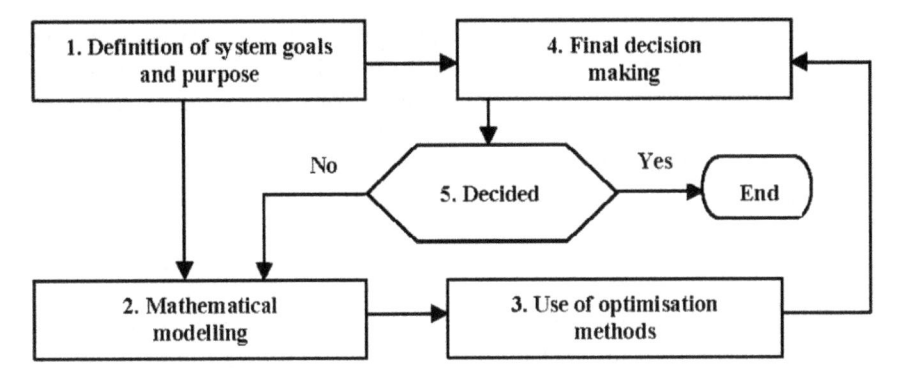

Fig. 4 The flow chart of competitiveness development and design of wastewater treatment plants for water quality optimization

a complex system, capable to solve using only some of the mathematical models based on MCDA and optimisation supported by advanced computational platforms and technology (Fig. 4).

The methodology of factor analysis using multi-criteria optimisation methods has been verified in practice on many complex systems in the field of water management systems. Some of the successful applications of this methodology include, among others, showing:

1. Multi-criteria optimisation of "Gornjak" water management system in the Mlava river basin: the analysis considered the position and size of several potential accumulations and multiple variant solutions for their use [107],
2. Multi-criteria optimisation of Foča-Goražde hydropower system on the Drina, involving the selection of the newest solution for the position of structures, which would meet the specific requirements for protection and preservation of a given space and environment [108],
3. Multi-criteria optimisation of the irrigation system in the Morava region, divided into nine drainage areas, which were separately analysed in terms of the optimal solution for drainage during the period of high groundwater levels and irrigation during the dry season [109].

These systems are continually updated using advanced computational platforms employing multiple layers of cyber security protocols.

4.3 A Novel Approach Towards Water Quality Assessment

Presently, the negative changes in water environment often exceed the existing monitoring and pollution control methods of natural waters condition, as diversity of pollutants provokes unpredictable response to the water quality and behaviour of water biota. There are documented cases of examples of negative changes without any substantial change in hydro-chemical and hydro-biological indicators of water quality. To adequately establish properties of aquatic systems, new approaches are needed to assess the condition of natural waters as a dynamic chemical-biological system. Recently, many researchers and practitioners, while elaborating and applying special hydrodynamic, physical-mathematic and biological methods, have obtained vital results by estimation and prognostication of physical and biological processes of waters self-purification. However, there are insufficient studies concerning methods of efficiency estimation, as well as quantitative data of pollutants chemical transformations in water.

Water quality and its self-purification capacity are influenced by many factors, one of them being the "redox state" of environment [110]. Due to oxidative and reducing nature in natural aquatic systems, redox processes are in the dynamic equilibrium with respect to biotic and abiotic compositions. Development of various micro-organisms is only possible in the environment with the redox potential within certain limits. To adequately estimate the ecological condition of natural waters, it is necessary to use the diverse research methods including the study of physical, chemical and biological processes occurring in the aquatic ecosystems, treated as an "open" dynamic system. The chemical processes in water environment connected with the oxidation of certain substances or the reduction of others, occur catalytically and under the influence of solar irradiation. The aquatic environment should be considered as an open oxidation–reduction system, and, therefore, explanation of the assessment criteria of water condition is based on a kinetic approach, the essence of which consists of the kinetics of the organic substances appearance and transformations in water, as a result of the interactions with the component elements of waters. At the same time, this approach implies that the oxidation–reduction processes' mechanisms in an aquatic medium occur with the involvement of dissolved oxygen and provoke the formation of its active intermediate forms. One of the most important oxidative agents of natural water is dissolved oxygen, which under the normal conditions is kinetically inert and needs activation either by means of photoinitiation or by complexation with transitional metal ion in the reduced state. The second important components of oxidative equivalents in nature is hydrogen peroxide, which is a product of two-electron reduction of oxygen. Furthermore, H_2O_2 causes oxidation–reduction processes much more readily than O_2. With oxygen and hydrogen peroxide participation in the self-purification chemical processes, active intermediates are formed, including free radicals with high reactivity. To mathematically represent total contribution of all oxidative agents with respect to the effective speed of the polluting substance transformation, characterizing self-purification ability of water medium, a rate constant $k(P_m)$ regarding the pollutant P_m, can be represented as follows:

$$k(p_m) = \sum k_i(\vec{\phi}) = k_{bio} + k_{cat} + k_{photo} + k_{radical} + k_{hydrolitic} \qquad (4)$$

where ϕ is multidimensional set of significant parameters of the environment, and subscripts bio, cat, photo, radical, hydrolytic correspond to transformation pathways of pollutants, such as, microbiologic, catalytic, photochemical, radical and hydrolysis, respectively. In the absence of any evidence of cross-influence of various pathways giving non-additive contribution to the overall transformation, such contribution is not considered. Thus, the system practically represents a closed cycle between the photosynthesis, chemical and microbiological destruction of the organic matter in biosphere.

Hydrogen peroxide plays a significant role in the oxidation processes, with several metal ions and complexes producing free radicals during their interactions, with hydroxyl radicals being the most active among them. Stationery concentration of OH radicals is determined by the rate of their initiation, W_i and the parameter characterizing the contents in OH radical "traps" in water, viz., inhibitors of radical oxidation processes:

$$[OH] = W_i / \Sigma k_i [S_i] \quad (5)$$

where k_i is rate constant of OH radicals' interaction with S_i "traps". The higher the stationery concentration of OH radicals in water; the higher the contribution of radical processes in transformations of polluting substances. Considering the dynamic parameters of an aquatic system, the redox state of natural waters can be assessed. Thus, an aquatic system is considered as being in oxidant state, once hydrogen peroxide accumulation rate in water is higher than the rate of the reducers formation in this system, i.e. the stationery concentration of H_2O_2 exceeds the stationery concentration of DH_2. In case of inverse inequality, water is in the reducing state. Thus, the water quality and its self-purification capacity are directly affected by the "redox" state. This can be explained by the fact that physical–chemical and biochemical processes occur in water environment, depending on the presence of oxidants and reducers, and, obviously, on their variations. This dynamic was determined by catalytic oxidation–reduction processes in water environment and the involvement of the transition metal ions and coordination compounds.

The negative redox potential of natural water may be typical only for groundwater from mountain springs or melt water, where there are low-state metals valence (Fe^{2+}, Mn^{2+}, Mo^{4+}, V^{4+}, U^{4+}). The oxidized state is determined by the presence of hydrogen peroxide, whereas the reducing state is determined by the presence of substances that can be readily oxidized with H_2O_2. The nature of the reducing substances is not clear, but a key role by nediol and thiolic substances is involved which come into the water medium as a result of pollution or the metabolism of living organisms. Catalytic oxidation of the reducing thiolic substances such as cysteine, thiourea, and glutathione with hydrogen peroxide occur in the water environment as a result of metabolic processes of hydrobionts, resulting from mass development of blue-green algae or anthropogenic pollution. Identification and classification of thiol compounds could clarify the management of natural waters quality and monitoring of the pathways of aquatic pollutants.

4.4 Contamination Remediation Strategies—Conventional and Novel

To maintain a high quality of water, contamination from polluted water needs to be remediated using advanced contamination remediation strategies to have an acceptable water quality. Remediation is a physical, chemical or biological process which enables removal of contaminants or to change the toxic state of material into nontoxic. Conventional physico-chemical and biological remediation methods for contaminant removal from water are precipitation and coagulation, distillation, adsorption, ion exchange, membrane water treatment (reverse osmosis, electrodialysis membrane treatment), catalytic processes (hydrogenation of nitrate, photocatalytic method, electrocatalytic oxidation), bioremediation (phytoremediation, vegetated filter strips, biologically active carbon filtration), magnetic separation and disinfection [111]. Precipitation and coagulation method for contaminant removal from water or wastewater is the change in material form from dissolved state into solid particle so that these impurities can be removed in subsequent solid/liquid separation processes [112, 113]. This technique is used to remove ionic components from water reducing the solubility. It can be used for the removal of metallic cations, anions (fluoride, cyanide and phosphate), organic molecules (phenols and aromatic amines), detergents and oily emulsions [113]. Distillation is the most common separation technique in which the mixed components in water are separated by the application of heat, based on the differences in boiling points of the individual components. During the adsorption process, dissolved contaminants adhere to the porous surface of the solid particles [111]. This remediation method is widely used for the treatment of wastewater, groundwater and drinking water to remove heavy metals, metalloids, halogens, pharmaceuticals, organic pollutants, etc. [114]. Ion exchange is a water treatment technique used to remove hardness and various contaminants (i.e. nitrate, arsenic, cobalt, DOC) in water [115]. During this process, an ion in water or wastewater is replaced with similarly charged ion attached to an immobile solid particle (naturally occurring inorganic zeolites or synthetically produced organic resins) [116]. During membrane treatment, water impurities are separated due to the existence of a gradient across the membrane. Based on the specific gradient type, there exit pressure- and non-pressure driven membrane processes [117]. Reverse osmosis is a pressure-driven process where a semi-permeable membrane removes water contaminants (i.e. dissolved salts, organic materials, bacteria) due to size exclusion, charge exclusion and physical–chemical interactions between solute, solvent and membrane [118]. Catalytic processes for water purification are generally described by the use of a combination of oxidation (O_3, H_2O_2), irradiation, electron and catalyst as the means of generating hydroxyl (·OH), superoxide ($O_2\cdot^-$) and hydroperoxyl ($HO_2\cdot$) radicals, etc. at ambient temperature and pressure [119]. Nitrate hydrogenation is a catalytic process by which hydrogen gas (H_2) is used to reduce nitrate to a potentially more useful or benign form of nitrogen, spontaneously releasing energy [120]. Bioremediation is a process that uses microorganisms to degrade contaminants in water or wastewater. It is a low-cost, low maintenance, environmentally friendly and sustainable approach for the cleanup of polluted water [121]. Phytoremediation, one of biotechnological methods for water treatment, is considered to be a versatile and sustainable natural treatment technique that uses the inherent capacity of certain plant species for the removal of contaminants from the polluted water environment [122]. Magnetic separation method for water remediation is based on the separation of magnetic solid particles from polluted water using permanent magnets. However, recent progress in this field showed the removal of non-magnetic water pollutants such as virus, algae, dissolved pollutants, etc. by the magnetic separation technique [123]. Disinfection treatment is applied to ensure microbiological quality. During this process, disinfectants such as chlorine, ozone, UV radiation, etc. are used to kill pathogenic bacteria in water to make it safe for users [124]. One of the recent methods for contamination remediation is nanophotocatalysis [125], using meta-materials to detect and capture organic materials, volatile organic compounds and microplastics. It is worth mentioning that all aforementioned water remediation systems do not sufficiently mitigate some of the emerging contaminants, which did not exist in sufficient quantities a decade or so ago. No one of these technologies can independently solve all the environmental problems. Therefore, the development of advanced and smart remediation techniques is one of the main environmental challenges facing the world today.

5 Conclusions—Observations, Analysis, Policy Initiatives and Future Recommendations

Good water quality is critically important for our health, quality of life and survival. Legislatively, many nations are debating that universal access to clean water should be treated as a fundamental right for human society. Since industrialization is part of a broader societal transformation and associated social changes that arise due to technological modernization, it is crucial to use a paradigm shift and convergence of technologies including state-of-the-art contamination monitoring and remediation strategies along with computational platforms and streaming data analytics to monitor water quality index. Since there are many water quality indexes, it is important to define a UWQI based on a maximum allowed limit for each contaminant that can be present in the water for uniformity. An outline of flow chart and logical steps for assessing UWQI is outlined in this investigation. The model used here is for drinking water only, although the modality is equally applicable to all other processed water functions and are based on input parameters. Hence, water contamination remediation strategies are categorized as per desired water process function, as a management strategy to remediate only those contaminants that are not needed for a desired function. Since the composition of contaminants varies from region to region, it is technological and economically beneficial to develop remediation strategies that are region specific and remove selected contaminants that may be present in water and in quantities specified by planned application. Such a distributed and decentralized water purification system is likely to be more technologically efficient, economical and is likely to serve a better purpose. Furthermore, it is critical to observe that there are numerous national security concerns emanating from the growing crisis of global freshwater scarcity. Many of the earth's freshwater ecosystems are being critically depleted and are used unsustainably to support growing residential and industrial demands. This results in ecological destabilization, thereby directly impacting current political, economic and social landscapes. Additionally, the basic need for clean drinking water depends on sources of water, exposome, and contamination remediation strategies. Hence it is necessary to conduct exposure research using the latest research methodologies. Use of exposure research using wide-spectrum bioinformatics, biomarkers research and the latest epidemiological tools is relevant to developing policy considerations. Considering current COVID-19 pandemic environment, detection of traces of nCOV-SARS2 in water discharge is alarming. It is, therefore, critical to use wide-spectrum disease surveillance tools, as part of integrative strategy for monitoring Universal Water Quality Index. Water, in the most natural environment with its living and non-living functionalities, should be subject to environmental protection. Global water resources are neither protected, nor managed sustainably due to the inadequate infrastructure and characterization mechanisms of water quality in many parts of the world. Upgrades and structural improvement in this regard are urgently required throughout the world. Advancements in water quality guidelines should reflect on water quality for sustainability required for global water resources management. As a part of ongoing investigation, a decentralized source of water, including atmospheric harvesting shows a promising future. Lastly, the policy recommendations must consist of strict guidance of unprocessed discharge from industries in addition to a robust lifecycle analysis, data-driven policies and study of impacts of microplastics in the aquatic environment, which is necessary for safe and sustainable marine environment, food chain supply, sustainable ecosystem—all concurrent with the benefits of the benefit of the latest technological innovations.

Acknowledgements Part of the project was funded through a NATO Science for Peace and Security grant.

References

1. Vaseashta A (2014) Emerging sensor technologies for monitoring water quality. In: Applications of nanomaterials for water quality. Future science book series, pp 66–84. https://doi.org/10.4155/ebo.13.208
2. Harhay MO (2011) Water stress and water scarcity: a global problem. Am J Public Health 101(8):1348–1349. https://doi.org/10.2105/AJPH.2011.300277
3. Biswas A, Tortajada C (2019) Water quality management: a globally neglected issue. Int J Water Resour Dev 35(6):913–916. https://doi.org/10.1080/07900627.2019.1670506
4. Water for The Future the West Bank and Gaza Strip, Israel, And Jordan (1999) Report by committee on sustainable water supplies for the middle east. Israel academy of sciences and

humanities. Palestine academy for science and technology. Royal Scientific Society, Jordan. U.S. National Academy of Sciences
5. Narayan R (2010) Use of Nanomaterials in water purification. Mater Today 13(6):44–46. https://doi.org/10.1016/S1369-7021(10)70108-5
6. Ritabrata R (2019) An introduction to water quality analysis. Int Res J Eng Technol 6:201–205
7. US Statues. "Interstate Quarantine Act of 1893," Chap. 114. US Statutes at Large, 27, February 15, 1893:449
8. PHS (Public Health Service) (1914) Final report of public health, service commission, bacteriological standard for drinking: water. Repr J Am Water Work Assoc 2(1):65–73. Treasury Department (1913) Pure Drinking Water for Passengers in Interstate Traffic Amendment to the Interstate Quarantine Regulations No. 6. Treasury Department Memo, 25 (1913) Records of the US public health service, record group 90. National Archives, Washington, DC
9. Phelps EB (1915a) Discussion following HE Jordan's 'The standards for drinking water used on interstate carriers'. In: Proceedings of the 8th annual convention of the indiana sanitary and water supply association, Indianapolis, IN, p 42; and Phelps EB (1915b) Enclosure in letter to the surgeon general of 3 March 1915, detailing his remarks at the February 1915 meeting of the Indiana sanitary and water supply association. Records of the US Public Health Service, Record Group 90, National Archives, Washington, DC
10. USPHS 956. Drinking Water Standards of 1962. USPHS. "Drinking Water Standards." Fed Reg, March 6, 1962:2152–2155
11. USPHS (1970a) Community water supply study: analysis of national survey findings. NTIS Pb214982. Springfield, VA: USPHS
12. USPHS (1970b) Community water supply study: significance of national findings. NTIS PB215198/BE. Springfield, VA: USPHS
13. USC (United States Congress) (1974) Safe drinking water act of 1974, US Codes of law, Title 42, Section 300g-1 (b) (4). (D). United States Congress, Washington, DC
14. The Safe Drinking Water Act of 1974. Public Law 93–523. 16 Dec 1974
15. Horton RK (1965) An index—number system for rating water quality. J Water Pollut Control Fed 37(3):300–305
16. Brown RM, McClelland NI, Deininger RA, Landwehr JM (1973) Validating The WQI" The paper presented at the national meeting of the American society of civil engineers on water resources engineering, Washington, DC
17. Prati L, Pavanell R, Pesarin F (1971) Assessment of surface water quality by a single index of pollution. Water Res 5:741–751
18. Walski TM, Parker FL (1974) Consumer's water quality index. J Environ Eng Div 100:593–611
19. Bascaron M (1979) Establishment of a methodology for the determination of water quality. Bull Inf Medio Ambient 9:30–51
20. Bhargava DS (1983) Use of a water quality index for river classification and zoning of Ganga River. Environ Pollut B 6:51–67
21. Dinius SH (1987) Design of water quality index. Water Resour Bull 23:823–843
22. Smith DG (1990) A better water quality indexing system for rivers and streams. Am J Water Resour 24:1237–1244
23. Wepener V, Euler N, Vuren JHJ, duPreez HH, Kohler A (1992) The development of an aquatic toxicity index as a tool in the operational management of water quality in the Olifants River (Kruger National Park). Koedoe 35:1–9
24. Cude CG (2001) Oregon water quality index: a tool for evaluating water quality management effectiveness. J Am Water Resour Assoc 37(1):125–137
25. Sargaonkar A, Deshpande V (2003) Development of an overall index of pollution for surface water based on a general classification scheme in Indian context. Environ Monit Assess 89:43–67
26. Almeida C, González SO, Mallea M et al (2012) A recreational water quality index using chemical, physical and microbiological parameters. Environ Sci Pollut Res 19:3400–3411. https://doi.org/10.1007/s11356-012-0865-5

27. Shivani Tyagi RC, Dubey RB, Ahamad F (2020) Multivariate statistical analysis of river ganga water at Rishikesh and Haridwar, India. Anal Chem Lett 10(2):195–213. https://doi.org/10.1080/22297928.2020.1756405
28. Boyacioglu H (2007) Development of a water quality index based on a European classification scheme. Water SA 33:101–106
29. Abbasi T, Abbasi SA (2012) Water quality indices, Elsevier BV Amsterdam, Netherlands 384
30. CCME (2001) "Canadian Water Quality Index 1.0" Technical report and user's manual, Gatineau, QC: Canadian council of ministries of the environment, Canadian environmental quality guidelines, Water Quality Index Technical Subcommittee, pp 1–5
31. Water Quality in British Columbia (2004). https://www.neef.ca/uploads/library/7920_Phippen2005_WQAttainment.pdf
32. Terrado M, Barcelo D, Tauler R, Borrell E, Campos SD (2010) Surface-water-quality indices for the analysis of data generated by automated sampling networks. Trends Anal Chem 29(1):40–52. https://doi.org/10.12691/ajwr-1-3-3
33. Boyacioglu H (2007) Development of a water quality index based on a European classification scheme. Water SA 33(1):101–106. WOS:000244250700013. https://doi.org/10.4314/wsa.v33i1.47882
34. World Health Organization (WHO) (2011) Guidelines for drinking-water quality, 4th edn. WHO, Geneva. https://www.who.int/water_sanitation_health/publications/2011/dwq_guidelines/en
35. Bala GS, Rao GVR, Raju PARK, Raju MJ (2017) Water quality index: a tool for evaluation of surface water quality. Int J Civ Eng Technol (IJCIET) 8(10):814–821, Article ID: IJCIET_08_10_086
36. Vaseashta A (2014) Advanced sciences convergence-based methods for surveillance of emerging trends in science, technology, and intelligence. Foresight 16(1):17–36. https://doi.org/10.1108/FS-10-2012-0074
37. Nikinmaa N (2014) An Introduction to aquatic toxicology. Academic Press. https://doi.org/10.1016/C2012-0-07948-3
38. Millennium Development Goals. https://www.undp.org/content/undp/en/home/sdgoverview/mdg_goals.html
39. Sustainable Development Goals. https://www.undp.org/content/undp/en/home/sustainable-development-goals.html
40. Unami K, Kawachi T (2003) Universal optimization of water quality management strategy. Adv Water Resour 26:465–472
41. Dennis KK, Marder E, Balshaw DM, Cui Y, Lynes MA, Patti, Barr (2017) Biomonitoring in the era of the exposome. Environ Health Perspect 125(4):502–510 Epub 2016 Jul 6
42. Wild CP (2005) Complementing the genome with an "exposome": the outstanding challenge of environmental exposure measurement in molecular epidemiology. Cancer Epidemiol Biomarkers Prev 14(8):1847–1850
43. Vermeulen R, Schymanski E, Barabási A, Miller G (2020) The exposome and health: where chemistry meets biology. Science 367(6476)392–396. https://doi.org/10.1126/science.aay3164
44. Lioy PJ, Rappaport SM (2011) Exposure science and the exposome: an opportunity for coherence in the environmental health sciences [Editorial]. Environ Health Perspectives 119:A466–A467. https://doi.org/10.1289/ehp.1104387
45. Stingone JA, Buck Louis GM, Nakayama SF, Vermeulen RC, Kwok RK, Cui Y, Teitelbaum SL (2017) Toward greater implementation of the exposome research paradigm within environmental epidemiology. Annual Rev Public Health 20(38):315–327. https://doi.org/10.1146/annurev-publhealth-082516-012750 Epub 2017 Jan 6
46. Wild CP, Scalbert A, Herceg Z (2013) Measuring the exposome: a powerful basis for evaluating environmental exposures and cancer risk. Environ Mol Mutagen 54(7):480–499. https://doi.org/10.1002/em.21777
47. Vrijheid M, Slama R, Robinson O, Chatzi L, Coen M, van den Hazel P, Nieuwenhuijsen MJ (2014) The human early-life exposome (HELIX): project rationale and design. Environ Health Perspectives 122:535–544. https://doi.org/10.1289/ehp.1307204

48. Robinson O, Vrijheid M (2015) The pregnancy exposome. Current Environ Health Rep 2(2):204–213. https://doi.org/10.1007/s40572-0150043-2
49. Baldwin J, Noorali S, Vaseashta A (2019) Wide spectrum bio-threats identification and classification. Chemical, Biological, Radiological and Nuclear (CBRN) Defense—Modernizing the Future Fight: Accelerate & Adapt. Wilmington, DE
50. Sarigiannis DA, Karakitsios SP, Handakas E, Papadaki K, Chapizanis D, Gotti A (2018) Informatics and data analytics to support exposome-based discovery: Part 1—assessment of external and internal exposure. In: Lytras M, Papadopoulou P (eds) Applying big data analytics in bioinformatics and medicine. IGI Global, Hershey, PA, pp 115–144. https://doi.org/10.4018/978-1-5225-2607-0.ch006
51. Sarigiannis DA (2019) The exposome paradigm in environmental health. In: Papadopoulou P, Marouli C, Misseyanni A (eds) Environmental exposures and human health challenges. IGI Global, Hershey, PA, pp 1–29. https://doi.org/10.4018/978-1-5225-7635-8.ch001
52. Sarigiannis DA, Gotti A, Handakas E, Karakitsios SP (2018) Informatics and data analytics to support exposome-based discovery: Part 2—computational exposure biology. In: Lytras M, Papadopoulou P (eds) Applying big data analytics in bioinformatics and medicine. IGI Global, Hershey, PA, pp 145–187. https://doi.org/10.4018/978-1-5225-2607-0.ch007
53. Water—Mapping, measuring, and mitigating global water challenges (2020) Water resources institute report
54. Fawell J, Nieuwenhuijsen MJ (2003) Contaminants in drinking water. Br Med Bull 68:199–208. https://doi.org/10.1093/bmb/ldg027
55. Bitton G (2014) Microbiology of drinking water: production and distribution. Wiley, Hoboken, NJ, USA. https://doi.org/10.1002/9781118743942
56. Villanueva CM, Kogevinas M, Cordier S, Templeton MR, Vermeulen R, Nuckols JR, Nieuwenhuijsen MJ, Levallois P (2014) Assessing exposure and health consequences of chemicals in drinking water: current state of knowledge and research needs. Environ Health Perspect 122:213–221. https://doi.org/10.1289/ehp.1206229
57. Demir V, Ergin S (2013) Occurrence and assessment of chemical contaminants in drinking water in Tuncely, Turkey. J Chem 238374. https://doi.org/10.1155/2013/238374
58. Alidadi H, Tavakoly Sany S, Zarif Garaati Oftadeh B, Mohamad, Shamszade, Fakhari (2019) Health risk assessments of arsenic and toxic heavy metal exposure in drinking water in northeast Iran. Environ Health Prev Med 24:59. https://doi.org/10.1186/s12199-019-0812-x
59. Carvalho FP, Fajgelj A (2013) Radioactivity in drinking water: routine monitoring and emergency response. Water Air Soil Pollut 224:1597. https://doi.org/10.1007/s11270-013-1597-y
60. Altikulac A, Turhan S, Gumus H (2015) The natural and artificial radionuclides in drinking water samples and consequent population doses. J Radiat Res Appl Sci 8:578–582. https://doi.org/10.1016/j.jrras.2015.06.007
61. Zehringer M (2019) Monitoring of natural radioactivity in drinking water and food with emphasis on alpha-emitting radionuclides: Ionizing and non-ionizing radiation. In: Osibote A (ed) Intech Open, London, pp 1–27. https://doi.org/10.5772/intechopen.90166
62. Lebreton LCM, van der Zwet J, Damsteeg JW, Slat B, Andrady A, Reisser J (2017) River plastic emissions to the world's oceans. Nat Commun 8:15611. https://doi.org/10.1038/ncomms15611
63. Lambert S, Wagner M (2018) Microplastics are contaminants of emerging concern in freshwater environments: an overview. In: Wagner M, Lambert S (eds) Freshwater microplastics: emerging environmental contaminants. The handbook of environmental chemistry, vol 58, pp 1–24. Springer Open. https://doi.org/10.1007/978-3-319-61615-5.
64. Eriksen M, Lebreton LCM, Carson HS, Thiel M, Moore CJ, Borerro JC (2014) Plastic pollution in the world's oceans: more than 5 trillion plastic pieces weighing over 250,000 tons afloat at sea. PLoS ONE 9(12). https://doi.org/10.1371/journal.pone.0111913
65. Vaseashta A, Ivanov V, Dekhtyar Y, Bolgen N (2020) Nanomaterials based technologies for identification and mitigation of environmental nano/microplastics, NP2020–011, Nanoposter 2020–9th virtual nanotechnology conference. The International Nanoscience Community

66. Wu C, Zhang K, Xiong X (2018) Microplastic pollution in inland waters focusing on Asia. In: Wagner M, Lambert S (eds) Freshwater microplastics: emerging environmental contaminants. The Handbook of Environmental Chemistry, **58**, 85–100. Springer Open. DOI: https://doi.org/10.1007/978-3-319-61615-5_5.
67. Crittenden JC, Trussell RR, Hand DW, Howe KJ, Tchobanoglous G (2012) MWH's water treatment: principles and design, 3rd edn. Wiley, Hoboken, N.J. https://doi.org/10.1002/9781118131473
68. Rist S, Hartmann NB (2018) Aquatic ecotoxicity of microplastics and nanoplastics: lessons learned from engineered nanomaterials. In: Wagner M, Lambert S (eds) Freshwater microplastics: emerging environmental contaminants, the handbook of environmental chemistry 2018, vol 58, pp 25–49. Springer Open. https://doi.org/10.1007/978-3-319-61615-5_2.
69. Ziccardi LM, Edgington A, Hentz K, Kulacki KJ, Driscoll KS (2016) Microplastics as vectors for bioaccumulation of hydrophobic organic chemicals in the marine environment: a state-of-the-science review. Environ Toxicol Chem 35:1667–1676. https://doi.org/10.1002/etc.3461
70. Moore CJ, Lattin GL, Zellers AF (2011) Quantidades e tipos de plásticos provenientes de dois rios urbanos que escoam para águas costeiras e praias do Sul da Califórnia. Revista de Gestão Costeira Integrada 11(1), Páginas 65–73. https://doi.org/10.5894/rgci194.
71. Barboza LGA, Vethaakd AD, Lavorante BRBO, Lundebye AK, Guilhermino L (2018) Marine microplastic debris: an emerging issue for food security, food safety and human health. Mar Pollut Bull 133:336–348. https://doi.org/10.1016/j.marpolbul.2018.05.047
72. Pfaller JB, Goforth KM, Gil MA, Savoca MS, Lohmann KJ (2020) Odors from marine plastic debris elicit foraging behavior in sea turtle. Curr Biol 30, PR213-R214. https://doi.org/10.1016/j.cub.2020.01.071
73. Bottrell ZLR, Beaumont N, Dorrington T, Steinke M, Thompson RC, Lindeque PK (2019) Bioavailability and effects of microplastics on marine zooplankton: a review. Environ Pollut 245:98–110. https://doi.org/10.1016/j.envpol.2018.10.065
74. Deng Y, Zhang Y (2017) Response to uptake of microplastics and related health effects. Sci Rep 7:46687
75. Bing W, Xiaomei W, Su L, Zhizhi W, Ling C (2019) Size-dependent effects of polystyrene microplastics on cytotoxicity and efflux pump inhibition in human Caco-2 cells. Chemosphere 221:333–341. https://doi.org/10.1016/j.chemosphere.2019.01.056
76. Wright SL, Kelly FJ (2017) Plastic and human health: a micro issue? Environ Sci Technol 51:6634–6647. https://doi.org/10.1021/acs.est.7b00423
77. Rist S, Almroth BC, Hartmann NB, Karlsson TM (2018) A critical perspective on early communications concerning human health aspects of microplastics. Sci Total Environ 626:720–726. https://doi.org/10.1016/j.scitotenv.2018.01.092
78. Vaseashta A (ed) (2015) Life cycle analysis of nanoparticles: risk, assessment, and sustainability. DEStech Publishers, Inc. www.destechpub.com/product/life-cycle-analysis-of-nanoparticles
79. Vaseashta A (2011) Technological advances in industrial water safety and security. In: Atimtay A, Sikdar S (eds) Security of industrial water supply and management. NATO science for peace and security series c: environmental security. Springer, Dordrecht. https://doi.org/10.1007/978-94-007-1805-0_4
80. Gatti A, Montanari S, Vaseashta A (2015) Nanopathology—risk assessment of mysterious cryptogenic diseases. In: Book: life cycle analysis of nanoparticles—risk, assessment, and sustainability. DEStech Publishers, PA, USA
81. Dris R, Gasperi J, Rocher V, Saad M, Renault N, Tassin B (2015) Environ Chem 12(5):592–599. https://doi.org/10.1071/EN14167
82. Lucas PJ, Cabral Ch, Colford JM Jr (2011) Dissemination of drinking water contamination data to consumers: a systematic review of impact on consumer behaviors. PLoS ONE 6. https://doi.org/10.1371/journal.pone.0021098
83. Zanchett G, Oliveira-Filho EC (2013) Cyanobacteria and cyanotoxins: from impacts on aquatic ecosystems and human health to anticarcinogenic effects. Toxins 5:1896–1917. https://doi.org/10.3390/toxins5101896

84. Canu IG, Laurent O, Pires N, Laurier D, Dublineau I (2011) Health effects of naturally radioactive water ingestion: the need for enhanced studies. Environ Health Perspect 119:1676–1680. https://doi.org/10.1289/ehp.1003224
85. Clewell HJ, Tan YM, Campbell JL, Andersen ME (2008) Quantitative interpretation of human biomonitoring data. Toxicol Appl Pharmacol 231(1):122–133. https://doi.org/10.1016/j.taap.2008.04.021
86. Bonnefoi MS, Belanger SE, Devlin DJ, Doerrer NG, Embry MR, van der Fukushima S, Laan JW (2010) Human and environmental health challenges for the next decade (2010–2020). Crit Rev Toxicol 40(10):893–911. https://doi.org/10.3109/10408444.2010.506640
87. Sobus JR, Tan YM, Pleil JD, Sheldon LS (2011) A biomonitoring framework to support exposure and risk assessments. Sci Total Environ 409(22):4875–4884
88. Gruden K, Hren M, Herman A, Blejec A, Albrecht T, Selbig J, Jeras M (2012) A "crossomics" study analyzing variability of different components in peripheral blood of healthy Caucasoid individuals. PLoS One, 7(1):e28761. https://doi.org/10.1371/journal.pone.0028761
89. Larkin A, Hystad P (2017) Towards personal exposures: how technology is changing air pollution and health research. Curr Environ Health Rep 4(4):463–471. https://doi.org/10.1007/s40572-017-0163-y
90. Steckling N, Gotti A, Bose-O'Reilly S, Chapizanis D, Costopoulou D, De Vocht F, Sarigiannis (2018) Biomarkers of exposure in environment-wide association studies - Opportunities to decode the exposome using human biomonitoring data. Environ Res 164:597–624. https://doi.org/10.1016/j.envres.2018.02.041 Epub 2018
91. de Nazelle A, Seto E, Donaire-Gonzalez D, Mendez M, Matamala J, Nieuwenhuijsen MJ, Jerrett M (2013) Improving estimates of air pollution exposure through ubiquitous sensing technologies. Environ Pollut 176:2–99
92. Nieuwenhuijsen MJ, Donaire-Gonzalez D, Foraster M, Martinez D, Cisneros A (2014) Using personal sensors to assess the exposome and acute health effects. Int J Environ Res Public Health 11(8):7805–7819. https://doi.org/10.3390/ijerph110807805
93. Nieuwenhuijsen MJ, Donaire-Gonzalez D, Rivas I, de Castro M, Cirach M, Hoek G, Sunyer J (2015) Variability in and agreement between modeled and personal continuously measured black carbon levels using novel smartphone and sensor technologies. Environ Sci Technol 49(5):2977–2982. https://doi.org/10.1021/es505362x
94. Loh M, Sarigiannis D, Gotti A, Karakitsios S, Pronk A, Kuijpers E, Cherrie JW (2017) How sensors might help define the external exposome. Int J Environ Res Public Health 14(4):434. https://doi.org/10.3390/ijerph14040434
95. Vaseashta A (2019) Cyber-physical systems—nanomaterial sensors based unmanned aerial platforms for real-time monitoring and analysis. In: Proceedings of the 4th international conference on nanotechnologies and biomedical engineering. Chisinau, Moldova
96. Lytras MD, Papadopoulou P (2018) Applying big data analytics in bioinformatics and medicine. IGI Global, Hershey, PA, pp 1–465. https://doi.org/10.4018/978-1-5225-2607-0
97. Spruit M, Lammertink M (2018) Effective and efficient business intelligence dashboard design: gestalt theory in dutch long-term and chronic healthcare. In: Lytras M, Papadopoulou P (eds) Applying big data analytics in bioinformatics and medicine. IGI Global, Hershey, PA, pp 243–271. https://doi.org/10.4018/978-1-5225-2607-0.ch010
98. Manrai AK, Cui Y, Bushel PR, Hall M, Karakitsios S, Mattingly C, Patel CJ (2017) Informatics and data analytics to support exposome-based discovery for public health. Annual Rev Public Health 38:279–294. https://doi.org/10.1146/annurev-publhealth-082516-012737
99. Adelman I, Moriz CT (1967) Society politics and economic development—quantitative approach, Baltimore
100. Kantorovič AB (1966) Matematičeskie problembi optimalonogo planirovania [Mathematical problems of optimal planning]. Novosibirsk
101. Child D (1970) The essentials of factor analysis. London
102. Stojanović R (1986) Optimalna strategija privrednog razvoja–faktorska analiza u izboru strategije razvoja, optimal strategy of economic development–factor analysis in the selection of development strategy, Savremena administracija. Belgrade, pp 151–210

103. Stener RE (1986) Multiple criteria optimization: theory, computation and application. Wiley, New York
104. Opricović S (1986) Višekriterijumska optimizacija [Multi-criteria Optimization], Naučna knjiga, Belgrade
105. Opricović S, Hajdarević D (1993) Primena faktorske analize u višekriterijumskoj optimizaciji [Use of Factor Analysis in Multi-criteria Optimization], SYM-OP-IS, Belgrade, pp 391–394
106. Milenković M (2012) Primena višekriterijumske optimizacije pri izboru scenarija regionalnog razvoja [Use of Multi-criteria Optimization in the Selection of a Regional Development Scenario], Proceedings, University of Niš, Faculty of Economy
107. Studija višenamenskog vodoprivrednog sistema "Gornjak" u slivu Mlave [A Study of "Gornjak" Multi-purpose Water Management System in the Mlava drainage basin], (1988), Faculty of Civil Engineering, Belgrade
108. Studija aktuelizacije idejnog rešenja hidroenergetskog iskorišćenja reke Drine na potezu Foča-Goražde [A Study of a Conceptual Solution for Hydropower Utilization of the Drina River at the Foča-Goražde Section] (1990) Energoprojekt, Belgrade
109. Generalno rešenje regionalnog hidrosistema snabdevanje vodom Morave–sistem navodnjavanja [General Solution for the Regional Hydropower Water Supply System of the Morava River – Irrigation System] (1995) Jaroslav Černi Institute, Belgrade
110. Matilainen A, Sillanpää M (2010) Removal of natural organic matter from drinking water by advanced oxidation processes. Chemosphere 80:351–365. https://doi.org/10.1016/j.chemosphere.2010.04.067
111. Sharma S, Bhattacharya A (2017) Drinking water contamination and treatment techniques. Appl Water Sci 7:1043–1067. https://doi.org/10.1007/s13201-016-0455-7
112. Jiang J-Q (2015) The role of coagulation in water treatment. Curr Opin Chem Eng 8:36–44. https://doi.org/10.1016/j.coche.2015.01.008
113. Wang LK, Vaccari DA, Li Y, Shammas NK (2005) Chemical precipitation. In: Wang LK, Hung YT, Shammas NK (eds) Physicochemical treatment processes. Handbook of environmental engineering, vol 3. Humana Press, Totowa, pp 141–197. https://doi.org/10.1385/1-59259-820-x:141
114. Bonilla-Petriciolet A, Mendoza-Castillo DI, Reynel-Avila HE (2017) Adsorption processes for water treatment and purification. Springer, Cham. https://doi.org/10.1007/978-3-319-58136-1
115. Amini A, Kim Y, Zhang J, Boyer T, Zhnag Q (2015) Environmental and economic sustainability of ion exchange drinking water treatment for organics removal. J Clean Prod 104:413–421. https://doi.org/10.1016/j.jclepro.2015.05.056
116. Akpor OB, Muchie M (2010) Remediation of heavy metals in drinking water and wastewater treatment systems: processes and applications. Int J Phys Sci 12:1807–1817
117. Madsen HT (2014) Membrane filtration in water treatment–removal of micropollutants. In: Sogaard EG (ed) Chemistry of advanced environmental purification processes of water. Fundamentals and applications. Elsevier, Amsterdam, pp 199–248. https://doi.org/10.1016/B978-0-444-53178-0.00006-7
118. Malaeb L, Ayoub GM (2011) Reverse osmosis technology for water treatment: state of the art review. Desalination 267:1–8. https://doi.org/10.1016/j.desal.2010.09.001
119. Dapeng L, Jiuhui Q (2009) The progress of catalytic technologies in water purification: a review. J Environ Sci 21:713–719. https://doi.org/10.1016/S1001-0742(08)62329-3
120. Wei L, Liu DJ, Rosales BA, Evans JW, Vela J (2020) Mild and selective hydrogenation of nitrate to ammonia in the absence of noble metals. ACS Catal 10:3618–3628. https://doi.org/10.1021/acscatal.9b05338
121. Chakraborty S, Sikder J, Mukherjee D, Mandal MK, Arockiasamy DL (2013) Bioremediation of water: a sustainable approach. In: Piemonte V, Falco MD, Basile A (eds) Sustainable development in chemical engineering innovative technologies. Wiley, Hoboken, pp 241–266. https://doi.org/10.1002/9781118629703.ch10
122. Shyamala S, Manikandan NA, Pakshirajan K, Tang VT, Rene ER, Park H-S, Behera ShK (2019) Phytoremediation of nitrate contaminated water using ornamental plants. J Water Supply Res Technol AQUA 68:731–743. https://doi.org/10.2166/aqua.2019.111

123. Karapinar N (2003) Magnetic separation: an alternative method to the treatment of wastewater. Eur J Miner Process Environ Prot 3:215–223
124. Collivignarelli MC, Abba A, Benigna I, Sorlini S, Torretta V (2018) Overview of the main disinfection processes for wastewater and drinking water treatment plants. Sustainability 10:86. https://doi.org/10.3390/su10010086
125. Chen Y (2020) Photodegradation of pharmaceutical waste by nano-materials as photocatalysts. In: Singh et al (eds) Nano-materials as photocatalysts for degradation of environmental pollutants—challenges and possibilities. Elsevier, NY, pp 143–152. https://doi.org/10.1016/B978-0-12-818598-8.00008-0

Ashok Vaseashta (M'79-SM'90) received Ph.D. in Materials Science and Engineering (minor in Electrical Engineering) from Virginia Polytechnic Institute and State University, Blacksburg, Virginia, USA. He is Executive Director of Research for International Clean Water Institute in VA, USA, Chaired Professor of Nanotechnology at the Academy of Sciences of Moldova and Professor, Nanotechnology and Biomedical Engineering at the Faculty of Mechanical Engineering, Transport and Aeronautics at the Riga Technical University. Prior to his current position, he served as Vice Provost for Research at the Molecular Science Research Center in Orangeburg, South Carolina. He served as visiting professor at the 3 Nano-SAE Research Centre, University of Bucharest, Romania and visiting scientist at the Helen and Martin Kimmel Center of Nanoscale Science at the Weizmann Institute of Science, Israel. He served the U.S. Department of State in two rotations, as strategic S&T advisor and U.S. diplomat. His research interests span nanotechnology, environmental/ecological science, and safety and security. His research on nanotechnology has been on improving the understanding, design, and performance of nanofibers and sensors/detectors, mainly for applications such as wearable electronics, target drug delivery, detection of biomarkers and toxicity of nano and xenobiotic materials. In the security arena, he has worked on counterterrorism, countering unconventional warfare and hybrid threats, critical-Infrastructure protection, biosecurity, dual-use research concerns, and mitigating hybrid threats. He has authored over 250 research publications, edited/authored eight books on nanotechnology, and presented many keynotes and invited lectures worldwide. He serves on the editorial board of several highly reputed international journals. He is an active member of several national and international professional organizations. He is a fellow of the American Physical Society, Institute of Nanotechnology, and the New York Academy of Sciences. He has earned several other fellowships and awards for his meritorious service including 2004/2005 Distinguished Artist and Scholar award.

Gor Gevorgyan was born in the capital city of Yerevan, Armenia, in 1984. He has been received the B.S. and M.S. degrees in biology from the Yerevan State University, Armenia, in 2006 and 2008, accordingly, and has been awarded Ph.D. degree in biology by the resolution of the Council of National Academy of Sciences of Armenia in 2011. Since 2011, he has been a Senior Scientific Researcher in the Scientific Center of Zoology and Hydroecology, National Academy of Sciences of Armenia, and since 2020, he has been Head of the Scientific Group of Applied Hydroecology in the Center. He is the author of more than 50 scientific publications. His research interests include aquatic and soil ecology, aquatic biodiversity, water and soil quality, water and soil resources management, limnology, environmental impact assessment, climate change and its consequences, fish farming.

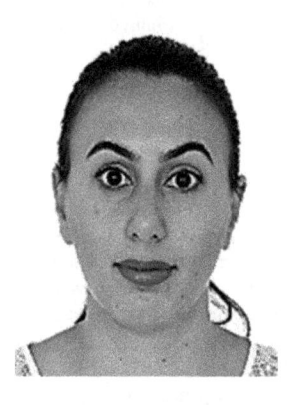

Doğa Kavaz received her B.Sc. degree from the Department of Chemistry, Hacettepe University in 2006. She received the M.Sc. degree in Biochemistry (Chemistry) and Ph.D. degree in Nanotechnology (Chemistry) from Hacettepe University, in 2009 and 2013. During her Ph.D., she went to Harvard University as a Researcher for 1 year with a scholarship from TUBITAK. She took part in a joint project with various universities and developed specialized nanoparticles for different applications. She has over 30 indexed publications. She is currently working as Associate Professor and Researcher of Environmental Research Centre in Cyprus International University. Her research interests include the development and characterization of biomedical materials to be used in medical purposes and synthesis of sustainable nanomaterials using green chemical methods. It also includes the application of these materials for Contaminants Detection, water treatment, soil quality improvement, and sustainable building materials.

Ognyan Ivanov was born in Haskovo, Bulgaria. He received the B.S. and M.S. degrees in solid state physics from the Physical Faculty of the Sofia University in 1982 and the Ph.D. degree in acoustoelectric effects from Institute of Solid State Physics, Bulgarian Academy of Sciences, Sofia, in 1989. From 1986 to 1994, he was Assistant Professor in the Institute of Solid State Physics (ISSP), Sofia. Since 2000, he has been an Associated Professor in the Institute of ISSP. From 2009 to 2010 he was lecturer in University of Architecture, Civil Engineering and Geodesy, Sofia. From 2012 is Member of The Scientific Council of Institute of Solid State Physics and Head of Electromagnetic Sensors Department. From 2019 is Director of GIS – Transfer center # 6 Electro-magnetic sensors, Bulgarian Academy of Sciences, He is the author of two books, more than 90 articles, and participant in 50 scientific conferences. Author in six patents. His research interests are in the field of acoustoelectric and optical and electron properties of solids, electromagnetic field – matter interactions and development of sensors on that basis. Member of Optical society of Amerika and different scientific, organization and editorial boards. Dr Ivanov has 6 awards from Bulgarian and international exhibitions and competitions.. He is a participant or leader in 16 projects of ISSP with outside organizations. He is a team leader of the Bulgarian team that works on development of new types of sensors under the European project COUNTERFOG—*Security programme.*

Mohammad Jawaid done PhD (Polymer Composites) from Universiti Sains Malaysia, Georgetown, Penang, Malaysia. Presently working as Senior Fellow (Professor) at Biocomposite Technology Laboratory, Institute of Tropical Forestry and Forest Products (INTROP), Universiti Putra Malaysia, Serdang, Selangor, Malaysia, and also has been Visiting Professor at the Department of Chemical Engineering, College of Engineering, King Saud University, Riyadh, Saudi Arabia since June 2013. His area of research interests includes hybrid composites, lignocellulosic reinforced/filled polymer composites, advance materials: graphene/nanoclay/ fire retardant, modification and treatment of lignocellulosic fibres and solid wood, biopolymers and biopolymers for packaging applications, nanocomposites and nanocellulose fibres, and polymer blends. So far, he has published 40 books, 65 book chapters, more than 350 peer-reviewed international journal papers, Presently, he is supervising 12 Ph.D. students and 6 Master's students in the fields of hybrid composites, green composites, nanocomposites, natural fiber-reinforced composites, nanocellulose, etc. Dr. Mohammad Jawaid received Excellent Academic Award in Category of International Grant-Universiti Putra Malaysia-2018 and also Excellent Academic Staff Award in industry High Impact Network Award. He is also Winner of Newton-Ungku Omar Coordination Fund: UK-Malaysia Research and Innovation Bridges Competition 2015. Recently he recognized with Fellow and Charted Scientist Award from Institute of Materials, Minerals and Mining, UK. He is also life member of Asian Polymer Association, and Malaysian Society for Engineering and Technology. He has professional membership of American Chemical Society and Society for polymers Engineers, USA.

Dejan Vasović received the B.S. and M.S. degrees in Environmental Protection Engineering from University of Niš, Faculty of Occupational Safety, Niš, Serbia, in 2006 and 2011, respectively, and the Ph.D. degree in the same field from University of Niš, Faculty of Occupational Safety, Niš, Serbia, in 2016. Since 2017, he works as an Assistant Professor at the Faculty of Occupational Safety, Environmental Protection Department. Currently, he is the Vice - Chair of the Department of Environmental and Occupational Quality Management. He has authored three international book chapters and more than twenty papers in journals with impact factor. His research interests include environmental protection, water resources management, system standards, emergency management, natural hazards. He is also a reserve officer from 2008, with the rank of lieutenant. Mr. Vasović is the member of the European Society of Safety Engineers (ESSE) and Balkan Environmental Association (B.EN.A.).

Water Resources, Nature of Contaminants, Impact on Health and Water Quality

Adrian Lucian Cococeanu and Teodor Eugen Man

Abstract Water, like energy, is an essential component of almost all human activities. The water supply is vital for feeding the population, for producing material goods (raising living standards) and for maintaining the integrity of the natural systems on which life on earth depends. Water use is defined as any social or economic unit that needs water of a certain quality to carry out its activity, which satisfies this need through a unitary set of constructions and installations through which the water supply is made, the use and wastewater disposal. The need for drinking water sources dates back to prehistoric times and it is an indispensable resource that we must protect and exploit rationally in a sustainable way to ensure quantity and quality for the future generations.

Keywords Water resources · Natural water circuit · Sustainable water use · Groundwater · Surface water

1 Brief History and Water Directive

The need for maintaining drinking water sources, dates back to prehistoric times. From the beginning, human settlements are connected or have been near the sources of water. The Euphrates, the Indus, the Ganges, the Tiber, the Yan-Tse are the main courses along which the first human civilizations appeared. According to archeological evidence about the existence of centralized water supply systems, during the Nippur civilization, in Samaria about 5000 years ago an arched drain was found with stones fixed by descending feathers, the water being collected through fountains and

---

A. L. Cococeanu
Civil Engineering Faculty, Department of Hydrotechnical Engineering, Politehnica University of Timişoara, Iuliu Podlipny Street, no. 23C, 300703 Timişoara, Romania
e-mail: adrian.cococeanu@student.upt.ro

T. E. Man (✉)
Civil Engineering Faculty, Department of Hydrotechnical Engineering, Politehnica University of Timişoara, Ulpia Traiana Street, no. 80/3, 300771 Timişoara, Romania
e-mail: eugen.man@upt.ro

© Springer Nature Switzerland AG 2021
A. Vaseashta and C. Maftei (eds.), *Water Safety, Security and Sustainability*,
Advanced Sciences and Technologies for Security Applications,
https://doi.org/10.1007/978-3-030-76008-3_5

cisterns. In the 16th–17th centuries, with the appearance of capitalized manufactures, water supply development techniques were imposed. Thus, in the 18th–19th centuries, the first centralized water supply systems were built in cities in England, Germany, France and Russia.

Industrial development in various advanced countries has led to a high demand for water and a higher quality, respectively, which has led to the promotion and concerns related to the development of water supply systems, thus developing techniques for using coagulation-flocculant reagents for intensification of settling, rapid filtration, chlorine disinfection, etc. Hydrology has developed as an empirical science through the way in which the mathematical solution of flows interacts, on the one hand and the evaluation and physical observations of the behavior of aquifers and groundwater basins, on the other hand.

The worldwide interest in water resources management and the recognition of hydrogeology as a discipline appeared in 1960 with the publication of a large number of hydrogeology books and international journals. In 1956, the International Association of Hydrogeologists was founded and the establishment of the International Hydrological Decade of UNESCO, the forerunner of the International Hydrological Program, was established.

Water sources are a vulnerable and limited renewable natural resource and represent a natural heritage that must be rationally protected and exploited. In the conditions of global warming, the sustainable management of water resources, the preservation of the self-regulation capacity and the support of the aquatic ecosystems is very important. The European Union's Water Framework Directive is the cornerstone of Europe's water policy history. It establishes a common framework for the sustainable and integrated management of all water bodies (groundwater, inland surface waters, transitional waters and coastal waters) and requires that all impact factors and economic implications be taken into account. In this respect, the Directive requires the establishment of a program of measures to improve the quality of water quality [1–3].

The new strategy for monitoring and characterizing water quality is based on a new concept of integrated water monitoring that involves a triple integration:

- of investigation areas at hydrographic basin level: surface waters in natural regime (rivers, lakes, transitional waters/brackish waters, coastal marine waters), artificial surface waters or waters with strongly anthropogenic modified regime, groundwater, protected areas, effluents;
- of the investigation media: water, sediments to which the biological components are integrated;
- of the monitored elements/components: biological, hydro morphological and physico-chemical (qualitative and quantitative).

2 The Water Framework Directive

The aims of the water framework directives are to prevent further deterioration, to protect and improve the condition of aquatic ecosystems and, in terms of water requirements, terrestrial ecosystems and wetlands directly dependent on aquatic ecosystems [1, 2];

- to promote the sustainable use of water based on a long-term protection of available water resources;
- the objective is the advanced protection and among others the improvement of the aquatic environment through specific measures for the progressive reduction of discharges, emissions or losses of priority substances and the cessation or gradual cessation of discharges, emissions or losses of dangerous priority substances;
- progressive reduction of groundwater pollution and prevention of subsequent pollution.

In addition, the Water Framework Directive contributes to:

- providing a drinking water in sufficient quantities, of good quality, from surface and groundwater as needed, for a sustainable, rational and equitable use;
- protection of territorial waters and marine waters;
- achieving the objectives of relevant international agreements, including those aimed at preventing and eliminating pollution of the marine environment.

The Water Framework Directive imposes a number of obligations on EU Member States, classified in terms of (a): planning; (b): adoption of regulations; (c): monitoring; (d): consultation and (e): reporting. In practical terms, the Directive requires [1, 2]:

- a wider range of tools for monitoring and classifying waters, in order to assess their ecological status;
- a system for authorization and registration of water samples and accumulations to protect the ecological status of water resources;
- an official system of planning at basin level and application of appropriate measures to limit pollution diffusion in water.

The following benefits are provided:

Environmental benefits:

- improving the protection and general improvement of the quality of the aquatic environment; promoting more efficient ways of using water in order to reduce environmental pressures on the aquatic environment;
- ensuring an efficient and sustainable management of the aquatic environment.

Social benefits:

- increasing the opportunities for involvement and influence of aquatic management by all social actors involved;

- improving the quality of available information about the aquatic environment and its management;
- ensuring the protection of the aquatic environment for the purpose of sustainable development and the provision of ecological services, including the aspects related to favoring the development of the recreational potential.

Economic benefits:

- ensures a fair cost-effectiveness approach, which will lead to a real basis for water use prices;
- favors the achievement of the balance between social, economic and environmental needs by defining environmental objectives;
- increasing the opportunities for involvement and influence of aquatic management by all social actors involved;
- improving the quality of available information about the aquatic environment and its management;
- ensuring the protection of the aquatic environment for the purpose of sustainable development and the provision of ecological services, including the aspects related to favoring the development of the recreational potential.

2.1 The Natural Water Circuits

Water is a precious commodity and is indispensable for life and constitutes more than 60% of living material. Through the cooler, water vapor, present in the mantle of gas that surrounds the earth, condenses for thousands of years. Since then, the amount of water is constant and represents a reserve of 1350 million km^3 (1 billion m^3) but 99.7% of this water is salty or stagnant. Accessible water from rivers and groundwater represents 0.3% of the total reserve.

In the natural cycle, water is regenerated in two levels,

- during evaporation, water vapor is pure, does not contain mineral salts or impurities
- on the occasion of filtration in soils: biological activity and the capacity of soils to filter restoring the clarity of water [4–7].

Filtered water does not have the same composition as water vapor in contact with the soil, some minerals dissolve in water, while water vapor does not contain minerals.

2.1.1 Evaporation: The Transformation of Water from a Liquid State into a Gaseous or Vapor State

Evaporation is the process by which water transforms from a liquid state to a gaseous or vapor state. Evaporation is the main way in which liquid water returns to the general water circuit in the form of vapor in the atmosphere [4–7]. Oceans, seas, lakes and rivers provide about 90% of the atmosphere's moisture through the process of evaporation, and the remaining 10% comes from plant perspiration. Heat is provided by the sun to evaporate. Energy is needed to break the bonds that hold water molecules together, so water evaporates easily at boiling point (100 °C) but evaporates much harder at freezing point. When the relative humidity in the air is 100% (at the "saturation" state), evaporation can no longer continue. The evaporation process absorbs heat from the environment, so water that evaporates from your skin will cool down.

Evaporation from the oceans is the main means by which water reaches the atmosphere. The large surface area of the oceans (over 70% of the Earth's surface is covered by oceans) allows large-scale evaporation to occur. On a global scale, the amount of water evaporated is almost equal to the amount of water that falls to the ground in the form of precipitation. However, this varies geographically. Above the oceans the amount of water in the form of vapor is more common than precipitation, while on the continent's evaporation is exceeded by precipitation. The largest amount of water that evaporates from the oceans falls back into them in the form of precipitation. Only 10% of the water evaporated from the oceans is transported above the earth and falls in the form of precipitation. Once evaporated, one molecule of water remains approx. 10 days in the air.

2.1.2 Water Storage in the Atmosphere in the Form of Vapors, Clouds and Moisture

Although the atmosphere is not a large reservoir of water, it is a means by which water can move the globe from one place to another. There is always water in the atmosphere. Clouds are the most visible form of atmospheric water, but clean air also contains water—water in particles too small to be seen. At any time, the volume of water in the atmosphere is approx. 12900 km^3. If all the water in the atmosphere fell at once, it could cover the earth with a layer of 2.5 cm of water [4–8].

2.1.3 Condensation: The Process by Which Water Is Transformed from a Gaseous State to a Liquid State

Condensation is the process by which water vapor in the air is transformed into liquid water. Condensation is important for the water circuit because it forms clouds. They can produce precipitation, which is the main way water returns to Earth. Condensation is the opposite of evaporation. Clouds form in the atmosphere because air containing water vapor rises and cools. The sun heats the air in the immediate vicinity of the Earth's surface, it becomes lighter and rises where temperatures are lower. As the air cools, the condensation process intensifies, favoring the formation of clouds [4–9].

2.1.4 Precipitation: The Release of Water from the Clouds

Precipitation is water released from the clouds in the form of rain, sleet, snow or hail. Precipitation is the main way in which atmospheric water returns to earth. Most of the precipitation is in the form of rain. Clouds contain water vapor and small drops of water, which are too small to fall in the form of precipitation but are large enough to form visible clouds. Water is constantly evaporating and condensing in the atmosphere. Most of the condensed water in the clouds does not fall in the form of precipitation, due to the rising currents that support the clouds. To produce precipitation, the very fine drops of water must first condense and combine to produce a drop large and heavy enough to fall from the cloud in the form of precipitation. It takes millions of particles of water to produce a single drop of rain [4–9].

2.1.5 Water Storage in Ice, Glaciers and Snow

Water stored for long periods in the form of ice, snow and glaciers are an integral part of the water circuit. Most of the Earth's ice masses, almost 90%, are in Antarctica, while the Greenland ice sheet contains 10% of the Earth's total ice mass. On a global scale, the climate is constantly changing, although this change is not happening so quickly that it is being noticed by people [9, 10].

2.1.6 Leakage of Water Resulting from Melting Snow

All over the world, the flow resulting from melting snow is an important factor in the movement of water around the globe. Where there are colder climates much of the spring runoff and river flows come from melting ice and snow. Rapid snowmelt can trigger in addition to floods and landslides and debris movements. The leakage caused by melting snow varies depending on the season and also from year to year. The lack of water accumulated in the form of a layer of snow in winter can reduce the amount of water for the rest of the year. This can affect the amount of water in downstream storage lakes, which in turn can affect the amount of water available for irrigation and water supply to the population [7–10].

2.1.7 Surface Runoff: Water Resulting from Precipitation Flowing on the Surface of the Soil to Rivers

Probably a lot of people think that precipitation that falls on the surface of the earth, drains, flows into rivers and then flows into the oceans. It is actually more complicated because rivers also lose or accumulate water from the soil. However, much of the water in rivers comes directly from rainfall runoff, defined as surface runoff. Usually some of the rainfall seeps into the soil, but when rainfall falls on a saturated or impermeable soil, it will turn into runoff on the slope. Water will flow along small formations of the hydrographic network, to then reach the larger rivers. The interaction between precipitation and surface runoff varies with time and geographical area, as well as all components of the water circuit. Only a third of the amount of precipitation that falls to ground level reaches watercourses and rivers, only to then return to the oceans. The other two thirds of the precipitation evaporate, is lost by evapotranspiration or infiltrates into groundwater. Surface runoff can also be used by people for their own needs [7–10].

2.1.8 Water Flow Through Riverbeds: The Movement of Water in a River

The flow of water through the riverbed refers to the amount of water that flows into a river, stream or creek. Rivers are important not only for people, but also for life everywhere. Rivers are used for water supply, irrigation, to produce electricity, to transport wastewater, to transport goods and to obtain food. Rivers play a crucial role for all plant and animal species. Rivers help keep groundwater aquifers filled with water by infiltrating water through their riverbeds. And of course, the oceans retain their water because the rivers feed them constantly. The river basin is the surface on which all the water from precipitation and runoff gathers towards an end point. The size of the river basin depends on the closure point—the entire surface of the land from which water is drained and drains to this closure point represents the river basin corresponding to that point. Watersheds are important because the flow and water quality of a river are affected by human-induced or non-man-made phenomena, which occur in the river basin. in a minute. Of course, the main influence on the flow is exerted by the flow of precipitation in the river basin. The size of a river depends largely on the size of its own basin [7–10].

2.1.9 Freshwater Storage

A part of the water circuit that is obviously essential for the existence of life on Earth is the existence of fresh water at ground level. Surface waters include streams, ponds, lakes, artificial lakes and freshwater swamps. The amount of water in rivers and lakes is constantly changing due to changes in inlet and outlet flows. Inlet flows come from precipitation, surface and underground runoff, tributary flows.

Outflows from lakes and rivers include evaporation and discharge into groundwater (aquifer). Surface water is also used by people to meet their needs. The amount and location of surface water changes in time and space, either naturally or through human intervention. Life can occur even in the desert if there is a source of surface or groundwater. At the same time, groundwater exists due to the downward movement of surface water to groundwater aquifers. Fresh water is relatively rare on the Earth's surface [7–10].

2.1.10 Infiltration: The Vertical Movement of Water from the Soil Surface into the Soil or Porous Rock

Everywhere in the world, a quantity of water that falls on the surface of the soil in the form of rain and snow seeps into the subsoil and rock. The amount of infiltrated water depends on a number of factors. Some infiltrated water will remain in the first layer of the soil surface, from where it can enter a watercourse through its banks. Part of the amount of water can infiltrate deeper by refilling underground aquifers. Water can travel considerable distances or remain in the groundwater reservoir for long periods of time, returning to the surface or discharging into other bodies of water such as rivers and oceans. As precipitation seeps into the soil, an unsaturated zone and a saturated zone generally form. In the unsaturated area there is a certain amount of water present in the small holes in the soil, but it is not saturated. In the upper part of the unsaturated area, the soil has cracks created by the roots of the plants where precipitation can infiltrate. Water from this area of the soil is used by plants. Below the unsaturated zone there is a saturated zone where water completely fills the gaps between the rock and the soil particles [7–10].

2.1.11 Groundwater Discharge (Aquifer)

You can see water everywhere every day, in lakes and rivers, in the form of ice, rain and snow. There are also immense amounts of water that cannot be seen—water that is lying and moving underground. People have been using groundwater for thousands of years and continue to use it today, mainly as drinking water and for irrigation. Life on Earth depends on groundwater as well as surface water. Some of the precipitation that falls on the earth seeps into the soil and becomes groundwater. Once in the ground, some of this water circulates near the earth's surface and comes to the surface very quickly in the riverbeds, but due to gravity, much of this water continues to flow deeper into the earth. and the speed of groundwater movement are determined by the multitude of characteristics of aquifers and impermeable layers (dense rock through which water penetrates with difficulty) from the earth. The movement of groundwater depends on the permeability (how easy or difficult water can move) and the porosity (volume of holes in the material) of surface rocks. If the rock allows water to move relatively easily through it then groundwater can travel significant distances in a matter of days. But groundwater can also drain vertically into deep aquifers, from where it needs thousands of years to return to the surface [7–10].

2.1.12 The Spring: The Place Where Groundwater Comes to the Surface of the Earth

A spring is the result of water coming to the surface of the earth from an aquifer filled to the brim. The size of the springs varies from small springs that flow only after heavy rainfall, to huge lakes where millions of liters flow per day. The springs can form in any kind of rock, but they are found mainly in limestone and dolomite, which they are easily eroded and can be dissolved by rains that become acidic. As the rock dissolves and crumbles, spaces can form that allow water to flow. If the runoff is horizontal, water can reach the surface of the earth, resulting in a spring. Spring water is usually clear [7–10].

2.1.13 Evapotranspiration: The Movement of Water Vapor from the Leaves of Plants into the Atmosphere

Evapotranspiration is the process by which moisture is transported by plants from the roots to the small pores on the dorsal side of the leaves, where it turns into vapor and is removed into the atmosphere. Evapotranspiration is the evaporation of water from the leaves of plants. Studies have shown that about 10% of the moisture in the atmosphere is released by plants through evapotranspiration. During the growing season, a leaf will sweat much more water than its weight. Atmospheric factors influencing evapotranspiration The amount of water that plants sweat varies greatly geographically and over time (Fig. 1).

There are a number of factors that determine the amount of water that evaporates:

- Temperature: The amounts of evapotranspirated water increase with temperature, especially during the growing season, when the air is warmer.

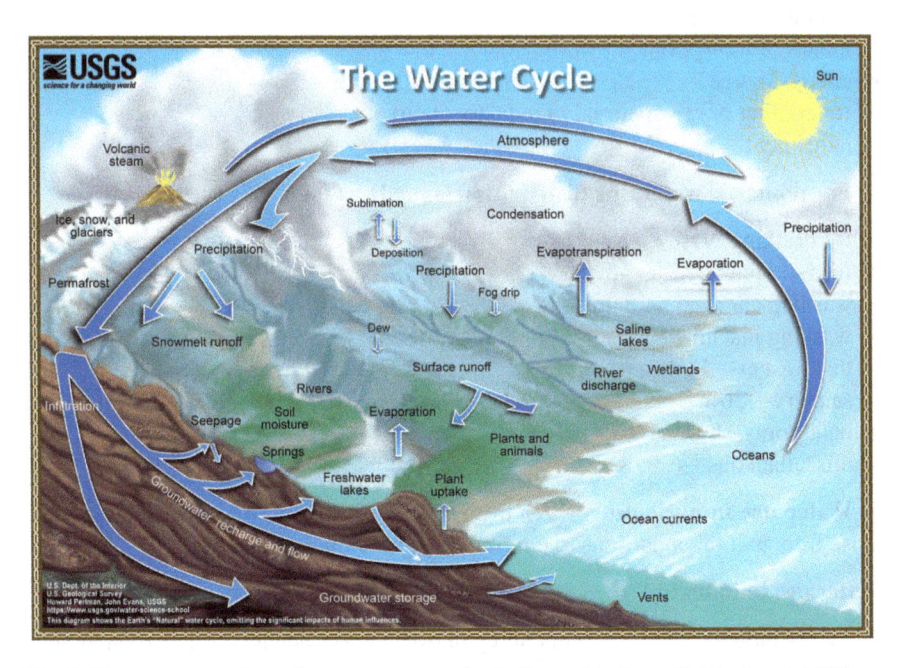

Fig. 1 The water cycle (https://www.usgs.gov—*Credit* Howard Perlman, Public domain) [11]

- Relative humidity: As the relative humidity of the air around the plant increases, the amount of evapotranspirated water decreases. It is easier for water to evaporate in drier air than in the wettest.
- Wind and air movement: A faster movement of air around the plant will lead to more abundant evapotranspiration.
- Plant type: Plants remove water by evapotranspiration in different amounts. Some plants that grow in arid regions, such as cacti, conserve precious water, sweating less than other plants [7–10].

2.2 Water Uses

Water, like energy, is an essential component of almost all human activities. The water supply is vital for feeding the population, for producing material goods (raising living standards) and for maintaining the integrity of the natural systems on which life on earth depends. Water use is defined as any social or economic unit that needs water of a certain quality to carry out its activity, which satisfies this need through a unitary set of constructions and installations through which the water supply is made, the use and wastewater disposal [3].

Water uses:

- human and animal food about 10%
- irrigation in agriculture approximately 50%
- natural resource of fish or aquaculture
- means of transport
- energy source
- for cooling or raw material in industry about 50%
- agreement

For domestic water, the distribution is made as follows:

- for drinking—1%
- for food preparation—6%
- washing cars and gardens—6%
- washing machine—12%
- dishwasher—10%
- sanitary—20%
- bathrooms, showers—39%
- miscellaneous—6%

Water use in agriculture—for irrigation is the supply of water in addition to those from natural conditions to ensure high agricultural production [3].

Water use in industry

- energy industry as a source of energy in hydropower plants, for the production of steam necessary for the operation of thermoelectric and nuclear power plants

- in manufacturing processes: as a hydraulic agent for the transport of materials, for ore processing, in the chemical industry, as a raw material in the food industry
- as a cooling or heating agent to ensure the development of a technological process in good conditions [3].

2.3 Water Resources Types

From the administrative point of view, the waters are classified as follows:

- **international waters**; are the waters on which our state is riparian with other states, those which enter or cross national borders, as well as those on which the interests of foreign states have been recognized by international treaties and conventions;
- **territorial waters (inland maritime)**; are the waters comprised between the part of the shore of the country to the sea, the extent and delimitation of which is established by national law;
- **national waters**; are the inland waterways, rivers, canals and lakes, as well as the waters of the border rivers and streams from the Romanian shore to the border line established by international treaties and conventions.

According to the criterion of objective location and destination, the waters are classified as follows:

- freshwater resources—surface and groundwater;
- water for the population—fresh water necessary for the life and ambiance of human settlements;
- drinking water—surface or groundwater, which, naturally or after physicochemical and/ or microbiological treatment, can be drunk;
- domestic wastewater;
- water for industry;
- industrial wastewater;
- water for irrigation—from surface water sources;
- Drying water.

The establishment of the quality status of the different water categories is made only on the basis of the quality indicators correlated with the different uses of the water, from the legislation in force [3, 5, 6].

2.4 Groundwater

Groundwater is the water inside the earth's crust that circulates in rocks through cracks and pores and accumulates in the form of aquifers. The aquifers (permeable rocks soaked in water) overlap layers of impermeable rocks made of clays, marls, crystalline and porous rocks. Groundwater arises from precipitation that seeps into the ground, or water infiltrates from the bed of running and stagnant water (rivers, streams, lakes), this penetration of water through the permeable layers will be stopped by impermeable rock that plays the role of a channel this system of canals can be superimposed [12–16].

Groundwater categories

The groundwater origin differs, but usually is formed by precipitation that infiltrates deep into the earth's crust (muddy waters). Sometimes they are caught between two impermeable layers and are called captive aquifers. When a groundwater table or a deep aquifer is intersected by a valley, water comes to the surface and flows. The place where an underground water appears is called a spring.

According to the water drainage regime, the following are classified:

- springs with continuous flow
- springs with intermittent flow (springs, geysers)

  Outbreaks—occur in karst regions and give a large amount of water
  Geysers—are intermittent hot springs in volcanic regions
  The sources are classified:

- by water temperature:

  – cold—with water temperatures equal to or lower than the average air temperature
  – warm—with water temperatures higher than average air temperatures

- according to the amount of dissolved mineral substances

  – fresh water—salt content less than 1 g/l
  – mineral waters containing up to 5 g/l
  – brines—waters with mineralization higher than 50 g/l

Causes that endanger the quality of groundwater Human activities can negatively influence the quality (by air, soil or surface water pollution) and the amount of groundwater (by irrational use of water). The natural causes are primarily drought due to the small amount, or lack of precipitates [12–16].

2.5 Surface Water

Surface waters are waters that run on the surface of the soil. Surface waters are produced by the general drainage of rain or the appearance of groundwater on the surface. Once the surface water appears, it follows the road that offers minimum resistance and can be flowing, as in the case of rivers and streams, or stagnant as in the case of lakes or ponds. The quality of rivers and streams varies depending on seasonal flows and can change. significantly due to the precipitation and runoff they receive. Lakes and ponds generally have less sediment than rivers and yet are subject to greater impact in terms of microbiological activity.

Surface water characteristics

Surface water differs in many characteristics: its flow and variations (in flowing ones), temperature, concentration and nature of dissolved or suspended substances, biological and microbiological content, each body of liquid water with its bed and its living beings being a different ecosystem. Surface waters have a special role both for the behaviors of nature and for human life. Rivers contribute to the drilling of relief influences the formation of climate and vegetation, soils. The quality of natural waters is determined by their physical, chemical and biological characteristics. Next, only those indicators will be described that allow, on the one hand, the general characterization of the water and on the other hand determine the water treatability and the choice of the corresponding technological treatment flow [12–14, 17] (Table 1).

3 Conclusion

Water managers and planners are slowly beginning to change their perspective and perceptions about how best to meet human needs for water; they are shifting from a focus on building supply infrastructure to improving their understanding of how water is used and how those uses can best be met. There is growing interest on the part of water managers around the world to implement these approaches to lessen pressures on increasingly scarce water resources, reduce the adverse ecological effects of human withdrawals of water, and improve long-term sustainable water use.

Table 1 Substances in natural waters

Provenance	Positive ions	Negative ions	Colloids	Suspensions	Gases
Contact of water with ores, soils and rocks	Calcium ($Ca+2$) Iron ($Fe+2$) Magnesium ($Mg+2$) Manganese ($Mn+2$) Potassium ($K+$) Sodium ($Na+$) Zinc ($Zn+2$)	Bicarbonates (HCO_3^-) Carbonates (CO_3^{2-}) Chlorides (Cl^-) Fluoride (F^-) Nitrate (NO_3^-) Phosphate (PO_4^{3-}) Hydroxide (OH^-) Sulfate (SO_4^{2-}) Silica (H_3SiO_4) Borates ($H_2BO_3^-$)	Clay silica (SiO_2) Ferric oxide (Fe_2O_3) Aluminum oxide (Al_2O_3) Manganese dioxide (MnO_2)	Clay mud sand and other inorganic soils	Carbon dioxide (CO_2)
From the atmosphere through the rain	Hydrogen ($H+$)	Bicarbonate (HCO_3^-) Chlorides (Cl) Sulfate (SO_4)		Dust pollen	Carbon dioxide (CO_2) Nitrogen (N_2) Oxygen (O_2) Sulfur dioxide (SO_2)
Decomposition of organic matter in the environment	Ammonia (NH_3) Hydrogen (H) Sodium (Na)	Bicarbonate (HCO_3^-) Chlorides (Cl^-) Acid sulfides (HSAcid sulfides (HS^-) Nitrate (NO_3) Nitrite (NO_2^-) Organic radicals	Colored vegetable matter, organic residues	Topsoil, organic residues	Carbon dioxide (CO_2) Nitrogen (N_2) Oxygen (O_2) Hydrogen sulfide (H_2S) Ammonia (NH_3) Hydrogen (H_2) Methane (CH_4)
Living organisms from the environment			Bacteria algae viruses	Algae, diatoms, fish tiny organisms	Carbon dioxide (CO_2) Ammonia (NH_3) Methane (CH_4)

(continued)

Table 1 (continued)

Provenance	Positive ions	Negative ions	Colloids	Suspensions	Gases
From industry, agriculture and other human activities	Inorganic ions, heavy metals	Inorganic ions, organic molecules, dyes	Inorganic, organic, colored solids, chlorinated organic compounds, bacteria, worms, viruses	Clay, silt, coarse sand or other inorganic solids, organic compounds, petroleum, corrosive compounds	Chlorine (Cl_2) Sulfur dioxide (SO_2)

Note Substances in natural waters data from Rojanschi et al. [5]

References

1. Framework Directive 2000/60/EC 102. http://www.mmediu.ro/beta/domenii/managementul-apelor-2/managementul-resurselor-de-apa/cooperarea-internationala-in-domeniul-apelor/directiva-cadru-pentru-apa/
2. Directive 2007/60/EC of the European Parliament and of the Council of 23 October 2007 on the assessment and management of floods
3. Leal Filho W, Sümer V (2015) Sustainable water use and management
4. http://elearning.masterprof.ro/
5. Rojanschi V, Bran F, Diaconu G (2002) Engineering and environmental protection. Economic Publishing House, Bucuresti
6. Florescu C, Mirel I, Carabeţ A, Stäniloiu C (2015) Watersupply (Alimentari cu apa). www.ct.upt.ro/users/ConstantinFlorescu/index.htm
7. Church TM (1996) An underground route for the water cycle. Nature 380(6575):579–580
8. Schlesinger WH, Jasechko S (2014) Transpiration in the global water cycle. Agric For Meteorol 189–190(2014):115–117
9. Vladimirescu I (1984) The basics of technical hydrology (Bazele hidrologiei tehnice). Technical Publishing House, Bucureşti, p 83
10. Wilby RL (1995) Contemporary hydrology. Departament of Geography, University of Derby, UK
11. https://www.usgs.gov
12. Cococeanu A, Man ET, Florescu C (2015) Modern technologies for the 3rd Millennium, Water source assessment for drinking water. Assessment/optimization for water treatment technologies, Oradea. https://arhicon.uoradea.ro/ro/cercetare/tmpm-iii-isi-proceedings
13. Cococeanu A, Man ET, Vlaicu I (2016) Sustainable development in rural area through projects—basic infrastructure—water and sewerage systems in the timiş County, Managemnt Agricol, editura Agroprint Timisoara 2016, seria I, vol XVIII(1). ISSN 1453-1410, E-ISSN 2069-2307
14. Cococeanu A, Man ET, Florescu C, Beilicci R, Vlaicu I (2019) Interferences past–present–future. In: The water supply of Timisoara Municipality from Underground Water Sources. Sci Bull Politeh Univ Timişoara 64(78)(2):25–32. ISSN 2601-8020
15. Mike Edmunds W, Shand P (2008) Natural groundwater quality
16. Wagner Brian J, Gorelick Steven M (1987) Optimal groundwater quality management under parameter uncertainty. Water Resour Res 23:1162–1174
17. Ouyang Y, Nkedi-Kizza P, Wu QT, Shinde D, Huang CH (2006) Assessment of seasonal variations in surface water quality. Water Res 40(20):3800–3810

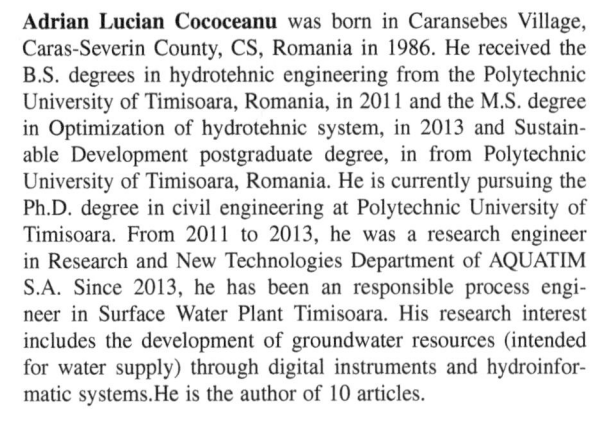

Adrian Lucian Cococeanu was born in Caransebes Village, Caras-Severin County, CS, Romania in 1986. He received the B.S. degrees in hydrotehnic engineering from the Polytechnic University of Timisoara, Romania, in 2011 and the M.S. degree in Optimization of hydrotehnic system, in 2013 and Sustainable Development postgraduate degree, in from Polytechnic University of Timisoara, Romania. He is currently pursuing the Ph.D. degree in civil engineering at Polytechnic University of Timisoara. From 2011 to 2013, he was a research engineer in Research and New Technologies Department of AQUATIM S.A. Since 2013, he has been an responsible process engineer in Surface Water Plant Timisoara. His research interest includes the development of groundwater resources (intended for water supply) through digital instruments and hydroinformatic systems.He is the author of 10 articles.

Teodor Eugen Man is affilitaed with the Polytechnic University of Timişoara. Currently he is Professor in the field of Civil Engineering and Installations, having completed 16 doctoral theses and in internship currently having 10 Ph.D. students. Leading positions held in UPT: Vice-Rector (1996–2000), Member of the Senate (1996–2000, 2000–2001), Dean of the Faculty of Hydrotechnics (2008–2011), Director of the Department of Hydrotechnics (2011–2012), Vice-Dean of the Faculty of Constructions and member of the Management Office of the Teachers' Council (2012–2014), Director of the Research Center in Hydrotechnics and Environmental Protection (2009–2018). He has developed a number of 23 volumes of courses, laboratory guides, monographs, textbooks, of which 6 as sole author and 10 as first author, over 300 published scientific papers of which 17 sole author, 145 first author, over 50 published abroad and over 30 articles listed ISI or ISI Proceedings, 3 patents, 8 innovations, 116 research and development contracts (grants), of which 70 was responsible/project manager, technical expertise: 70, verification projects: 242. He is a member of prestigious professional associations in the country and abroad, he was a visiting professor at Artois University, IUT Bethune in France, also for 14 years he taught specialized courses at the Faculty of Environmental Protection, University of Oradea. Prof. Man's CV was also presented in the "Dictionary of Romanian Personalities" (Contemporary Biographies—ed. 2012, 2017). *Member of the editorial board*: the Scientific and Technical Bulletin of UPT, Department of Hydrotechnics; The scientific and technical bulletin of T.U. "Ghe. Asachi" Iaşi; Annals of the University of Oradea, Faculty of Environmental Protection and Faculty of Constructions and Architecture; Scientific Annals of Ovidius University Constanta; Revista Hidrotehnica, A.N. "Apele Romane", Bucharest. *Professional, cultural, educational associations*: Associate member of the Bucharest Academy of Agricultural and Forestry Sciences (ASAS); Member of the General Association of Romanian Engineers (AGIR); Founding Member of the Bucharest Association of Land Improvements and Agricultural Constructions (AIFCR); Member of the Romanian National Society for Soil Science Bucharest (SNRSS); Member of the National Committee for Irrigation and Drainage (CNRID) Bucharest, affiliated to the International Commission of Irrigation and Drainage, New Delhi, India (ICID).

Contamination Detection and Mitigation

Nanomaterials and Their Role in Removing Contaminants from Wastewater—A Critical Review

Violeta-Carolina Niculescu, Marius Gheorghe Miricioiu, and Roxana-Elena Ionete

Abstract Removal of contaminants from wastewater has become an important research area because the amount of available drinking water in the world continues to decline due to rising demand and/or long periods of drought. Furthermore, the chemical and petrochemical industry generates a wide variety of highly toxic residues. Treatment of wastewater is a controversial field in terms of environment protection. In this chapter, several nanomaterials, which impart their unique properties, will be discussed. Among nanomaterials, carbon nanotubes (CNTs) are a form of carbon allotrope with a graphite-like structure, displaying various adsorption characteristics, as a result of the diameter, internal geometry, physical and chemical properties or the obtaining method. Contaminants removal using CNTs needs further research, only limited studies being available and more practical applications are needed to confirm the results. Several other adsorbent nanomaterials have been reported in literature. Among them, mesoporous materials have large surface areas and narrow pore size distribution, ranging from 20 to 100 Å, being suitable for liquid phase reactions because they favor the diffusion of the reactants to the active site. The adsorbents can be very effective for adsorption of several types of contaminants, such as heavy metals and different types of dyes. Recently, advanced research targeted the wastewater treatment by using nano catalysts, nano photocatalysts or membranes. The purpose of this chapter was to accomplish a comprehensive overview on the use of nanomaterials in wastewater treatment. The renewed interest in the environment pollution has led to the development of effective models describing the performances of these technologies.

Keywords Adsorption · Catalysis · Nanomaterial · Wastewater

V.-C. Niculescu (✉) · M. G. Miricioiu · R.-E. Ionete
National Research and Development Institute for Cryogenic and Isotopic Technologies – ICSI Ramnicu Valcea, 4th Uzinei Street, PO Raureni, Box 7, 240050 Ramnicu Valcea, Romania
e-mail: violeta.niculescu@icsi.ro

M. G. Miricioiu
e-mail: marius.miricioiu@icsi.ro

R.-E. Ionete
e-mail: roxana.ionete@icsi.ro

© Springer Nature Switzerland AG 2021
A. Vaseashta and C. Maftei (eds.), *Water Safety, Security and Sustainability*, Advanced Sciences and Technologies for Security Applications, https://doi.org/10.1007/978-3-030-76008-3_6

1 Introduction

Safe drinking water and sanitation facilities are essential for human health and well-being, but also a basic necessity for crops and animals. Since water is the key of the body's metabolic processes, once contaminated with toxic compounds, it can cause serious diseases such as cholera, diarrhea, dysentery, hepatitis and typhoid. Up to 1 million people die every year due to diarrhea caused by drinking water or poor hygiene [1]. Due to rapid urbanization and climate change, it will become a challenge to consume water with proper quality, for general needs and even for agriculture. In 2017, according to World Health Organization, 1400 million people use water sources located 30 min away from them, 206 million takes more than 30 min to collect water, 435 million uses groundwater and 144 million uses untreated water from surface sources [1]. Therefore, to achieve high quality standards, wastewater treatment must be performed.

A combination of several factors lead to water pollution, such as the discharge of effluents from various sources (industrial, domestic), intensive use of pesticides in agricultural activities, and poor isolation of landfills. [2]. According to the European Environment Agency, the main contaminants present in soil due to commercial and industrial activities are heavy metals (31%), mineral oil (20%), polycyclic aromatic hydrocarbons (16%), aromatic hydrocarbons (13%), chlorinated hydrocarbons (13%) and others (7%) [3, 4]. The identification of appropriate technological advances and gaps that map with the contaminants removal from wastewater, must be a priority for correctly informing the scientists regarding the application of the best standards, resulting the safe use of water. Nevertheless, only a fraction of the wastewater streams are collected and appropriately treated by various methods (Fig. 1), i.e. primary (settling treatment), secondary (biological methods for reduction of organic compounds) and tertiary (stringent methods for reduction of nutrients) [5]. In Serbia, for example, 48% of the population collects sewage, but does not treat it before evacuation, while at the opposite pole, in Netherlands, 99% of the population applies strict treatments.

During the last decades, several unconventional and conventional technologies were applied for wastewater treatment, taking into account the pollution types, sources and levels. One of the classical technologies, intensively applied, is based on sand filters. Its main advantage consists in the complex filtration, resulting in the removal of solids, such as chemical (e.g. nitrite, nitrate, heavy metals and pesticides) and biological contaminants [6]. Two types of sand filters were used, namely slow and rapid sand filter introduced in various tank designs, depending on the water demands. The filtration rate for the rapid filter is up to 50 times greater than in case of slow filter, due to the particles size, which were significantly higher, ~up to 11 times. Another conventional technology used for many years for wastewater treatment is based on the use of biological activated carbon (BAC) filter, consisting of activated carbon covered by a biofilm [6]. The purification mechanism involves both adsorption and biodegradation [7]. A different filter used only for wastewater treatment, through its aeration configuration, is the biological aerated filter (BAF) [8, 9]. Also,

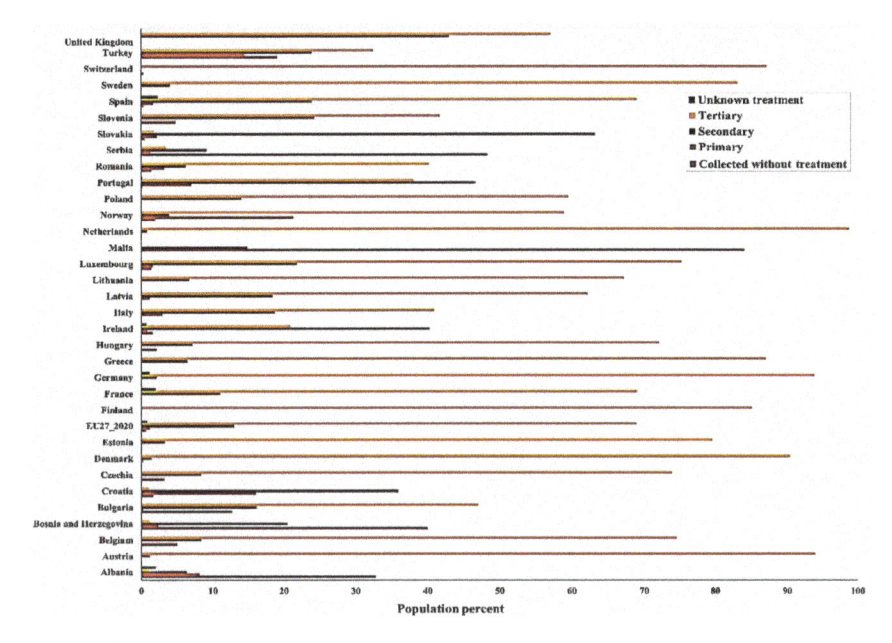

Fig. 1 Collected urban wastewater and types of the treatments

membrane filtration and chemical treatments by using chlorination, UV irradiation and permanganate oxidation have been and are still used, but, in some cases, these treatments do not reach the required standards for wastewater [10–13]. Nowadays, new highly efficient treatments with low-cost operation and environmentally friendly composition are used for human and environment.

In order to have an overview of the efficiency of removing contaminants, including metals, the treatment processes (Fig. 2) were structured in four main categories (nanofiltration, adsorption, reverse osmosis and ion exchange) [14]. It has been observed that conventional adsorption technology has the highest tendency to remove As and Hg, but lower efficiency for Ni removal. Reverse osmosis has a higher efficiency for removing Ni and Zn.

An important aspect of these technologies is the fabrication and operation cost. Taking into consideration the higher cost per treated water volume but lower operation cost, the nanofiltration remains a promising solution for water treatments, being intensively studied in order to obtain cheaper materials with higher efficiency and environmentally friendly properties [14]. The interest in using nanomaterials for wastewater treatment arises from their superior characteristics, such as high surface area, high surface free energy, tunable pores or reactive sites [15]. Therefore, various nanomaterials have been used in different wastewater treatment methods, involving adsorption, photocatalysis or membranes.

The aim of this chapter is to accomplish a comprehensive overview of results obtained in the past years on the use of nanomaterials in wastewater treatment. The renewed interest in the environment pollution has led to the development of

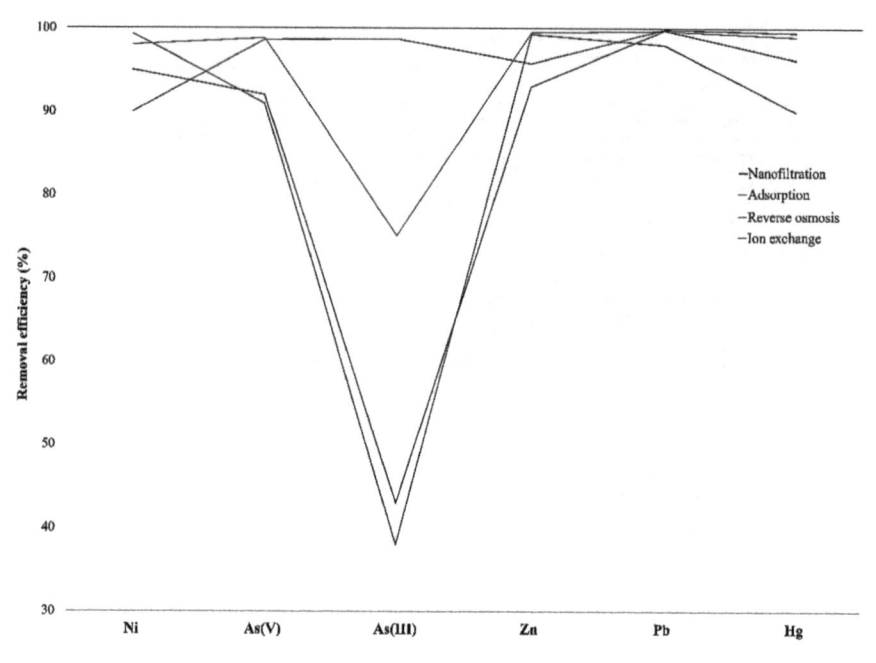

Fig. 2 Technologies efficiency for wastewater treatment

effective models, which equally well describe the performances of these technologies. For wastewater treatment, the nanomaterials can be considered easily adaptable approaches, but some concerns need to be addressed: their limitations, advantages, disadvantages and future perspectives. Moreover, their health risk must be evaluated by the research communities, being responsible for generating suitable regulation to surpass this concern.

2 Nanotechnology for Wastewater Treatment

The necessity of clean water is increasing worldwide due to the freshwater diminishing resources, caused by population increase, extended droughts, climate changes and strict water quality regulation [15, 16]. Nanotechnology has proved to be one of the most suitable methods for wastewater treatment. It can appropriately mitigate many of the water quality problems by using various functional nanoparticles and/or nanofibers [17]. Nanotechnology uses "materials with any external dimension in the nanoscale (around 1–100 nm) or having internal structure or surface structure in the nanoscale" (Fig. 3) [18, 19].

Nanomaterials have significantly improved physical, chemical and biological characteristics resulted from their structure and high surface area [21, 22]. These unique properties (Table 1) were studied for implementation in wastewater treatment

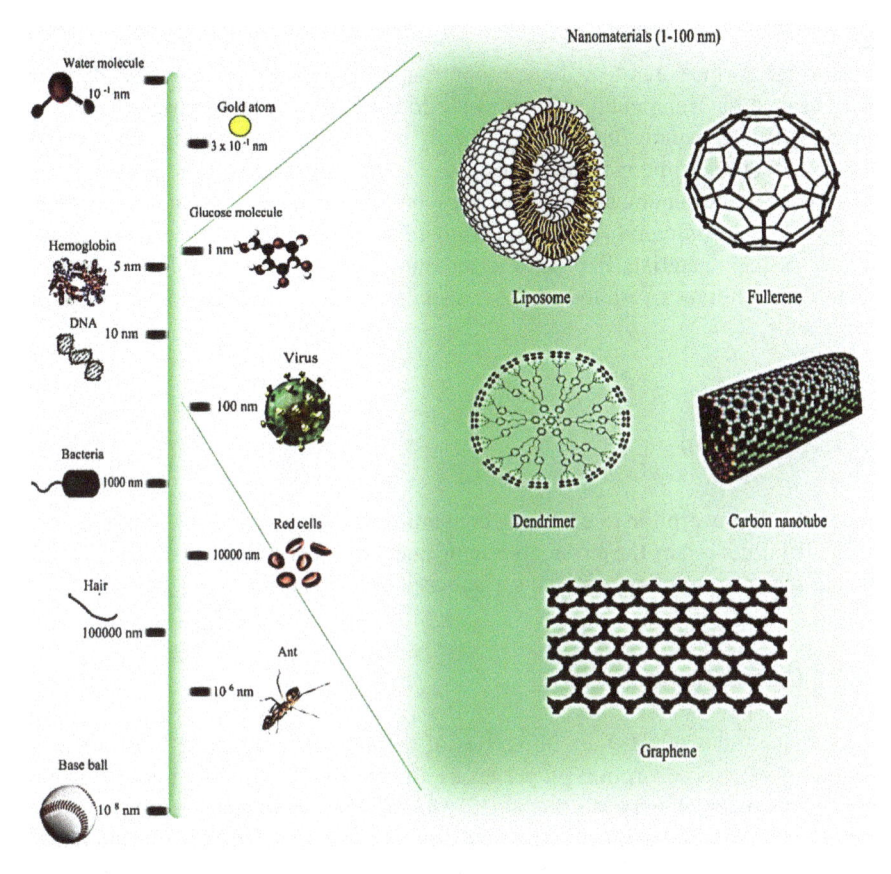

Fig. 3 A size comparison of nanoparticle with other larger-sized materials [20]

Table 1 Potential applications of nanomaterials in wastewater treatment

Treatment method	Nanomaterial	Nanomaterial improved characteristics
Adsorption	Carbon nanotubes, metal oxides, nanofibers, metal-organic frameworks	High specific surface area, selective adsorption sites, tunable pores, easy reuse, etc.
Photocatalysis	Nano-TiO_2, silica derivates	High stability and selectivity, low toxicity and costs, etc.
Membranes	Zeolitic, polymeric, mixed matrix membranes	High permeability and selectivity, hydrophilicity, low toxicity, mechanical and chemical stability, etc.

[23].

The high surface area-to-volume proportion of nanomaterials improves the reactivity against environmental contaminants. In the context of wastewater treatment and remediation, nanotechnology can supply both water quality and quantity, with low-costs and real-time measurements [24, 25]. Energy preservation results in cost savings due to the nanomaterials small sizes, but the total usage cost of the nanotechnology must be compared with other commercial techniques [26]. The development of various nanomaterials like nano adsorbents, nano catalysts, zeolites or nanostructured membranes resulted in toxic metals removal or organic and inorganic compounds.

2.1 Adsorption

Generally, the adsorption of emerging contaminants on the surface of nanomaterials is mainly influenced by the physical structure and chemical properties of the material, such as the pore structure, specific surface area or surface functional groups.

2.1.1 Carbon Nanotubes

Carbon nanotubes (CNTs) are allotropes of carbon with a graphite-like structure, exhibiting various adsorption properties as a result of the chirality, internal geometry and diameter, or synthesis method [27–33]. Carbon nanotubes are single-walled nanotubes (SWNT), having an internal diameter of about 1 nm [34, 35] and multi-walled nanotubes (MWNT), formed by a number of concentric tubes or laminated graphene layers [35–37]. Multi-walled carbon nanotubes can be obtained from single-walled CNTs by using supplementary chemical processing methods, in order to improve the contact area by several times and the amount of active sites for adsorption [38]. Table 2 gives an overview of the applications where CNTs have been used for the removal of emerging contaminants from water.

Specific surface area has an important influence on the adsorption performance of CNTs and it mainly depends on the presence of single- or multi-walled structures. For example, when SWNT were used, tetracycline was removed from wastewater with a 92% efficiency, while MWNT removed only 16% [44]. The adsorption coefficient (K_d) values of SWCNTs, MWCNTs were almost 1500 and 1100 respectively [44]. The sorption data of tetracycline on MWCNTs were evaluated using the Langmuir model, the maximum adsorption capacity being 269.5 mg/g and the efficiency 99.8% [42].

There are only few studies that compare the behavior of single- and multi-walled carbon nanotubes, the majority revealing better performance for the single-walled carbon nanotubes. Also, contradictory results were obtained by using the same carbon nanotubes for removal the same contaminant [45, 46]. For example, the removal of sulfamethoxazole from aqueous solutions was tested under various conditions [45,

Table 2 Adsorption of some emerging contaminants on carbon nanotubes

Nanomaterial	Contaminant	Treatment conditions	Maximum adsorption − q_m (mg/g)/coefficient − k_f (mmol^{-1-n}L^{n}kg^{-1})	References
MWCNT	Norfloxacin	T = 30 °C; pH = 7	q_m = 89	[39]
MWCNT	Sulfamethoxxazole	pH = 7	q_m = 46	[40]
MWCNT	Sulfamethoxxazole	pH = 6	k_f = 510	[41]
MWCNT	Tetracycline	T = 20 °C	q_m = 270	[42]
MWCNT	Sulfonamides	T = 25 °C	k_f = 352–2815	[43]
MWCNT	Chloramphenicol	T = 25 °C	k_f = 570–618	[43]
MWCNT	Non-antibiotic pharmaceuticals	T = 25 °C	k_f = 318–1521	[43]
MWCNT	Tetracycline	pH = 5	k_f = 240	[29, 44]
KOH-activated MWCNT	Sulfamethoxxazole	pH = 6	k_f = 2300	[29, 44]
KOH-activated MWCNT	Tetracycline	pH = 6	k_f = 800	[29, 44]
SWCNT	Tetracycline	pH = 5	k_f = 1150	[29, 44]
KOH-activated MWCNT	Sulfamethoxxazole	pH = 6	k_f = 5200	[29, 44]
KOH-activated MWCNT	Tetracycline	pH = 6	k_f = 1400	[29, 44]

46]. Some authors reported that, from the multiple factors that can be varied, such as pH, adsorbent dosage or adsorbate concentration, the effect of pH affects adsorption capacity most strongly [45]. Others reported that, in the same conditions, the adsorption capacity was mostly influenced by adsorbent quantity or initial concentration of the adsorbate [46].

The adsorption capacity of CNTs can be improved by functionalizing them with other reactive nanomaterials, which is an area of ongoing investigation. For example, zero valent iron (nZVI) was immobilized on the surface of the CNTs to remove the diazo dye Direct Red 23 from aqueous solution [47]. The emerging contaminants removal by adsorption on CNTs still needs further research, only limited studies being available and more experimental proof being needed in order to sustain the reported trends.

2.1.2 Metal-Organic Framework (MOF) Nanomaterials

The adsorption properties of some MOFs are summarized in Table 3. Zeolitic imidazole framework (ZIF)-magnetic graphene oxide exhibited high adsorption efficiency

Table 3 Adsorption of some emerging contaminants on MOFs

Nanomaterial	Contaminant	Treatment conditions	Maximum adsorption – q_m (mg/g)	References
Cr(III) terephtalat-MIL101	Dimetridazole	T = 25 °C; pH = 6	$q_m = 185$	[48]
Cr(III) terephtalat-MIL101	Metronidazole	T = 25 °C; pH = 6	$q_m = 188$	[48]
Cr(III) terephtalat-MIL101	Naproxen	T = 25 °C; pH = 7	$q_m = 156$	[49]
Cr(III) terephtalat-MIL101	Ketoprofen	T = 25 °C; pH = 7	$q_m = 80$	[49]
Zeolitic imidazole framework-magnetic graphene oxide	Benzotriazole	T = 40 °C	$q_m = 300$	[50]
Zeolitic imidazole framework-8	1H-benzotriazole	T = 30 °C	$q_m = 299$	[51]
Zeolitic imidazole framework-8	5-tolyltriazole	T = 30 °C	$q_m = 397$	[51]
Metal organic framework-porous carbon	Ibuprofen	T = 25 °C; pH = 5	$q_m = 320$	[52]
Metal organic framework-porous carbon	Diclofenac solution	T = 25 °C; pH = 5	$q_m = 400$	[52]

against benzotriazole (300 mg/g) [50]. The ZIF-8 adsorption capacity for 1H–benzotriazole and 5-tolyltriazole was better evaluated by pseudo-second-order kinetics, fitting the Langmuir adsorption model with an adsorption capacity of 298.5 and 396.8 mg/g, respectively [51].

Various mechanisms were proposed for MOFs adsorption of pollutants from wastewater, such as Lewis acid–base interactions, electrostatic interactions, π–π interactions or H-bonding [50]. For example, it was reported that the adsorption of nitroimidazole antibiotics on MOFs was achieved by H-bonding between the –NO_2 group from nitroimidazole and –NH_2 from the modified MOFs [48]. One of the most important parameters that influence MOFs adsorption capacity is the pH. The 1H–benzotriazole and 5-tolyltriazole adsorption on ZIF-8 slightly decreased with the pH increasing [51]. The ZIFs negatively charged with magnetic reduced graphene oxide displayed rather stable adsorption for benzotriazole at pH varying between 4 and 9 [50]. Once the pH increased to 10, adsorption decreased due to the inhibition of electrostatic adsorption by the negatively-charged species graphene oxide [50].

Another MOF, MIL-101, was used for the saccharin adsorption from wastewater, displaying stable adsorption capacity at pH ranging from 3 to 7, which was attributed to the electrostatic interaction of positively charged MOF with negatively charged deprotonated form of saccharin and the stable H-bonding between the NH_2

function on urea-MIL-101 and saccharin anion [53]. Two MOFs composites, MIL-101/chitosan (MIL-101/CS) and MIL-101/sodium alginate (MIL-101/SA) were used for the adsorption of benzoic acid (BEN), IBP, and ketoprofen (KET), exhibiting similar variation of the pH-dependent adsorption, reaching a maximum adsorption at pH around 4, due to the influence of pKa-dependent electrostatic interaction [54]. Urea-modified MIL-101 manifested a decrease in the adsorption capacity one the pH was increased, as a result of the electrostatic interaction between the positive surface charge on MIL-101 and the negatively charged oxygen from the $-NO_2$ group of the nitroimidazole antibiotics [48].

2.1.3 Mesoporous Silica

Mesoporous silica materials (such as SBA-3, SBA-15, MCM-41 or MCM-48) gained intensive interest as potential adsorbents over the last years, due to their high surface area, tunable, ordered and uniform pores, high pore volume, thermal and mechanical stability and option for functionalization [55, 56]. As a consequence, they have been applied as adsorbents for organic dyes [57–59], heavy metals from wastewater [60, 61], polycyclic aromatic hydrocarbons [62], as well as other organic contaminants [63]. The adsorption capacity of several mesoporous silica materials for various dyes is summarized in Table 4.

As noted, SBA-15 manifested a significantly higher adsorption efficiency than MCM-48, due to its larger pore size (5.27 nm vs. 3.0 nm), allowing dye molecules to easily diffuse from SBA-15 surface to pores [58, 67]. Furthermore, the mesoporous silica adsorption capacity is dependent on the functional group, initially having a negative surface charge due to the Si—OH groups. In this respect, in order to improve the adsorption processes, the mesoporous silica surface was functionalized with groups suitable for adsorption of specific compounds. Various functionalized mesoporous silica materials were used for adsorbing dyes (Table 4). For example, mesoporous silica functionalized with amino or carboxylic groups have been used for adsorption of acidic and basic dyes, with good selectivity and rapid adsorption rate due to the high surface area and to the strong electrostatic interactions [68, 69]. The adsorption of Remazol Red dye by MCM-41-NH_2 reached an efficiency of 98.2%, higher than that obtained using Fe(III)/Cr(III) hydroxide (9%) or various carbon-based adsorbents [68, 69]. Mesoporous silica materials can be easily protonated in water, resulting in charging their surface which can interact with other ions in solution. As a consequence, mesoporous silica could be applied as efficient adsorbents for the removal of various organic contaminants [73].

Also, in the case of using mesoporous silica as adsorbent, the pH controls the amplitude of the electrostatic charges shared by the ionized contaminants molecules. In general, low pH will increase the rate removal of an anionic dye, while that of a cation dye will decrease [74–76]. For example, the removal of cationic methylene blue dye using 3-aminopropyl triethoxysilane-mesoporous silica was increased once the pH increased, a maximum adsorption capacity (66 mg/g) being achieved at pH equal to 7 [77]. The capacity of dimethyldecylamine-mesoporous silica for removing

Table 4 Adsorption of various dyes by functionalized or not mesoporous silica

Nanomaterial/textural properties (surface area – m^2/g; pore diameter-nm)		Functional group	Contaminant	Removal efficiency (R%)/Equilibrium adsorption-Q_e (mmol/q)	References
SBA	1435	–	Orange G—anionic dye	79/0.35	[64]
	1005; 21.6	Ethylenediamine Aminopropyl penta-ethylene-hexamine	Acid Blue 113—anionic dye	83.6/–	[65]
	1290; 6.8		Acid Green 28—anionic dye	95.9/–	[65]
	659; 5.2	–	Methylene Blue—cationic dye	99.1/0.15	[58]
	659; 5.2	–	Janus Green B—cationic dye	–/0.13	[58]
	707; 1.9	Hyperbranched-polyglycerol	Methylene Blue—cationic dye	–/<0.5	[66]
MCM	1149; 3.0	–	Methylene Blue—cationic dye	–/0.04	[67]
	1222; 5.6	–	Methylene Blue—cationic dye	–/0.07	[67]
	1647; 3.8	–	Methylene Blue—cationic dye	–/0.14	[67]
	215; 1.8	– NH_2	Remazol Red—anionic dye	99.1/–	[68]
	774; 2.5	– NH_2	Acid Blue 25—anionic dye	–/0.60	[69]

(continued)

Table 4 (continued)

Nanomaterial/textural properties (surface area – m²/g; pore diameter-nm)		Functional group	Contaminant	Removal efficiency (R%)/Equilibrium adsorption-Q_e (mmol/q)	References
	754; 2.5	– COOH	Methylene Blue—cationic dye	– /0.30	[69]
Other mesoporous silica	285; 19.0	Dimethyldecylamine	Dark Yellow GG—anionic dye	– /1.71	[70]
	285; 19.0	– NH_2	Red Violet X-2R—anionic dye	– /0.90	[71]
	313	– NH_2	Acid Black 1—anionic	83.0/–	[72]
	516; 4.6	Dimethyldecylamine	Dark Yellow GG—anionic dye	– /1.96	[71]
	516; 4.6	Dimethyldecylamine	Red Violet X-2R—anionic dye	– /1.96	[71]

sulphonated azo dye from wastewater (0.3 mg/g) was slowly increased by decreasing the pH under 4. A higher pH conducted to the adsorption capacity decrease, probably due to the deprotonation of the surface groups and to the protonation of the acidic functional groups of the dye, resulting an electrostatic repulsion between the mesoporous silica and the adsorbate [70]. The maximum adsorption capacities of the acidic dyes were reported in solutions with a pH varying from 2 to 6, while for the cationic dyes, the optimum pH varied between 7 and 11 [78].

Heavy metal ions (Pb, Zn, Cd or Cr) are known as emerging contaminants in a water source. Their direct or indirect release as a by-product from different industries in wastewater stream is a part of water pollution. Ni-SBA-15 and Ni-MCM-41 obtained through co-condensation were used as adsorbents for removing Ni^{+2} from wastewater with an adsorption rate and capacity up to 95% [79]. Determination of Pb^{+2} traces in wastewater was achieved using mesoporous silica functionalized with Pb(II) [80]. The bifunctional modified Al^{+3}/Ti^{+4}-MCM-41 was used to remove Cd^{+2} ions from wastewater [81]. Additionally, the M41S and SBA series were preferred for removal of Cr(VI) from wastewater, due to their unique pore structure [82]. Functionalization of SBA-15 and MCM-41 with amino groups conducted to effective removal of heavy metals such as Pb^{+2}, Cd^{+2}, Cu^{+2}, Ni^{+2} and Cr^{+2}, reaching high adsorption rate (around 99%) [83]. Also, SBA-15 functionalized with two types of functional groups, propyl-trimethylammonium and propyl-ammonium were obtained for nitrates removal from wastewater [84]. The adsorption capacity was influenced by the nature of the functional group and, also, by the synthesis method. It can be concluded that the selectivity and capacity of metal ions adsorption are affected by mesoporous silica obtaining method, functional groups and pH [85].

2.2 Photocatalysis

An advanced oxidation process for removing trace contaminants is the photocatalytic oxidation. The most studied photocatalysts consist of either metal oxides (such as TiO_2 and ZnO) or carbon nanotubes and graphene oxides combined with metal oxides (such as TiO_2, Cu_2O Co_3O_4 and $ZnFe_2O_4$). TiO_2 nanoparticles are used effectively for the photocatalysis of wastewater pollutants like benzenes, polychlorinated biphenyls (PCBs) or chlorinated alkanes [86]. Also, the microcystins were removed from wastewater by photocatalysis using TiO_2 nanoparticles in a "falling film" reactor [87].

Organic contaminants can be efficiently removed by doping TiO_2 with noble metals, due to the hydroxyl radical appearance [88]. For example, nano-TiO_2 doped with noble metal were applied in the methylene blue removal in the visible-light domain [89]. Al_2O_3 was deposited onto nanoporous TiO_2 and it was effectively used for the total organic removal from wastewater [90]. Similarly, significant results were obtained with photocatalysts derived from mesoporous silica, for example combining TiO_2/Al-MCM-41 and TiO_2/Al-SBA-15 for the of phenolic compounds removal from wastewater [91]. Carbon nanotubes were applied as reinforced photocatalytic

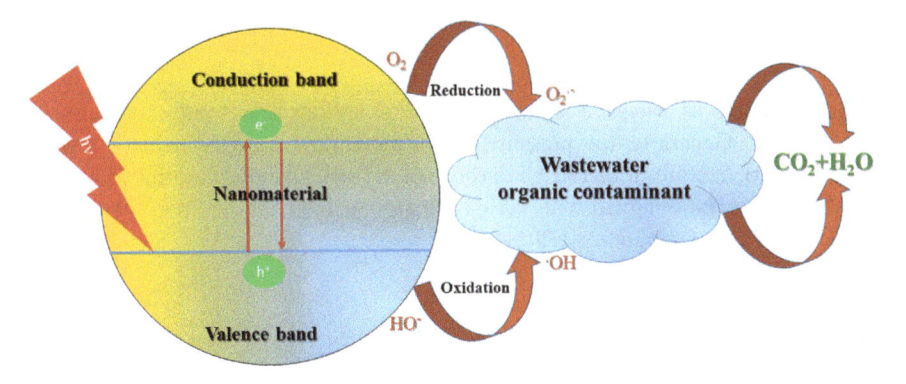

Fig. 4 Representation of photocatalytic mechanism

composite materials along with TiO_2 or ZnO, in order to improve their total surface area, defects or electrical conductivity, affecting the overall photocatalytic activity [92]. A mechanism of action for enhancing photocatalytic activity proposed the involvement of band gap or energy gap defined as energy intervals (no electrons exist) between the valence and conduction bands (Fig. 4).

The valence band consists in the highest energy state with electrons, whereas the conduction band is the lowest energy band without electrons [93]. Photons raised from various light sources can be exposed to a nanocatalyst, the vibration band electrons being excited and moving to the conduction band. In this manner a vacancy or hole appears in the vibration band. The holes react with water molecules or hydroxyl groups, resulting in hydroxyl radicals (·OH) that directly oxidize the pollutants on the carbon nanotubes surface. On the other hand, the excited electrons moved to conduction band form hydroxyl radicals, which interact with oxygen molecules, resulting superoxide radical ions ($O_2 -·$) that rapidly attacks and oxidizes the target contaminant.

The photocatalysis can be influenced by various parameters, such as light radiation, the type and nature of semiconductor, temperature, pH, as well as contaminant concentration [94]. Although photocatalysis efficiency is increased when UV light is used, various nanomaterials had been tested using visible light for photodegradation of pharmaceuticals and organic dyes [95]. The pH can affect the band edge position, and the surface charge of the nanocatalyst particles. In photocatalysis, the effect of pH is correlated with the catalyst surface charge, as well as with the ionic form of the substrate [96]. The photocatalysis can be improved if an oxidant is added to the reaction. This is captured on the catalyst surface, reducing the hole-electron recombination and promoting the formation of hydroxyl ions. For example, in the photocatalytic oxidation of sulfamethoxazole, hydrogen peroxide was used as oxidant agent which can absorb light, thus resulting the charge separation [97]. Regardless of the light type, introduction of photocatalysts in wastewater treatment can conduct to a decrease in energy requirement.

2.3 Membranes

Membranes act as physical barriers allowing various ions and molecules to pass through. Generally, the pressure-driven membrane process includes reverse osmosis (RO), nanofiltration (NF), microfiltration (MF) and ultrafiltration (UF). The membranes can be obtained with various shapes such as hallow fiber, tubular and spiral, with various separation efficacy.

2.3.1 Zeolite Membranes

Precise nanoscale crystal of 2D zeolites and obtaining of zeolite nanosheets with appropriate mechanical stability received great attention in the last years. Briefly, zeolite membranes are obtained using similar methods as for graphene and MOF nanosheets [98]. Various studies have been achieved in order to obtain a well dispersed suspension of nanosheets via exfoliation method, but their morphology and structure and were affected. Due to these disadvantages, only few studies achieved the rational design and obtaining of 2D membranes, based on pristine 2D zeolite nanosheets. For example, zeolite nanosheets with uniform thickness ($\cong 3.5$ nm) were prepared [99]. However, in order to produce well-characterized membrane microstructures, the focus should remain on the preferred orientation, designed interfaces and grain boundary control, with emphasis on reproducibility and stability under multicomponent contaminants mixtures. This can be accomplished by incorporating 2D zeolite nanosheets in an appropriate polymer matrix resulting mixed matrix membranes (MMMs).

Zeolite membranes were used as a substitute to polymeric membranes for desalination of complex wastewaters containing organic solvents or radioactive compounds, as well as in the situation when high temperature operation is required [100]. A preparation method of hydroxysodalite nano porous zeolite membranes on mullite support was reported and the membranes were used in desalination by pervaporation technique, studying the effect of various operation conditions such as feed pressure, temperature or rate on water flow. It was concluded that increased pressure, feed rate and temperature linearly influenced the wastewater flow.

2.3.2 Mixed Matrix Membranes (MMMs)

The aim of developing these membranes was to combine the advantageous properties of the two types of polymeric and ceramic membranes and increasing the overall process efficiency. Apart from the wastewater treatment, the MMMs have revolutionized other areas where separation or purification is important, such as gas separation [101]. Several researchers defined four types of MMMs, based on the membrane structure and filler location in the membrane structure (Fig. 5), namely conventional

Fig. 5 Illustration of MMMs types: **a** conventional nanocomposite, **b** thin film nanocomposite, **c** TFC with nanocomposite substrate, **d** surface located nanocomposite

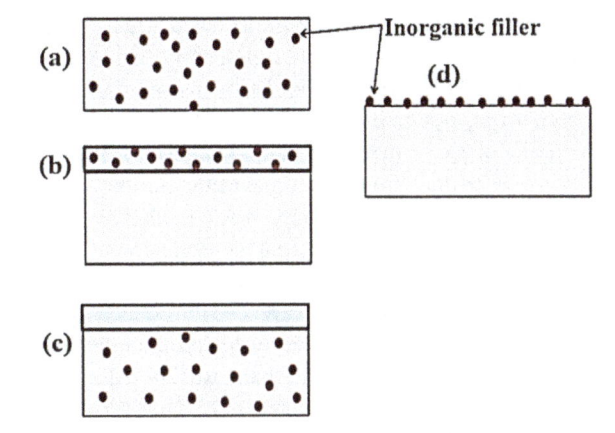

nanocomposite, thin film nanocomposite, thin film composite with nanocomposite substrate, and surface located nanocomposite [102].

Various MMMs contain inorganic fillers which attach to the support materials by covalent bonds, hydrogen bonds or van der Waals forces. These inorganic fillers can be obtained through sol gel, photothermal synthesis, thermal plasma synthesis, inert gas condensation, flame synthesis, low-temperature reactive synthesis, pulsed laser ablation, spark, mechanical alloying/milling, electrodeposition and so on [103]. Inorganic fillers contribute to obtain the MMMs desired properties. In the water treatment, these fillers have been incorporated for various purposes: disinfection [104], selectivity improvement [103] or to surpass membrane fouling [105]. Examples of inorganic fillers can be zeolite [106], TiO_2 [107], silica [108] or carbon nanotubes [109]. Figure 6 offers an illustration of various inorganic fillers for MMMs used in water treatment [110].

Fig. 6 Different types of inorganic fillers used for MMMs

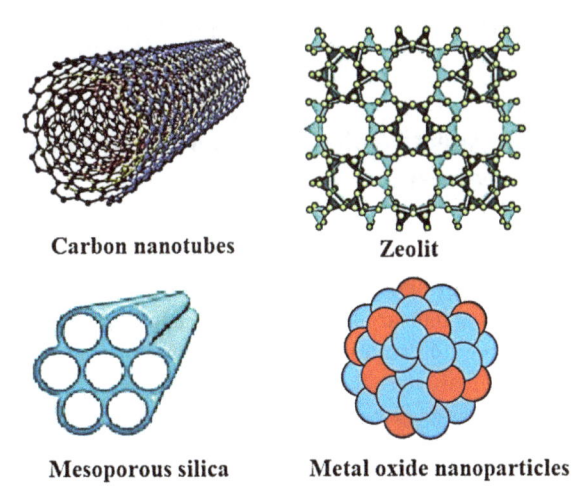

Carbon nanotubes are currently considered as vital for water treatment, especially for desalination, being able to significantly decrease the cost and energy consumption [109]. MMMs can be obtained also by introducing organic fillers such as cyclodextrin, polypyrrole, polyaniline or chitosan beads into substrate matrix, mainly through blending or phase inversion [111, 112]. The advantage of organic fillers consists in having functional groups that makes them more suitable than the inorganic ones.

A nanocomposite membrane was obtained by blending polyaniline nanofibers in polysulfone polymer, resulting a membrane with good permeability and antifouling characteristic, resulting the water flow increasing up to 1.6 times [112]. Polyaniline nanospheres and oligomers were also introduced into polysulfone matrix, increasing the water flow from 1.7 to 4 times higher than the neat polymeric membranes [113].

The β-cyclodextrin polyurethane was mixed into polysulfone matrix for removal of Cd^{+2} contaminants from water [111]. The permeability of the obtained membranes increased up to 489 Lm^2/h, due to the appearance of wider pores on the surface, higher hydrophilicity and better pores inter-connectivity. The disadvantage was that β-cyclodextrin reduced the membrane strength due to the macro-voids appeared in the structure, resulting a lower mechanic stability [111].

Recent development was achieved by using hybrid fillers to obtain MMMs. Such membranes consist in two different fillers introduced in a continuous phase to accomplish a targeted purpose or to improve the overall process efficiency. For example, the combination of Fe(II)-Fe(III) oxide and polyaniline was introduced into polyethersulfone matrix, resulting a removal of 85% for Cu(II) from wastewater [114]. An antifouling MMM was prepared by Fe_2O_3 nanoparticles and multiwalled carbon nanotube inclusion into polyvinyldene fluoride, speeding the degradation of contaminants such as cyclohexanoic and humic acid [115]. The Fe_2O_3 nanoparticles improved the membrane hydrophilicity but caused the decrease of surface porosity. Reduced graphene oxide/polythiophene (rGO/PTh) were immersed into polyether sulfone matrix, designing an antifouling membrane with high permeability [116]. Despite the observed advantages of hybrid fillers, they could also affect the membrane efficiency, pore blockage being frequently observed [114].

3 Prospective of Nanomaterials Application in Wastewater Treatment

The key issue of nanotechnology introduction in wastewater treatment consists in the possibility of finding nanomaterials in high quantities at low costs. Scaling up these materials at industrial level remains a major milestone in nanotechnology application for wastewater treatment. Also, the nanomaterials characteristics (for example, high surface areas, size, shape or dimensions), their interaction with other contaminants than the targeted ones or with living beings are not fully elucidated and further research needs to be achieved. Environmental fate and toxicity of nanomaterials towards humans are still not fully explored.

The nanomaterials stability (oxidative, photochemical, biological or hydrolytic) in environment needs to be studied. Up to now, it was demonstrated that carbon nanotubes or TiO_2 nanoparticles are very toxic for humans. Many nanomaterials have carcinogenic effect and obstruct the normal cellular roles of lungs or immune system. In order to use nanomaterials in wastewater treatment systems, efficient methods need to be developed, being able to prevent the nanomaterials passing through the treated. Also, cost-benefit evaluation needs to be approached in order to evaluate the nanotechnology application for wastewater treatment.

The nanostructured membranes can be used for the degradation of various organic and inorganic contaminants. To improve their performances, it will be necessary a better understanding of the nanocomposites membranes formation. In this respect, the priority concern in the real field wastewater treatment must be directed towards the pattern of the nanoparticles within the matrix, as well as toward the changes in their structures and properties.

Mixed matrix membranes are claimed to be efficient in terms of efficiency, permeability and selectivity; however, some difficulties were identified, restricting their wider applications. The drawbacks include the discovery of compatible nanoparticles, complexity of the synthesis, high cost, morphology control, as well as structural defects. Furthermore, the introduction of inorganic particles into an organic membrane for wastewater treatment presents a potential hazard to environment and human health, a milestone that must be addressed in the near future. Despite this, it is considered that MMMs have great potential, their successful and competitive application requiring a combined effort to solve the identified drawbacks in order to compete with the classical purification technologies. This chapter aimed to provide a systematic review and a critical bibliometric analysis on nanomaterials and techniques (such as adsorption, photocatalysis or membrane technology) that can be applied for the removal of various classes of contaminants from wastewater.

4 Conclusions

This study intended to emphasize the use of nanomaterials for removing pollutants from wastewater by adsorption, catalysis or membrane processes. While many studies approached the endocrine disrupting chemicals removal, it also must be mentioned the increased interest in pharmaceuticals and personal care products. Both the adsorption and catalysis processes showed great potential for removing pollutants from wastewater. For the adsorption technology, carbon nanotubes and mesoporous silica have attracted an increased interest, the proposed mechanisms including hydrophobic effect, hydrogen bonding, covalent bonding, π-π interactions or electrostatic interaction. In the last years, metal organic frameworks nanomaterials were studied for removing pollutants from wastewater. Among the nanomaterials used for photocatalysis, TiO_2 was, by far, the most studied. Membrane technology has efficiently replaced conventional water treatment. The idea of hybrid or mixed matrix membranes has risen, combining characteristics of polymeric and ceramic membranes by introducing inorganic particles as fillers in an organic polymer matrix, improving the efficiency, permeability and selectivity.

Acknowledgements This research was financially supported by the Romanian Ministry of Education and Research, NUCLEU Program-Financing Contract no. 9 N/2019, under Project PN 19 11 03 01 "Studies on the obtaining and improvement of the acido-basic properties of the nanoporous catalytic materials for application in wastes valorisation".

References

1. https://www.who.int/news-room/fact-sheets/detail/drinking-water. Accessed 05 Nov 2020
2. Hasan HA, Abdullah SRS, Kofli NT, Yeoh SY (2016) Interaction of environmental factors on simultaneous biosorption of lead and manganese ions by locally isolated Bacillus cereus. J Ind Eng Chem 37:295–305
3. https://www.eea.europa.eu/data-and-maps/figures/main-contaminants-at-industrial-and-commercial-sites-affecting-soil-in-europe-as-of-total. Accessed 02 Nov 2020
4. Badea SL, Geana EI, Niculescu VC, Ionete RE (2020) Recent progresses in analytical GC and LC mass spectrometric based-methods for the detection of emerging chlorinated and brominated contaminants and their transformation products in aquatic environment. Sci Total Environ 722:
5. https://www.eea.europa.eu/data-and-maps/daviz/urban-waste-water-treatment-in-europe#tab-chart_1_filters=%7B%22rowFilters%22%3A%7B%7D%3B%22columnFilters%22%3A%7B%7D%3B%22sortFilter%22%3A%5B%22collected_without_treatment%22%5D%7D. Accessed 9 Nov 2020
6. Hasan HA, Muhammad MH, Ismail NI (2020) A review of biological drinking water treatment technologies for contaminants removal from polluted water resources. J Water Process Eng 33:
7. Li Z, Dvorak B, Li X (2012) Removing 17b-estradiol from drinking water in a biologically active carbon (BAC) reactor modified from a granular activated carbon (GAC) reactor. Water Res 46:2828–2836
8. Hasan HA, Abdullah SRS, Kamarudin SK, Kofli NT (2013) On-off control of aeration time in the simultaneous removal of ammonia and manganese using a biological aerated filter system. Process Saf Environ Prot 91:415–422
9. Han M, Zhao ZW, Gao W, Tian Y, Cui FY (2016) Effective combination of permanganate composite chemicals (PPC) and biological aerated filter (BAF) to pre-treat polluted drinking water source. Desalin Water Treat 57:28240–28249
10. He S, Wang J, Ye L, Zhang Y, Yu J (2014) Removal of diclofenac from surface water by electron beam irradiation combined with a biological aerated filter. Radiat Phys Chem 105:104–108
11. Ricardo AR, Carvalho V, Velizarov S, Crespo JG, Reis MA (2012) Kinetics of nitrate and perchlorate removal and biofilm stratification in an ion exchange membrane bioreactor. Water Res 46:4556–4568
12. Wang S, Ma X, Liu Y, Yi X, Du G, Li J (2020) Fate of antibiotics, antibiotic-resistant bacteria, and cell-free antibiotic-resistant genes in full-scale membrane bioreactor wastewater treatment plants. Bioresour Technol 302:
13. Schijven JF, Berg HHJL, Colin M, Dullemont Y, Hijnen WAM, Magic-Knezev A, Oorthuizen WA, Wubbels G (2013) A mathematical model for removal of human pathogenic viruses and bacteria by slow sand filtration under variable operational conditions. Water Res 47:2592–2602
14. Bolisetty S, Peydayesh M, Mezzenga R (2019) Sustainable technologies for water purification from heavy metals: review and analysis. Chem Soc Rev 48:463–487

15. Lee EJ, Schwab KJ (2005) Deficiencies in drinking water distribution systems in developing countries. J Water Health 3:109–127
16. Moe CL, Rheingans RD (2006) Global challenges in water, sanitation and health. J Water Health 4:41–58
17. Savage N, Diallo MS (2005) Nanomaterials and water purification: opportunities and challenges. J Nanoparticle Res 7:331–342
18. Masciangioli T, Zhang WX (2003) Peer reviewed: environmental technologies at the nanoscale. Environ Sci Technol 37:102A–108A
19. Eijkel JCT, van den Berg A (2005) Nanofluidics: what is it and what can we expect from it? Microfluid Nanofluidics 1:249–267
20. Panneerselvam S, Choi S (2014) Nanoinformatics: emerging databases and available tools. Int J Mol Sci 15:7158–7182
21. Rickerby DG, Morrison M (2007) Nanotechnology and the environment: a European perspective. Sci Technol Adv Mater 8:19–24
22. Vaseashta A, Vaclavikova M, Vaseashta S, Gallios G, Roy P, Pummakarnchana O (2007) Nanostructures in environmental pollution detection, monitoring, and remediation. Sci Technol Adv Mater 8:47–59
23. Qu X, Alvarez PJJ, Li Q (2013) Applications of nanotechnology in water and wastewater treatment. Water Res 47:3931–3946
24. Riu J, Maroto A, Rius FX (2006) Nanosensors in environmental analysis. Talanta 69:288–301
25. Theron J, Walker JA, Cloete TE (2008) Nanotechnology and water treatment: applications and emerging opportunities. Crit Rev Microbiol 34:43–69
26. Crane RA, Scott TB (2012) Nanoscale zero-valent iron: future prospects for an emerging water treatment technology. J Hazard Mater 211–212:112–125
27. Rodriguez O, Peralta-Hernandez JM, Goonetilleke A, Bandalac ER (2017) Treatment technologies for emerging contaminants in water: a review. Chem Eng J 323:361–380
28. Ahmed MJ, Theydan V (2012) Adsorption of cephalexin onto activated carbons from Albizia lebbeck seed pods by microwave-induced KOH and K_2CO_3 activations. Chem Eng J 211–212:200–207
29. Ji L, Shao Y, Xu Z, Zheng S, Zhu D (2010) Adsorption of monoaromatic compounds and pharmaceutical antibiotics on carbon nanotubes activated by KOH etching. Environ Sci Technol 44:6429–6436
30. Kim HJ, Choi K, Baek Y, Kim D, Shim J, Yoon J, Lee J (2014) High performance reverse osmosis CNT/polyamide nanocomposite membrane by controlled interfacial interactions. ACS Appl Mater Interfaces 6(4):1–15
31. Singh RK, Patel KD, Kim JJ, Kim TH, Kim JH, Shin US, Lee EJ, Knowles JC, Kim HW (2014) Multifunctional hybrid nanocarrier: magnetic CNTs unsheathed with mesoporous silica for drug delivery and imaging system. ACS Appl Mater Interfaces 6(4):2201–2208
32. Wu D, Pan B, Wu M, Peng H, Zhang D, Xing B (2012) Coadsorption of Cu and sulfamethoxazole on hydroxylized and graphitized carbon nanotubes. Sci Total Environ 427–428:247–252
33. Zhang D, Pan B, Zhang H, Ning P, Xing B (2010) Contribution of different sulfamethoxazole species to their overall adsorption on functionalized carbon nanotubes. Environ Sci Technol 44:3806–3811
34. Lara IV, Zanella I, Fagan SB (2014) Functionalization of carbon nanotube by carboxyl group under radial deformation. Chem Phys 428:117–120
35. Ren X, Chen C, Nagatsu M, Wang X (2011) Carbon nanotubes as adsorbents in environmental pollution management: A review. Chem Eng J 170:395–410
36. Gupta VK, Agarwal S, Saleh TA (2011) Chromium removal by combining the magnetic properties of iron oxide with adsorption properties of carbon nanotubes. Water Res 45:2207–2212
37. Kim H, Hwang YS, Sharma VK (2014) Adsorption of antibiotics and iopromide onto single-walled and multi-walled carbon nanotubes. Chem Eng J 255:23–27

38. Cho HH, Huang H, Schwab K (2011) Effects of solution chemistry on the adsorption of ibuprofen and triclosan onto carbon nanotubes. Langmuir 27:12960–12967
39. Yang W, Lu Y, Zheng F, Xue X, Li N, Liu D (2012) Adsorption behavior and mechanisms of norfloxacin onto porous resins and carbon nanotube. Chem Eng J 179:112–118
40. Tian Y, Gao B, Morales VL, Chen H, Wang Y, Li H (2013) Chemosphere removal of sulfamethoxazole and sulfapyridine by carbon nanotubes in fixed-bed columns. Chemosphere 90:2597–2605
41. Ji L, Chen W, Zheng S, Xu Z, Zhu D (2009) Adsorption of sulfonamide antibiotics to multiwalled carbon nanotubes. Langmuir 25:11608–11613
42. Zhang L, Song X, Liu X, Yang L, Pan F, Lv J (2011) Studies on the removal of tetracycline by multi-walled carbon nanotubes. Chem Eng J 178:26–33
43. Zhao H, Liu X, Cao Z, Zhan Y, Shi X, Yang Y, Zhou J, Xu J (2016) Adsorption behavior and mechanism of chloramphenicols, sulfonamides, and non-antibiotic pharmaceuticals on multi-walled carbon nanotubes. J Hazard Mater 310:235–245
44. Ji L, Chen W, Bi J, Zheng S, Xu Z, Zhu D, Alvarez PJ (2010) Adsorption of tetracycline on single-walled and multi-walled carbon nanotubes as affected by aqueous solution chemistry. Environ Toxicol Chem 29:2713–2719
45. Zhang S, Shao T, Kose HS, Karanfil T (2010) Adsorption of aromatic chemicals by carbonaceous adsorbents: a comparative study on granular activated carbon, activated carbon fiber and carbon nanotubes. Environ Sci Technol 12:1–10
46. Zhang D, Pan B, Wu M, Wang B, Zhang H, Peng H, Wu D, Ning P (2011) Adsorption of sulfamethoxazole on functionalized carbon nanotubes as affected by cations and anions. Environ Pollut 159:2616–2621
47. Sohrabi MR, Mansouriieh N, Khosravi M, Zolghadr M (2015) Removal of diazo dye Direct Red 23 from aqueous solution using zero-valent iron nanoparticles immobilized on multiwalled carbon nanotubes. Water Sci Technol 71:1367–1374
48. Seo PW, Khan NA, Jhung SH (2017) Removal of nitroimidazole antibiotics from water by adsorption over metal–organic frameworks modified with urea or melamine. Chem Eng J 315:92–100
49. Song JY, Jhung SH (2017) Adsorption of pharmaceuticals and personal care products over metal-organic frameworks functionalized with hydroxyl groups: quantitative analyses of H-bonding in adsorption. Chem Eng J 322:366–374
50. Andrew Lin KY, Der Lee W (2016) Self-assembled magnetic graphene supported ZIF-67 as a recoverable and efficient adsorbent for benzotriazole. Chem Eng J 284:1017–1027
51. Jiang JQ, Yang CX, Yan XP (2013) Zeolitic imidazolate framework-8 for fast adsorption and removal of benzotriazoles from aqueous solution. ACS Appl Mater Interfaces 5:9837–9842
52. Bhadra BN, Ahmed I, Kim S, Jhung SH (2017) Adsorptive removal of ibuprofen and diclofenac from water using metal-organic framework-derived porous carbon. Chem Eng J 314:50–58
53. Seo PW, Khan NA, Hasan Z, Jhung SH (2016) Adsorptive removal of artificial sweeteners from water using metal-organic frameworks functionalized with urea or melamine. ACS Appl Mater Interfaces 8:29799–29807
54. Zhuo N, Lan Y, Yang W, Yang Z, Li X, Zhou X, Liu Y, Shen J, Zhang X (2017) Adsorption of three selected pharmaceuticals and personal care products (PPCPs) onto MIL-101(Cr)/natural polymer composite beads. Sep Purif Technol 177:272–280
55. Yan Z, Li G, Mu L, Tao S (2006) Pyridine-functionalized mesoporous silica as an efficient adsorbent for the removal of acid dyestuffs. J Mater Chem 16:1717–1725
56. Niculescu V, Miricioiu M, Geana I, Ionete RE, Paun N, Parvulescu V (2019) Silica mesoporous materials – an efficient sorbent for wine polyphenols separation. Rev Chem 70:1513–1517
57. Monash P, Pugazhenthi G (2009) Removal of crystal violet dye from aqueous solution using calcined and uncalcined mixed clay adsorbents. Adsorption 45:94–104
58. Huang CH, Chang KP, Ou HD, Chiang YC, Wang CF (2011) Adsorption of cationic dyes onto mesoporous silica. Micropor Mesopor Mater 14:102–109

59. Boukoussa B, Hamacha R, Morsli A, Bengueddach A (2013) Adsorption of yellow dye on calcined or uncalcined Al-MCM-41 mesoporous materials. Arabian J Chem 10:S2160–S2169
60. Aguado J, Arsuaga JM, Arencibia A, Lindo M, Gascon V (2009) Aqueous heavy metals removal by adsorption on amine-functionalized mesoporous silica. J Hazard Mater 163:213–221
61. Shahbazi A, Younesi H, Badiei A (2014) Functionalized nanostructured silica by tetradentate-amine chelating ligand as efficient heavy metals adsorbent: Applications to industrial effluent treatment. Korean J Chem Eng 9:1598–1607
62. Vidal CB, Barros AL, Moura CP, De Lima ACA, Dias FS, Vasconcellos LCG, Fechine P, Nascimento RF (2011) Adsorption of polycyclic aromatic hydrocarbons from aqueous solutions by modified periodic mesoporous organosilica. J Colloid Interface Sci 357:466–473
63. Niculescu V, Iordache M, Miricioiu M, Asimopolos L (2018) Phosphorus removal from wastewater in the presence of mesoporous materials. Prog Cryogen Isotopes Sep 21:29–42
64. Anbia M, Hariri SA, Ashrafizadeh SN (2010) Adsorptive removal of anionic dyes by modified nanoporous silica SBA-3. Appl Surf Sci 256:3228–3233
65. Anbia M, Salehi S (2012) Removal of acid dyes from aqueous media by adsorption onto amino-functionalized nanoporous silica SBA-3. Dyes Pigments 94:1–9
66. Chen Z, Zhou L, Zhang F, Yu C, Wei Z (2012) Multicarboxylic hyperbranched polyglycerol modified SBA-15 for the adsorption of cationic dyes and copper ions from aqueous media. Appl Surf Sci 258:5291–5298
67. Wang S, Li H (2006) Structure directed reversible adsorption of organic dye on mesoporous silica in aqueous solution. Micropor Mesopor Mater 97:21–26
68. Santos DO, de Lourdes Nascimento Santos M, Costa JAS, de Jesus RA, Navickiene S, Sussuchi EM, Mesquita ME (2013) Investigating the potential of functionalized MCM-41 on adsorption of Remazol Red dye. Environ Sci Pollut Res 7:5028–5035
69. Ho KY, Mckay G, Yeung KL (2003) Selective adsorbents from ordered mesoporous silica. Langmuir 19:3019–3024
70. Yang H, Feng Q (2010) Characterization of pore-expanded amino-functionalized mesoporous silicas directly synthesized with dimethyldecylamine and its application for decolorization of sulphonated azo dyes. J Hazard Mater 180:106–114
71. Yang H, Feng Q (2010) Direct synthesis of pore-expanded amino-functionalized mesoporous silicas with dimethyldecylamine and the effect of expander dosage on their characterization and decolorization of sulphonated azo dyes. Micropor Mesopor Mater 135:124–130
72. Mahmoodi NM, Khorramfar S, Najafi F (2011) Amine-functionalized silica nanoparticle: preparation, characterization and anionic dye removal ability. Desalination 279:61–68
73. Walcarius A, Mercier L (2010) Mesoporous organosilica adsorbents: nanoengineered materials for removal of organic and inorganic pollutants. J Mater Chem 20:4478–4511
74. Khaled A, Nemr AE, Sikaily A, Abdelwahab O (2009) Removal of Direct N Blue-106 from artificial textile dye effluent using activated carbon from orange peel: adsorption isotherm and kinetic studies. J Hazard Mater 165:100–110
75. Salleh MAM, Mahmoud DK, Karim WAWA, Idris A (2011) Cationic and anionic dye adsorption by agricultural solid wastes: a comprehensive review. Desalination 280:1–13
76. Yan H, Li H, Yang H, Li A, Cheng R (2013) Removal of various cationic dyes from aqueous solutions using a kind of fully biodegradable magnetic composite microsphere. Chem Eng J 223:402–411
77. Karim AH, Jalil AA, Triwahyono S, Sidik SM, Kamarudin NHN, Jusoh R et al (2012) Amino modified mesostructured silica nanoparticles for efficient adsorption of Methylene Blue. J Colloid Interface Sci 386:307–314
78. Bharathi K, Ramesh S (2013) Removal of dyes using agricultural waste as low-cost adsorbents: a review. Appl Water Sci 3:773–790
79. Rong H, Zhihong W, Lei T, Yi Z, Weiming L, Da X, Whei C, Youwen T (2018) Design and fabrication of highly ordered ion imprinted SBA-15 and MCM-41 mesoporous organosilicas for efficient removal of Ni2 + from different properties of wastewaters. Micropor Mesopor Mater 257:212–221

80. Cui H, Shu Liu YL, Zhang JF, Zhou Q, Zhong R, Yang ML, Hou XF (2017) Novel Pb(II) ion-imprinted materials based on bis-pyrazolyl functionalized mesoporous silica for the selective removal of Pb(II) in water samples. Micropor Mesopor Mater 241:165–177
81. Kulamani P, Krushna GM, Suresh KD (2012) Adsorption of toxic metal ion Cr(VI) from aqueous state by TiO2-MCM-41: equilibrium and kinetic studies. J Hazard Mater 241:395–403
82. Gu XX, Luo LL, Wu J, Zhong SX, Chen JR (2012) Progress of mesoporous materials for the adsorption of Cr(VI) in wastewater. Adv Mater Res 550:2129–2133
83. Wu YQ, Wei JW, Wang DQ (2013) Removal of heavy metals in water by functionalized mesoporous silica materials: a review. Adv Mater Res 785:693–696
84. Dioum A, Hamoudi S (2014) Mono- and quaternary-ammonium functionalized mesoporous silica materials for nitrate adsorptive removal from water and wastewaters. J Porous Mater 21:685–690
85. Dindar MH, Yaftian MR, Rostamnia S (2015) Potential of functionalized SBA-15 mesoporous materials for decontamination of water solutions from Cr(VI), As(V) and Hg(II) ions. J Environ Chem Eng 3:986–995
86. Kabra K, Chaudhary R, Sawhney RL (2004) Treatment of hazardous organic and inorganic compounds through aqueousphase photocatalysis: a review. Ind Eng Chem Res 43:7683–7696
87. Shephard GS, Stockenstrom S, De Villiers D, Engelbrecht WJ, Wessels GFS (2002) Degradation of microcystin toxins in a falling film photocatalytic reactor with immobilized titaniumdioxide catalyst. Water Res 36:140–146
88. Han X, Kuang Q, Jin M, Xie Z, Zheng L (2009) Synthesis of titania nanosheets with a high percentage of exposed (001) facets and related photocatalytic properties. J Am Chem Soc 131:3152–3153
89. Wu L, Yu JC, Fu X (2006) Characterization and photocatalytic mechanism of nanosized CdS coupled TiO_2 nanocrystals under visible light irradiation. J Mol Catal A Chem 244:25–32
90. Sun D, Meng TT, Loong TH, Hwa TJ (2004) Removal of natural organic matter from water using a nano-structured photocatalyst coupled with filtration membrane. Water Sci Technol 49:103–110
91. Phanikrishna Sharma MV, Durga Kumari V, Subrahmanyam M (2008) Photocatalytic degradation of isoproturon herbicide over TiO_2/Al-MCM-41 composite systems using solar light. Chemosphere 72:644–651
92. Di Paola A, García-Lopez E, Marcì G, Palmisano L (2012) A survey of photocatalytic materials for environmental remediation. J Hazard Mater 211:3–29
93. Das R (2017) Carbon nanotube in water treatment. Nanohybrid catalyst based on carbon nanotube. Carbon nanostructures. Springer, Cham, pp 23–54. https://doi.org/10.1007/978-3-319-58151-4_2
94. Kocí K, Obalova L, Lacny Z (2008) Photocatalytic reduction of CO_2 over TiO_2 based catalysts. Chem Pap 62:1–9
95. Pastrana-Martínez LM, Morales-Torres S, Likodimos V, Figueiredo JL, Faria JL, Falaras P, Silva AMT (2012) Advanced nanostructured photocatalysts based on reduced graphene oxide-TiO_2 composites for degradation of diphenhydramine pharmaceutical and methyl orange dye. Appl Catal B Environ 123–124:241–256
96. Zúñiga-Benítez H, Aristizábal-Ciro C, Peñuela GA (2016) Heterogeneous photocatalytic degradation of the endocrine-disrupting chemical Benzophenone-3: parameters optimization and by-products identification. J Environ Manag 167:246–258
97. Kaniou S, Pitarakis K, Barlagianni I, Poulios I (2005) Photocatalytic oxidation of sulfamethazine. Chemosphere 60:372–380
98. Rehman F, Thebo KH, Aamir M, Akhtar J (2020) Chapter 8—nanomembranes for water treatment. Amrane A, Rajendran S, Nguyen TA, Assadi AA, Sharoba AM (eds) In micro and nano technologies, nanotechnology in the beverage industry. Elsevier, pp 207–240. ISBN 9780128199411. https://doi.org/10.1016/B978-0-12-819941-1.00008-0
99. Rangnekar N, Shete M, Agrawal KV, Topus B, Kumar P, Guo Q, Ismail I, Alyoubi A, Basahel S, Narasimharao K, Macosko CW, Mkhoyan A, Al-Thabaiti S, Stottrup B, Tsapatsis M (2015)

2D zeolite coatings: Langmuir-Schaefer deposition of 3 nm thick MFI Zeolite nanosheets. Angew Chem Int Ed 54:6571–6575
100. Kazemimoghadam M (2010) New nanopore zeolite membranes for water treatment. Desalination 251:176–180
101. Miricioiu MG, Iacob C, Nechifor G, Niculescu VC (2019) High selective mixed membranes based on mesoporous MCM-41 and MCM-41-NH_2 particles in a polysulfone matrix. Front Chem 7:1–10
102. Yin J, Zhu G, Deng B (2013) Multi-walled carbon nanotubes (MWNTs)/polysulfone (PSU) mixed matrix hollow fiber membranes for enhanced water treatment. J Membr Sci 437:237–248
103. Taurozzi JS, Arul H, Bosak VZ, Burban AF, Voice TC, Bruening ML, Tarabara VV (2008) Effect of filler incorporation route on the properties of polysulfone–silver nanocomposite membranes of different porosities. J Membr Sci 325:58–68
104. Kim SH, Kwak SY, Sohn BH, Park TH (2003) Design of TiO_2 nanoparticle self-assembled aromatic polyamide thin-film-composite (TFC) membrane as an approach to solve biofouling problem. J Membr Sci 211:157–165
105. Li JH, Xu YY, Zhu LP, Wang JH, Du CH (2009) Fabrication and characterization of a novel TiO_2 nanoparticle selfassembly membrane with improved fouling resistance. J Membr Sci 326:659–666
106. Ma N, Wei J, Liao R, Tang CY (2012) Zeolite-polyamide thin film nanocomposite membranes: Towards enhanced performance for forward osmosis. J Membr Sci 405:149–157
107. Rahimpour A, Jahanshahi M, Mollahosseini A, Rajaeian B (2012) Structural and performance properties of UV-assisted TiO_2 deposited nano-composite PVDF/SPES membranes. Desalination 285:31–38
108. Jadav GL, Singh PS (2009) Synthesis of novel silica-polyamide nanocomposite membrane with enhanced properties. J Membr Sci 328:257–267
109. Kim HJ, Choi K, Baek Y, Kim DG, Shim J, Yoon J, Lee JC (2014) High-performance reverse osmosis CNT/polyamide nanocomposite membrane by controlled interfacial interactions. ACS Appl Mat Interf 6:2819–2829
110. Savage N, Diallo MS (2005) Nanomaterials and water purification: opportunities and challenges. J Nanopart Res 7:331–342
111. Adams FV, Nxumalo EN, Krause RW, Hoek EM, Mamba BB (2012) Preparation and characterization of polysulfone/β-cyclodextrin polyurethane composite nanofiltration membranes. J Membr Sci 405:291–299
112. Fan Z, Wang Z, Duan M, Wang J, Wang S (2008) Preparation and characterization of polyaniline/polysulfone nanocomposite ultrafiltration membrane. J Membr Sci 310:402–408
113. Zhao S, Wang Z, Wei X, Tian X, Wang J, Yang S, Wang S (2011) Comparison study of the effect of PVP and PANI nanofibers additives on membrane formation mechanism, structure and performance. J Membr Sci 385:110–122
114. Daraei P, Madaeni SS, Ghaemi N, Salehi E, Khadivi MA, Moradian R, Astinchap B (2012) Novel polyethersulfone nanocomposite membrane prepared by PANI/Fe_3O_4 nanoparticles with enhanced performance for Cu(II) removal from water. J Membr Sci 415–416:250–259
115. Alpatova A, Meshref M, McPhedran KN, Gamal El-Din M (2015) Composite polyvinylidene fluoride (PVDF) membrane impregnated with Fe_2O_3 nanoparticles and multiwalled carbon nanotubes for catalytic degradation of organic contaminants. J Membr Sci 490:227–235
116. Saf AO, Akin I, Zor E, Bingol H (2015) Preparation of a novel PSF membrane containing rGO/PTh and its physical properties and membrane performance. RSC Adv 5:42422–42429

Violeta Carolina Niculescu was born in Râmnicu Valcea, Vâlcea, Romania in 1979. She received the B.S. and M.S. degrees in chemistry and environment quality from the Chemistry Faculty, University of Craiova, Romania, in 2002 and 2003 respectively, and the Ph.D. degree in chemistry from "Ilie Murgulescu" Physical-Chemistry Institute, Romania Academy, Bucharest, Romania, in 2012. From 2003 to 2016, she was a Scientific Researcher with the Research-Development and Technological Transfer Laboratory from National Research and Development Institute for Cryogenics and Isotopic Technologies – ICSI Râmnicu Vâlcea, Romania. Since 2016, she has been a Senior Scientific Researcher with the ICSI ANALYTICS Department. She is the author of two international book chapters, more than 70 scientific articles in ISI or other international databases indexed journals, more than 120 scientific papers presented at national or international conferences/congresses, 4 awards and distinctions, more than 25 national or international projects as project leader and/or scientific officer. Her research interests include life sciences, with main interest in synthesis, characterization and applications of new biological-active compounds and new porous and non-porous materials used as adsorbents/catalysts for gas separation and wastewater treatment. Her current research is directed toward improving the quality of life, security and environment. Dr. Niculescu's awards include the Golden Medal at the Belgian and International Trade Fair for Technological Innovation "Eureka!" – INNOVA 2012, 15–17 November 2012, Brussels, Belgium; Diploma of Excellence and Silver Medal at the International Workshop for Research, Innovation and Inventions – PRO INVENT 2011, IXth Edition, Cluj-Napoca, Romania; Silver Medal at the International Workshop for Inventions, Scientific Research and New Technologies – Inventika 2011, XVth Edition, 2011, Bucharest, Romania and Diploma of Excellence and Silver Medal at The International Exhibition of Research, Innovation and Inventions–PRO INVENT 2014, XIIth Edition, 19-21 Mar. 2014, Cluj-Napoca, RO.

Marius G. Miricioiu was born in Râmnicu Vâlcea, Vâlcea, Romania in 1985. He received the B.S., M.S. and Ph.D. degrees in chemical engineering from Politehnica University of Bucharest, Romania, in 2009, 2011 and 2017, respectively. He is currently pursuing the postdoctoral studies in chemical engineering at Politehnica University of Bucharest, Romania. From 2013 to 2017, he was a Research Assistant with the National Research and Development Institute for Cryogenics and Isotopic Technologies, Râmnicu Vâlcea, Vâlcea, Romania. Since 2017, he is 3rd degree researcher with the same institute. He is the author of more than 40 articles. His research interest includes pollutants reduction through unconventional/conventional techniques (wastewater treatments and CO_2 reduction). He is reviewer for MDPI journals.

Roxana Elena Ionete was born in Râmnicu Vâlcea city, Vâlcea county, Romania in 1971. She received the B.S. and MD degree in mechanical engineering, from Transilvania University of Braşov, Romania, in 1994 and the Ph.D. degree in engineering sciences from the Technical University of Construction of Bucharest, Romania, in 2003. She is a Senior Researcher, head of department, with the ICSI Analytics, National Research and Development Institute for Cryogenics and Isotopic Technologies - ICSI, Râmnicu Vâlcea, Romania. She has published over 50 articles in ISI journals, more than ten books and books chapters, and holds five patents. Her research activities have been rewarded with more than 10 prices and medals at national and international conferences. Her research interests include theoretical and experimental studies in the fields of stable isotope applications for environmental/hydrology and health. She has dedicated a lot of interest to the study of isotopic exchange processes in the interrelationships between abiotic factors and living systems, with direct applications in environmental science, health and agriculture/food safety and security.

Polymeric Nanocomposite Membranes for Water Filtration

Jnyana Ranjan Mishra, Sukanya Pradhan, Smita Mohanty, and Sanjay K. Nayak

Abstract The entire world is facing with a severe dilemma of water pollution which is mainly concerned with the chemical and biological contaminants that have endangered the quality of drinking water. The Membrane separation technology is vastly acknowledged as an advanced process for water filtration and purification, having the potential to alleviate global matter of contention of freshwater scarcity. Various nanofillers have been used by many in water purification. This methodology has been accepted as feasible and active to produce a multifunctional nanocomposite membrane i.e. membranes with higher performance along with their synergistic effects of organic polymer matrix and inorganic nanomaterials for water and wastewater treatment. Recently, inorganic nanofillers such as inorganic metal oxides (SiO_2, TiO_2, ZnO, Ag, etc.), carbon based (CNT, GO) and mixed nanoparticles (SiO_2-TiO_2, GO-TiO_2, GO-SiO_2, etc.) have been extensively used to prepare polyvinylidene fluoride (PVDF) nanocomposite membranes with desired properties for water purification. This review aims to highlight the performance of nanofiller incorporated PVDF membranes and the novel strategies explored for fabrication of the PVDF nanocomposite membrane to cater to the current requirements and expectations of water purification performances.

Keywords PVDF · Ultrafiltration · Purification · Membrane · Nanocomposite · Nanofiller

1 Introduction

Currently, accessing pure drinking water appears to be a major concern in both developed and developing countries due to worldwide population growth, global climate change and deterioration in drinking water quality. For this reason, people are getting affected by a variety of waterborne diseases or heavy metal poisoning. Hence there

---

J. R. Mishra · S. Pradhan · S. Mohanty (✉) · S. K. Nayak
Laboratory for Advanced Research in Polymeric Materials (LARPM), School for Advanced Research in Polymers (SARP), Central Institute of Petrochemicals Engineering & Technology (CIPET), Bhubaneswar, Odisha, India

© Springer Nature Switzerland AG 2021
A. Vaseashta and C. Maftei (eds.), *Water Safety, Security and Sustainability*, Advanced Sciences and Technologies for Security Applications, https://doi.org/10.1007/978-3-030-76008-3_7

is an urgent need for a new and improved water treatment technology with higher efficiency and more sustainability towards water purification. The emerging nanomaterial and nanotechnology have provided innumerable opportunities to develop high-performance polymeric membranes which can impart tailored properties such as durability, high flux, hydrophilic and fouling resistance, self-cleaning, photo catalytic, photo degradation and competency to remove most toxic metals like lead (Pb), arsenic (As) etc. Organic polymers, mostly polysulfone (PSF) [1–12] poly(ether sulfone) (PES), polyacrylonitrile (PAN) [13], polyamide [14–17], polyimide, cellulose acetate membrane [18–25] polylactic acid [26, 27], Poly vinyl alcohol (PVA) [28], poly-(vinylidenefluoride) (PVDF) [29–32] and polytetrafluoroethylene (PTFE) are widely used in the field of water filtration membrane [33].

Currently, many researchers consider PVDF as a foremost polymer for the development of membranes for water purification, thanks to its high mechanical strength [34] durability, thermal stability [35] and ease of casting via nonconventional techniques, such as non-solvent induced phase separation (NIPS) process or thermally induced phase separation (TIPS) [36–41]. Moreover, some membrane manufacturers like Asahi Kasei Chemicals, GE, Merck Millipore, Koch Membrane Systems, Hyflux and Siemens Water Technologies etc., have also developed products with PVDF membrane for the purpose of water purification. However, PVDF is vulnerable to various organic contaminations such as proteins and exhibits significantly low water flux due to its hydrophobic nature [42]. In order to upgrade and enhance the performance of PVDF membranes various attempts have been made for the modification of PVDF membrane such as physical blending [43–46], surface modification [47] and chemical grafting [48].

Progress in nanotechnology research offers a promising resolution to develop the next generation water purification membranes which can deliver high performance and inexpensive water purification solution [49]. The novel membranes composed of inorganic nanomaterials as additives, demonstrate their great potential in new water treatment technology with desired state-of-the-art performance with additional functions, such as antibiotic resistance, fouling resistance, and catalytic reactivity [50]. However, the smaller size, high surface energy and incompatibility of the nanoparticles with PVDF lead to their agglomeration during mixing in polymer solution. This, consequently, causes heterogeneous dispersion of nanoparticles in the membranes. Non-uniform dispersion of the nanoparticles alters the performance of the membrane due to change in microstructure, pore sizes, surface topography and fouling resistance of the membrane. At present, organic-inorganic hybrid membranes formed by dispersing nanoparticles using mechanical stirring or sonication seem less effective. Typically, various methods, such as (1) surface modification of nanoparticle [51–53], (2) in situ preparation of nanoparticles through sol-gel process [54], (3) addition of a third component to increase the interaction between nanoparticles, are major practices for minimizing agglomeration of nanoparticles [55].

1.1 Nanoparticles in Polymeric Membrane Systems

In water treatment membrane processes, various range of polymers such as polyamide, cellulose triacetate, poly(vinylidene fluoride) (PVDF), poly(ether sulfone) (PES), poly(ether imide) (PEI) employing Forward Osmosis (FO), Reverse Osmosis (RO), ultra-filtration (UF), nano-filtration (NF), etc., have been extensively used. It should be noted that, whatever process may appear as essential; all the membranes used in such techniques are required to perform with the highest water permeability with the greatest foulant rejection. These requirements of membrane technology combined with the current advances in nanoscience have led to the discovery of a new research trend that involves the utilization of nanomaterials as hydrophilic additives in the polymeric membrane matrix'.

Nano-additives are organic/inorganic nano-sized particles which can effectively perform as a hydrophilic additive for polymeric membranes for water filtration. This material exhibits excellent performance characteristics with reference to large surface area to volume ratio, high hydrophilicity, porosity and contaminants isolation efficiency. Hence, to accomplish further perfections in water filtration, employing nanomaterials as an additive has become a new potential alternative for macromolecular pore formers for separation of water contaminants. This has led to the great demand amongst the researcher's community for using nanocomposite in water treatment polymeric membranes. Much research has been demonstrated in the past regarding the inclusion of nanoparticles into the design and development of novel materials. However, these nanomaterials possess limitations such as a high-pressure requirement and surface pore blockage when used at higher concentrations, etc. and hence, there is an utmost necessity for developing a cost-effective and reliable technology for water filtration.

1.2 Fabrication Methods of Membrane

The techniques employed to fabricate polymer membranes depends mainly on the selection of base polymer and the structure-properties of the desirable membrane. The most popular methods of fabricating polymeric membranes are electrospinning, track-etching, interfacial polymerization, stretching, and phase inversion.

1.2.1. *Phase inversion* is a demixing approach that uses the homogenous polymer solution converted into solid using different ways such as immersion precipitation, thermally induced phase separation, vapor induced etc. Out of all the methods described above, thermally induced phase separation and immersion precipitation are most generally employed to fabricate polymeric membranes in a different range of morphologies.

1.2.2 *Immersion precipitation* method is employed with a polymer solution casted onto a suitable support and then submerged in a coagulation bath that contains a nonsolvent. Many researchers have developed a wide variety of techniques that facilitate the customization of the pore structure of the membrane, cross-section morphology, including by varying a variety of parameters when immersion precipitation occurs, such as bath temperature, immersion time, base polymer, solvents and non-solvents used, additives etc.

1.2.3 *Evaporation-induced phase separation* is a simple technique of preparing membranes in which a polymer solution of desired viscosity is prepared with a solvent/non solvent. Subsequently, the prepared polymer solution is casted onto a flat porous substrate, employing the doctor blade method. The choice of solvents depends majorly upon boiling points that have a direct influence on the morphology of films. PVC, PS, PVDF, and PVAc microporous membranes have all been created by employing a variety of organic solvents and demonstrated the effect of solvents on pore size, shape and surface morphology.

1.2.4 *Interfacial polymerization (IP)* is another significant technique for commercially fabricating thin-film composite (TFC), nanofiltration, and reverse osmosis membranes. The morphology and the barrier membrane layer composition can be altered by the following factors such as reaction time, solvent type, monomer concentration, and subsequent treatment etc.

1.2.5 *Stretching* is the most common technique employed to fabricate microporous membranes. In this method there is no use of solvent, however the temperature of the polymer is increased beyond its melting point and then extruded to create thin sheets which then stretched to form porosity. This method deals with extremely crystalline polymers, with the crystalline areas providing strength and the amorphous areas allowing porosity. There are two phases involved in the stretching process, cold stretching and hot stretching. Cold stretching is employed for the micropores nucleation of the polymeric precursor film while hot stretching is then used to increasing the porosity and overall finish of the membrane.

1.2.6 *Electrospinning* is relatively new and is used to fabricate porous membranes for a number of uses, desalination, and filtration among them. High potential is applied between the collector and the polymer solution droplet and electrostatic potential is raised to desired value to overcome the droplet's surface tension and a charged liquid jet is obtained. Such fibrous membranes are unique in that the fiber and morphology of nano/microfibres' aspect ratios (length/diameter) may be controlled by varying the level of electric potential applied, the fabrication environment, solution flow rate, and viscosity. The morphology and diameter of the fiber controls the electrospun mats' surface morphology, pore size distribution, hydrophobicity, and porosity.

2 Polymeric Nanocomposite Membranes

The polymeric membranes due to its low cost, pore size range, configuration flexibility and scalability holds a great choice amongst the researchers and industrial communities. The combination of nanomaterials with a polymer matrix results in the formation of nanocomposite membranes with novel or improved properties. The nanomaterials such as graphene, graphene oxide (GO), reduced graphene oxide (RGO), CNTs, CNFs, ZnO, nano-ferrite etc. have gained particular interest as additives to polymeric membranes for water filtration and purification owing to unique properties such as like high surface area, low cost, fast adsorption kinetics, thermal stability, chemical stability, and mechanical stability. The following are the various membranes incorporated with the discussed nanomaterials which improve the mechanical, chemical and thermal properties of the membranes along with the water purification performances.

2.1 PVDF/Carbon-Based Nanocomposite Membrane

Currently peer reviewed publications have perceived the remarkable developments towards the use of carbon based nanoparticles, mainly carbon nanotube (CNTs) and graphene oxide (GO) in nanocomposite membrane for water purification [56, 57]. CNTs and GO are promising nanomaterials for the development of the next generation membranes which exhibit high flux, high selectivity, and low fouling capabilities. The water molecules could easily pass through CNTs which closely resembles the transportation of water molecule through polymeric membrane [58–62]. Similarly, a great effort has been devoted to developing GO incorporated membrane in industrial quantities. It has also been observed that GO incorporated polymeric membranes exhibited a higher transport rate of water and higher selectivity towards substances with various sizes [63, 64].

2.2 PVDF/CNT Membrane

The high aspect ratios along with smooth hydrophobic walls and the suitable inner pore diameter of CNTs in PVDF/CNT membrane lead to easier transportation of water molecule through the membrane and have remarkably improved the parameters required for augmenting the water purification capabilities. Furthermore, the operating temperature in membrane distillation (MD) process is significantly reduced and at same time the water fluxes increase with the addition of CNT in PVDF membrane [63].

Madaeni et al. [64] evolved a novel PVDF based microfiltration (MF) membrane by depositing multiwalled carbon nanotubes (MWCNTs) on the surface of PVDF membrane. The membrane then was coated with polydimethylsiloxane (PDMS) which showed super-hydrophobicity with excellent antibiofouling property. The super-hydrophobic surface developed by the PDMS coating reduced the interaction of the water molecule with the membrane surface and also declined the fouling affinity. Gethard et al. [65] demonstrated that in the PVDF/CNTs membranes, CNT afford an additional pathway for the transport of solute through the membrane as shown in Fig. 1. The PVDF/CNTs water purification membranes resulting a higher permeation of vapor flux for a wide range of salt concentration, even at a relatively lower temperature.

Silva et al. [66] prepared flat-sheet direct contact membrane distillation (DCMD), using MWCNT/PVDF membranes via phase inversion process. The report conveyed that surface chemistry of MWCNT, content of MWCNT loading, addition of PVP etc. significantly affect the performance of the membrane. MWCNT/PVDF membranes could be a good option for DCMD as the membrane contains large sponge-like pores with higher pore density which results a in a greater water flux and salt rejection completely as shown in Fig. 2.

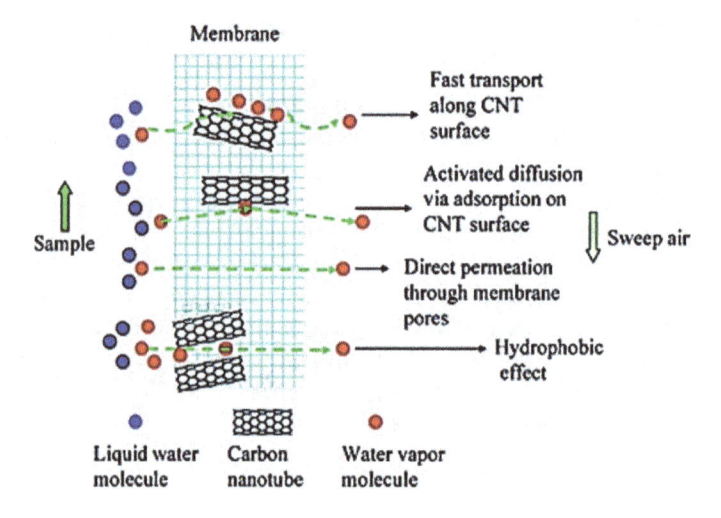

Fig. 1 Mechanism of membrane distillation of CNTs-PVDF membranes [65]

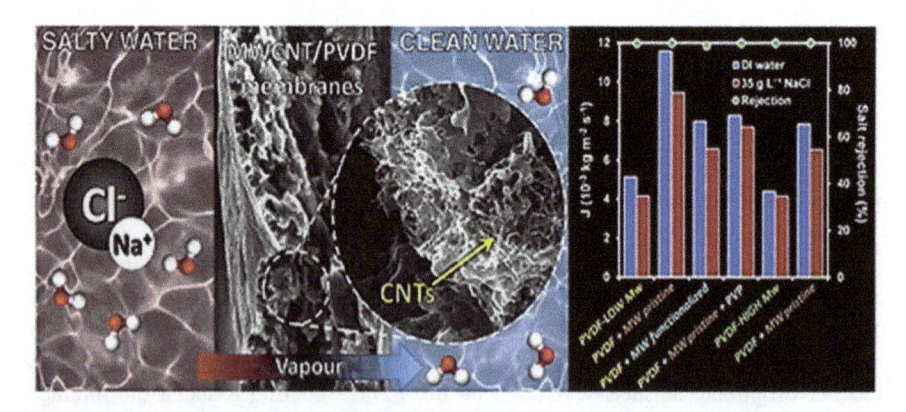

Fig. 2 Salt rejection by MWCNT/PVDF membrane [66]

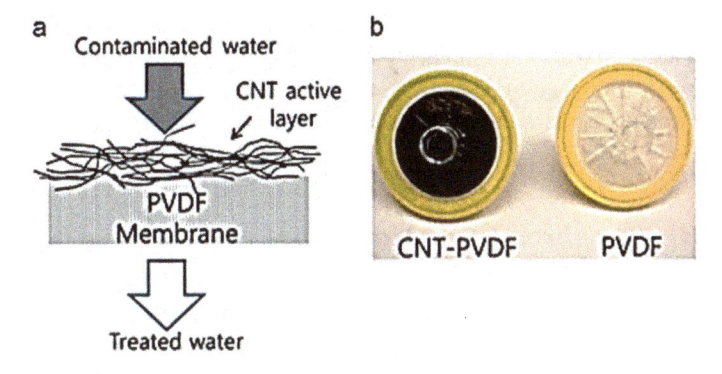

Fig. 3 PVDF membrane with MWCNT layer on top

PVDF membranes casted via phase inversion process generally forms finger-like pores which consequently reduces mechanical strength. CNT is an appealing material to improve the antifouling as well as mechanical properties. The membranes containing CNTs decrease the deposition of organic contaminates and reduces the membrane fouling. The fouling affinity of membrane drops because of uniform porous layers on the surface of the membrane which opposes the confession of organic layers to form fouling on the membrane surface. In a study by Ajmani et al. [60], it was found that CNT layered membranes have the anti-fouling ability for longer period. PVDF membranes blended with MWCNT of larger diameter were more effective in controlling fouling on membrane surface and also tripled the time to form noticeable fouling. Further, the study also found that the structural difference of CNT layers is also a major causative factor for antifouling properties.

PVDF membrane with SWCNT and MWCNT (as shown in Fig. 3) layers on top can be used remove of pharmaceuticals and personal care products (PPCP) like triclosan (TCS), acetaminophen (AAP) and ibuprofen (IBU) effectively from water and established as a viable technology for water treatment [67].

The consequence of modified MWCNT on the various performances of PVDF membrane was reported by Ma et al. [68]. Authors prepared a hybrid ultrafiltration membrane based on PVDF membrane via phase inversion with incorporation of oxidized MWCNT and pristine MWCNT varying from 0.2 to 2 wt% in the PVDF casting dope. PVDF membrane containing 1 wt% of oxidized MWCNTs have a water flux which is 11 times higher than the virgin membrane. Similarly, BSA rejection also increases to the tune of 22.2% as compared to pure PVDF membrane. Authors explained that the higher BSA rejection and permeability of the developed membrane is mainly because of the presence of hydrophilic oxygen-containing groups on the surface of oxidized MWCNT. In another study, Zhao et al. [69] prepared a PVDF membrane by adding various wt% of hyper branched poly (amine-ester) (HPAE) functionalized MWCNTs (MWNTHPAE). The presence of MWNTHPAE on membrane surface promoted denser and more stable hydration layer which in turn resulted in higher hydrophilicity. The hydrophilic group in MWNTHPAE formed H-bonding interaction with water molecules, as a result of which the tendency of protein adsorption was significantly failed. Also, the flux recovery of 2 wt. incorporated PVDF membrane increased to 95.7% as compared to 82% for PVDF.

2.3 PVDF/Graphene Oxide Membrane

The introduction of GO to the polymer casting solution remarkably improved the properties of the membrane mainly because of the presence of functional groups such as carboxyl, carbonyl, hydroxyl groups etc., hence became a smart choice as a nanofiller for the fabrication membranes. The mechanical properties, hydrophilicity, foul resistance, water, flux recovery properties improved with the addition of GO due to the increase in pore channel width in PVDF membrane.

GO incorporated PVDF membrane was fabricated by Wang et al. [70] by the immersion phase inversion method and effects of GO on mechanical strength and permeation properties along with anti-fouling behavior of PVDF membrane were studied. The GO blended PVDF membranes show more surface hydrophilicity due to the hydrophilic character of GO. Results showed an increase in tensile strength by 123% and permeability by 94.4% in case of GO blended PVDF membrane. Further, pure water flux recovery ratio found to have higher than the unblended PVDF membranes. Moreover, hydrophilicity was confirmed from decreased contact angle which in turn entails improved anti-fouling ability of the membrane. Changa et al. [71] followed immersion precipitation phase inversion process in order to fabricate a series of GO and PVP incorporated PVDF UF membranes and investigate the combined effects of PVP and GO on the performance of the membrane.

The presence of GO in PVDF/GO/PVP membrane significantly increases the hydrophilicity of the surface and anti-fouling performance. Further, PVP forms H-bonding with GO (Fig. 4), and acts as an efficient pore forming agent which quickens the rate of phase inversion and also causes a rise in in the roughness of the surface. Similar results were also reported by Zhao et al. [63, 72] for PVDF/GO nanocomposite membranes.

Zhao et al. [72] prepared PVDF/GO nanosheet UF membrane using immersion precipitation phase inversion process to inspect the effect of GO content on the hydrophilicity, morphology and foul resistance properties. FTIR study indicated the existence of large number of hydroxyl groups. These hydroxyl groups attached to the surface of GO nanosheet increases the hydrophilicity. The SEM analysis of the developed membrane evidenced an increased porosity with higher mean pore size and sub-structure of figure like pores. AFM images indicated a significant improvement in the anti-fouling behavior of the membrane surface by the addition of GO nanosheet. The pure water flux and water permeation flux increased by 79% and 99% respectively with the addition of 2 wt% of GO nanosheet.

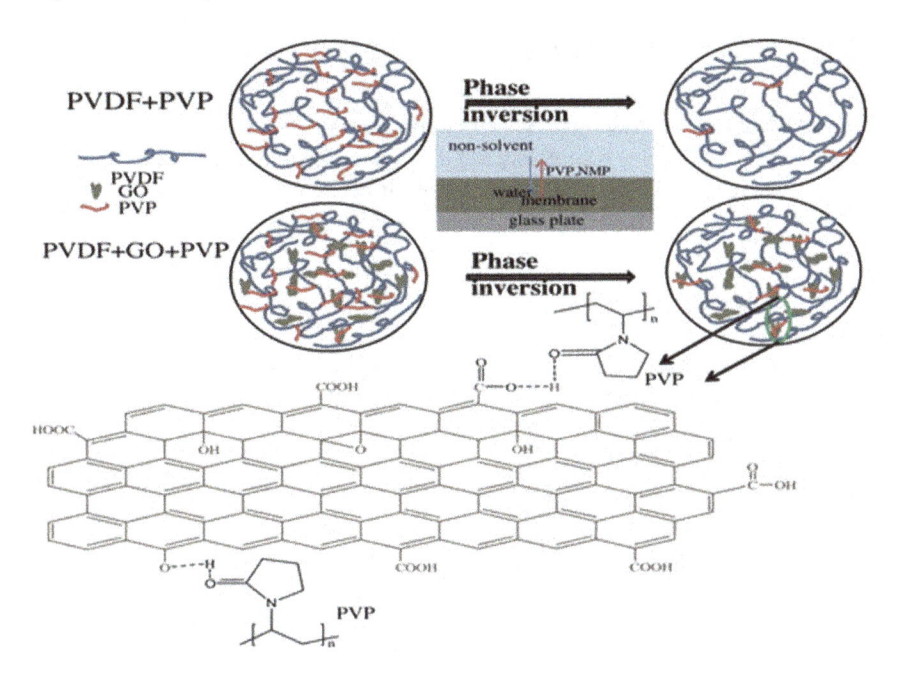

Fig. 4 PVDF/GO/PVP membrane formation mechanism [71]

3 PVDF/Inorganic Metal Oxide Nanocomposite Membrane

The use of nanoparticles in the PVDF matrix increases the mechanical strength of the membrane along with the enhancement in a number of properties like contact angle, hydrophilicity, antibacterial property, and many more. It was also known that specific elements have attraction for particular contaminates which may be organic, inorganic or heavy metal. As different regions of the world are affected by different contaminates, the use of specific nanoparticles has become very important. The following are the details of specificity of the different nanoparticles.

3.1 PVDF/Inorganic Metal Oxide Nanocomposite Membrane

The use of nanoparticles in the PVDF matrix increases the mechanical strength of the membrane along with the enhancement in a number of properties like contact angle, hydrophilicity, antibacterial property, and many more. It was also known that specific elements have attraction for particular contaminates which may be organic, inorganic or heavy metal. As different regions of the world are affected by different contaminates, the use of specific nanoparticles has become very important. The following are the details of specificity of the different nanoparticles.

3.2 PVDF/ZnO Membrane

Zinc is the last transition element in the first row and is in +2 oxidation state in ZnO, so although its d-orbital is completely filled, its 4s-orbital remains vacant. It thereby has the capacity to accommodate electrons, so it behaves a somewhat like a Lewis acid. Zinc Oxide (ZnO) as a nanoparticle provides various promising properties which include:

- Has antibacterial nature [73]
- Resistant to ultraviolet [74, 75]
- Cheap as compared to TiO_2 and Al_2O_3 nanoparticles [76]
- Has bactericidal properties
- Good adsorption capacity for gases like H_2, CO and CO_2 [77–81].
- Has high catalytic activity.
- Can be easily implanted in a membrane which leads to
  - Increase in physical and chemical properties
  - Increase in hydrophilicity and mechanical properties.

Taking into account the above listed properties of ZnO nanoparticles, various studies have been steered to improve the properties of PVDF membrane of which, a few are listed in the table. For example, Liang et al. [76] reported the development of a PVDF membrane blending it with ZnO nanoparticle manufactured by NIPS method [82–85] so as to obtain a novel anti irreversible fouling membrane with different dosages (6.7–26.7%of PVDF weight). The results showed that with the addition of 6.7% (wt% of PVDF) content of ZnO nanoparticle the water permeability of the membrane doubled. This is due to the increase in hydrophilicity (decrease in contact angle) of the membrane. There was also an increase in the flux recovery (up to 100%) in comparison to the unmodified membrane (78% and continued to decline). The modified membranes had better morphology in comparison to the unmodified membrane as a number of small pores of size 20–30 nm were formed in the modified membrane. However, the results also showed that the increase in the amount of ZnO nanoparticles also reduced properties like hydrophilicity, water permeability, flux recovery.

Hong et al. [86] also studied the effects of addition of ZnO nanoparticles into PVDF microfiltration membrane by preparing membranes through phase inversion method using dimethyl acetamide (DMAc) as solvent and varying the ZnO nanoparticle content 0.001–1(wt%). Their results showed an increase in the viscosity of the modified membrane dopes with respect to an increase in the content of ZnO nanoparticle. This might be due to the intensified interactive force between the PVDF molecule which was caused due to the specific area and surface energy of ZnO nanoparticles which are higher. An increment in the mean pore size (7.02×10^{-8} m to 7.98×10^{-8} m) as well as porosity (71.30–75.16%) was found as the ZnO nanoparticle content increased. However, when the ZnO content was more than 0.005%, the viscosity of the dopes increased significantly with a minute change in hydrophilicity leading to decrease in the porosity and mean pore size. Further, maximum pure water flux was found when the ZnO nanoparticle content was 0.005% because of greater hydrophilicity, pore size and porosity. The highest flux recovery observed was highest for the membrane having 0.01% ZnO nanoparticle content (67.12%). The BSA adsorbed on the surface of the membrane could be easily removed than that on the unmodified membrane due to increased hydrophilicity and decrease in roughness of the modified PVDF/ZnO membrane. In addition, an increase in the tensile strength, elongation at break, enthalpy of fusion and melting temperature were seen with increase in ZnO nanoparticle content from 0 to 0.01%. However, when the ZnO nanoparticle content exceeded 0.01% there was a decrease in the properties recorded.

Zhang et al. [87] prepared PVDF/ZnO nanoparticles membrane by two different methods for their application in the removal of copper ions. In the first method, the PVDF membrane was first prepared using phase inversion process then the membrane was treated with potassium hyper manganate and Cetyltrimethyl ammonium bromide (PVDF-CTAB) and another membrane was treated with potassium hyper manganate and Sodium dodecyl benzene sulphonic acid (PVDF-SDBS). Both PVDF-CTAB and PVDF-SDBS were immersed into ultrasonically processed ZnO suspension for 4 h. and then washed. In another method, induced phase inversion process with varying contents of ZnO nanoparticles (1–5 wt%) was utilized to prepare the membranes. Their results showed that the value of water flux increased with the increase in ZnO nanoparticle content (285–465 L/m^{2}h). It may be due to the increase in hydrophilicity as evident from the decrease in contact angle from (62° to 50°). In addition, the contact angle decreased till 3 wt% of ZnO nanoparticle content then increased above 3 wt% of ZnO and also a maximum water flux was found at the same content of ZnO nanoparticle. Therefore 3% of ZnO nanoparticle content was found optimum for attaining good hydrophilicity and permeability [80]. Moreover, there was a drop in the values of BSA adsorption capacity with the rise in ZnO nanoparticle content which indicated the increase in anti-fouling ability of the membrane. The results of the Cu^{2+} ions done by maintaining a pH of 6 and for an equilibrium time of 120 min showed that PVDF/ZnO membrane had adsorbed 87.4 μg/cm^2 of Cu^{2+} ions which equivalent to 9 times more than that of unmodified membrane.

Khan et al. [88] prepared copper doped zinc oxide nanoparticles (ZnO: Cu) by using co-precipitation technique by employing two routes of synthesis and suggested its efficiency as antimicrobial, antifungal properties for cosmetics, pharmaceuticals as well as for water filtration applications. The results also revealed that Cu-doped ZnO nanoparticles have shown the prodigious ability of photocatalytic activity by disintegrating the organic dye. Figure 5 depicts the SEM images of pure ZnO nanoparticles at different magnifications. Several groups have reported the efficiency of the combination of PVDF and ZnO which is represented in Table 1.

From the above types of membranes, the optimized concentration of ZnO nanoparticles lies in between 0.001 and 3 weight percentage. Although it depends on other factors, but from all the above theoretical observations it was found that ZnO is a novel material at its nano level for water purification.

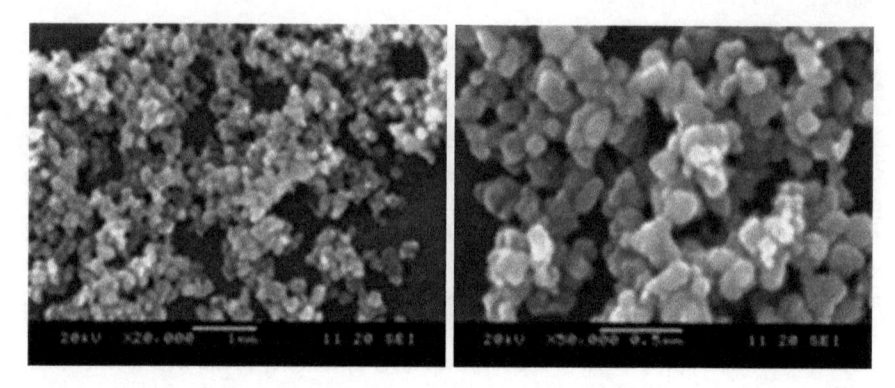

Fig. 5 SEM images of pure ZnO nanoparticles at different magnifications

Table 1 Examples of some of the works reported in the area of ZnO_2-modified PVDF membrane

S. No	Membrane	ZnO content	Remarks	References
1	PVDF/ZnO	6.7–26.7% (% weight of PVDF)	Flux recovery was 100% whilst for unmodified membrane, it was 78% Mechanical strength doubled when ZnO content was around 20.7–26.7% as compared to unmodified membrane	[76]
2	PVDF/ZnO	0.001–1 (wt%)	0.005 wt% presence of ZnO nanoparticle in membrane lead to highest pure water flux, porosity largest pore size and least surface roughness	[86]
3	PVDF/ZnO (CTAB, SDBS)	1–5 (wt%)	Two method of membrane preparation were followed PVDF/ZnO membrane with 3 wt% content of nanoparticle was found to have best surface hydrophilicity and it was capable of absorbing 9 times more amount of Cu^{2+} ions than unmodified membrane	[87]

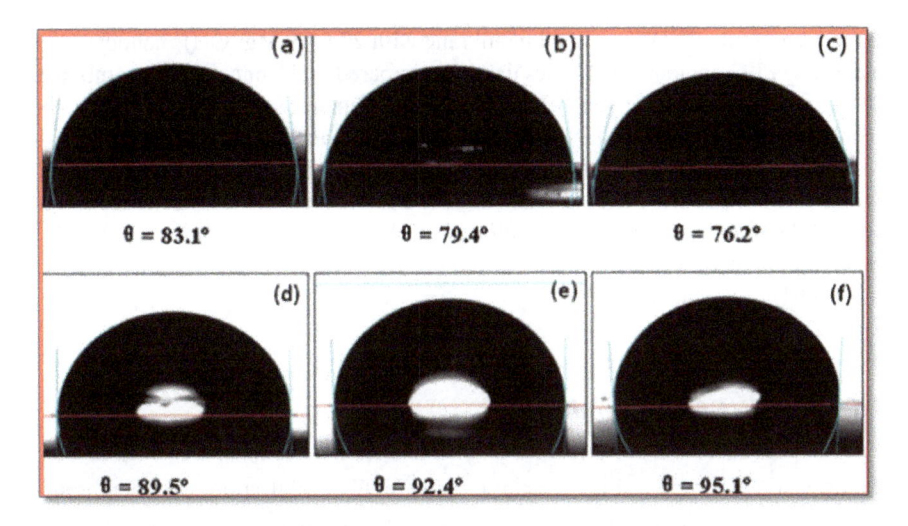

Fig. 6 Membranes with 0 wt% (**a** and **d**), 10 wt% (**b** and **e**), and 20 wt% (**c** and **f**) $Mg(OH)_2$ nanoparticles (**a–c** for PEG = 5 wt%; **d–f** for PEG = 10 wt%) [90] and their water contact angles

3.3 PVDF/Mg (OH)$_2$ Membrane

Dong et al. in two different studies [89, 90], have studied the antibacterial and antifouling properties of PVDF membrane modified with $Mg(OH)_2$ nanoparticles. The examined modified membranes varied in PEG concentration (5–10 wt%) and $Mg(OH)_2$ content (10–20 wt%). $Mg(OH)_2$ nanoparticles were blended well in the PVDF membrane matrix as concluded from residual weight test carried out on the modified membrane. In addition, no biofilm formation was found on the surface of modified membrane as compared to unmodified membrane. This may be explained by the low adsorption of *E.coli* and BSA owing to the addition of $Mg(OH)_2$ nanoparticles content to the membrane. An increase in hydrophilicity with increase in $Mg(OH)_2$ nanoparticles content at PEG 5 wt% content was found this can be explained by the generation of large number of –OH groups due to addition of $Mg(OH)_2$ nanoparticles to the membrane. However, addition of 10 wt% PEG content led to generation of larger pores on the membrane surface resulting in smoother surface and high contact angle even after addition of $Mg(OH)_2$ nanoparticles. Moreover, introduction of $Mg(OH)_2$ nanoparticles to membrane having 10 wt% PEG leads to a decrease in porosity as compared to addition of $Mg(OH)_2$ nanoparticles to membranes with 5 wt% PEG proving the effect of varied PEG content to the dispersion of $Mg(OH)_2$ nanoparticles in the membrane. In addition, it was found that modified membranes having 10 wt% PEG had lower permeability as compared to modified membranes with 5 wt% PEG this can be explained by the entrapment of $Mg(OH)_2$ nanoparticles in the large pores generated due to the excess content of PEG in the membrane.

The entrapment leads to smoothening of the surface and hence simultaneously increases the water contact angle. This is the cause of the decrease in the permeability of the membrane. PVDF modified membrane with 20 wt% Mg $(OH)_2$ nanoparticles and 5 wt% PEG content had 4 times the flux compared to the unmodified membrane. The contact angle data given below gives us the change in hydrophilicity of the membrane with respect to the change in PEG and nanocontainer. Biofilm generated during filtration contained detached dead bodies of *E.coli* and bacteria due to reaction with $Mg(OH)_2$ nanoparticles [89, 91] un-adsorbed bacteria due to increased hydrophilicity of the membrane (addition of –OH groups) and un-trapped *E.coli* (due to small pore size). Further the rate of biofilm formation decreases in the presence of $Mg(OH)_2$ nanoparticles.

3.4 PVDF/Fe_3O_4 Membrane

Huang et al. [92] prepared Fe_3O_4 incorporated PVDF ultrafiltration (UF) membranes with and without the application of parallel magnetic field. Their results showed that in both the cases (i.e. with and without applied magnetic field), there was an increase in pure water flux with the increase in Fe_3O_4 content (0–60 wt%) which sharply increases when the Fe_3O_4 content was over 60 wt%. In addition, there was no significant difference observed in pore dimensions and surface porosity of the membrane [92, 93]. The trend of pure water flux may be caused due to the presence of Fe_3O_4 nanoparticles on the surface of the membrane leading to disturbance in the normal phase inversion process [94]. A 23–35% difference was found in pure water flux of PVDF-Fe_3O_4 membranes (i.e. with and without application of magnetic field). The increased amount of Fe_3O_4 (40–65 wt%) and the magnetic field applied at the time of membrane casting, "lamellar macro voids" were generated which were responsible for the increase in pure water flux and also a decline in resistance to permeation. The rejection value for the membranes (i.e. with and without application magnetic field) was found around 93% when the Fe_3O_4 content was 65 wt%. This may be due to the reduction in the size of pores on the surface with increase in Fe_3O_4 content [95]. Both the PVDF-Fe_3O_4 membranes (i.e. with and without application magnetic field) were found to have good compaction resistance ability. Unmodified PVDF membrane and PVDF-Fe_3O_4 membrane prepared without applying magnetic field with Fe_3O_4 content below 65 wt% showed same tensile strength whilst PVDF-Fe_3O_4 membrane prepared by applying magnetic field with Fe_3O_4 content below 75 wt% had higher tensile strength than unmodified PVDF membrane. The magnetic field applied during the preparation of the membrane forced the Fe_3O_4 particles to align in the direction of applied magnetic field therefore an increase in tensile strength was seen in the direction of magnetic field whereas the tensile strength was low in the cross direction.

Maet et al. [96] studied the formation and the characteristics of PVDF ultrafiltration membrane by the addition of ferrous chloride. The membranes were made via immersion precipitation method with different doses of ferrous chloride ($FeCl_2.4H_2O$) keeping PVDF content constant. The resulted membrane had a rufous color which became deeper with positive increase in the content of ferrous chloride.

Also, it was found that there was an increase in the amount of 0.3 μm diameter particulates with increase of ferrous chloride content. EDX analysis done on the membrane showed the presence of ferric oxide on the surface of the membrane. X-Ray photoelectron spectroscopy analysis done on the membrane showed the presence of iron oxyhydroxide (FeOOH) in the membrane. FeOOH is a type of catalyst similar in properties to ozone used in the removal of organic pollutants in water. The membrane permeability and antifouling property is also increased due to greater hydrophilic nature of FeOOH resulted due to the existence of highly dense surface hydroxyl group. Further with increase in ferrous chloride content in the membrane, an increase in pure water flux (42.35–328.60 $Lm^{-2}h^{-1}$) was observed with increase in the thickness of the membrane (42–80 μm). The ferrous chloride membranes had higher compaction resistant when compared to membranes made of other additives like LiCl, $CaCl_2$, $MgCl_2$ due to the higher F/F_0 values (0.88–1.28). The increase in compaction resistance can be supported by the presence of the 0.3 μm diameter particulates in the interior surfaces of the membrane leading to lowering amount of squeezing which also benefits the water permeation ability. Stability test was done to check the flushing out of derivatives of ferrous chloride from the membrane and it was found that the derivatives were intact when tested with water, but the membrane failed miserably with acid wash test. The color change of the membrane was supported by the fact that Fe^{2+} got oxidized to Fe^{3+} by the oxygen present in the air contacting the solution during the preparation and also due to the H_2O crystals added along with Ferrous chloride.

3.5 PVDF/Al_2O_3 Membrane

Aluminum based nano powder has come to concentration of the scientific community working in the area of water purification technology due to its easy availability, enough hydrophilicity, good mechanical strength and high stability. Besides these advantages there is also a major problem associated with Al_2O_3 which is generally regarding dispersion of these nanoparticles. The dispersion of Al_2O_3 nanoparticle is not perfect. Generally, people prefer to use gamma Al_2O_3. First of all, gamma-Al_2O_3 is porous in nature. Also, these nanoparticles have high surface activity, high adsorptive ability etc. With the addition of these nanoparticles, the surface gets enriched with oxygen atoms. Even a slightly acidic condition converts these oxygen atoms into hydroxyl groups. The presence of these hydroxyl groups may lead to enhancement in hydrophilicity of the membrane. Simultaneously, the increase in hydrophilicity [97–99] increases the antifouling property [97, 98, 100, 101] the membrane.

In this process of preparation [99] of the complex suspension, the basic gamma Al_2O_3 particles need be added in to the PVDF solution before the addition of acid. The addition of the acid is necessary for the mechanism of formation of the suspension. The first step involves the abstraction of proton leading to the formation of the unsaturated system. Then the addition of acid leads to the formation of a carbocation center which means a positive center where the oxygen atom can interact and make bond with. If the acid is not added to the solution then the unsaturation will remain like this and there will be side reaction of water molecules with the unsaturated sides. The above facts are confirmed by the IR data. Let us consider the IR spectra shown by the membrane prepared without the addition of acid. The spectrum shows a peak at 1640 cm^{-1} which is due to the presence of carbon double bond. A peak also appears at 3000 cm^{-1} which indicates the presence of OH group which is introduced by the interaction of water molecule with the unsaturated side. But when acid is added, the unsaturated side is lost resulting in the absence of the peak near 1640 cm^{-1} and also at 3300 cm^{-1}. So, the treatment with acid is undoubtedly a necessary step in this process.

The surface hydrophilicity of the membrane was obtained via the measurement of contact angle. As the concentration of the alumina increases the contact angle decreases. For example, as we add alumina to the above solution by 0, 1, 2, 3 wt%, the contact angle shows values near 91, 88, 82 and 79. Thus the membrane having 3% of alumina has highest hydrophilicity. And by looking at the SEM image of different concentration membrane, it was found that the increase in concentration of the alumina increases the porosity of the membrane. When we came to the field of filtration performance, the membrane having concentration of alumina 2 wt% showed highest pure water flux rate i.e. 134.4 L/m^2, also this membrane has the highest BSA rejection of 93.4%. The membranes having 0 and 1 wt% have lower performance due to lower hydrophilicity in the first case and agglomeration of particle in the second case. Also, the contact angle difference between the membrane having concentration 2 and 3 wt% is very small. Another most important property of the membrane is the antifouling property. It is a very important but complicated one. It depends upon a number of factors but depends a lot on the flux rate. The membrane having high flux rate will have lower membrane fouling than membrane having lower one [102].

3.6 PVDF/TiO$_2$ Membrane

Unification of TiO$_2$ nanoparticles with PVDF membrane has garnered greater attention due to its distinctive physicochemical performance and superior chemical stability and properties. Using TiO$_2$ as a nanofiller, properties like permeability, hydrophilicity, removal of contaminants under UV irradiation can be tailored and also self-cleaning and anti-fouling properties can be achieved in polymeric membranes.

TiO$_2$ exists in the form of anatase and rutile which could mainly be differentiated by their structural properties. Anatase TiO$_2$ nanoparticles are smaller in size and form smaller pores membrane surface and more apertures as compared to larger size rutile TiO$_2$ which in turn results higher permeability and anti-fouling performance.

For sure, the desired properties PVDF/TiO$_2$ nanocomposite membrane enhanced because of the interaction between TiO$_2$ and PVDF. The mechanical strength of PVDF/TiO$_2$ nanocomposite membrame was apparently high which atoms in PVDF (shown Fig. 7) are mainly due to the Vander Waal interaction of Ti4+ ions of TiO$_2$ nanoparticle with fluorine (F).

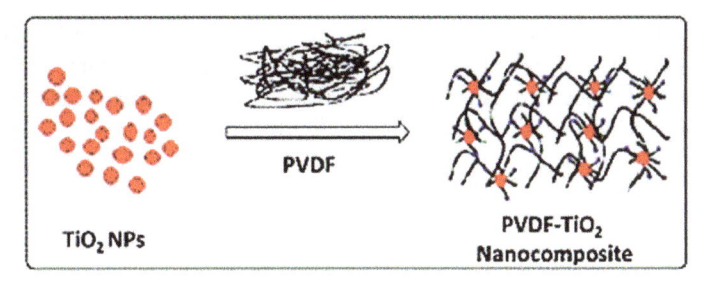

Fig. 7 Formation of PVDF/TiO2 nanocomposite [106]

However, the main problem lies in the fixation of nano TiO_2 on the surface of the PVDF membrane. To fix the nano TiO_2 and for its immobilization two methods are often used; in one case the PVDF and the nanoparticle holder like polyacrylic acid are first merged and then the membrane is treated with nano solution. In the second case the particle holder and the nanoparticles are first sonicated and then the membrane is treated with the solution. The grafting yield and also the dispersion of the particles are higher in later.

It was successfully recognized that under UV irradiation, the anti-fouling properties of the membrane can be boosted significantly due to the photocatalytic behavior of TiO_2 nanoparticles [103–105]. In the presence of UV irradiation, nano TiO_2 particles engender strong oxidizing agents which preclude deposition of various organic compounds, proteins, fats etc. Similarly, Mohd Yatim et al. [106] reported that the resulted super hydrophilicity due to TiO_2 nanoparticle in PVDF membrane allowed to attain high water flux under UV irradiation. Besides the photocatalytic activity [107–109], low cost, biocompatibility, non-toxicity, environmentally friendliness, high chemical resistivity, optical stability and it has also high quantum yield [110–113].

Tavakolmoghadam et al. [114] prepared PVDF/TiO_2 membrane using mixed solvents triethyl phosphate (TEP) and dimethylacetamide (DMAc) through immersion precipitation method and studied various properties of the membrane. The water contact angle (WCH) decreased from 70.53° to 60.5° which indicated higher hydrophilicity of PVDF membrane with 0.25 wt% TiO_2 of membrane than neat PVDF. The improved anti-fouling properties thanks to higher hydrophilicity also caused a 96.85% flux recovery ratio (FRR) in PVDF/TiO_2 membrane.

Méricq et al. [115] adopted phase inversion method to cast PVDF membranes trapped TiO_2 nanoparticle and investigated the effect of TiO_2 concentration. They reported that at intermediate content of TiO_2 of about 25 wt%, optimum water permeability of greater than 150 L h^{-1}/m^{-2} was observed due to high hydrophilicity of the membrane and due to the formation of finger-like macro voids in membrane. The membrane permeability decreased at higher concentration of TiO_2 due agglomeration of TiO_2 which blocked the pores of the membrane. At the same time lower concentration of TiO_2 had no impact on the performance of the neat PVDF membrane.

Hydrothermally prepared TiO_2 nanofibers in alkali medium used to fabricate PVDF/TiO_2 composite membrane via a facile wet chemical process was reported

[116] which have prodigious potential applications in water purification because of its high mechanical, anti-fouling, permeability performance. The membrane exhibited 95% of original water flux restoration with the aid of solar irradiation and admirable anti-fouling performance.

3.7 PVDF/SiO$_2$ Membrane

Huang et al. [117] prepared PVDF/silica (SiO$_2$) hybrid membranes through phase inversion by a tetraethoxysilane (TEOS) sol–gel process. The study was highlighted with improvement in various properties like mechanical and thermal properties, porosity, hydrophilicity, anti-fouling properties, BSA rejection rate etc. Also, when TEOS concentration exceeded 20%, the PVDF crystalline structure changed from β phase to α phase. The preparation of SiO$_2$ nanoparticle enforced PVDF hollow fiber membranes via an immersion precipitation method was reported by Hashim et al. [118] A remarkable improvement in pure water flux was observed when the SiO$_2$ particles were washed out from the membrane using either 20 wt% NaOH solution at 70 °C or HF acid at room temperature.

Wu et al. [119] designed a unique methodology to PVDF/SiO$_2$ membranes modified nanoparticles. The authors demonstrated the addition of siloxane and PEG functionalized SiO$_2$ nanoparticles to enhance antifouling properties. During PVDF/SiO$_2$ membrane casting via phase inversion process, siloxane chains or PEG chains in modified SiO$_2$ spontaneously drifted to polymer–water interface.

Siloxanes protect the membrane surface from the deposition of foulants whereas PEG shields the adsorption of protein and cell adhesion. The functionalized SiO$_2$ nanoparticle prevents the particle aggregation during phase inversion as depicted in Fig. 8. As a result, PEG or siloxane chain could sufficiently attach to the skin-layer resulting in the enrichment of polysiloxane or PEG on the surface. The test experimental results revealed a 33 and 27% improvement in flux recovery ratio with the addition of 2 wt% SiO$_2$–COOH and SiO$_2$–COOH–PEG nanoparticles respectively along with long term anti-fouling performance. However, at higher concentration of SiO$_2$, a declining tendency of the above properties was observed due to the agglomeration of SiO$_2$ nanoparticles. The low surface energy, thermal stability, strong shear resistance, free rotation of silicon-oxygen backbone in PDMS make it an efficient foul releasing material.

Wang et al. [120] PVDF/PDMS modified SiO$_2$ (SiO$_2$-g-PDMS) hybrid membrane via non solvent induced phase separation(NIPS) was prepared. SiO$_2$-g-PDMS nanoparticles have both hydrophilic core and hydrophobic shell, which will be capable to improve the foul resistant and separation performance. The results indicated that water flux recovery ratio (FRR-W) of the PVDF/SiO$_2$-g-PDMS hybrid membrane with bovine serum albumin (BSA) and humic acid (HA) increased due to the increase in fouling repulsion as well as the fouling release performance of the membrane.

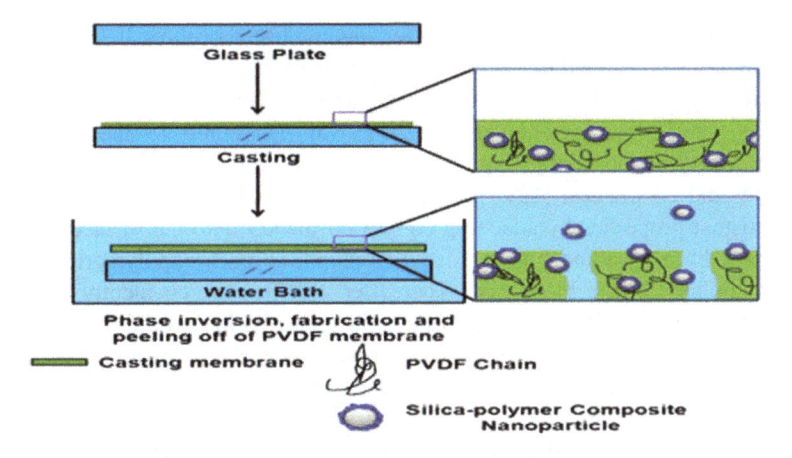

Fig. 8 Schematic diagram of surface enrichment of surface modified silica nanoparticles during phase inversion [119]

3.8 PVDF/Ag Membrane

Silver nanoparticles (AgNPs) display incomparable robust inhibitory and biocidal performance among various biocidal materials against a variety of microorganism. Recently, substantial work has been done to assimilate AgNPs into polymer nanocomposite membrane surface by either in situ or ex-situ process. Besides the well-known antifouling [100] properties of silver (Ag) nanoparticle, the hydrophilic nature of the membrane raises its importance. Ag is stable and exists in water without any interaction. Its electronic configuration shows that it is very much inert towards water molecule, so it exits in water without any chemical reaction with water molecules. But Ag atom lacks one electron to fulfill its d-orbital. So, there is always a secondary interaction which makes Ag a hydrophilic species. The interaction between the Ag and membrane is purely secondary, but this interaction is sufficient enough to resist the force of water which comes with pressure. The concentration of the Ag nanoparticles affects the properties of the membrane to a greater extent. Membranes containing different percentage of Ag nanoparticles were taken and it was found that the flux rate increases [100] with increase in concentration of Ag nanoparticles. When the Ag concentration increases to 1.5 weight percent the flux rate increases almost three times than the flux rate of the PVDF membrane without Ag nanoparticles. The BSA rejection percentage value also decreased by the addition of Ag nanoparticles. There is an interaction between the BSA solution and membrane surface [100]. This interaction is mainly electrostatic. At lower pH value, the solutions exist as cationic species and the membrane is in anionic state. So, the rate of rejection becomes high. But by the addition of Ag nanoparticles the membrane behaves like a cationic one. So, the attractive force between the membrane and solution decreases to some extent, which affects the rejection rate. Contact angle also varies directly with the concentration of Ag nanoparticles. The antibacterial property [100–102] of the Ag nanoparticles was confirmed by the halo test. From the test it could be concluded that the inhibition zone of E. coli bacteria increases with the increase in concentration of the Ag nanoparticles. The sample containing no Ag nanoparticles [100] showed higher bacterial growth. This confirmed the antibacterial property of Ag nanoparticles (Fig. 9).

Kwak and co-workers [121] prepared PVDF nanocomposite membranes with foul resistant properties by attaching AgNPs covalently onto the surface of PVDF membrane by a thiol end-functionalized amphiphilic block copolymeric linker to form a thiolated PVDF membrane. Thereafter, AgNPs were covalently embedded to the thiolated PVDF membrane surface via impregnation process, as illustrated in Fig. 10. The resulting PVDF/Ag nanocomposite membrane exhibited admirable foul resistant property which can be an attractive high-efficiency membrane for water purification applications.

Li et al. [122] prepared the poly(vinylidene fluoride)-g-poly(acrylic acid) (PVDF-g-PAA) membranes. Then, the grafted membrane was immersed in a silver nitrate

Fig. 9 Schematic illustration of the synthesis of SiO_2-g-PDMS nanoparticles [120]

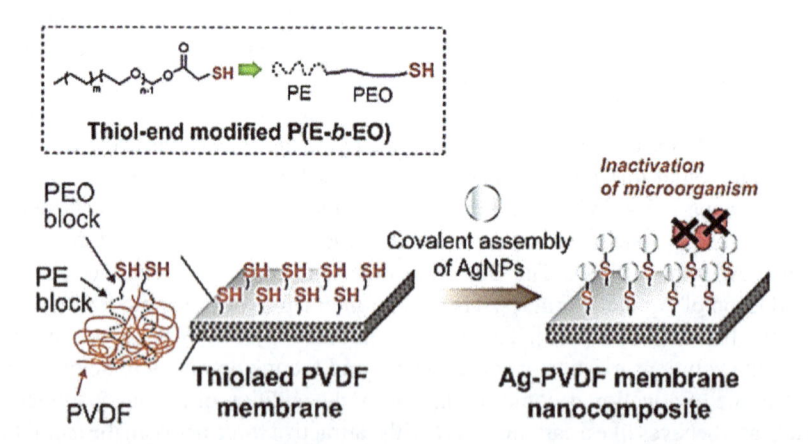

Fig. 10 Immobilization of AgNPs to the PVDF membrane mediated by a thiol-end functional amphiphilic block copolymer linker [121]

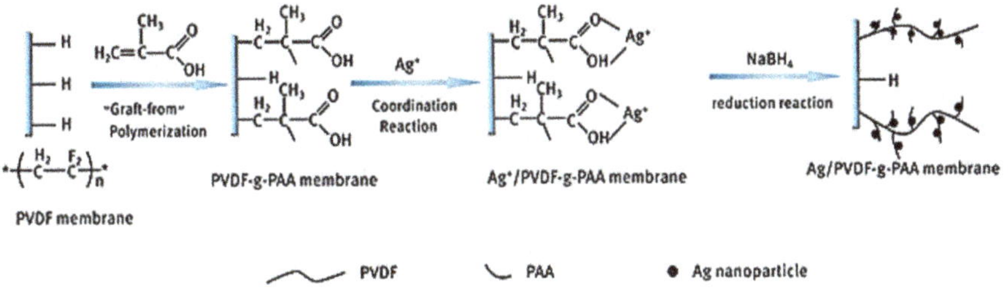

Fig. 11 Silver modification of PVDF membrane [122]

solution as result of which Ag$^+$ ions bonded with the carboxyl group of PAA coordinately to form a Ag nanoparticle followed by reduction of Ag+ ions with NaBH$_4$ to get immobilized PVDF-g-PAA membrane, as shown in Fig. 11.

4 PVDF/Hybrid Nanoparticles Membrane

To develop multiple properties in the membrane and in order to improve its applicability, researchers have experimented with different nanocomposites for the modification of the membranes. By the use of these nanocomposites the deficit of single nano based membrane was mitigated. The article mentioned below gives us some idea about different successful combinations of the nanoparticles.

Alpatova et al. [123] in their study they examined the performance and various properties like physiochemical properties, permeability of the membrane which is a composite of PVDF-Fe$_2$O$_3$ and MWCNT. Their results showed that there was a drop in the contact angle (<90°) of all the membranes which were modified whereas the unmodified membrane had a contact angle (95°) which was related to the enhanced hydrophilicity of the modified membranes. In addition, this decrease in contact angle was caused by the addition of Fe$_2$O$_3$ not MWCNT. The highest number of pores were observed for the modified membrane having 0.5 wt% Fe$_2$O$_3$ content. The MWCNT helped in even-spreading of the Fe$_2$O$_3$ content throughout the membrane. The modified membranes were found to have an increased permeate flux along with the rise in trans membrane pressure (0.17–1.02 mPa). This might be attributed to the diffusive effect of nanopores in MWCNTs [123]. A further rise in the permeate flux was noticed with an increase in the content of Fe$_2$O$_3$ up to 1% which was then followed by decline at 2%. The modified membrane with 1% content of Fe$_2$O$_3$ was found to be the optimum due to its better surface morphology, hydrophilicity, and clean water flux than other combinations. The results of the stability test showed that only 1.9% of iron was leached out after continuous filtration of 9 h.

Lijunet al. [124] developed a PVDF/Fe^{3+}-TiO$_2$ membrane and studied about its characterization and catalytic activity to decompose H$_2$O$_2$. Their study showed that with an optimum decomposition rate of H$_2$O$_2$ was found when ratio of Fe^{3+}/TiO$_2$ was 1:9 in the membrane. In addition, with increase in the Fe^{3+} content in the membrane there was an increase in the decolorization rate of Orange IV. It was also found that addition of 21% of Fe^{3+}-TiO_2 in the membrane led to the highest decomposition rate of H_2O_2. The presence of high amount of Fe^{3+} content increased the catalytic activity of the membrane rather than the concentration of TiO_2. The increase in Fe^{3+}/TiO_2 content in the membrane also decreased the contact angle thereby increasing the hydrophilicity of the membrane because of the hydroxyl radical.

Figure 12 [123] shows the variation in SEM image with different composition of MWCNT (M) and Fe_2O_3 (F) nanoparticles. H. Bai and co-workers [125] prepared a flat round PVDF membrane which was modified with TiO_2/ZnO composite photocatalyst by atomic layer deposition (ALD) and light driven to study its various properties like hydrophilicity, permeability, antifouling. Their results showed that roughness values of the membrane rose with the increase in ZnO content (108.8–126.0 mm). The elemental mapping results showed that Ti and Zn particles were uniformly spread all over the surface of the membrane as well as on the pore walls. The mean pore size of the original PVDF membrane decreased after the deposition with 200 ALD cycles of TiO_2 and ZnO respectively. Because of the deposition of TiO_2 and ZnO layer on the PVDF membrane there was an improvement in hydrophilicity [126]. A fall of 82.6% in water contact angle was noted after visible light irradiation of the membrane which was modified (TiO_2:ZnO = 1:3) for 30 min, indicating the photo-induced super-hydrophilicity of the membrane [126]. The pure water flux rate were measured in dark and visible light irradiation condition; it was found that in dark condition the maximum rise in pure water flux was 0.017 cm/s for modified membrane (TiO_2:ZnO = 1:3; ZnO content 0.923 μmol/cm^2) and in visible light irradiation there was a 33.5% increase as compared to that of unmodified membrane. These changes are due to enhanced hydrophilicity (decreased contact angle) owing to the deposition layer of TiO_2/ZnO. In addition, 80% removal efficiency of Methylene blue was observed for TiO_2/ZnO composite modified membrane. Modified membrane (TiO_2:ZnO = 1:3) had a rate constant up to ten times higher than unmodified membrane (0.0075–0.11 min^{-1}) indicating that degradation of methylene blue was performed by photo catalysis [123]. The TiO_2/ZnO composite membrane had high total organic carbon removal rate even after five times recycle proving the reusability of the modified membrane. Under visible light irradiation the total organic carbon removal of *Hyaluronic acid* was up to 89.2% for the modified membrane (TiO_2:ZnO = 1:3) and 72.1% in dark displaying combined effect of rejection along with photo degradation. There was an increase in total organic carbon removal of

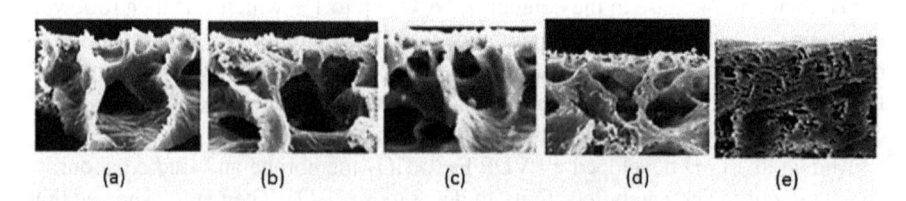

Fig. 12 SEM cross-sectional view images of fabricated membranes: **a** M-0-F 0 membrane, **b** M-0.2 F 0.2 membrane, **c** M-0.2-F 0.5 membrane, **d** M-0.2-F 1 membrane, **e** M-0.2-F 2 membrane

Hyaluronic acid with increase in trans-membrane pressure (0.05–0.125 bar) in dark condition. Whilst under visible light irradiation at 0.1 bar trans membrane pressure, an excellent total organic carbon removal of Hyaluronic acid was observed by modified composite ($TiO_2:ZnO = 1:3$) membrane.

Li et al. [127] in their study investigated about the visible light response activity and surface hydrophilicity of Ag/TiO_2 modified PVDF membrane. They used blending/photo reduction combined method for casting PVDF-Ag/TiO_2 membranes and the composition of the membrane were confirmed by the XPS and XRD tests. The FTIR spectra displayed an increase in amount of hydroxyl group on the surface of Ag/TiO_2 modified membrane in comparison to pristine PVDF membrane; this also supports the decrease of contact angle and increase in hydrophilicity of the Ag/TiO_2 modified PVDF membrane. Moreover, the Ag/TiO_2 modified PVDF membrane maintained a high flux of 600 L/m^2 even after three filtration cycles in comparison with pristine PVDF membrane having a flux of 150 L/m^2 h. Further visible light activity response was investigated by checking the amount of methylene blue degradation Ag/TiO_2 modified PVDF membrane which displayed 51% higher degradation of methylene blue in comparison to the TiO_2 modified membrane and pristine PVDF membrane. Their results also showed that Ag/TiO_2 modified PVDF membranes displayed high bactericidal effect [128] on *E. coli* and *S. Aureus* in comparison to the pristine PVDF membrane and the effect kept on increasing with increase in amount of Ag nanoparticles.

A novel PVDF nanocomposite ultrafiltration membrane comprising rGO/TiO_2 (a mixture rGO and TiO_2) has been fabricated using the well-known phase inversion method [129]. It could be clearly understood from the report that the agglomeration of TiO_2 significantly reduced with the use of rGO in the membrane. Also, hydrophilicity, foul resistant properties also increased. The increase in the hydrophilicity and permeability of the membranes could be explained by the presence hydrophilic oxygenated groups on rGO/TiO_2. The membranes also showed remarkably enhanced performance at a rGO/TiO_2 loading of 0.05 wt%. At 0.05 wt% rGO/TiO_2 content, optimum foul resistance was observed and also the membrane developed a rise in pure water flux by 54.9% when compared with neat PVDF membrane.

In another study it was reported that the synergistic effect of GO/TiO_2 hybrid nano mixture in PVDF membrane improved the superior flux recovery ratio, self-cleaning, photocatalytic activity and antifouling function [130]. The GO/TiO_2-PVDF membrane exhibited considerably 50–70% enhancement in photodegradation efficiency with two fold higher water flux ion than the virgin PVDF membrane with 92.5% BSA reject. A related study was also reported by Xu et al. [130] Membranes for enhanced water treatment [131] having Photocatalytic activity with the photo inactivation of bacteria [132–141] by the use of nanocomposites was also studied (Figs. 13 and 14).

The possible mechanism for the photodegradation of BSA was detailed elsewhere [142, 143]., as explained in Fig. 15. Under the UV irradiation, TiO_2 was excited and it generated photo-electrons (e^-) and holes (h^+). Oxygen molecule (O_2) captured the electrons on the surface of TiO_2 and produced O_2^-, HO_2^-, H_2O_2 and hydroxyl radical ($^\bullet OH$). The $^\bullet OH$ radical also resulted due to the interaction of H^+ with H_2O

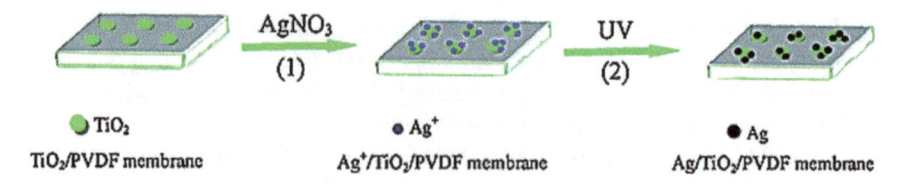

Fig. 13 Preparation of Ag/TiO$_2$/PVDF membrane

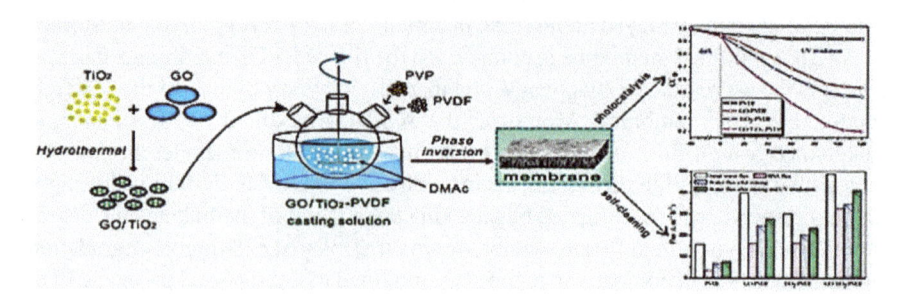

Fig. 14 PVDF/GO/TiO$_2$ membrane preparation [130]

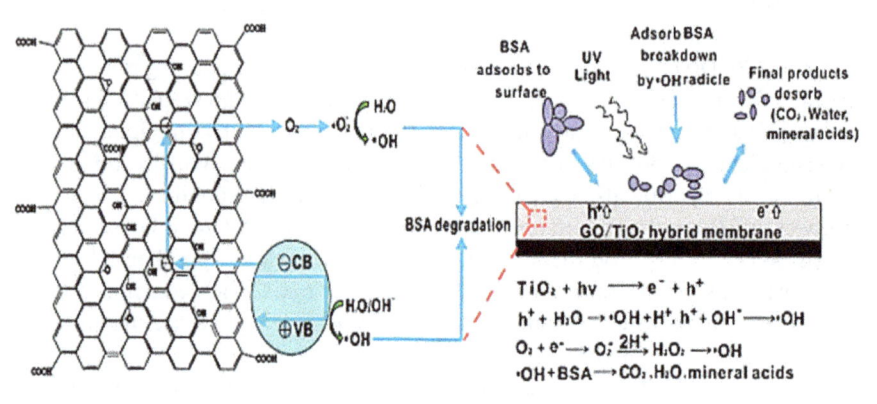

Fig. 15 Possible mechanism of Photo-catalysis and process of GO/TiO2-PVDF membranes [142, 143]

and OH$^-$. The •OH is capable of degrading many complex organic compounds. Damodar et al. [142] worked on the UV photodegradation of BSA. In the study a FRR of 98% was reported for BSA filtration by a UV irradiation for 30 min using a PVDF/TiO$_2$ membrane. Photodegradation of BSA was also studied by Yang et al. [143] at 365 nm UV irradiation and it was suggested that BSA degrades in two steps: first into tiny fragments, which then in turn into small inorganic molecules as they undergo mineralization. This effective photocatalytic property is undoubtedly very instrumental in diminishing fouling of the membrane because of the oxidation of foulants by photocatalysis thereby guaranteeing the fact that if we compare the

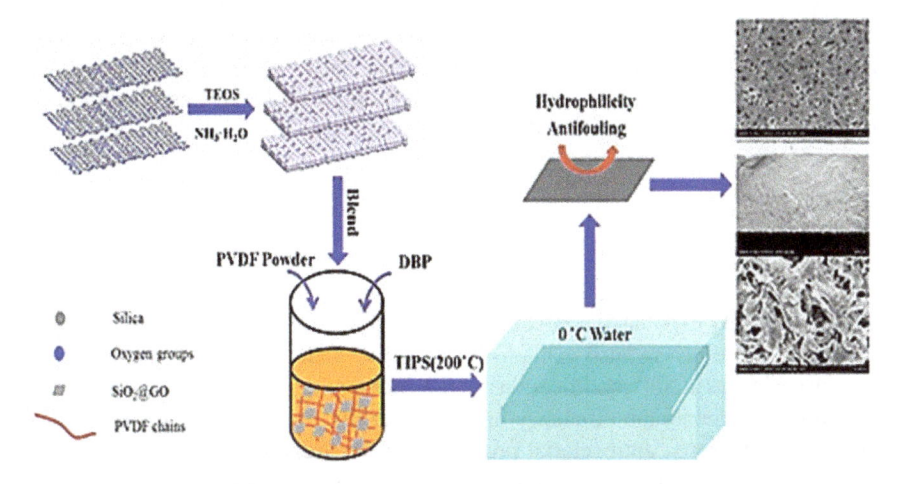

Fig. 16 Preparation of PVDF/SiO$_2$@GO membranes [144]

GO/TiO$_2$ modified membranes with the traditional ones [144] then the former can sustain high permeate flux for definitely a longer period of time (Fig. 16).

Li et al. [145] follows thermally induced phase separation (TIPS) technique for preparing SiO$_2$@GO modified PVDF nanohybrid membrane. How the nanoparticles affect the performances of the membrane was investigated and the result declared the increase in membrane performances of the modified membrane. The properties of the modified membrane were studied, and the results gave the idea that the modification leads to increase in hydrophilicity, reduction in permeation flux, increase in bubhaine serum albumin(BSA) rejection and improved anti-fouling ability of membranes.

It is not only the flux increases after the use of GO@SiO$_2$ nanohybrid particles but also the antifouling properties of the membranes enhance. Its preparation involves the introduction of the "three-dimensional structure" nanohybrid particle (GO@SiO$_2$) along with polyvinylpyrrolidone (PVP) to polyvinylidene fluoride (PVDF) casting solution. GO@SiO$_2$ nanohybrid particles were prepared through in situ hydrolysis and condensation of precursor tetraethoxysilane (TEOS) on the surfaces of graphene oxide (GO) nanosheets. The nanohybrid membranes were prepared through phase inversion technique by an immersion precipitation method. Compared with PVDF/PVP, PVDF/SiO$_2$/PVP and PVDF/GO/PVP membranes, PVDF/GO@SiO$_2$/PVP membrane exhibited remarkable performance in hydrophilicity, mechanical force and also in surface roughness. The rejection rate of bovine serum albumin (BSA) surpassed 78.5% under high water flux of 1232 L/(m^{2}h). Also, the highest flux recovery rate of 77.5% for BSA and 82.1% for sodium alginate was observed for the membrane. The antifouling properties of the nanohybrid membrane were further studied. This was done using our own foulant probes with the help of an Atomic Force Microscope (AFM) and thereby determining the adhesion forces of foulant and membrane. From the study it was concluded that adhesion forces between the membrane with GO@SiO$_2$ and the foulants were very weaker compare to the membrane without these nanocomposites. Thus, the observed results implied that the new nanohybrid membrane had altogether the best properties and antifouling proficiency. And it undoubtedly delivers an advanced method for improving the performance of PVDF membranes [146].

5 Conclusion

The above study explains the effect of different types of nanoparticles on the PVDF membrane. The review also explains the selective and also the versatile nature of nanoparticles used for modification of the membrane. The use of nanoparticles increases the membrane efficiency and the usability by increasing hydrophilicity, antifouling property, antibacterial property, tensile strength etc. It was also found that specific nanoparticle is efficient for removing specific types of impurities. The variation of the membrane property with respect to the change in concentration of nanoparticles is also explained in this article. The optimization values of different nanoparticles are also given. The Contact angle is decreasing with the increase in concentration of pore forming agent and also the nanoparticles. The pore structure, size and uniformity of pore distribution are also controlled by the use of nanoparticles. Overall, the concentration of PVDF, pore forming agent, solvent of dope solution for making membrane, the selection of linker and the nanoparticle are the deciding factors for membrane efficiency and selectivity.

References

1. Geise GM, Lee H-S, Miller DJ, Freeman BD, McGrath JE, Paul DR (2010) Water purification by membranes: the role of polymer science. J Polym Sci Part B Polym Phys 48(15):1685–1718
2. Ghosh AK, Hoek EMV (2009) Impacts of support membrane structure and chemistry on polyamide–polysulfone interfacial composite membranes. J Membr Sci 336(1–2):140–148
3. Hegab HM, Zou L (2015) Graphene oxide-assisted membranes: fabrication and potential applications in desalination and water purification. J Membr Sci 484:95–106
4. Li X, Janke A, Formanek P, Fery A, Stamm M, Tripathi BP (2018) One pot preparation of polysulfone-amino functionalized SiO_2 nanoparticle ultrafiltration membranes for water purification. J Environ Chem Eng 6(4):4598–4604
5. McCloskey BD, Park HB, Ju H, Rowe BW, Miller DJ, Freeman BD (2012) A bioinspired fouling-resistant surface modification for water purification membranes. J Membr Sci 413:82–90
6. Zodrow K, Brunet L, Mahendra S, Li D, Zhang A, Li Q, Alvarez PJ (2009) Polysulfone ultrafiltration membranes impregnated with silver nanoparticles show improved biofouling resistance and virus removal. Water Res 43(3):715–723
7. Werber JR, Osuji CO, Elimelech M (2016) Materials for next-generation desalination and water purification membranes. Nat Rev Mater 1(5):1–15
8. Esfahani MR, Tyler JL, Stretz HA, Wells MJM (2015) Effects of a dual nanofiller, nano-TiO_2 and MWCNT, for polysulfone-based nanocomposite membranes for water purification. Desalination 372:47–56

9. Ramanan SN, Shahkaramipour N, Tran T, Zhu L, Venna SR, Lim C-K, Singh A, Prasad PN, Lin H (2018) Self-cleaning membranes for water purification by co-deposition of photo-mobile 4, 4′-azodianiline and bio-adhesive polydopamine. J Membr Sci 554:164–174
10. Yin J, Kim E-S, Yang J, Deng B (2012) Fabrication of a novel thin-film nanocomposite (TFN) membrane containing MCM-41 silica nanoparticles (NPs) for water purification. J Membr Sci 423:238–246
11. Yoon K, Hsiao BS, Chu B (2009) Formation of functional polyethersulfone electrospun membrane for water purification by mixed solvent and oxidation processes. Polymer 50(13):2893–2899
12. Zhang L, Shan C, Jiang X, Li X, Liangmin Yu (2018) High hydrophilic antifouling membrane modified with capsaicin-mimic moieties via microwave assistance (MWA) for efficient water purification. Chem Eng J 338:688–699
13. Yoon K, Hsiao BS, Chu B (2009) High flux ultrafiltration nanofibrous membranes based on polyacrylonitrile electrospun scaffolds and crosslinked polyvinyl alcohol coating. J Membr Sci 338(1–2):145–152
14. Cadotte J, Forester R, Kim M, Petersen R, Stocker T (1988) Nanofiltration membranes broaden the use of membrane separation technology. Desalination 70(1–3):77–88
15. Kang G-d, Cao Y-m (2012) Development of antifouling reverse osmosis membranes for water treatment: a review. Water Res 46(3):584–600
16. Lee SY, Kim HJ, Patel R, Im SJ, Kim JH, Min BR (2007) Silver nanoparticles immobilized on thin film composite polyamide membrane: characterization, nanofiltration, antifouling properties. Polym Adv Technol 18(7):562–568
17. Tan Z, Chen S, Peng X, Zhang L, Gao C (2018) Polyamide membranes with nanoscale turing structures for water purification. Science 360(6388):518–521
18. Nouri M, Marjani A, Tajdari M, Heidary F, Salimi M (2018) Preparation of cellulose acetate membrane coated by PVA/Fe3O4 nanocomposite thin film: an in situ procedure. Colloid Polym Sci 296(7):1213–1223
19. Bhanthumnavin W, Wanichapichart P, Taweepreeda W, Sirijarukula S, Paosawatyanyong B (2016) Surface modification of bacterial cellulose membrane by oxygen plasma treatment. Surf Coat Technol 306:272–278
20. Chou W-L, Da-Guang Yu, Yang M-C (2005) The preparation and characterization of silver-loading cellulose acetate hollow fiber membrane for water treatment. Polym Adv Technol 16(8):600–607
21. Reuvers AJ, Van den Berg JWA, Smolders C (1987) Formation of membranes by means of immersion precipitation: part I. A model to describe mass transfer during immersion precipitation. J Membr Sci 34(1):45–65
22. Rahimpour A, Madaeni S (2007) Polyethersulfone (PES)/cellulose acetate phthalate (CAP) blend ultrafiltration membranes: preparation, morphology, performance and antifouling properties. J Membr Sci 305(1–2):299–312
23. Saljoughi E, Sadrzadeh M, Mohammadi T (2009) Effect of preparation variables on morphology and pure water permeation flux through asymmetric cellulose acetate membranes. J Membr Sci 326(2):627–634
24. Kusworo TD, Wibowo AI, Harjanto GD, Yudisthira AD, Iswanto FB (2014) Cellulose acetate membrane with improved perm-selectivity through modification dope composition and solvent evaporation for water softening. Res J Appl Sci Eng Technol 7(18):3852–3859
25. Worthley CH, Constantopoulos KT, Ginic-Markovic M, Pillar RJ, Matisons JG, Clarke S (2011) Surface modification of commercial cellulose acetate membranes using surface-initiated polymerization of 2-hydroxyethyl methacrylate to improve membrane surface biofouling resistance. J Membr Sci 385:30–39
26. Gao A, Liu F, Xue L (2014) Preparation and evaluation of heparin-immobilized poly (lactic acid)(PLA) membrane for hemodialysis. J Membr Sci 452:390–399
27. Ma H, WenXin S, Tai Z, Sun D, Yan X, Liu B, Xue Q (2012) "Preparation and cytocompatibility of polylactic acid/hydroxyapatite/graphene oxide nanocomposite fibrous membrane. Chin Sci Bull 57(23):3051–3058

28. Svang-Ariyaskul A, Huang RYM, Douglas PL, Pal R, Feng X, Chen P, Liu L (2006) Blended chitosan and polyvinyl alcohol membranes for the pervaporation dehydration of isopropanol. J Membr Sci 280(1–2):815–823
29. Zhu L-P, Jing-Zhen Yu, You-Yi X, Xi Zhen-Yu, Zhu B-K (2009) Surface modification of PVDF porous membranes via poly (DOPA) coating and heparin immobilization. Colloids Surf B 69(1):152–155
30. Liu F, Awanis Hashim N, Liu Y, Moghareh Abed MR, Li K (2011) Progress in the production and modification of PVDF membranes. J Membr Sci 375(1–2):1–27
31. Zhang M, Nguyen QT, Ping Z (2009) Hydrophilic modification of poly (vinylidene fluoride) microporous membrane. J Membr Sci 327(1–2):78–86
32. Khayet M, Matsuura T (2001) Preparation and characterization of polyvinylidene fluoride membranes for membrane distillation. Ind Eng Chem Res 40(24):5710–5718
33. Tomaszewska M (1996) Preparation and properties of flat-sheet membranes from poly(vinylidene fluoride) for membrane distillation. Desalination 104:1–11
34. Sukitpaneenit P, Chung TS (2009) Molecular elucidation of morphology and mechanical properties of PVDF hollow fiber membranes from aspects of phase inversion, crystallization and rheology. J Membr Sci 340(1–2):192–205
35. Zhai Y, Wang N, Mao X, Si Y, Jianyong Yu, Al-Deyab SS, El-Newehy M, Ding B (2014) Sandwich-structured PVdF/PMIA/PVdF nanofibrous separators with robust mechanical strength and thermal stability for lithium ion batteries. J Mater Chem A2(35):14511–14518
36. Hou D, Dai G, Wang J, Fan H, Zhang L, Luan Z (2012) Preparation and characterization of PVDF/nonwoven fabric flat-sheet composite membranes for desalination through direct contact membrane distillation. Sep Purif Technol 101:1–10
37. Wang YK, Chung TS, Gryta M (2008) Hydrophobic PVDF hollow fiber membranes with narrow pore size distribution and ultra-thin skin for the fresh water production through membrane distillation. Chem Eng Sci 63:2587–2594
38. Bonyadi S, Chung T-S (2009) Highly porous and macrovoid-free PVDF hollow fiber membranes for membrane distillation by a solvent-dope solution co-extrusion approach. J Membr Sci 331(1–2):66–74
39. Bonyadi S, Chung TS (2007) Flux enhancement in membrane distillation by fabrication of dual layer hydrophilic–hydrophobic hollow fiber membranes. J Membr Sci 306(1–2):134–146
40. Edwie F, Teoh MM, Chung T-S (2012) Effects of additives on dual-layer hydrophobic–hydrophilic PVDF hollow fiber membranes for membrane distillation and continuous performance. Chem Eng Sci 68(1):567–578
41. Fan H, Peng Y (2012) Application of PVDF membranes in desalination and comparison of the VMD and DCMD processes. Chem Eng Sci 79:94–102
42. Cao X, Ma J, Shi X, Ren Z (2006) Effect of TiO_2 nanoparticle size on the performance of PVDF membrane. Appl Surf Sci 253(4):2003–2010
43. Dong C, He G, Li H, Zhao R, Han Y, Deng Y (2012) Antifouling enhancement of poly (vinylidene fluoride) microfiltration membrane by adding Mg (OH)2 nanoparticles. J Membr Sci 387:40–47
44. Yu LY, Xu ZL, Shen HM, Yang H (2009) Preparation and characterization of PVDF–SiO_2 composite hollow fiber UF membrane by sol-gel method. J Membr Sci 337:257–265
45. Oh SJ, Kim N, Lee YT (2009) Preparation and characterization of PVDF/TiO_2 organic–inorganic composite membranes for fouling resistance improvement. J Membr Sci 345(1–2):13–20
46. Wang P, Ma J, Wang Z, Shi F, Liu Q (2012) Enhanced separation performance of PVDF/PVP-g-MMT nanocomposite ultrafiltration membrane based on the NVP-grafted polymerization modification of montmorillonite (MMT). Langmuir 28(10):4776–4786
47. Wei X, Wang Z, Wang J, Wang S (2012) A novel method of surface modification to polysulfone ultrafiltration membrane by preadsorption of citric acid or sodium bisulfite. Membr Water Treat 3(1):35–49
48. Chiang Y-C, Chang Y, Higuchi A, Chen W-Y, Ruaan R-C (2009) Sulfobetaine-grafted poly (vinylidene fluoride) ultrafiltration membranes exhibit excellent antifouling property. J Membr Sci 339(1–2):151–159

49. Qu X, Brame J, Li Q, Alvarez PJJ (2013) Nanotechnology for a safe and sustainable water supply: enabling integrated water treatment and reuse. Acc Chem Res 46(3):834–843
50. Ying Y, Ying W, Li Q, Meng D, Ren G, Yan R, Peng X (2017) Recent advances of nanomaterial-based membrane for water purification. Appl Mater Today 7:144–158
51. Razmjou A, Mansouri J, Chen V (2011) The effects of mechanical and chemical modification of TiO2 nanoparticles on the surface chemistry, structure and fouling performance of PES ultrafiltration membranes. J Membr Sci 378(1–2):73–84
52. Shi F, Ma Y, Ma J, Wang P, Sun W (2013) Preparation and characterization of PVDF/TiO2 hybrid membranes with ionic liquid modified nano-TiO2 particles. J Membr Sci 427:259–269
53. Zhang S, Wang R, Zhang S, Li G, Zhang Y (2014) Treatment of wastewater containing oil using phosphorylated silica nanotubes (PSNTs)/polyvinylidene fluoride (PVDF) composite membrane. Desalination 332(1):109–116
54. Yu L-Y, Shen H-M, Xu Z-L (2009) PVDF–TiO$_2$ composite hollow fiber ultrafiltration membranes prepared by TiO2 sol–gel method and blending method. J Appl Polym Sci 113(3):1763–1772
55. Chae S-R, Hotze EM, Wiesner MR (2014) Possible applications of fullerene nanomaterials in water treatment and reuse. Nanotechnology applications for clean water. William Andrew Publishing, pp 329–338
56. Mubarak NM, Sahu JN, Abdullah EC, Jayakumar NS (2014) Removal of heavy metals from wastewater using carbon nanotubes. Sep Purif Rev 43(4):311–338
57. Zhang X, Pan B, Yang K, Zhang D, Hou J (2010) Adsorption of sulfamethoxazole on different types of carbon nanotubes in comparison to other natural adsorbents. J Environ Sci Health Part A 45(12):1625–1634
58. Lu C, Su F (2007) Adsorption of natural organic matter by carbon nanotubes. Sep Purif Technol 58(1):113–121
59. Wu H, Tang B, Wu P (2010) Novel ultrafiltration membranes prepared from a multi-walled carbon nanotubes/polymer composite. J Membr Sci 362(1–2):374–383
60. Ajmani GS, Goodwin D, Marsh K, Howard Fairbrother D, Schwab KJ, Jacangelo JG, Huang H (2012) Modification of low pressure membranes with carbon nanotube layers for fouling control. Water Res 46(17):5645–5654
61. Nair RR, Wu HA, Jayaram PN, Grigorieva IV, Geim AK (2012) Unimpeded permeation of water through helium-leak–tight graphene-based membranes. Science 335(6067):442–444
62. Zhao C, Xu X, Chen J, Yang F (2013) Effect of graphene oxide concentration on the morphologies and antifouling properties of PVDF ultrafiltration membranes. J Environ Chem Eng 1(3):349–354
63. Goh PS, Ismail AF, Ng BC (2013) Carbon nanotubes for desalination: performance evaluation and current hurdles. Desalination 308:2–14
64. Madaeni SS, Zinadini S, Vatanpour V (2013) Preparation of superhydrophobic nanofiltration membrane by embedding multiwalled carbon nanotube and polydimethylsiloxane in pores of microfiltration membrane. Sep Purif Technol 111:98–107
65. Gethard K, Sae-Khow O, Mitra S (2011) Water desalination using carbon-nanotube-enhanced membrane distillation. ACS Appl Mater Interfaces 3(2):110–114
66. Silva TLS, Morales-Torres S, Figueiredo JL, Silva AMT (2015) Multi-walled carbon nanotube/PVDF blended membranes with sponge-and finger-like pores for direct contact membrane distillation. Desalination 357:233–245
67. Wang Y, Zhu J, Huang H, Cho H-H (2015) Carbon nanotube composite membranes for microfiltration of pharmaceuticals and personal care products: Capabilities and potential mechanisms. J Membr Sci 479:165–174
68. Ma J, Zhao Y, Zhiwei X, Min C, Zhou B, Li Y, Li B, Niu J (2013) Role of oxygen-containing groups on MWCNTs in enhanced separation and permeability performance for PVDF hybrid ultrafiltration membranes. Desalination 320:1–9
69. Zhao X, Ma J, Wang Z, Wen G, Jiang J, Shi F, Sheng L (2012) Hyperbranched-polymer functionalized multi-walled carbon nanotubes for poly (vinylidene fluoride) membranes: from dispersion to blended fouling-control membrane. Desalination 303:29–38

70. Wang Z, Hairong Yu, Xia J, Zhang F, Li F, Xia Y, Li Y (2012) Novel GO-blended PVDF ultrafiltration membranes. Desalination 299:50–54
71. Chang X, Wang Z, Quan S, Yanchao X, Jiang Z, Shao L (2014) Exploring the synergetic effects of graphene oxide (GO) and polyvinylpyrrodione (PVP) on poly (vinylylidenefluoride)(PVDF) ultrafiltration membrane performance. Appl Surf Sci 316:537–548
72. Zhao C, Xiaochen X, Chen J, Wang G, Yang F (2014) Highly effective antifouling performance of PVDF/graphene oxide composite membrane in membrane bioreactor (MBR) system. Desalination 340:59–66
73. Xu T, Xie CS (2003) Tetrapod-like nano-particle ZnO/acrylic resin composite and its multi-function property. Prog Org Coat 46(4):297–301
74. Vigneshwaran N, Kumar S, Kathe AA, Varadarajan PV, Prasad V (2006) Functional finishing of cotton fabrics using zinc oxide–soluble starch nanocomposites. Nanotechnology 17(20):5087
75. Zhang C-H, Huang Y-D, Yuan W-J, Zhang J-N (2011) UV aging resistance properties of PBO fiber coated with nano-ZnO hybrid sizing. J Appl Polym Sci 120(4):2468–2476
76. Liang S, Xiao K, Mo, Y, Huang, X (2012) A novel ZnO nanoparticle blended polyvinylidene fluoride membrane for anti-irreversible fouling. J Memb Sci 394–395, 184–192
77. Enesca A, Isac L, Duta A (2013) Hybrid structure comprised of SnO_2, ZnO and Cu_2S thin film semiconductors with controlled optoelectric and photocatalytic properties. Thin Solid Films 542:31–37
78. Wöll C (2007) The chemistry and physics of zinc oxide surfaces. Prog Surf Sci 82(2–3):55–120
79. Wang L, Zheng Y, Li X, Dong W, Tang W, Chen B, Li C, Li X, Zhang T, Xu W (2011) Nanostructured porous ZnO film with enhanced photocatalytic activity. Thin Solid Films 519(16):5673–5678
80. Zhou X, Guo X, Ding W, Chen Y (2008) Superhydrophobic or superhydrophilic surfaces regulated by micro-nano structured ZnO powders. Appl Surf Sci 255(5):3371–3374
81. Kwak G, Seol M, Tak Y, Yong K (2009) Superhydrophobic ZnO nanowire surface: chemical modification and effects of UV irradiation. J Phys Chem C 113(28):12085–12089
82. Bottino A, Camera-Roda G, Capannelli G, Munari S (1991) The formation of microporous polyvinylidene difluoride membranes by phase separation. J Membr Sci 57(1):1–20
83. Xin Y, Fujimoto T, Uyama H (2012) Facile fabrication of polycarbonate monolith by non-solvent induced phase separation method. Polymer 53(14):2847–2853
84. Susanto H, Ulbricht M (2009) Characteristics, performance and stability of polyethersulfone ultrafiltration membranes prepared by phase separation method using different macromolecular additives. J Membr Sci 327(1–2):125–135
85. Tang E, Cheng G, Ma X (2006) Preparation of nano-ZnO/PMMA composite particles via grafting of the copolymer onto the surface of zinc oxide nanoparticles. Powder Technol 161(3):209–214
86. Hong J, He Y (2012) Effects of nano sized zinc oxide on the performance of PVDF microfiltration membranes. Desalination 302:71–79
87. Zhang X, Wang Y, Liu Y, Junli X, Han Y, Xinxin X (2014) Preparation, performances of PVDF/ZnO hybrid membranes and their applications in the removal of copper ions. Appl Surf Sci 316:333–340
88. Khan SA, Noreen F, Kanwal S, Hussain G (2017) Comparative synthesis, characterization of Cu-doped ZnO nanoparticles and their antioxidant, antibacterial, antifungal and photocatalytic dye degradation activities. Dig J Nanomater Biostruct 12(3):877–889
89. Dong C, Cairney J, Sun Q, Maddan OL, He G, Deng Y (2010) Investigation of Mg (OH) 2 nanoparticles as an antibacterial agent. J Nanoparticle Res 12(6):2101–2109
90. Dong C, He G, Li H, Zhao R, Han Y, Deng Y (2012) Antifouling enhancement of poly (vinylidene fluoride) microfiltration membrane by adding Mg (OH) 2 nanoparticles. J Membr Sci 387:40–47
91. Dong C, Song D, Cairney J, Maddan OL, He G, Deng Y (2011) Antibacterial study of Mg (OH) 2 nanoplatelets. Mater Res Bull 46(4):576–582

92. Huang Z-Q, Chen K, Li S-N, Yin X-T, Zhang Z, Xu H-T (2008) "Effect of ferrosoferric oxide content on the performances of polysulfone–ferrosoferric oxide ultrafiltration membranes. J Membr Sci 315(1–2):164–171
93. Genne I, Kuypers S, Leysen R (1996) Effect of the addition of ZrO2 to polysulfone based UF membranes. J Membr Sci 113(2):343–350
94. Zhang Y, Li H, Lin J, Li R, Liang X (2006) Preparation and characterization of zirconium oxide particles filled acrylonitrile-methyl acrylate-sodium sulfonate acrylate copolymer hybrid membranes. Desalination 192(1–3):198–206
95. Wu C, Zhang S, Liu C, Yang D, Jian X (2008) Preparation, characterization and performance of thermal stable poly (phthalazinone ether amide) UF membranes. J Membr Sci 311(1–2):360–370
96. Ma J, Wang Z, Pan M, Guo Y (2009) A study on the multifunction of ferrous chloride in the formation of poly (vinylidene fluoride) ultrafiltration membranes. J Membr Sci 341(1–2):214–224
97. Yan L, Hong S, Li M, Li YS (2009) Application of the Al2O3–PVDF nanocomposite tubular ultrafiltration (UF) membrane for oily wastewater treatment and its antifouling research. Sep Purif Technol 66(2):347–352
98. Yan L, Li YS, Xiang CB (2005) Preparation of poly (vinylidene fluoride)(pvdf) ultrafiltration membrane modified by nano-sized alumina (Al_2O_3) and its antifouling research. Polymer 46(18):7701–7706
99. Liu F, Moghareh Abed MR, Li K (2011) Preparation and characterization of poly (vinylidene fluoride)(PVDF) based ultrafiltration membranes using nano γ-Al2O3. J Membr Sci 366(1–2):97–103
100. Li X, Pang R, Li J, Sun X, Shen J, Han W, Wang L (2013) In situ formation of Ag nanoparticles in PVDF ultrafiltration membrane to mitigate organic and bacterial fouling. Desalination 324:48–56
101. Shi H, Liu F, Xue L, Huafeng L, Zhou Q (2014) Enhancing antibacterial performances of PVDF hollow fibers by embedding Ag-loaded zeolites on the membrane outer layer via co-extruding technique. Compos Sci Technol 96:1–6
102. Shi H, Liu F, Xue L (2013) Fabrication and characterization of antibacterial PVDF hollow fibre membrane by doping Ag-loaded zeolites. J Membr Sci 437:205–215
103. Choi H, Sofranko AC, Dionysiou DD (2006) Nanocrystalline TiO2 photocatalytic membranes with a hierarchical mesoporous multilayer structure: synthesis, characterization, and multifunction. Adv Funct Mater 16(8):1067–1074
104. Madaeni SS, Zinadini S, Vatanpour V (2011) A new approach to improve antifouling property of PVDF membrane using in situ polymerization of PAA functionalized TiO_2 nanoparticles. J Membr Sci 380(1–2):155–162
105. Oh SJ, Kim N, Lee YT (2009) Preparation and characterization of PVDF/TiO2 organic–inorganic composite membranes for fouling resistance improvement. J Membr Sci 345(1–2):13-20
106. Mohd Yatim NS, Boon Seng O (2016) Performance of chemically modified TiO2-poly (vinylidene fluoride) DCMD for nutrient isolation and its antifouling properties. J Membr Sci Res 2(4):163–168
107. Farner Budarz J, Turolla A, Piasecki AF, Bottero J-Y, Antonelli M, Wiesner MR (2017) Influence of aqueous inorganic anions on the reactivity of nanoparticles in TiO_2 photocatalysis. Langmuir 33(11):2770–2779
108. Long M, Brame J, Qin F, Bao J, Li Q, Alvarez PJJ (2017) Phosphate changes effect of humic acids on TiO2 photocatalysis: from inhibition to mitigation of electron–hole recombination. Environ Sci Technol 51(1):514–521
109. Vequizo JJM, Matsunaga H, Ishiku T, Kamimura S, Ohno T, Yamakata A (2017) Trapping-induced enhancement of photocatalytic activity on brookite TiO2 powders: comparison with anatase and rutile TiO2 powders. ACS Catal 7(4):2644–2651
110. Lee JH, Kang M, Choung S-J, Ogino K, Miyata S, Kim M-S, Park J-Y, Kim J-B (2004) The preparation of TiO2 nanometer photocatalyst film by a hydrothermal method and its sterilization performance for Giardia lamblia. Water Res 38(3):713–719

111. Matthews RW (1987) Solar-electric water purification using photocatalytic oxidation with TiO2 as a stationary phase. Solar Energy 38(6):405–413
112. Necula BS, Lidy EF, Sebastian AJ, Zaat IA, Jurek D (2009) In vitro antibacterial activity of porous TiO2–Ag composite layers against methicillin-resistant Staphylococcus aureus. Acta Biomater 5(9):3573–3580
113. Gao M, Zhu L, Ong WL, Wang J, Ho GW (2015) Structural design of TiO 2-based photocatalyst for H 2 production and degradation applications. Catal Sci Technol 5(10):4703–4726
114. Tavakolmoghadam M, Mohammadi T, Hemmati M (2016) Preparation and characterization of PVDF/TiO2 composite ultrafiltration membranes using mixed solvents. Membr Water Treat 7(5):377–401
115. Méricq J-P, Mendret J, Brosillon S, Faur CJCES (2015) High performance PVDF-TiO2 membranes for water treatment. Chem Eng Sci 123:283–291
116. Zhang W, Zhang Y, Fan R, Lewis R (2016) A facile TiO_2/PVDF composite membrane synthesis and their application in water purification. J Nanoparticle Res 18(1):31
117. Huang X, Zhang J, Wang W, Liu Y, Zhang Z, Li L, Fan W (2015) Effects of PVDF/SiO2 hybrid ultrafiltration membranes by sol–gel method for the concentration of fennel oil in herbal water extract. RSC Adv 5(24):18258–18266
118. Hashim NA, Liu Y, Li K (2011) Preparation of PVDF hollow fiber membranes using SiO2 particles: the effect of acid and alkali treatment on the membrane performances. Ind Eng Chem Res 50(5):3035–3040
119. Wu H, Mansouri J, Chen V (2013) Silica nanoparticles as carriers of antifouling ligands for PVDF ultrafiltration membranes. J Membr Sci 433:135–151
120. Wang H, Zhao X, He C (2016) Enhanced antifouling performance of hybrid PVDF ultrafiltration membrane with the dual-mode SiO_2-g-PDMS nanoparticles. Sep Purif Technol 166:1–8
121. Park SY, Chung JW, Chae YK, Kwak S-Y (2013) Amphiphilic thiol functional linker mediated sustainable anti-biofouling ultrafiltration nanocomposite comprising a silver nanoparticles and poly (vinylidene fluoride) membrane. ACS Appl Mater Interfaces 5(21):10705–10714
122. Li J-H, Shao X-S, Zhou Q, Li M-Z, Zhang Q-Q (2013) The double effects of silver nanoparticles on the PVDF membrane: surface hydrophilicity and antifouling performance. Appl Surf Sci 265:663–670
123. Alpatova A, Meshref M, McPhedran KN, El-Din MG (2015) Composite polyvinylidene fluoride (PVDF) membrane impregnated with Fe2O3 nanoparticles and multiwalled carbon nanotubes for catalytic degradation of organic contaminants. J Membr Sci 490:227–235
124. Lijun X, Li Z, Jingyuan C, Junchi L, Ye T, Yingjie Z (2013) Characteristics and preparation of PVDF catalytic membrane modified by nano-TiO_2/Fe^{3+}. Open Mater Sci J 7(1)
125. Bai H, Liu Z, Sun DD (2012) A hierarchically structured and multifunctional membrane for water treatment. Appl Catal B Environ 111:571–577
126. Takeuchi M, Sakamoto K, Martra G, Coluccia S, Anpo M (2005) Mechanism of photoinduced superhydrophilicity on the TiO2 photocatalyst surface. J Phys Chem B 109(32):15422–15428
127. Li J-H, Yan B-F, Shao X-S, Wang S-S, Tian H-Y, Zhang Q-Q (2015) Influence of Ag/TiO2 nanoparticle on the surface hydrophilicity and visible-light response activity of polyvinylidene fluoride membrane. Appl Surf Sci 324:82–89
128. Yu B, Leung KM, Guo Q, Lau WM, Yang J (2011) Synthesis of Ag–TiO2 composite nano thin film for antimicrobial application. Nanotechnology 22(11):115603
129. Safarpour M, Khataee A, Vatanpour V (2014) Preparation of a novel polyvinylidene fluoride (PVDF) ultrafiltration membrane modified with reduced graphene oxide/titanium dioxide (TiO2) nanocomposite with enhanced hydrophilicity and antifouling properties. Ind Eng Chem Res 53(34):13370–13382
130. Xu Z, Tengfei W, Shi J, Teng K, Wang W, Ma M, Li J, Qian X, Li C, Fan J (2016) Photocatalytic antifouling PVDF ultrafiltration membranes based on synergy of graphene oxide and TiO_2 for water treatment. J Membr Sci 520:281–293
131. Kim E-S, Hwang G, El-Din MG, Liu Y (2012) Development of nanosilver and multi-walled carbon nanotubes thin-film nanocomposite membrane for enhanced water treatment. J Membr Sci 394:37–48

132. Tahir K, Ahmad A, Li B, Nazir S, Ullah Khan A, Nasir T, Khan ZUH, Naz R, Raza M (2016) Visible light photo catalytic inactivation of bacteria and photo degradation of methylene blue with Ag/TiO$_2$ nanocomposite prepared by a novel method. J Photochem Photobiol B Biol 162:189–198
133. Hirakawa T, Kamat PV (2004) Photoinduced electron storage and surface plasmon modulation in Ag@ TiO$_2$ clusters. Langmuir 20(14):5645–5647
134. Takai A, Kamat PV (2011) Capture, store, and discharge. Shuttling photogenerated electrons across TiO$_2$–silver interface. ACS nano 5(9):7369–7376
135. Yang D, Sun Y, Tong Z, Tian Y, Li Y, Jiang Z (2015) Synthesis of Ag/TiO2 nanotube heterojunction with improved visible-light photocatalytic performance inspired by bioadhesion. J Phys Chem C 119(11):5827–5835
136. Yao T, Shi L, Wang H, Wang F, Wu J, Zhang X, Sun J, Cui T (2016) A simple method for the preparation of TiO$_2$/Ag-AgCl@ polypyrrole composite and its enhanced visible-light photocatalytic activity. Chem Asian J 11(1):141–147
137. Prakash J, Kumar P, Harris RA, Swart C, Neethling JH, Janse van Vuuren A, Swart HC (2016) Synthesis, characterization and multifunctional properties of plasmonic Ag–TiO$_2$ nanocomposites. Nanotechnology 27(35):355707
138. Bian J, Yang Q, Fazal R, Li X, Sun N, Jing L (2016) Accepting excited high-energy-level electrons and catalyzing H2 evolution of dual-functional Ag-TiO2 modifier for promoting visible-light photocatalytic activities of nanosized oxides. J Phys Chem C 120(22):11831–11836
139. Chen J-J, Wu JCS, Wu PC, Tsai DP (2011) Plasmonic photocatalyst for H2 evolution in photocatalytic water splitting. J Phys Chem C 115(1):210–216
140. Ren R, Wen Z, Cui S, Hou Y, Guo X, Chen J (2015) Controllable synthesis and tunable photocatalytic properties of Ti^3+-doped TiO$_2$. Sci Rep 5(1):1–11
141. Kulkarni RM, Malladi RS, Hanagadakar MS, Doddamani MR, Bhat UK (2016) Ag-TiO$_2$ nanoparticles for photocatalytic degradation of lomefloxacin. Desalination Water Treat 57(34):16111–16118
142. Damodar RA, You S-J, Chou H-H (2009) Study the self-cleaning, antibacterial and photocatalytic properties of TiO$_2$ entrapped PVDF membranes. J Hazardous Mater 172(2–3):1321–1328
143. Yang GQ, Li ZH, Lin ZX, Wang XX, Liu P, Fu XZ (2005) Investigation of TiO2 photocatalytic degradation of bovine serum albumin. Guang pu xue yu guang pu fen xi Guang pu 25(8):1309–1311
144. Li Z-K, Lang W-Z, Miao W, Yan X, Guo Y-J (2016) Preparation and properties of PVDF/SiO$_2$@ GO nanohybrid membranes via thermally induced phase separation method. J Membr Sci 511:151–161
145. Xu Z, Zhang J, Shan M, Li Y, Li B, Niu J, Zhou B, Qian X (2014) Organosilane-functionalized graphene oxide for enhanced antifouling and mechanical properties of polyvinylidene fluoride ultrafiltration membranes. J Membr Sci 458:1–13
146. Zhu Z, Jiang J, Wang X, Huo X, Yawei X, Li Q, Wang L (2017) Improving the hydrophilic and antifouling properties of polyvinylidene fluoride membrane by incorporation of novel nanohybrid GO@ SiO$_2$ particles. Chem Eng J 314:266–276

Jnyana Ranjan Mishra presently works as a lecturer in Chemistry in Government Polytechnic Kalahandi, Odisha. He has completed his M. Phil degree from Utkal university, Odisha and joined as a research fellow in the Laboratory for Advanced Research in Polymeric Materials (LARPM), the R&D wing of Central Institute of Petrochemicals Engineering & Technology (CIPET) in 2016. He has also registered and continuing for Ph.D. programme under Utkal University from 2017. His area of research interests includes the synthesis of multifunctional polymeric membrane and its modification to induce various advanced properties. The modification specifically includes the functionalization as well as hybridization of multiple nanoparticles and their incorporation into various polymers for the formulation of different polymer based membranes with outstanding properties.

Sukanya Pradhan, Dr., is currently working as a Research scientist at LARPM- Central Institute of Petrochemicals Engineering & Technology (CIPET), Ministry of Chemicals and Fertilizers, Department of Chemicals and Petrochemicals, Govt. of India She has received her Ph.D. in Polymers and plastics technology from Anna University, India in 2019. She has published over 20 papers in peer reviewed journals, book chapters and international conferences proceedings. Her research interest includes Polymer synthesis, Polymer characterization coatings and adhesive, Polymer composites, Biodegradable polymers, feminine hygiene. She has presented many invited and contributed talks at international conferences. She has over 5 years of teaching experience at undergraduate, graduate and post-graduate levels. She has active participation in applied innovative teaching methods to encourage student learning objectives. She is also a life member of various professional bodies (Indian Science Congress Association, Orissa Chemical Society).

Smita Mohanty, Dr., is the Principal scientist at School for Advanced Research in Polymers (SARP):LARPM-Central Institute of Petrochemicals Engineering & Technology (CIPET), Ministry of Chemicals and Fertilizers, Department of Chemicals and Petrochemicals, Govt. of India. She has over 15 years of experience, with research interests in polymers, coatings, adhesives, paints and composites, biomaterials, and functionalization of polymeric materials. She has published over 160 articles in international peer reviewed journals and is the winner of the National Awards by the Department of Chemicals and Petrochemicals, Govt. of India, and Young Scientist Award by Department of Science and Technology (DST), Govt. of India. She was a recipient of Outstanding Research faculty 2017–18 in Materials Science discipline through Scopus by Careers 360 and has 7 Indian patents to her credit.

Sanjay K. Nayak, Prof. (Dr.) is the Director General of Central Institute of Petrochemicals Engineering & Technology (CIPET), Ministry of Chemicals and Fertilizers, Department of Chemicals and Petrochemicals, Govt. of India. He has been known for his pioneering work on Ecofriendly Composites, Nanocomposites, Bio Polymers and value addition of Plastics from E-waste. He is the author of more than 450 Research Papers, 30 Books/Book Chapters and has more than 50 Patents & Registered Product Designs to his credit. He is also the recipient of 'Distinguished Scientist Award' in appreciation to his contribution to Polymer Science & Technology by Asian Polymer Association (APA), Honoris Causa (Honorary Doctorate) conferred by Utkal University (2017): for Nation Building and Contribution towards Science and Technology. 'Commendable Faculty Award' for Research in Material Science & High Citation Index (as per Scopus) and National Awards for Technology Innovations in Petrochemicals and Downstream Plastics Processing Industry by Govt. of India. He has been also recognized as 2% of Global Scientists by Stanford University, USA.

Electrospun Nanomaterials: Applications in Water Contamination Remediation

Nimet Bölgen and Ashok Vaseashta

Abstract Recent growth in industrialization and globalization has resulted in excessive water-stress in our water resources. Pollution in water pollution comes from anthropogenic activities and level of pollution and emerging contaminants have become an immediate challenge to the living organisms. Water resources now consist of metals, inorganics, pathogens, returned pharmaceuticals, dyes, pesticides, toxic industrial chemicals, organic pollutants, and other new and emerging contaminants, which need to be sufficiently treated before disposal back into the natural body of water. Several membrane-based water purification technologies have attracted attention and interest in the last decade. One of the methodologies that is applied commonly is by using membranes. Among various membranes, electrospun nanofibrous membranes offer unique properties, such as high and tunable porosity, high surface area, tunability in the composition and structure, functionalization ability and good mechanical behaviour, to be used in water purification compared to conventional membranes. This chapter reviews the past achievements and recent research trends of electrospun nanofibrous materials employed for water remediation.

Keywords Electrospinning · Nanofibers · Membranes · Wastewater · Water purification

---

N. Bölgen (✉)
Engineering Faculty, Chemical Engineering Department, Mersin University, 33343 Mersin, Turkey
e-mail: nimet@mersin.edu.tr

A. Vaseashta
International Clean Water Institute, Manassas, VA, USA

Institute of Electronic Engineering and Nanotechnologies "D. Ghitu", ASM, Chisinau, Moldova

Institute of Biomedical and Nanotechnologies, Riga Technical University, Riga, Latvia

A. Vaseashta
e-mail: prof.vaseashta@ieee.org

© Springer Nature Switzerland AG 2021
A. Vaseashta and C. Maftei (eds.), *Water Safety, Security and Sustainability*,
Advanced Sciences and Technologies for Security Applications,
https://doi.org/10.1007/978-3-030-76008-3_8

1 Introduction

Pollution of freshwater resources is a major challenge that humanity is facing in the twenty-first century [1]. Water contamination is caused by disposal of wastewater from industrial wastes, untreated domestic sewage, agricultural activities and environmental influences [2]. In developing and industrialized countries, introduction of contaminants such as chemicals, pesticides, particulates, or biological materials into water sources threatens human health and natural ecological system. The progressive water pollution and hence the scarcity of clean water supply constitute a major threat that increases public health and environmental concerns. Every year millions of people, the majority of which is children, die from sicknesses transmitted through waterborne bacteria and enteric viruses due to insufficient sanitation [3, 4]. In this regard, significant research is directed into the development of novel approaches to disinfect and decontaminate water. A necessity of water purification technology, due to population growth, increasing industrialization and agricultural activities, is attracting immense interest in the development of more efficient and cost-effective systems that can be used in treatment of polluted water.

Water treatment processes are conducted for removing contaminants in water. Several water purification processes such as sedimentation, adsorption, precipitation, disinfection, aerobic and anaerobic digestion chemical oxidation and membrane filtration have attracted the interest of researchers in the past [5–7]. Among these technologies, membrane-based systems gained a great deal of attention of both academic and industries, having advantages such as high separation efficacy, scalability, high selectivity, low power consumption, and high stability [6]. Membrane technologies can be categorized according to separation principles and membrane properties such as microfiltration, ultrafiltration, nanofiltration, reverse osmosis, forward osmosis, and pressure retarded osmosis [7–11]. Membranes play a key role in the membrane-based contamination separation processes. Polymeric membranes are the most widely used materials for water filtration applications due to their favorable physical and chemical properties. Polymeric membranes can be fabricated by different techniques such as solvent casting, phase inversion, stretching, track-etching, interfacial polymerization, layer-by-layer deposition, and electrospinning [12]. Among them, electrospinning is the most versatile, efficient and simple technique, which has been applied to produce nanofibrous membranes. Electrospun nanofibers have unique and interesting characteristics, such as small diameters with excellent pore connectivity, high surface area to volume ratio, high orientation of nanofibers, good mechanical properties, good water permeability, and have garnered much consideration for water treatment applications [13]. Electrospun nanofibers were employed for various water applications, based on their extraordinary permeability, selectivity, and mechanical stability. Furthermore, it is feasible to fabricate multifunctional membranes by embedding functional materials into nanofibers or loading target groups on the electrospun nanofiber surface. With this perception, this chapter encompasses a review of the recent progress in fabrication and modification of electrospun nanofibers and their application in water purification.

2 Nanofiber Fabrication by Electrospinning Technique

Electrospinning is a versatile technique that has known to produce nanofibers with diameters at nanoscale [14–16]. A typical diagram of the electrospinning setup is shown in Fig. 1. Mainly there are three essential parts in a simple electrospinning system: a high voltage power supply, a syringe connected to a needle (spinneret) and a metal collector. In the electrospinning process, polymeric nanofibers are prepared by applying a high voltage to create an electrically charged jet of polymer solution from a spinneret. The voltage is applied gradually from a variable source, which causes charging of the solution.

When the applied voltage overcomes the surface tension of the polymer solution at a critical voltage, a Taylor cone appears and the polymer jet undergoes an elongation process, which allows the jet to reach the collector as a fiber. Before reaching the collector screen, the solvent evaporates, and the polymer solidifies and is collected as an interconnected web of small fibers [17].

2.1 Effects of Different Parameters on Electrospinning

There are several intrinsic, operational and surrounding process parameters affecting the size and morphology of nanofibers and the resultant membrane pore size such as viscosity, surface tension and conductivity of the solution, solvent type, polymer concentration, applied voltage, working distance between spinneret and collector, spinneret radius, and ambient temperature and humidity.

In electrospinning, process optimization entails the selection of appropriate polymer-solvent and process conditions in order to control the morphology and structure of the resulting nanofibers. The selection of the polymer depends mostly on the end-application. High molecular weight polymers are selected for electrospinning

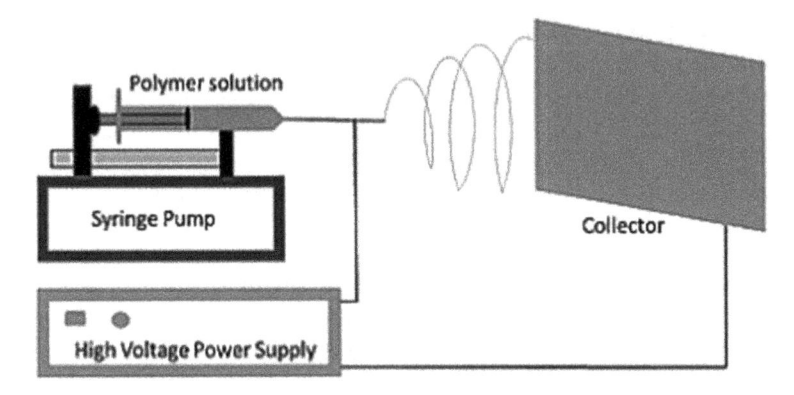

Fig. 1 Diagram of electrospinning set-up

as they provide sufficient chain entanglement during elongation of the polymer jet. Solvents that dissolve the polymer need to have good conductivity and less surface tension to facilitate successful fabrication of nanofibers [18]. Volatility of the solvent also affects the successful production of nanofibers. Solvents with too high volatility cause blockage of the needle due to rapid solvent evaporation, while those with too low volatility will prevent fiber drying after reaching the collector. For a given pair of polymer-solvent, there exists a threshold of appropriate polymer concentration to produce homogeneously distributed nanofibers without bead formation. Bead formation starts to occur on nanofibers, when the viscosity of the polymer solution is too low, which causes heterogeneity of the nanofibrous structure. Increase in polymer concentration results in fabrication of nanofibers without bead formation. Similarly, solutions containing higher molecular weight polymer results in higher viscosity and higher chain entanglement, hence the formation of beads is prevented. In addition, as the polymer concentration increases, jet elongation becomes more difficult and slow, which results in increased nanofiber diameter and pore size.

It is revealed that solution surface tension plays critical role in determining the range of concentrations that influence the morphologies of electrospun nanofibers [19]. High surface tension causes resistance to the flow of polymer solution and stretching of the jet. High voltage should be applied to overcome the surface tension of the solution for the initiation of the jet. High surface tension may also inhibit the electrospinning process due to the instability of the jet and cause the formation of beaded fibers [20]. Although by lowering the surface tension of the spinning solution, electrospinning can occur at lower electrical field, not all solvents with low surface tension will always be suitable for electrospinning. A polymer-solvent system with high conductivity and low surface tension is mostly suitable for the successful fabrication of nanofibers in order to adjust the overall surface tension of the solution.

Electrical conductivity of a polymer-solvent system significantly affects the spinnability and it is dependent on the polymer and solvent type, and as well as the existence of the ionizable salts [21]. The increase of the electrical conductivity of the solution promotes high electric charge, thereby inducing stretching of a polymer jet. It has been found that an increase in conductivity leads to a significant decrease in the diameter of the electrospun nanofibers. It was observed that the jet radius was inversely proportional to the cube root of the electrical conductivity of the solution. On the other hand, low conductivity of the solution causes weak electrical force, thereby resulting in insufficient elongation of a jet to produce uniform nanofibers, and bead formation may occur due to jet break up, forming droplets rather than fibers [22]. With the use of ionic salts, the uniformity of the fibers is increased and bead-free fibers are obtained. However, although an increase in the conductivity enhances electrospinning, its excessive increase can cause instability of the Taylor cone, which consequently generates beads. Basically, in order to obtain electrospun nanofibers composed of uniform and bead-free fibers the upper and lower boundaries of the conductivity of the polymer solution should be optimized to an appropriate value [23].

In the electrospinning process, one of the key parameters is the magnitude of the applied voltage to the solution. The electric field created between the syringe needle tip and the collector induces the necessary charges on the solution to initiate electrospinning [24]. The necessary charges are transferred to the polymer solution, after attainment of threshold voltage, jet is ejected from the needle and attracted to the collector, forming the fibers. The critical voltage to generate jets varies for each polymer-solvent system and is dependent on the polymer solution concentration/viscosity. Some researchers have suggested that the formation of thinner nanofibers is favored under a higher applied voltage, which was due to the increase in the electrostatic repulsive force on the fluid jet which provides the narrowing of the fiber diameter [23]. Other authors have reported that higher applied voltage can eject more polymer solution which facilitates the formation of a larger diameter fiber [25]. Thus, voltage influences fiber diameter and an optimum voltage is required to obtain nanofibers with decreased diameters.

An optimum flow rate of the polymer solution from the syringe is necessary as it influences the jet velocity and solution transfer. The flow rate must provide sufficient time for the solvent evaporation [26]. High flow rate may result in beaded nanofibers due to unavailability of sufficient evaporation of solvent before reaching the collector. It has been observed that high flow rates were associated with large Taylor cone, resulting in increased fiber diameter and pore size.

Another parameter that affects the fibers diameters and morphology is tip-to-collector distance. An optimum distance is required for the fibers to dry before reaching the collector. Bead formation occurs if the distance is too high or too low. The effect of tip-to-collector distance on morphology is minor, while the effect of other parameters is moderate [25].

Besides intrinsic and operational parameters, ambient temperature and humidity affect the morphology of electrospun nanofibers. It has been reported that at higher surrounding temperature nanofibers with smoother surfaces and smaller diameter can be obtained due to the reduction of surface tension and viscosity [27]. On the other hand, increase of temperature may accelerate the rate of evaporation of the solvent, which solidify the fibers and terminate the electrical stretching of polymeric jet prematurely. Therefore, the influence of temperature should be considered when optimizing the working conditions of electrospinning. Humidity is another parameter for the successful fabrication of nanofibers. It has been reported that at lower relative humidity, the nanofibers possessed thinner diameters [28].

The design of the spinneret and the collector affects the morphology and structure of the obtained nanofibers. To fabricate nanofibers with desired physical functions, a variety of production methods were developed by modifying the electrospinning apparatus. Multiple spinnerets enable fabricating multi-component nanofibers and enhance productivity [29]. Electrospinning setup with coaxial spinneret can generate nanofibers with core-sheath configuration [30]. Triaxial electrospinning method is used for fabrication of multilayered nanofibers with desirable features [31]. Alignment of nanofibers can be achieved by using a rotating cylindrical collector [32]. Rotating disk having sharp edge, rotating mandrel, and conductive grid can also be used as collector which enables to fabricate tailor-made nanofibers [33]. Furthermore, threedimensional nanofibrous matrices can be produced by controlling the design of collectors [34]. The development of new electrospinning apparatus designs continues in the research to achieve fiber orientation and threedimensional structure.

3 Electrospun Nanofibers for Water Treatment

3.1 Polymeric Nanofibrous Membranes

A wide variety of electrospun polymeric nanofibers have been investigated for the removal of organic, inorganic contaminants, and biologic agents from water. Both natural and synthetic polymers are used to prepare nanofibrous membranes. Synthetic polymers such as polyacrylonitrile (PAN), polyethersulfone (PES), polyvinyl alcohol (PVA), polyurethane (PU), polyvinylidene fluoride (PVDF); and natural polymers such as chitosan and cellulose were electrospun to be used for water treatment applications [35, 36].

3.1.1 Synthetic Polymers Based Nanofibers

Polyacrylonitrile—Kampalononwat et al. fabricated PAN nanofibers by electrospinning and modified them to obtain aminated PAN, which they evaluated its chelating property with metal ions Cu(II), Ag(I), Fe(II), and Pb(II) [37]. They observed that the initial pH, concentration of the metal solution, and contact time affected the amount of the metal ions adsorbed on the modified nanofibers. In addition, it was demonstrated that the regeneration of nanofibrous materials could be achieved by using hydrochloride acid solution. Zhao et al. produced branched polyethyleneimine grafted electrospun PAN membrane for Cr(VI) adsorption [38]. They demonstrated that the fiber diameter affected the adsorption capacity of the nanofibrous membranes. The decrease in fiber diameter could increase the adsorption capacity of the membranes. The electrostatic interaction and reduction mechanism provided an adsorption capacity of $q_m = 637.46$ mg/g. The membraned could be regenerated and recycled five cycles. Oxime grafted PAN was prepared by Hider et al. by combining electrospinning and chemical grafting [39]. The adsorption studies of dyes methylene blue, rhodamine B, and safranin T were performed. The adsorption capacity (q_m at equilibrium time) was 102.15 mg/g, 118.34 mg/g, and 221.24 mg/g for methylene blue, safranin T, and rhodamine B, respectively.

Polyethersulfone—Yoon et al. proposed electrospun PES membranes for water filtration applications [40]. Mechanical properties of the membrane were improved by using mixed solvents containing dimethylformamide and N-methyl-pyrrolidinone, and the hydrophilicity of the membrane was increased by an oxidation treatment with ammonium persulfate. The flux performance of the membranes was in the range of microfiltration.

Polyvinyl alcohol—Liu et al. electrospun PVA directly on a polyethylene terephthalate (PET) support and immersed the PVA membrane and PET support together in a glutaraldehyde solution in order to crosslink the matrix [41]. The filtration properties of the matrix were compared with Millipore GSWP 0.22 μm membrane. It was demonstrated that the PVA matrices' rejection ratio of polycarboxylate microsphere particles was higher than Millipore GSWP 0.22 μm membrane. A cake layer formed on electrospun PVA matrices during particle rejection test, however, despite cake formation permeates flux was higher compared to the Millipore membrane (Fig. 2). Therefore, PVA matrix was suggested to be used in bacteria or microorganism removal from wastewater. Mahanta and Valiyaveettill fabricated PVA nanofibers and modified surface hydroxyl groups of PVA with mercaptopropionic acid, mercaptosuccinic acid, and lysine, to obtain functional groups such as thiol and amine on the surface [42]. The membrane was proposed for extracting nanoparticles from wastewater. Silver and gold nanoparticles were used as model particles in adsorption studies. It was shown that the amine and thiol-modified nanofibers provided 90% extraction efficiency for silver and gold nanoparticles.

Polyvinylidene fluoride—Li et al. prepared a PVDF membrane by electrospinning [43]. Tetrabutylammonium chloride was added to the electrospinning solution which provided the formation of a branched-like structure. The microfiltration properties of the membrane were investigated. The diameter of the trunk fibers was between 100–500 nm, where the diameter of the fibers of branches was between 5–100 nm. It was demonstrated that the PVDF membrane with tetrabutylammonium chloride showed a retention ratio of 99.9% against 0.3 μm polystyrene particles. High flux rates at low pressure was achieved by using the tree like electrospun membrane. In another study, PVDF membranes were fabricated by electrospinning, the obtained membrane was heat treated, the surface of the membrane was exposed to plasma, and subsequently graft copolymerized with methacrylic acid [44]. The water filtration properties of the fabricated membrane were compared with a commercial PVDF membrane. The filtration performance of the produced membrane was better than the commercial product. It was demonstrated that with surface modification, electrospun nanofibers could be fabricated to have smaller pores while having high flux performance.

Polyurethane—Lev et al. produced PU electrospun membranes and their filtration efficiency with a commercial PVDF membrane with a similar pore size [45]. Bacterial removal efficiency of the produced samples was better compared to the commercial membrane. The semi-pilot scale experiments were also performed, which confirmed that the electrospun membrane had potential to be used in microbial filtration, with *Escherichia coli*. Fang et al. fabricated N-substituted PU electrospun membrane for self-cleaning and oil/water separation which has potential to be used in wastewater treatment [46].

3.1.2 Natural Polymers Based Nanofibers

Chitosan—Min et al. prepared electrospun chitosan nanofibers for removal of arsenate. Firstly, chitosan was electrospun with polyethylene oxide [47]. Addition of

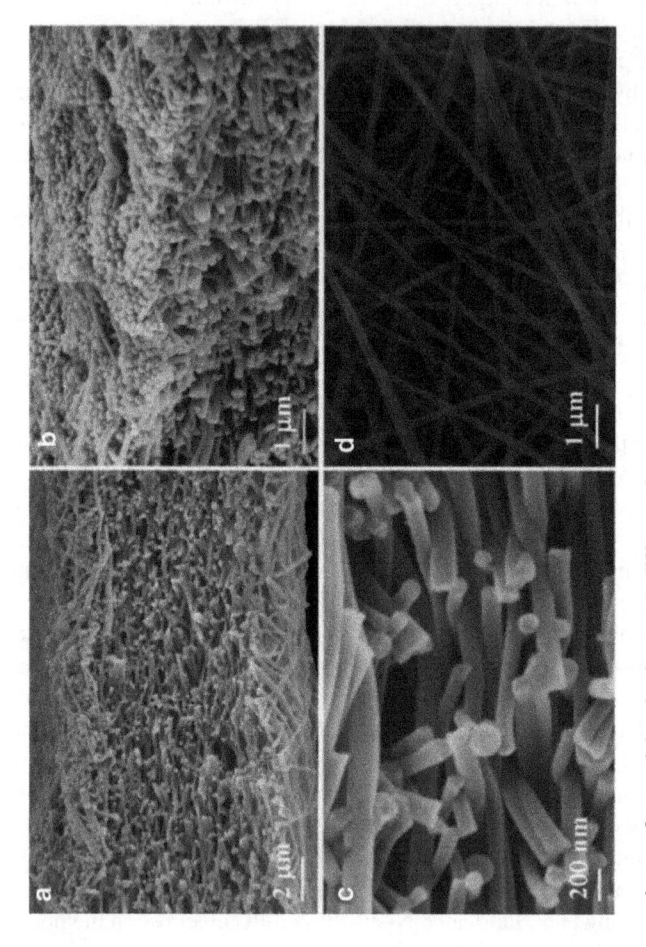

Fig. 2 SEM images of the membrane after particle rejection test in different views: **a** cross-section; **b** top layers of cross-section; **c** middle part of cross-section; **d** bottom surface (From Liu Y, Wang R, Ma H, Hsiao BS, Chu B (2013) High-flux microfiltration filters based on electrospun polyvinyl alcohol nanofibrous membranes. Polymer 54:548–556, copyright Elsevier)

polyethylene oxide enhanced the spinnability of the solution. After the nanofibers were formed polyethylene oxide was removed by washing with water. The adsorption capacity of the chitosan nanofibers was found to be 30.8 mg/g. The results demonstrated the potential use of electrospun chiton nanofibers as adsorbent for arsenic removal from water.

Cellulose—Tian et al. fabricated cellulose acetate nanofibers by electrospinning from its solution, subsequently the electrospun membranes were modified by using methacrylic acid [48]. Polymethyl methacrylate modified cellulose acetate membrane was evaluated for its efficiency in adsorption of heavy metal ions (Hg^{+2}, Cu^{+2}, Cd^{+2}) in water. It was demonstrated that the membrane had high and selective adsorptivity for Hg^{+2} and could be regenerated by ethylenedinitrilo tetraacetic acid solution.

3.2 Ceramic Nanofibers

Ceramic nanofibers have unique physical and chemical properties, therefore they are applicable in a wide variety of applications such as adsorption, photocatalysis, optoelectronics, nanoelectronics, and tissue engineering. There are several ways to produce ceramic nanofibers including template mediated growth of nanofibers, chemical vapour deposition, hydrothermal technique, laser mediated nanofiber synthesis, and electrospinning. Here, electrospinning has advantages such as simplicity, low cost and capability to produce continuous fibers in the range of 50–1000 nm or greater [49]. However, electrospinning is largely limited to the fabrication of nanofibers of synthetic and natural polymers, since the polymer solution or melt provide suitable rheological properties required for electrospinning. Ceramics are usually blended with a carrier polymer to be spinnable. The fabrication of ceramic nanofibers includes electrospinning of metal salts and a matrix polymer as a blend template, and calcination of the composite nanofibers to remove organic components.

Malwal et al. fabricated electrospun polyethylene oxide/nickel acetate nanofibers, and subsequently calcinated them to obtain pure nickel oxide nanofibers [50] (Fig. 3). The photocatalytic activity of the nanofibers was evaluated with a model dye under visible light irradiation. The nickel oxide nanofibers efficiently degraded the model dye Congo red following the pseudo-first order reaction. Peng et al. prepared porous hollow γ-Al_2O_3 nanofibers by single-capillary electrospinning of $Al(NO_3)_3$/PAN precursor solution, followed by sintering [51] (Fig. 4). The prepared hollow nanofibers were suggested as adsorbents to eliminate dyes from an aqueous solution.

Fig. 3 FE-SEM images of **a** polyethylene oxide/nickel acetate nanofibers and **b** nickel oxide nanofibers obtained after calcination at 500 °C (From Malwala D, Gopinath P (2015) Fabrication and characterization of poly(ethylene oxide) templated nickel oxide nanofibers for dye degradation. Environ Sci Nano 2:78–85, copyright Royal Society of Chemistry.)

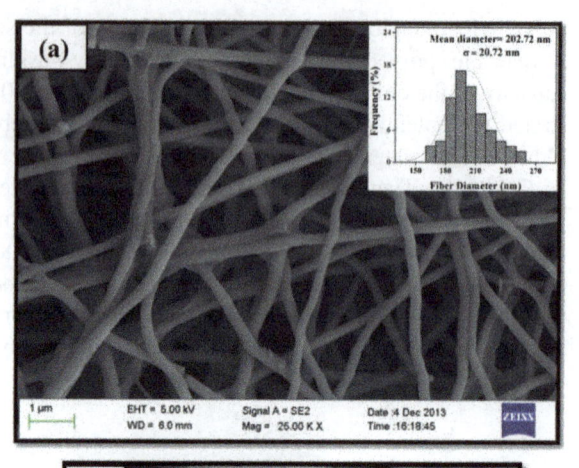

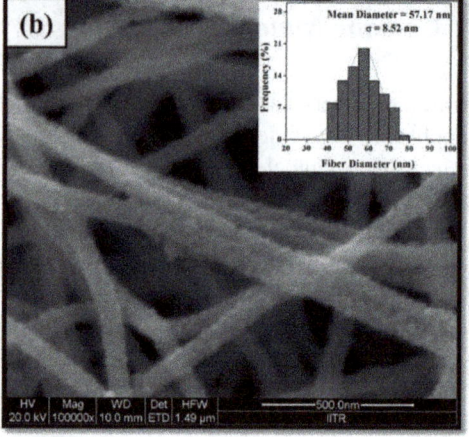

3.3 Composite Nanofibers

In recent years, electrospun composite materials attracted the interest of researchers, because the advantages of several types of materials can be combined in a matrice, which provides improved water purification.

Nie et al. fabricated PAN-CuS composite nanofibers with CuS nanoparticles distributed on the surface of nanofibers by hydrothermal method [52] (Fig. 5). It was demonstrated that the nanofibers showed catalytic activity to degrade methylene blue in the presence of hydrogen peroxide and had potential to be used in dye removal from wastewater. In another study, polyamide 6 nanofibers were modified with titanium dioxide nanoparticles [53]. The photocatalytic activity of the membranes was evaluated. The membranes had efficient photocatalytic activity under UV light irradiation, and in addition the membranes were successful in removing *Escherichia*

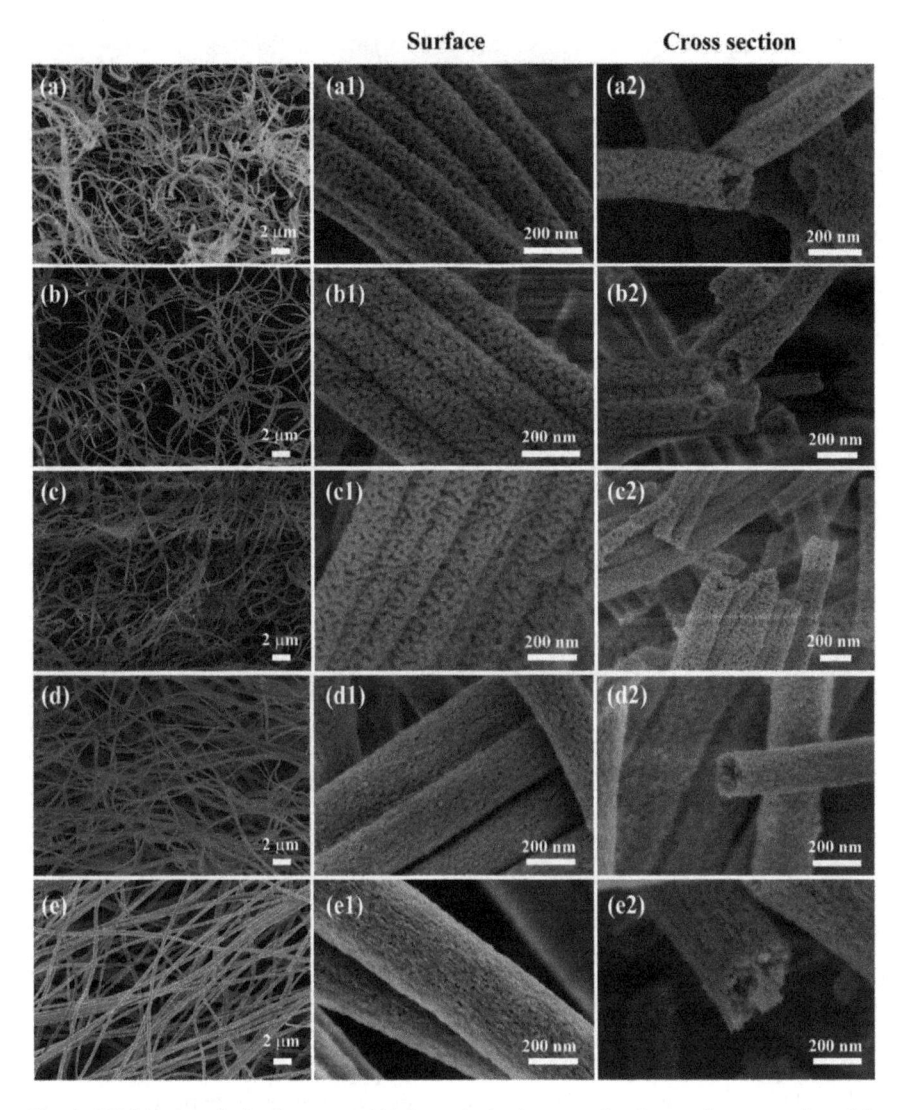

Fig. 4 SEM low-resolution images and high-resolution images of surface and cross-section of the porous hollow γ-Al_2O_3 nanofibers sintered at 800 °C with different weight ratios of $Al(NO_3)_3$/PAN to PAN **a** 1:10, **b** 2:10, **c** 3:10, **d** 5:10, and **e** 10:10 (From Peng C, Zhang J, Xiong Z, Zhao B, Liu P (2015) Fabrication of porous hollow γ-Al_2O_3 nanofibers by facile electrospinning and its application for water remediation. Micropor Mesopor Mat 215:133–142, copyright Elsevier.)

coli from water. Taha et al. produced cellulose acetate/silica composite nanofibers by combining sol gel with electrospinning [54].

The adsorption capacity of the membrane was evaluated. The maximum adsorption capacity of the membrane for Cr(VI) was estimated to be 19.46 mg/g. Parekh et al. developed silver nanoparticles coated electrospun PAN nanofibers for water filtration

Fig. 5 TEM images of PAN/CuS composite nanofibers after recycling five times with different magnifications (From Nie G, Li Z, Lu X, Lei J, Zhang C, Wang C (2013) Fabrication of polyacrylonitrile/CuS composite nanofibers and their recycled application in catalysis for dye degradation. Appl Surf Sci 284: 595–600, copyright Elsevier.)

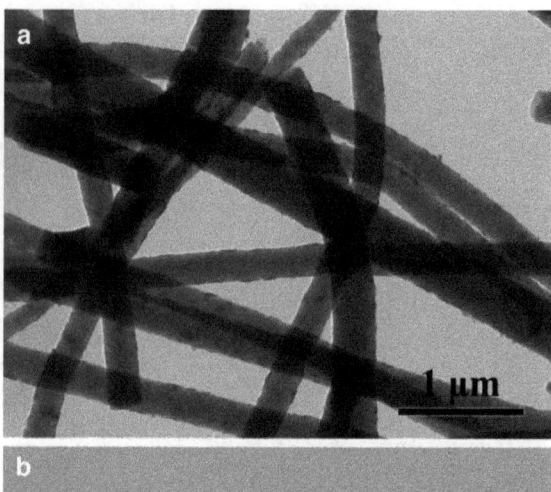

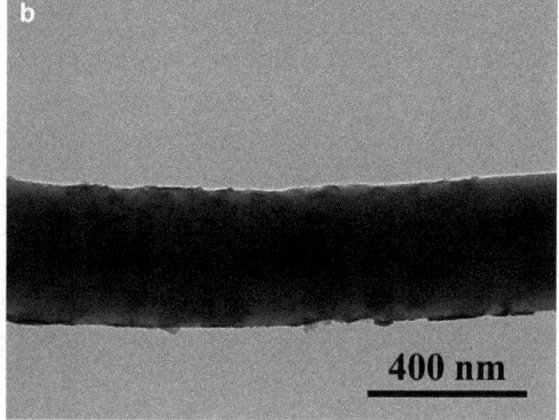

[55]. The membrane was applied as antimicrobial membrane with 100% reduction of Gram negative bacteria. In another study, hydrophilic graphene oxide was incorporated in PVDF membrane via electrospinning [56]. The composite membranes showed high water flux and incorporation of graphene oxide improved antifouling ability of the membrane, therefore the membrane was suggested to be used as water treatment membrane applications.

4 Conclusion

Over the last decades, electrospinning became a globally recognized simple method to fabricate nanofibrous membranes. High surface area to volume ratio, high porosity, good mechanical properties, ability to be functionalized or modified, make them attractive for a wide variety of applications. In terms of water filtration purposes, electrospun membranes can perform as both adsorption and filtration membranes. Both synthetic and natural polymers and ceramics can be used for preparation of electrospun nanofibrous membranes for removal of contaminants from wastewater. In this review different polymeric, ceramic or composite electrospun membranes used in water treatment applications was summarized. In the last few decays, modification and functionalization of electrospun nanofibers by pre or post-treatment methods and combination of nanofibers with functional nanoparticles have been explored. However further research is required to prepare. electrospun nanofibrous membranes on industrial scale with reusable properties.

Conflict of Interest The authors declare no conflict of interest for this work.

References

1. Seckler D, Barker R, Amarasinghe U (1999) Water scarcity in the twenty-first century. Int J Water Res Dev 15:29–42
2. Dodds WK, Perkin JS, Gerken JE (2013) Human impact on freshwater ecosystem services: a global perspective. Environ Sci Technol 47:9061–9068
3. Shannon MA, Bohn PW, Elimelech M, Georgiadis JG, Mariñas BJ, Mayes AM (2008) Science and technology for water purification in the coming decades. Nature 452:301–310
4. Montgomery MA, Elimelech M (2007) Water and sanitation in developing countries: including health in the equation. Environ Sci Technol 41:17–24
5. Li Q, Mahendra S, Lyon DY, Brunet L, Liga MV, Li D, Alvarez PJJ (2008) Antimicrobial nanomaterials for water disinfection and microbial control: potential applications and implications. Water Res 42:4591–4602
6. Ulbricht M (2006) Advanced functional polymer membrane. Polymer 47:2217–2262
7. Bazargan AM, Keyanpour-Rad M, Hesari FA, Esmaeilpour MG (2011) A study on the microfiltration behavior of self-supporting electrospun nanofibrous membrane in water using an optical particle counter. Desalination 265:148–152
8. Shukla R, Cheryan M (2002) Performance of ultrafiltration membranes in ethanol–water solutions: effect of membrane conditioning. J Membr Sci 198:75–85
9. Sundarrajan S, Balamurugan R, Kaur S, Ramakrishna S (2013) Potential of engineered electrospun nanofiber membranes for nanofiltration applications. Dry Technol 31:163–169
10. Fujioka T, Khan SJ, McDonald JA, Roux A, Poussade Y, Drewes JE, Nghiem LD (2013) N-nitrosamine rejection by nanofiltration and reverse osmosis membranes: the importance of membrane characteristics. Desalination 316:67–75
11. Hoover LA, Schiffman JD, Elimelech M (2013) Nanofibers in thin-film composite membrane support layers: enabling expanded application of forward and pressure retarded osmosis. Desalination 308:73–81
12. Lalia BS, Kochkodan V, Hashaikeh R, Hilal N (2013) A review on membrane fabrication: structure, properties and performance relationship. Desalination 326:77–95
13. Balamurugan R, Sundarrajan S, Ramakrishna S (2011) Recent trends in nanofibrous membranes and their suitability for air and water filtrations. Membranes 1:232–248
14. Huang ZM, Zhang YZ, Kotaki M, Ramakrishna S (2003) A review on polymer nanofibers by electrospinning and their applications in nanocomposites. Compos Sci Technol 63:2223–2253
15. Burger C, Hsiao BS, Chu B (2006) Nanofibrous materials and their applications. Annu Rev Mater Res 36:333–368
16. Bölgen N, Menceloğlu YZ, Acatay K, Vargel I, Pişkin E (2005) In vitro and in vivo degradation of non-woven materials made of poly(epsilon-caprolactone) nanofibers prepared by electrospinning under different conditions. J Biomater Sci Polym Ed 16(12):1537–1555

17. Piskin E, Bölgen N, Egri S, Isoglu IA (2007) Nanomedicine (Lond) 2(4):441–457
18. Bhardwaj N, Kundu SC (2010) Electrospinning: a fascinating fiber fabrication technique. Biotechnol Adv 28:325–347
19. Deitzel JM, Kleinmeyer J, Harris D, Tan NCB (2001) The effect of processing variables on the morphology of electrospun nanofibers and textiles. Polymer 42:261–272
20. Hohman MM, Shin M, Rutledge G, Brenner MP (2001) Electrospinning and electrically forced jets. II. Applications. Phys Fluids 13:2221–2236
21. Subbiah T, Bhat GS, Tock RW, Parameswaran S, Ramkumar SS (2005) Electrospinning of nanofibers. J Appl Polym Sci 96:557–569
22. Zong X, Kim K, Fang D, Ran S, Hsiao BS, Chu B (2002) Structure and process relationship of electrospun bioadsorbable nanofiber membrane. Polymer 439:4403–4412
23. Wongchitphimon S, Wang R, Jiraratananon R (2011) Surface modification of polyvinylidene fluoride-co-hexafluoropropylene (PVDF–HFP) hollow fiber membrane for membrane gas absorption. J Membr Sci 381:183–191
24. Hu J, Wang X, Ding B, Lin J, Yu J, Sun G (2011) One-step electro-spinning/nettingtechnique for controllably preparing polyurethane nano-fiber/net. Macromol Rapid Comm 32:1729–1734
25. Zhang C, Yuan X, Wu L, Han Y, Sheng J (2005) Study on morphology of electrospun poly (vinyl alcohol) mats. Eur Polym J 41:423–432
26. Yuan XY, Zhang YY, Dong CH, Sheng J (2004) Morphology of ultrafine polysulfone fibers prepared by electrospinning. Polym Int 53:1704–1710
27. Yang GZ, Li HP, Yang JH, Wan J, Yu DG (2017) Influence of working temperature on the formation of electrospun polymer nanofibers. Nanoscale Res Lett 12:55–65
28. Wang X, Ding B, Yu J, Yang J (2011) Large-scale fabrication of two-dimensional spider-web-like gelatin nano-nets via electro-netting. Colloid Surf B 86:345–352
29. SalehHudin HS, Mohamad EN, Mahadi WNL, Afifi AM (2018) Multiple-jet electrospinning methods for nanofiber processing: a review. Mater Manuf Process 33:479–798
30. Moghe AK, Gupta BS (2008) Co-axial electrospinning for nanofiber structures: preparation and applications. Polym Rev 48:353–377
31. Khalf A, Singarapu K, Madihally SV (2015) Influence of solvent characteristics in triaxial electrospun fiber formation. React Funct Polym 90:36–46
32. Matthews JA, Wnek GE, Simpson DG, Bowlin GL (2002) Electrospinning of collagen nanofibers. Biomacromol 3:232–238
33. Park S, Park K, Yoon H, Son JG, Min T, Kim GH (2007) Apparatus for preparing electrospun nanofibers: designing an electrospinning process for nanofiber fabrication. Polym Int 56:1361–1366
34. Shah Hosseini N, Bölgen N, Khenoussi N, Yılmaz N, Yetkin D, Hekmati A, Schacher L, Adolphe D (2018) Novel 3D electrospun polyamide scaffolds prepared by 3D printed collectors and their interaction with chondrocytes. Int J Polym Mater PO 67:143–150
35. Liao Y, Loh CH, Tian M, Wang R, Fane AG (2018) Progress in electrospun polymeric nanofibrous membranes for water treatment: fabrication, modification and applications. Prog Polym Sci 77:69–94
36. Feng C, Khulbe KC, Matsuura T, Gopal R, Kaur S, Ramakrishna S, Khayet M (2008) Production of drinking water from saline water by air-gap membrane distillation using polyvinylidene fluoride nanofiber membrane. J Membr Sci 311:1–6
37. Kampalanonwat P, Supaphol P (2010) Preparation and adsorption behavior of aminated electrospun polyacrylonitrile nanofiber mats for heavy metal ion removal. ACS Appl Mater Interfaces 2:3619–3627
38. Zhao R, Li X, Sun B, Li Y, Li Y, Yang R, Wang C (2017) Branched polyethylenimine grafted electrospun polyacrylonitrile fiber membrane: a novel and effective adsorbent for Cr(VI) remediation in wastewater. J Mater Chem A 5:1133–1144
39. Haider S, Binagag FF, Haider A, Al-Masry WA (2014) Electrospun oxime-grafted-polyacrylonitrile nanofiber membrane and its application to the adsorption of dyes. J Polym Res 21(371):1–13

40. Yoon K, Hsiao BS, Chu B (2009) Formation of functional polyethersulfone electrospun membrane for water purification by mixed solvent and oxidation processes. Polymer 50:2893–2899
41. Liu Y, Wang R, Ma H, Hsiao BS, Chu B (2013) High-flux microfiltration filters based on electrospun polyvinylalcohol nanofibrous membranes. Polymer 54:548–556
42. Mahanta N, Valiyaveettil S (2011) Surface modified electrospun poly(vinyl alcohol) membranes for extracting nanoparticles from water. Nanoscale 3:4625–4631
43. Li Z, Kang W, Zhao H, Hu M, Wei N, Qiu J, Cheng B (2016) A novel polyvinylidene fluoride tree-like nanofiber membrane for microfiltration. Nanomaterials 6(152):1–11
44. Kaur S, Ma Z, Gopal R, Singh G, Ramakrishna S, Matsuura T (2007) Plasma-induced graft copolymerization of poly(methacrylic acid) on electrospun poly(vinylidene fluoride) nanofiber membrane. Langmuir 23:13085–13092
45. Lev J, Holba M, Došek M, Kalhotka L, Mikula P, Kimmer D (2014) A novel electrospun polyurethane nanofibre membrane–production parameters and suitability for wastewater (WW) treatment. Water Sci Technol 69(7):1496–1501
46. Fang W, Liu L, Li T, Dang Z, Qiao C, Xu J, Wang Y (2016) Electrospun N-substituted polyurethane membranes with self-healing ability for self-cleaning and oil/water separation. Chem Eur J 22:878–883
47. Min LL, Yuan ZH, Zhong LB, Liu Q, Wua RX, Zheng YM (2015) Preparation of chitosan based electrospun nanofiber membrane and its adsorptive removal of arsenate from aqueous solution. Chem Eng J 267:132–141
48. Tiana Y, Wub M, Liua R, Li Y, Wanga D, Tana J, Wue R, Huang Y (2011) Electrospun membrane of cellulose acetate for heavy metal ion adsorption in water treatment. Carbohydr Polym 83:743–748
49. Malwal D, Gopinath P (2016) Fabrication and applications of ceramic nanofibers in water remediation: a review. Crit Rev Env Sci Tec 46:500–534
50. Malwala D, Gopinath P (2015) Fabrication and characterization of poly(ethylene oxide) templated nickel oxide nanofibers for dye degradation. Environ Sci Nano 2:78–85
51. Peng C, Zhang J, Xiong Z, Zhao B, Liu P (2015) Fabrication of porous hollow γ-Al_2O_3 nanofibers by facile electrospinning and its application for water remediation. Micropor Mesopor Mat 215:133–142
52. Nie G, Li Z, Lu X, Lei J, Zhang C, Wang C (2013) Fabrication of polyacrylonitrile/CuS composite nanofibers and their recycled application in catalysis for dye degradation. Appl Surf Sci 284:595–600
53. Blanco M, Monteserín C, Angulo A, Pérez-Márquez A, Maudes J, Murillo N, Aranzabe E, Ruiz-Rubio L, Vilas JL (2019) TiO_2-doped electrospun nanofibrous membrane for photocatalytic water treatment. Polymers 11(747):1–11
54. Taha AA, Wu Y, Wang H, Li F (2012) Preparation and application of functionalized cellulose acetate/silica composite nanofibrous membrane via electrospinning for Cr(VI) ion removal from aqueous solution. J Environ Manag 112:10–16
55. Parekh SA, David RN, Bannuru KKR, Krishnaswamy L, Baji A (2018) Electrospun silver coated polyacrylonitrile membranes for water filtration applications. Membranes 8(59):1–11
56. Janga W, Yuna J, Jeonb K, Byun H (2015) PVdF/graphene oxide hybrid membrane via electrospinning for water treatment application. RSC Adv 5(58):46711–46717

Nimet Bölgen was born in Adana, Turkey, in 1979. She received the B.S. degree in Chemical Engineering from Ankara University, Ankara, in 2002, and the M.S degree in Chemical Engineering from Hacettepe University, Ankara, in 2004. She received Ph.D. degree from Bioengineering Department of Hacettepe University, Ankara, in 2008. From 2009 to 2014, she was an Assistant Professor, and from 2014 to 2020 she was an Associate Professor with the Chemical Engineering Department of Mersin University, Mersin. Since 2020, she has been a Professor with the Chemical Engineering Department of Mersin University, Mersin. She is the author of 10 book chapters, 38 articles and more than 100 proceedings. Her research interest includes tissue engineering, biomaterials, controlled drug release systems, natural and synthetic polymers, cryogels and electrospun nanofibers. She is the Editorial Board Member of the journal Biomaterials and Biomechanics in Bioengineering. Prof. Bölgen is the member of Tissue Engineering and Regenerative Medicine International Society.

Ashok Vaseashta (M'79-SM'90) received Ph.D. in Materials Science and Engineering (minor in Electrical Engineering) from Virginia Polytechnic Institute and State University, Blacksburg, Virginia, USA. He is Executive Director of Research for International Clean Water Institute in VA, USA, Chaired Professor of Nanotechnology at the Academy of Sciences of Moldova and Professor, Nanotechnology and Biomedical Engineering at the Faculty of Mechanical Engineering, Transport and Aeronautics at the Riga Technical University. Prior to his current position, he served as Vice Provost for Research at the Molecular Science Research Center in Orangeburg, South Carolina. He served as visiting professor at the 3 Nano-SAE Research Centre, University of Bucharest, Romania and visiting scientist at the Helen and Martin Kimmel Center of Nanoscale Science at the Weizmann Institute of Science, Israel. He served the U.S. Department of State in two rotations, as strategic S&T advisor and U.S. diplomat. His research interests span nanotechnology, environmental/ecological science, and safety and security. His research on nanotechnology has been on improving the understanding, design, and performance of nanofibers and sensors/detectors, mainly for applications such as wearable electronics, target drug delivery, detection of biomarkers and toxicity of nano and xenobiotic materials. In the security arena, he has worked on counterterrorism, countering unconventional warfare and hybrid threats, critical-Infrastructure protection, biosecurity, dual-use research concerns, and mitigating hybrid threats. He has authored over 250 research publications, edited/authored eight books on nanotechnology, and presented many keynotes and invited lectures worldwide. He serves on the editorial board of several highly reputed international journals. He is an active member of several national and international professional organizations. He is a fellow of the American Physical Society, Institute of Nanotechnology, and the New York Academy of Sciences. He has earned several other fellowships and awards for his meritorious service including 2004/2005 Distinguished Artist and Scholar award.

Water Treatment by Green Coagulants—Nature at Rescue

Manoj Kumar Karnena and Vara Saritha

Abstract Water purification and treatment are significant challenges for environmental engineers to several extents, continually increasing pressure to provide safe drinking water to the consumers. Nevertheless, water treatment with chemicals is minimized as they are toxic to the environment and humans. One of the fundamental steps for water treatment for consumption is water clarification by Coagulation and flocculation, eliminating colloidal particles, impurities, algae, etc. The sustainability in water treatment is only possible by the usage of natural materials combined with innate technologies. In these lines, the present book chapter will appraise the efficiency of natural coagulants' in treating surface water. Several researchers tested the natural coagulants for water treatment; but, various natural coagulants and their coagulation efficiency were not presented. Thus, an attempt is made to achieve a comprehensive account of widely used different plant-based coagulants, to understand their properties and efficiencies in treating water. The approaches mentioned in the book chapter will develop a knowledge database on available coagulants and their best utility techniques that can be adopted at any given time.

Keywords Natural coagulants · Water treatment · Turbidity · Coagulation · Flocculation

1 Introduction

Water is a vital resource for biological activities and human beings; strengthening the profiles of lakes, rivers, oceans, and forests becomes a segment in the hydrological cycle marking its significance in developing social life and ecosystems [126]. In recent years be wary about the resources are becoming a crucial mission progressively. The global community alerted the WHO (World Health Organization) and the UN (United Nations) regarding water scarcity and unrestrained disposal of pollutants into surface water. When measured as a parameter affecting human lives, it is

---

M. K. Karnena (✉) · V. Saritha
Department of Environmental Science, GITAM Institute of Science, (Deemed to Be) University, Visakhapatnam 530045, India

© Springer Nature Switzerland AG 2021
A. Vaseashta and C. Maftei (eds.), *Water Safety, Security and Sustainability*, Advanced Sciences and Technologies for Security Applications,
https://doi.org/10.1007/978-3-030-76008-3_9

undoubtedly one of humankind's intricate development's critical factors. It is crucial to connect human health, education, and poverty, whose implications are essential for humans' development [14]. Also, in the challenge of Millennium Development Goals, water is an intersecting aspect. Availability and reuse of water are not a new concern; the human community is worried about water stress for many years. [32]. Intended essential for life, present inappropriate water use, including industrialization, urbanization, increasing population, and climate change, enhanced these reserves [15]. The scarcity of water affects all community sectors, including social and economic, resulting in a threat to natural resource sustainability [60]. The water's purity level for consumption is crucial as it directly affects individual health [75]. Sowmeyan et al. [124], reported that about two million people use unhealthy aqua, resulting in diarrhea besides several diseases caused due to water, specifically in developing countries. Regardless of advancements in water treatment technology, one of the significant challenges confronted by many developing countries is the deficiency of safe and clean water for their citizens [104]. Safe and pure water requirements triggered detailed and comprehensive research for treating water and wastewater [77]. Treating water to make it potable is a question of essential significance, together for emerging and advanced countries. The trait of aqua in developing countries generally is unsatisfactory and unsafe for wellbeing. In contrast, in developed countries, water decontamination involves the application of chemicals, knowing it is not safe on health upon long-standing vulnerability is questionable [82, 87].

Many communities in most of the third world countries depend to some extent on surface water sources for domestic water supply. Water drawn from rivers contains high turbidity, particularly in the rainy season. Many studies have reported that seasonal variation in turbidity is a significant problem in treating surface water [85, 93]. Generally, surface waters contain fine suspended particles, sediments, and natural organic materials, which cause turbidity. Chemically turbidity of water is due to the existence of negatively charged elements in the colloidal structure, hence positively charged agents called coagulants are used to form complexes in a process called Coagulation [88, 101]. Coagulation is a crucial process in treating surface water, removing dissolved chemical species, and turbidity. Coagulation is simple, economical, and dependable consuming less energy, which is practiced commonly. Furthermore, it is a proven process for removing suspended, colloidal, and dissolved particles by aggregating macro and micro particulates to large size, which eventually settle by sedimentation. Coagulation is also applicable for removing many other impurities like micro-pollutants, organic compounds, color and oils, and fats [67].

Conventional Coagulation was achieved using chemical-based coagulants like alum (Aluminium sulfate) and ferric chloride. Alum is a broadly used coagulant accepted by many treatment units worldwide due to its performance, availability, management, and cost-effectiveness. The efficiencies of chemical-coagulants [47], nevertheless, drawbacks related to their usage include their ineffectiveness in waters with lower-temperature [63], costs associated with the acquisition, generation of larger volumes of sludge, alteration of treated water pH and harmful impacts on humans. Shreds of proof relating to alum coagulants to Alzheimer's disorder in humans are also reported [51].

Thus, it becomes necessary to search for feasible alternatives that are safe for humans and the environment [62]. Many studies have explored and evaluated cleaner and eco-friendly coagulants originating from natural sources like plants, animals, and microorganisms. Advantages of natural coagulants over chemical coagulants include the production of lower sludge volume (20–30% than that of alum), biodegradability, economical, does not alter pH of treated water, require lower doses, and safe for humans [58, 130]. Technologies using natural, locally available coagulants entails water treatment practices that are environmentally friendly, culturally suitable, and reduce the external dependency of mechanisms and equipment.

Natural coagulants from plant origin are derived from leaves, seeds, bark or sap, fruit extract, and roots [67]. Some widely used natural coagulants that have reported to have good coagulation ability include *Moringa olefiera* [74, 97], starch [106], *Strychnos potatorum* [114] and cactus [3, 8]. Accepted coagulants like tannins have also been applied for water treatment, along with wastes of fruits and numerous legumes and vegetables [29, 30].

Studies to improve Coagulation and flocculation are being carried out during recent years, focusing on developing novel coagulants incorporating natural organic polymers and metal coagulants [137]. The greater efficiency of natural organic polymers is attributed to their greater molecular weight conferring better aggregation properties. Hence, growth in the size and molecular weight of coagulants can be a means towards additional improvement, which could be achieved by composite coagulants [92]. Thus, these novel composite coagulants will combine the positive aspects of respective individual components resulting in enhanced coagulant stability, being effective in broader pH range, requiring low dosage, and forming denser flocs virtuous settling properties. Furthermore, these coagulants simplify treatment procedure and improve overall cost-effectiveness and reduce the extent of toxicity because of non-reacted monomers in polymer due to interactions among metal species and natural coagulant complexation reactions [128].

2 History of Coagulation—Development of Coagulation technology over the years

The history of coagulation tracks back to three thousand years. Around 2000 BC Egyptians, used smeared almond as coagulant for clarification of river water. During 77 AD Romans got familiar with alum which was used as a coagulant. Alum as coagulant for water treatment was used in England by 1757. During 1881 alum was adopted as coagulant formally for treating public water supplies [49, 70, 132].

Use of Ferric Chloride and Aluminium Sulphate for modern water treatment was initiated about a decade ago after which scientific studies on Coagulation were started. Widely known rule of Schulze-Hardy rule was put forth explaining mechanisms of

Coagulation [10, 49, 64]. Concept of particle collision that formed the foundation for knowing variations in particle number during the course of flocculation was developed by Smoluchowski [123]. Mattson was the first to report aluminium and iron hydrolysis products, which was widely accepted after 30 years [84]. Influence of pH and different anions on formation of floc was studied by Black and Colleague. Afterwards, the focus of research in Coagulation was directed towards better floc production through mechanical methods, along with exploration for superior coagulant aids like silicates, limestone and bentonite [18]. In 1940s, Derjaguin, Landau, Verwey and Overbeek independently proposed the DLVO theory of colloid stability colloid interaction based on Vander Waals attraction and electrical interaction that result in repulsion or attraction which depends on surroundings [40, 129].

Novel coagulation concept was later established by Langelier and Ludwig [78], illustrating two mechanisms for exclusion of colloidal particles as:

- Compression of the double layer, procedure that allows particles to overcome the repulsive forces resulting in agglomeration and precipitation, and
- Enmeshment of precipitate during which physical enmeshment of small particles takes place by metal precipitate as they form and settle.

Both of these mechanisms were theoretically explained by LaMar and Healey 1963 and they have also suggested terms Coagulation and flocculation. Micro- electrophoresis has introduced in developing concept of Coagulation for studying destabilization of colloidal that has allowed quantification of zeta potential on particles of colloid during 1960s, along with study of stoichiometric relationship among coagulant dosages which are mandatory for neutralizing colloids along with concentration of these particles in water. Additionally, these studies too indicated influence of chemical factors like ionic strength, properties of colloids for the removal and pH on colloidal particle surface potential.

Investigations during 1970s highlighted that mechanism of Coagulation and performance of Coagulation are always related to water quality standards. After the identification of organic compounds like haloforms and halogenated natural organic matter (NOM) in water after treated, principal focus of coagulation/flocculation was to be focussed on to remove NOM. Stringent drinking water standards have made many water treatment plants to adopt granular activated carbon (GAC) filter for effective removal of organic compounds from water [111].

During 1980s scientific studies on Coagulation for removal of NOM assessed on various surface water sources focussed on various factors like coagulant type, coagulant dosage, addition of chemicals, pH, re-stabilization zones and water quality characteristics [61]. Apart from NOM other precursors like humic substances, microorganisms like Legionella, Giardia, Cryptosporidium and different viruses have still remained as the main impurities of concern during 1990s. Removal of these impurities can be achieved by treatment process which includes Coagulation, filtration and disinfection. To achieve anticipated results coagulation is an indispensable process to be adopted [39, 89].

During the later years it is understood that Coagulation is also influenced by upstream and downstream treatment processes like pre-oxidation process in upstream and settling, filtration, adsorption (activated carbon) and disinfection at downstream respectively. At the same time pre-ozonation effect on attainment and efficiency of Coagulation on different raw surface waters with diverse amounts of NOM along with the effect of Coagulation on adsorption with activated carbon have been taken up which still are under research and development.

Moreover, studies focussing on improved Coagulation through adding excess coagulant dose, reducing pH or through preparation of effective coagulants are being evaluated targeting elimination of NOM and enhancing performance of Coagulation. Mathematical models are also developed for studying coagulant dose, removal efficiency of NOM and also for describing coagulation physical aspects [26, 34].

3 Theory of Coagulation

Coagulation essentially consists of destabilization and aggregation of stable small colloidal contaminations to larger particles known as floc, that are effectively removed by physico-chemical processes like rapid mixing, slow mixing, separation of solid-liquid stages in sedimentation and filtration. Colloidal impurities generally include viruses, bacteria, protozoans, color producing substances, inorganic solids and organic matter.

Conventional method of removing these impurities include a combination of the following steps:

- **Coagulation**—includes addition of coagulants and rapid stirring in raw water for complete mixing of the coagulant;
- **Flocculation**—includes slow stirring of water after rapid stirring in order to assist the particles to flocculate, so that they gain weight and settle down;
- **Sedimentation**—after coagulation and flocculation certain time is provided for the flocs to settle;
- **Filtration**—includes passing the treated water through the filters for removal of all the suspended matters.

In principle, Coagulation de-stabilizes the colloidal particles by charge neutralization after dosing of the coagulant, whereas in flocculation aggregation the colloids (with sizes μm and sub-μm) take place to form floc with mm size. Although coagulation and initial stage of flocculation happen quite swiftly, in actual process, there is a slight dissimilarity. Hence, terms either "coagulation" or "flocculation" can be adopted for describing complete process.

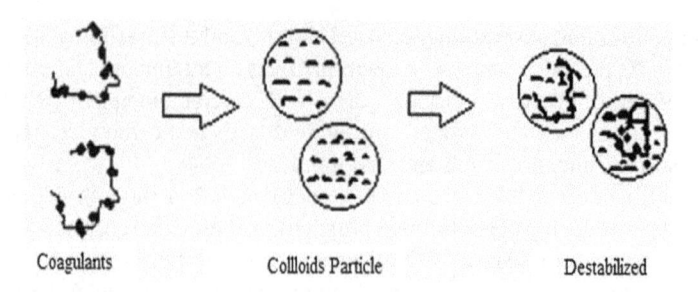

Fig. 1 Double layer compression (Redrawn with permission from Duan et al. [46])

4 Coagulation Mechanisms

As identified by O'Melia and Dempsey, Coagulation of colloidal particles can be attained in four mechanisms, which are: compression of double layer, neutralization of charge, sweep Coagulation and inter-particle bridging [33].

4.1 Double Layer Compression

Classical mechanism used to define particle destabilization; double layer compression is accomplished through addition of an electrolyte (coagulant) into the suspension of colloids. Under stable conditions the colloidal particles exist in high stability where they are unable to get close to each other owing to the presence of thick double layer. The added electrolyte brings about the change in the ionic strength of the double layer around the colloid resulting in destabilization of the colloid [46, 125]. Compression of double layer has been established to be significant destabilization mechanism (Fig. 1).

4.2 Charge Neutralization

Colloids under normal conditions possess negative charge. The charge neutralization process includes adding of metals/polymers having cations that neutralize negative charges of colloids subsequently resulting in surface charge neutralization. Thus, efficiency of the mechanism is intensely reliant on coagulant dosage, since restabilization of particle can be evident after exceeding optimum dosage (Fig. 2) [20, 29, 125].

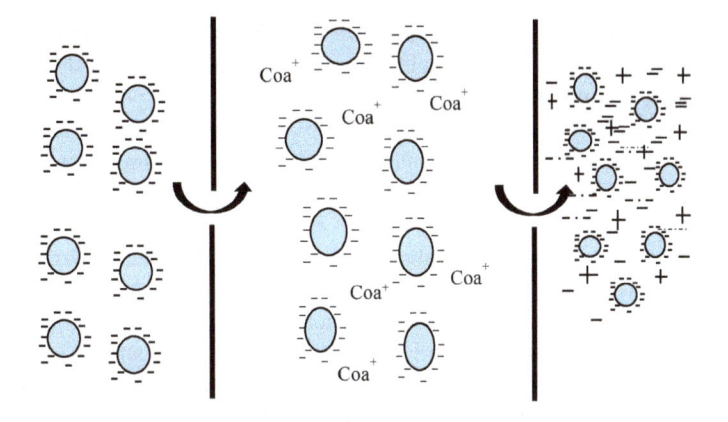

Coa⁺ = Coagulants with net +ve charge

Fig. 2 Charge neutralization (Redrawn with permission from Choy et al. [29])

Fig. 3 Sweep Flocculation (Redrawn with permission from Duan and Gregory [45])

4.3 Sweep Coagulation

The process of forming coagulant precipitates through addition of huge quantity of coagulant is the chief process leading to sweep Coagulation (Fig. 3). The formed coagulant precipitates enmesh the colloidal particles thus successfully removing the colloidal suspension [122]. In comparison to charge neutralization sweep coagulation could result in enhanced Coagulation [45].

4.4 Inter-particle Bridging

The long-chain polymers or polyelectrolytes form bridge between them binding numerous colloids together. Higher bridging competence is achieved with larger molecular weight coagulants owing to their extended polymeric chains. The governing factor for bridging is availability of adequate vacant surface on particles so that polymer chain segments can attach [20, 29, 108].

Polymer branches that are formed may cause restabilization due to particle destabilization. In the process of inter-particle bridging, elements contour mesh-like matrix comprising of colloids that are destabilized and branches of polymer when they come together. This floc may entrain smaller particles as it begins to settle, which is stated as sweep floc that include other coagulation mechanisms in order to form the initial floc. Sweep floc is predominate mechanism of Coagulation in most of the conventional water treatment plants [7].

The mechanism of sweep coagulation is associated with metal coagulants like Al^{3+} and Fe^{3+}. These metals tend to produce huge quantity of metal hydroxide when added in larger concentrations. The formed amorphous hydroxide sweeps particles of colloid on its way towards downward for settling [109]. Significant particle removal is observed with sweep coagulation in comparison to only destabilization by charge neutralization. This can be partly owed to significantly enhanced aggregation rate, due to the increased concentration of solid. The open structure of hydroxide precipitates results in a large efficient volume concentration even with a small mass improving the possibility of adsorbing additional particles. Amirthatajah et al. 1990 illustrated that coagulation mechanisms are governed by pH and coagulant dose.

5 Factors Governing Coagulation

Factors that govern Coagulation include temperature, pH, alkalinity, concentration and composition of ions, type of coagulant used and dosage of coagulant [57, 102]. Characteristics of floc obtained from chemical coagulation using alum was compiled from different studies and anticipated that these principles are same for natural coagulants owing to the similar mechanisms of Coagulation. Depending on type and nature of natural coagulant, chief mechanisms of Coagulation important for turbidity removal is neutralization of charge and bridging, which are also governed by pH and coagulant dose. Hence, in order to understand complete prospective of natural coagulants optimization studies are required.

5.1 Temperature

One of the important parameters that governs water treated by coagulation-flocculation is temperature. At lower temperatures the process of Coagulation and flocculation is affected due to alteration in solubility of coagulant, increase in viscosity of water and obstructing the kinetics of hydrolysis reactions and particle flocculation. Furthermore, lower temperatures require elevated coagulant dosage, inclusion of filter aids, extended flocculation durations and lesser flotation, sedimentation and filtration rates, precisely upon use of alum.

5.2 pH

The most significant parameter affecting the surface charge of colloids, charge on functional groups of natural organic matter, charge on the coagulant species in dissolved phase, solubility of coagulant and surface charge of floc particles is pH, which is important for effective Coagulation.

5.3 Alkalinity

pH reduction during the coagulation process is limited by the alkalinity of raw water. At lower alkalinity concentrations coagulants tend to consume entire alkalinity available resulting in reduction of pH for efficient treatment. On the other hand, presence of high concentration alkalinity might require higher coagulant dose in order to reduce the pH towards favourable coagulation conditions. pH adjustment can be taken up by adding acid in case of high alkalinity, but this process might prove difficult for water services which deal with variations in influent of both alkalinity and coagulant dosage.

5.4 Type of Coagulants

Owing to the numerous interrelated parameters in the coagulation and flocculation process it is utmost vital to choose the precise coagulant, considering the quality of water entering the water treatment plant. Usually when common coagulants do not meet the required standards of treatment then coagulant aids are used. Hence, the type of coagulant selected should fit its use under critical conditions, so that the treated water can be of high quality for the user. Moreover, increase in sludge volume and inappropriate dewatering of sludge is due to the type of coagulant applied, which can be reduced by using a suitable coagulant [50].

5.5 Coagulant Dosage

Coagulant dosage is a significant parameter for the process of Coagulation and flocculation since it requires utmost care and appropriate control for achieving desired quality of treated water. In general, scanty coagulant dosage will result in substandard quality of treated water, due to the lack of process optimization. Diverse quality of impurities possessed by diverse water samples will need optimization of coagulant dose along with the type of coagulant used. Therefore, determining optimum coagulant dosage is required for achieving utmost turbidity removal at least cost of treatment. Charge neutralization and bridging are the only two mechanisms of the four coagulation mechanisms which are negatively affected by dosage of coagulant because of the stoichiometric relationship.

Coagulant dosages can fall under three categories like under dosing, optimum dosing and over dosing. Upon exceeding optimum coagulant dosage particle restabilization resulting in reduced turbidity removal efficacies. While surplus cationic coagulant adsorption in colloidal particles of neutral nature would result in reversing of charge which consequently mark repulsion of particles hampering the agglomeration of particles. Moreover, crowded coagulants of solution might further limit availability of adsorption sites for bridging of particles since the surface of the coagulant will be completely covered [20]. Hence, adding higher coagulant doses might not improve the coagulant process. On the other hand, coagulant dose lesser than required would result in inadequate and inefficient coagulation process as majority of colloidal particles are left in suspension.

6 Types of Coagulants

Coagulants can be either natural or chemical-based, natural coagulants are recognized for use in conventional water purification as apparent from citation in numerous ancient records [22, 44]. While alum has globally spread as coagulant intended to treat public water supply during 19th century [68]. China is world's primitive user of alum in treatment of water, which was adopted by other nations. Metal coagulants like ferric salts have reigned since 1880 in the United States but efficiency of chemical coagulants was doubtful under some working conditions like low temperature which caused admittance of polymerized aluminium coagulants (PACl). PACl reaped mounting market due to its greater efficiency and lower alkalinity consumption [20].

Infringement of chemical coagulants have ceased down the traditional water treatment methods employing natural coagulants which were restricted to rural and developing countries. This was marked as the commencement of paradigm change towards dependency on chemical coagulants for treatment of turbid waters. This shift has gradually resulted to dormancy in development of natural coagulants, making traditional methods obsolete.

6.1 Chemical Coagulants

The most widely accepted coagulants in water and wastewater treatment are Aluminum [Al (III)] and Iron [Fe (III)] based coagulants and organic coagulants (Fig. 4). Furthermore, composite coagulants are being explored in the recent years [25]. Metal salts of aluminium and iron form the metal coagulants which include Aluminum sulfate (Alum), Aluminum chloride, Acidified Aluminum Sulfate (Acid

Flow Chart

- **Metal Coagulants**
  - Al based
  - Fe based
- **Organic Coagulants**
  - Cationic
  - Anionic
  - Non-Ionic
- **Polymerized Inorganic Coagulants**
  - Al or Fe (III) based polymeric inorganic coagulants
  - Al and Fe based blended polymeric inorganic coagulants
  - Si based polymeric inorganic coagulants
  - Composite Coagulants
  - Composite Inorganic - Organic Coagulants

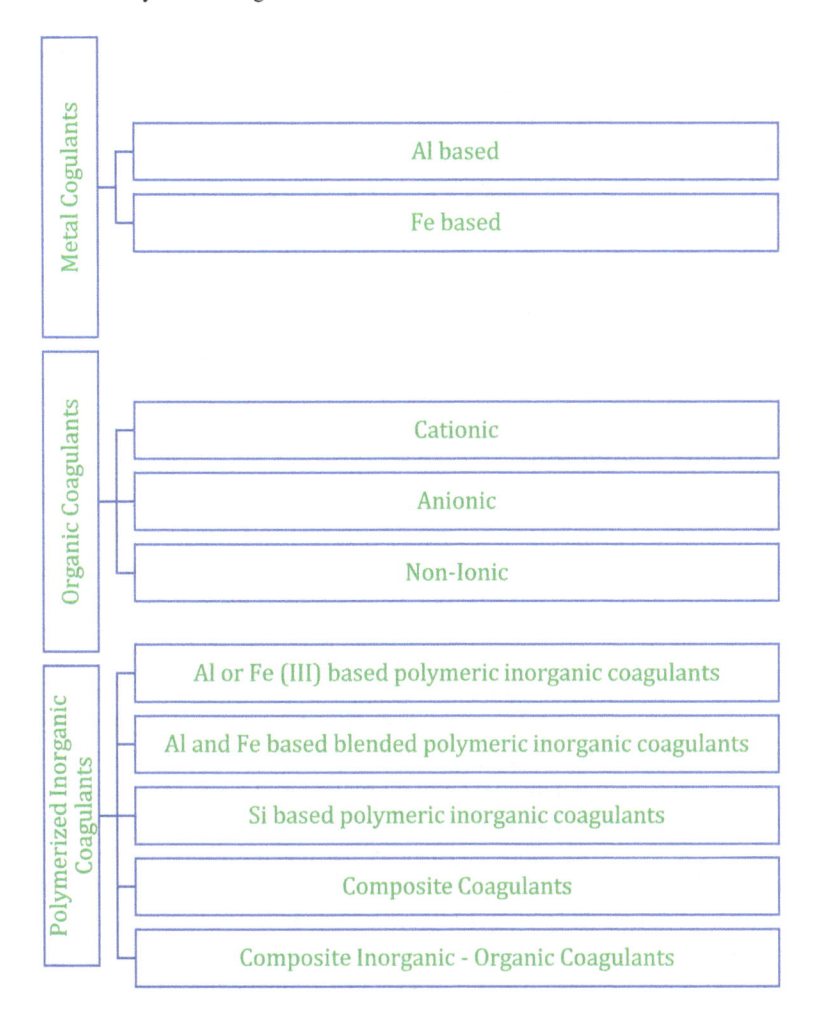

Fig. 4 Flow chart of different coagulants

Alum), Sodium Aluminate, Ferrous Sulfate, Ferric Sulfate (Alum), Ferric Chloride and Chlorinated Ferrous Sulfate. Coagulants are described as follows:

6.1.1 Aluminium Based Metal Coagulants

Widely adopted aluminium based metal coagulant is alum ($Al_2(SO_4)_3$), others being aluminium chloride ($AlCl_3$). Use of alum, however, can leave comparatively high aluminium residues in the treated water precisely under conditions of low temperatures and low pH causing health effects and other difficulties in distribution system like impulsive flocculation. Though it can be prevented by controlling pH, but then natural organic matter and turbidity will also be affected [28, 51].

6.1.2 Ferric-Based Metal Coagulants

Commonly used ferric salt-based coagulants are ferric chloride ($FeCl_3$) and ferric sulphate ($Fe_2(SO_4)_3$) [23]. Ideal pH range for ferric-based coagulants is reported to be 4.5 to 6 with 29 to 70% removal of NOM in terms of dissolved organic carbon [1, 21, 23].

6.2 Organic Coagulants

Utilized as primary coagulants or coagulant aids for drinking water treatment, organic polymers are polymers with high-molecular-weight generally known as polyelectrolytes. When only organic polymers are used these may be effective in particle destabilization but result in production of poor-quality floc. Furthermore, organic polymers do not possess any disinfection by-product precursor removal [48]. Polymer preparations can contain contaminants originating from manufacturing process which include residual monomers, reaction by-products, which might have a potential negative impact on health of humans. Furthermore, polymers containing such contaminants can undergo reaction with chemicals added for treating water forming undesirable secondary products. These concerns have made countries like Switzerland and Japan to adopt legislations barring utilization of such organic (synthetic) polymers in treatment of potable water. Stringent limits on use of organic polymers have been established by United States, France and Germany [2, 81].

Classification of organic polymers is based on the electric charge on their macro ions that dissociate in water. These are Cationic, Anionic and Non-ionic. Among these cationic polymers are generally used as primary coagulants, whereas anionic and non-ionic polymers are stated as coagulant aids or flocculants.

6.2.1 Cationic Polyelectrolyte

Example of this class include Poly dimethyldiallyl ammonium chloride produces ions which are positively charged upon dissolution in water. These are generally utilized as primary coagulants/coagulant aids to metal coagulants owing to the fact that suspended and colloidal particles present in water are usually negatively charged.

6.2.2 Anionic Polyelectrolyte

Example of this class include Copolymer acrylamide/Trimethylaminoethyl acrylate, which produce negatively charged ions upon dissolution in water. These polymers are generally used in blend with metal coagulants for removing positively charged solids.

6.2.3 Non-ionic Polyelectrolyte

Example of this class include Non-ionic polyacrylamide, which have balanced ions, releasing both cations and anions upon dissolution in water.

6.3 Polymerized Inorganic Coagulants

The development and use of polymerized inorganic coagulants trace back to 1980s. Since than these coagulants have proved to be promising with their greater performance in treating water and wastewater. Their performance can be owed to their wider working pH range, lower sensitivity towards water temperature and moreover these require quite small dosages to attain the same treatment competence and also result in very fewer residual concentrations of metal-ions. These characteristics are accredited to the existence of wide range of polymeric species consisting great molecular weights. In general polymers possess higher cationic charge that increase their surface activity and capacity of charge-neutralization, making them more competitive, enhancing the rate of colloidal charge neutralization, floc development and settlement, cost reduction, production of less sludge over conventional coagulants [56, 71]. Generally used inorganic polymeric coagulants are Al and Fe (III) based, a few examples are as follows:

6.3.1 Aluminum Chlorohydrate

Probably the first ploymerized coagulant that was developed in United Kingdom during 1950s. Used both as sludge conditioner and as primary coagulant it has a theoretical formula $Al_2(OH)_5Cl$.

6.3.2 Al or Fe (III) Based Polymeric Inorganic Coagulants

Having wield application for water and wastewater treatment these include Polyferric chloride and Polyaluminum chloride, which are hydrolysed products of Al and Fe (III) [38, 79]. Other hydrolysed products from aluminum chloride with aluminum sulphate and ferric sulphate are Polyaluminum chloride Sulphate and Polyferric Sulphate respectively [56, 71].

6.3.3 Al and Fe (III) Based Blended Polymeric Inorganic Coagulants

Blended inorganic coagulant prepared from solution of Al and Fe which is hydrolyzed with alkaline materials is polyaluminumferric chloride [54].

6.3.4 Si Based Polymeric Inorganic Coagulants

These are prepared from metal polymerized inorganic coagulants in combination with polysilicic acid [19, 54, 55, 91]. Ex: Polyferric Silicate chloride, Polyaluminum Silicate chloride, and Poly-aluminum-ferric Silicate chloride, etc.

6.3.5 Composites Coagulants

When cationic polyelectrolyte organic polymers are injected into metal polymerized inorganic coagulants, these form composite coagulants. Ex: PACl with PDMDAAC, PFCl with PDMDAAC [133].

6.3.6 Composite Inorganic-Organic Coagulants

Latest innovation in coagulants is preparing composite coagulants by combining organic polymers to inorganic polymerized flocculants forming inorganic-organic coagulants to possess the advantages of both components.

6.4 Drawbacks of Chemical Coagulants

It can be understood that 21st century have been driven by chemical coagulants for water and wastewater treatment primarily because of comfort of availability, management, use, storage and mixing, being economical etc. [83, 118, 125]. Though these have been widely applicable, their usage comes with limitations and is being probed upon imparting brown colour to equipment by iron salts and impacts on environment due to higher concentrations of residual aluminium is in treated water, wastewater and sludge [37, 83, 136].

Harmful effects of using chemical coagulants upon human health were published in 1960s [121]. Moreover biochemical, neuropathological and epidemiological studies advocate [13, 86, 103, 110, 131] and senile dementia due to residual aluminium in treated waters. Results from several epidemiological and clinical observation reported at least 70% positive correlation between Alzheimer's disease and aluminium residue in drinking water. Nevertheless, conflicting results were also reported owing to absence of significant indication in several cases [134]. Therefore, a decisive relation between aluminium and development of Alzheimer's disease cannot be determined. However, in order to minimize risk over sustained period of consumption, threshold concentration of aluminium in treated water was observed and stated to be 0.02 mg/L [42], 0.1 mg/L [86, 110] and proposed aluminium concentration by WHO is 0.2 mg/L [121].

Limitations of their usage include their ineffectiveness at lower temperatures, harmful human health effects, generation of large volumes of sludge and pH variations in treated water [30, 116]. Environments like extreme pH and temperature result in production of very sensitive and fragile flocs leading to poor sedimentation and these flocs might undergo rupture upon any physical force. Generation of huge hydrous oxide natured sludge that is non-biodegradable is additional chief drawback of using metal coagulants. It is also reported that sludge obtained from using alum is hard to dewater. Also, alum sludge can cause uptake of phosphorous from plants through absorption of inorganic phosphorus leading to aluminium phytotoxicity [41]. Owing to these drawbacks of chemical coagulants, an urgent need for considering other potential alternatives fitting in the objective of green chemistry is required for treating water, which can minimize environmental damages and protect wellbeing of human population.

6.5 Advantage of Using Natural Coagulants

Advantages of natural coagulants are evident; they are economical, will not change pH of treated water [135], highly biodegradable, require low dose of coagulant, omit pH adjustment as these will not consume water alkalinity like alum [98], low/negligible toxicological risk, harmless to human health and aquatic life, subsequently lesser impact on environment [73].

Additionally, these are renewable in nature and hence have negligible net impact on global warming [117], does not cause secondary pollution [109], generally possess several surface charges increasing the efficacy of coagulation process [52], they are non-corrosive, eliminating concern of pipe line corrosion [30]. Furthermore, they produce lesser sludge volumes, nearly five times lesser, which can be treated biologically and disposed of as soil conditioners as the sludge possess higher nutritional value and is non-toxic [94]. This will reduce treatment and handling costs of sludge, making natural coagulants better sustainable option [31].

All the above-mentioned advantages are specifically amplified if the coagulant derived plant is indigenous or if these plants are cultivated, harvested and processed locally [52]. Moreover, in the present stage of climate change, environmental degradation and depleting natural resources, adaptation of natural coagulants, satisfying the requirements of green technology is vital effort towards sustainable development initiatives [88, 135].

7 Natural Coagulants

Due to environmental concerns worldwide, searching for alternatives to the chemical coagulants for water treatment has become compulsory as they produce toxic substances during their extraction and purification process. Thus, several researchers are searching for and exploring natural coagulants to treat water [36, 37, 116]. Usage of natural coagulants for water purification and wastewater is well-known through human history from ancient times to date [94, 112]. Natural coagulant usage has been dated for more than 200 years in countries like Africa, India, and China for removing turbidity [9]. In India, it is apparent as the Sanskrit writing dating back to 400 AD, also from the Old Testament and Romans records dating back to 77 AD [44]. Natural coagulants can be categorized into cationic, anionic, or non-ionic, also called polyelectrolytes [135]. Examples include [109]:

- Natural Cationic polymers: cationic starches and chitosan
- Natural Anionic polymers: modified lignin sulfonates and sulphated polysaccharides
- Natural Non-ionic polymers: cellulose derivatives and starch [65].

The mechanisms exhibited by the natural coagulants include adsorption and charge neutralization, and adsorption, and inter particulate bridging. Adsorption and neutralization results while suspended particles adsorb on oppositely charged ions; interparticle bridging happens once chains of polysaccharide coagulants adhere to several particles making particles bound to a coagulant and this can be attributed to long-chained structures with great molecular weights, which greatly enhance the number of unoccupied adsorption sites. It is understood that precisely these two mechanisms provide fundamental principles to internal mechanisms of plant-based coagulants. Moreover, electrolytes in the aqueous conditions can enable these coagulants' coagulants, as the electrostatic repulsion between particles is low [66].

Natural coagulants based on the origin can be grouped into three categories: plant-derived, animal-derived microorganism derived (Fig. 5).

- **Animal-derived**: While examples of animal sources include chitosan from the shells of crustaceans [22] and isinglass from shredded fish bladders [16]. Chitosan, natural polysaccharide possesses beneficial features like the competence of adsorbing metal ions due to amino groups, biocompatibility, hydrophilicity, non-toxic, biodegradability, and linear cationic polymer with high molecular weight.
- **Microorganism derived**: examples of microorganisms derived are microbial polysaccharides and alginate. Alginates are known for their characteristics like gelling, keeping water viscosifying and stabilizing properties [80] and Oladoja,

Fig. 5 Flow chart of different categories of natural coagulants

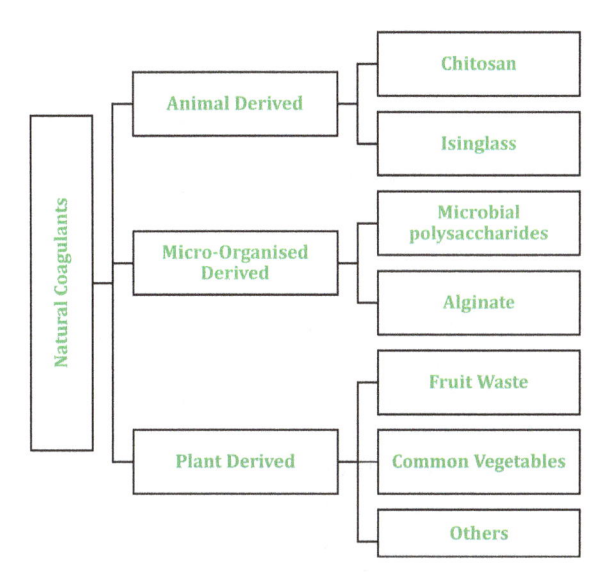

in his studies he reported that high flocculating activity due to the existence of polysaccharides and protein with neutral sugar, an amino sugar, and uranic acids [95, 98].

- **Plant-derived**: The availability of coagulants from plants is more significant than that from animal sources, suggesting that plant-derived coagulants are impending alternatives to chemical coagulants. Hence these have gained attention over the years [29]. These advantages are economical and efficient in treating waters with low to medium turbidity (50–500NTU).

7.1 Classification of Plant-Derived Coagulants

Sources of plant-derived coagulants are vast; these may be directly taken from various parts of plants or even waste after processing plant parts, leading to a disarray in the compilation. From the literature, plant-derived coagulants can be classified as originating from vegetables and legumes, fruit waste, and others (Fig. 6).

7.1.1 Common Vegetables

Moringa Olifera: is native north Indian species belonging to family Moringacea, it is a deciduous tree, and its coagulative properties were identified in the 18th century. The seeds contain positive ion proteins, which act as an antibacterial flocculant and used for water clarification. Amagloh and Benang [6], compared alum with moringa and revealed that 12 g/1000 ml of *Moringa olifera* powder showed analogous turbidity removal to alum. Pritchard et al. [105] reported that using *Moringa olifera*, 84%

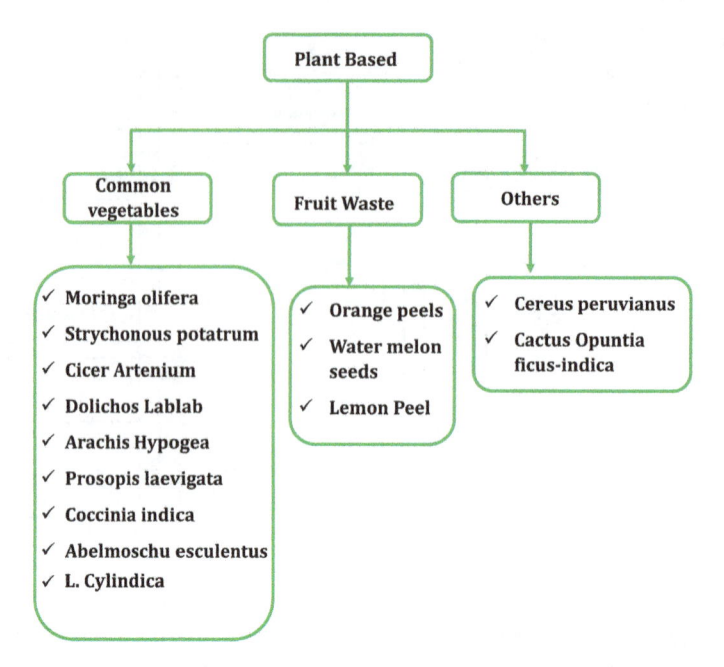

Fig. 6 Coagulants derived from plants

reduction in turbidity was achieved with an initial turbidity of 146 NTU. *Moringa oleifera* is a flocculant that is available naturally around regions like India, Africa, Madagascar, and Arabia; six of fourteen Moringa species are available abundantly. Developing countries like India and Sudan are using these seeds from generations for treating the water, and it is not new for them [59]. Also, comparison studies of these species were done with aluminum sulfate and ferric chloride in turbid water (160 NTU), and this species showed a reduction of 97% turbidity. In contrast, remaining coagulants like aluminum sulfate and ferric chloride showed 99 and 98%. Color removal using these coagulants was studied by Pritchard et al. [105], and achieved 83% reduction in colour by *Moringa oleifera*, 88% by alum and 93% by ferric sulfate. Gaikwad and Munavalli [53], reported that at lower turbidity (35NTU) by using lesser concentrations 50 mg/l, they achieved 90.6% removal. In comparison with other synthetic coagulants, the turbidity removal or performance is less. Hence, *Moringa oleifera* is competent enough and can be used as an alternative to the synthetic coagulants.

Strychnos potatorum: (Nirmali) is a traditional medicinal tree and available in central parts of India, Burma, and Srilanka. It is a medium-sized tree; materials form this tress produce alkaloids like acetyldaibolin and diabolin used for the treatment of epilepsy and urolithiasis. Niramli seeds contain alkaloids like nonvaccine, strychnine, etc., whereas bark has campest Erol, sitosterol, and isomotiol. The seed powder consists of polyelectrolytes of anionic charge and destabilizes the impurities in water by intermolecular bonding. Earlier studies revealed that extracts from the seed comprised of lipids, carbohydrates, and alkaloids containing carboxyl groups and alcohol groups. The carboxyl and alcohol groups are responsible for Coagulation and enhance the intermolecular bonding. Galactan and galactomannan are polysaccharides extracted from the seeds of *Strychnos potatorum* that have a capability in reducing the turbidity of the solution up to 80%. Galactomannans consist of 1,4-linked dmannopyranosyl with terminal ends d-galactopyranosyls. Even though many researchers used nirmali seeds for the removal of turbidity, the exact mechanism that has been associated with the Coagulation was not explained clearly. Some of the researchers summarised that presence of alcohol chains in the seeds is helping for adsorption. In bench-scale studies, *Strychnos potatorum* seeds showed good adsorption, and direct filtration is acting as an effective method for treating low turbidity water [53, 96].

Cicer arietinum: belongs to the family of Fabaceae and commonly called as chickpea, Choubey et al. [27], reported that 95.89% of turbidity reduction had been observed in comparison with alum in his study. These seeds have antibacterial properties and help in the removal of microorganisms from water. *Cicer arietinum* is more effective in removing high turbidity of water by maintaining the experimental setting three minutes high mixing, followed by twelve minutes slow mixing with 100 mg/l dosage. These plant materials are available cheaply in all parts of Asia. Contrarily, these are biodegradable and safe for consumption by humans [9].

Dolichos lablab: belongs to Fabaceae family and is commonly called Hyacinth bean. Shilpaa et al. [119], reported 99.14% of turbidity reduction with an initial NTU of 500 by treating with seeds of *Dolichos lablab*. Asrafuzzaman et al. [9], by their study stated that lower dosages of the coagulants showed high efficiency in removing the turbidity of the water, however increasing the dosage of the coagulants decreased turbidity removal of the solution. The most effective dose of hyacinth bean for turbidity removal is 200 mg/500 ml and showed a 68% removal at 60 min settling time [27].

Arachis hypogea: belong to the family of Fabaceae are known as peanut. Birima et al. [17], used the seeds of Arachis hypogea for turbidity removal and achieved 92% efficiency by extracting the peanuts with 6 mol/NaCl with initial NTU 200 turbidity using 200 mg/l dose. However, peanut extract with distilled water showed only a 31.5% reduction. The proteins present inside the peanut are responsible for the coagulation activity, and they found that KCl and $NaNO_3$ are also suitable solvents to extract the proteins in the coagulants, which are useful for coagulation activity.

Prosopis laevigata: commonly known as honey mesquite and belongs to Fabaceae, these trees can grow up to 13 meters. *Prosopis laevigata* have similar characteristics to the *Prosopis galactomannans* [24]. The gums of these seeds are generally used as a coagulant. Torres and Carpinteyro-Urban [127], used these seeds and concluded that galactomannans present in the seeds act as a coagulating agent for treating the water and wastewater.

Coccinia Indica: Patale and Pandya [99], used the *C. indica* fruit extract and stated that mucilage of these substances showed high efficiency in removing turbidity in comparison to low turbidity. Turbidity removal with an initial 100 NTU was obtained 94% removal, whereas the lowest turbidity 10 NTU got 82% removal.

Abelmoschus esculentus: Abelmoschus esculentus, commonly called Okra, belongs to the family Malvaceae. Mishra et al. [120], used okra seeds powder as a coagulant and compared the turbidity removal with alum. Okra seeds showed similar results to alum in clarification, 72% turbidity removal was achieved by using lesser concentrations of the coagulant in their study. Raji et al. [107], conducted similar investigations and used optimum dosage of 300 mg/l and reduced turbidity from 745 NTU to 11 NTU at neutral pH and concluded that using optimum dosage will give 99% efficiency.

Luffa cylindrica: also called sponge guard, consists of light-coloured black seeds. These plants are using traditional medicine as it contains triterpenoids and a-tocopherol acts as an anti-inflammatory drug [11]. *Luffa cylindrica* belongs to the family Cucurbitaceae. Sowmeyan et al. [124], used the extracts of the sponge guard and removed turbidity up to 85%. Extracts of the plant have not been specified by the authors and postulated that the plant extract might be seeds or whole fruits [4].

7.1.2 Fruit Waste as Coagulants

Citrus sinensis: are commonly called as oranges and belong to the family of Rutaceae. The peels of these fruits consist of carboxyl and hydroxyl functional groups along with substances like pectin and cellulose. The presence of these groups in the peels is acting as an adsorbent material in removing the organic substance from the water [115]. Pathak et al. [100], used the peel powder of orange and achieved 88.4% turbidity removal by using 60 mg/l dosage. They concluded that orange peel powder has the potential to reduce the turbidity. Klancnik et al. [76], used the peel powder and removed the colour up to 99.7%, and the efficiency of these substances is compared with the activated carbon treatment.

Citrus limon: belongs to Rutaceae family and locally called as lemon, and these fruits are available all over the world. Peels of these fruits have been used as a coagulant from ancient days and theses are non-corrosive and non-toxic to water treatment units and reduce sludge during the treatment. Dollah et al. [43], used the peel powder of Citrus limon and reduced the turbidity to 89% by using a coagulant dose of 60 mg/l.

Musa: The scientific name of the banana is Musa and is available around the word as edible fruit. The peels of this fruit are used for various purposes like composting and cosmetics preparation. These substances are biodegradable and will not cause any type of pollution. The Musa peels have high adsorption capacities due to the presence of lignocellulose. Mokhtar et al. [90], used banana peel powder as a coagulant and reduced the turbidity of water with 88% efficiency. Similar studies were conducted by John et al. [72], and achieved 95% turbidity removal. The stem juice of the Musa was used as a coagulant by Alwi et al. [5], and reported 98.5% efficiency in removing turbidity with 90 ml/l dosage.

Citrullus lanatus: is a plant species belonging to the Cucurbitaceae family and commonly known as watermelon. More than a thousand varieties are available and cultivable around the world. Datti et al. [35], used the seed powder of melon and achieved 75% turbidity removal with initial turbidity 58.7 NTU. Similar studies were conducted by Sekar et al. [113], and treated turbid waters with watermelon seed powder and obtained results within limits given by world health organization.

7.1.3 Others

Cactus Opuntia ficus-indica: belongs to cacti species, and it consists of mucilaginous cells and stores mucilage in both peel and pulp. These plant species are available in semi-arid regions of the world, and it can grow up to five meters. Many researchers [69, 88] studied the characteristics of opuntia, and mucilage is used as a coagulant. Owing to its availability and low cost, it has gained more importance in treating water and wastewater. Bandala et al. [12], converted the mucilage into powder and used as an adsorbent for coagulation technique and achieved excellent results.

Cereus peruvianus: belongs to the family of Cactaceae and commonly known as Peruvian Apple Cactus. These plants can grow up to 10 meters and is widely used as an ornamental plant. De Souza et al. [37], extracted mucilage from the Peruvian and used for the water clarification and achieved 95% efficiency in turbidity removal.

8 Conclusions

Nature's diversity and stability provide numerous solutions; so that species on the earth are distributed equally, and no other species are exhausted. Thus, knowledge of more materials having coagulating properties might provide options even in the abnormal or worst conditions leading to resilience, one of the primary sustainability principles. The present book chapter presents the properties and efficiency of natural coagulants classified under plant-based, fruit waste, etc. Furthermore, an in-depth understanding of the above said coagulants concerning their optimizing process parameters is essential in future research and exploring more natural resources having coagulating properties for treating water and wastewater.

References

1. Abbaszadegan M, Mayer BK, Ryu H, Nwachuku N (2007) Efficacy of removal of CCL viruses under enhanced coagulation conditions. Environ Sci Technol 41(3):971–977
2. Aizawa T (1990) Problems with introducing synthetic polyelectrolyte coagulants into the water purification process. Water supply 8:27–35
3. Al-Saati NHA, Hwaidi EH, Jassam SH (2016) Comparing cactus (Opuntia spp.) and alum as coagulants for water treatment at Al-Mashroo canal: a case study. Int J Environ Sci Technol 13(12):2875–2882
4. Altınışık A, Gür E, Seki Y (2010) A natural sorbent, Luffa cylindrica for the removal of a model basic dye. J Hazard Mater 179(1–3):658–664

5. Alwi H, Idris J, Musa M, Ku Hamid KH (2013) A preliminary study of banana stem juice as a plant-based coagulant for treatment of spent coolant wastewater. J Chem 2013
6. Amagloh FK, Benang A (2009) Effectiveness of Moringa oleifera seed as coagulant for water purification
7. Amirtharajah A, O'melia CR (1990) Coagulation processes: destabilization, mixing, and flocculation. McGraw-Hill, Inc., USA, p 1194
8. Antov MG, Šćiban MB, Prodanović JM (2012) Evaluation of the efficiency of natural coagulant obtained by ultrafiltration of common bean seed extract in water turbidity removal. Ecol Eng 49:48–52
9. Asrafuzzaman M, Fakhruddin ANM, Hossain M (2011) Reduction of turbidity of water using locally available natural coagulants. ISRN Microbiol 2011
10. Austen PT, Wilber FA (1884) Annual report of the State Geologist of New Jersey
11. Azeez MA, Bello OS, Adedeji AO (2013) Traditional and medicinal uses of Luffa cylindrica: a review. J Med Plants 1(5):102–111
12. Bandala ER, Tiro JB, Lujan M, Camargo FJ, Sanchez-Salas JL, Reyna S, Torres LG (2013) Petrochemical effluent treatment using natural coagulants and an aerobic biofilter. Adv Environ Res 2(3):229–243
13. Banks WA, Niehoff ML, Drago D, Zatta P (2006) Aluminum complexing enhances amyloid β protein penetration of blood–brain barrier. Brain Res 1116(1):215–221
14. Beltrán-Heredia J, Sánchez-Martín J, Gómez-Muñoz MC (2010) New coagulant agents from tannin extracts: preliminary optimisation studies. Chem Eng J 162(3):1019–1025
15. Benetti AD (2008) Water reuse: issues, technologies, and applications. Engenharia Sanitaria e Ambiental 13(3):247–248
16. Biggs S (2015) Polymeric flocculants. In: Encyclopedia of surface and colloid science. CRC Press, pp 5918–5934
17. Birima AH, Hammad HA, Desa MNM, Muda ZC (2013) Extraction of natural coagulant from peanut seeds for treatment of turbid water. In: IOP conference series: earth and environmental science, vol 1. IOP Publishing Ltd., pp 1–4
18. Black AP (1934) Coagulation with iron compounds. J (Am Water Works Assoc) 26(11):1713–1718
19. Boisvert JP, To TC, Berrak A, Jolicoeur C (1997) Phosphate adsorption in flocculation processes of aluminium sulphate and poly-aluminium-silicate-sulphate. Water Res 31(8):1939–1946
20. Bolto B, Gregory J (2007) Organic polyelectrolytes in water treatment. Water Res 41(11):2301–2324
21. Bond T, Goslan EH, Parsons SA, Jefferson B (2010) Disinfection by-product formation of natural organic matter surrogates and treatment by coagulation, MIEX® and nanofiltration. Water Res 44(5):1645–1653
22. Bratby J (2006) Coagulation and flocculation in water and wastewater treatment, 2nd edn. IWA Publishing, London
23. Budd GC, Hess AF, Shorney-Darby H, Neemann JJ, Spencer CM, Bellamy JD, Hargette PH (2004) Coagulation applications for new treatment goals. J-Am Water Works Assoc 96(2):102–113
24. Chaires-Martínez L, Salazar-Montoya JA, Ramos-Ramírez EG (2008) Physicochemical and functional characterization of the galactomannan obtained from mesquite seeds (Prosopis pallida). Eur Food Res Technol 227(6):1669
25. Chen Z, Fan B, Peng X, Zhang Z, Fan J, Luan Z (2006) Evaluation of Al30 polynuclear species in polyaluminum solutions as coagulant for water treatment. Chemosphere 64(6):912–918
26. Cheng RC, Krasner SW, Green JF, Wattier KL (1995) Enhanced coagulation: a preliminary evaluation. J-Am Water Works Assoc 87(2):91–103
27. Choubey S, Rajput SK, Bapat KN (2012) Comparison of efficiency of some natural coagulants-bioremediation. Int J Emerg Technol Adv Eng 2(10):429–434
28. Chow CW, van Leeuwen JA, Fabris R, Drikas M (2009) Optimised coagulation using aluminium sulfate for the removal of dissolved organic carbon. Desalination 245(1–3):120–134

29. Choy SY, Prasad KMN, Wu TY, Ramanan RN (2015) A review on common vegetables and legumes as promising plant-based natural coagulants in water clarification. Int J Environ Sci Technol 12(1):367–390
30. Choy SY, Prasad KMN, Wu TY, Raghunandan ME, Ramanan RN (2014) Utilization of plant-based natural coagulants as future alternatives towards sustainable water clarification. J Environ Sci 26(11):2178–2189
31. Choy SY, Prasad KN, Wu TY, Raghunandan ME, Ramanan RN (2016) Performance of conventional starches as natural coagulants for turbidity removal. Ecol Eng 94:352–364
32. Cisneros BJ, Rose JB (eds) (2009) Urban water security: managing risks: UNESCO-IHP, vol 5. CRC Press
33. Crittenden JC, Trussel RR, Hand DW, Howe KJ, Tchobanoglous G (2005) Coagulation, mixing and flocculation. Water Treat Princi Design 2
34. Crozes G, White P, Marshall M (1995) Enhanced coagulation: its effect on NOM removal and chemical costs. J-Am Water Works Assoc 87(1):78–89
35. Datti Y, Barau SS, Nura T (2019) Chemical compositions and the phytochemical constituents of the seed of Sesamum Indicum grown at Katsina State, Northern Nigeria. Fudma J Sci 3(4):201–205. Issn: 2616-1370
36. de Souza MTF, Ambrosio E, de Almeida CA, de Souza Freitas TKF, Santos LB, de Cinque Almeida V, Garcia JC (2014) The use of a natural coagulant (Opuntia ficus-indica) in the removal for organic materials of textile effluents. Environ Monit Assess 186(8):5261–5271
37. de Souza MTF, de Almeida CA, Ambrosio E, Santos LB, de Souza Freitas TKF, Manholer DD, ..., Garcia JC (2016) Extraction and use of Cereus peruvianus cactus mucilage in the treatment of textile effluents. J Taiwan Inst Chem Eng 67:174–183
38. Dempsey BA, Ganho RM, O'Melia CR (1984) The coagulation of humic substances by means of aluminum salts. J-Am Water Works Assoc 76(4):141–150
39. Dempsey BA, Sheu H, Ahmed TT, Mentink J (1985) Polyaluminum chloride and alum coagulation of clay-fulvic acid suspensions. J-Am Water Works Assoc 77(3):74–80
40. Deryagin BV, Landau LD (1941) Theory of stability of strongly charged lyophobic sols and adhesion of strongly charged particles in electrolytic solutions. J Exp Theor Phys 11:802
41. Dharmappa HB, Hasia A, Hagare P (1997) Water treatment plant residuals management. Water Sci Technol 35(8):45–56
42. Doll R (1993) Alzheimer's disease and environmental aluminium. Age Ageing 22(2):138–153
43. Dollah Z, Abdullah ARC, Hashim NM, Albar A, Badrealam S, Zaki ZM (2019) Citrus fruit peel waste as a source of natural coagulant for water turbidity removal. In: Journal of physics: conference series, vol 1349, no 1. IOP Publishing, p 012011
44. Dorea CC (2006) Use of Moringa spp. seeds for coagulation: a review of a sustainable option. Water Sci Technol Water Supply 6(1):219–227
45. Duan J, Gregory J (2003) Coagulation by hydrolysing metal salts. Adv Coll Interface Sci 100:475–502
46. Duan J, Niu A, Shi D, Wilson F, Graham NJD (2009) Factors affecting the coagulation of seawater by ferric chloride. Desalinat Water Treat 11(1–3):173–183
47. Edzwald JK (1993) Coagulation in drinking water treatment: particles, organics and coagulants. Water Sci Technol 27(11):21–35
48. Edzwald JK, Becker WC, Wattier KL (1985) Surrogate parameters for monitoring organic matter and THM precursors. J-Am Water Works Assoc 77(4):122–132
49. Faust SD, Aly OM (1998) Chemistry of water treatment. CRC Press
50. Fazeli M, Safari M, Ghobaee T (2014) Selecting the optimal coagulant in turbidity removal and organic carbon of surface water using AHP. Bull Environ Pharmacol Life Sci 3:78–88
51. Flaten TP (2001) Aluminium as a risk factor in Alzheimer's disease, with emphasis on drinking water. Brain Res Bull 55(2):187–196
52. Freitas TKFS, Oliveira VM, De Souza MTF, Geraldino HCL, Almeida VC, Fávaro SL, Garcia JC (2015) Optimization of coagulation-flocculation process for treatment of industrial textile wastewater using okra (A. esculentus) mucilage as natural coagulant. Ind Crops Prod 76:538–544

53. Gaikwad VT, Munavalli GR (2019) Turbidity removal by conventional and ballasted coagulation with natural coagulants. Appl Water Sci 9(5):130
54. Gao BY, Hahn HH, Hoffmann E (2002) Evaluation of aluminum-silicate polymer composite as a coagulant for water treatment. Water Res 36(14):3573–3581
55. Gao BY, Yue QY, Wang BJ (2006) Properties and coagulation performance of coagulant poly-aluminum-ferric-silicate-chloride in water and wastewater treatment. J Environ Sci Health, Part A 41(7):1281–1292
56. Gao B, Yue Q (2005) Effect of $SO_4 2-/Al^{3+}$ ratio and $OH-/Al^{3+}$ value on the characterization of coagulant poly-aluminum-chloride-sulfate (PACS) and its coagulation performance in water treatment. Chemosphere 61(4):579–584
57. Geng Y (2006) Application of flocs analysis for coagulation optimization at the Split Lake water treatment plant
58. Ghebremichael KA, Gunaratna KR, Henriksson H, Brumer H, Dalhammar G (2005) A simple purification and activity assay of the coagulant protein from Moringa oleifera seed. Water Res 39(11):2338–2344
59. Gottsch E (1992) Purification of turbid surface water by plants in Ethiopia. Walia 1992(14):23–28
60. Guarino AS (2017) The economic implications of global water scarcity. Res Econom Manage 2(1):51
61. Gullick RW (2003) AWWA's source water protection committee outlines how to maintain the highest quality source water. J-Am Water Works Assoc 95(11):36–42
62. Gunaratna KR, Garcia B, Andersson S, Dalhammar G (2007) Screening and evaluation of natural coagulants for water treatment. Water Sci Technol Water Supply 7(5–6):19–25
63. Haaroff J, Cleasby JL (1988) Comparing aluminum and iron coagulants for in-line filtration of cold waters. J Am Water Works Assoc 80(4):168–175
64. Hardy WB (1900) A preliminary investigation of the conditions which determine the stability of irreversible hydrosols. Proc R Soc London 66(424–433):110–125
65. Hassan MAA, Hui LS, Noor ZZ (2009) Removal of boron from industrial wastewater by chitosan via chemical precipitation. J Chem Nat Res Eng 4:1–11
66. Huang C, Chen Y (1996) Coagulation of colloidal particles in water by chitosan. J Chem Technol Biotechnol Int Res Process Environ Clean Technol 66(3):227–232
67. Jadhav MV, Mahajan YS (2013) Investigation of the performance of chitosan as a coagulant for flocculation of local clay suspensions of different turbidities. KSCE J Civ Eng 17(2):328–334
68. Jahn SAA (2001) Drinking water from Chinese rivers: challenges of clarification. J Water Supply Res Technol—AQUA, 50(1):15–27
69. Jeon JR, Kim EJ, Kim YM, Murugesan K, Kim JH, Chang YS (2009) Use of grape seed and its natural polyphenol extracts as a natural organic coagulant for removal of cationic dyes. Chemosphere 77(8):1090–1098
70. Jiang JQ (2001) Development of coagulation theory and pre-polymerized coagulants for water treatment. Sep Purif Methods 30(1):127–141
71. Jiang JQ, Graham NJ (1998) Preparation and characterisation of an optimal polyferric sulphate (PFS) as a coagulant for water treatment. J Chem Technol Biotechnol Int Res Process Environ Clean Technol 73(4):351–358
72. John B, Baig U, Fathima N, Asthana S, Sirisha D (2017) Removal of turbidity of water by banana peel using adsorption technology. J Chem Pharmaceut Res 9(4):65–68
73. Kakoi B, Kaluli JW, Ndiba P, Thiong'o G (2016) Banana pith as a natural coagulant for polluted river water. Ecol Eng 95:699–705
74. Katayon S, Noor MMM, Asma M, Ghani LA, Thamer AM, Azni I, Suleyman AM (2006) Effects of storage conditions of Moringa oleifera seeds on its performance in coagulation. Biores Technol 97(13):1455–1460
75. Kingsely OJ, Nnaji JC, Ugwu BI (2017) Biodisinfection and coagulant properties of mixed Garcinia kola and Carica papaya seeds extract for water treatment. Chem Sci Int J 1–9
76. Klančnik M (2014) Coagulation and adsorption treatment of printing ink wastewater. Acta graphica: znanstveni časopis za tiskarstvo i grafičke komunikacije 25(3–4):73–82

77. Kumar P, Rouphael Y, Cardarelli M, Colla G (2017) Vegetable grafting as a tool to improve drought resistance and water use efficiency. Front Plant Sci 8:1130
78. Langelier WF, Ludwig HF (1949) Mechanism of flocculation in the clarification of turbid waters. J (Am Water Works Assoc) 41(2):163–181
79. Leprince A, Fiessinger F, Bottero JY (1984) Polymerized iron chloride: an improved inorganic coagulant. J-Am Water Works Assoc 76(10):93–97
80. Mabrouk ME (2014) Production of bioflocculant by the marine actinomycete Nocardiopsis aegyptia sp. nov. Life Sci J 11:27–35
81. Mallevialle J, Bruchet A, Fiessinger F (1984) How safe are organic polymers in water treatment? J-Am Water Works Assoc 76(6):87–93
82. Martyn CN, Coggon DN, Inskip H, Lacey RF, Young WF (1997) Aluminum concentrations in drinking water and risk of Alzheimer's disease. Epidemiology 281–286
83. Matilainen A, Vepsäläinen M, Sillanpää M (2010) Natural organic matter removal by coagulation during drinking water treatment: a review. Adv Coll Interface Sci 159(2):189–197
84. Mattson S (1928) The electrokinetic and chemical behavior of the alumino-silicates. Soil Sci 25(4):289–312
85. McConnachie GL, Folkard GK, Mtawali MA, Sutherland JP (1999) Field trials of appropriate hydraulic flocculation processes. Water Res 33(6):1425–1434
86. McLachlan DRC (1995) Aluminium and the risk for Alzheimer's disease. Environmetrics 6(3):233–275
87. Miller RG, Kopfler FC, Kelty KC, Stober JA, Ulmer NS (1984) The occurrence of aluminum in drinking water. J-Am Water Works Assoc 76(1):84–91
88. Miller SM, Fugate EJ, Craver VO, Smith JA, Zimmerman JB (2008) Toward understanding the efficacy and mechanism of Opuntia spp. as a natural coagulant for potential application in water treatment. Environ Sci Technol 42(12):4274–4279
89. Miltner MJ, Nolan SA, Summers RS (1994) Evolution of enhanced coagulation for DBP coagulation: critical issues in water and wastewater treatment. In: Proceeding of the 1994 national conference on environmental engineering, ASCE, Boulder, CO
90. Mokhtar NM, Priyatharishini M, Kristanti RA (2019) Study on the effectiveness of banana peel coagulant in turbidity reduction of synthetic wastewater. Int J Eng Technol Sci 6(1):82–90
91. Moussas PA, Zouboulis AI (2008) A study on the properties and coagulation behaviour of modified inorganic polymeric coagulant—polyferric silicate sulphate (PFSiS). Sep Purif Technol 63(2):475–483
92. Moussas PA, Tzoupanos ND, Zouboulis AI (2011) Advances in coagulation/flocculation field: Al-and Fe-based composite coagulation reagents. Desalinat Water Treat 33(1–3):140–146
93. Muyibi SA, Okuofu CA (1995) Coagulation of low turbidity surface waters with Moringa oleifera seeds. Int J Environ Stud 48(3–4):263–273
94. Ndabigengesere A, Narasiah KS, Talbot BG (1995) Active agents and mechanism of coagulation of turbid waters using Moringa oleifera. Water Res 29(2):703–710
95. Nwodo UU, Green E, Mabinya LV, Okaiyeto K, Rumbold K, Obi LC, Okoh AI (2014) Bioflocculant production by a consortium of Streptomyces and Cellulomonas species and media optimization via surface response model. Colloids Surf, B 116:257–264
96. Ociński D, Jacukowicz-Sobala I, Kociołek-Balawejder E (2016) Alginate beads containing water treatment residuals for arsenic removal from water—formation and adsorption studies. Environ Sci Pollut Res 23(24):24527–24539
97. Ogunsina BS, Radha C, Indrani D (2011) Quality characteristics of bread and cookies enriched with debittered Moringa oleifera seed flour. Int J Food Sci Nutr 62(2):185–194
98. Oladoja NA (2015) Headway on natural polymeric coagulants in water and wastewater treatment operations. J Water Process Eng 6:174–192
99. Patale V, Pandya J (2012) Mucilage extract of Coccinia indica fruit as coagulant-flocculent for turbid water treatment. Asian J Plant Sci Res 2(4):442–445
100. Pathak PD, Mandavgane SA, Kulkarni BD (2015) Fruit peel waste as a novel low-cost bio adsorbent. Rev Chem Eng 31(4):361–381

101. Peavy HS, Rowe DR, Techobanoglous G (1985) Water quality: defination, characteristics, and perspectives. Environ Eng 11–43
102. Pernitsky DJ, Edzwald JK (2006) Selection of alum and polyaluminum coagulants: principles and applications. J Water Supply Res Technol—AQUA 55(2):121–141
103. Polizzi S, Pira E, Ferrara M, Bugiani M, Papaleo A, Albera R, Palmi S (2002) Neurotoxic effects of aluminium among foundry workers and Alzheimer's disease. Neurotoxicology 23(6):761–774
104. Pritchard M, Craven T, Mkandawire T, Edmondson AS, O'neill JG (2010) A comparison between Moringa oleifera and chemical coagulants in the purification of drinking water– an alternative sustainable solution for developing countries. Phys Chem Earth, Parts A/B/C 35(13–14):798–805
105. Pritchard M, Mkandawire T, Edmondson A, O'neill JG, Kululanga G (2009) Potential of using plant extracts for purification of shallow well water in Malawi. Phys Chem Earth, Parts A/B/C 34(13–16):799–805
106. Qudsieh IY, l-Razi AF, Kabbashi NA, Mirghani MES, Fandi KG, Alam MZ, ..., Nasef MM (2008) Preparation and characterization of a new coagulant based on the sago starch biopolymer and its application in water turbidity removal. J Appl Polymer Sci 109(5):3140–3147
107. Raji YO, Abubakar L, Giwa SO, Giwa A (2016) Assessment of coagulation efficiency of okra seed extract for surface water treatment. Int J Sci Eng Res 6(2):1–7
108. Rasteiro MG, Garcia FAP, Ferreira P, Blanco A, Negro C, Antunes E (2008) The use of LDS as a tool to evaluate flocculation mechanisms. Chem Eng Process 47(8):1323–1332
109. Renault F, Sancey B, Badot PM, Crini G (2009) Chitosan for coagulation/flocculation processes–an eco-friendly approach. Eur Polymer J 45(5):1337–1348
110. Rondeau V, Commenges D, Jacqmin-Gadda H, Dartigues JF (2000) Relation between aluminum concentrations in drinking water and Alzheimer's disease: an 8-year follow-up study. Am J Epidemiol 152(1):59–66
111. Rook JJ (1974) Formation of haloforms during chlorination of natural waters
112. Sanghi R, Bhatttacharya B, Singh V (2002) Cassia angustifolia seed gum as an effective natural coagulant for decolourisation of dye solutions. Green Chem 4(3):252–254
113. Sekar M, Sutharesan N, Mashi DA, Shaiful MH, Shazni M, Wei KM, Abdullah MS (2014) Comparative evaluation of antimicrobial properties of red and yellow watermelon seeds. Int J Curr Pharmaceut Res 6(3):35–37
114. Sen AK, Bulusu KR (1962) Effectiveness of nirmali seed as coagulant and coagulant aid. Indian J Environ Health 4:233–244
115. Shah PD, Kavathia S (2015) Development and application of hybrid materials in coagulation and flocculation of wastewater. J Environ Res Develop 9(4):1218–1224
116. Shamsnejati S, Chaibakhsh N, Pendashteh AR, Hayeripour S (2015) Mucilaginous seed of Ocimum basilicum as a natural coagulant for textile wastewater treatment. Ind Crops Prod 69:40–47
117. Sharma BR, Dhuldhoya NC, Merchant UC (2006) Flocculants—an ecofriendly approach. J Polym Environ 14(2):195–202
118. Sher F, Malik A, Liu H (2013) Industrial polymer effluent treatment by chemical coagulation and flocculation. J Environ Chem Eng 1(4):684–689
119. Shilpaa B, Akankshaa K, Girish P (2012) Evaluation of cactus and hyacinth bean peels as natural coagulants. Int J Chem Environ Eng 3(3)
120. Mishra S, Singh S, Srivastava R (2017) Okra seeds: an efficient coagulant. Int J Res Appl Sci Eng 5(VI)
121. Simate GS, Iyuke SE, Ndlovu S, Heydenrych M, Walubita LF (2012) Human health effects of residual carbon nanotubes and traditional water treatment chemicals in drinking water. Environ Int 39(1):38–49
122. Sincero AP, Sincero GA (2002) Physical-chemical treatment of water and wastewater. CRC Press

123. Smoluchowski MV (1917) An experiment on mathematical theorization of coagulation kinetics of the colloidal solutions. Zeitschrift fur physikalisch Chemie 92:129–168
124. Sowmeyan R, Santhosh J, Latha R (2011) Effectiveness of herbs in community water treatment. Int Res J Biochem Bioinf 1(11):297–303
125. Teh CY, Budiman PM, Shak KPY, Wu TY (2016) Recent advancement of coagulation–flocculation and its application in wastewater treatment. Ind Eng Chem Res 55(16):4363–4389
126. Theodoro JP, Lenz GF, Zara RF, Bergamasco R (2013) Coagulants and natural polymers: perspectives for the treatment of water. Plastic Polymer Technol 2(3):55–62
127. Torres LG, Carpinteyro-Urban SL (2012) Use of Prosopis laevigata seed gum and Opuntia ficus-indica mucilage for the treatment of municipal wastewaters by coagulation-flocculation. Nat Resour 3(2):2012
128. Tzoupanos ND, Zouboulis AI (2008) Coagulation-flocculation processes in water/wastewater treatment: the application of new generation of chemical reagents. In: 6th IASME/WSEAS international conference on heat transfer, thermal engineering and environment (HTE'08), August 20th–22nd, Rhodes, Greece, pp 309–317
129. Verwey EJW, Overbeek JTG, Van Nes K (1948) Theory of the stability of lyophobic colloids: the interaction of sol particles having an electric double layer. Elsevier Publishing Company
130. Vijayaraghavan G, Shanthakumar S (2015) Efficacy of alginate extracted from marine brown algae (Sargassum sp.) as a coagulant for removal of direct blue2 dye from aqueous solution. Global Nest J 17(4):716–726
131. Walton JR (2013) Aluminum involvement in the progression of Alzheimer's disease. J Alzheimers Dis 35(1):7–43
132. Wang Q, Yang Z (2016) Industrial water pollution, water environment treatment, and health risks in China. Environ Pollut 218:358–365
133. Wang Y, Gao BY, Yue QY, Wei JC, Zhou WZ (2006) Novel composite flocculent ployferric chloride-polydimethyldiallylammonium chloride (PFC-PDMDAAC): its characterization and flocculation efficiency. Water Pract Technol 1(3)
134. Wettstein A, Aeppli J, Gautschi K, Peters M (1991) Failure to find a relationship between mnestic skills of octogenarians and aluminum in drinking water. Int Arch Occup Environ Health 63(2):97–103
135. Yin CY (2010) Emerging usage of plant-based coagulants for water and wastewater treatment. Process Biochem 45(9):1437–1444
136. Zhu G, Zheng H, Zhang Z, Tshukudu T, Zhang P, Xiang X (2011) Characterization and coagulation–flocculation behavior of polymeric aluminum ferric sulfate (PAFS). Chem Eng J 178:50–59
137. Zouboulis AI, Moussas PA, Tzoupanos ND (2009) Coagulation-flocculation processes applied in water or wastewater treatment. In: Industrial waste: environmental impact, disposal and treatment. Nova Science Publishers, Inc., Hauppauge, pp 289–324

Manoj Kumar Karnena received the B.S. degree in Environmental Science and the M.S. degree in Environmental science from GITAM (Deemed to be) University, India. Later, he pursued his Ph.D. in Water and Wastewater treatment at the same institution. He was the university topper and gold medalist in bachelors and post-graduation. He holds an Environmental Law degree from National Law University, Delhi, and Industrial Safety Degree from Annamalai University, Tamilnadu. Since 2019, he has been working as a teaching assistant with the Department of Environmental Science, GITAM Institute of Science, GITAM (Deemed to be) University, India. His greatest strengths are his research and communication skills. He started teaching and working in technical work in 2019 in various backgrounds. Manoj Kumar Karnena has written more than 50 papers since he started writing in 2017. He has written several book chapters and books on the topic within his research interest and holds two patents. His research interest is in Water and Wastewater Treatment, Natural Coagulants, Nanocatalysis, Storage Technologies, and membrane technology. Manoj Kumar Karnena had won several engineering designs and environmental protection awards in water and wastewater treatment.

Vara Saritha is presently Associate Professor in the Department of Environmental Science, GITAM (Deemed to be University), Visakhapatnam and has been teaching for about 16 years. After completion of her graduation and master's from Andhra University, she pursued her M.Tech (Environmental Management) and Ph.D. in Environmental Science and Technology from JNTU, Hyderabad. She has published around eighty national and international research publications and nearly 8 book chapters to her credit. She has supervised doctoral students. Her research interests include areas like water and wastewater treatment, environmental toxicology, epidemiological studies and corrosion studies.

Application of the Systems Approach and System Standards in Water Safety Plan Development and Implementation

Dejan Vasović, Goran Janaćković, and Ashok Vaseashta

Abstract Since the emergence of the first organized social communities, there has always been a need for continuous and safe provision of basic life necessities such as food and water. Nowadays, access to safe and sustainable drinking water and other sanitation services is vital for a healthy and dignified life. Even the sustainability ratio of any modern society is often connected to availability, affordability and sustainability of water-related services. On the other hand, as a result of different intentional and unintentional stressors affecting water services, contemporary science is increasingly focused on the safety of all sub-systems related to the transformation of water resources, a part of the environment, into an adequate product for end users such as households and industries. Within such a system, both qualitative and quantitative issues should be identified and properly managed. In this sense, the systems approach allows a comprehensive view of a water-system as an entirety, rather than a mere sum of its building blocks. Internal and external application of the systems approach should be considered for the purpose of water safety plan development. Internal application pertains to the drinking water supply system observed as the technical entity for collection, transformation and distribution of drinking water, where the elements of the process are taken into consideration. External application is focused on the context of a broader social environment that governs the system, regarding both normal operations and emergency events. Such stakeholder analysis includes

---

D. Vasović (✉) · G. Janaćković
Faculty of Occupational Safety in Niš, University of Niš, Čarnojevića 10a, Niš, Serbia
e-mail: dejan.vasovic@znrfak.ni.ac.rs

G. Janaćković
e-mail: janackovic.goran@gmail.com

A. Vaseashta
International Clean Water Institute, Manassas, VA, USA

Institute of Electronic Engineering and Nanotechnologies "D. Ghitu", ASM, Chisinau, Moldova

Institute of Biomedical and Nanotechnologies, Riga Technical University, Riga, Latvia

A. Vaseashta
e-mail: prof.vaseashta@ieee.org

© Springer Nature Switzerland AG 2021
A. Vaseashta and C. Maftei (eds.), *Water Safety, Security and Sustainability*,
Advanced Sciences and Technologies for Security Applications,
https://doi.org/10.1007/978-3-030-76008-3_10

end-user requirements and expectations, as well as economic, technical, organizational and educational capacities of a local government, sustainability of the chosen solution, and hazard and threat identification. Threats should be analysed both at the micro and macro levels, where influencing factors are identified and described in more detail. In addition to the application of the systems approach, in the context of the development and management of water safety plans, it is necessary to consider the provisions of the EN 15975 standard, which is one of the aims of this paper. A brief analysis of typical examples of the use of such plans is also provided. The main conclusion refers to the improvements generated by the implementation of the various water safety plans.

Keywords Water safety · Security · Plans · Systems approach · EN 15975-1 · EN 15975-2

1 Introduction

Modern conditions in which the human community exists imply a high level of interaction with various media of the environment in which the community lives, including water, air, land, flora, fauna and environmental resources. In a way, the human community changes or influences the environment in which it exists, while simultaneously transforming itself by responding to the pressures to which it is exposed from the environment. The analysis of the environment and the state of environmental protection, as well as a critical review of the previous experiences of institutions dealing with this issue, identify the need for innovation in the field of environmental safety and security management [1]. Today, all the natural resources, including water, are increasingly endangered. It is necessary to acknowledge the fact that changes in the environment due to anthropogenic influence are often irreversible and that science and practice today focus on the further development of humanity towards the development of a culture of prevention instead of a culture of response and remediation. The environment is characterized by great variability and heterogeneity in time and space, which is the result of a constant influence of complex ecological conditions. Many interdependent processes take place in the environment and change in any of its part affects the entire environment. The consequences of environmental changes can be positive, but also far-reaching and negative. Modern, large and rapid development of technology over time places a burden on the environment and manifests as harmful effects, creating a problem of how to preserve the quality of all environmental media [2].

In such conditions, a special problem arises from the management of those ecosystem components that have the character of resources that support the development of social communities that use and modify them but ultimately reject them (in the form of solid waste, wastewater etc.). This indicates that one of the biggest challenges facing the scientific community in the 21st century is the management of water resources. Water endangerment can be singled out as the most serious challenge because water is inextricably linked to food production, energy production and aquatic and terrestrial environmental media functions. Uneven quantitative and qualitative temporal and spatial distribution of water, somewhat exacerbated by global climate change, and increasing accidental pollution, caused by direct or indirect discharge of hazardous and harmful substances into aquatic environments, are the catalysts for more efficient and effective action of the international community and national services [3–5]. The basis for undertaking a number of activities begins with the harmonization of legislative acts, institutional organization in the field of water and environmental protection, provision of financial resources, etc. and leads to specific plans for undertaking preventive protection measures [6]. The said activities in the field of water management are universally applicable in all segments of the environment.

2 Background

The development of science during the 19th and the beginning of the 20th century went in the direction of specialization. The loss of the idea of the whole and the non-reliance on the results from other scientific disciplines led to the enormous effort and the results from the related fields remaining unused [7]. The complexity of modern systems has been increasing, with environmental systems becoming particularly complex. The reasons for the increased complexity are both internal and external. The former is caused by the existence and interaction of different elements and stakeholders, while the latter are caused by the relations of the observed system with other systems in the environment. The analysis of complex systems is based on understanding the problem, defining the priorities and identifying the most important interactions. In complex systems, the problem of control stands out as one of the basic problems [8]. Wiener points out that cybernetics provides a general methodological basis for studying the control of complex systems, incorporating analytical research methods from various scientific fields [7]. According to Angyal, the system is a new methodological instrument for studying complex phenomena, i.e. the whole, whereby an element becomes a part of the system based on the value (contribution) it has for the system [9]. There are important elements that are equal for all systems [10, 11], so in reality, this parallelism, transferred to science, enables the formulation of universally applicable principles.

Environmental decision-making has significant socio-cultural dimensions, especially with regard to limited available resources, so the dimension of using available knowledge has become even more important [12, 13]. Effective involvement of stakeholders in the planning and decision-making process is crucial for obtaining meaningful solutions. The participatory process requires, among other things, defining the concepts, setting the goals, formulating the models and applying them to decision-making, as well as evaluating the outputs and the outcomes [14]. Indicators can be used to express the current state of a system. They can be used to monitor the state of a system or to assess the success of the defined policies, for example, security aspects [15, 16]. Hierarchical organization of performance indicators and a selection of key performance indicators based on the application of multi-criteria analysis and group decision-making is useful when analysing the characteristics of a system or the consequences of adverse events [17]. Likewise, environmental regulatory decision-making requires the application of proper models and their detailed evaluation during all stages of development, from problem identification, through definition of a conceptual model and the implementation of a specific model, to its verification and validation [18, 19].

To fulfil the aim of this paper, it is necessary to define the precise causal relationship between the terms system, systems approach, system standards, water resources, water safety, water security, water risk, water scarcity, water stress, sustainability, natural hazards, anthropogenic hazards, and response activities by the society [20–22].

The following terminology is considered:

- *system*—a set of purposefully and functionally connected elements, whose connections are based on specific principles and laws, on the basis of which the whole acquires the properties that are not characteristic of its individual elements.
- *systems approach*—the systems approach to management does not deal with parts of organizations but with the organization as a single system, composed of interconnected parts. An organization is viewed as a complex dynamic system consisting of several interconnected and dependent subsystems and elements. The systems approach is a modern scientific approach to the phenomenon of particular organization, which is defined as a complex system dominated by technical, organizational and institutional subsystems. These standards usually contain two crucial parts—the part that refers to terms and definitions and the part that refers to requirements. Compliance, verification, monitoring and reporting are often defined by a set of supporting or complementary standards.
- *system standards*—standards that define the structure of the entire management system, whether or not the management pertains to the field of quality, environment, occupational safety, information security, risks, etc.
- *water resources*—exclusively the water that has a significant use value for end-users (households, industry, public sector, agriculture). Water as a resource is always quantitatively smaller than the total water present in an area. In terms of safe water supply, water resources are those that are affordable, of appropriate quality, that can be used over a long period of time, but also those that are exposed to the lowest possible risks of any kind.
- *water safety is broadly defined as the acceptable availability of adequate quality and quantity of water necessary for the smooth functioning of the economy, fulfilment of health needs, meeting of the needs of households and aquatic and terrestrial ecosystems, and consideration of the threats of actions by people who intend to 'disrupt' the desired water quality and quantity status.*
- *water security*—the capacity of a population to safeguard sustainable access to adequate quantities and acceptable quality of water for sustaining livelihoods, human well-being, and socio-economic development, for ensuring protection

against water-borne pollution and water-related disasters, and for preserving ecosystems in a climate of peace and political stability [23].
- *water risk* determines the ratio of probability and the consequence that a certain region can be subjected to the harmful effects of water (which is one of the basic aspects of water management issues). Water risk needs to be distinguished from the risk of exposure to surface or groundwater pollution.
- *water scarcity*, in the narrowest sense, refers to the lack of water in terms of quantity, i.e. volume or volume flow intended for the needs of water supply or insufficient quality of available water. Modern research popularizes an approach in which the concept of water scarcity is defined as the ratio of the amount of water intended for human needs in relation to the available drinking water in a particular region.
- *water stress* refers to the availability of water resources intended to meet the needs of society but also of the environment in which a particular social community exists, i.e. to meet the water needs of all users in the society while guaranteeing minimum volume flows required to preserve aquatic populations.
- *sustainability* implies equal distribution of resources, not only spatially between users in a given period, but also temporally between different generations of users over time. The idea is to use resources in such a way that they are readily available to all generations, without any fear that some future generation will fare worse than the current ones. In order to achieve sustainability, it is necessary to reconsider the notion of what we consider basic needs at the level of society. Interpretations of needs vary widely from one region, one city, and even one person to another.
- *natural hazard*—a natural process or phenomenon that may cause loss of life, injury or other health impacts, property damage, loss of livelihoods and services, social and economic disruption, or environmental damage [24].
- *anthropogenic hazard*—a hazard derived from the processes induced entirely or predominantly by human activities and choices. This term does not include the occurrence or risk of armed conflicts and other situations of social instability or tension that are subject to international humanitarian law and national legislation [25].
- *response activities*—the way society can respond to and improve the existing state of water supply, which it can do in the domains of risk identification, assessment and management. The creation of concrete plans and their implementation is a concrete example of society's reaction.

The scientific and historical data, as well as the results of statistical analyses, indicate that there is a constant, ever-present, increasing trend of risks to which the water/resources/user system is exposed in terms of frequency, types or intensity of hazards and pressures, whether anthropogenic or environmental, intentional or unintentional. The evaluation of modern challenges, risks and hazards has to rely on the multi-criteria analysis, based on the systems approach, aimed at predicting new hazards and their consequences, and with an emphasis on the measures that directly or indirectly reduce the risk or consequences [17]. In this sense, the purpose of responsible and preventive behaviour in such a system is reflected in the identification of the level of danger of different hazards and the ability to manage them [26].

3 Models and Methods—Systems Approach and System Standards

Three basic methodological approaches are characteristic of the research of systems: the observational approach, the analytical approach, and the systems approach. The observational approach consists of a systematic observation of a phenomenon and the recording of events, critical analysis of the recorded facts, formulation of hypotheses about the observed phenomenon, and testing of the hypotheses by further observations. The analytical approach was spawned from the development of technical tools that allowed researchers to switch from passive observers to active experimenters. The systems approach, as a modern approach to scientific research, is based on a comprehensive analysis of a problem and a research of goals and alternatives in order to indicate the possible directions of actions for solving the problem. This methodological approach emphasizes the analysis of relationships within the system and the analysis of relationships of the system with its environment, followed by the identification of alternative strategies for achieving system objectives. The systems approach allows a comprehensive view of a water-system as an entirety, rather than a simple sum of its building blocks.

Systems thinking is a framework for perceiving mutual relations, setting patterns of change, and understanding the reality at a higher level. Instead of thinking about individual events, the emphasis is on thinking about patterns and the structure. The subject of reflection at this level is the observation of events in the context in which they take place, focusing on the interconnectedness of events, discovering their causes and predicting future events [27]. Consideration of water supply systems at different levels requires different complexities at different system levels, which implies that the same phenomenon has different characteristics at different levels of observation. The basis of the systems approach is the application of systems analysis. The systems approach concentrates on the interaction between elements by examining the effects of interactions and compares the behaviour of the model and the real system. The knowledge base consists of less rigorous models in relation to the analytical approach, but these models are useful in the decision making and acting, as well as in the analysis of nonlinear interactions. Systems analysis as a methodological procedure is a process of learning about the system, understanding and interpreting it by applying logical and cognitive procedures. As a general methodology of system research, systems analysis consists of five basic phases: formulating the research problem, determining the goals, determining the limitations, defining a system, and assessing the system. Applied to environmental systems, within the previously mentioned phases, information about the researched system is collected and used. The main functions of system analysis are model formation, information collection and information synthesis.

Effective environmental management must take into account complex interactions in the observed systems [28]. Adequate analysis of complex systems can significantly improve policy definition, governance and decision making. This often requires large investments, a lot of time and significant stakeholder participation. Success factors are defined in order to analyse the realized modelling outcomes, which depends on project management, stakeholder engagement, model characteristics and possibilities of their application, but also on a number of contextual factors [29]. The general systems theory is based on the laws of dependencies, differences and changes. It uses the systems approach as a methodological approach to scientific research. In a comprehensive analysis of the problem, the research of goals and alternatives points to possible actions to solve the selected problem. The systems approach implies the acceptance of superiority of the whole over its parts. It concentrates on the interaction between the elements and examines the effective interactions and the emphasis on global perception. The facts are proved by comparing the functioning (or the behaviour) of the model to the real system.

The model as an abstraction of a researched system contains its essential properties. It is a simplified description of a system, and the process of creating a model is called modelling. The model of the system is created during systems analysis and it enables experimentation. Models and systems are equal in terms of their structures, functions or behaviours. This allows the study of the model to gain new knowledge about the system. Models can be classified according to different criteria. According to the relations between the variables in models, there are models describing causal connections and correlation models. The correlation models define static relations between variables and can be used for trend forecasts. The causal models describe the relationship between cause and effect in a system and they are made for prediction or control purposes. According to the number of elements in relation to the original system, there are isomorphic and homomorphic models. The isomorphic models have the same number of elements as a real system, with each element of the real system corresponding to one particular element of the model and vice versa. In contrast, the homomorphic models are realized with a smaller number of elements compared to the original system, so that one element of the model describes more elements of the real system. According to their purpose, there are research models and decision-making models. The research models are used to study real systems—for their description, explanation, understanding, prediction, as well as behaviour management. The decision-making models are used to prepare, execute and convey decisions. Finally, according to the modelling method, there are abstract models and physical models. The abstract models are abstractions of real systems. They formalize knowledge and form an abstract representation based on observations and experiments. Abstractions can be symbolic or verbal. In the symbolic models, mathematical, logical and graphical symbols are used to describe the system. The physical models are real physical objects similar to the analysed system (smaller copies of the original, of the same physical nature), with the same variables and parameters, described by the same units.

The analysis of water systems requires the application of different models. Simulation models, inventory models, graphical and network models are just some of them [12]. A good model must include a precisely defined purpose and evaluation procedure, supported by performance indicators and evaluation assumptions [30]. Different scenarios can be defined for each event. A scenario based on an assessment of extreme events is important for analysing the possibilities of implementing plans in practice [31]. A lack of resources (human, organizational or technical) can be crucial for the inability to implement seemingly well-defined water safety plans. Therefore, special emphasis is given to training and capacity building at all levels of decision making and response, as well as capacity evaluation [32]. Regulatory style affects regulatory objectives, and inadequate style can lead to the absence of expected outcomes. Different risk management strategies can be applied to avoid that, for example, the participatory approach. Hasan et al. describe the application of the regulatory ladder in the context of water safety plans, emphasizing the importance of education and development of organizational skills [33]. The experience from other countries and organizations in the implementation of water safety plans can help during the identification of the main hazards [33, 34]. At the organizational level, important dimensions of water safety planning include leadership commitment, possession of adequate technical knowledge and competent persons, governance, and exchange of previous experience through cooperation and collaboration [35].

4 Water Safety/Security/Sustainability Plan Development

The Water Safety Plan represents a risk management tool in the public water supply system developed and introduced by the World Health Organization (WHO) in 2004. It is based on a detailed analysis of the entire water supply system from source to tap (similar to the HACCP (Hazard Analysis and Critical Control Points) approach in the production of safe food, often referred to as *"from the field to the dining table"*), and its main purpose is to ensure the continuity of safe water supply, whether in qualitative or in quantitative terms. Nowadays, this initiative is accepted worldwide, with a tendency to increase the number of countries developing water safety plans. Figure 1 shows the global status of water safety plans policy development.

The implementation of the water safety plan implies a systematic risk analysis that can be applied to any water supply system, large or small, public or private, whether it is a lake as a source of water supply or an underground source [37]. For the purpose of this paper, the water safety plan development and implementation system are shown in Fig. 2. The problem of water safety plan development is observed at the micro and macro levels.

At the macro level, the main stakeholders are classified by levels, identifying the local level, the regional level and the national level. Organizations and inter-organizational actors are observed at the micro level. This primarily refers to organizational management, technical services and executors in individual organizations. Likewise, established management systems are very important in achieving

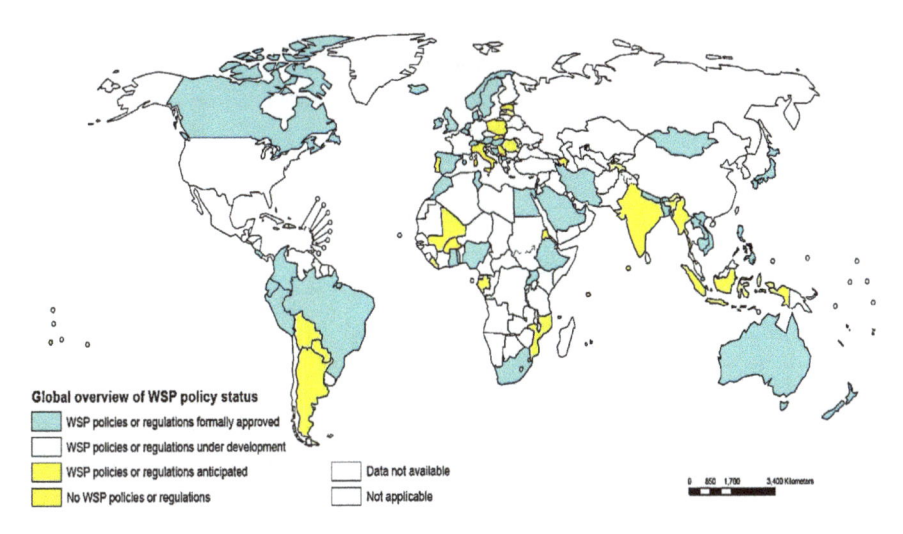

Fig. 1 Water safety plans—global status [36]

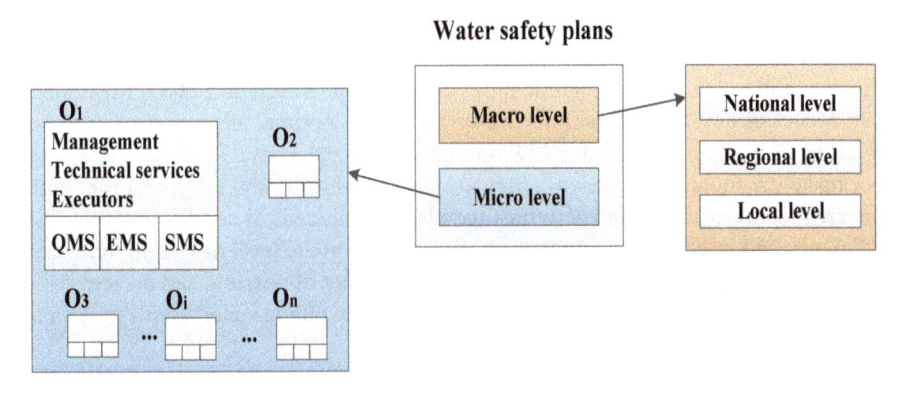

Fig. 2 Water safety plan development process

the set goals. These are the safety management system (SMS), the environmental management system (EMS), and the quality management system (QMS) [38].

The development of the plan and the corresponding description of the water system is based on the defined goals or targets. These targets are primarily of a health and safety nature. Based on the objectives, the selected team develops a plan identifying all the key elements, stakeholders and risks. The next step is the implementation of the plan. The success of the plan implementation is constantly monitored, the situation in the field is monitored and then the risks are assessed. All identified irregularities and risks, as well as opportunities, are used to improve the existing plan. The cycle is shown in Fig. 3.

Proper management of the system, guided by goals, leads to the fulfilment of its purpose. Therefore, it is important to monitor the indicators and to obtain efficient

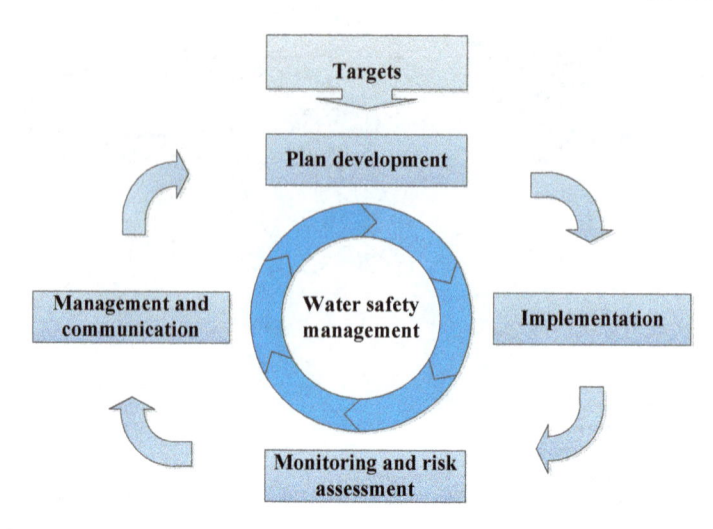

Fig. 3 Continual water safety management process

and effective communication among different stakeholders involved in its implementation [39]. Due to intensive automation of the monitoring process, which made the data collection process significantly advanced and simplified, the most important aspect of any modern monitoring system is the optimal allocation of sensors, in terms of both time and space [40]. On the other hand, data collection alone does not make the monitoring system efficient. Data processing and interpretation, with the creation of metadata and activities based on them, leads to certain changes. The efficiency and effectiveness of the protection system at the level of one public water supply company is based on an appropriate assessment of possible and present risks to water quality and quantity. Risk management in the public water supply system is a dynamic process that includes phases related to:

- risk identification.
- consideration of legal and other requirements in the subject area.
- risk assessment and control, i.e. taking measures to minimize the risk.
- measurement and monitoring of the effects of implemented activities; and
- measures to continuously improve the risk management system.

The pillar of an efficient risk management system in the water supply system is represented by standardized procedures for managing such risks, based on the requirements of EN 15975-1 and EN 15975-2 standards, developed and published by The European Committee for Standardization CEN – Comité Européen de Normalisation. These European standards describe the general principles of water supply management in crisis situations, including precautions and measures during a crisis. The standards address all the entities and stakeholders who share responsibility in the provision of safe drinking water throughout the entire supply chain from the source to the point of use [41, 42]. A similar approach to risk management is defined by the World Health Organization, but it is defined in the form of an instruction and not in the form of an international standard [43, 44]. These European standards describe good practice principles of drinking water supply management in the event of a crisis, including preparatory and follow-up measures as well as the principles of a risk management approach to improve the integrity of the drinking water supply system. Additionally, these standards provide an adequate terminological basis for the development of risk management systems in the field of public water supply as well as guidelines for the implementation of water safety plans, of which the following definitions are the most important:

- drinking water supplier—body responsible for delivering drinking water;
- drinking water supply system integrity—existence of drinking water supply system suitable to meet the specified quality, quantity, continuity and pressure targets in accordance with legal/regulatory requirements and the drinking water supplier's objectives;
- hazard—biological, chemical, physical or radiological agent in, or condition of water, with the potential to cause harm to public health;
- hazardous event—event that introduces hazards to, or fails to remove them from, the drinking water supply system;
- risk—combination of the likelihood of a hazardous event and the severity of consequences, if the hazard occurs in the drinking water supply system;
- crisis—event or situation with the potential to seriously affect a drinking water supplier that may require other organizational structures and possibly more than the usual means of operation to respond to an emergency;
- crisis management—special kind of organizational capability designed to guide a drinking water supplier through a crisis, outside the organization of normal operations;
- verification routine confirmation, through the provision of objective evidence, that the drinking water supply system is delivering water in accordance with the set objectives and that the risk management approach is effective [41, 42].

According to EN 15975-2, all drinking water supply systems are faced with risks, which need to be controlled adequately in order to ensure appropriate risk control throughout different management activities. A consistent and systematic risk management approach allows the drinking water supplier to analyze and to compare risks that may occur in the elements of the drinking water supply chain (technical failures, natural hazards, disasters or malicious attacks). The risk management approach defined by this standard is shown in Fig. 4, while the risk continuum is shown in Fig. 5.

Positive experience in applying these standards indicates that risk factors will never be fully identified, evaluated and made predictable. Moreover, the nature of the public water supply process is such that an increasing number of public water supply systems are in increasing contact with hazards from the environment (partly due to climate change). Therefore, the subjects in the water supply system will always be vulnerable (sensitive) to pressures (hazards) to some extent. In that sense, the purpose of responsible, preventive behaviour in the water supply system is reflected

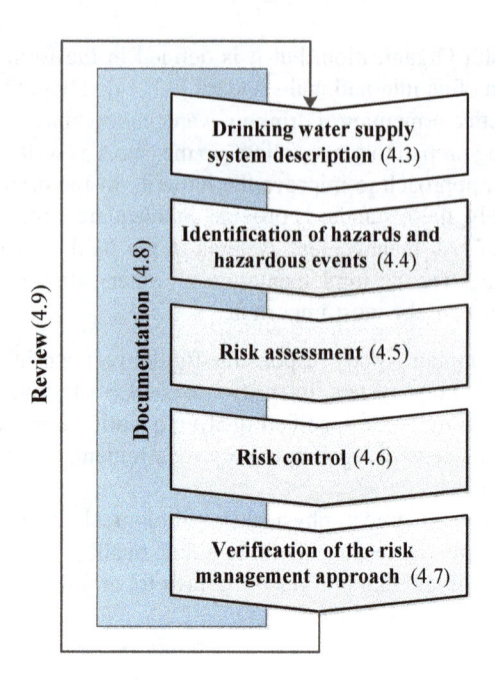

Fig. 4 Risk management process [42]

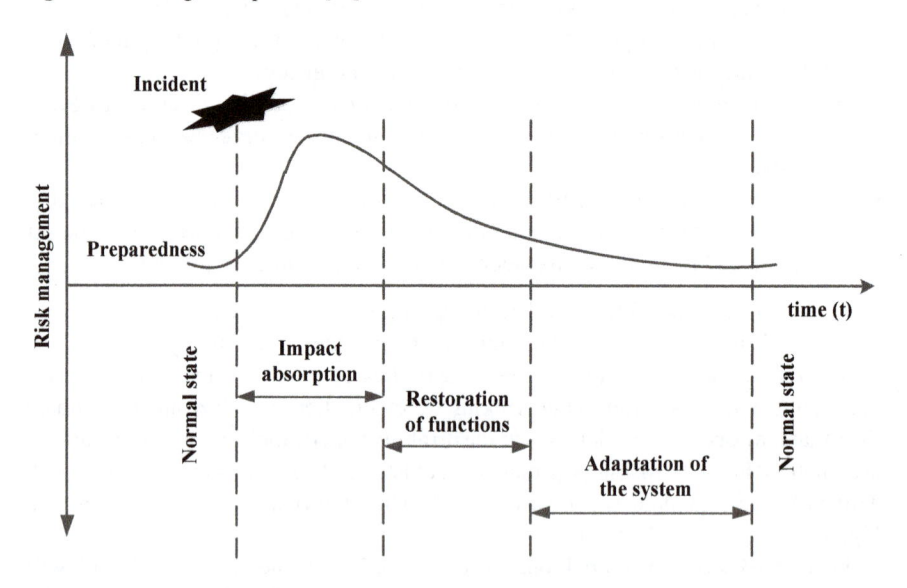

Fig. 5 Risk management continuum

Application of the Systems Approach ... 255

in the identification of the level of vulnerability of the water supply system to various hazards and the improvement of capabilities, i.e. the capacity to manage them. Functionally, the risk management capacity in the public water supply system can be understood as the ability of the system to achieve the desired goals in either regular or emergency situations. The application of the systems approach enables the analysis of influential components in the water supply system, while the role of the EN 15975 standard is the standardization of risk management processes and activities.

5 The Experience of Serbia

In the Republic of Serbia, according to the Law on communal activities (or public utilities) from [45], the supply of drinking water represents a system of interconnected activities related to the capture, purification, processing, storage and delivery of potable water through the public water supply network to the measuring instrument of the consumer, including the measuring instrument. Currently, all water supply networks are state owned.

The data for 2018 shows that in the Republic of Serbia, through the public water supply systems of the local self-government centres, 654 million m^3 per year are captured, i.e., about 21 m^3/s. About 65% of abstracted water (424 million m^3–13 m^3/s) is sent for consumption, of which about 48% (317 million m^3 per year–10 m^3/s) for household consumption, and about 17% (108 million m^3 per year)–3 m^3/s) for consumption of industry and institutions. Households consume about 74.5%, the industrial sector about 10.6%, and other users about 14.9% of the total supplied water. Table 1 shows the potable water balance in the Republic of Serbia [46].

In the Republic of Serbia, 146 municipal companies take care of water supply to 171 local self-government units (Niš, Belgrade, Novi Sad and another 168 cities and municipalities), i.e., city centres. 87.9% of the total population (6 945 000) is

Table 1 Potable water balance—Republic of Serbia

No	Indicator	Republic office for statistics	
		*10^6 m^3	%
1.	Abstracted water	654	100
2.	Water sent to the network	424	64
3.	Billed water		
4.	Households	317	75
5.	Industry	45	10
6.	Other users	62	15
7.	Water losses—unbilled water	230	36

connected to centralized water supply systems. Water supply systems for urbanized settlements consist of 59 drinking water treatment plants (projected capacity of about 910 million m^3 per year and operating capacity of about 610 million m^3 per year) and about 28,000 kilometres of water supply network (average 238 km per enterprise). Plants as well as the network need reconstruction or improvement of treatment capacity and technology. Potable water sector structure as well as the sector sustainability assessment are shown on Figs. 6 and 7, respectively [47].

Water supply is mostly continuous, except in cases of technologic accidents, severe pollution, extreme hydrological impacts or in cities with a chronic shortage of drinking water, both in quantity and quality. Having that in mind, there are numerous

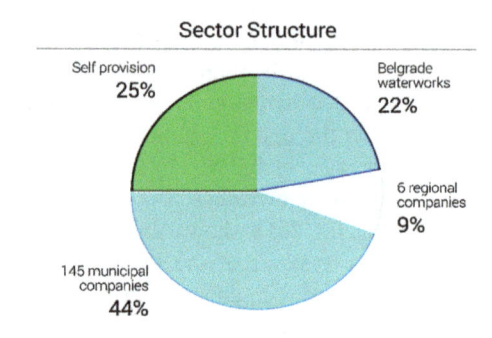

Fig. 6 Sector structure

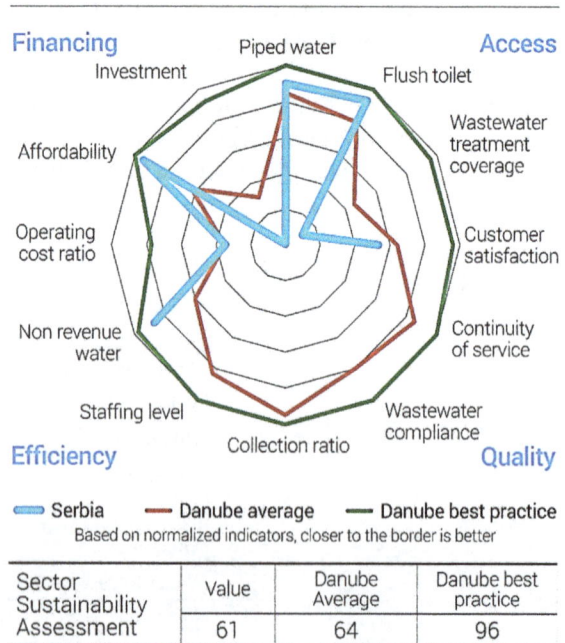

Fig. 7 Sector sustainability assessment

Sector Sustainability Assessment	Value	Danube Average	Danube best practice
	61	64	96

practices aimed at achieving safe water supply, specified in the national overall water management strategy, the laws and the by-laws, although the Republic of Serbia did not formally adopt the water safety plan methodology. As defined by Milenkovic et al., Serbia has been an active participant in the enactment and implementation of documents, declarations, strategies, and other documents dedicated to environmental quality preservation and sustainable development [48].

Currently, the draft version of the Law on Water Intended for Human Use has been created in Serbia and is being harmonized with the relevant EU directives. The relevant legislation of the European Union has been transposed into the proposed legal solution—The Drinking Water Directive—as well as its amended version from 2015 and the proposal of the new European Drinking Water Directive. The new directive aims to improve consumers' access to updated water quality data by harmonizing standards for the materials in contact with drinking water, access to water and access to information regarding water quality. Once the Law is enacted, it will be the legal basis for the adoption of a water safety plan. This is highly relevant, because the risk to safe water supply in many parts of Serbia is significant, from the source to the tap of the consumers.

At this point, the monitoring of drinking water quality is carried out on the basis of a set of laws that cannot be used for quick identification and timely response to inadequate water quality, particularly at the source point.

6 Conclusions

The availability of sufficient quantities of healthy and safe water today is of immeasurable importance and it implies a continuous supply of drinking water, which provides sufficient quantities according to the criterion of intended use. The level of risk to human life and health, as well as to the economy and the environment, is conditioned by the quality of the water at the source, the efficiency of treatment in order to obtain drinking water and the integrity of the distribution system.

Based on these facts, it can be easily concluded that, for the safe functioning of the entire system, it is necessary to perform a comprehensive risk assessment and then set control mechanisms, thus managing the assessed risk. Risk assessment and management must cover all phases of water supply, starting from water intake to the end users. In order to control risks to human health, monitoring programs should ensure that measures are in place through the water supply chain and to consider the information about the water bodies from which the water used for human consumption is taken. Consequently, in order to properly reduce the risks derived from hazardous events, it is necessary to introduce suitable protection mechanisms—water safety plans. Drinking water resources represent a common good that needs to be preserved in a sustainable way.

Future research should focus on particular determinants that affect the activities in public water supply systems, on training models, on the interconnection of authorized entities and on the strengthening of the role of academic institutions. In addition, the experiences of the countries or organizations that have defined, implemented and monitored the effects of the water safety plans must be taken into account.

7 Recommendations

Due to the strategic importance of supplying the population with safe and good drinking water, all water supply organizations must create proper risk prevention plans. Risk control involves the selection and application of risk mitigation options to preserve the integrity of the water supply system. Given the significant role of drinking water suppliers in protecting public health, priority should be given to control measures that offer a high degree of process safety and operational stability to control risks that would otherwise have severe consequences for public health. The identification of potential risk control measures should follow the prioritization of identified risks, while the selection of appropriate measures should be based on standard specifications or procedures that the water supplier can support. The risk control measures applied may be preventive or reactive to mitigate the risk. The development and implementation of the water safety plans in the Republic of Serbia would provide multiple benefits, such as avoidance of incidents affecting the health of the population, improved compliance with legal and other requirements, greater consumer trust, as well as cost-effectiveness and investment planning.

Acknowledgements The presented research is supported by the Ministry of Education, Science and Technological Development of the Republic of Serbia, No. 451-03-9/2021-14/200148, as well as by the Serbian Academy of Sciences and Arts, Niš Branch, through the project "Integral approach to storm water management in urban catchment areas in the South-East Serbia region", No: O-15-18.

References

1. Nastasiuc L, Bogdevici O, Aureliu O, Culighin E, Sidorenko A, Vaseashta A (2016) Monitoring water contaminants: a case study for the Republic of Moldova. Pol J Environ Stud 25(1):221–230. https://doi.org/10.15244/pjoes/58888
2. Nikolic V, Vasovic D (2015) Tailor made education: environmental vs. energy security and sustainable development paradigm. In: Caleta D, Radovic V (eds) Comprehensive approach as "Sine Qua Non for Critical Infrastructure Protection & Managing Terrorism Threats to Critical Infrastructure Challenges for South Eastern Europe". NATO SPS series D: information and communication security. IOS Press, Amsterdam, Berlin, Tokyo, Washington
3. The Hyogo Framework for Action 2005–2015. United Nations Office for Disaster Risk Reduction
4. The Sendai Framework for Disaster Risk Reduction 2015–2030. United Nations Office for Disaster Risk Reduction
5. Law on Environmental Protection (2018) Official Gazette of the Republic of Serbia, No 95/2018
6. ISO 31000: 2018 (2018) Risk management—requirements. ISO, Geneva
7. Wiener N (1948) Cybernetics—or control and communication in the animal and the machine. The MIT Press, Cambridge

8. Ashby WR (1956) An introduction to cybernetics. Chapman and Hall, London
9. Angyal A (1941) Foundations for a science of personality. The Commonwealth Fund, New York
10. Bertalanffy L von (1950) The theory of open systems in physics and biology. Science III:22–29
11. Bertalanffy L (1968) General system theory (foundations, development, applications). Penguin Books, London, p 1968
12. Votruba L, Kos Z, Nachazel K, Patera A, Zeman V (1988) Analysis of water resource systems. Elsevier, Amsterdam
13. van Kerkhoff L, Pilbeam V (2017) Understanding socio-cultural dimensions of environmental decision-making: a knowledge governance approach. Environ Sci Policy 73:29–37
14. Voinov A, Kolagani N, McCall MK, Glynn PD, Kragt ME, Ostermann FO, Pierce SA, Ramu P (2016) Modelling with stakeholders—next generation. Environ Model Softw 77:196–220
15. Malenovic-Nikolic J, Vasovic D, Janackovic G, Milosevic L, Ilic Krstic I (2018) Realisation of the goals of sustainable development based on application of energy indicators in environmental engineering. J Environ Prot Ecol 19(1):216–225
16. Janaćkovic G, Malenovic-Nikolic J, Vasovic D (2016) On efficiency and security of energy systems. Commun Dependabality Qual Manag Int J 19(4):30–39
17. Janackovic G, Vasovic D, Malenovic-Nikolic J, Musicki S, Vranjanac Z (2018) Vulnerability assessment of municipality areas to natural disasters based on group fuzzy analytic hierarchy process. J Environ Prot Ecol 19(4):1526–1535
18. National Research Council (2007) Models in environmental regulatory decision making. National Academies Press, Washington (DC)
19. Savic S, Stankovic M, Janackovic G (2020) Theory of systems and risk. Akademska misao, Belgrade (in Serbian; in press)
20. Vasovic D (2016) Hybrid model of environmental capacity management. Doctoral dissertation, University of Nis, Faculty of Occupational Safety in Nis, Nis, Serbia
21. Janackovic G (2015) Models for integrated safety system management based on collaborative work. Doctoral dissertation, University of Nis, Faculty of Occupational Safety in Nis, Nis, Serbia
22. Vaseashta A (2011) Technological advances in industrial water safety and security. In: Atimtay A, Sikdar S (eds) Security of industrial water supply and management. NATO science for peace and security series C: environmental security. Springer, Dordrecht. https://doi.org/10.1007/978-94-0071805-0_4
23. Herriet B et al (2013) Water security and the global water agenda—a UN water analytical brief. United Nations University, Institute for Water, Environment and Health, Hamilton, Canada
24. UNISDR Terminology on Disaster Risk Reduction (2009) United Nations International Strategy for Disaster Reduction (UNISDR), Geneva, Switzerland
25. United Nations Office for Disaster Risk Reduction (UNDRR) (2017) United Nations International Strategy for Disaster Reduction (UNISDR), Geneva, Switzerland
26. Vasovic D, Janackovic G, Malenovic Nikolic J, Musicki S, Markovic S (2018) Multimodality in the field of resource protection. J Environ Prot Ecol 19(4):1519–1525
27. Anderson V, Johnson L (2000) System thinking basics, from concepts to causal loops. MA Pegasus Communications, Waltham
28. Vasovic D, Janackovic G, Malenovic Nikolic J, Milosevic L, Musicki S (2018) Promoting reflective practice in resource protection area: a step to forecast outcomes in uncertainty. J Environ Prot Ecol 19(3):1320–1329
29. Merritt WS, Fu B, Ticehurst JL, El Sawah S, Vigiak O, Roberts AM, Dyer F, Pollino CA, Guillaume JHA, Croke BFW, Jakeman AJ (2017) Realizing modelling outcomes: a synthesis of success factors and their use in a retrospective analysis of 15 Australian water resource projects. Environ Model Softw 94:63–72
30. Crout N, Kokkonen T, Jakeman AJ, Norton JP, Newham LTH, Anderson R, Assaf H, Crokeg BFW, Gaber N, Gibbons J, Holzworth D, Mysiak J, Reichl J, Seppelt R, Wagener T, Whitfield P (2008) Good modeling practice. In: Jakeman AJ, Voinov AA, Rizzoli AE, Chen SH (eds) Chap. 2. Environmental modelling, software and decision support: state of the art and new perspectives. Elsevier, Amsterdam, The Netherlands, pp 15–32

31. Neto FF, Gómez-Martín MB (2020) Water safety plan integrated to the land use and occupation measures: proposals for Caraguatatuba-SP, Brazil. Land Use Policy 97(104732):1–8
32. Ferrero G, Setty K, Rickert B, George S, Rinehold A, DeFrance J, Bartram J (2019) Capacity building and training approaches for water safety plans: a comprehensive literature review. Int J Hyg Environ Health 222(4):615–627
33. Hasan H, Parker A, Pollard SJT (2020) Whither regulation, risk and water safety plans? Case studies from Malaysia and from England and Wales. Sci Total Environ. Article no. 142868. https://doi.org/10.1016/j.scitotenv.2020.142868
34. Li H, Smith CD, Cohen A, Wang L, Li Z, Zhang X, Zhong G, Zhang R (2020) Implementation of water safety plans in China: 2004–2018. Int J Hyg Environ Health 223:106–115
35. Roeger A, Tavares AF (2018) Water safety plans by utilities: a review of research on implementation. Util Policy 53:15–24
36. World Health Organization (2017) Global status report on water safety plans: a review of proactive risk assessment and risk management practices to ensure the safety of drinking-water. World Health Organization, Geneva, Switzerland
37. Gevorgyan GA, Mamyan AS, Hambaryan LR, Khudaverdyan SK, Vaseashta A (2016) Environmental risk assessment of heavy metal pollution in armenian river ecosystems: case study of Lake Sevan and Debed river catchment basins. Pol J Environ Stud 25(6):2387–2399. https://doi.org/10.15244/pjoes/63734
38. Vasovic D, Malenovic Nikolic J, Janackovic G (2016) Evaluation and assessment model for environmental management under the Seveso III, IPPC/IED and Water framework directive. J Environ Prot Ecol 17(1):356–365
39. Janackovic G, Savic S, Stankovic M (2013) Selection and ranking of occupational safety indicators based on fuzzy AHP: case study in road construction companies. S Afr J Ind Eng 24(3):175–189
40. Mukherjee R, Diwekar UM, Vaseashta A (2017) Optimal sensor placement with mitigation strategy for water network systems under uncertainty. Comput Chem Eng 103:91–102. ISSN 0098-1354. https://doi.org/10.1016/j.compchemeng.2017.03.014
41. European Committee for Standardization—CEN (2011) EN 15975-1: security of drinking water supply—guidelines for risk and crisis management—part 1: crisis management. Brussels, Belgium
42. European Committee for Standardization—CEN (2013) EN 15975-2: security of drinking water supply—guidelines for risk and crisis management—part 2: risk management. Brussels, Belgium
43. World Health Organization—WHO (2017) Safely managed drinking water—thematic report on drinking water 2017. Geneva, Switzerland
44. World Health Organization & International Water Association (2009) Water safety plan manual: step-by-step risk management for drinking-water suppliers. Geneva, Switzerland
45. Law on Public Utilities (2018) Official Gazette of the Republic of Serbia, No 95/2018
46. Annual report on public utilities in Republic of Serbia (2018) Ministry of the Construction, Transport and Infrastructure. Belgrade
47. Danube Water Program—Water and Wastewater Services in the Danube Region (2015) The World Bank, Washington, USA
48. Milenkovic M, Vaseashta A, Vasovic D (2021) Strategic planning of regional sustainable development using factor analysis method. Pol J Environ Stud 30(1):1–7. https://doi.org/10.15244/pjoes/124752. ISSN 1230-1485 (in press)

Dejan Vasović received the B.S. and M.S. degrees in Environmental Protection Engineering from University of Niš, Faculty of Occupational Safety, Niš, Serbia, in 2006 and 2011, respectively, and the Ph.D. degree in the same field from University of Niš, Faculty of Occupational Safety, Niš, Serbia, in 2016. Since 2017, he works as an Assistant Professor at the University of Niš, Faculty of Occupational Safety, Environmental Protection Department. Currently, he is the Vice-Chair of the Department of Environmental and Occupational Quality Management. He has authored three international book chapters and more than twenty papers in journals with impact factor. His research interests include environmental protection, water resources management, system standards, emergency management, natural hazards. He is also a reserve officer from 2008, with the rank of lieutenant. Mr. Vasović is the member of the European Society of Safety Engineers (ESSE) and Balkan Environmental Association (B.EN.A.).

Goran Janaćković received the B.S. and M.S. degrees in Computer Engineering from University of Niš, Faculty of Electronic Engineering, Niš, Serbia, in 2000 and 2004, respectively, and the Ph.D. degree in Safety Engineering from University of Niš, Faculty of Occupational Safety, Niš, Serbia, in 2015.

Since 2016, he has been an Assistant Professor with the University of Niš, Faculty of Occupational Safety. Currently, he is the Chair of the Department of System Research in Safety and Risk. He has authored three books on system and risk management, and more than twenty papers in journals with impact factor. His research interests include systems theory, risk and emergency management, safety information systems, and multicriteria analysis. Mr. Janaćković is the member of the European Society of Safety Engineers (ESSE) and Balkan Environmental Association (B.EN.A.).

Ashok Vaseashta (M'79-SM'90) received Ph.D. in Materials Science and Engineering (minor in Electrical Engineering) from Virginia Polytechnic Institute and State University, Blacksburg, Virginia, USA. He is Executive Director of Research for International Clean Water Institute in VA, USA, Chaired Professor of Nanotechnology at the Academy of Sciences of Moldova and Professor, Nanotechnology and Biomedical Engineering at the Faculty of Mechanical Engineering, Transport and Aeronautics at the Riga Technical University. Prior to his current position, he served as Vice Provost for Research at the Molecular Science Research Center in Orangeburg, South Carolina. He served as visiting professor at the 3 Nano-SAE Research Centre, University of Bucharest, Romania and visiting scientist at the Helen and Martin Kimmel Center of Nanoscale Science at the Weizmann Institute of Science, Israel. He served the U.S. Department of State in two rotations, as strategic S&T

advisor and U.S. diplomat. His research interests span nanotechnology, environmental/ecological science, and safety and security. His research on nanotechnology has been on improving the understanding, design, and performance of nanofibers and sensors/detectors, mainly for applications such as wearable electronics, target drug delivery, detection of biomarkers and toxicity of nano and xenobiotic materials. In the security arena, he has worked on counterterrorism, countering unconventional warfare and hybrid threats, critical-Infrastructure protection, biosecurity, dual-use research concerns, and mitigating hybrid threats. He has authored over 250 research publications, edited/authored eight books on nanotechnology, and presented many keynotes and invited lectures worldwide. He serves on the editorial board of several highly reputed international journals. He is an active member of several national and international professional organizations. He is a fellow of the American Physical Society, Institute of Nanotechnology, and the New York Academy of Sciences. He has earned several other fellowships and awards for his meritorious service including 2004/2005 Distinguished Artist and Scholar award.

Chitosan–Gelatin Cryogels as Bio-Sorbents for Removal of Dyes from Aqueous Solutions

Didem Demir, Nimet Bölgen, and Ashok Vaseashta

Abstract Dye effluents of wastewater of textile and other industries, particularly reactive azo dyes, are serious safety and health problem for the environment and living organisms. This study demonstrates the results of the removal of Procion Red MX-5B (PR MX-5B) from an aqueous solution by using cryogels prepared from natural polymers. A polymeric cryogel based on chitosan and gelatin was synthesized by cryogelation via crosslinking reaction of polymers at cryogenic conditions. Chitosan-gelatin cryogel (CH-GEL) crosslinked with glutaraldehyde was utilized for adsorption of PR MX-5B dye from aqueous solutions. The synthesized cryogel was chemically characterized by using Fourier Transform Infrared Spectroscopy (FTIR). The scanning electron microscopy (SEM) was used for the morphological characterization of cryogels. The fabricated cryogel has a highly interconnected and open macroporous structure. The water uptake properties of the cryogel in distilled water was determined. The parameters affecting the adsorption of dyes, such as pH of dye solution, initial concentration of dye and adsorbent dosage were studied. The maximum adsorption capacity of the CH-GEL cryogel for the removal of PR MX-5B in 50 mg/L of dye solution was 64.56 mg/g. The macroporous CH-GEL cryogels, due to their highly interconnected porous structure, can be effectively used as adsorbents to remove the PR MX-5B reactive dye from wastewater sources. As a result, an

---

D. Demir · N. Bölgen (✉)
Chemical Engineering Department, Mersin University, Engineering Faculty, 33343 Mersin, Turkey
e-mail: nimet@mersin.edu.tr

D. Demir
e-mail: didemdemir@mersin.edu.tr

A. Vaseashta
International Clean Water Institute, Manassas, VA, USA

Institute of Electronic Engineering and Nanotechnologies "D. Ghitu", ASM, Chisinau, Moldova

Institute of Biomedical and Nanotechnologies, Riga Technical University, Riga, Latvia

A. Vaseashta
e-mail: prof.vaseashta@ieee.org

© Springer Nature Switzerland AG 2021
A. Vaseashta and C. Maftei (eds.), *Water Safety, Security and Sustainability*,
Advanced Sciences and Technologies for Security Applications,
https://doi.org/10.1007/978-3-030-76008-3_11

eco-friendly biosorbent was fabricated for the removal of toxic dye molecules from aqueous solutions.

Keywords Wastewater · Dye removal · Adsorption · Biosorbent · Chitosan · Gelatin · Cryogel

1 Introduction

Pollution by the discharge of raw waste of industries in water resources remains one of the major health and safety problems that humanity and other life forms in our world are facing today. The wastewater discharged from textile, printing dyeing and papermaking industries contains a large number of substances such as dyes, color residues, catalytic chemicals, auxiliary chemicals and cleaning solvents [12, 20, 23]. These pollutants cause environmental and health problems due to their toxic, nondegradable, stable, mutagenic, allergenic and carcinogenic characteristics [6, 23]. Among all these substances, dyes are the least desired materials [3].

Dyes used in the textile industry have been classified as nitro, azo, acid, basic, direct, sulphur and reactive dyes according to their chemical structures and industrial applications [18]. Procion Red MX-5B (PR MX-5B, $C_{19}H_{10}C_{12}N_6Na_2O_7S_2$) is a typical reactive mono-azo dye, also known as Reactive red 2 and is commonly used in coloring cellulosic fibers, nylon, silk and wools. The biggest problem of this reactive dye is that only 60–70% of the dye reacts with the material during the dyeing process and the remaining is discharged into the aquatic environment [10]. The accumulation of the dyeing substance in water causes mutagenic, cytotoxic, genotoxic and carcinogenic effects on aquatic organisms due to decreasing light permeability of water which causes negative effects on photosynthetic activities. Furthermore, giving this wastewater to the environment may cause great damages to the human organs including kidneys, liver and brain [22]. Due to the serious health and environmental impact due to the PR MX-5B, it is mandatory to treat wastewater containing dyes before being discharged into the environment and water treatment facilities.

There are many physical, chemical and biological techniques for removing dyes from wastewater including coagulation and flocculation, membrane separation, electrocoagulation and adsorption. Adsorption has been found to be an efficient and economical process for the removal of pollutants such as dyes from wastewater.

The adsorption of dyes on natural biopolymers, such as chitin, chitosan, alginate, cellulose, lignin and gelatin, is possible even at low concentrations. These biosorbents are versatile, eco-friendly, biocompatible, biodegradable. In addition, their functional groups can make complexes with a wide range of molecules, including phenolic compounds, dyes and metal ions [2]. The adsorption of different types of dyes on various biosorbents have been studied and these studies demonstrated that these biosorbents are efficient biomaterials that have an extremely high affinity for many classes of dyes [1, 5, 11, 17].

The aim of this chapter is to report an investigation of the PR MX-5B adsorption capacity of chitosan-gelatin (CH-GEL) cryogels. CH-GEL cryogels were firstly fabricated by cryogelation technique and characterized by using FTIR, SEM and swelling measurement analysis. The effects of pH, dye concentration and adsorbent dosage on the rate of adsorption of PR MX-5B were studied in a batch system. The adsorption capacity of samples was calculated using experimental data.

2 Materials and Methods

2.1 Materials

Low molecular weight chitosan was obtained from Sigma Aldrich, USA. Gelatin for microbiology, glutaraldehyde (25%, v/v) as a crosslinking agent and acetic acid (100%, v/v) as a solvent were purchased from Merck, Germany. PR MX-5B from DyStar UK Limited was supplied by the Chemical Engineering Department of Mersin University. An amount of 1 g of dye was weighed and dissolved in 1 L of distilled water to be used as a stock solution. Solutions at the desired concentrations were prepared by diluting the stock solution.

2.2 Production of CH-GEL Cryogel

The ratio between chitosan and gelatin was fixed at 1:1 (wt.: wt.). Total polymer concentration was 2%, w/v. Calculated amounts of chitosan and gelatin were dissolved in an aqueous acetic acid solution (6%, v/v). After obtaining a homogeneous mixture, glutaraldehyde (3%, v/v) was added quickly to the polymer solution. The mixture was immediately poured into plastic syringes (2 mL) and placed into a refrigerated bath circulator (WiseCircu WCR-P6, Daihan, Korea) for 3 h at $-12\ °C$, then incubated at $-16\ °C$ for 24 h in the freezer. After that, the cryogels were thawed at room temperature and washed several times with distilled water. Before carrying out the characterization and adsorption experiments, produced samples were freeze-dried.

2.3 Characterization Studies of CH-GEL Cryogel

Morphology of the cryogel sample before adsorption studies was examined using SEM (Quanta 400F Field Emission Supra 55, Zeiss, Germany) at an accelerating voltage of 5 kV. The images were taken at $80000\times$ magnification. The FTIR spectra of the CH-GEL cryogel before adsorption studies was determined by using FTIR spectrometer (Perkin Elmer, FT-IR/FIR/NIR Spectrometer Frontier, ATR, USA). The spectral measurements were recorded in the wavenumber range of 400–4000 cm^{-1}.

2.4 Measurement of the Equilibrium Swelling Ratio

The swelling ratio values of the synthesized cryogels in distilled water were determined by gravimetric measurements at room temperature. Firstly, the freeze-dried cryogels were cut into samples of 3 cm diameter and 3 cm height. The weight of each sample before swelling (W_0) and after swelling (W_t) was measured at different time intervals using an analytical balance (Schimadzu, ATX 224, Japan). The measurements continued until the weight of swollen samples reached a constant value. All measurements were done in triplicate. The percentage of swelling ratio (SR%, w/w) of the samples was calculated by the following equation.

$$SR\%, \text{w/w} = [(W_t - W_0)/W_0] * 100 \tag{1}$$

2.5 Adsorption Experiments

Adsorption studies were performed to determine the PR MX-5B dye adsorption capacity of the synthesized CH-GEL cryogel samples in a batch system. Prior to adsorption experiments, all samples were cut into discs with dimensions of 9 mm in diameter and 3 mm in height sizes. 0.05 g cryogel was immersed into 50 mL of the PR MX-5B dye solution of the desired initial concentration in 100 mL Erlenmeyer flasks. The flasks were placed in a shaking water bath at an agitation speed of 125 rpm at room temperature for 120 min. After specified time intervals, 3 mL of the dye solution was taken and analyzed by UV-Vis spectrophotometer (Chebious, Optimum-One UV-Vis, Italy). The dye concentration was analyzed at a wavelength of 538 nm. The effect of various parameters such as initial pH, initial dye concentration and adsorbent concentration on the equilibrium adsorption capacity was investigated. The parameters applied for each variable are summarized in Table 1. Adsorption experiments were performed in duplicate.

The amount of PR MX-5B dye adsorbed per unit weight of CH-GEL cryogel at equilibrium (q_e, mg/g) was calculated using the following equation:

Table 1 Experimental variables and parameters of batch adsorption studies

Variables	Experimental parameters		
	pH	Dosage, g/L	Dye, mg/L
Effect of pH	3–11	1	50
Effect of dye concentration	3	1	25–100
Effect of adsorbent dosage	3	0.25–1.25	50

$$q_e = \frac{(C_0 - C_e) \times V}{m} \qquad (2)$$

where C_0 and C_e are the initial and the final concentrations of dye in aqueous solution (mg/L), respectively. V is the volume of the dye solution (L) and m is the weight of the swollen cryogel sample (g).

In addition, the color removal was calculated by the mass balance equation given below:

$$\text{Color removal}, \% = \frac{C_0 - C_e}{C_0} \times 100 \qquad (3)$$

3 Results and Discussion

3.1 Chemical Structure and Morphology of Cryogel

The chemical structure and morphology of CH-GEL cryogel are shown in Fig. 1. The FTIR spectrum of cryogel (Fig. 1a) showed the pattern of components of gelatin and chitosan and confirmed that the polymers were successfully crosslinked. The broadband near 3287 cm^{-1} is related to Amide A band of gelatin and, –NH and –OH vibrations of chitosan. The peaks located at 1640, 1551 and 1406, and 1241 cm^{-1} are assigned to Amide I, Amide II and Amide III peaks, respectively, present in both chitosan and gelatin polymers [21]. There are two main mechanisms in the chemical formation of CH-GEL composites: crosslinking reaction between chitosan and glutaraldehyde, and electrostatic interaction between chitosan and gelatin. The crosslinking reaction of chitosan with glutaraldehyde is carried out with a Schiff base

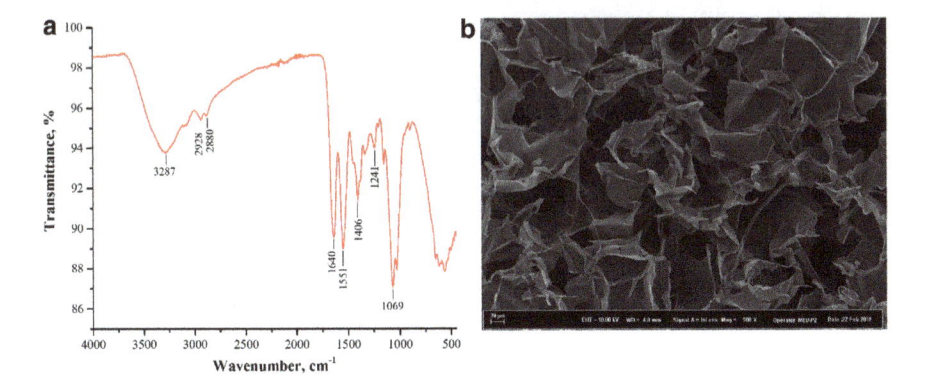

Fig. 1 FTIR spectrum **a** and SEM image **b** of CH-GEL cryogel

formed between the amine group of chitosan and the aldehyde group of glutaraldehyde. The aldehyde groups (C = O stretching) of glutaraldehyde generally appear between 1700 and 1750 cm^{-1} [21]. The absence of this band indicates the absence of free aldehyde groups, which proves that glutaraldehyde completely reacted with chitosan [13]. In addition, the FTIR spectra revealed the formation of electrostatic interaction between the amino groups of chitosan and the carboxyl groups of gelatin [15].

SEM analysis of produced cryogels was also carried out, and the SEM image is shown in Fig. 1b. The cryogel has an interconnected pore structure with pore sizes ranging from 20 to 100 μm. The produced cryogel is a mega porous material that can be used especially for the adsorption of large dye molecules [4].

3.2 The Equilibrium Swelling Ratio of Cryogel

The swelling capacity of the prepared CH-GEL cryogel in distilled water is shown in Fig. 2. First, the scaffold starts to uptake a high amount of distilled water, and the rate of swelling increases very fast. After a certain period of time, the water uptake becomes constant, and the CH-GEL cryogels achieve their equilibrium swelling capacity. The equilibrium swelling percentage of 6382.63 ± 40.29% was obtained at the end of 160 min. This high swelling value indicates that high amounts of water can be retained in superabsorbent cryogel which provides high absorption capacity of the material (Hakam 2015). The water molecules enter into the cryogel network and create an outward pressure by providing a rapid swelling which is a perfect condition for the adsorption of dye molecules.

Fig. 2 Swelling behaviour of the prepared CH-GEL cryogels in distilled water, insect shows the digital image of the swollen cryogels at different time intervals

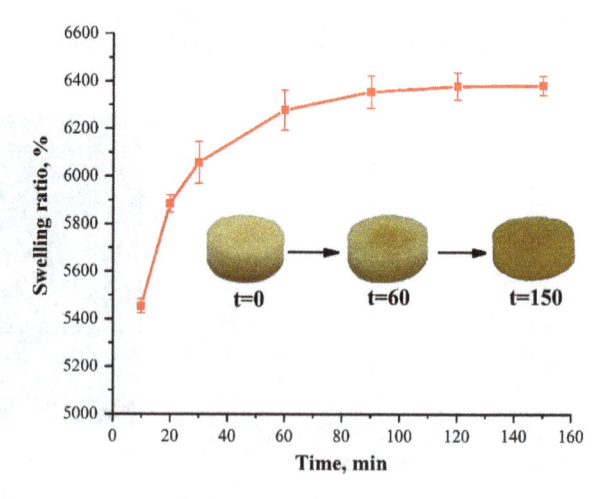

3.3 The Equilibrium Adsorption Studies

The effects of parameters, such as initial pH, initial dye concentration and adsorbent dosage, on adsorption of PR MX-5B on CH-GEL cryogels were investigated in a batch system.

3.3.1 Effect of Initial pH

The adsorption experiments were conducted with the pH of dye solution ranging from 3 to 11. The effect of initial pH on the adsorption capacity and color removal ability of CH-GEL cryogel are shown in Fig. 3a and b, respectively. The experiments were carried out at fixed adsorbent dosage (1 g/L), initial dye concentration (50 mg/L), room temperature (25 °C) and constant agitation speed (125 rpm). The pH of the solution is an important parameter for adsorption of dyes because it affects the surface charge and/or surface characteristics of the adsorbent, the degree of ionization and the dissociation of functional groups on active sites of the adsorbent [7]. From Fig. 3b, it was observed that the maximum equilibrium PR MX-5B adsorption capacity (q_{max}) was at pH 3 and q_{max} value decreased with increasing pH of the dye solution. In other words, the adsorption of dye on cryogel decreased with an increase in pH. The highest color removal (98.55%) was observed at pH 3, which can be due to the ionic interaction between the dye and polysaccharides in acidic environments [16, 19]. Therefore, it is very important to determine the zero point charge of adsorbent to understand the adsorption mechanism in detail. In a study by He et al., the different structures of chitosan-gelatin microspheres in acidic and alkaline media were reported. When the pH is less than 6, the electrostatic interaction between the anionic group of dye and the protonated group of chitosan-gelatin complex increases. Thus, ionic force occurs between these groups, resulting in an increase in dye adsorption [8].

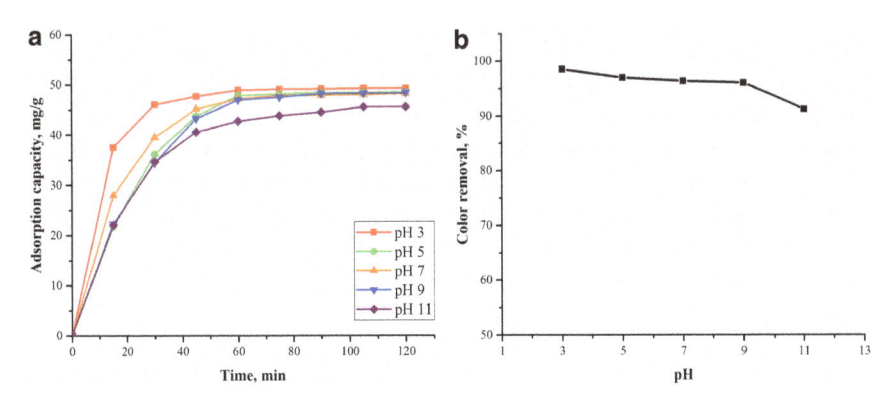

Fig. 3 Effect of pH on adsorption capacity **a** and color removal **b** of PR MX-5B on CH-GEL cryogel

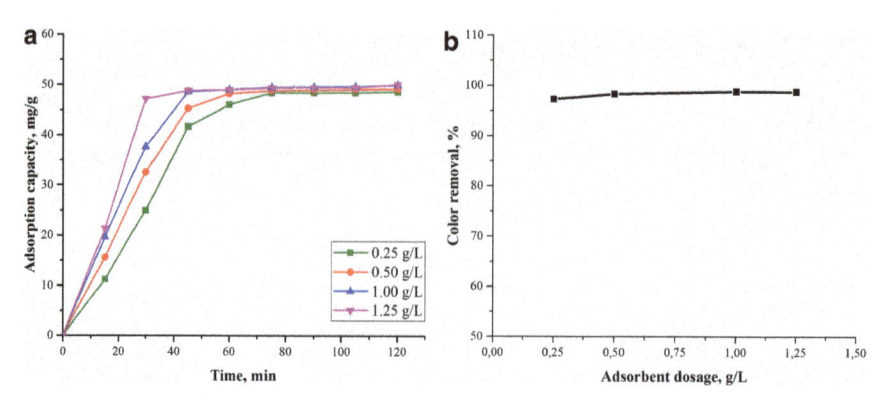

Fig. 4 Effect of adsorbent dosage on adsorption capacity **a** and color removal **b** of PR MX-5B on CH-GEL cryogel

3.3.2 Effect of Adsorbent Dosage

Batch system experiments were conducted at various CH-GEL cryogel dosages ranged from 0.25 to 1.0 g/L while the other experimental parameters were constant. The effect of the amount of adsorbent, CH-GEL cryogel, on the PR MX-5B adsorption is shown in Fig. 4. A rapid initial adsorption of PR MX-5B took place within the first 60 min, after which the adsorption reached equilibrium at 80 min. This is because of the easy adsorption of dye molecules onto the surface of polymers due to their ability and the available sites on gelatin and chitosan surfaces [9]. The increase in adsorbent dosage showed no significant changes in removal efficiency. Further adsorption experiments were performed with the fixed adsorbent dosage of 0.5 g/L to better clarify the adsorption process using a lower mass of adsorbent.

3.3.3 Effect of Dye Concentration

The effect of initial dyestuff concentration on the adsorption rate was examined in the 25, 50, 75 and 100 mg/L concentration range at 25 °C, 125 rpm agitation speed, pH 3 and 0.5 g/L adsorbent dosage. The adsorption capacity and percentage of removal of color at different dye concentrations are shown in Fig. 5. The percentage of color removal and adsorption capacity decreased with increasing initial dye concentration. The reason for the decrease in color removal is that there are no unoccupied active sites on the adsorbent at high dye concentrations. Above optimum dye concentration, more active sites will be required for adsorption [14].

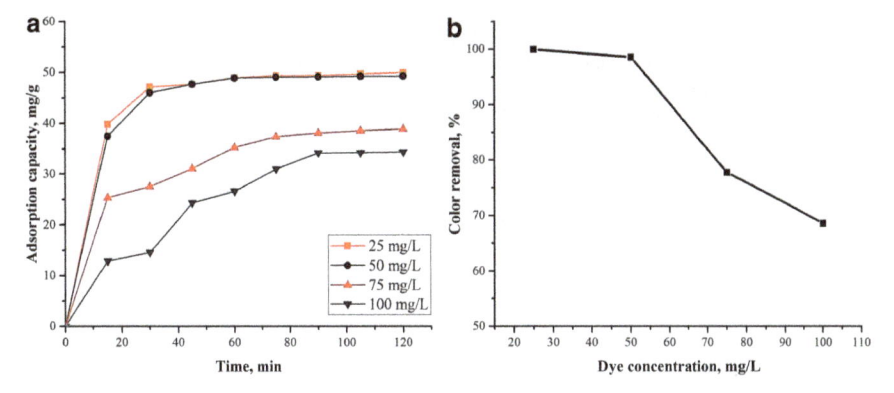

Fig. 5 Effect of initial dye concentration on adsorption capacity **a** and color removal **b** of PR MX-5B on CH-GEL cryogel

4 Conclusions

Natural polymer based cryogels for dye removal from aqueous solutions were prepared by using glutaraldehyde as a crosslinking agent at cryogenic conditions. The cryogels were successfully synthesized in monolithic shape and characterized with chemical, physical, morphological and gravimetric analyzes. Preliminary results of characterization studies suggest that the cryogel prepared has an interconnected mega porous structure, which is suitable for the adsorption of large molecule dyes. The adsorption of PR MX-5B dye on CH-GEL cryogel has been optimized in terms of pH of initial dye solution, amount of the adsorbent and concentration of the dye solution. The percentage of color removal and adsorption efficiency were calculated for each adsorption studies. The dye removal is pH and dye concentration dependent and found to be maximum at pH 3 and 25 mg/L dye concentration. The high adsorption capacity of CH-GEL cryogel samples for the removal of PR MX-5B dye from water solutions showed the potential cryogels in water purification applications. Cryogels as superabsorbent materials would be very effective for the adsorption of dye effluents from wastewater to improve overall safety and human health. Further investigations are currently ongoing on dyes and returned pharmaceuticals in wastewater streams.

Conflict of Interest Authors express no conflict of interest in this work.

References

1. Asadi S, Eris S, Azizian S (2018) Alginate-based hydrogel beads as a biocompatible and efficient adsorbent for dye removal from aqueous solutions. ACS Omega 3:15140–15148
2. Crini G (2006) Non-conventional low-cost adsorbents for dye removal: a review. Bioresour Technol 97:1061–1085

3. Das A, Pal A, Saha S, Maji SK (2009) Behaviour of fixed-bed column for the adsorption of malachite green on surfactant-modified alumina. J Environ Sci Heal—Part A Toxic/Hazardous Subst Environ Eng 44:265–272
4. Esquerdo VM, Cadaval TRS, Dotto GL, Pinto LAA (2014) Chitosan scaffold as an alternative adsorbent for the removal of hazardous food dyes from aqueous solutions. J Colloid Interface Sci 424:7–15
5. Gioiella L, Altobelli R, De Luna MS, Filippone G (2016) Chitosan-based hydrogel for dye removal from aqueous solutions: optimization of the preparation procedure. AIP Conf Proc 1736
6. Gita S, Hussan A, Choudhury TG (2017) Impact of textile dyes waste on aquatic environments and its treatment. Environ Ecol 35:2349–2353
7. Hadi AG (2014) Removal of cationic dye from aqueous solutions using chitosan. Indian J Appl Res 4:4–6
8. He B, Xue H (2015) Adsorption behaviors of acid dye by amphoteric chitosan/gelatin composite microspheres. Water Qual Res J Canada 50:314–325
9. Labidi A, Salaberria AM, Fernandes SCM, Labidi J, Abderrabba M (2019) Functional chitosan derivative and chitin as decolorization materials for methylene blue and methyl orange from aqueous solution. Materials (Basel) 12
10. Maas R, Chaudhari S (2005) Adsorption and biological decolourization of azo dye reactive red 2 in semicontinuous anaerobic reactors. Process Biochem 40:699–705
11. Mabel MM, Sundararaman TR, Parthasarathy N, Rajkumar J (2019) Chitin beads from Peneaus sp. shells as a biosorbent for methylene blue dye removal. Polish J Environ Stud 28:2253–2259
12. Madhav S, Ahamad A, Singh P, Mishra PK (2018) A review of textile industry: wet processing, environmental impacts, and effluent treatment methods. Environ Qual Manag 27:31–41
13. Nieto-Suárez M, López-Quintela MA, Lazzari M (2016) Preparation and characterization of crosslinked chitosan/gelatin scaffolds by ice segregation induced self-assembly. Carbohydr Polym 141:175–183
14. Pathania D, Sharma S, Singh P (2017) Removal of methylene blue by adsorption onto activated carbon developed from Ficus carica bast. Arab J Chem 10:1445–1451
15. Qiao C, Ma X, Zhang J, Yao J (2017) Molecular interactions in gelatin/chitosan composite films. Food Chem 235:45–50
16. Rajendra Sukhadeorao D (2019) Chitosan formulations: chemistry, characteristics and contextual adsorption in unambiguous modernization of S&T. Hystersis of Composites. IntechOpen, 1–15
17. Saber-Samandari S, Saber-Samandari S, Joneidi-Yekta H, Mohseni M (2017) Adsorption of anionic and cationic dyes from aqueous solution using gelatin-based magnetic nanocomposite beads comprising carboxylic acid functionalized carbon nanotube. Chem Eng J 308:1133–1144
18. Benkhaya S, El Harfi S, El Harfi A (2017) Classifications, properties and applications of textile dyes: a review. Appl J Envir Eng Sci 3:311–320
19. Sánchez-Duarte RG, Sánchez-Machado DI, López-Cervantes J, Correa-Murrieta MA (2012) Adsorption of allura red dye by cross-linked chitosan from shrimp waste. Water Sci Technol 65:618–623
20. Siddique K, Rizwan M, Shahid MJ, Ali S, Ahmad R, Rizvi H (2017) Textile wastewater treatment options: a critical review. enhancing cleanup of environmental pollutants. Springer International Publishing, pp 183–207
21. Staroszczyk H, Sztuka K, Wolska J, Wojtasz-Pająk A, Kołodziejska I (2014) Interactions of fish gelatin and chitosan in uncrosslinked and crosslinked with EDC films: FT-IR study. Spectrochim Acta A Mol Biomol Spectrosc 117:707–712
22. Tekoglu O, Ozdemir C (2010) Wastewater of textile industry and its treatment processes. BALWOIS 1–11
23. Wang C, Li J, Wang L, Sun X, Huang J (2009) Adsorption of dye from wastewater by zeolites synthesized from fly ash: kinetic and equilibrium studies. Chinese J Chem Eng 17:513–521

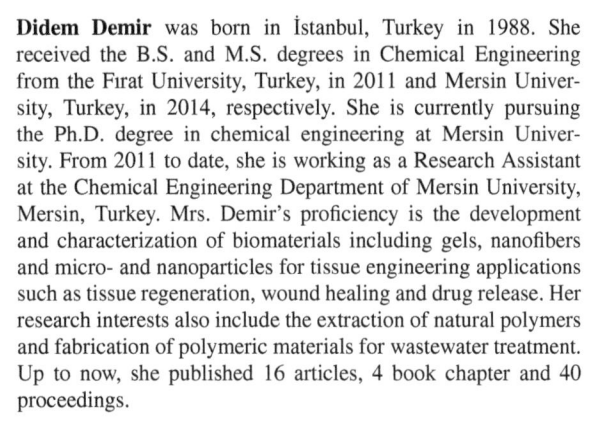

Didem Demir was born in İstanbul, Turkey in 1988. She received the B.S. and M.S. degrees in Chemical Engineering from the Fırat University, Turkey, in 2011 and Mersin University, Turkey, in 2014, respectively. She is currently pursuing the Ph.D. degree in chemical engineering at Mersin University. From 2011 to date, she is working as a Research Assistant at the Chemical Engineering Department of Mersin University, Mersin, Turkey. Mrs. Demir's proficiency is the development and characterization of biomaterials including gels, nanofibers and micro- and nanoparticles for tissue engineering applications such as tissue regeneration, wound healing and drug release. Her research interests also include the extraction of natural polymers and fabrication of polymeric materials for wastewater treatment. Up to now, she published 16 articles, 4 book chapter and 40 proceedings.

Nimet Bölgen was born in Adana, Turkey, in 1979. She received the B.S. degree in Chemical Engineering from Ankara University, Ankara, in 2002, and the M.S degree in Chemical Engineering from Hacettepe University, Ankara, in 2004. She received Ph.D. degree from Bioengineering Department of Hacettepe University, Ankara, in 2008. From 2009 to 2014, she was an Assistant Professor, and from 2014 to 2020 she was an Associate Professor with the Chemical Engineering Department of Mersin University, Mersin. Since 2020, she has been a Professor with the Chemical Engineering Department of Mersin University, Mersin. She is the author of 10 book chapters, 38 articles and more than 100 proceedings. Her research interest includes tissue engineering, biomaterials, controlled drug release systems, natural and synthetic polymers, cryogels and electrospun nanofibers. She is the Editorial Board Member of the journal Biomaterials and Biomechanics in Bioengineering. Prof. Bölgen is the member of Tissue Engineering and Regenerative Medicine International Society.

Ashok Vaseashta (M'79–SM'90) received Ph.D. in Materials Science and Engineering (minor in Electrical Engineering) from Virginia Polytechnic Institute and State University, Blacksburg, Virginia, USA. He is Executive Director of Research for International Clean Water Institute in VA, USA, Chaired Professor of Nanotechnology at the Academy of Sciences of Moldova and Professor, Nanotechnology and Biomedical Engineering at the Faculty of Mechanical Engineering, Transport and Aeronautics at the Riga Technical University. Prior to his current position, he served as Vice Provost for Research at the Molecular Science Research Center in Orangeburg, South Carolina. He served as visiting professor at the 3 Nano-SAE Research Centre, University of Bucharest, Romania and visiting scientist at the Helen and Martin Kimmel Center of Nanoscale Science at the Weizmann Institute of Science, Israel. He served the U.S. Department of State in two rotations, as strategic S&T

advisor and U.S. diplomat. His research interests span nanotechnology, environmental/ecological science, and safety and security. His research on nanotechnology has been on improving the understanding, design, and performance of nanofibers and sensors/detectors, mainly for applications such as wearable electronics, target drug delivery, detection of biomarkers and toxicity of nano and xenobiotic materials. In the security arena, he has worked on counterterrorism, countering unconventional warfare and hybrid threats, critical-Infrastructure protection, biosecurity, dual-use research concerns, and mitigating hybrid threats. He has authored over 250 research publications, edited/authored eight books on nanotechnology, and presented many keynotes and invited lectures worldwide. He serves on the editorial board of several highly reputed international journals. He is an active member of several national and international professional organizations. He is a fellow of the American Physical Society, Institute of Nanotechnology, and the New York Academy of Sciences. He has earned several other fellowships and awards for his meritorious service including 2004/2005 Distinguished Artist and Scholar award.

Macroporous Cryogels for Water Purification

Didem Demir, Ashok Vaseashta, and Nimet Bölgen

Abstract Cryogels are polymeric matrices with 3D architecture fabricated by cryotropic gelation at subzero temperatures via co-polymerization of monomers or crosslinking of linear polymers. The main characteristic feature of these materials that distinguishes them from other gels is their continuous and highly porous structure resulting from cryopolymerization that occurs in a two-phase environment. Highly porous polymeric cryogels can be produced in various shapes and sizes. There are many application areas of cryogels due to their individual properties. The applications of supermacroporous cryogels in the biomedical and biotechnological fields have been well reported in the literature. However, studies on the use of these unique gels in wastewater treatment as a solid adsorbent have started to attract attention in the last few years. The cryogels are suitable adsorbents for the removal of non-degradable organic and synthetic toxic compounds in wastewater. In addition, these cryogel matrices have superior properties including high surface area, high porosity, spongy structure, and physical stability during swelling conditions compared to other gel materials available. This chapter highlights the details of cryotropic gelation, process parameters effective on characteristics of cryogels and utilization areas of these materials specifically in water purification applications.

Keywords Polymer · Cryogel · Wastewater · Water pollution · Water purification

---

D. Demir · N. Bölgen (✉)
Chemical Engineering Department, Engineering Faculty, Mersin University, 33343 Mersin, Turkey
e-mail: nimet@mersin.edu.tr

D. Demir
e-mail: didemdemir@mersin.edu.tr

A. Vaseashta
Int'l Clean Water Institute, Manassas, VA, USA

Biomedical Engineering and Nano Technologies Institute, Riga Technical University, Riga, Latvia

A. Vaseashta
e-mail: prof.vaseashta@ieee.org

1 Introduction

Water pollution has become one of the most serious problems with the rapid growth of the industry and population in the world. Although about 71% of the world's surface is water-covered, only less than 1% of water on the surface of the earth is available for drinking, as reported by many investigators [1] The use of pesticides, antibiotics, solvents, petrochemicals, dyes and some other organic pollutants in urban, industrial and agricultural applications result in the release of toxic and non-degradable compounds into the rivers, lakes, coastal waters, seas and oceans [2]. These compounds have several adverse impacts on living organisms and environment, even at very low concentrations. Therefore, these pollutants should be treated and removed from wastewater before entering the environment.

For many years, different pretreatment methods have been extensively studied for the purification of wastewater including chemical coagulation, electro-coagulation, ozonation, membrane filtration and adsorption [3]. Compared with the other methods, adsorption by solid adsorbents is a suitable technique which is easily designed, highly efficient, and economic [4]. Many synthetic, natural and waste materials have been used as adsorbents due to their porosity, pore structure, nature of their adsorbing surfaces and economic viability [5].

In the last few years, cryogels (cryotropic hydrogel), having great flexibility, rapid reversible transformation in 3D microstructures in response to external stimuli and outstanding swellability in aqueous media than traditionally prepared hydrogel, mechanical strength and macroporous structure have gained great importance and interest in water treatment applications. Furthermore, high adsorption capacity for heavy metals, dyestuffs, antibiotics, and microorganisms, coupled with high mechanical stability under swelling/deswelling conditions, biocompatibility, adequate cellular migration, ingrowth, regenerative capability, antimicrobial properties and low cost are some of the essential features for using cryogels as a suitable adsorbent for wastewater purification. Furthermore, the polymeric surfaces of the cryogels may be modified with specific affinity ligands or functional nanoparticles to increase the binding capacity [6].

Cryogels can be fabricated with a process known as cryotropic gelation (cryogelation) that has been applied to prepare interconnected supermacroporous 3D cryogel matrices [7]. Various monomers or linear polymers have been used to prepare polymeric cryogels for application in many different areas. The natural or synthetic polymers such as chitosan, gelatin, starch, silk, poly (vinyl alcohol), poly (lactic acid), poly (ethylene glycol) diacrylate, etc. or their composites can be used to obtain crosslinked cryogels via chemical or physical crosslinking approaches during cryogelation process [8]. Cryogels can also be prepared by copolymerization of monomers including acrylamide, N-isopropyl acrylamide and N-vinyl-caprolactam with the presence of an initiator and crosslinker [9]. The type and concentration of the polymer/monomer, amount of crosslinking agent and cryogelation temperature affect the physical, chemical and morphological properties of cryogels which result in the adsorption behavior of the material. This chapter summarizes the fabrication method of cryogels, the effective factors on properties of the synthesized cryogels, application areas of these polymeric materials prepared in different shapes and the recent studies of the supermacroporous cryogels for water purification applications.

2 Fabrication of Cryogels

Cryogels are 3D polymeric network structures which are generally fabricated at subzero temperatures by crosslinking polymerization of monomers or by crosslinking of polymers [8]. Different production techniques such as salt leaching, gas foaming, freeze drying and cryogelation are used in the fabrication of porous materials [10]. Compared to other methods, cryogelation also known as cryogelling and cryotropic gelation is the least time requiring, easiest and effective method. To understand their characteristics, this section provides the typical cryogelation process and unique properties of the synthesized cryogels.

Cryogelation is a simple technique used for the production of cryogels with a macroporous morphology up to 200 μm in pore diameter [11]. The process includes three main steps: (I): phase separation with ice crystal formation, (II): crosslinking polymerization of monomers or crosslinking of polymers, and (III): thawing of the ice crystals at room temperature, to form an open and highly interconnected polymer network [12, 13]. Typically, the monomers and polymer precursors, or polymers are first homogeneously dissolved in an appropriate solvent to form an initial solution. This mixture is then filled in a mold such as a tubular column or plastic syringe. The gel solution is then frozen to subzero temperatures (down to $-40\ °C$) below the freezing point of the selected solvent. During freezing, a large volume of the solvent is crystallized and the ice crystals in micrometer-size act as porogens (pore forming agents) that provide the formation of the final porous structure. Meanwhile, the crosslinking reaction of the polymer proceeds at the concentrated liquid microphase [14]. After a suitable incubation time for gelling is completed, the frozen gel is allowed to thaw at room temperature. The ice crystals melt and leave large and continuous interconnected pores in melted areas [9]. The resulting material possesses a macroporous, spongy, compressible and mechanically stable structure [15]. The fabrication steps of a cryogel with an interconnected porosity using cryogelation technique are illustrated in Fig. 1.

3 Factors Effecting the Characteristic Properties of Cryogels

The composition (type of solvent/polymer/monomer/crosslinker etc.) and concentration of initial solution, freezing temperature and incubation time are the main key parameters affecting the characteristic properties of the synthesized cryogels [16].

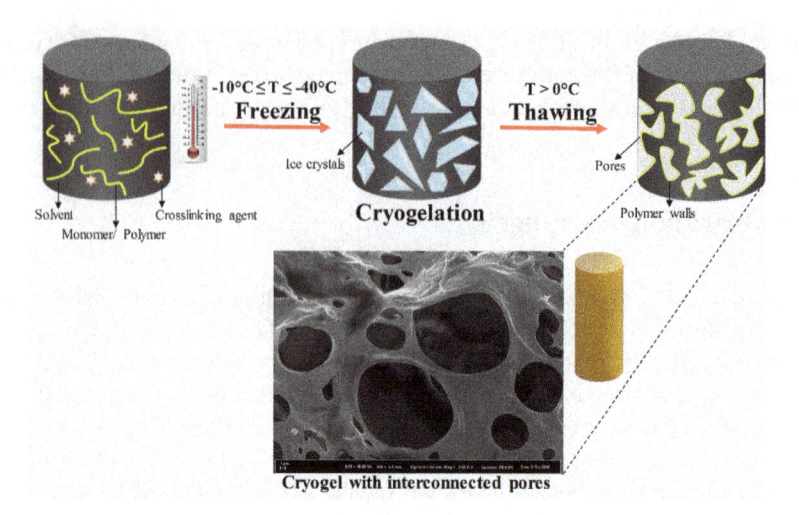

Fig. 1 The overall scheme of the processing steps of a cryogel with interconnected porosity by cryogelation technique

During cryogelation process, changing any of these fabrication parameters will result in a change in morphology of the cryogel such as polymer wall thickness, pore size, pore structure and porosity [17].

3.1 The Content of Initial Solution

Cryogels can be produced from various types of monomers and their derivatives in the presence of an initiator and crosslinker (Table 1). Ammonium persulfate (APS) and potassium persulfate (KPS) are water soluble initiators which are generally used to initiate the polymerization reactions. Ultraviolet (UV) and electron-beam (EB)

Table 1 Monomers used for the fabrication of polymeric cryogels

Monomer	Crosslinker	Solvent	Temperature	References
Acrylamide (AAm)	MBAA	Water	−13 °C	[18]
N-isopropylacrylamide (NIPAM)	MBAA	Water	−18 °C	[19]
2-hydroxyethyl methacrylate (HEMA)	MBAA	Water	−16 °C	[20]
N-Vinylcaprolactam (VCL)	MBAA	Dimethyl sulfoxide containing water	−12 °C	[21]
Lauryl methacrylate (LMA)	Divinylbenzene	Acetic acid	−18 °C	[22]

Table 2 Synthetic and natural polymers for the fabrication of cryogels

Polymer	Crosslinking	Solvent	Temperature	References
Chitosan	Glutaraldehyde	Acetic acid	−16 °C	[27]
Collagen	Dialdehyde starch	Water	−15 °C	[28]
Poly (ethylene glycol) diacrylate	UV light	Water	−40 °C	[29]
Poly (vinyl alcohol) Starch	Freeze-thawing cycles	Water	−20 °C	[30]
Poly (vinyl alcohol) Gelatin	Glutaraldehyde	Water	−12 °C	[31]

radiations can also be used as initiators [8]. N,N'-methylenebisacrylamide (MBAA) is the commonly used crosslinking agent to reinforce the structure.

Synthetic and natural polymers can also be used for preparation of the cryogels via crosslinking during cryogelation, in the presence of a crosslinker (Table 2) [8]. Crosslinking by chemical agents (glutaraldehyde, genipin, heparin, etc.) or physical crosslinking by freeze-thawing cycles are common methods to produce a cryogel matrix. Polymers are chosen for cryogel fabrication due to their distinctive features such as mechanical stability, chemical variability, compatibility, degradability, non-toxicity, manufacturing facility and cost-effectiveness. Cryogels based on both natural and synthetic polymers and their composites have been used in other applications including scaffolds in tissue engineering, chromatographic materials in biotechnology, adsorbents for water treatment, carriers for immobilization of biomolecules and cells [23, 24, 25, 26].

3.2 Effect of the Initial Concentration

The concentration of monomers/polymers in the initial solution can influence the structural properties of the synthesized cryogels. In general, as the initial concentration increases, the cryogel exhibits a smaller pore size due to the formation of thicker walls, resulting in a rigid and less spongy structure. In general, the starting monomer/polymer takes place in non-frozen phase and its concentration affects the size of the non-frozen phase, which results in the formation of the final polymeric wall of the cryogel. As a result, a larger non-frozen phase, with a less porous structure and with thick pore walls will be formed [32, 33]. In addition, increased polymer concentration in pore walls leads to increased mechanical strength of the cryogel [34]. For example, Bruns et al. fabricated injectable ligand pre-functionalized polyethylene glycol (PEG) cryogels and studied the effect of polymer concentration on the microstructure of PEG cryogels. Increase in polymer concentration has been shown to increase the cryogel stiffness [35]. Similarly, Dispinar et al. reported that altering the initial precursor concentration caused a change on the mechanical properties of the cryogels [36].

3.3 Cryogelation Temperature and Duration

Cryogelation, a polymerization process, is intended to form a crosslinked and macroporous gel network at sub-zero temperatures. The basic difference between cryogels and hydrogels is the reaction temperature. While hydrogels require temperatures at or above room temperature, cryogels require lower temperatures, especially below the freezing point of the appropriate solvent for polymer/monomer [37]. The reaction proceeds around ice crystals formed at temperatures below zero and the interconnected macroporous network of cryogel separated by crosslinked polymer walls is formed by melting the ice crystals at room temperature, after completion of the reaction. Therefore, the cryogelation temperature and duration have a considerable effect on the porous structure of the cryogel, such as shape, size and density of the pores [17]. Polymerization at low freezing temperatures causes the formation of smaller and numerous ice crystals and hence cryogel with smaller pore sizes are fabricated [38]. In addition, a temperature that is not low enough causes the solution to remain in a non-frozen state, which leads to the formation of an undesirable structure [39]. It is reported that a temperature range between -10 and $-21\ °C$ and a freezing time of 24 h provides an optimal interconnected ice crystal structure in the macrometer range [40].

4 Application Areas of Supermacroporous Cryogels

Several cryogels have been fabricated so far for many different applications. These cryogels are formed in different shapes and sizes, such as monoliths, disks, membranes and spherical beads. Depending on the interconnected porosity, flexibility and spongy structure of the cryogels, they have the potential to be used in tissue engineering, biomedical, bio-separation, bio-catalysis, water treatment and protein purification applications, as shown in Fig. 2.

5 Cryogels Used in Water Purification Applications

Water pollution is an ever increasing global concern that adversely impacts the health of billions of people worldwide. In addition to human health issues related to waterborne contaminants, there are additional concerns as they relate to toxicity to wildlife, long-term ecosystem impact and economic consequences. The contaminated water can come from a variety of sources that include non-degradable products such as

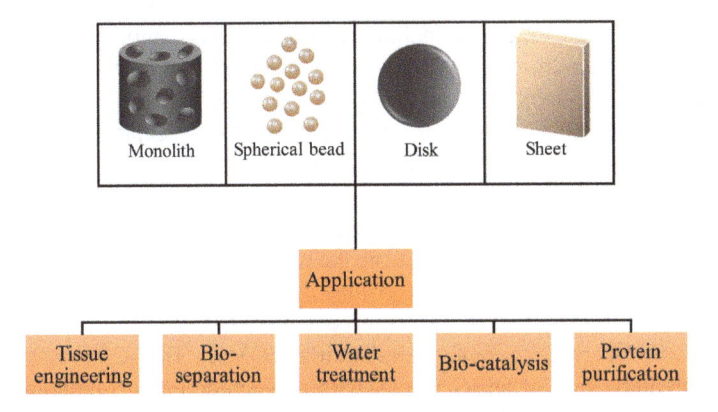

Fig. 2 Application areas of supermacroporous cryogels in various shapes and sizes

pesticides, phenolic chemicals, surfactants, heavy metals, highly toxic chemicals and biological pollutants [41]. Therefore, in order to make wastewater safe, it must be treated using state-of-the-art filtration systems and contaminants detection systems with wide-spectrum detection capability down to ppb (parts per billion). In this context, many approaches have been developed and reported in related literature for the purification of wastewater by using different functional adsorbents. Among them, macroporous cryogels are novel biosorbents suitable for different kinds of contaminated water sources with their enhanced adsorption capability. The most important feature that distinguishes cryogels from hydrogels as adsorbents, is their controllable mechanical properties during swelling in water, which affect their use for different cycles in water treatment processes [42]. Also, the cryogels have additional advantages over other adsorbents such as high swelling capability, flexibility and short diffusion path due to the interconnected large pores in their 3D networks. So far, only a very limited literature exists on the applications of cryogels for water treatment applications. The main applications of cryogels as adsorbents will be discussed according to the types of pollutants: (a) antibiotics; (b) metallic contaminants and (c) dyestuffs.

5.1 Antibiotics

Water pollution caused by residues of drugs and pharmaceuticals is a rapidly growing problem. The pharmaceutical products are often discharged into water resources during or after their manufacturing and often even untreated, depending upon prevailing regulations. Furthermore, disposal of unused pharmaceuticals by individuals, unregulated companies, private hospitals and even from households has significantly increased [43], resulting in alarming levels of Estrogens, painkillers and antibiotics in water [44]. Additionally, many other returned pharmaceuticals are typically found in water and include anti-cancer drugs, antidepressants, severe acute respiratory syndrome (SARS) virus, nonsteroidal anti-inflammatory drugs, antianxiety agents, betablockers and analgesics [45], including recently discovered nCOV-SARS2, a.k.a. COVID-19, novel coronavirus. In particular, antibiotics pose a special problem that can lead to the development and spread of antibiotics resistant bacteria and genes which can have serious effects on the ecological system [46]. To overcome this problem, macroporous cryogels can be efficient adsorptive materials for the removal of antibiotic effluents from the wastewaters. A study by Erşan et al. is one of the first studies that aims to determine the adsorption characteristics of tetracycline (TC) antibiotic on tannin based cryogels (TAB CRG), in which tannin was obtained from Quercus macranthera gall extract (QMG). TAB CRG was prepared by cryopolymerization of QMG extract with acrylamide (AAm) monomer. N,N'-methylene bisacryl amide (MBAA) was used as a crosslinker for immobilization of tannin in elastic and spongy matrices of cryogels. Adsorption of TC on cryogels was studied in a batch system by optimization the experimental parameters including pH, adsorbent dosage, initial TC concentration and temperature to achieve the maximum yield of TC removal (Fig. 3).

A rapid, exothermic and spontaneous adsorption process has been observed and reached equilibrium within 150 min. The monolayer capacity (Q_{max}, mg/g) value

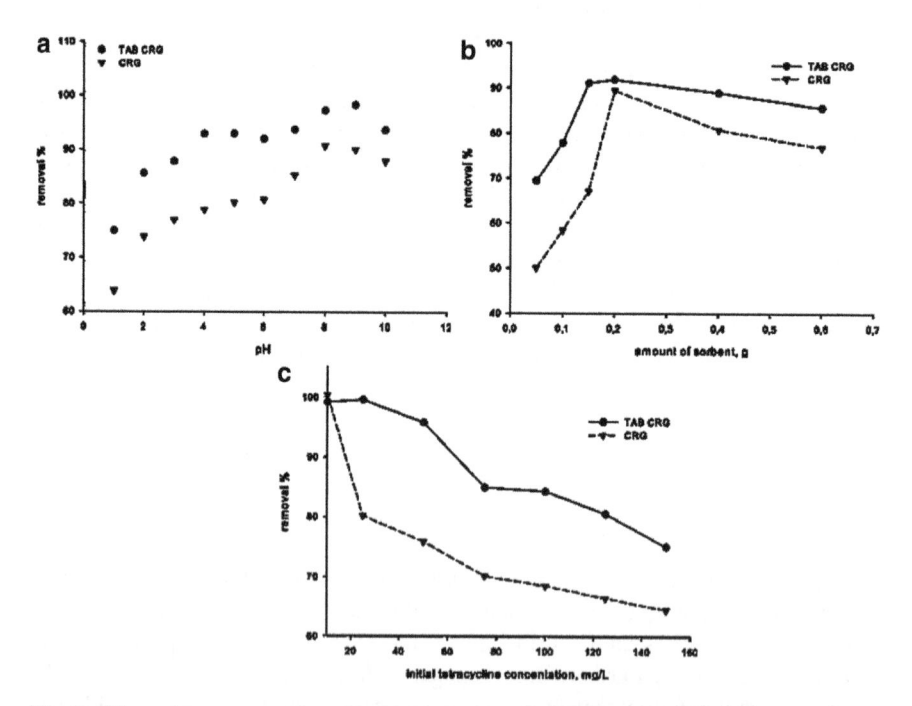

Fig. 3 Effect of the parameters on the adsorption TC to TAB CRG and CRG cryogels. **a** Effect of pH, **b** Effect of sorbent concentration and **c** Effect of initial antibiotic concentration. Figures reproduced from Erşan et al. [47] with permission from Elsevier

was about 65 mg/g for TAB CRG. The adsorption capacity of TAB CRG was found to be comparable with other conventional adsorbents [47].

5.2 Metallic Contaminants

Numerous elements fall into the category of metallic pollutants in wastewater such as cadmium (Cd), chromium (Cr), copper (Cu), nickel (Ni), arsenic (As), lead (Pb) and zinc (Zn). These heavy metals can enter into the water sources through mining activities, industries bearing heavy metals and household applications [48]. It is very important to treat metal-contaminated wastewater prior to its discharge because there are many adverse health effects on the environment and living organisms, mainly human beings. Cryogels are attractive sorbents for heavy metals due to the high chemical reactivity of functional groups (hydroxyls, carboxyls and amines) of polymers (polysaccharides, e.g. chitin, chitosan, starch, alginate and gelatin) which increase the adsorption capacity of metal ions [49]. Additionally, due to the high water retention capacity of cryogels, the diffusion of water containing metallic contaminants into the cryogel is also an effective factor that increases the adsorption efficiency [50].

In a study by Erdem et al., the adsorption efficiency of Cu (II) ions on novel macroporous cryogels based on jeffamine (a commercial PEG-containing di-amino terminated triblock copolymers) with various molecular weights (RC 600, RC 900 and RC 2003) was investigated. The cryogels were synthesized at two steps: (I) formation of schiff base cryogels with the combination of jeffamine and glutaraldehyde at $-18\ ^\circ$C, and (II) reduction of the cryogels by using $NaBH_4$ (as a reducing agent) and NaOH (for deprotonation) to create cryogels having additional active functional groups. The adsorption capacity of the reduced cryogels was examined as a function of dosage, initial concentration, pH, contact time and temperature. The surface morphology of the cryogels before and after adsorption of Cu (II) ions was analyzed using a scanning electron microscope as shown in Fig. 4a. An interconnected macroporous structure was observed before adsorption. Many irregular small particles were formed on the surface of cryogels after the adsorption of Cu (II) ions and confirmed by EDX analysis (Fig. 4b). The decrease in the percentage of adsorbed Cu (II) ions by increasing the molecular weight of jeffamine was attributed to the decrease in the number of active functional groups as the molecular weight increased. RC 600 revealed higher adsorption capacity for Cu (II) ions compared to RC 900 and RC 2003 due to the higher amine content [51].

5.3 Dyestuffs

The development of the industries such as paper, textile and food industry positively affect the economic development in the world. However, with the development and

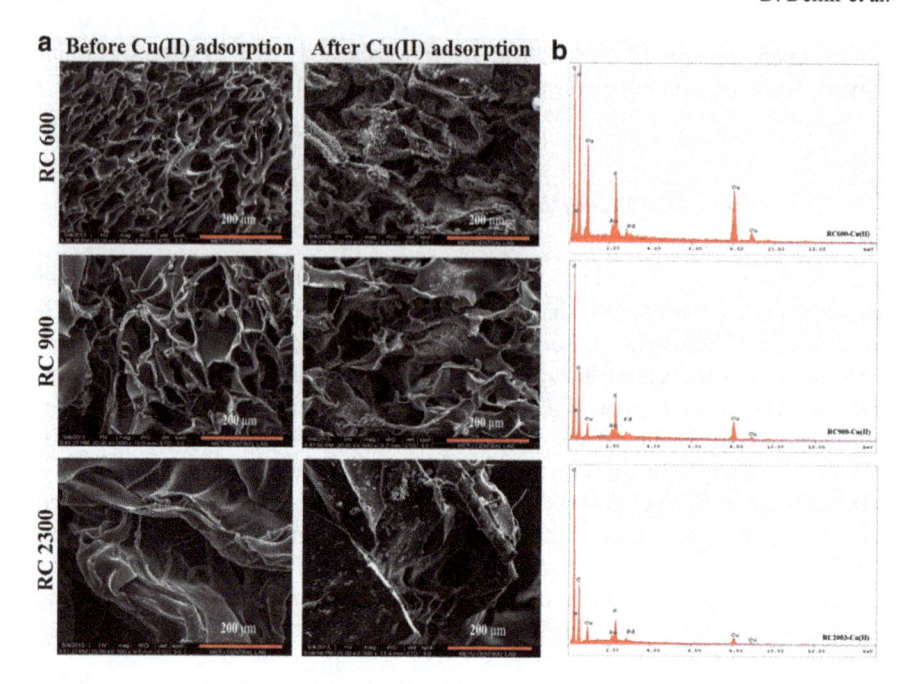

Fig. 4 a Structure of cryogels (RC 600, RC 900 and RC 2003) before and after Cu (II) adsorption. **b** EDX Spectra of cryogels after Cu (II) adsorption. Redesigned from Erdem [51] with permission from Elsevier

growth of these industries, the amount of contaminants, especially dyes, given to the waters is increasing [52]. Various types of synthetic and natural dyes are added during dyeing process and after finishing this process tons of non-degradable color effluents are discharged into the water systems [53]. Dyestuffs remain in water cause dangerous results, particularly for the aquatic ecosystem. Even just a little amount of dye pollutes large volumes of water, decreases the oxygen concentration in water body, blocks the sunlight, inhibits photosynthesis and damages biological activities in aquatic life [54]. Therefore, it is absolutely necessary to eliminate the dye effluents from water sources. The dye adsorption on cryogels made from various types of synthetic and natural polymers have been studied previously. Dobritoiu and Patachia studied the methylene blue (MB) adsorption on PVA based cryogels fabricated by repeated freezing–thawing cycles.

They used three types of biopolymers to synthesize PVA/scleroglucan, PVA/microfibers of cellulose and PVA/zein cryogels. The removal capacity of MB was evaluated according to the effects of the initial dye concentration, contact time and the composition of materials, then the adsorption kinetic and thermodynamic parameters were determined. During adsorption, the cryogels changed their color from white to blue and the amount of adsorbed dye into the cryogels increased with the increasing amount of the initial dye concentration in aqueous solutions as shown in Fig. 5. They concluded that adsorption was related to the type and morphology

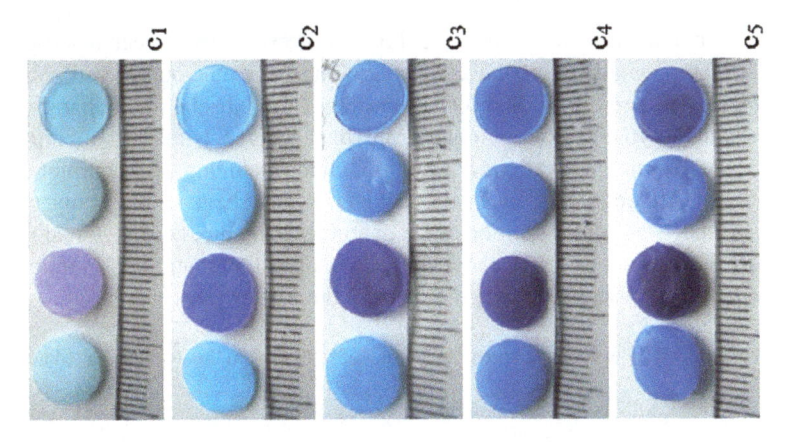

Fig. 5 The digital camera image of cryogels balanced in methylene blue solutions with different initial dye concentrations (C_1: 8.00×10^{-6}, C_2: 2.40×10^{-5}, C_3: 4.00×10^{-5}, C_4: 5.60×10^{-5} and C_5: 8.00×10^{-5} mol/L). Adapted from the figures of Dobritoiu and Patachia [55] with permission from Elsevier

of the cryogels and the highest sorption efficiency (9.5279 mg/g) was obtained for PVA/scleroglucan cryogels [55].

Studies on the adsorption of dyestuffs on cryogenic materials continued to increase in the following years. For instance, Sahiner et al. reported the removal of some organic dyes such as methylene blue (MB), rhodamine 6G (R6G) and methyl orange (MO) by bare poly(4-vinylpyridine) (p(4-VP)) cryogel and graphene oxide (GO) embedded p(4-VP)/GO cryogels. The p(4-VP) cryogels and different amounts of GO added p(4-VP)/GO cryogel composites were synthesized via free radical polymerization at -18 °C. The samples were suitable for the adsorption and separation of MB, R6G and MO toxic organic dyes and their mixtures as a column filter material. The adsorption efficiency of cryogels varied according to the chemical structure and surface charge of the dye and cryogel type used [19].

6 Conclusions and Future Perspective

During the past several decades, cryogels have been very useful as biomaterials in tissue engineering and biomedical applications. However, the recent use of polymeric cryogels for water purification applications has attracted great interest from many researchers. The research is ongoing and shows significant potential for contamination adsorbents for water purification. Highly interconnected large macropores and sponge-like morphology, appropriate mechanical and chemical stability are the main characteristic features of these materials. Within the framework of water contamination remediation, this chapter describes fabrication methods of cryogels, effect of several parameters that improve the properties of the synthesized materials and nature of contaminants that can be treated from wastewater using such macroporous cryogels. Having highly interconnected, continuous large macropores with a spongy morphology and appropriate stability in swollen form, are the main characteristic features of the cryogels. These unique properties of cryogels make them attractive matrices for adsorption of various types of pollutants including dyes, antibiotics and heavy metals. Although the idea of using cryogels as adsorbents is interesting, studies need to be improved by the addition of nanoparticles, nanotubes or functional groups to impart the features of cryogels such as mechanical strength and antimicrobial nature against bacteria and viruses that exist in the wastewater sources.

Conflict of Interest The authors express no conflict of interest with the work presented here.

References

1. Singh NB, Nagpal G, Agrawal S, Rachna (2018) Water purification by using adsorbents: a review. Environ Technol Innov 11:187–240
2. Inyinbor Adejumoke A, Adebesin Babatunde O, Oluyori Abimbola P, Adelani-Akande Tabitha A, Dada Adewumi O, Oreofe Toyin A (2018) Water pollution: effects, prevention, and climatic impact. In: Glavan M (eds) Water challenges of an urbanizing world. IntechOpen
3. Nouri Alavijeh H, Sadeghi M, Rajaeieh M, Moheb A, Sadani M, Ismail A (2017) Integrated ultrafiltration membranes and chemical coagulation for treatment of baker's yeast wastewater. J Membr Sci Technol 07:1–9
4. Jiuhui QU (2008) Research progress of novel adsorption processes in water purification: a review. J Environ Sci 20:1–13
5. Rashed MN (2013) Adsorption technique for the removal of organic pollutants from water and wastewater. In: Mohamed Nageeb R (eds) Organic pollutants—monitoring, risk and treatment. IntechOpen
6. Yao K, Yun J, Shen S, Wang L, He X, Yu X (2006) Characterization of a novel continuous supermacroporous monolithic cryogel embedded with nanoparticles for protein chromatography. J Chromatogr A 1109:103–110
7. Saylan Y, Denizli A (2019) Supermacroporous composite cryogels in biomedical applications. Gels 5:1–20
8. Okay O, Lozinsky VI (2014) Polymeric cryogels macroporous gels with remarkable properties. Springer International Publishing
9. Bencherif SA, Braschler TM, Renaud P (2013) Advances in the design of macroporous polymer scaffolds for potential applications in dentistry. J Periodontal Implant Sci 43:251–261
10. Memic A, Colombani T, Eggermont LJ, Rezaeeyazdi M, Steingold J, Rogers ZJ, Navare KJ, Mohammed HS, Bencherif SA (2019) Latest advances in Cryogel technology for biomedical applications. Adv Ther 2:1800114
11. Savina IN, Ingavle GC, Cundy AB, Mikhalovsky SV (2016) A simple method for the production of large volume 3D macroporous hydrogels for advanced biotechnological, medical and environmental applications. Sci Rep 6:1–9
12. Bölgen N, Plieva F, Galaev IY, Mattiasson B, Pişkin E (2007) Cryogelation for preparation of novel biodegradable tissue-engineering scaffolds. J Biomater Sci Polym Ed 18:1165–1179
13. Kumar A, Mishra R, Reinwald Y, Bhat S (2010) Cryogels: freezing unveiled by thawing. Mater Today 13:42–44
14. Henderson TMA, Ladewig K, Haylock DN, McLean KM, O'Connor AJ (2013) Cryogels for biomedical applications. J Mater Chem B 1:2682–2695

15. Hixon KR, Lu T, Sell SA (2017) A comprehensive review of cryogels and their roles in tissue engineering applications. Acta Biomater 62:29–41
16. Bölgen N, Yang Y, Korkusuz P, Güzel E, El Haj AJ, Pişkin E (2008) Three-dimensional ingrowth of bone cells within biodegradable cryogel scaffolds in bioreactors at different regimes. Tissue Eng Part A 14:1743–1750
17. Bakhshpour M, Idil N, Perçin I, Denizli A (2019) Biomedical applications of polymeric cryogels. Appl Sci 9:553
18. Kahveci MU, Beyazkilic Z, Yagci Y (2010) Polyacrylamide cryogels by photoinitiated free radical polymerization. J Polym Sci, Part A: Polym Chem 48:4989–4994
19. Sahiner N, Yildiz S, Sagbas S (2018) Graphene oxide embedded P(4-VP) cryogel composites for fast dye removal/separations. Polym Compos 39:1694–1703
20. Santos T, Brito A, Boto R, Sousa P, Almeida P, Cruz C, Tomaz C (2018) Influenza DNA vaccine purification using pHEMA cryogel support. Sep Purif Technol 206:192–198
21. Hwang Y, Zhang C, Varghese S (2010) Poly(ethylene glycol) cryogels as potential cell scaffolds: Effect of polymerization conditions on cryogel microstructure and properties. J Mater Chem 20:345–351
22. Chen X, Sui W, Ren D, Ding Y, Zhu X, Chen Z (2016) Synthesis of hydrophobic polymeric cryogels with supermacroporous structure. Macromol Mater Eng 301:659–664
23. Aslıyüce S, Denizli A (2017) Design of PHEMA cryogel as bioreactor matrices for biological cyanide degradation from wastewater. Hacettepe J Biol Chem 45:639–645
24. Demir D, Bölgen N (2017) Synthesis and characterization of injectable chitosan cryogel microsphere scaffolds. Int J Polym Mater Polym Biomater 66:686–696
25. Lozinsky VI, Galaev IY, Plieva FM, Savina IN, Jungvid H, Mattiasson B (2003) Polymeric cryogels as promising materials of biotechnological interest. Trends Biotechnol 21:445–451
26. Yeşilova E, Osman B, Kara A, Tümay Özer E (2018) Molecularly imprinted particle embedded composite cryogel for selective tetracycline adsorption. Sep Purif Technol 200:155–163
27. Demir D, Özdemir S, Yalçın MS, Bölgen N (2019) Chitosan cryogel microspheres decorated with silver nanoparticles as injectable and antimicrobial scaffolds. Int J Polym Mater Polym Biomater 1–9
28. Mu C, Liu F, Cheng Q, Li H, Wu B, Zhang G, Lin W (2010) Collagen cryogel cross-linked by dialdehyde starch. Macromol Mater Eng 295:100–107
29. Madaghiele M, Salvatore L, Demitri C, Sannino A (2018) Fast synthesis of poly(ethylene glycol) diacrylate cryogels via UV irradiation. Mater Lett 218:305–308
30. Bagri LP, Bajpai J, Bajpai AK (2011) Evaluation of starch based cryogels as potential biomaterials for. Bull Mater Sci 34:1739–1748
31. Ceylan S, Göktürk D, Bölgen N (2016) Effect of crosslinking methods on the structure and biocompatibility of polyvinyl alcohol/gelatin cryogels. Biomed Mater Eng 27:327–340
32. Kirsebom H, Topgaard D, Galaev IY, Mattiasson B (2010) Modulating the porosity of cryogels by influencing the nonfrozen liquid phase through the addition of inert solutes. Langmuir 26:16129–16133
33. Srivastava A, Jain E, Kumar A (2007) The physical characterization of supermacroporous poly(N-isopropylacrylamide) cryogel: mechanical strength and swelling/de-swelling kinetics. Mater Sci Eng, A 464:93–100
34. Plieva FM, Galaev IY, Mattiasson B (2007) Macroporous gels prepared at subzero temperatures as novel materials for chromatography of particulate-containing fluids and cell culture applications. J Sep Sci 30:1657–1671
35. Bruns J, McBride-Gagyi S, Zustiak SP (2018) Injectable and cell-adhesive polyethylene glycol cryogel scaffolds: independent control of cryogel microstructure and composition. Macromol Mater Eng 303:1–15
36. Dispinar T, Van Camp W, De Cock LJ, De Geest BG, Du Prez FE (2012) Redox-responsive degradable PEG cryogels as potential cell scaffolds in tissue engineering. Macromol Biosci 12:383–394
37. Butun S, Demirci S, Yasar AO, Sagbas S, Aktas N, Sahiner N (2017) 0D, 1D, 2D, and 3D soft and hard templates for catalysis. Stud Surf Sci Catal 177:317–357

38. Svec F (2010) Porous polymer monoliths: amazingly wide variety of techniques enabling their preparation. J Chromatogr A 1217:902–924
39. Plieva F, Huiting X, Galaev IY, Bergenståhl B, Mattiasson B (2006) Macroporous elastic polyacrylamide gels prepared at subzero temperatures: control of porous structure. J Mater Chem 16:4065–4073
40. Bäcker A, Göppert B, Sturm S, Abaffy P, Sollich T, Gruhl FJ (2016) Impact of adjustable cryogel properties on the performance of prostate cancer cells in 3D. Springerplus 5:1–12
41. Andrabi SM, Tiwari J, Singh S, Sarkar J, Verma N, Kumar A (2016) Supermacroporous hybrid polymeric cryogels for efficient removal of metallic contaminants and microbes from water. Int J Polym Mater Polym Biomater 65:636–645
42. Atta AM, Ezzat AO, Al-Hussain SA, Al-Lohedan HA, Tawfeek AM, Hashem AI (2018) New crosslinked poly (ionic liquid) cryogels for fast removal of methylene blue from waste water. React Funct Polym 131:420–429
43. Pruden A, Larsson DGJ, Amézquita A, Collignon P, Brandt KK (2013) Management of Options for Reducing the Release of Antibiotics. Environ Health Perspect 878:878–885
44. Vaseashta A (2013) Emerging sensor technologies for monitoring water quality, Future Medicine. © Future Science Ltd. https://doi.org/10.4155/ebo.13.208
45. Miazek K, Brozek-Pluska B (2019) Effect of PHRs and PCPs on microalgal growth, metabolism and microalgae-based bioremediation processes: a review. Int J Mol Sci 20:2492
46. Li B, Zhang T (2010) Biodegradation and adsorption of antibiotics in the activated sludge process. Environ Sci Technol 44:3468–3473
47. Erşan M, Bağda E, Bağda E (2013) Investigation of kinetic and thermodynamic characteristics of removal of tetracycline with sponge like, tannin based cryogels. Colloids Surf B Biointerfaces 104:75–82
48. Barakat MA (2011) New trends in removing heavy metals from industrial wastewater. Arab J Chem 4:361–377
49. Vandenbossche M, Jimenez M, Casetta M, Traisnel M (2015) Remediation of heavy metals by biomolecules: a review. Crit Rev Environ Sci Technol 45:1644–1704
50. Gunatilake SK (2015) Methods of removing heavy metals from industrial wastewater. J Multidiscip Eng Sci Stud Ind Wastewater 1:13–18
51. Erdem A, Ngwabebhoh FA, Yildiz U (2017) Novel macroporous cryogels with enhanced adsorption capability for the removal of Cu(II) ions from aqueous phase: modelling, kinetics and recovery studies. J Environ Chem Eng 5:1269–1280
52. Yaseen DA, Scholz M (2019) Textile dye wastewater characteristics and constituents of synthetic effluents: a critical review. Springer, Berlin Heidelberg
53. Mittal A, Mittal J, Malviya A, Kaur D, Gupta VK (2010) Adsorption of hazardous dye crystal violet from wastewater by waste materials. J Colloid Interf Sci 343:463–473
54. Gita S, Hussan A, Choudhury TG (2017) Impact of textile dyes waste on aquatic environments and its treatment. Environ Ecol 35:2349–2353
55. Dobritoiu R, Patachia S (2013) A study of dyes sorption on biobased cryogels. Appl Surf Sci 285:56–64

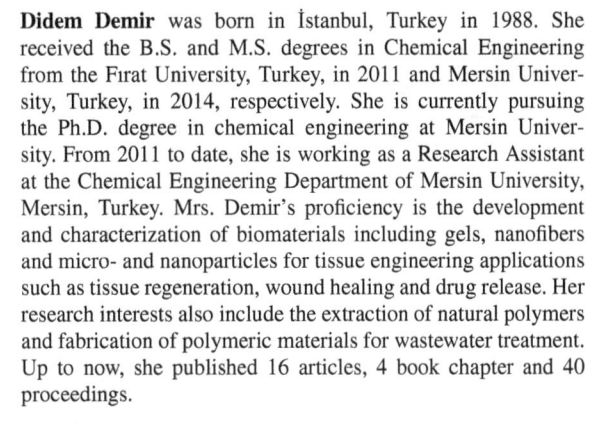

Didem Demir was born in İstanbul, Turkey in 1988. She received the B.S. and M.S. degrees in Chemical Engineering from the Fırat University, Turkey, in 2011 and Mersin University, Turkey, in 2014, respectively. She is currently pursuing the Ph.D. degree in chemical engineering at Mersin University. From 2011 to date, she is working as a Research Assistant at the Chemical Engineering Department of Mersin University, Mersin, Turkey. Mrs. Demir's proficiency is the development and characterization of biomaterials including gels, nanofibers and micro- and nanoparticles for tissue engineering applications such as tissue regeneration, wound healing and drug release. Her research interests also include the extraction of natural polymers and fabrication of polymeric materials for wastewater treatment. Up to now, she published 16 articles, 4 book chapter and 40 proceedings.

Ashok Vaseashta (M'79-SM'90) received Ph.D. in Materials Science and Engineering (minor in Electrical Engineering) from Virginia Polytechnic Institute and State University, Blacksburg, Virginia, USA. He is Executive Director of Research for International Clean Water Institute in VA, USA, Chaired Professor of Nanotechnology at the Academy of Sciences of Moldova and Professor, Nanotechnology and Biomedical Engineering at the Faculty of Mechanical Engineering, Transport and Aeronautics at the Riga Technical University. Prior to his current position, he served as Vice Provost for Research at the Molecular Science Research Center in Orangeburg, South Carolina. He served as visiting professor at the 3 Nano-SAE Research Centre, University of Bucharest, Romania and visiting scientist at the Helen and Martin Kimmel Center of Nanoscale Science at the Weizmann Institute of Science, Israel. He served the U.S. Department of State in two rotations, as strategic S&T advisor and U.S. diplomat. His research interests span nanotechnology, environmental/ecological science, and safety and security. His research on nanotechnology has been on improving the understanding, design, and performance of nanofibers and sensors/detectors, mainly for applications such as wearable electronics, target drug delivery, detection of biomarkers and toxicity of nano and xenobiotic materials. In the security arena, he has worked on counterterrorism, countering unconventional warfare and hybrid threats, critical-Infrastructure protection, biosecurity, dual-use research concerns, and mitigating hybrid threats. He has authored over 250 research publications, edited/authored eight books on nanotechnology, and presented many keynotes and invited lectures worldwide. He serves on the editorial board of several highly reputed international journals. He is an active member of several national and international professional organizations. He is a fellow of the American Physical Society, Institute of Nanotechnology, and the New York Academy of Sciences. He has earned several other fellowships and awards for his meritorious service including 2004/2005 Distinguished Artist and Scholar award.

Nimet Bölgen was born in Adana, Turkey, in 1979. She received the B.S. degree in Chemical Engineering from Ankara University, Ankara, in 2002, and the M.S degree in Chemical Engineering from Hacettepe University, Ankara, in 2004. She received Ph.D. degree from Bioengineering Department of Hacettepe University, Ankara, in 2008. From 2009 to 2014, she was an Assistant Professor, and from 2014 to 2020 she was an Associate Professor with the Chemical Engineering Department of Mersin University, Mersin. Since 2020, she has been a Professor with the Chemical Engineering Department of Mersin University, Mersin. She is the author of 10 book chapters, 38 articles and more than 100 proceedings. Her research interest includes tissue engineering, biomaterials, controlled drug release systems, natural and synthetic polymers, cryogels and electrospun nanofibers. She is the Editorial Board Member of the journal Biomaterials and Biomechanics in Bioengineering. Prof. Bölgen is the member of Tissue Engineering and Regenerative Medicine International Society.

Materials and Processes for Treatment of Microbiological Pollution in Water

Marwa Alazzawi and Hilal Turkoglu Sasmazel

Abstract Clean and safe water is vital for the life and health of human beings. However, there are still millions of people around the world with inadequate clean water sources. Microbiological pollution is one of the most concerned water pollutants and is the crucial cause of waterborne diseases like diarrhea, resulting in about two million deaths annually due to severe dehydration (WHO in Guidelines for drinking-water quality. Incorporating the first addendum, WHO, Geneva, 2017). It is critical to develop methods using advanced materials and process to mitigate contaminants from water resources. Production of safe water usually involves disinfection and decontamination processes. Conventional disinfection process, such as chlorination, is challenged by the formation of disinfection by-products. Furthermore, the presence of emerging pathogenic, that resist conventional water treatment techniques, raised the crucial necessity for emerging materials and techniques for treating water from microbiological pollution (Shannon et al. in Nature 452:301–310, 2008). This chapter describes bacterial, viral, and protozoal microbiological pollution in water supplies and the application of emerging materials and techniques to eliminate such contaminations.

Keywords Water · De-contamination · Purification · Emerging materials

---

M. Alazzawi
Department of Biomedical Engineering, Al Nahrain University, Al Jadriya Bridge, Baghdad 64074, Iraq

H. Turkoglu Sasmazel (✉)
Department of Metallurgical and Materials Engineering, Atilim University, Incek, 06830 Golbasi, Ankara, Turkey
e-mail: hilal.sasmazel@atilim.edu.tr

Fig. 1 Diagram of fecal-oral disease cycle

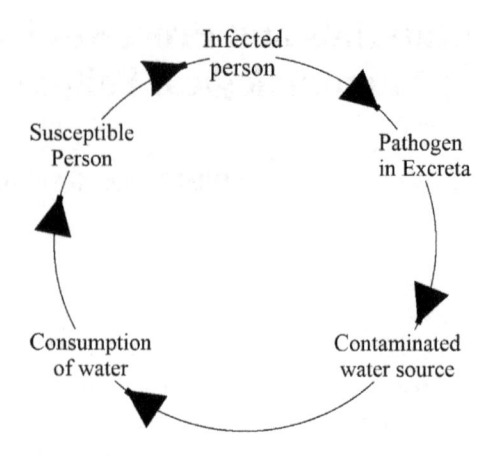

Infected person → Pathogen in Excreta → Contaminated water source → Consumption of water → Susceptible Person → Infected person

1 Introduction: Water Pollution and Microbiological Pollution

Generally, water pollution is the contamination of water resources, such as lakes, rivers, oceans, aquifers, and groundwater, with chemical, biological and/or radioactive substances that pose chronic or acute health risks to exposed personnel [2]. Water microbiological pollution refers to the inclusion or growth of pathogenic microorganisms such as bacteria, viruses, and protozoan parasites into the water used by humans. Usually, this type of pollution is transmitted via a fecal-oral route from animal or human excreta. Figure 1 demonstrates the fecal-oral route.

Microbiological pollution has several properties that distinguish it from other water pollutions [35], such as:

- Causing acute diseases in addition to chronic health problems.
- Some of the contamination can grow in the environment.
- Multiplying inside their host after infection.
- No cumulative effect, unlike many chemical agents.
- Many factors affect diseases caused by microbiological contamination for example the dose, virulence of the pathogen, and the immune status of the individual.

2 Types of Microbiological Water Pollution

Infectious diseases caused by pathogenic bacteria, viruses, and protozoa are the most common health risk associated with drinking water [27]. Most waterborne pathogens gain access to water supplies via human or animal wastes. Generally, Ingestion of waterborne pathogens initiates infection in the gastrointestinal tract. On the other hand, inhalation and direct contact are also routes of transmission of pathogens into the water, leading to contamination of the respiratory tract, skin and even brain [35].

2.1 Bacterial Water Pollution

Bacteria represent a large family of prokaryotic microorganisms with the size of approximately 1 μm and shapes ranging from spheres to rods and spirals [4]. Common waterborne bacterial pathogens are listed according to the disease it transmits, focusing on bacteria's general information and the characterization of the disease it causes.

2.1.1 Cholera

Cholera is a diarrheal disease caused by the bacterium *Vibrio cholera*, mostly transmitted by water or food that has been contaminated by human or animal wastes. *Vibrios* are small curved-shaped Gram-negative rods. *Vibrios* are facultative anaerobes capable of both fermentative and respiratory metabolism. Several *Vibrios* species can infect humans, but *Vibrio cholera* is the most important of those species. *Vibrio cholera* is a very diverse bacterial species with 200 serotypes. Only two of these serotypes (O1 and O139) are involved in true cholera [3].

The incubation period for cholera is one to three days. The disease characterization is acute and very intense diarrhea that can exceed 1 L\h. Cholera patient feels thirsty, has muscular pains and general weakness. In the absence of treatment, the mortality of patients infected with cholera is 50% due to circulatory collapse and dehydration with cyanosis [34].

2.1.2 Salmonellosis

Salmonellosis is a bacterium caused by the *Salmonella* pathogen. *Salmonella* spp. is a member of the family *Enterobaderiaceae*, which are Gram-negative motile straight rods. At present, there are more than 2,500 *Salmonella* species in the world, with new ones emerging every year. *Salmonella* genus contains two species: *Salmonella enterica* and *Salmonella bongoni*. Some species of *Salmonella* spp. show host specificity; as an example, *Salmonella typhi* and *Salmonella paratyphi* are generally restricted to human beings. On the other hand, a wide range of species like *Salmonella typhimurium* and *Salmonella enteritidis* infect humans and animals [14].

There are two types of Salmonellosis caused by the *Samonella* pathogen: Typhoid and paratyphoid fever caused by *Salmonella typhi* and *Salmonella paratyphi*, respectively, and gastroenteritis caused by non-typhoidal species. Symptoms of gastroenteritis infection are diarrhea that lasts for three to five days, fever, and abdominal pain. Typhoid fever is a more severe disease than gastroenteritis and can be fatal. Usually, the waterborne pathogen causes typhoid fever [15].

2.1.3 Shigellosis or Bacillary Dysentery

Shigella spp. pathogen causes serious intestine disease that is called Shigellosis or the severe form of Shigellosis Bacillary dysentery. *Shigella* spp. is a Gram-negative non-motile rod-shaped member of the *Enterobacteriaceae* family. *Shigella* divided into four species: *Shigella dysenteria, Shigella flexneri, Shigella boydii,* and *Shigella sonnei.*

Shigella spp. is a pathogen with a low infectious dose (<10 bacteria) that spread through fecal-oral route by ingestion of contaminated food, drinking contaminated water or person-to-person contact. The Shigellosis incubation period is one to four days with symptoms of watery diarrhea, fever and fatigue. Then, *Shigella* spp. pathogens destroy the colonic mucosa that leads to inflammatory reactions and ulceration followed by bloody or mucoid diarrhea [9].

2.1.4 Pathogenic Escherichia coli Strains

Pathogenic *Escherichia coli* is Gram-negative rod-shaped elective anaerobic bacteria that belong to the genus *Shigella*, within the family *Enterobacteriaceae*. *Escherichia coli* pathogens are transmitted through the fecal-oral route by ingesting contaminated water, food, and contact with an infected animal or people. Pathogenic *Escherichia coli* is divided into six main strains: enteropathogenic (EPEC), enterotoxigenic (ETEC, namely O148), enteroaggregative (EAggEC), entero-invasive (EIEC, namely O124), entero-hemorrhagic (EHEC, namely O157) and diffusely adherent (DAED) [9].

Pathogenic *Escherichia coli* can cause a wide spectrum of intestinal and extra-intestinal diseases such as urinary tract infections and meningitis. *Escherichia coli* serotypes O148, O157, and O124 that are transmitted particularly from consumption of contaminated water can cause watery diarrhea with abdominal cramps and nausea. Usually, most patients recover after ten days. In few cases, the disease may become life-threatening, especially in infants and aged patients [35].

2.1.5 Emerging Pathogen

Sharma et al. defined the emerging waterborne pathogens as "any new, reemerging or drug-resistant infection whose incidence in humans has increased within the past two decades or whose incidence threatens to increase in the future" [30]. The emerging pathogens outlined here are examples of bacterial pathogens transmitted through contaminated drinking water, and there are no microbiological indicators of their presence.

Helicobacter pylori is a Gram-negative spiral-shaped bacterium classified as *Campylobacter pylori*. At least fourteen species of *Helicobacter* exist, but only *Helicobacter pylori* recognized as a pathogen for humans. The main route of transmission is orally by person-to-person contact within the families, although consumption of contaminated drinking water has been suggested as a source of infection. Mostly, *Helicobacter Pylori* infection is initiated during childhood and remains with no symptoms. If the infection is not treated, it may lead to complications such as gastric cancer [16].

Mycobacterium avium Complex is a group of pathogenic species including *Mycobacterium avium* and *Mycobacterium intercellular,* which is part of the non-tuberculous *Mycobacterium* species. *Mycobacterium avium complex* aerobic rod-shaped bacterial pathogen has the ability to survive and proliferate in the water at temperature up to 50 °C and wide pH range. In addition to that, this *Mycobacterium* pathogen is highly resistant to the used water treatment techniques such as chlorine and chemical disinfectants. The *Mycobacterium avium complex* pathogens can cause a wide range of infections involving the respiratory and gastrointestinal tract, with the possibility of spreading to other locations in the body [35].

2.2 Viral Water Pollution

Viruses are microscopic infectious pathogens that invade living host and infect their bodies by replication inside living cells of an organism. Hundreds of viruses infect humans, and some of them are released to the environment by excretion which is called enteric viruses. The route of transmission of enteric viruses indicates fecal-oral transmission, with humans are considered to be the only source, except for *hepatitis E* viruses.

The most important viruses transferred to humans by ingestion of contaminated water, are members of six families, including RNA virus families such as *Picornaviridae, Calicivridae, Hepeviridae, Reoviridae, Astroviridae,* and the *Adenoviridae* within the family of DNA viruses. These viruses cause either asymptomatic infections or a wide range of symptoms involving mild to severe gastroenteritis and infectious hepatitis [35].

2.3 Protozoan Water Pollution

During the last decade, protozoan parasites such as *Acanthamoeba* spp., *Cryptosporidium parvum, Cyclospora cayetanensis, Entamoeba histolytica, Naeglenis fowleri, Toxoplasma gondil, Isospora belli, Giardia intestinalis,* and *Microsporidia*, were documented as the most common cause of waterborne outbreaks even when the water quality met the requirement. These protozoan parasites are micro-sized pathogens that are transmitted to humans through the digestion of contaminated

Table 1 Viruses and protozoan parasites pathogens associated with water contamination [27]

Pathogen	Health significance	Associated disease
Viruses		
Adenovirus	High	Gastroenteritis
Enteroviruses	High	Gastroenteritis
Hepatitis A virus	High	Hepatitis
Hepatitis E virus	High	Infectious hepatitis
Rotavirus	High	Gastroenteritis
Sapoviruses	High	Acute viral gastroenteritis
Pathogen		
Astroviruses	High	Diarrhea
Norovirus	High	Gastroenteritis
Protozoa		
Acanthamoeba spp.	High	Amoebic meningoencephalitis, keratitis, encephalitis
Cryptosporidium parvum	High	Cryptosporidiosism diarrhea
Cryptosporidium cayetanensis	High	Diarrhea
Entamoeba histolytica	High	Amoeba dysentery
Giardia intestinalis	High	Diarrhea
Naegleria fowleri	High	Infection of the brain
Toxoplasma gondii	High	Toxoplasmosis, miscarriage, birth defect

water. Table 1 lists some common viruses and protozoan parasites related to the water contamination, the associated disease and health significance [27].

3 Methodology Used in Microbiological Water Pollution Treatment

The treatment of contaminated water with pathogenic microorganisms is essential for a healthy human life. There is a diverse range of water microbiological treatment methodologies, including, disinfection, coagulation, flocculation, sedimentation, and filtration. These methods could be used individually or in combination to achieve microbiological water treatment. Here, we highlight some of the microbiological water treatment methods that are used in either piped water supplies treatment or point-of-use treatment.

3.1 Disinfection

Disinfection is the most widely used method for water treatment from microbiological contamination. This method includes chlorine-based disinfectants, such as chlorine or chlorine dioxide, besides ozone, some other oxidant, and strong acids and bases [28]. The mechanisms by which disinfectants deactivate pathogenic microorganisms include:

- Interference with cellular metabolic processes of the pathogen.
- Impairment of the cellular function of the pathogen by destruction of major constituents, for example, cell wall.
- Inhabitation of the growth of the pathogen by blocking the synthesis of cellular constituent for example DNA.

Chlorine and chlorine-based disinfectants are strong oxidants that have been used to treat water since a long time ago. Unfortunately, chlorine-based disinfectants tend to generate toxic by-products such as trihalomethanes and other halogenated hydrocarbons. Therefore, to reduce the health risk from these toxic by-products, various post-treatment methods have been used such as filtration. On the other hand, alternative disinfection chlorine-free methods such as ultraviolet and ozonation are used to treat contaminated water, notwithstanding the historical use of sunlight to disinfect water. It is also important to recognize that some pathogenic microorganisms are more resistant to disinfection than others [29]. Table 2 summarizes different disinfection methods in terms of relative cost and effectiveness.

Table 2 Comparison of different disinfection methods for microbiological disinfection of water [5]

Method	Relative cost	Effectiveness
Chlorine (various forms)	Low	Effective with most but not all microbes
		Chlorine residuals provide post-treatment control
		Chlorination by-products can be hazardous
Bromine/iodine	Moderate	Better for point-of-use applications
		Toxicity issues if misused
Oxidants (O_3, H_2O_2, UV)	Moderate high	Highly effective against microbes
		Require post-sterilization treatment for downstream storage and use
		Require electrical power or storage of hazardous oxidants

3.2 Coagulation, Flocculation, and Sedimentation

Coagulation-flocculation is the critical first and most important chemical technique in removing pathogenic microorganisms in water. The coagulation principle is to reduce of the repulsive electrical potential between particles by adding coagulants that promote destabilization and agglomeration of the contaminated particles into clots [10]. Different types of chemical coagulants are used in this conventional water treatment such as aluminum chloride and ferric chloride. However, aluminum-based and ferric-based inorganic coagulants salts have drawbacks, such as toxic waste and high levels of residuals in treated water that are associated with cancer, Alzheimer's disease, and other neurological illnesses. This has led to increasing the interest in organic-based polymer coagulants that are biodegradable and non-toxic [22]. Usually, coagulation is followed by flocculation. Flocculation is a gentle stirring technique that encourages clots to agglomerate into larger particles called flocs. Floc size increases until it reached the optimum size to be settled and removed by sedimentation or filtration. Figure 2 demonstrates coagulation-flocculation technique. Sedimentation is a physical water treatment process to remove suspended particles after settling down. Many techniques are used in sedimentation such as gravity, centrifugal acceleration or electromagnetism [8].

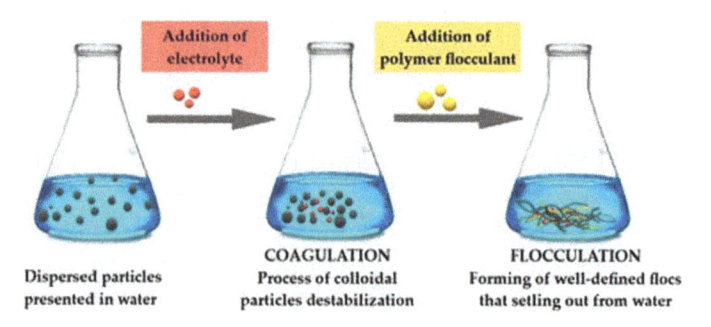

Fig. 2 Coagulation-flocculation water treatment techniques illustrations [23]

Fig. 3 TiO$_2$ structures according to the structural dimensionality and expected property [1]

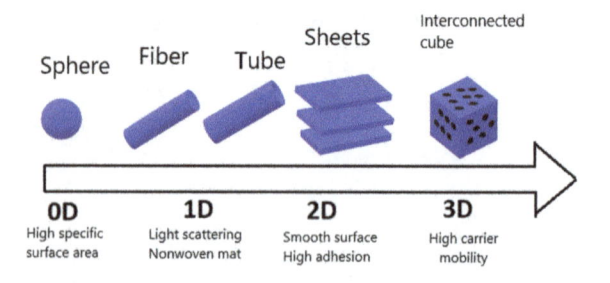

3.3 Filtration

The filtration method removes particulate contamination according to their size. Filter removal capability depends on pore size and technical aspects. One or combinations of the following techniques are used in the filtration method: mechanical, sedimentation, flocculation, adsorption, straining, and/or biological metabolism [5]. The least expensive and most common in high-volume water treatment filtration method is sand filters. However, pre-filtration and post-filtration treatment is needed to facilitate the complete removal of pathogenic microorganisms that is small enough to transit sand filters. The development of membrane filtration technology introduces a wide spectrum of more capable filtration methods. A micro-filtration membrane is capable of filtering particles down to 0.01 μm diameter. The ultra-filtration method utilizes semipermeable membranes with pressure to filter particles down to 500 Da in molecular weight, leaving water and small molecular solute. Nano-filtration membrane is a very small pore size membrane, typically a few nanometers in diameter which can filter particles down to 100 Da in molecular weight. Nano-filtration membrane offers the ability to filter a wide spectrum of contamination such as viruses. Reverse osmosis offers greater selectivity than any other type of membrane filtration [26].

Membrane filtration methods provide high-quality pure water, but as well as any other filtration methods offer challenges. These filtration methods are high-priced relative to other treatment methods. Also, it can be damaged easily; even minor damage can degrade membrane performance considerably. Furthermore, membrane filtration performance can suffer from clotting, bio-fouling, and scale formation. Accordingly, membrane filtration methods have a limited lifetime [5].

4 Materials Used in Microbiological Water Pollution Treatment

Microbiological water pollution is caused by the emerging of pathogenic bacteria, viruses, and protozoan. Some of the water treatment techniques mentioned including filtration, disinfection and coagulation. A broad range of materials involved in microbiological water treatment. Here we mention some of the recent emerging materials and nanomaterials in this field.

Silver has been known to have antibacterial properties for ages. Silver, especially ionic silver Ag^+ and silver nanoparticles AgNP, have been shown to have antibacterial properties against a range of both Gram-negative and Gram-positive bacteria [32]. In addition to that, some researchers have demonstrated the biocidal action of silver nanoparticles against some common types of viruses [11].

Numerous studies have been conducted on the efficiency of ionic silver [generated from silver salts (silver nitrate and silver chloride)] for water disinfection against a wide range of bacteria. Usually, ionic silver is used as a secondary disinfectant to replace or reduce the level of chlorine. Also, silver is commonly used in conjunction with copper as an ionization process for water treatment. When copper ions are dissolved in water, they oxidized immediately into Cu^{+2} ions which search for particles of opposite polarity such as bacteria and viruses. The copper ions will disturb cell wall permeability and cause silver ions to penetrate the core of microorganisms, causing them to eventually die. This technique is generally used for hospital hot water systems and treatment of swimming pool water [18].

Silver nanoparticles in conjunction with filtration are currently being extensively used as household point-of-use drinking water disinfection techniques. A wide range of filtration membranes utilized for silver nanoparticles coating includes polyurethane foams [13], polystyrene resin beads [24], fiberglass [25], titania [20] and activated carbon composite supporting magnetite [31]. Even though ionic silver and silver nanoparticles have shown effectiveness against bacteria pathogen, their effectiveness is limited against viruses and protozoa. In addition to that, silver has a toxic effect on mammalian cells [33].

Titanium dioxide (TiO_2) is the most widely used photocatalyst-mediated water disinfection. Photocatalysis can be defined as the chemical reaction that may occur when a light source interacts with the surface of semiconductor material. The concept of water purification using Titanium dioxide as photocatalysis is the utilization of photon energy to oxidize water molecules for the formation of active radicals, which are toxic to microorganisms [36].

Although the TiO_2 photocatalyst system for water disinfection has advantages in a fast response and good photo-efficiency even in weak light, it is inactive in visible light. In recent years, many studies on TiO_2 to maximize photocatalyst activity and widen the range of photon absorption band have been conducted. Many of these studies have focused on modifying the morphology of TiO_2, which can exist in a wide range such as zero-dimensional TiO_2 spheres, one-dimensional with TiO_2 fibers, rods, tubes, two-dimensional with TiO_2 nanosheets, and three-dimensional with interconnected architecture [17]. Figure 3 shows various dimensions of TiO_2. In addition to that, the introduction of metal, such as Pt, Pd, Au, Ag, and Fe, onto TiO_2 surfaces to increase the activity under the visible irradiation and enhance the photocatalytic activity [37].

Carbon nanotubes—Owing to their remarkable antimicrobial activity, Carbon nanotubes (CNTs) have received notable attention in water treatment from pathogenic microorganisms in recent years. Direct contact between microorganisms and CNTs, which leads to deterioration of pathogens cell wall and cytoplasmic membrane, is the mechanism of action of CNTs against microorganisms [21].

CNTs are mainly divided into either single-walled (SWCNT) or multi-walled (MWCNT), which differ in the graphene sheet arrangement. Several studies have explored that CNTs can be incorporated with polymer or nanomaterials to produce a membrane that is effective in bacteria and virus's inactivation. However, challenges exist for using CNTs in water treatment such as toxicity and cost-effectiveness [19].

Zero valent iron (ZVI) has great potential in biological contamination water treatment due to its chemical, electronic, magnetic, mechanical, catalytic and optical properties. However, ZVI still suffers from several drawbacks including the activity decreases in time, aggregation of ZVI particles and storage difficulties due to high reactivity. Some studies have explored methods to overcome these drawbacks for example stabilizing ZVI on starch or sodium carboxymethyl cellulose to increase the stability [7] or incorporating another metal in the structure of ZVI to increase the effective surface area [12]. Another study has established that ZVI sand filtration can remove biological contamination, including bacteria and viruses [6].

5 Discussion

Problems of unsafe water and inadequate decontamination method are most acute in developing countries. However, people in industrialized countries can also get sick from contaminated water. Microbiological pollution is one of the crucial issues that contribute to water safety and sustainability. In this regard, many technologies are adapted in water treatment from microbiological pollution. Current water treatment technologies involve a combination of physical, chemical, and biological processes to remove microbiological pollutants such as bacteria, viruses, and protozoan parasites from water. However, these treatment techniques have advantages and disadvantages.

Disinfection with chlorine and chlorine-based disinfectants is effective with most microbiological pollutions. Besides, it is a relatively low-cost technique. However, chlorine and chlorine-based disinfectant generate toxic by-products which contribute to post-treatment methods, such as filtration to reduce health risks. On the other hand, chlorine-free disinfectants such as UV and ozonation are used to treat contaminated water with no toxic by-products.

Coagulation-flocculation is a conventional water chemical pre-treatment technique, usually followed by sedimentation or filtration to eliminate the suspended particles. Different types of coagulants and flocculants are used in this technique. However, some of them have drawbacks such as toxic residuals. Therefore, organic-based polymer coagulants and flocculants are introduced to decrease the toxic effect.

Filtration is the most effective and most expensive technique in water treatment. Microorganism removal depends on filter pore size, the material used, and technical aspects. The main drawback of filtration is the limited life-time due to damaging and clotting.

Emerging materials such as silver, zero-valent iron, carbon nanotubes, and titanium dioxide are introduced in water treatment techniques, owing to their antimicrobial properties. Most of them are used in filtration techniques for example, silver nanoparticle is currently being extensively used as household point-of-use drinking water disinfection techniques. A wide range of filtration membranes utilized for silver nanoparticle coating includes polyurethane foams, polystyrene resin beads, fiberglass, titania, and activated carbon composite supporting magnetite. Furthermore, titanium dioxide is the most widely used as photocatalyst-mediated water disinfection. Although, these materials suffer from drawbacks such as activity decrease in time, toxicity, cost-effectiveness, limited effectiveness against microorganisms. In this regard, many studies improve these materials to attain the best effectiveness.

6 Conclusion and Future Directions

The dramatic need for high water quality is clear. Consequently, there is a large effort to conceive better and cheaper techniques for purifying drinking water concerning microbiological contamination. Moreover, emerging materials were introduced to different water treatment techniques to overcome drawbacks of the conventional ways. Many studies are needed in the field of water treatment to produce the highest water quality with cost-effectiveness. Improvement of the existing materials, a combination of the treatment techniques, and combination of the materials are favored in modern water treatments.

References

1. Akira J, Kazuya N, Tsuyoshi O, Manivannan A, Donald T (2013) Recent aspects of photocatalytic technologies for solar fuels, self-Cleaning, and environmental cleanup. Electrochem Soc Interface 22:51–56
2. Alrumman S, El-kott A, Keshk S (2016) Water pollution: source and treatment. Am J Environ Eng 6:88–89
3. Ashbolt NJ (2004) Microbial contamination of drinking water and disease outcomes in developing regions. Toxicology 198(1–3):229–238
4. Cabral JPS (2010) Water microbiology: bacterial pathogens and water. Int J Environ Res Public Health 7:3657–3703
5. Cahoon LB (2019) Water purification: treatment of microbial contamination. In: Ahuja S (ed) Advances in water purification techniques. Elsevier, pp 385–395
6. Chopyk J, Kulkarni P, Nasko DJ, Bradshaw R, Kniel KE, Chiu P, Sharma M, Sapkota AR (2019) Zero-valent iron sand filtration reduces concentrations of virus-like particles and modifies virome community composition in reclaimed water used for agricultural irrigation. BMC Res Notes 12:223
7. Devi P, Dalai AK (2019) Effects of carboxymethyl cellulose grafting on stability and reactivity of zerovalent iron in water systems. J Clean Prod 229:65–74
8. Domopoulou A, Gudulas K, Papastergiadis E, Karayannis V (2015) Coagulation/flocculation/sedimentation applied to marble processing wastewater treatment. Mod Appl Sci 9(6):137–144
9. Garcia-Aljaro C, Momba M, Muniesa M (2017) Pathogenic members of *Escherichia coli* and *Shigella* spp. shigellosis. Global Water Pathogen Project (GWPP)
10. Gologan D, Popescu E (2016) The optimization of coagulation-flocculation process in a conventional treatment water plant. IOSR J Environ Sci Toxicol Food Technol 10:2319–2399
11. De Gusseme B, Sintubin L, Baert L, Thibo E, Hennebel T, Vermeulen G, Uyttendaele M, Verstraete W, Boon N (2010) Silver for disinfection of water contaminated with viruses. Appl Environ Microbiol 76(4):1082–1087
12. Hasemzadeh G, Momenpour M, Omidi F, Hosseini MR, Ahani M, Barzegari A (2014) Applications of nanomaterials in water treatment and environmental remediation. Front Environ Sci Eng 8:471–482

13. Jain P, Pradeep T (2005) Potential of silver nanoparticles-coated polyurethane foam as an antibacterial water filter. Biotechnol Bioeng 90(1):59–63
14. Kim S (2010) Salmonella serovars from foodborne and waterborne diseases in Korea, 1998–2007: total isolates decreasing verses rare serovars emerging. J Korean Med Sci 25(12):1693–1699
15. Kit WH, Mohtar DA, Hui LM, Sattar ZM, Abdul Halim AN, Muhamad NS, Saidin SF, Abdul Majid AM (2011) Salmonellosis: the diseases, treatment, prevention and drug resistance. Webmed Central Bacteriol 2(12):1–12
16. Kusters JG, Van Vliet AH, Kuipers EJ (2006) Pathogenesis of helicobacter pylori infection. Clin Microbiol Rev 19(3):449–490
17. Lee SY, Park SJ (2013) TiO_2 photocatalyst for water treatment applications. J Ind Eng Chem 19(6):1761–1769
18. Lin YE, Stout JE, Yu VL (2011) Controlling *Legionella* in hospital drinking water: an evidence-based review of disinfection methods. Infect Control Hosp Epidemiol 32(2):166–173
19. Liu D, Mao Y, Ding L (2018) Carbon nanotubes as antimicrobial agents for water disinfection and pathogen control. J Water Health 16(2):171–180
20. Liu J, Wang Z, Luo Z, Bashir S (2012) Effective bactericidal performance of silver-decorated titania nano-composites. Dalton Trans 42:2158–2166
21. Maksimova YG (2019) Microorganisms and carbon nanotubes: interaction and applications (review). Appl Biochem Microbiol 55:1–12
22. Matilainen A, Vepsalainen M, Sillanpaa M (2010) Natural organic matter removal by coagulation during drinking water treatment: a review. Adv Colloid Interface Sci 159(2):189–197
23. Maćczak P, Kaczmarek H, Ziegler-Borowska M (2020) Recent achievements in polymer bio-based flocculants for water treatment. Materials 13:3951–3992
24. Mthombeni NH, Mpenyana-Monyatsi L, Onyango MS, Momba MNB (2012) Breakthrough analysis for water disinfection using silver nanoparticles coated resin beads in fixed-bed column. J Hazard Mater 217–218:133–140
25. Nangmenyi G, Xao W, Mehrabi S, Mintz E, Economy J (2009) Bactericidal activity of Ag nanoparticles-impregnated fiberglass for water disinfection. J Water Health 7(4):657–663
26. Pendergast M, Hoek E (2011) A review of water treatment membrane nanotechnologies. Energy Environ Sci 4:1946–1971
27. Ramirez-Castillo FY, Loera-Muro A, Jacques M, Garneau P, Avelar-Gonzalez FJ, Harel J, Guerrero-Barrera AL (2015) Waterborne pathogens: detection methods and challenges. Pathogens 4(2):307–334
28. Schoenen D (2001) Requirements for catchment, treatment, and Surveillance of drinking water to avoid the transmittance of pathogenic bacterial, viral, and parasitic organisms. Acta Hydrochim Hydrobiol 29(4):2–7
29. Shannon M, Bohn P, Georgiadis J, Marinas B, Mayes A (2008) Science and technology for water purification in the coming decades. Nature 452:301–310
30. Sharma S, Sachdeva P, Virdi JS (2003) Emerging water-borne pathogens. Appl Microbiol Biotechnol 61(5–6):424–428
31. Valušová E, Vandžurová A, Pristaš P, Antalik M, Javorský P (2012) Water treatment using activated carbon supporting silver and magnetite. Water Sci Technol 66:2772–2778
32. Wijnhoven S, Pejinenburg W, Herberts C, Hagens W, Oomen A, Heugens E, Roszek B, Bisschops J, Gosens L, Van de Meent D, Dekkers S, De Jong W, Van Zijverden M, Sips A, Geertsma R (2009) Nano-silver—a review of available data and knowledge gaps in human and environmental risk assessment. Nanotoxicology 3(2):109–138
33. World Health Organization (1993) Guidelines for drinking-water quality, 2nd edn. Health Criteria Other Support Info., WHO, Geneva, Switzerland, pp 338–343
34. World Health Organization (2008) Guidelines for drinking-water quality, 3rd edn. Incorporating the First Addendum, WHO, Geneva
35. World Health Organization (2017) Guidelines for drinking-water quality, 4th edn. Incorporating the First Addendum, WHO, Geneva

36. Wu MJ, Bak T, O'Doherty PJ, Moffitt MC, Nowotony J, Bailey TD, Kersaitis C (2014) Photocatalysis of titanium dioxide for water disinfection: challenges and future perspectives. Int J Photochem 2014:1–9
37. Zar M, Varghese S, Nair S (2012) Photocatalytic water treatment by titanium dioxide: recent updates. Catalysts 2:572–601

Marwa Alazzawi, Mrs. was born in Baghdad, Iraq in 1986. She received the B.S. degree in biomedical engineering from the Jordanian university of science and technology, Jordan, in 2010 and the M.S. degree in biomedical engineering from Al-Nahrain university, Iraq in 2013. She is currently pursuing the Ph.D. degree in biomedical engineering at Al-Nahrain university, Iraq. Her research interest includes tissue engineering, fabrication on nano- and micro-structured scaffold using electrospinning, biomechanics and human gait cycle.

Hilal Turkoglu Sasmazel, Prof. Dr. was born in TURKEY in 1976. She received B.S. degree in Chemical Engineering from Hacettepe University, Ankara, in 1998 and M.S. and Ph.D. degrees in Bioengineering from Hacettepe University, Ankara, TURKEY in 2001 and 2007 respectively. Since 2019, she has been a full professor with the Metallurgical and Materials Engineering Department, Ankara, Atilim University. She is the author of more than 10 e-book chapters and more than 40 SCI-indexed articles, and 2 patents. She is involved in more than 20 national and international projects as a leader. Her research interests include biomaterials, nanomaterials, tissue engineering, cell culture, polymeric and composite materials, surface modifications by plasma and wet chemistry and electrospinning process. She is the founder and director of the Polymer/Composite Materials Laboratory, Biocompatibility of Biomaterials Laboratory and Materials Antibacterial Property Testing Laboratory from the Atilim University. She is Turkey Branch President of Executive Council of ModTech Professional Association Iasi, Romania, since April 2017. Her awards include TUBITAK-2214 Research Scholarship in 2001, TUBITAK-2211 Doctoral Scholarship in 2005, Young Women in Science in 2009.

Methods and Characteristics of Conventional Water Treatment Technologies

Adrian Lucian Cococeanu and Teodor Eugen Man

Abstract Water in its natural state is never absolutely pure. In nature, water may be consumed by humans, if it satisfies certain condition of drinking for it to be considered clean, healthy i.e. free of micro-organisms, parasites or substances, that by its number or concentration could pose a potential danger to human health. Water quality is governed by its physical and chemical characteristics, which in turn provides the information that we need to determine the water treatment processes to ensure that the minimum required quality of water is met. Each and every source of water has its own characteristics and we need to identify a suitable water treatment methodology that is particular to removing the source specific contaminants. This chapter deals with some particular treatment solutions trough analyses and pilot study, to determine the best water treatment solutions.

Keywords Water characteristics · Water treatment · Purification · Disinfection process · Pilot plant

1 Introduction

Water in its natural state is never absolutely pure. The water in nature may be consumed by humans if it satisfies the condition of drinking water that is healthy and clean, free of micro-organisms, parasites or substances by number or concentration that could pose a potential danger to human health. Water can influence the health of the population by its quality. Natural water source intended for drinking water must ensure that a required quantity of water all year that meets certain quality

---

A. L. Cococeanu · T. E. Man (✉)
Civil Engineering Faculty-Department of Hydrotechnical Engineering, Politehnica University of Timişoara, Iuliu Podlipny Street, no. 23C, 300703 Timişoara, Romania
e-mail: eugen.man@upt.ro; eugen@zavoi.com; eugen.man@upt.ro

A. L. Cococeanu
e-mail: adrian.cococeanu@student.upt.ro; cococeanu.adrian@gmail.com

© Springer Nature Switzerland AG 2021
A. Vaseashta and C. Maftei (eds.), *Water Safety, Security and Sustainability*,
Advanced Sciences and Technologies for Security Applications,
https://doi.org/10.1007/978-3-030-76008-3_14

standards. Water intended for human consumption must meet certain characteristics, namely: organoleptic characteristics: taste, smell, color and respective physical characteristics, physico-chemical, radioactive, biological and microbiological.

1.1 Potable Water Characteristics and Conventional Elimination Methods

The chemical and physical quality of water may affect its acceptability to consumers. Turbidity, color, taste, and odor, whether of natural or other origin, affect consumer perceptions and avourur. In extreme cases, consumers may avoid aesthetically unacceptable but otherwise safe supplies in avour of more pleasant but less wholesome sources of drinking-water. It is therefore wise to be aware of consumer perceptions and to take into account both health-related guidelines and aesthetic criteria when assessing drinking-water supplies. The combined perception of substances detected by the senses of taste and smell is often called "taste". Taste problems in drinking-water supplies are often the largest single cause of consumer complaints. Changes in the normal taste of a public water supply may signal changes in the quality of the raw water source or deficiencies in the treatment process. Water should be free of taste and odor that would be objectionable to the majority of consumers.

1.1.1 Smell

In case of groundwater, smell is due either to natural causes: salts, hydrogen sulphide, carbon dioxide; or due to foreign influences: viz. the infiltration of contaminated surface water, liquid fuels, etc. In surface water case may be due to living microorganisms: algae, protozoa; or due to decomposition of plant and animal organisms; or due substances of industrial origin: phenol, cresols, copper, zinc, aluminum. It highlights some specific smells: the smell of rotten eggs: indicating the presence of hydrogen sulfide, the strong smell of fish: indicates the presence of diatoms, musty smell of earth: it is given by the yellow-blue algae, the smell of grass: is given by green and blue algae, etc. In practice, conventional methods are adopted for the elimination of smell, such as: aeration; oxidation; coagulation—flocculation, decantation, filtration; activated carbon adsorption [1, 2].

1.1.2 Taste

In the case of groundwater, a peculiar taste is due either to natural causes: dissolving salts, such as, $CaCO_3$, $MgCO_3$, $CaSO_4$, Na_2SO_4, $CaHCO_3$, $MgHCO_3$, or due presence of specific metals Fe, Cu, Mn, Zn, or foreign influences, viz. the infiltration of surface water, contaminated liquid fuels. In the case of surface water, it may be due to algae growth and other plants either because the products resulting from the processes of metabolism and/or the ingress of solid or liquid waste. It highlights some different types of taste: salted taste—indicate the presence of sodium chloride, the bitter taste—indicates the presence of sulphate and magnesium chloride, bland taste or sweet—is given by the calcium sulphate, the sour taste—is given by the ferric chloride, acid taste—is given by the carbon dioxide, metallic taste—the waters with a pH of 3.0–5.5. In practice, conventional methods are adopted for the elimination of taste, such as: aeration, oxidation; coagulation—flocculation, decantation, filtration; activated carbon adsorption; ion exchange, reverse osmosis [3, 4, 1, 5]

1.1.3 Color

Coloring of water sources is caused by dissolved mineral substances such as compounds of iron, e.g. brick red, or compounds of manganese, e.g. black and tan, or due to solutes of organic nature such as chlorophyll from the leaves fresh, e.g. yellowish green; Xanthophyll of withered leaves, e.g. yellow; humic acids, e.g. reddish brown or by industrial waste water discharged. In the apparent color case due to the presence of suspended matter, the recommended method of elimination is water filtration, and in true colors case due to the presence of dissolved or colloidal substances, the recommended methods of elimination are coagulation, flocculation, decantation, filtration [4, 1, 2].

1.1.4 Turbidity

It is characterized by lack of transparency of the waters that contains suspended and colloidal materials of mineral origin, such as sand, silt, clay or colloids suspended solids and organic origin: decomposition of plant and animal organisms, humic acids, bacteria, algae. In practice, conventional methods are adopted for the removing of turbidity, such as coagulation, flocculation, settling, and filtration [3, 1, 6].

1.1.5 Electrical Conductivity

The electrical conductivity of natural waters depends on the quantity of dissolved substances, the nature of the ions, the total concentration and of the temperature at which the measurement is taken. It is used to control the constancy of chemical water quality. Electrical conductivity of natural waters is between 166 and 1,666 μs/cm [3, 6].

1.1.6 Water Reaction or Water pH

pH represents the concentration of the hydrogen ions present in a liter of water. Largely determined by biological and chemical processes, the value ranges as, 6.5 < pH < 8.5 for biological processes that take place under normal conditions, and for 6.5 > pH > 8.5, the biological cycle is unsustainable. The recommended water treatments are: oxidation, aeration, and coagulation. For pH < 6.5, water has a corrosive action on materials with which they come in contact, hence conventional methods for adjusting pH from the specialized theory, such as treatment with lime-wash; treatment with sodium hydroxide; treatment with sulfuric acid or a mineral acid are used [3, 1, 6].

1.1.7 Water Hardness

It is given by all of the soluble salts of calcium and magnesium and are characterized by temporary hardness, viz.: $CaHCO_3$, $MgHCO_3$ and permanent hardness is caused by salts viz.: $CaCl_2$, $CaSO_4$, $MgCl_2$, $MgSO_4$. Natural waters can be: soft water: 0–60 mg/L, and may have adverse effects on the cardiovascular system; hard water in the range of 120–300 mg/L. Hardness is not dangerous in terms of physiological characteristics, since humans need some minerals to stay healthy, and the World Health Organization (WHO) states that drinking-water may be a contributor of calcium and magnesium in the diet and could be important for those who are marginal for calcium and magnesium intake. However, very rough waters: >300G may prevents absorption of important elements for the body (iodine) and digesting heavy food. Conventional methods for controlling hardness are to balance adjustment of lime-carbonic acid solution or basic ion exchange [1, 2].

1.1.8 Chemical Oxidation Ability

It represents the content of organic matter of natural waters that coming from: the decomposition of living organisms in water, soil washing basin, such as humic substances, discharge of domestic and industrial wastewater. Organic substances contain the following elements: carbon, oxygen, nitrogen, phosphorus, and sulfur. The organic substances in the water determine the degree of contamination of water by following indicators, such as, CBO-5 (biochemical consumption of Oxygen), Chemical consumption of oxygen (CCO) coupled with biodegradable organic matter, CCO/CBO (the degree of biodegradability), TOC (total organic carbon) and the degree of organic contamination, expressed in mg C/L. The organic substances present in the water results in bad taste and smell, provides nutritional support for microorganisms in the distribution system, makes the water treatment processes for drinking water difficult, and renders them toxic and carcinogenic. Conventional methods for the removal of organic substances oxidation; coagulation, flocculation; activated carbon adsorption [1, 2].

1.1.9 Ammonium, NH_{4+}

Natural levels in groundwaters are usually below 0.2 mg/L of ammonia. Higher natural contents (up to 3 mg/L) are found in strata rich in humic substances or iron or in forests. Surface waters may contain up to 12 mg/L. Ammonia may be present in drinking-water as a result of disinfection with chloramines. Furthermore, Ammonia, in groundwater may come from the reduction of nitrites by the bacteria autotrophic, reduction of nitrates sands containing ferrous ions, protein degradation and organic matter nitrogen from animal waste and plant in soil, discharges of human and/or animal manure, industrial discharges. Ammonia in surface water comes from rain and snow, reducing nitrites or nitrates from waste animal or vegetable contained in soil organic matter decomposition nitrogen in algae, plants or bed vegetable from the bottom of rivers, discharges of human and/or animal manure, industrial discharges. The conventional method for removing ammonia from natural waters is the chlorine oxidation at break-point followed by activated carbon filtration [4, 1, 2].

Nitrites, NO_{2-}

They are unstable compounds that through complete oxidation turn into nitrates that are stable compounds. Nitrites indicate pollution with ammonia or ammonia compounds, arising from bacterial oxidation of ammonia or nitrates due to the reduction. Excess of nitrates in drinking water can cause hypotension in case of adults and in case of newborn children methemoglobinemia newborn, gastric cancer. Conventional methods for removing nitrites from natural waters are: chemical oxidation; anionic ion exchangers; reverse Osmosis; electrodialysis [1, 2].

Nitrate, NO_3

In natural waters they come from precipitation by nitrogen oxides in the atmosphere produced by nitrogen oxides, combustion of fossil fuels, ashes of vegetation burned, the bacterial nitrification of ammonia through microorganisms nitrosomonas and notrosococcus of nitrification nitrite by nitrobacteria, form springs, following the dissolution of nitrate content in the rocks at sea depth of soil erosion containing nitrate in wastewater discharges containing nitrates, agricultural, chemical and natural fertilizers. Conventional methods for removing nitrates from natural waters are: anionic ion exchangers; reverse osmosis;electrodialysis; mixing with a source of water that does not contain nitrate [1, 6].

Iron and Manganese

Iron and manganese are found in natural waters or in the form of soluble salts, insoluble salts or in the form complexes with colloidal organic substances. The iron in water favors the development ferobacteria. In concentrations over 0.3 mg Fe^{2+}/L, it renders water a metallic taste, sweet and sour, and is cloudy yellow, but in contact with air becomes red. Ferruginous waters is not harmful to the human body and it is recommended for people suffering from anemia, constituting a tonic for the body. Manganese concentrations over 0.2 mg Mn^{2+}/L in the presence of oxygen precipitates to form a dark brown deposit. Manganese is not toxic, and it is considered essential for life since the human need ~1.5–5 mg / day. Conventional methods for removing iron and manganese from natural waters are: aeration, chemical oxidation, filtration, alkalizing, coagulation; cation exchanger; reverse osmosis, ultrafiltration, electrodialysis [1, 6].

Gases Contained in the Water

Gases contained in water: nitrogen gas, N_2, oxygen, O_2, carbon dioxide, CO_2 is present in all water which is in contact with the ambient air. Ammonia, NH_3, methane, CH_4 resulting from the anaerobic biological decomposition of organic matter [1, 6].

Bacteria

Banal bacteria are expressed by the total number of germs. They have no influence on the human body and their damaging action is due to a large amount of toxins obtained from a large number of ordinary bacteria, so that the drinking water limit the number of ordinary bacteria. Coliform bacteria: *E. coli, Escherichia coli, Streptococcus faecalis, perfrigens Clostridiums* is fecal pollution indicator that lives in human and animal intestines. *Saprophytes* and pathogenic bacteria are the largest source of bacteriological pollution of water in waste of human and animal feces that can contain germs Pathogens [1, 6]. The principal risks to human health associated with community water supplies are microbiological. The approach assumes that health authorities will be aware of other specific sources of risk in each region, such as chemical contamination, and will include these in the monitoring scheme. The parameters recommended for the minimum monitoring of community supplies are those that best establish the hygienic state of the water and thus the risk of waterborne infection.

It is necessary to make physico-chemical and microbiological determinations for the water sources to establish the treatment process. The sources must be counted both from the point of view of capacities and exploited sustainably without prejudicing the next generations but also qualitatively to ensure a good water quality to ensure standards. In counting these aspects, it is necessary to calibrate from an economic point of view, the use of operating treatment costs to ensure the need for water at a quality that is regulated by the standards in force to be economically sustainable for the population and economic agencies providing such services.

2 Usual Water Treatment Lines

It is necessary to treat the water every time a parameter is higher in quantity than the norms in force. Depending on the quality of the water intended for human consumption, the treatment plants will be built from a set of more or less complex modules, the sequence can go from a single disinfection stage for a good quality groundwater, to an assembly that it includes about ten stages for a poor quality surface water (category A3). Directive 75–440 / EEC of 16 June 1975 fixes three qualities A1, A2, A3 of surface waters, these being associated with the following three types of treatments, viz. T1, T2, T3:

- T1: simple physical treatment with disinfection
- T2: normal treatment, chemical and disinfection
- T3: physical treatment, advanced chemistry, refining and disinfection [3, 7, 8].

2.1 Simple Disinfection

Common water disinfection methods include UV, chemicals such as chlorine, unscented bleach and chloramines, distillation, ozonation and, of course during times of crisis, boiling. The most common sanitizing treatment for wells is shock chlorination. Often, a strong chlorine bleach (sodium hypochlorite) solution is used in the well and throughout the distribution piping. Unscented household chlorine bleach can be used to sanitize wells. Chlorine bleach is a sodium hypochlorite (NaOCl) solution containing approximately three to six percent available chlorine. However, certain precautions and safety equipment must be used, such as information about the well, including the volume of the water in the well; depth of the well; static water level; and size of casing. The vast majority of good quality groundwater is distributed after a simple disinfection, as shown in Fig. 1.

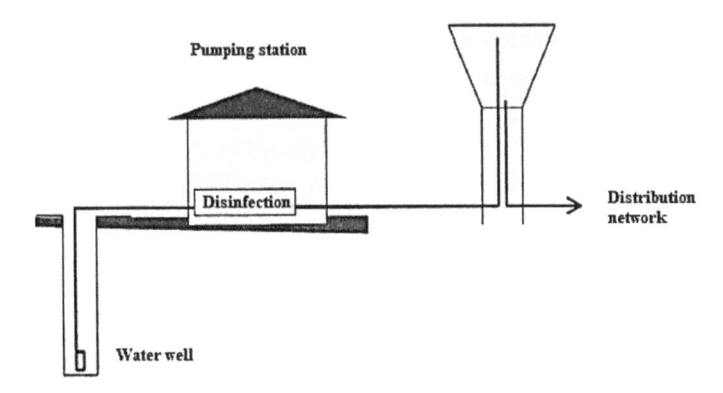

Fig. 1 Simple disinfection

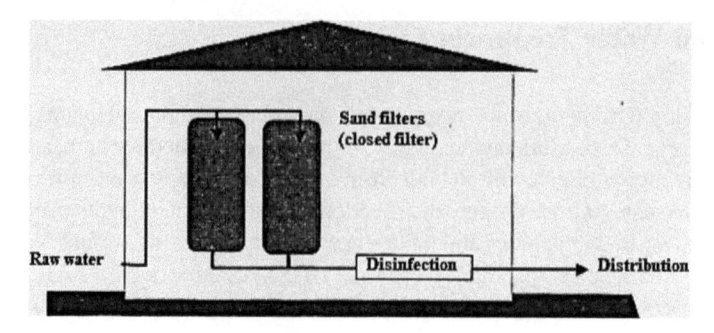

Fig. 2 Physical treatment—closed sand filters

2.2 Treatment Type T1: Simple Physical Treatment with Disinfection

This treatment is mainly for good quality surface waters without a large variation in turbidity; also valid for all groundwater, mainly karstic waters with low turbidity. These waters have characteristics close to those of the previous class but have concentrations in suspended matter that require prior physical treatment. Coagulation is done before the filter (coagulation in the filter). In general, closed type filters are used which have small dimensions and which allow important filtration speeds.

Currently, a good alternative to this treatment is the use of ultrafiltration membranes, followed by disinfection. This channel allows the treatment of waters with turbidities of up to 250 NTU. In addition, it admits less "clean" water from a bacteriological point of view because this process is an effective barrier to germs [8] (Fig. 2).

2.3 Treatment Type T2: Normal Treatment, Chemical and Disinfection

The waters for which this treatment is used are surface waters, rich in suspension and organic matter and whose content of micropollutants exceeds the drinking water standards. This treatment can also be applied to waters rich in iron and manganese. The role of pre-oxidation stage is the oxidation of ammoniacal nitrogen (if the oxidant is chlorine), dissolved iron and manganese, with subsequent formation of precipitates, also elimination of algae and protection of settlers' tanks. Pre-oxidation helps to improve coagulation-flocculation process. Purification stage is achieved through the process of coagulation-flocculation in usual way with iron or aluminum salts, decanting the flakes formed in the previous stage, thus eliminating the suspensions and turbidity. For final clarification filtration process is accomplished through the sand filter (single layer), or sand+coal (double layer). Final stage represents the

Fig. 3 Water treatment—pre-oxidation-clarification-disinfection

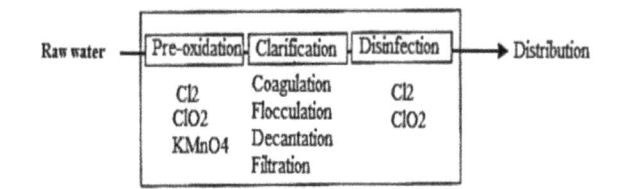

Raw water → [Pre-oxidation (Cl2, ClO2, KMnO4) | Clarification (Coagulation, Flocculation, Decantation, Filtration) | Disinfection (Cl2, ClO2)] → Distribution

disinfection which has the role of elimination of pathogens. Some installations use ozone as the final disinfectant. This variant tends to disappear because the action of ozone on the residual organic matter generates biodegradable compounds, favoring bacterial revival in the network [8] (Fig. 3).

2.4 Treatment Type T3: Physical Treatment, Advanced Chemistry, Refining and Disinfection

This treatment is as complete as possible and is applicable to surface waters, rich in organic matter, color, micropollutants. The dies of the installations will be more or less complex depending on the type of water and the parameters that must be corrected. They will include the steps presented in the previous case but will be completed with a refining step. This refining stage will include the absorption on granular activated carbon (CAG) or granular activated powder (CAP) (this case is reserved for installations that include a purification step). Other stage is oxidation of ozone (or ozone/hydrogen peroxide, if the water is rich in pesticides) followed by an absorption of granular activated carbon and it is necessary in restoring the chalco-carbon balance, which can be reduced to a simple pH correction but can also resort to more complex techniques. The integration of membrane technologies in a complete surface water treatment chain tends to develop. The use in the treatment of refining, ultrafiltration, or even nano-filtration are being studied for some installations, with good results [8].

2.5 Pilot Plant Research Stations

Pilot plant research station checks a theoretical question about a particular phenomenon at a certain reaction or behavior of a device. Any pilot plant research station must have the following characteristics:

- STATES—the measuring range, the overload of the system, the safety limit
- METROLOGICAL—accuracy, correctness, loyalty, sensitivity, reproducibility
- TECHNICAL—stability, inertia, generality, reliability
- DYNAMIC—static, dynamic regime.

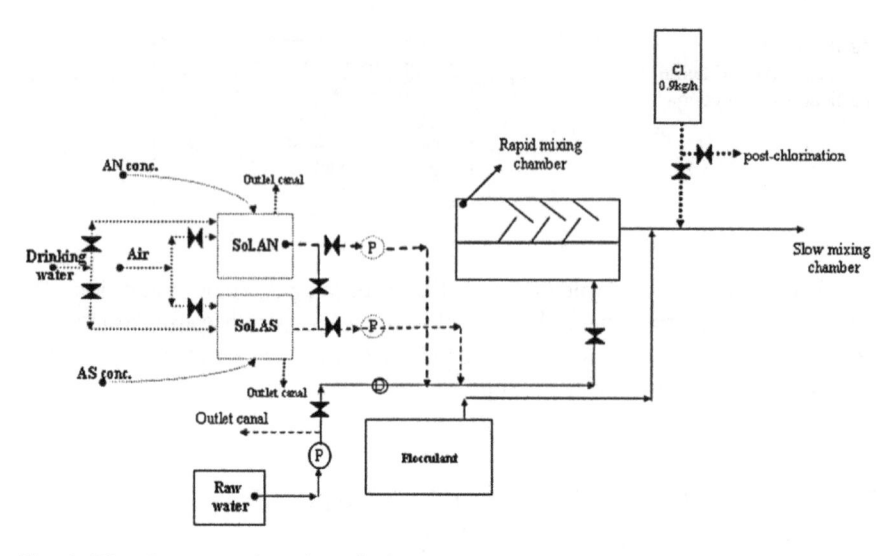

Fig. 4 Pilot plant research station—Surface water treatment plant. *Source* Aquatim S.A (with permission)

Figures 4, 5 and 6 show schematics of the technological flow of station's pilot of surface water treatment STA Bega with which they have conducted studies and have integrated the curves dosage of aluminum sulphate, a sodium aluminate, have been tested different reagents and additives, different methods of disinfection, filtration tests for washing of filters.

The purpose and importance of evaluation methods is:

- Verification, identification systems/processes/operations/equipment that can be optimized
- Selection, optimization methods or best application of technologies.
- The valuation methods are:
    - Theoretical: specialty literature, technical books, operating instructions
    - Statistics: specific databases (quality, technological data, etc.)
    - Technical laboratory facilities, low tonnage installations (Stations Pilot).

Factors that influencing the methods of evaluation:

- Statistics: existence/non-existence of databases, situation, accuracy of existing data, etc.
- Technique: especially the precision, accuracy equipment, measurements, etc.

The choice of methods of evaluation is done based on evaluation/analysis of existing data implementation methods evaluation is done through the development of research methodology/Optimization of the System/process/operations/equipment analyzed.

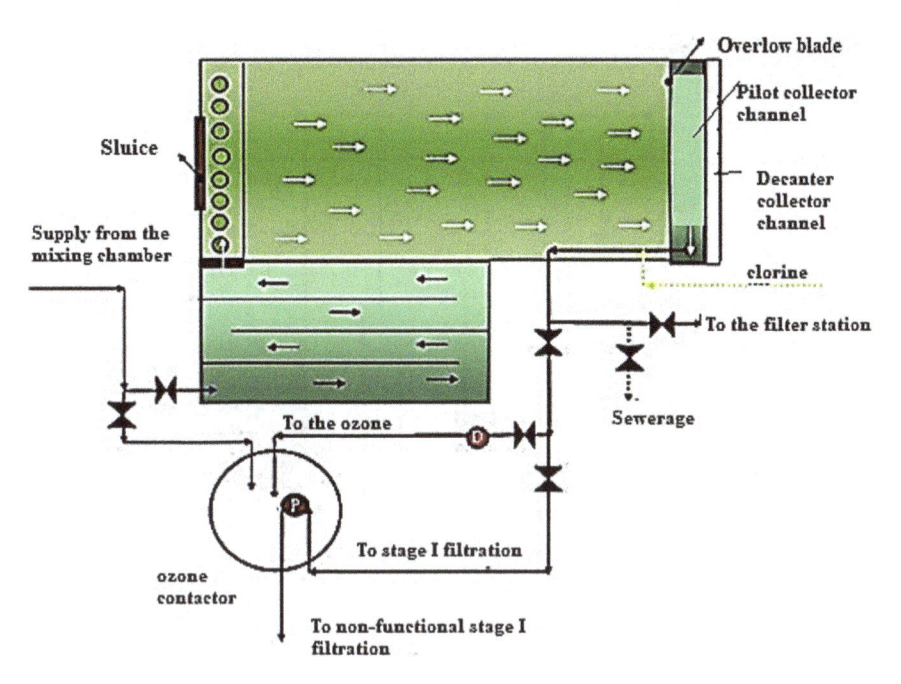

Fig. 5 Pilot plant research station—Surface water treatment plant. *Source* Aquatim S.A. (with permission) [9]

3 Potable Water Treatment Process

3.1 Pretreatment

In water treatment processes, pretreatment is the first phase and helps the next technological processes and respectively protects the installations against corrosion and abrasion. Purification and possible refining steps (ozone, activated carbon) are intended only for colloids and dissolved substances. On the other hand, modern decanters and filters work in optimal conditions if they are overloaded with coarse suspended matter and aquatic plants. Therefor, the treatment plant must be protected by adequate pre-treatment [8].

3.1.1 Water Intake

The intake system is to reliably deliver an adequate quantity of water for the best available quality. Reliable intake systems are costly and may represent as much as 20% or higher of the total water treatment plant investment. An intake system must possess a high degree of reliability and be able to supply the quantity of water demanded by a water utility under the most adverse conditions. For large capacity of water

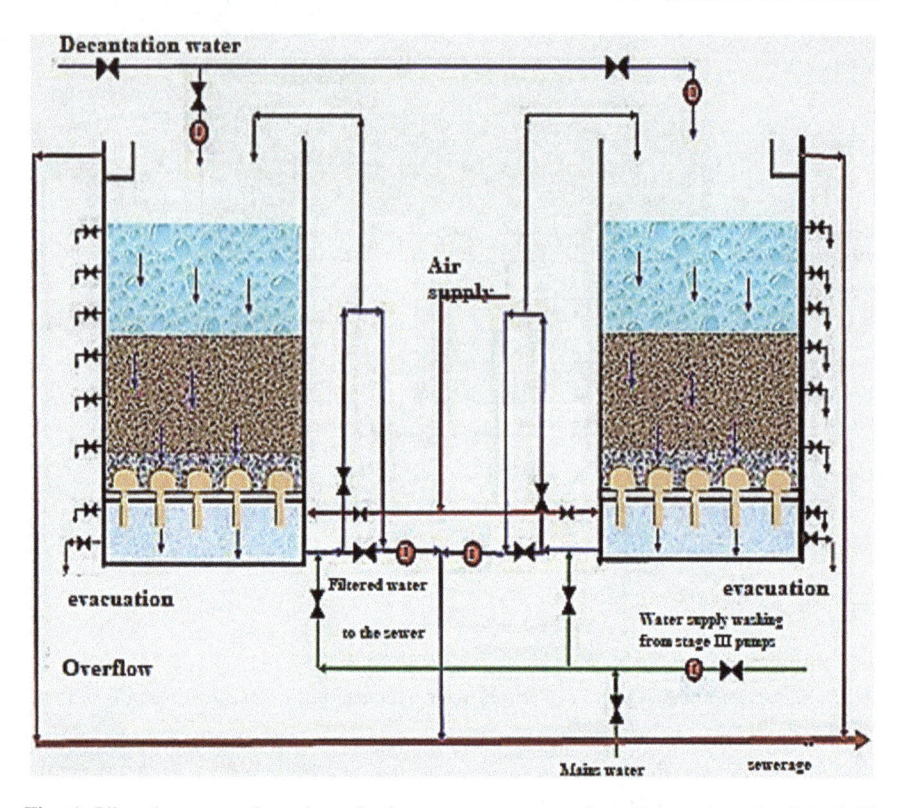

Fig. 6 Pilot plant research station—Surface water treatment plant. *Source* Aquatim S.A. (with permission) [9]

treatment plant, detailed survey including hydraulic calculation, mapping, pumping Station, source of power supply and other scope must be done for designing intake plant. Each intake system presents unique challenges. River capturing is dependent on the case depending on the hydrological regime, the navigation possibilities, the nature of the rivers and the upstream activities [8].

Capture from stagnant water (lake, barrier lake) is possible to execute a water intake from tower with several tiers in order to change the sampling height depending on the variations of the water level, the seasonal phenomena of thermal stratification, reverse flow, etc. In general, the sampling quota must provide for the possibility to be chosen in such a way that the water content of the suspended matter, colloidal matter, iron or manganese, seedling, is as low as possible [8].

Raw water storage is storing water in the raw water tank has several roles. For safety, it can allow the power supply to be stopped for a day, a week or even a month, depending on the capacity of storage, when the source is polluted. Other role is purifying that allows a mechanical purification and biological (self-purification). They generally improve water quality at least for certain parameters (turbidity, suspended matter, NH_4+, unwanted bacteria). Reservoirs must be protected against possible degradation of water quality, in particular, eutrophication trends must be carefully studied and combated. These conditions must be created and/or maintained: at least 5 m deep, a water circulation (by agitation or by the arrival and departure of water from different angles) and avoiding the proliferation of algae [8].

3.2 Pre-treatment at Raw Water Intake

The first possible treatment is screening with the aim to remove large debris that could impede the implementation of subsequent treatment stages.

3.2.1 Grills

It consists in the use of grills that allow the elimination of large bodies, floating on the surface or in water. The grills are iron bars between 1 and 2 cm thick, with gaps from 1 to 100 m. Coarse fixed grill are with bars having gaps between 5 and 10 cm and are generally completely immersed. Fine mechanical grill with space between the bars, between 3 and 25 mm, are either with automatic or manual washing and with straight or curved bars [8].

3.2.2 Screens

This operation consists of passing water through perforated screens made of metal (steel, stainless steel), or plastics (nylon, perlon, polyethylene) stretched on frames; these panels being mounted on rotating strips or drums having a dihedral or semi-cylindrical shape, in order to increase the filtration surface.

Depending on the size of the stitches, we distinguish:

- Macrosites—metal sieves, with an opening between 0.15 and 12 mm, intended to retain certain suspended, floating or semi—floating materials, vegetable or animal remains, insects, algae, grasses, etc.
- Microsites—metal or plastic sieves with meshes smaller than 100 microns, generally of the order of 30 microns, intended to retain clogged materials of very small dimensions (especially phyto and zooplankton) [8].

3.2.3 Petroleum Products Separation

The goal is to separate the free-floating oil products to avoid entrainment in the treated water. Generally, siphon walls are used to make a dam in the way of oil products collected on the surface, which can then be evacuated through a fixed or mobile spillway [8].

3.2.4 Desanding

It aims to extract gravel, sand, and other more or less fine mineral particles from raw water, in order to avoid deposits on channels and pipes, to protect pumps and other devices against abrasions and to avoid overloading the next treatment steps. The de-sanding equipment is generally rectangular of the corridor type, there is also the separation of the sand by cyclone (hydrocyclone on the discharge of the pumps), but there is the risk of wear by abrasion [8].

3.2.5 Presettling

This pre-treatment phase is an operation that precedes the clarification of surface waters. the goal is to completely remove fine sands and mud as much as possible. The clarification yield of a pre-decanter (as removal of suspended matter) varies between 50 and 60%, without reagents; it can reach 75–98% with an adequate dose of coagulant and/or flocculant. Like decanters, pre-decanters are rectangular or circular in shape [8].

3.2.6 Aeration

This pre-treatment phase serves to elimination of excess gases (H_2S, CO_2), in the case of groundwater and enrichment of water with dissolved oxygen [3, 8].

3.3 Oxidation

An oxidant is a compound capable of fixing electrons (e), being the opposite of a reducer that tends to give up electrons. This phenomenon is represented by the following reaction Eq. 1.

$$\textbf{Oxidant} + n\,e- \leftrightarrow \textbf{Reducer 1} + \textbf{Oxidant 2}. \qquad (1)$$

Depending on the abilities to give up and capture electrons, the reaction takes place mainly in one direction or the other.

A good oxidizer should possess the following properties:

- oxidation and/or disinfection power as high as possible
- a non-disturbing conjugate reducer (non-toxic, easy to remove)
- the weakest ability to generate toxic compounds through side effects or parasites.

Oxidation is a process that can be used in different stages of treatment. At the beginning of the process, allowing: oxidation of mineral compounds such as iron, manganese, ammonium; oxidation of organic pollutants; elimination of taste and smell; inhibition of algae development in the clarification stage; improvement of the coagulation stage; and flocculation, (a): in the middle of the process, to increase biodegradability before filtration on activated carbon, or (b): towards the end of the treatment, to ensure water disinfection, before distribution There is no ideal oxidant, the choice being made according to the water to be treated and the compounds to be removed. The main disinfectants used in water treatment for drinking are: Chlorine, chlorine dioxide; ozone; mono chloramines; ultraviolet radiation; potassium permanganate [3, 8].

3.4 Coagulation in the Water Treatment Process

Most surface waters contain a large amount of impurities which, depending on the size of the particles dispersed in the water, can be: decantable, suspended, colloidal or dissolved. The turbidity and color of a water are mainly caused by the presence in the water of very small particles called colloidal particles. These particles can remain suspended in water for a long time because their specific gravity is very close to that of water.

The colloidal particles in solution are very stable because films with electric charges of the same kind are formed around them, which cause the particles to repel each other. To accelerate the settling process of these particles, chemical reagents are used, which by dissolving them in water produce ions of the opposite sign to the colloidal particles, the partial neutralization of these charges leading to the agglomeration of colloids into flakes, larger and heavier aggregates. In the literature, this process of treating water with chemical reagents is called coagulation-flocculation.

- Coagulation consists, first of all in the destabilization of the colloidal particle and then in the agglomeration of the particles in small aggregates or flakes.
- Flocculation (L. flocculus—tuft, clump), is the process of enlarging and uniformizing the flakes under the influence of a slow agitation. In conclusion, chemical coagulation, in water treatment, is a complex physico-chemical process, which has the effect of removing colloidal particles, pollutants and microorganisms from water, by destabilizing them with chemical reagents, then agglomeration into small aggregates and finally, removing by sedimentation.

Therefore, chemical coagulation is a process that involves three phases, namely: coagulation, flocculation and sedimentation [3, 10].

3.5 Settling Process

The decantation achieves the gravitational separation of the insoluble matter and in particular of the flocks generated in the previous coagulation-flocculation stages. The settling process takes place in the so-called settlers which are of different types.

Static decanters that can be with vertical flow: radial, conical and horizontal flow: longitudinal. Decanters with suspension layer: pulsator, accelerator, decanters with sludge recirculation, decanters with ballasted flocs, lamellar decanters also combined decanters: decanter with slurry and slats, decanter with recirculation and slats [3, 8].

3.6 Filtration

Filtration is a process that allows the separation of solid–liquid when passing through a porous medium. Rapid filtration allows to improve the quality of water from decantation by reducing turbidity. Another role is loosening, as it is performed when the filtration through sand is followed by a filtration on granular activated carbon having the effect of eliminating taste and color. As filter material can be used: quartz sand, granular activated carbon, anthracite, marble, volcanic tuff.

The factors that influence the filtering process are granulometry of the filter medium, filtering speed, the height of the filter layer, the height of the water layer. The pressure drop in a filter is a measure of the resistance to the passage of water, to the passage of the porous medium. It is caused by the materials or flakes deposited in the spaces between the granules. Time function at the exit of the filters is characterized by three stages after washing the filters, the turbidity is relatively high, it decreases rapidly, this is the maturation phase. During operation the turbidity remains relatively constant and at the end of the filtration cycle the turbidity increases. When the limit turbidity is reached or when the pressure drop is maximum, the washing of the filters is triggered. The washing of filters can be done with water under pressure, expansion-free washing with simultaneous use of air and water or washing with air and water in two successive phases. The filters are of two types: slow filters and fast filters that can be opened or closed. Depending on the adjustment mode, the filters can be classified into: constant flow and progressive charge filters, constant flow filters and clogging compensation or decreasing flow filters [3, 11, 12].

3.7 Membrane Techniques

A membrane is a thin wall, liquid or solid, which puts a resistance to the transfer of different constituents of a fluid. The principle of membrane technology is the fact that it works without the addition of chemicals, with a relatively low energy use and easy and well-arranged process conductions. The membrane separation process is based on the presence of semi permeable membranes and hence it has the property of selective separation of molecular or ionic chemical aspects. The separation is done under the influence of a driving force that can be:

- the difference in chemical potential (dialysis)
- the difference of electric potential (electrodialysis)

- pressure (piezo dialysis, filtration on semipermeable membranes).

Semipermeable membranes are used to treat water for human consumption. These are porous, water-permeable materials that retain all species larger than pore size. Depending on the retention power obtained (separation threshold), different processes can be distinguished: microfiltration, ultrafiltration, nanofiltration, reverse osmosis. Classification of membranes are filters that eliminate larger compounds than pores. They are divided by the nature organic: based on cellulose derivatives or synthetic polymers or minerals: they consist of alumina, titanium oxide, carbon, zirconium. In general, mineral membranes have a very high chemical inertia, being resistant to high temperatures, and high pressures and are very resistant to mechanics, which facilitates their washing. Divided by structure homogeneous: the membrane consists of a single layer of cup or regular material (size between 50 and 200 μm), asymmetric: the membrane consists of a very fine surface (0.1–1 μm) or compound: resembles an asymmetric membrane without the dermis and its support being of the same material. Also divided by configuration—there are two types of membranes, flat or in the form of cavity fibers. The arrangement of the flat membranes is done in plate modules. The fluid circulates between two adjacent membranes plates and at the exit, the permeate is recovered. These modules are almost compact but easily removable, which facilitates the replacement / washing of the membranes. Frequently, the membranes are rolled in spirals, the fluid circulating from the outside to the center of the spiral. Water circulation is important, but it is less compact. Cavity fiber arrangement—the fibers are grouped into bundles. A module comprises one or more beams. This configuration is very compact and allows reverse washing.

Filtering modes can be front or tangential filtering. Front filtering this is a classic filter. The slurry to be treated is transported perpendicular to the membrane, at which the suspended matter accumulates during the process. Under the impact of this accumulation, the filter flow decreases. To avoid this flow decrease, it is necessary to regenerate the membrane by countercurrent washing or the use of tangential filtration. Tangential filtration can be designed as a technique to direct the filtrate suspensions parallel to the filter medium. They allow the separation of the permeate from the concentrate. This tangential circulation limits the accumulation of particles on the filter medium. Reverse washing across the membrane remains necessary for membrane regeneration [8, 13].

3.7.1 Microfiltration

It can be frontal or tangential and allows the separation of micron particles (suspended matter, colloids) and particles. The pores of the ultrafiltration membranes have diameters between 0.1 and 10 μm, the most used being those of 0.2 μm. Microfiltration guarantees an absolute bacterial sterilization, but also a viral safety. Under these conditions, the addition of oxidant downstream of microfiltration is indispensable to ensure viral disinfection and to maintain a residual disinfectant in the network.

The solution can be used in general, for waters with low turbidity but which are occasionally very heavy (karst waters) [8, 13].

3.7.2 Ultrafiltration

It is generally preceded by a pretreatment consisting of sand filtration and/or microfiltration. Each ultrafiltration membrane is characterized by the separation threshold, in other words, the molar mass corresponding to a retina of 90% of a molecular species. Ultrafiltration is tangential, made starting from asymmetric or compound membranes. The porosity of the membranes varies between 0.001 and 0.1 μm. The membranes used in Europe for ultrafiltration have a porosity of 0.01 μm. In this case, the viral particles are retained, being a technology capable of providing sterile water. The final disinfection is performed only as a safety measure to avoid possible recontamination in the network. Instead, ultrafiltration has no action on pesticides, nitrates, iron, manganese, sulfate ions. Ultrafiltration is perfectly adapted to the purification of groundwater in order to eliminate mainly particles (suspended matter or microorganisms) [8, 13].

3.7.3 Nanofiltration

It can be preceded by a microfiltration and/or ultrafiltration. The membranes have a porosity between 0.003 and 0.001 μm. The advantage of nanofiltration over other tangential filtration processes is not only to ensure microbial sterility (ultrafiltration), but it is also to reduce dissolved organic pollution and micropollutants in significant proportions. Thus, reductions in trihalomethanes and pesticides lead to processes with interesting particularizations. This technique also allows water softening (elimination of calcium and magnesium).

Nanofiltration therefore allows the treatment of highly mineralized surface waters, whose softening is pursued, ensuring in a single stage the elimination of organic matter and hardness. Parameters that influence membrane filtration are numerous and are related to the nature of the membrane: structure, thickness, porosity. The nature of the suspensions: loaded, clogging power, deformable particles or presence of dissolved gases influence membrane filtration and operating parameters, such as, passage speed (0–6 m/s), transmembrane pressure (0–5 bar for microfiltration and ultrafiltration, 5–10 bar for nanofiltration. All these variables influence the flow of filtered water. [8, 13]

3.8 Activated Carbon Absorption

Absorption is a physical or chemical process during which an absorbent is fixed on the surface of an absorbent, at the interference between two phases, solid–liquid.

The absorbent power of activated carbon is used in the secondary or tertiary stage of water treatment lines, in the purpose of retaining pollutants not eliminated by classical methods. Also, absorption is a process commonly used in water treatment to remove humic substances, color, taste and odor, phenols, cresols and a number of non-biodegradable toxic substances (trihalomethanes). Activated carbon is a major component used in water treatment, either in the form of CAG granules or in the form of CAP powder. This is the material that, keeping all the proportions, is the most economical to remove a very large amount of pollutants.

The use of carbon as a biological support is a fairly recent technique [8]. The activated carbon in water treatment has two properties, viz. adsorbent power (physico-chemical treatment) and bacterial support. Two variants are currently used: activated carbon powder (CAP) and granular activated carbon (CAG). The porosity of coal allows the adsorption of a significant number of molecules and therefore to eliminate many unwanted substances present in water like organic matter, organic micropollutants, residual oxidants and some substances that impress the taste water.

3.8.1 Granulated Activated Carbon, CAG

Granular activated carbon is used in filter layers, usually fed continuously on the top. The charcoal granules located at the top of the filters and are therefore the first saturated. In double-layered filters, associated with sand, fed according to the technologies and objectives pursued, either through the upper part or through the lower part. Like all filter materials, granular activated carbon is characterized by granulometry, effective size (TE), uniformity coefficient (optimum between 1.5 and 1.8), friability (loss in % of TE after 750 or 1500 strokes), bulk density, humidity (on delivery must be less than 5% by weight), ash content (less than 10%).

3.8.2 Granular Activated Carbon in Filtration in the First Stage

If the carbon filter is right after the decanter, we have a filtration in the first stage and must fulfill two roles: actual filtration (elimination of turbidity) and adsorption. This use has often been mentioned for the conversion of a fast sand filtration plant into an activated carbon adsorption filtration, without major changes in the structure of the plant. TE will be around 0.9–1 mm, analogous to sand filters, for filtration lives of the order of 10 m h. If the coal is renewable, the friability must be very poor to limit mechanical wear. Washings are thus frequent (comparable to those of the sand filter) and because a significant disturbance of the CAG layers that will not be reclassified [8].

3.8.3 Double-Layer Filtering

Double-layered filtration is a controversial alternative in the case of modernizing existing installations. By applying this solution, new equipment and constructions are no longer needed. In fact, the existing filter will not allow the use of significant heights of activated carbon and it will not be particularly efficient in absorption. The lower layer is therefore constructed of sand (TE ~ 0.6–0.7 mm) and the upper layer of CAG. The selection of materials, nature, density, TE and friability are chosen taking into account the hydraulic requirements: total load loss, maximum allowed costs for construction works existing, minimum fluidization rates—very different for the two materials—in order to obtain a net separation of the two layers, after washing and limiting the loss of materials during washing. In relation to sand filters, there is an improvement in water quality by eliminating organic matter and in times of crisis limiting the consequences of certain accidental pollution [11, 12, 14].

3.8.4 Granular Activated Carbon in Filtration in the Second Stage

It is a common step, namely the installation of granulated activated carbon in a second filtration stage, in order to keep in this stage, the role of refining and diminishing natural organic matter and/or the role of eliminating accidentally appeared micropollutants. Clarified by the processes of decantation and filtration on sand of suspended materials, and the elimination of organic matter is achieved by adsorption and by biological activities that take place on coal. One can also seek to promote the biological activity of nitrified bacteria, trying to turn ammonia into nitrites in the case of ammonium-rich waters. The actual size can be chosen smaller for the first stage 0.55 or 0.65 mm. A filter needs to retain suspended matter, so it is washed less: the washing time is determined by the need to remove biomass that dissolves on the filter. The efficiency of this carbon filter is increased by a previous oxidation of the water with a strong oxidant (ex: ozone). The organic molecules attacked by oxidation will be able to adsorb and biodegrade more easily. Pesticides without saturating the CAG very quickly. Some CAG filters are used to remove traces of residual oxidant, after a disinfection by strong chlorination. CAG is manifested by the so-called catalytic action. Second-stage filtration is the most adopted technical solution for obtaining good water quality [11, 12].

3.8.5 Activated Carbon Powder CAP

It is a coal with reduced granulation: 10–15 microns, which is infected starting from a slip with a concentration of 10% CAP in the decanter, even on channels that include only filtration on coarse filters. The applied doses are determined by flocculation tests (Jar test) and vary from 5 to 50 g/m^3 depending on the case. It is used effectively against the temporary pollution of raw water (with organic matter, taste-generating components, and micropollutants). Currently, one of the ways to improve the treatment of water for human consumption is to develop refining treatments, combining membranes with CAP.

3.9 Ion Exchange Resins

Cation exchange capacity (CEC) is a fundamental soil property used to predict plant nutrient availability and retention in the soil. It is the potential of available nutrient supply, not a direct measurement of available nutrients. Resin structure are typically either gel or macroporous. Ion exchange resins are solid materials, obtained by copolymerization of styrene and acrylates with divinylbenzene (DVB), which ensures the crosslinking of linear polymer structures, which are ~5 mm in diameter and chemically inert. Ion exchange is much faster with macroporous resins.

The polymeric structure is grafted with active groups carrying permutable ions, which define the four types of resins:

- carboxyl group—weakly cationic
- sulfonic grouping—strongly cationic
- tertiary anime grouping—weakly anionic
- Quaternary ammonium group—strongly anionic.

The resin is electrically neutral, each active position being neutralized by a counter ion from the aqueous solution that hydrates the resin. This counter ion can be exchanged with a counter ion from the percolating solution, when passing it through the resin layer. Depending on the resin the counter ions can be H^+, Na^+, OH^-, Cl^-, with total and useful exchange capacity of resin that has a defined number of positions, interchangeable with mobile ions, which define the total exchange capacity, expressed in meq/100 g of soil. Values over 25 meq/100 g soil are found with heavy clay soils, organic, or muck soils. The total capacity depends on the following parameters: concentrations and types of ions to be fixed, temperature, flows, type of regenerant, regeneration conditions.

Moisturizing capacity—Water is retained in the porosity of the resins. The amount of water retained conditions the ion exchange rate. The hydration capacity is expressed in grams of water retained in 100 g of dry resin. Granulometry—conditions the flow rate of the water to be treated and the reaction speed. Large granules favor important flows while small granules increase the reaction speed [15].

Ion exchange mechanisms types of reactions—The ion exchange reaction can be reversible (softening type) or irreversible the reversible reaction—example in the case of a cationic resin, follows as follows in Eq. 2.

$$R - NA_2 + Ca^{2+} \leftrightarrow R - CA + 2NA^+ \qquad (2)$$

In this balance equation, the process from left to right corresponds to the exchange phase (softening) and when it flows from right to left it corresponds to the resin regeneration phase. Irreversible reaction this reaction takes place from the moment a strong acid is fixed by a strong basic anion exchanger. The corresponding reaction can be represented as follows in Eq. 3:

$$R - OH + HCl \rightarrow R - Cl + H_2O \tag{3}$$

3.9.1 Disinfection

Disinfection of drinking water is the final stage of water treatment for human consumption before distribution. It allows the elimination of all pathogenic microorganisms from the water, with the observation that there may be common germs because disinfection is a sterilization. The processes applied for disinfection can be chemical and physical. Disinfectants used in this stage of treatment have effects like bactericidal to destroy bacteria, virulicidal to destroy viruses and remaining effect meaning that the disinfectant remains at a residual dose in the distribution network, which guarantees the biological quality of the water. The residual disinfectant has a bacteriostatic effect, against bacterial revival and a bactericidal effect against minor and point pollution that may occur in the network. Some disinfectants have a very short lifespan and have a residual effect. Disinfection can be chemical in nature (chlorine, monochloramine, chlorine dioxide, ozone) or physical in nature (ultraviolet radiation) [3, 1].

Chlorine

Several products are susceptible to the supply of chlorine necessary for disinfection (chlorine gas, sodium hypochlorite) and is more effective than hypochlorite.
Conditions for good chlorine disinfection:

- pH < 8
- bactericidal power: $0.1 < Cl_2 < 0.2$ mg/l, contact time 10–15 min
- virucidal power: $0,3 < Cl_2 < 0.5$ mg/l, contact time: 30–45 min
- Optimal chlorine efficiency occurs at 4 < pH < 6.

Free chlorine has good stability (but is rapidly reduced by reducing compounds such as iron, manganese, ammonia or organic compounds). It is remnant which allows to ensure the additional protection of the water during the transfer or to the consumers. Chlorine has the disadvantage of forming organochlorine compounds leading to a deterioration of the organoleptic properties of the treated water and in some cases, generating incompatibilities with the standards [3, 1].

Monochloramines

They produce results from the reaction between chlorine and ammonia. Their disinfecting power is clearly lower than chlorine.
Conditions for a good disinfection:

- bactericidal power: 1 mg/l, contact time 15 to 20 min
- virulicidal power: 10–20 mg/l, contact time: a few hours the interest in monochloramins lies in their great stability over time.

This allows long-distance water supply with reasonable starting treatment doses. Despite these advantages, monochloramines are very rarely used as disinfectants because there is a risk of forming dichloramines which gives the water unbearable odors, comparable to those of pool water [3, 1].

Chlorine Dioxide

It is a water-soluble gas. Chlorine dioxide has more important bactericidal and virucidal power than chlorine and its efficiency remain constant over a wide pH range between 6 and 10.
Conditions for a good disinfection:

- bactericidal power: $0.1 < ClO_2 < 0.2$ mg/l, contact time 5–10 min
- virucidal power: $0.3 < ClO_2 < 0.5$ mg/l, contact time: 30 min.

Chlorine dioxide is a stable compound but is slightly less residual than chlorine. Chlorine dioxide forms fewer taste-generating compounds compared to chlorine. Chlorine dioxide does not generate trihalomethanes [3].

Ozone

Ozone is a very strong bactericidal and virulicidal agent, superior to chlorine. Conditions for a good disinfection:

- bactericidal power: 0.1 up to 0.2 mg/l, contact time 1 la 2 min
- virucidal power: 0.3 up to 0.5 mg/l, contact time: 4 min.

Ozone is very little remaining. Residual ozone disappears the faster the higher the temperature and pH. The stability of ozone in water is 10–20 min. Ozone is a very strong oxidizer, but rarely used alone in disinfection because it is residual and increases the biodegradability of organic matter, increasing the risk of bacterial revival in the network [3].

Ultraviolet Radiation

Chemical disinfection processes, such as chlorine or ozone, are the most common. However, as a purely physical process, it uses the properties of ultraviolet radiation, developed particularly in mountainous regions and in regions where the water captured is of very good quality and a small and well-maintained distribution network. Ultraviolet radiation is part of the radiation located outside the visible spectrum, with a wavelength between 100 and 400 nm Ultraviolet radiation with a wavelength from 200 to 280 nm are the only ones that have germicidal properties: they destroy bacteria, viruses, fungi, algae, yeast, the maximum effect is obtained at 253.7 nm. The disinfection of a water with ultraviolet radiation consists in the application on the water of a certain light intensity, for a period of time.

4 Conclusions

Water in its natural state is never absolutely pure. The water in nature may be consumed by humans if it satisfies the condition of drinking water that is healthy and clean, free of micro-organisms, parasites or substances by number or concentration could pose a potential danger to human health. Water characteristics, the chemical and physical quality of water provide the information that we need to determinate the water treatment process to ensure quantity and quality of water. Every source of water that we identify to use for water alimentation is particular and has his own mark that need a particular treatment solution trough analyses and pilot study to determine the best water treatment solutions. The evaluation methods are theoretical based to specialty literature, technical books, operating instructions, also statistics based on specific databases and technical laboratory facilities. The choice of methods of evaluation is done based on evaluation/analysis of existing data implementation methods evaluation is done through the development of research methodology. Is important that the determined sources must be counted both from the point of view of capacities and exploited sustainably without prejudicing the next generations but also qualitatively to ensure a good water that respects the standards in force. Counting these aspects, they also need to be calibrated from an economic point of view, exploitation operating treatment costs to ensure the need for water at a quality regulated by the standards in force to be economically sustainable for the population and economic agencies for which these services are provided.

References

1. Cococeanu A (2015) Modern technologies for the 3rd millennium, water sourse assesment for drinking water.Assesment/optimization for water treatment technologies, Oradea
2. Frederick W (1990) Pontius, 4th edn. Water Quality and Treatment, American Water Works Association
3. Binnie C, Kimber M, Smethurst G (2002) Basic water treatment (3rd ed). London: Thomas Telford Ltd. pp 5–89
4. Haiduc I (1996) Chimia mediului ambiant. Editura Univarsității Babeş-Bolyai, Cluj-Napoca, Controlul calității apei, pp 25–40
5. Vigneswaran S, Visvanathan C (1995) Water treatment processes: simple options. CRC Press, Boca Raton, Florida, pp 4–10
6. Părăuşanu V (1982) Tehnologii chimice. Editura Ştiinţificăşi Enciclopedică, Bucureşti, pp 5–10
7. Florescu C, Mirel I, Carabeţ A, Stăniloiu C (2015) Alimentări cu apă, pp 4–8. www.ct.upt.ro/users/ConstantinFlorescu/index.htm
8. Jumanca V (1996) Instalaii de captare, tratare şi epurare a apelor. Universitatea din Bacău, pp 10–25
9. Technical archive of Aquatim SA, Timisoara
10. Ianculescu O, Ianculescu D (2002) Procesul de coagulare–floculare în tratarea apei de alimentare. Optimizarea camerelor din staţiile de tratare, Editura Matrix Rom, Bucureşti, pp 3–8
11. Florescu C (2010) Tratarea apelor pentru potabilizare prin utilizarea filtrelor rapide cu straturi multiple. Editura Eurostampa, pp 55–70
12. Florescu C, Mirel I (2010) Filtrarea rapidă cu straturi multiple pentru procesele de limpezire din staţiile de tartare. Editura Matrix Rom Bucureşti, pp 20–30
13. Sagle A, Freeman B (2021) Fundamentals of membranes for water treatment. University of Texas at Austin, pp 5–10. https://www.twdb.texas.gov/publications/reports/numbered_reports/doc/R363/C6.pdf. Accessed 19 Mar 2021
14. Giurconiu M, Mirel I, Carabeţ A, Chivereanu D, Florescu C, Stăniloiu C (2002) Construcţii şi instalaţii hidroedilitare. Editura de Vest, Timişoara, pp 5–15
15. Ionescu T (1964) Schimbători de ioni. Editura Tehnică Bucureşti, pp 4–8

Adrian Cococeanu was born in Caransebes Village, Caras-Severin County, CS, Romania in 1986. He received the B.S. degrees in hydrotehnic engineering from the Polytechnic University of Timisoara, Romania, in 2011 and the M.S. degree in Optimization of hydrotehnic system, in 2013 and Sustainable Development postgraduate degree, in from Polytechnic University of Timisoara, Romania. He is currently pursuing the Ph.D. degree in civil engineering at Polytechnic University of Timisoara. From 2011 to 2013, he was a research engineer in Research and New Technologies Department of AQUATIM S.A. Since 2013, he has been an responsible process engineer in Surface Water Plant Timisoara. His research interest includes the development of groundwater resources (intended for water supply) through digital instruments and hydroinformatic systems.He is the author of 10 articles.

Teodor Eugen Man is affilitaed with the Polytechnic University of Timișoara. Currently he is Professor in the field of Civil Engineering and Installations, having completed 16 doctoral theses and in internship currently having 10 PhD students. Leading positions held in UPT: Vice-Rector (1996-2000), Member of the Senate (1996-2000, 2000-2001), Dean of the Faculty of Hydrotechnics (2008-2011), Director of the Department of Hydrotechnics (2011-2012), Vice-Dean of the Faculty of Constructions and member of the Management Office of the Teachers' Council (2012-2014), Director of the Research Center in Hydrotechnics and Environmental Protection (2009-2018). He has developed a number of 23 volumes of courses, laboratory guides, monographs, textbooks, of which 6 as sole author and 10 as first author, over 300 published scientific papers of which 17 sole author, 145 first author, over 50 published abroad and over 30 articles listed ISI or ISI Proceedings, 3 patents, 8 innovations, 116 research and development contracts (grants), of which 70 was responsible / project manager, technical expertise: 70, verification projects: 242. He is a member of prestigious professional associations in the country and abroad, he was a visiting professor at Artois University, IUT Bethune in France, also for 14 years he taught specialized courses at the Faculty of Environmental Protection, University of Oradea. Prof. Man's CV was also presented in the "Dictionary of Romanian Personalities" (Contemporary Biographies - ed. 2012, 2017). Member of the editorial board: the Scientific and Technical Bulletin of UPT, Department of Hydrotechnics; The scientific and technical bulletin of T.U. "Ghe. Asachi" Iași; Annals of the University of Oradea, Faculty of Environmental Protection and Faculty of Constructions and Architecture; Scientific Annals of Ovidius University Constanta; Revista Hidrotehnica, A.N. "Apele Romane", Bucharest. Professional, cultural, educational associations: Associate member of the Bucharest Academy of Agricultural and Forestry Sciences (ASAS); Member of the General Association of Romanian Engineers (AGIR); Founding Member of the Bucharest Association of Land Improvements and Agricultural Constructions (AIFCR); Member of the Romanian National Society for Soil Science Bucharest (SNRSS); Member of the National Committee for Irrigation and Drainage (CNRID) Bucharest, affiliated to the International Commission of Irrigation and Drainage, New Delhi, India (ICID).

Water Security

Environmental Forensic Tools for Water Resources

Ilija Brčeski and Ashok Vaseashta

Abstract Environmental forensics consists of a set of defensible scientific methods to address histories and sources of contamination in environment and involves the reconstruction of past environmental events, such as the timing, types and amounts, and sources of chemical releases to the environment. Interrogations necessitating environmental forensic applications usually relate to understanding of the extent, duration, and responsibility for environmental contamination sites in a regulatory and/or legal context. These approaches are also integral to due diligence of environmental aspects related to mergers, acquisitions and remediation cost recovery. Techniques such as chemical fingerprinting, chemical fate and transport modeling, hydrogeological investigation, and reconstructing operational histories, among others are at the heart of many investigations. These and newer techniques, such as multivariate receptor statistical modeling, continue to evolve and have become more sophisticated over time, as have the types of problems to which they are applied. Scenarios in which environmental forensics have been applied have ranged from remote Arctic environments to urban sediments. In both extreme scenarios, the chemical condition of the environment—i.e., the background or baseline—is a central part of any investigation. It is upon this background that additional contamination from one or several responsible methods may be juxtaposed. The types of problems to which environmental forensic techniques are commonly applied include: identifying and quantifying contributions from various sources to contaminated sites, distinguishing natural background and diffuse anthropogenic background from specific pollution sources, differentiating specific sources of petroleum and natural gas, delineating time frames

---

I. Brčeski (✉)
Faculty of Chemistry, University of Belgrade, Belgrade, Serbia
e-mail: ibrceski@chem.bg.ac.rs

European Academy of Science and Arts, Salzburg, Austria

A. Vaseashta
International Clean Water Institute, Manassas, VA, USA
e-mail: prof.vaseashta@ieee.org

Institute of Electronic Engineering and Nanotechnologies "D. Ghitu", ASM, Chisinau, Moldova

Institute of Biomedical and Nanotechnologies, Riga Technical University, Riga, Latvia

of releases, reconstructing historical concentrations and pathways of releases for dose reconstruction in toxic torts, and conducting causal analysis to determine associations between observed conditions and potential sources. A detailed analysis with examples is presented in this chapter to shed more light on this evolving subject.

Keywords Fingerprinting · Forensics · Water · Environment · Contaminants · Smart water

1 Introduction

Since the ancient time, the humans have engaged in the impermissible activities in the environment, due to survival, growth and industrialization. Activities, such as burning, deforestation, ore extraction, generation and uncontrolled disposal of all kinds of waste, conventional or unconventional wars, and even the recent technical advances are all part of the growth and development of civilization. Energy is crucial for economic growth and hence several of these activities that cause environmental contamination, are necessary for the human survival. While, dependence of technology is necessary for societal evolution and growth, the resulting contamination is natural, primarily as a byproduct. The motivation to produce more and use more is driven by capitalist economy, where profit and comfort are the leading indicators of success, which always comes with waste generation and environmental contamination. By promoting education, awareness and technological tools, it is necessary to realize improvements to identify and mitigate urban and regional environmental pollution globally. In addition to growth related introduction of contaminants, there are other instances, such as trade competition, infringement of trademark, intentional introduction of contaminants, nefarious intention to introduce harmful and unauthorized use of resources. All of such activities require a forensic examination of contaminants, additives, tracer elements or to identify contaminants for proper identification. Due to incremental awareness and recent advances in testing methodologies, some of the environmental contamination may not have been fully identified due to lack of earlier awareness, characterization tools and determination of toxicity of several contaminants and xenobiotic mixtures. In fact, the importance of the toxicological study of mixtures cannot be understated since these toxicants can enter the body via ingestion of contaminated water or consumption of foods or crops. The tools that are currently available are ever more powerful and demonstrate capability to assess a large volume of waste and also a large variety of contaminants, whether the contaminants are in pure form or mixed with other contaminants.

Pollution impacts our health and safety on a daily basis and in many ways. The use of unauthorized chemicals in oil blending compromises the quality of the air we breathe; mercury and other metals illegally released from unapproved mining into rivers and the sea endangers aquatic ecosystems and water supplies; as well as illegal dumping of waste at landfill sites contaminates the soil where food is grown. As a consequence, unauthorized pollution sources threaten environmental sustainability, public health, safety and the quality of life. In addition to sustainable development, it is mandatory to develop international standards and treaties, such as declarations, conventions, and agreements concerning internationally accepted legal regulations, defining the norms guided by environmental ethics and codifying the penal policy. The key message of these regulations reads "the polluter pays the damage caused". Thus, the polluter represents the criminal entity, has to be identified, and the damage needs to be established. The international organization engaged in prevention of criminal activities have organized Interpol Environmental Crime Groups, as a special team in this field, which is designed for the protection of natural resources, prevention of pollution and wildlife preservation. As with any other type of environmental crime, pollution crime is primarily driven by a high-reward, low-risk business model, where criminals exploit regional inequalities such as labor costs, weak environmental legislation and law enforcement capacity. Transnational and cross-border offences are also linked to pollution crimes, requiring a coordinated law enforcement response, both at the national level (inter-agency cooperation) and internationally.

At the core of these regulations, is identification of contaminants by tools of environmental forensics. For identification of the criminal activities and trace it back to the polluter, forensic techniques are used based on a nexus of scientific disciplines. In this regard, a new environmental field—practically a new scientific discipline has been formed and is called "Environmental Forensics", which is the application of defensible scientific methods to address questions related to release histories and sources of contamination in the environment. Unlike the classic forensics, the crime is not territorially restricted here, it often covers the borders of a number of countries and the damage can be caused to millions of individuals, creating many interested parties and complex legal activities, often at the international level. In addition, organizations, such as the International Society of Environmental Forensics has grown out of the need for a platform to present scientific investigations that address environmental contamination subjected to law, public debate, or formal argumentation as well as the evaluation of the basic science that serves as underpinnings to those activities.

In this Chapter, we will attempt, in an inadmissibly simplified manner, to outline only a part of possible reconstructions of past environmental events, such as the timing, types and amounts, and sources of chemical releases to the environment, for water resources. The forensic techniques explored in this chapter include:

- Water quality and shallow water table modeling,
- Hydrologic Modeling for shallow groundwaters and wetlands,
- Sediment transport modeling of streams and rivers
- Tracing pollutants in watersheds
- Wetland drainage and Sediment age dating.
- Microbial Fingerprinting, RNA, DNA, PLFA
- Fingerprinting for trademarks.

2 Water—Statistics and Characteristics

If the Earth is perceived as a living organism, not only through the philosophers' eyes, but also through the eyes of the naturalists, then, certainly water stream can is seen analogous to bloodstream. By observation, large and small bodies of water, such as rivers, reservoirs and capillary network are perceived, in the same modality. Thus, water represents an exceptional and hardly (being a slow process), a renewable natural resource. It is quite certain that in the 21st century, it represents a major strategic resource in the world. Even though after the 1970s energy crisis, it was believed that energy would be the problem of the next century, however, it was proved that the Earth was significantly richer in energy than it was first estimated. Fearing far greater dependence of the rich countries on the undeveloped part of the world, energy saving and developing new energy resources have become an imperative. Nuclear energy also contributes to energy stability. While the world was consumed with energy related issues, water consumption has been growing at great rate and remained unnoticed until recently. Growth in industry requires enormous quantities of water. The increase in number of population and their gravitation toward the urban centers, the use of washing agents and the comfort of living boost have significantly increased quantities of communal wastewaters, thus the concept of so-called renewable resource has been on the rise. The distribution of this water is provided in Table 1 [1].

As global water pollution reaches to an alarming level, a review of products produced by chemical industries reveal that many compounds with great stability have been synthesized, which made waste waters treatment much more difficult and costly. This is concurrent with a significant increase in the use of plastics, throughout the world, in cosmetics, washing products, micro- and nano- plastics—as an emerging contaminant in water, additives and pesticides, degreasers, etc.

In our Water globe, slightly over 70% of its surface is under water and it is estimated that the water is about 1.5 billion km^3, which renders it as the largest part, although practically useless without expensive desalination treatment plants. This is followed by crystal water which is within rocks and minerals composition. Only a small portion of total water is available for use, as surface fresh waters and ground waters. During normal course of water consumption, it sustains/undergoes

Table 1 Water distribution in nature [1]

Where found	Quantity	
	Million km^3	%
Global ocean	1250	83.51
Crystal water	230	15.45
Ice	15.50	1.007
Surface fresh waters	0.25	0.015
Ground waters	0.25	0.015
Atmosphere	0.15	0.008

Table 2 Example of water consumption [1]

Use	Unit	Quantity (L)
Drinking, cooking, washing	Person/day	20–30
Taking a bath in the tub		200–300
Car washing		200–300
Washing of streets, parks	m^2	1–2
Brewery	1 L of beer	5–8
Dairy plant	1 L of processed milk	3–6
Sugar refinery	100 kg of sugar beet	1,500–2000
Slaughterhouse/Abattoir	1 slaughtered ox	5.500
Paper mill	1 kg of fine paper	400–600

changes and is categorized as inappropriate for re-use without a prior treatment. A typical consumption consists of a sequence of steps/factors, some of which consume enormous quantities of water, with respect to the resources. There is quite a lot data on water consumption for several activities, and some of the examples are provided in Table 2. Some industries consume a very large amount of water, such as semiconductor processing, thermal power plant, nuclear power plants, agriculture and dye industries. It is worth noting that some of the most populated regions around the world, are physically located near the large river banks. Many of these rivers span international boundaries. Due to trans-boundaries treaties, legal procedures and other issues related to water contamination are often complex and sometimes difficult to assess due to dilution factor.

It is evident from the Table 2, that the quantities of water available for regular use by general population are rather small and it is critical to maintain certain water quality that is free of the pollutants. With the notion of global warming and associated climate changes, it is argued that the temperature of the earth will increase resulting in reduction in access to the usable water resources, such as decreased surface waters quantity, drastic drop in ground waters level, icebergs melting, etc. The national legislature, international associations, professional organizations, non-government sector and other interested parties in preserving water as the resource, have often questioned the water quality, however the source of introducing water pollution remains indirect or unnoticed from the main parameters. For several chemicals, it is critical to assess the source and their origin, using scientific discipline, viz. the environmental forensics. Some of these techniques are rather expensive and complicated, requiring highly specialized persons, those who possess several years' experience. The methodologies should be able to detect not only the pollutant, but also the polluter. Some of these techniques use large amounts of water to study the trends, such as rate, level and flow, even during previous periods, to be able to conform to local regulations and adjudicate the case accurately and expeditiously, to preserve one of the most precious and miraculous molecules the Nature has created—the water molecule.

2.1 Some Peculiar Characteristics of Water

Water is a rather unique molecule. To be able to characterize it, it is essential to know its peculiar characteristics. The most peculiar characteristics of water is that the scientist cannot agree on "what is water". For the chemists, "the molecule of a defined and stable structure", the physicists, "matter with anomalous characteristics", the biologists "the origin of life is a secret, but the condition is not...", the physical-chemists, "a miracle molecule", the technologists, "raw material, solvent, catalyst, energy and transport fluid, etc.", the meteorologists, "climate creator", the geologists, "it dissolves everything, water and ice are creators of the Earth", the agriculturists and nutritionists, "food", the politicians, "the owners of the drinking water will be the masters of the world", and the like. Water, indeed, as a molecule is one of the most exotic molecules designed by the Nature and certainly, the most peculiar molecule encountered by the mankind. Its properties, as per their peculiarity can only be compared to liquid helium. Although it represents a simple chemical compound with a corresponding formula of H_2O (11.19 mass% of hydrogen and 88.81 mass% of oxygen), if we consider these isotopes, many different combinations are possible, of which there are 91 stable forms. That means that the complex mixture of nine compounds display quite different properties. The complexity is further authenticated since there are a number of theories on liquid–water structure. There is a series of unexpected physical-chemical properties, anomalies, with respect to its structure or mass, as shown in Table 3.

Water also shows other anomalies, of which temperature and density anomalies are maybe the most important ones for the living world. In the temperature interval of 30-40 °C certain changes are noticed, for example, at 35 °C water has the least heat conductivity, and at temperature from 0 to 35 °C the biggest drop of electronic water polarization is noticed under the effect of the exterior field (optimum temperature for living organisms is within these limits). Another anomaly is observed in the temperature interval of 55–60 °C. At 55 °C there are minimum changes in electronic polarization of water molecules under the effect of the adiabatic pressure, with minimum changes of dielectric permeability depending on the pressure and significantly increased sound transmission. The density anomaly reveals that the ice density is smaller than the liquid water, namely that maximum density is at 4 °C. This anomaly plays an enormous role in nature, since during colder periods, water cools down to 4^0 C and sinks, being heavier and pushes ice up and also serves as thermal insulation, protecting sea life beneath.

2.2 Usual Chemical Water Composition and Pollutants

Chemical composition of natural waters is the result of their interaction with its environment [2]. During water and air interaction, if air is polluted it may contain organic and inorganic substances, dust particles, soot and even radioactive particles.

Table 3 Overview of some water properties [1]

Property/characteristics	Comparison to the "normal" liquids	Importance
Physical state	Liquid unlike H_2S, H_2Se, H_2Te	Exceptional. Exists in 3 states. Without it—no liquid water on the Earth
Heat capacity	Rather great	Global temperature regulation, heat transporter
Latent evaporation heat	great	Important for atmospheric physics and as the equilibrium in condensation and evaporation processes
density	Anomaly at 4 °C	Anomalous dilation: Ice density less than water. Water surfaces freezing, however, not the denser deep water
Surface tension	Rather great	Important surface phenomena, forming drops in atmosphere, transport through bio-membranes
Dielectric constant	Rather great	Excellent solvent
Dissociation	Rather small	Neutral medium, with equally represented acid and alkaline properties, exceptional solvent
Heat conductivity	Rather great	Important for heat transport mechanisms and climate on the Earth

Often, there are acid forming gases (in addition to carbon-dioxide) which significantly change water pH value. A liter of rainwater falling from the height of 1 km rinses through 3.26×10^5 L of air [3, 4]. The next interaction which forms yet another composition of water is with soil. Additionally, salts dissolve due to absorption/adsorption and ion exchange. Hence, natural waters represent complex systems containing salt ions, compounds of organic compounds, minerals—both in the form of real solutions, and also as colloids, suspensions or emulsions. Besides, water-atmosphere and water-soil interactions, its chemical composition is also impacted by water-living world interaction, and in particular human activity. There are many other chemical reactions that take place occurring during these interactions. At the level of ions, interactions can be hydrolytic, redox, and ion exchange reactions, among others. At the molecular level, water forms several complexes due to different interactions, depending on the conditions such as temperature, pressure and geological specificity. The things get complicated by participation of aquatic living world, namely by microbiological transformation of compounds in water.

In natural waters the presence of several stable elements of the Periodic table are noticed and they are mostly found in the form of ions. Chlorine, sodium, magnesium, calcium potassium and sulfur ions are most prevalent. As regards gases, oxygen and carbon-dioxide are particularly important. About 90–95% of total salts contents consist of cations and anions formed of these elements. Nitrogen exists in the form of ammonia ion (or ammonia), nitrite or nitrate ion, and phosphorus in the form of phosphates. Due to the importance for the living world this group of compounds is termed as biogenic. The third large group are organic compounds and they represent humus, consisting of hydro-carbons, living world metabolites, and biogenic elements. Included there are also humins and humic acids. Humus acids are divided into humatolenic, humic and fulvic acids. Fulvic acids are also high mass molecular compounds of oxy carbonic acids and have a lower content of carbon atoms than the humic ones, with significantly greater acidity. They are much more prevalent than humic acids and mainly create complexation of metal ions by humic acids from water solutions, the elemental contents of which is smaller than 1 mg/L and represent microelements which may occur as cations (Li, Ba, Sr…) "heavy" metals (Ni, Cd, Cu…), anions (J, F, Br …) or radioactive metals.

Depending on the water resource, certain specificities also occur to help identify the timeline of mineralization of rivers, river beds and lakes. As an example, the lakes are characterized by significantly slower water replacement and increased evaporation. This gives rise to increased general mineralization, but certainly depends on lake type and inflow of "fresh" water. For accumulation lakes, chemical composition is mainly formed in the initial period of their emergence. This period is several years long following which their balanced state is formed, similar to the lakes. It has been emphasized that human activity significantly impacts natural waters chemical composition. From the aspect of quality, the impact is almost always negative. The effects of water chemical pollution can be summarized in the following manner:

- aesthetic: appearance, unpleasant smell
- temperature: usually of increased temperature
- oxygen-wise: reduction in oxygen
- toxic: chronic toxification of aquatic environments
- acidity/alkaline: pH regimes disruptions
- eutrophic: intensive development of certain organisms.

Chemical substances that pollute aquatic environments are classified in the group of "priority control chemicals" and possess the following properties: toxic in small concentrations, bio accumulative, persistent, and cancerous. The impact of drinking chemically contaminated water on human health has been studied with particular interest, particularly due to the emergence of highly toxic substances. It is believed that about 25,000 substances that are widely used, pollute the waters in some way or the other, of which some may exist even in traces. As an example of the cancerous substances is polychlorinated biphenyls. Typically, 1 kg pollutes 10^9 L of water (it is believed that about 10^6 tons of these compounds have been produced globally!) and chemical structure provides them with an exceptional stability in nature. Along with the complexity of natural chemical composition, as mixtures, a serious analytical issue presents in identification of such huge number of potential pollutants. Strategic approach to mitigating such issue is continuous and persistent monitoring of water resources quality for defining their natural chemical composition, as well as publishing the list of substances, particular pollutants, in accordance with regional and international standards. Water resources quality monitoring implies water and sediments chemical testing, testing of micro- and macrophytic vegetation, as well as microbiological testing of water, testing of benthos and radiological testing. In addition to the customary microbiological—parasitological specific assessments are also are required, e.g. the analyses of microorganisms that disintegrate phenols or mineral oils. Unlike chemical testing the biological testing provides information on water quality, for significantly longer period.

Correlation between biological and chemical testing is extremely necessary. Many compounds impact the living word growth and production, most often in terms of toxicity: metals (mercury, cadmium, zinc, copper), organic compounds (pesticides, PCB, phenols..), dissolved gases (chlorine, ammonia, methane...) or anions (cyanides, sulfides, sulfites, nitrites...). Classification and data processing provide the information on the possibility of using these water resources for human use. Although the physical parameters that may present the pollutants (non-reactive suspensions, granulometry, sound,...) in water are not considered, it is necessary to note that temperature increase may present a great cause of polluting chemical processes in natural waters. Given that the increase in mean natural temperature is evident, temperature may induce consequential impact on pollutants. Some of the reasons for such standpoint will be stated only perfunctorily: since the increase in temperature impacts a decrease in gases solubility, and hence, it impacts the oxygen dissolution and increases the lethal degree of aerobic organisms. Oxidative disintegration is made difficult/impossible, when anaerobic flora is stimulated and the disintegration products are reduction compounds of nitrogen (NH_3), sulfur (H_2S, mercaptans..), and carbon (CH_4). Furthermore, there is an increased evaporation and concentration, which has a negative impact on water ecosystems and waters quality. The release of the deposited methane, at the bottom, with low oxygen availability and with reduction conditions, means additional quality deterioration. These are good prerequisites for formation of large anoxic (oxygen free) areas in waters.

2.3 Applied Forensics Methods for Water

The emergence of gas chromatograph, first with flame ionization detector, then with detector with electrons capture, and later with mass spectrometer coupled with a number of other detectors, has provided completely novel possibilities of analytical determination of organic molecules, in particular and influenced by the development of almost new, novel yet complex, chemical analysis branch: Analytical Chemistry of Waters. There are many techniques developed for analytical determination, although many of them are also the variations of the following techniques: Fourier Transform Infra-red (FTIR) and Ultraviolet–Visible (UV/VIS) spectroscopy, atomic absorption/emission spectrometry (AA/ES), optical or with mass detection, X-ray fluorescence (XRF) spectrometry, ion chromatography, highly efficient liquid chromatography (LC) (spectroscopic, spectrometric detection...), neutron activation analysis, as well as other irreplaceable forensic technique: viz. isotopes mass spectrometry. The application of certain techniques depends on the type and concentration of determining molecule or ion, or on the goal: identification of certain chemical type, its concentration or origin. Often, the polluting substance disintegrates or transforms in water, thus identification of the product of its eventual path is extremely important or determining the time of origin, based on the product and is significant, especially when selecting the method. Processing and interpretation of the results obtained is particularly important and the current software solutions have a huge selection of options with a future in artificial intelligence (AI).

3 Specific Knowledge Skills for Environmental Forensics

Analytical chemistry has become an enormous field of knowledge that is literally increasing on a daily basis. Each of its branches assumes enormous proportions, especially with new experimental methodologies, data analytics and AI. This presents a special challenge for a forensic chemist, who almost invariably, works with samples that are most complex admixtures, since in nature, soil and water practically represent admixtures. Sorption soil characteristics for certain ions dissolved in water impact the water composition significantly. These sorption characteristics depend on soil composition. Basic knowledge of pedology is also necessary in this field. Knowledge of stability of compounds (organic and inorganic), as well as their redox stability is of great importance. Sampling their conservation and preparation for analysis, to a great extent, determines the result. The essential knowledge of the analytical techniques, their advantages and deficiencies will direct environmental forensic scientist to the correct results, regardless of the "user-friendly" software used. Interpretation of results, most often means inter-correlation of the values obtained, finding the meaning in order to make a correct decision, by using digital tools or software solutions. Often, a forensic scientist/chemist is also a court expert, since his/her judgment may have an exceptional legal or economic weight, the reason for which have been stated above, and also requires the assurance that is based on many years of experience. Hence, true forensic environmentalists are rare. That is compensated for by a team of experts, where sometimes, and really sometimes, the truth is forfeited for some inadequate compromises made regarding the truth.

3.1 Terrain

Information on area or the medium of an interest for the forensic environmentalist is of particular significance since forensic approach without detailed study area is impossible. All of this evolves from the surrounding area that ends up in water, surface or groundwater, which for a hydrologist may be different media, however for the forensic scientist, it is critical to draw their interconnection between the distribution of the potential pollutants or pollution source to be recorded in the area. All photographing techniques can be classified in the photogrammetry, as defined by scientific discipline, since as early as, circa 1867 [5], which can be deemed to be the first big step in the field of data processing. Photographs (as well as aerial photographs, satellite photographs), mostly serve for precise topographic mapping of terrain situation, pollution sources, potential locations of finding toxic waste or geological disruption in terrain [6]. Hydrological and pedological terrain maps are also of particular importance. The next big step in this field was made by emergence of the Geographic Information System (GIS). This system serves to present, process and analyze the geographic data from the ground by using digital tools. GIS may show base infrastructural data, which are taken as final and usually do not serve for further processing or modeling. Other data that GIS can process are specific (cadasters of waste waters discharge, mine tailings ponds, landfills, etc.). Special data bases in GIS technology relate to concrete activities: exact sampling locations, specific photographs, technical documentation that relates to infrastructure and similar. As GIS technology is connected with the Global Positioning System (GPS), the exact location determination is enabled automatically [7]. These data sets are most often formed by the researcher themselves in the field.

All these procedures described in principle are rather expensive, in particular the photogrammetric determination. Often there is a compromise, since for environmental forensic projects, it is long term expense, as compared to an sporadic event which yields quick results. Furthermore, with invention of drones as imaging/photographing devices, have facilitated imaging with enhanced speed of terrain/field activities, albeit on a narrow local level. The photographing devices may provide images in different specters. By conjoining software processing of digital data (photographs), the beginning (source) of a pollution (colored or fluorescent material from chemical industry) can be identified in surface waters, especially using devices such as hyperspectral imagers and light detection and ranging (LIDARs) devices. This can be an excellent forensic tool not only in "visible" (in ultraviolent/visible specter part), but also the thermal water pollution (in infrared specter). The data in infrared specter are irreplaceable for understanding thermal pollution, first of all due to rapid mixing of water in natural conditions and also due to the absence of typical "marker" that can be monitored in real-time, not only with thermal pollution, but also with chemical, in addition to location and time of occurrence and pollutant manner of distribution of the events taking place in the field. Recently devices for terrain/field measurements possessing inconceivable possibilities have emerged, which provide data in addition to usual terrain measurements (pH, conductivity, turbidity, electro-chemical oxygen or certain ions measuring by selective electrodes). There already exist terrain gas chromatographs with a series of advanced detectors (photo ionization, mass spectrometry, etc.), FTIR devices, and with bases and software for data processing which present an exceptionally powerful tool in this field [8].

3.2 Overview of Some of Most Widely Spread Pollutants and Techniques

3.2.1 Heavy Metals

About 80 elements of the Periodic table meet the basic conditions to be termed as metals. There are many divisions of metals as per their properties. There certainly is no other term in environmental chemistry that has such an indeterminate meaning, as "heavy metals". The term "heavy metals" relates to their density, however, lacks general consent, since the bottom density threshold starts from 3.5 g/cm^3 [9] and goes over 7 g/cm^3 (scandium has the density of 2.9 g/cm^3!). Metals expressed as and with non-metallic properties are listed in the heavy metals group, they are, viz. "semimetals"—such as arsenic, tellurium or non-metallic modification of selenium [10, 11]. However, the most commonly tacit, under the term "heavy metals" in the environmental chemistry, justifiably but not from the chemical aspect, are metallic ions (cations) that are extremely eco-toxic (Pb, Cd, Hg, As, Cu, Zn, Cr, Ni, Co...). Natural sources of heavy metals are the rocks containing them or which they leach from in the natural conditions, as well as volcanic eruptions. Anthropogenic emission results from industrial processes, mining, most mine concentrates for processing phases, metals, agricultural activities—in particular obtaining and use of phosphate or special fertilizers, production of batteries of all types and sizes, production of pigments and paints, metal packaging and material waste. Industrial wastewater contains significant quantities of metallic ions. Combustion of fossil fuels in processing also produce a large amount of heavy metals' emission in nature. Each of these emitters, both depending on its nature and also on the place of origin has its characteristic "fingerprints", identifiable by forensic scientist environmentalists.

It is important to emphasize that in natural conditions, metals ions and cations are encountered, which react not only with water but also with other numerous ionic types, forming complex compounds having various characteristics, such as solubility, sorption characteristics, and bioavailability. Solubility of metal compounds is of particular importance, as many nitrates of metals, some sulfates (except Pb and Ba) and many chlorides (but not of mercury and lead) are soluble. On the other hand, majority of hydroxides, carbonates, phosphates and sulfides are poorly soluble. It should be noted that in water there is no insoluble compound in the proper meaning of the word. Solubility will, to a great extent depends on concentration of hydrogen ion (pH), and that is usually the consequence of many hydrolytic reactions. Oxide-reductive medium will also impact the speciation of the metallic compounds The presence of organic substances (humic and fulvic acid) result in the formation of complex compounds where organic molecules appear as ligands. As stated earlier, water is in the continuous contact with soil and its composition and granulometry (grain size distribution), which makes the forensic scientist's job more complex. The majority of researchers in this field are physical chemists, namely analysts and may fully disregard the presence of microorganisms, which take part in biochemical redox processes and with their enzyme systems make transformations of metallic ions

compounds which are otherwise non-degradable in water. This property signifies that they accumulate in biota and through ecosystem they are transferred to plants, animals and human population, disrupting the biological equilibrium of any medium they find themselves in. There is a voluminous literature in this field. As stated earlier that classification is inconsistent and inadequate for forensic scientist environmentalists and also, reactions of metal with the environment are numerous and complex. Hence, one can conclude that metallic ion in water is important for at least, for the following two reasons, viz. metallic ions in water have an exceptional eco-toxic factor and serve as tracers/markers.

Eco-Toxic Factor of Metallic Ions

This might be the most significant and most sought after task assigned to the forensic scientist in this field. Forensic scientist is typically not a toxicologist and also that eco-toxicology is not necessarily human toxicology (e.g. boron, vanadium and molybdenum are important for the fauna, but not for the humans). Forensic scientist should determine inorganic and organic speciation of metals and leave the results interpretations to other experts in specialized fields. Organic and inorganic compounds of certain metals may have different toxicity. Inorganic arsenic compounds are extremely toxic, whereas its organic compounds (foodstuff from marine organisms/seafood) is significantly less toxic, while mercury organic compounds from marine organisms are extremely toxic. Toxicity of lead and cadmium compounds depends on their solubility, namely absorbability [11]. Oxidation ion state also has great impact on toxicity. Arsenic, as As(III)-ion is much more toxic than As(V)-ion [12], chrome as Cr(III) is practically non-toxic with respect to Cr(VI) [13], mercury as Hg(I) is practically insoluble and biologically less (hardly) available in respect to Hg(II). It is particularly interesting that some metals in the same oxidation state, depending on anions, show different toxicological effects, and as an example, zinc-chloride is much more toxic than zinc-sulfate [14].

As water is a universal solvent, water-soil/sediment-biological material interaction will give rise to certain degree of accumulation and bioaccumulation of metallic ions [15, 16]. Soil usually serves as sorbent/barrier for further spreading of metallic ions towards sensitive ground waters reservoirs, used for drinking water. Depending on the pollution degree and soil characteristics, ground waters pollution is significantly likely in the vicinity of large rivers, where the population density is great with significant hazardous effects [15, 17–19]. That means a complete image on aquatic eco-systems pollution must be acquired, if the forensic scientist should determine the concentrations of metallic ions in water, sediment and living organisms. Metallic ions will not even stay in water for long term, but will be sorbed on suspended particles, which originate from soil erosion and rock material disintegration and will separate as sediment (solid substance of mixed mineralogical composition), namely by bio-accumulation and will pass on to the living organisms. Thus, their concentration will practically be the highest at the bottom. This results in the sediment being a "memory card" of the aquatic eco-system. With the change of any of chemical factors (e.g. pH medium), there will be release of the bound ions of metals from the sediments and the living organisms consume greater concentrations, which ends up in the food chain and leads to intake in the human food chain. Several forensic assessments have demonstrated similar sorption characteristics of sediments [20, 21]. Realistically, this would mean that once an aquatic soil is contaminated with greater metallic ions concentration, the sorbed ions remain in the sediment practically permanently, regardless of the current concentration of free ion at the water surface, down to 50 cm, which is the usual depth for sampling. Water serves the role of an extraction agent—sometimes sediments contamination testing is performed by the sequential extraction method with aggressive solutions in order to determine the leaching degree [22]. In aquatic living organisms' heavy metals serve as biomarkers [11]. If a satisfactory correlation among heavy metals concentration in all three cases is found then it can be deemed that eco-toxicological case has been resolved.

Tracer

It is not a strictly defined word. As emphasized with "heavy metals" there is no strict definition of the notion or terminology "tracer". Some terms are often used without thinking about their complexity, namely multiple meaning or mostly literary convenience Within a broader context, it can be applied to represent as the "visualization of the flow/course or movement" and can be also be used as "an indicator of change of the assumed/expected state". If, by means of tracer, the pollution flow is modeled, the starting point will also be the pollution source, whereas its change in concentration will indicate the characteristics of the route the tracer passes through, which in this particular case is water. If tracer is a certain chemical type then it represents a simple real tracer. It can be also be used as the indicator of connection of the hydrological network (natural or created one), as well as monitoring of the underground flow rate [23]. For certain chemical types, it is important that they are well soluble, mobile and that they do not hydrolyze and that their sorptivity is low. Cations alkaline metals meet this condition [10], while for anion, the chlorides serve as an example.

If complex agglomerates, or several chemical types are of common origin and have permanent ratio in the admixture, then it can be said "with caution" that it represents the "fingerprint". These chemical compounds must have a unique composition, unlike the environment. Naturally, one isotope or isotopic ratio of one element can be monitored separately. Tracer can be mechanical particles of specific form or color which can be identified by a microscope (23), complex organic compounds or geological (mineralogical) composite. Meaning, water can "carry" certain sediment particles (or their specific soluble types) as geochemical tracer, and that by analyzing them, we learn about the origin (starting point of the occurrence), monitor the flow and rate and also interpret biological or geochemical substance cycle in the aquatic environment, with the knowledge of adoption mechanism by biota [16]. Tracing compounds must not always be perceived as a reliable "fingerprint". However, by their flow, i.e. interaction with the environment, certain components may possess different sorptivity, thus the quantitative ratio may change, whereas in geological material, among others, it may come to mechanical separation, and then both quantitative and qualitative ratio of certain types in the compound may change. In fact, a "SmartWater®" is a traceable liquid and forensic asset marking system (taggant) that is applied to items of value to identify thieves and deter theft. The liquid leaves a long lasting and unique identifier, whose presence is invisible except under an ultraviolet black light. Fingerprinting of conflict materials is a subject of much interest and Laser induced breakdown spectroscopy (LIBS) is used for ascertaining provenance for elemental fingerprinting.

Environment compromises of complex tracer compound in water, simply by the natural emission of several components. It is worth noting that mining is 4,000 years old, at the minimum, where practically the impact on water pollution by metals started and still is an active field to this day. Maximum emissions were in the period of extensive industrial revolution. While the activities and impact declined after the 1980s, the presence of heavy metals still persists and has been confirmed [24, 25] almost globally, in sediments of rivers. Some of these rivers originate from the same geographic location and represent an important water resource for about 1.4 billion of people [26]! By analysis of the river sediments at different depths the data on pollution history can be obtained, and even on flow rate of water [25, 27]. Chemostratigraphy is the study of the chemical variations within sedimentary sequences to determine stratigraphic relationships and is method of studying the presence of metallic ions of different masses at certain sediments depths. It is an excellent forensic tool for studying the past, namely the history of pollution (in the period of several years to a millennium), which comprises of both geochemical testing of matrices (identification of similar ones) and also their detailed chemical characteristics. A great number of samples, necessary for such research also provide data that are processed by multivariant statistical methods.

A significant progress has been made in the field of environmental forensic, where isotopes are studied as tracers. Not only do they have significantly and more reliable geological fingerprint, but also the substances of anthropogenetic origin having different isotopic values than those of geological origin. Hence, establishing the pollutant origin becomes considerably more certain. This scientific discipline, called "geochemical isotopic signature", is based on physical and physical-chemical studies (isotopes fractioning), geochemistry (atomic types and their ratios in soil), and radiology (radioactive decomposition of unstable isotopes). Isotopes are the atoms of a chemical element having the same number of protons in their nucleus and with different number of neutrons (atomic mass). They can be stable, without spontaneous radioactive disintegration, namely with insignificant radioactivity with enormously long period of semi-disintegration and unstable, while radioisotopes disintegrate spontaneously and have relatively short period of semi-disintegration. Both of them are used in tracer techniques. The total number of particles in nucleus is stated on the left-hand side above the symbol of the chemical element (^{56}Fe, ^{65}Cu..). Isotopic ratio is the permanent ratio of the isotope of one element in the nature and is characteristic for each element. The majority of the Periodic table elements represent an isotopic admixture. Although in chemical terms, it is considered that admixture components reactivity is identical, however, that is not the case. The reactants from the isotopic admixture of different mass react at different rate with other reactants, or biomolecules, and also govern different rates of physical phenomena, such as diffusion, evaporation, and their kinetic isotopic effect. The rate of direct and return reaction of certain isotopes is not the same ("equilibrium isotope effects"), which also causes isotopic fractioning. The rate of reaction and equilibrium concentrations are conditioned also by isotope basic physical characteristics, their origin, admixture characteristics, chemical properties (coordination number), as well as other factors impacting this fractioning [28], which is enrichment/depletion of isotope, as a consequence of the reaction (e.g. sedimentation, redox reaction or complexation with organic ligands, or mixing the reservoirs with different isotopic signature) and by calculating their ratio can clarify much on their interconnection or flows. Physical phenomena such as evaporation, sorption, and diffusion also cause fractioning, which can clarify much about both historical and present events. As many biological processes are controlled by the chemical reaction rate, they are prone to kinetic and equilibrium isotopic effect. Concentrations of certain isotopic fractions indicate transport of anthropogenic and natural pollutants in aquatic mediums and by way of rate of water evaporation from the natural water resources [28–31].

Stable Isotopes in Research

Lead isotopes are widely used as tracers in river sediments testing. Lead compounds are mostly insoluble, with a tendency to be sorbed on the river sediment. Different fraction per sediment particle size will have different lead concentration. Natural lead isotopic "signature" (isotopic admixture) depends on lead, uranium and thorium presence. Namely, Pb has four isotopes: ^{204}Pb, ^{206}Pb, ^{207}Pb and ^{208}Pb. Only ^{204}Pb (1.4%) is stable and does not have "parents", whereas ^{206}Pb (23.6%) is the offspring of ^{238}U disintegration, ^{207}Pb (22.6) originates from ^{235}U, and ^{208}Pb (52.3%) from ^{232}Th. Therefore, the ^{204}Pb concentration is permanent throughout the geological history of our planet. Even though, it is more difficult to determine due to small prevalence and the results are more difficult to interpret due to the absence of its radiological "cousin", its increased presence indicates to older mineralogical forms, whereas the younger ones are richer in other isotopes. Also, the relatively poor prevalence of lead isotope originating from uranium and thorium indicates to anthropogenic, and not geological origin. A great majority of lead is actually of anthropogenic origin and originates from ore extraction (for over thousands of years), its processing as far as metal, and in the recent history also by combustion of fossil fuels—coal in particular, from the gasoline engines (tetraethyl lead), production of batteries, paints and pesticides production, and waste incineration. Different origins will have different isotopic ratio. In river sediments, it is evident in studies of the geological and anthropogenic impact (pollution), thus it can be established with great certainty, whether the pollution originates, for example, from pre-industrial period, from mining activities (of geological type) or from combustion/incineration. Facilitated measurement of isotopic admixture, sorptivity and transport with river sediments, predictability of origin, possibility to determine it in all environmental segments, then the identifiable isotopic composition in respect to the origin, makes these isotopes as exceptional tracers. The possibility of sequential sediments extraction and their testing in any extraction agent, also provide the historical overview of pollution origination over a long time interval [25, 28, 32–35].

Zinc isotopes represent the admixture of ^{64}Zn, ^{66}Zn, ^{67}Zn, ^{68}Zn, ^{70}Zn. Variations in isotopic admixture in zinc are governed more strictly by rock material and industrial processes. Since, zinc is a biogenic element and reacts with organic molecules or is adsorbed by the organic material, and oxides of iron, Zinc isotopes can be used as the tracers for materials originating from mines and processes, including coal operated thermal power plants. Restriction of zinc isotope as tracer is that the isotopic admixture fractioning can happen only following certain processes, such as diffusion upon being deposited on materials. Yet, Zn isotopes has applications in water tracer testing, particularly being a bio-element.

Copper isotopes are the isotopic admixture of ^{63}Cu and ^{65}Cu. Copper, unlike zinc, is stable only in one oxidation state. This redox phenomenon plays a significant role in isotopic fractioning in copper. Harder soluble Cu(I) exists in geological material of sulfide types, whereas Cu(II) exists in aquatic solutions. This can be an excellent indicator for rivers pollution as the consequence of mining activities [36]. Many sulfide minerals are prone to redox reactions by hemolytic autotrophic microorganisms, thus the soluble Cu(II) is accumulated in the cell wall [37]. Sometimes the redox reaction in the medium or of microbe type, may also cause certain unclear issues since the subsequent dispersion in water and/from the sediments can result in different isotopic fractioning [25].

Chrome isotopes are the admixture of ^{50}Cr, ^{52}Cr, ^{53}Cr, ^{54}Cr. Chrome, similar to copper, emerges in two stable oxidation states—toxic and water-soluble Cr(VI), and significantly less toxic and significantly water-insoluble Cr(III). Redox processes (for example by reduction of Cr(VI) to Cr(III) by chemical or microbe reactions) result in isotopic fractioning, which can be an indicator of ground waters pollution (37), but also of bi-geochemical cycles (the ratio between insoluble and soluble chrome forms may provide the assessment on natural and anthropogenic pollution as well as the assessment of the pollution migration process, through equilibrium isotopic admixtures [38, 39].

Mercury isotopes represent the isotopic admixture of ^{196}Hg, ^{198}Hg, ^{199}Hg, ^{200}Hg, ^{201}Hg, ^{202}Hg, ^{204}Hg. Mercury is an unusual metal, not only because of different oxidation states, but also due to complex geochemical and biochemical cycle. Mercury is one of the priority polluters in the environment, with extremely eco-toxic characteristics. Many organic and inorganic reactions impact the isotopic fractioning of mercury isotopes (redox, methylations, absorption and adsorption, evaporation and condensation) which results in isotopic ratios variation and differentiation of natural and anthropogenic impacts on aqueous eco-systems. These isotopic variations can indicate to mercury pollution sources, pollution history, transport and transformations in aqueous systems [40–42].

Unstable Isotopes in Research

Unstable atomic nuclei, prone to disintegration during certain time period, namely the offspring of the parents—radioactive disintegration, are unstable isotopes. Radiology and radiochemistry deals with this subject in great details and for environmental forensics in waters, they can be used similar to stable isotopes, viz. source, course, distribution of the pollutant in aqueous environments (inclusive also of biogeochemical cycles), namely in sediments; monitoring of changes in time in the chemistry of large aqueous reservoirs (lakes, seas), such as precise time date recording of the changes occurred in eco-systems. Similar factors, as with stable isotopes, impact isotopic fractioning of these isotopes. Naturally, the heavier the isotopes, the difference in isotopic masses is smaller, rendering their determination more problematic [43].

Strontium isotopes represent the isotopic admixture of ^{84}Sr, ^{86}Sr, ^{87}Sr, ^{88}Sr. Only ^{87}Sr is radionuclide and is formed by beta emission of radioactive ^{87}Rb. The ratio of stable $^{86}Sr/^{88}Sr$ in nature is constant and amounts to 0.1194 and that ratio will appear in all inorganic or biogenic species. Other ratios will depend on the impact of the environment (inorganic-mineralogical, organic-biogenic) on their fractioning. The ration $^{87}Sr/^{86}Sr$ in surface waters of the global rivers is 0.703–0.730 (in North America is 0.710). Ground waters and lakes (water in the lakes represents the admixture of ground waters and river waters as well as rainwaters) have different ratio depending on geochemical characteristics (contribution of strontium from atmospheric precipitations is practically negligible), whereas the ratio $^{87}Sr/^{86}Sr$ in ocean waters is rather conservative (in the Pacific 0.709) [43, 44]. By careful selection and significant number of samples, Chemostratigraphy data of the strontium isotope ratio can be obtained and conclusions drawn on historical changes of water regimes [45]. Monitoring rivers flow and their pollution by the pollutants from agricultural activity, or as the indicator of bottled mineral water origin, with a ratio of $^{87}Sr/^{86}Sr$ considered to be an excellent and rather acceptable parameter [46, 47].

There are also many other isotopes and isotopic pairs that can be used as tracers or indicators of changes in waters. ^{234}Th that is the product of disintegration of ^{238}U and ^{7}Be which is formed in atmosphere as the result of proton and neutron originating from cosmic irradiation, interaction with nuclei of ^{14}N and ^{16}O isotopes [48, 49]. These unstable isotopes have the property of adsorbing on sediment particles (for example those suspended in the river), similar to heavy metals. This indicates that their ratio is a good model which can be used for correlation of sorbed heavy metals on particles. $^{234}Th/^{7}Be$ activity eliminates the variations of specific nuclide activity depending on the nature or particle size. Since the concentration of ^{234}Th depends on the presence of parents ^{238}U, the sorption of which depends on salinity, while ^{7}Be will come from the atmospheric precipitations, the interaction of the fresh and salty waters at the origin of large rivers can also be monitored [48].

A large group of isotopes is used as forensic tool, being the isotopes of rare earth elements (REE). This group of elements, the usual oxidation state of which is +3 has similar ion radii. However, a small difference in radius, but with the increase of atomic number, results in specificity in certain pairs, such as Sm/Nd ratio in various mediums [49]. Neodymium isotopic "signature" (unstable ^{143}Nd is similar to ^{87}Sr) is similar to strontium in many globally significant rivers. Similar to all metal ions, they also have affinity towards the particles suspended in rivers, which conjoin the oceans, creating isotopic signatures [50, 51]. In addition, there are lantana-cerium [49], lutetium-hafnium [52], rhenium-osmium (not all are REE series) [53, 54] as well as many other from several hundreds of atomic nuclei around us, with the stability period (semi-disintegration) which are multifold greater than the Space age, and those after semi-disintegration, which is measured in fractions of a second. All would have been great in forensics, had their concentrations in water and sediments been not extremely low, which for forensic scientist poses a serious issue.

Metal ions, their stable isotopes and radioisotopes, individually or in certain ratios, demonstrate the history of event in the nature. Since we have the capability to read this history, we might be able to reconstruct it into separate years, in terms of rivers flow rates, rate of evaporation, average temperatures, suspended or deposited aqueous sediments, which informs us about the natural interactions of water with the environment, and the remains of the biological non-disintegrable material on hydrological biogeochemical cycles. The cycles that would be established can predict our future. Departing from the customary cycles in the recent "industrial" ages, indicates a disruptive factor.

Instrumental Techniques

The work on implementation of instrumental techniques covers the following basic principles:

- working conditions: laboratory microclimate also including the manner of air flow and purification, personnel workload with the number of samples or working on different techniques,
- basic infrastructure: method of samples preservation, standard substances and reference materials,
- qualification of the technical personnel: education and experience (equally important) in certain technique, age and condition of instruments
- method of samples preparation: this is of exceptional importance and in addition to method of preparation it requires most often also extremely clean reagent medium for work,
- results reading and processing: requiring also certain knowledge of mathematics.

Instrumental techniques in this field are extremely expensive and may be practical in certain parts of the world than for the less developed part of the world. It should be emphasized that in forensic explorations, there is no routine procedure to look for the traces, however the conclusion must be unambiguous. Most widespread and basic techniques are the techniques of atomic absorption/emission spectrophotometry, AAS, (flame, graphite furnaces, hydride and cold steam techniques), with all modifications and procedures for increasing the sensitivity. They were preceded by UV-VIS spectrophotometric, which were preceded by volumetric methodologies, which is the analytical pre-history of metallic ions. Currently we are in the era of inductively coupled plasma and related numerous techniques of which two are dominant: optical emission spectroscopy and mass spectrometry with all possible variations (some similar or identical to AAS, for example hydride technique). Recently, use of Fourier Transform Infrared spectroscopy (FTIR) with Focal Plane Array (FPA) detectors and Raman spectroscopy, in conjunction with image processing software has been used to trace micro/nano plastics, not only to identify additives but also to trace it back to the source of origination. It should also be mentioned here the chromatographic techniques (such as ion chromatography), which may be coupled with some of the stated techniques in order to increase specificity, sensitivity and selectivity.

The major techniques for determining the ratio of stable metallic isotopes is the inductively coupled plasma with mass detection ICP-MS (it can also be quadrupole ICP-MS, ICP-QMS), and its upgrades, multi-collector inductively coupled plasma with mass spectrometry, MC-ICP-MS, namely magnetic sector MS ICP-MS, followed by sector field ICP-SFMS, which can be coupled with laser ablation, LA ICP-MS [55–57]. Furthermore, thermal ionizing mass spectrometry (TIMS) is used [58], as one of the prominent techniques. There are also other modifications, but the basis of all techniques is represented by metal ion mass spectrometry.

It is important to realize that for the water and sediments, the isotopic separation has been taking place over long time period of time and as a result of the transition of water from solid into liquid state, and then into gaseous state (and vice versa), the isotopic signatures so created indicate the origin of the spring water and its characteristic of underground water reservoirs [59, 60]. This has been done using earlier methods such as gamma spectrometry of isotopes, X-ray fluorescence, scanning electron microscopy, electron microprobe or wavelength dispersive spectroscopy [48, 61]. Finally, one of the most sensitive and maybe the most powerful techniques of the analytical metal determination in all environmental segments (and perhaps the most expensive), is a complex gamma spectrometric technique, in addition to an important technique called the neutron activation analysis [62].

3.2.2 Organic Substances

Hydrosphere consists of an enormous number of natural and synthesized organic compounds, all inclusive of sediments. Natural compounds have always been present and included in bio-geochemical cycles and transformations, but since they are classified in complex admixtures, our knowledge about them is insufficient. Practically nothing is known about millions of synthesized organic compounds or on their eco-chemical characteristics or eco-toxicity. This is due to the fact that the techniques for their determination are relatively new, new and emerging compounds continue to be developed due to rapid advances in technological innovations, and perhaps the most important being that most of the time the use of several materials commences without necessary testing. The classification itself or their sequence, as per priority list is complex and non-uniform, since the properties may be unknown, and it is difficult to establish any assessment criteria. This is true, specifically for nanomaterials and nano-biomaterials. With respect to their eco-toxicity, this is still in a learning curve in terms of their stability in natural conditions, induced oxidative stress, suppression of normal defense mechanisms, procoagulant activities and immune response for a human body. Except for some natural stable compounds, the characteristics are either largely unknown as regards to the admixture's complexity or very complex. However, with recent advances in characterization equipment, data analytics and artificial intelligence, the environmental forensics scientists will have significant tools in their arsenal to solve complex mysteries of tracing contaminants, interaction of contaminants with human body, and toxicokinetic to help improve quality of life.

Petroleum Hydrocarbons—Pollutants, Markers and Additives

Petroleum is an exceptionally complex product of natural processes about which the scientific debates are still ongoing. Hundreds or even thousands of compounds (mainly formed from carbon and hydrogen) are found in the admixture, termed as petroleum and represent "petroleum hydrocarbons". Although the classification is complicated, based on their molecular structure, its constituents can be classified into alkanes, alkenes and aromatic hydrocarbons. Alkanes in petroleum (paraffin) contain a single-sequence carbon-carbon bond (most often of linear or branching type) and are a dominant fraction in natural petroleum. Alkenes (olefins) (also linear and branching), are formed more during petroleum processing. Both of them are poorly water soluble. Aromatic compounds are of a ring-shaped structure, with different number of carbons in the ring, among which there are "conjugated" dual bonds. They are multiple water soluble compared to alkanes and alkenes. They can be of a single ring ("monoaromatic") and their representatives of significance are benzene and its basic derivatives, viz. toluene, ethyl benzene and xylenes (usually, forming the group "BTEX"), as well as other derivatized structures of benzene or substituted benzenes. If they are formed of a number of aromatic nuclei ("polynuclear"), then they are termed polynuclear aromatic hydrocarbons (PAHs). Polynuclear hydrocarbons possess a rather broad value for water solubility. Depending on the processing technology, certain compounds are grouped in admixtures of various physical properties such as the density or boiling point. The majority serve as fuel for different type engines or as solvents. Petrol is liquid and contains liquid alkanes (up to 12 carbon atoms) and alkenes (up to 7 carbon atoms) and monoaromatic compounds. The middle fraction, known as "diesel" and "kerosene", contain compounds with up to 24 carbon atoms of alkanes being dominant, and monoaromatic and poorly soluble polyaromatics hydrocarbons, whereas the rest may contain long chains hydrocarbons, insoluble polyaromatic compounds and other residues. Metallic ions can be found in fuels and, certainly, also in the compounds serving to boost their basic purpose characteristics—the additives.

If the presence of petroleum hydrocarbon is established, then in addition to determining its origin, also hydrocarbon type should be determined, where the material originates from, namely the owner. The chemical signature techniques are also represented by the "fingerprint" term. In fact, there are no universal protocols, or rules that can be applied to all cases, not even the universal method. It is often that the forensic scientists have to combine various analytical techniques, use different standard admixtures or finished products in the analyses, and compare peaks or to use the appropriate software, to obtain comparable values as per compound composition and signals intensity. The most commonly applied technique is gas chromatography, which separates the admixture into individual components and determine the components with certain detectors. Petrol provide different peak heights of the compounds present (C4–C12) and this chemical signature will depend on the refining type, degree of purity, year of production. Petrol for different use will have specific composition (for automobiles, jet fuel or high-octane). They will even differentiate the composition based on regions, as different regions have different regulations. In addition to aliphatic, they will also contain significant quantities of monoaromatic compounds (BTEX). These admixtures may be characteristic also for many solvents used for cleaning, most often the metal surfaces; however, they will contain more easy-to-evaporate components, so as not to leave traces on the surfaces. Diesel fuels, also possesses a specific signature depending on the origin and age, whereas those with longer chains, that remained upon distillation processes of fuels, such as motor, hydraulic, or transformer oils, will also have their own characteristic fingerprints [63].

A group of compounds representing biomarkers is of particular interest. It consists of sesquiterpanes (C25), diterpanes (C20), triterpanes (C30), steranes (C30), and hopanoid compounds (64). This becomes complicated when the petroleum hydrocarbons reach their natural conditions, i.e., in addition to sorptivity, some components are evaporative, water soluble and some prone to redox, chemical or biochemical-microbiological changes. In time the concentrations of evaporative components will decline on the account of harder-to-evaporate ones and so will the "signature", including the presence of light fractions, n-hexane and n-pentane. In the context of water, majority of hydrocarbon admixtures are poorly soluble. Therefore, water soluble fraction will be rather different from the original compound, whereas the insoluble one will be similar. In soluble phase the presence of BTEX should be expected as well as other monoaromatic substituents, if petrol is polluters, namely some phenanthrenes or anthracenes, if diesel is the cause of pollution. If chemical or microbiological redox processes are present, then changes in peaks will be perceived, first of all of alkane fraction [63]. Interactions with biological material, in addition to the fingerprint of the hydrocarbon fraction in water, will also leave the traces in the organisms, comparable not only to water but to the sediment in water. With a careful approach distribution of pollution through water sediment and organism can be monitored [64, 65].

Markers and Additives

In petroleum, hydrocarbons are of particular interest for the forensic scientist, primarily since they indicate to their origin and age. Markers indicating origin can be alkane, of isoprenoid type, such as pristane (C19) and phytane (C20) ratio. Their ratio depends on redox conditions, temperature and organic substance characteristics

[66, 67]. C17/pristane and C18/phytane ratios indicate the age. Namely, as petroleum hydrocarbons degradation is performed by microbiological oxidation, C17 and C18 hydrocarbons are more easily degraded in respect to pristane and phytane, the ratios will be changeable in time [64, 68]. Except on these molecules, biodegradable activities can also be observed on hopanoid compounds that also serve as markers, namely chronological indicators of degradation processes [69, 70]. As emphasized earlier, for forensic researches in water, BTEX group of compounds are particularly important, where ratios of certain components in the group may represent marker admixture and indicate the contamination origin and age [71].

Many substances are added to fuels as additives for various reasons, sometimes separately, sometimes in an admixture. The knowledge of their characteristics refers both to the manufacturer and also to hydrocarbon type, and its fingerprint may also indicate the time of getting into water. Some of additives have their degradation paths, where the degradation product can be found in ground waters. Some are added as stabilizers, oxygenates, and paints [67]. Complex "organic lead" compounds used to be added as anti-detonation agents (tetraethyl- and tetramethyl- lead and other alkyl lead compounds), but also many other compounds that prevent lead oxide depositing in engines (ethylene dibromide, ethylene dichloride) [67, 72]. In contact with water, their hydrolytic products (even though they hydrolyze slowly) are excellent forensic traces which indicate to pollution type and age. Other additives as well that are complex compounds of metallic ions may be good indicators, like nickel carbonyl, iron pentacarbonyl, and in particular, methylcyclopentadienyl manganese tricarbonyl (MMT) is indicative, that has not been added by all manufacturers, namely has different history of use [73]. Paints, that can also serve as good traces, are added for marking different types, namely different fuel quality. Usually they are of azo, diazo, or anthraquinone structures [74].

Of all additives, "oxygenate", as oxygen sources, increase the octane number, and therefore boost fuel performance when used in engines. As per their structure they are mostly alcohols or ethers, and methyl-*tert*-butyl ether (MTBE) stands out as particularly significant for forensics. After having perceived advantages of use, an abrupt production boost of this additive followed due to an increasingly greater application (especially after lead compound elimination). Detailed studies of its routes of degradation in aqueous medium and also based on the knowledge of these reactions kinetics and conclusions were made at the time of occurrence of pollution [75, 76]. The human health effects from MTBE at low environmental doses are unknown, however animal studies involving high doses of MTBE have shown effects such as skin and eye irritation. This is also of particular interest to the Bureau of Alcohol, Tobacco, Firearms and Explosives (ATF) agents for forensics. Additional characteristics, such as highly soluble benzene does not get absorbed and degradation is not rapid and qualifies as a good tracer for modeling movement in waters. Based on its detected concentrations in water or their absence, the contamination origin and type can be identified (fuel, pipeline, tanker) or that origin is from non-point source, such as urban atmosphere [77]. In order to identify pollution based on release/effusion of hydrocarbon material that does not contain MTBE, the forensic scientist must also have the field data available and good models of spreading through the environment.

One of key questions to be answered is whether the pollution resulted from effusion from several sources and in different time periods and which one is dominant. In addition to testing MTBE concentration, forensics must also have the knowledge of BTEX concentrations and ratios, as well as alkane fractions, and their degradation paths [78]. This is necessary, since the MTBE quantities in use, as well as the natural resources loaded with this compound, cause water pollution, and therefore also drinking water pollution [79, 80], since it is found in some water sources, mainly in urban areas with leaking underground gasoline storage tanks.

Polycyclic Aromatic Hydrocarbons (PAHs)

PAHs are a large group of compounds, most commonly formed of two or more rings which could also be substituted and are rather stable which makes them convenient for forensic testing, although they represent only a small percentage of total compounds present in the admixture. Sometimes three-membered heterocyclic dibenzothiophenes are classified in this group. Their origin might be from the geological period of petroleum forming under specific conditions (petrogenic), then from combustion or organic substance pyrolysis (pyrogenic) and under the effect of oxidation processes, microbiological-biochemical, or organic substance in the sediments (biogenic). Biogenic PAHs practically do not take part in pollution [81].

Three-membered alkylated rings are convenient for determining origin, since their ratio remains fixed during processing, and as their degradation kinetics stays similar during longer time periods as well. This ratio represents the "source", the initial admixture. This ratio is characteristic for determining certain natural admixture, which makes them specific and enables their individual identification [82]. Since they are mostly poorly soluble they are the most often found in sediments, and the ratio of three-membered PAHs does not change for decades and as such can be adopted by biota [82–84]. This indicates that the origination of the "signature", depends on the condition of obtaining, first of all of temperature, which indicate that we will have two specific "signatures": petrogenic and pyrogenic, as they were formed in two temperature regimes of different duration. Those that are of long-term lasting, petrogenic ones, will have lower content of less stable linear components (anthracene), and more stable non-linear multi-membered ring like structures whereas pyrogenic will have more unstable, linear components with smaller number of rings [85]. The forensic scientist can now differentiate the origin of PAHs, which were deposited in the sediment from atmospheric precipitations, by combustion in the vicinity of urban centers (anthropogenous) or originate from petroleum [86–88]. With the accelerated industrial development and demand in energy, as well as by generating large quantities of waste, the requirement for differentiating specific PAHS signatures has significantly increased, in particular with the large incinerators (gas or waste) hidden (by underground reservoirs, pipelines leakage) or visible (by effluence) of petroleum pollutants. In order to come across the correct data, the background contribution has to be separated [89, 90]. The significance of field observation, information on processes and the sampling method is of considerable importance to obtain the results.

Halogenated Organic Compounds

Organic halogen compounds are a large class of natural and synthetic chemicals that contain one or more halogens (fluorine, chlorine, bromine, or iodine) combined with carbon and other elements. The simplest organochlorine compound is chloromethane, also called methyl chloride (CH_3Cl). Other simple organo-halogens include bromomethane (CH_3Br), chloroform ($CHCl_3$), and carbon tetrachloride (CCl_4). Similarities with other halogenated compounds are greater than the differences. The chlorine-containing insecticide dichlorodiphenyltrichloroethane (DDT) is highly effective in killing disease-ridden mosquitoes, ticks, and fleas, but it is virtually nontoxic to mammals. The fluorine-containing pesticide "1080," or flouroacetic acid ($FCH_2 CO_2H$), is highly toxic and often lethal to all mammals. The industrial and combustion by-product dioxin is highly toxic to some animals but not to others; in humans, dioxin causes the skin disease chloracne. There are several known natural organo-halogens that have grown from ~30 in 1968 to nearly ~4000 during the early 2000s. Many are the same as synthetic chemicals. They are biosynthesized by marine organisms and hence pose difficulty in forensics.

Halogenated Solvents

During 1960s and 1970s, the production and use of halogenated solvents increased abruptly, due to their versatility, from being reagents in organic synthesis to cleaning and degreasing agents. The common chlorinated solvents are Trichlorethylene ($ClCH–CCl_2$), Perchloroethylene (tetrachlorethylene, $Cl_2C–CCl_2$), Methylene chloride (CH_2Cl_2), Carbon tetrachloride (CCl_4)), Chloroform ($CHCl_3$), 1,1,1-trichloroethane (methyl chloroform, $CH_3–CCl$) since chlorinated solvents are the most popular halogenated solvents. Their utility comprised not only of "large" industries but also small size enterprises, such as those producing metal objects, printing machines, cleaning clothes and many similar activities. A great part of rubber, paints, pesticides, resins or plastic masses industry, and even electronics, rests on these compounds. Also, the use in chemical laboratories for various extraction procedures and basic solvents is quite significant. One or more hydrogen atoms may be substituted in different compounds classes, most often in saturated (dichloromethane, trichloromethane, tetrachloromethane, 1,2-dichloroethane, 1,1,1-trichloroethane, non-saturated (trichloroethylene, tetrachloroethylene...) and aromatic hydrocarbons (chlorbenzene, 1,2-dichlorobenzene, 1,2,4-trichlorobenzene...). Practically, they are omnipresent around us. Although toxicity has not been precisely stated in any of cases previously, due to this property, they present a great hazard for the environment, in particular for the ground waters. Unlike the substances, they are always found in nature as petroleum compounds or metallic ions, halogenated solvents as purely synthetic compounds—completely strange to the environment. The nature has developed many mechanisms to minimize their hazardous impact, for example: solubility in water is small, sorbing on sediments, such as immobilization, pH dependent spreading, redox potential of the medium and mostly metallic ions are in non-soluble (less hazardous) form, and numerous population of microorganism is able to "eat" or transform natural organic molecules or by biosorption with extinct biomass to lessen metallic ions spreading. Also, there are many other mechanisms that form part of the bio-geological cycle.

In halogenated solvents none of this is applicable and many of their properties for which they are actually used are completely opposite to those that nature would have mitigated. Yet another, unpleasant albeit important fact is that, due to abundance presence in the environmental the methods and instrumentation for their detection have been developed much later after great damage has already been done [91]. Their great volatility may result in the false feeling that it is more difficult for them to get into ground waters. They can penetrate the soil and waters by gas diffusion, and if they evaporate, the chemistry of atmosphere "forms" rather toxic compounds that it "returns back" by precipitations back into the hydrosphere. Their density is greater than water density that will lead to penetration into ground waters aqueous layers and according to the physical chemistry laws, the concentration will accumulate. Low viscosity and possibility of drawing into the pores or filling the interphase spaces adds to pollution spreading. Small sorptivity to soil and sediment indicates that it will stay in the aqueous phase, and also extremely low biodegradability means that they can stay unchanged for long period of time in ground waters. Their solubility in water, although small, is thousands of times greater than maximum concentrations acceptable by the national and international standards for drinking waters [91–93]. This was observed in 1999 in an incident of leakage of about 2,100 T 1,2-dichloroethane (commonly known also as ethylene dichloride-EDC), leading to the destruction of petrochemical complex in a small town of Pancevo in Serbia.

Forensic testing of chlorinated solvents implies determining the sources and age of pollution. Additives testing, determining degradation products, as well as isotopic analysis form the basis for such researches. Many chlorinated solvents for the industrial application require additives to boost their stability (i.e. decrease in oxidation, hydrolysibility, etc.). Countries usually add the additives in various concentrations (or as different compounds) which facilitates source identification [94, 95]. Tens of compounds, organic and inorganic are used as additives with different properties. Naturally, the original sample is required for determining the type of chlorinated solvent in nature, for comparison.

The presence or absence of the degrading products, or their subsequent products is used to identify the source and time of pollution. In many cases, both of degrading products and also mutual ratios of chlorinated solvents have been described, as an example, when 1,1-dichloroethane is formed from trichloroethane and perchloro ethane, chloroethane is formed as degradation products of trichloroethane or 1,2-dichloroethane [95]. The period of semi-disintegration (disintegration half-life) in certain conditions of trichloroethane to 1,1-dichloroethane is about 4.5 years, and from 1,1-dichloretfane and trichloroethane ratio, with the knowledge of transformation ratios and transformation kinetics the time period of the presence of pollution by similar compounds can be calculated [95]. Chlorinated solvents remain concealed for long period in surfaces and ground waters as pollutants, in particular in the states where its use is not strictly controlled, as was the case, in mid twentieth century in the developed world [96].

Polychlorinated Biphenyls, PCBs

Polychlorinated biphenyls are synthetic compounds formed by catalytic chlorination of biphenyls at increased temperature and pressure. This way, under controlled conditions 209 congeners can be formed under this name. Their production started in 1929 and lasted uninterrupted, at least in the USA, practically until the 1980s, when they were proclaimed environmental pollutants [97]. They were produced and sold under numerous synonyms, by various manufacturers. Their use was rather widespread: in electricity transformers, additives to the hydraulic oils, as plasticizers, additives in paints, and pesticides industry, among others. Even nowadays in many countries of the Eastern part of the world there are thousands of transformer units containing polychlorinated biphenyls. In the 1960s, when initial research results were released, traces of PCBs could be detected in people and animals around the world—not only in heavily populated areas such as New York City, but also in remote areas as far as the Arctic. These findings of such widespread and persistent contamination contributed to the banning of the chemical in 1979. Because PCBs exist in sediments, scientists need to determine if it is better to dredge and remove contaminated sediments from waterways or if it is safer to leave the sediments in place and cover with clean sediments, allowing them to naturally biodegrade. The methods and instruments for congeners separation and their full identification were developed in the early 1990s [98], more than half a century after the initial use and great damage caused. Congeners ratio, and in particular carbon and chlorine ratio, which in fact is their "fingerprint" are important indicators of the time of formation and origin of polychlorinated biphenyls, from one or several sources [95, 99–101]. These organic synthetic products are currently in the environment or in units at least several decades old. Their "fingerprint" is complex with maximum number of 209 constituents (congener). Thus, the conclusions in the forensic terms may be made only after the careful perceiving of the condition in the field and by comparison of several techniques, inclusive of degradation products, although it concerns stable compounds. Same as halogenated solvents these compounds also demonstrate similar harmful properties in nature.

Trihalomethanes

Trihalomethanes (THMs) are a group of chemicals—chloroform, bromodichloromethane, dibromochloromethane, and bromoform—formed, along with other disinfection byproducts, when chlorine or other disinfectants used to control microbial contaminants in drinking water react with naturally occurring methane, often derived from other forms of organic and inorganic matter in water. These are usually broken up by chlorine or bromine used to disinfect water. THMs are part of a larger class of halogenated organic compounds, including others of water supply health concern such as halogenated acetic acids (HAA). THM are the most common halogenated organics and pose yet another challenge due to its potential presence in the drinking water. Additionally, organic substances present in the natural water (humic and fulvic acids, common macromolecule organic substances as colloids, methane) with chlorine will form a series of halogen based compounds. The presence of other halogenide ions, bromides in particular, will make the processes more complex due to inter-halogen reactions, however, the final results is the presence of halogenated organic compounds, most often methane halogen compounds [102–104]. Sometimes, to solve the issue of potential presence of micro-organisms some of which may be pathogen) due to accident situations (flooding sources, hydric diseases occurrence) the chlorination intensity is greater, namely large quantities of chlorine are added. Halogen organic elements concentrations are then extremely high and hence attention should be drawn to the fact that upon chlorination in regular circumstances water remain in pipes longer, and thus a small quantity of organic substance will produce these compounds. The most important role of the forensic scientist in this case should be determining the potential of THMs formation in the early stage of establishing the site as drinking water resource, and its closer and farther surroundings as well.

Pharmaceuticals

There is a growing concern about the occurrence of pharmaceuticals in water bodies and in drinking water. Pharmaceuticals get into the water supply via human excretion and by drugs being flushed down the toilet. Due to increased health concerns of ingesting returned pharmaceuticals, testing of returned pharmaceuticals in the waters has become a new tool in forensics. Many compounds used as medications possess the characteristics of chemical markers, since they possess characteristic mainly for the urban sewage waters, some are not easily biodegradable or their products are known, they do not evaporate easily, they may even be water soluble. This is further compounded recently by COVID19, as traces of nCOV-SARS2 have been detected in water. The source of pharmaceuticals in water is not just from manufacturing plants, as antibiotics and drugs are used in the livestock industry, and for streams receiving runoff from animal-feeding operations, pharmaceuticals such as acetaminophen, caffeine, cotinine, diphenhydramine, and carbamazepine, have been found. In fact, several compounds from this and other groups meet the requirements of steroids, alkylbenzenes, trialkyl amines, caffeine, antibiotics, and nicotine derivatives. In this regard, identifying characteristic of returned pharmaceuticals is considered as an excellent indicator of anthropogenic pollution, namely sewage networks or mixing with clean natural waters. Well established experiments of their concentrations monitoring for a longer period of time may indicate to the degree of use not only of medications but also narcotics or illegal substances [105, 106].

Micro Plastics

The quantity of plastic mass in the world has been increasing for quite some time. With omnipresence of plastics, the plastics is present almost everywhere. The term micro plastic implies the objects $\leq$ 5 mm of certain forms, or in the form of "microfibers", which is a very frequent case with extremely adverse impact not only on the surface waters, but on the sediment as well [106, 107]. Polyester, nylon, cellulose and acrylic fibers are much prevalent and can be used as the indicators of anthropogenic impact, as the markers contribution to certain water courses pollution, with regards to their form and physical characteristics. These testing can be perceived also as a model system in forensic procedures, since many laboratories and space are already contaminated with this material, therefore exceptional measures are required not to make the mistake originating from the basic contamination. The testing techniques in this case are different, most often based on microscopy (with or without polarized light, phase contrast—phase contrast microscopy (PCM), transmission electron microscopy(TEM), scanning electron microscopy (SEM)), whereas the plastic type is most often determined by infrared spectroscopy. The techniques of determining mineral fibers (asbestos) are similar.

3.2.3 Some Instrumental Techniques in This Field

Testing the presence of organic compounds in water and sediment is a rather difficult forensic issue. First, because concentration of potentially harmful single compounds should be determined and then all compounds in the admixture identified or their products in time under the impact of various factors and their interaction with the environment. This must be done for the admixtures that sometimes contain hundreds of similar compounds. Therefore, only some of widely applied techniques will be listed here. In the beginning it was impossible to have the analytical determining relate only to single compounds, but only to group ones, for example the presence of easily oxidizable organics substance, through consumption of potassium permanganate, more difficult oxidizable substances through chemical oxygen consumption (COD) or biological-microbiological oxidizing compounds (BOD) [108]. These volumetric redox techniques are simple and ingenious that they have been preserved until present, practically unchanged, as basic indicators of organic substance presence. Only on a later date a similar indicator—total organic carbon content (TOC), was added to the list.

The spectrophotometric techniques in the UV/VIS fields are used extensively. In fact, determining the total phenols and phenol compounds is practically colorimetric with excellent sensitivity and has also been retained until present in the majority of larger national standards within this field. Infrared spectroscopy, first without and then with mathematical signals processing, is also a technique for determining total hydrocarbons [109]; similar to it is scanning fluorescence spectroscopy [110]. However, for the development of forensic analytics, it is necessary to develop selective methods which would determine also small concentrations, "traces" of certain compounds. In order to determine them individually, they have to be separated. Thus, begins the development of the chromatographic techniques with apparently simple concept that in the liquid phase the components of a certain admixture are dissolved, then they pass over stationary (solid) (or liquid which is on the solid carrier) phase and those that possess greater affinity stay longer and in the end are eluted again by mobile phase. Depending on the elution phase and method, certain chromatography methods are differentiated, viz. Paper Chromatography, Thin Layer Chromatography,

Gas Chromatography (GC), High-Performance Liquid Chromatography (HPLC) or Gel Chromatography, then Ion Exchange Chromatography (IC), some techniques are based on biological affinity, Bio-Affinity Chromatography, and their numerous modifications [111].

Often it is necessary to prepare the sample prior to analysis, with an aim to perform certain rough separation, or to perform concentration, and most often it is one, other or both. Many physical-chemical techniques are used for preparation: extraction with different solvent, many distillation types, adsorption and various separation and elution on columns, and also some of the stated chromatography methods. Purge and trap techniques significantly reduce detection limits and facilitate the works with smaller volumes of water samples tested. Adsorption on the appropriate carrier, in addition to extraction, has been increasingly applied recently. The development of the stationary phase and the method of its packing have resulted in the emergence of columns that are rather efficient in components separation in many chromatographic techniques, in particular the emergence of capillary columns in gas chromatography. As with this technique volatile compounds are determined, temperature regulation is of particular significance. Parallel to this, is the technology of certain components detection, which has been developed on specific detectors. The first ones are, especially for gas chromatography, non-selective thermal conductivity (TC), and then flame ionization detector (FID), significantly more sensitive also with linear response in respect to concentration, is also practically non-selective. The first good selectivity was achieved with electron capture detector (ECD), and then with nitrogen- phosphorus detector (NPD).

All detectors, with significantly improved characteristics are in use even nowadays. The emergence of mass detectors, very specific ones, represented an enormous progress in environmental forensics. Separated molecules of a certain compounds on columns, under the effect of the beam of particles are fragmented into certain small parts, according to certain fragmentation rules, with defined mass. Depending on mass and the conditions in electromagnetic field in the detectors, they travel a certain path emitting a signal on the path travelled that is always the same for the same mass. Putting the signal fragments together, the molecule is reconstructed. The compound mass specter consists of individual fragments masses, namely their signals. Now chromatography seems simple: for samples input, separation column, detector emitting signal and signal writer, and quite naturally for the mobile phase, as well. The place for samples input can be fully controlled, same as column temperature. The mobile phase flow is under full control, and the columns contain highly selective layers. The detectors have many modifications with an aim of precises mass determination, or by subsequent fragmentation in cycles. Powerful computers have data bases for masses of many fragments and mathematical algorithms serve to connect masses signals and reconstruct many compounds. It is similar with other techniques, in that the HPLC has mostly optical detectors with different wavelengths, short columns bearing mobile phase pressures in thousands kPa, which serve for simultaneous separation of tens and sometimes hundreds of thermally unstable compounds.

The emergence of mass detectors of similar characteristics as in gas chromatography, transforms HPLC into one of the most reliable techniques for determining the medications and biological molecules in waters and in the environment generally.

Another aspect of mass spectrometry, also in organic molecules, is isotopic determining. Gas chromatography/combustion/isotope ratio mass spectrometry (GC/C/IRMS) techniques are based on the isotopic composition of different elements and can differentiate the origin or age of organic compounds [112], only with a microliter or a couple of microliters of prepared samples. As per future, a minimum of two predictions are likely to be realized. One of them in the near future is more intensive merging of separating techniques and the techniques of destructive or non-destructive detection under the supervision of the computers with huge databases of compounds characteristics. This will make a true hybrid analytics or data analytics for forensics. This will also result in great selectivity or possibility of confirming the results in the automated regime, as well as significant increase of the number of simultaneously determined parameters, accompanied by numerous and inconceivable implications. Secondly, significantly more difficult in a long time interval will be the dream of every analysts—the atom/molecule as the limit of chemical analysis. To paraphrase Robert Boyle (circa 1627-169) *"element as chemical analysis limit"*, namely, we are still analytically blind, both as regards to elements and also as regards to organic molecules. Realizing that the lower determination limit of the AAS CV or ICP OES Hydride technique can be ng/kg and the pollutant Hg (with relative atomic mass of 200.592), we can observe not less than 3×10^{12} atoms of Hg/kg. Similarly, realizing that the lower determination limit of the GC ECD technique can be μg/kg and the pollutant of certain PCBs (3,4,4′,5-Tetrachlorobiphenyl, $C_{12}H_6Cl_4$) (relative molecular mass of 292), we can observe not less than 2×10^{15} molecules per kg! The more we lower the determination limits, the greater achievement is created, and increasingly more rigorous conditions are required. Yet, we are quite far away from knowing what number of atoms and molecules can cause damage to the environment, most possibly significantly less than we can perceive now!

4 Conclusions and Future Directions

This chapter aims to raise awareness of environmental forensics with the objective that it becomes as a standalone course, due to its significance and importance. The book in its entirety discusses safety, security and sustainability of water and this chapter adds yet another dimension to support the three pillars as envisaged in the book. Scientific papers are usually strictly defined/restricted and are driven by the results to be interpreted in a conclusion to be drawn. Because of the time and current lifestyle, sometimes only the conclusion is read. It is generally deemed that all foretasted text is clear. A book is written to incite contemplations, and thus it never represents the final as conclusive part, but the beginning or the contents inside. Further, the chapters contain a broader overview of the issues discussed. The authors of the chapters are entitled not to be overly precise, or even to fantasize, but the goal is to provide an overview to the readers. To express the authors opinions, even though imprecise, and not always to hide themselves behind the quotations, a compromise must exist. Have the authors succeeded in that? We doubt it, however if at least one environmental chemist undertakes the hard and ungrateful job of a forensic scientist, then our intention has scored the point successfully. As future direction, we list below a course that may be adopted for future Environmental Forensics. This course will introduce students to the principles of environmental forensic science. The students will learn about the fate and transport of chemicals in the environment. The environmental "crime scene," where pollution first occurs, is often far away from the area most impacted by the pollution. Along the way, pollutants undergo changes as they interact with the environment. Linking a pollutant to its source involves understanding and quantifying these changes. The course may be divided into several sections, viz: what is Environmental Forensic Science?; environmental pollutants; toxicity of environmental contaminants; oil contamination; hydrocarbon finger-printing; fate of chemicals; environmental partitioning; biological transformations; contaminant transport by groundwater; environmental transport models; groundwater transport; transport of atmospheric pollutants; and atmospheric dispersion.

References

1. Gaćeša S, Klašnja M (1994) Water and wastewater technology. Yugoslav Brewers Association, Belgrade
2. Harrison RM (ed.) (2001) Pollution: causes, effects and control. Royal society of chemistry
3. Horne RA et al. (1977) Chemistry of our environment. Wiley
4. Veselinović DS et al (1995) Conditions and processes in the environment. University of Belgrade, Faculty of Physical Chemistry
5. Meydenbauer A (1867) Die Photometrographie. Wochenblatt des Architektenvereins zu Berlin 1(14):125–126
6. Brilis GM, Gerlach CL, van Waasbergen RJ (2000) Remote sensing tools assist in environmental forensics. Part I: traditional methods. Environ Forensics 1(2):63–67
7. Brilis GM et al (2001) Remote sensing tools assist in environmental forensics: Part II—Digital tools. Environ Forensics 2(3):223–229
8. Falbe AJ, Regitz M (1997) Römpp Lexikon Chemie. Neubearbeitete Auflage 10
9. Duffus JH (2002) "Heavy metals" a meaningless term? (IUPAC Technical Report). Pure Appl Chem 74(5):793–807
10. Goyer R (2004) Issue paper on the human health effects of metals. US Environmental Protection Agency
11. Kuivenhoven M, Mason K (2019) Arsenic (Arsine) Toxicity. In: Stat Pearls [Internet]. StatPearls Publishing
12. Sun H, Brocato J, Costa M (2015) Oral chromium exposure and toxicity. Curr Environ Health Rep 2(3):295–303
13. Verma A (2018) Forensic aspect of metal poisoning: a review. Int J Res Appl Sci Eng Technol (IJRASET) 6(I)
14. Brčeski I (2010) Faculty of chemistry. University of Belgrade, unpublished results
15. Hazrat A, Ezzat K, Ikram I (2019) Environmental chemistry and ecotoxicology of hazardous heavy metals: environmental persistence, toxicity, and bioaccumulation. J Chem 6730305. https://doi.org/10.1155/2019/6730305

16. Adesanya OO et al. (2020) Source identification and human health risk assessment of heavy metals in water sources around bitumen field in Ondo State, Nigeria. Environ Forensics 1–12
17. Woldeamanuale T, Hassen A (2017) Toxicity study of heavy metals pollutants and physicochemical characterization of effluents collected from different paint industries in addis Ababa, Ethiopia. J Forensic Sci Criminal Invest 5(5), 001
18. Sankhla MS et al (2016) Heavy metals contamination in water and their hazardous effect on human health-a review. Int J Curr Microbiol App Sci 5(10):759–766
19. Yunus K, Zuraidah MA, John A (2020) A review on the accumulation of heavy metals in coastal sediment of Peninsular Malaysia. Ecofeminism Clim Change
20. Alves RH, Rietzler AC (2015) Ecotoxicological evaluation of sediments applied to environmental forensic investigation. Brazil J Biol 75(4):886–893
21. Sakan S et al (2002) Water quality parameters of the Tisza River. J Environ Protect Ecol 3(4):828–33
22. Dimkić M, Keckarević D, Brčeski I, Pavlović D, Papić P, Duduković A, Vrvić MM (1990) The study of degradation of high concentration of pollutants in water on a natural model. Advanced Programme of the Second International ISEP Congress, Vienna (Austria), p 16
23. Pfendt P Faculty of chemistry. University of Belgrade, unpublished results
24. Meybeck M (2013) Heavy metal contamination in rivers across the globe: an indicator of complex interactions between societies and catchments. In: Proceedings f H04 understanding freshwater quality problems in a changing world 361: 3–16
25. Miller JR (2013) Forensic assessment of metal contaminated rivers in the 21st century using geochemical and isotopic tracers. Minerals 3(2):192–246
26. Mao G et al (2019) Spatiotemporal variability of heavy metals and identification of potential source tracers in the surface water of the Lhasa River basin. Environ Sci Pollut Res 26(8):7442–7452
27. Macklin MG, Klimek K (1992) Dispersal, storage and transformation of metal contaminated alluvium in the upper Vistula basin, southwest Poland. Appl Geogr 12(1):7–30
28. Wiederhold JG (2015) Metal stable isotope signatures as tracers in environmental geochemistry. Environ Sci Technol 49(5):2606–2624
29. Kelemen Z, Gillikin DP, Bouillon S (2019) Relationship between river water chemistry and shell chemistry of two tropical African freshwater bivalve species. Chem Geol 526:130–141
30. Morana C et al (2015) Biogeochemistry of a large and deep tropical lake (Lake Kivu, East Africa: insights from a stable isotope study covering an annual cycle. Biogeosciences 12(16):4953–4963
31. Kelemen Z et al (2017) Calibration of hydroclimate proxies in freshwater bivalve shells from Central and West Africa. Geochim Cosmochim Acta 208:41–62
32. Bird G (2011) Provenancing anthropogenic Pb within the fluvial environment: developments and challenges in the use of Pb isotopes. Environ Int 37(4):802–819
33. Passmore DG, Macklin MG (1994) Provenance of fine-grained alluvium and late Holocene land-use change in the Tyne basin, northern England. Geomorphology 9(2):127–142
34. Dunlap CE et al (2008) The persistence of lead from past gasoline emissions and mining drainage in a large riparian system: evidence from lead isotopes in the Sacramento River, California. Geochimica et Cosmochimica Acta 72(24):5935–5948
35. Leybourne MI, Cousens BL, Goodfellow WD (2009) Lead isotopes in ground and surface waters: fingerprinting heavy metal sources in mineral exploration. Geochem Explor Environ Anal 9(2):115–123
36. Kimball BE et al (2009) Copper isotope fractionation in acid mine drainage. Geochim Cosmochim Acta 73(5):1247–1263
37. Bullen TD, Walczyk T (2009) Environmental and biomedical applications of natural metal stable isotope variations. Elements 5(6):381–385
38. Weiss DJ, Rehkdmper M, Schoenberg R, McLaughlin M, Kirby J, Campbell PG, Arnold T, Chapman J, Peel K, Gioia AS (2008) Application of nontraditional stable-isotope systems to the study of sources and fate of metals in the environment. Environ Sci Technol 42(3): 655–664

39. Raddatz AL, Johnson TM, Mcling TL (2011) Cr stable isotopes in Snake River Plain aquifer groundwater: evidence for natural reduction of dissolved Cr (VI). Environ Sci Technol 45(2):502–507
40. Yin R, Feng X, Shi W (2010) Application of the stable-isotope system to the study of sources and fate of Hg in the environment: a review. Appl Geochem 25(10):1467–1477
41. Sonke JE et al (2010) Sedimentary mercury stable isotope records of atmospheric and riverine pollution from two major European heavy metal refineries. Chem Geol 279(3-4):90–100
42. Sherman LS, Blum JD (2013) Mercury stable isotopes in sediments and largemouth bass from Florida lakes, USA. Sci Total Environ 448:163–175
43. Banner JL (2004) Radiogenic isotopes: systematics and applications to earth surface processes and chemical stratigraphy. Earth Sci Rev 65(3-4):141–194
44. Nakano T (2016) Potential uses of stable isotope ratios of Sr, Nd, and Pb in geological materials for environmental studies. Proc Japan Acad Ser B 92(6):167–184
45. McArthur JM (1994) Recent trends in strontium isotope stratigraphy. Terra Nova 6(4):331–358
46. Négrel P, Petelet-Giraud E, Widory D (2004) Strontium isotope geochemistry of alluvial groundwater: a tracer for groundwater resources characterization. Hydrol Earth Syst Sci 8(5):959–972
47. Montgomery J, Evans JA, Wildman G (2006) 87Sr/86Sr isotope composition of bottled British mineral waters for environmental and forensic purposes. Appl Geochem 21(10):1626–1634
48. Feng H, Cochran JK, Hirschberg DJ (1999) 234Th and 7Be as tracers for transport and sources of particle-associated contaminants in the Hudson River Estuary. Sci Total Environ 237:401–418
49. Belyanin D et al. (2019) Sources and accumulation of ^{7}Be, ^{210}Pb and ^{137}Cs isotopes in the annual needles of larch and cedar in Novy Urengoy region (Arctic part of Western Siberia). In: E3S web of conferences. EDP Sciences, p 12002
50. Piepgras DJ, Wasserburg GJ (1980) Neodymium isotopic variations in seawater. Earth Planetary Sci Letters 50(1):128–138
51. Piepgras DJ, Wasserburg GJ, Dasch EJ (1979) The isotopic composition of Nd in different ocean masses. Earth Planetary Sci Letters 45(2):223–236
52. Vervoort JD et al. (1999) Relationships between Lu–Hf and Sm–Nd isotopic systems in the global sedimentary system. Earth Planetary Sci Lett 168(1–2):79–99
53. Colodner D, Sachs J, Ravizza G, Turekian K, Edmond J, Boyle E (1993) The geochemical cycle of rhenium: a reconnaissance. Earth Planetary Sci Letters 117(1–2):205–221
54. Levasseur S, Birck J-L, Allègre CJ (1998) Direct measurement of femtomoles of osmium and the ^{187}Os/^{186}Os ratio in seawater. Science 282(5387):272–274
55. Douthitt CB (2008) The evolution and applications of multicollector ICPMS (MC-ICPMS). Anal Bioanal Chem 390(2):437–440
56. Halliday AN et al. (1998) Applications of multiple collector-ICPMS to cosmochemistry, geochemistry, and paleoceanography. Geochimica et Cosmochimica Acta 62(6):919–940
57. Baxter DC, Rodushkin I, Engström E (2012) Isotope abundance ratio measurements by inductively coupled plasma-sector field mass spectrometry. J Anal At Spectrom 27(9):1355–1381
58. Fantle MS, Bullen TD (2009) Essentials of iron, chromium, and calcium isotope analysis of natural materials by thermal ionization mass spectrometry. Chem Geol 258(1–2):50–64
59. Criss RE, Fernandes SA, Winston WE (2001) Isotopic, geochemical and biological tracing of the source of an impacted karst spring, Weldon Spring, Missouri. Environ Forensics 2(1):99–103
60. Winograd IJ, Friedman I (1972) Deuterium as a tracer of regional ground-water flow, southern Great Basin, Nevada and California. Geol Soc Am Bull 83(12):3691–3708
61. Kennedy SK, Walker W, Forslund B (2002) Speciation and characterization of heavy metal-contaminated soils using computer-controlled scanning electron microscopy. Environ Forensics 3(2):131–143
62. Glascock MD (2006) An overview of neutron activation analysis. University of Missouri Research Reactor (MURR), Columbia, MO

63. Zemo DA, Bruya JE, Graf TE (1995) The application of petroleum hydrocarbon fingerprint characterization in site investigation and remediation. Groundwater Monitor Remediation 15(2):147–156
64. Morrison RD (2000) Critical review of environmental forensic techniques: Part II. Environ Forensics 1(4):175–195
65. Blumer M, Souza GT, Sass J (1970) Hydrocarbon pollution of edible shellfish by an oil spill. Mar Biol 5(3):195–202
66. Zakaria MP, et al. (2000) Oil pollution in the Straits of Malacca, Malaysia: application of molecular markers for source identification. Environ Sci Technol 34(7):1189–1196
67. Kaplan IR, Galperin Y, Lu ST, Lee RP (1997) Forensic environmental geochemistry: differentiation of fuel-types, their sources and release time. Org Geochem 27(5–6):289–317
68. BrasselL SC et al (1981) Specific acyclic isoprenoids as biological markers of methanogenic bacteria in marine sediments. Nature 290(5808):693–696
69. Prince RC, et al. (1994) 17. alpha.(H)-21. beta.(H)-hopane as a conserved internal marker for estimating the biodegradation of crude oil. Environ Sci Technol 28(1):142–145
70. Wang Z, Fingas M, Sergy G (1994) Study of 22-year-old Arrow oil samples using biomarker compounds by GC/MS. Environ Sci Technol 28(9):1733–1746
71. Odermatt JR (1994) Natural chromatographic separation of benzene, toluene, ethylbenzene and xylenes (BTEX compounds) in a gasoline contaminated ground water aquifer. Org Geochem 21(10-11):1141–1150
72. Hurst RW, Davis TE, Chinn BD (1996) Peer reviewed: the lead fingerprints of gasoline contamination. Environ Sci Technol 30(7):304A–307A
73. Zayed J, Hong B (1999) L'espérance, Gilles. Characterization of manganese-containing particles collected from the exhaust emissions of automobiles running with MMT additive. Environ Sci Technol 33(19):3341–3346
74. Youngless TL et al. (1985) Mass spectral characterization of petroleum dyes, tracers, and additives. Anal Chem 57(9):1894–1902
75. Suflita JM, Mormile MR (1993) Anaerobic biodegradation of known and potential gasoline oxygenates in the terrestrial subsurface. Environ Sci Technol 27(5):976–978
76. Steffan RJ et al (1997) Biodegradation of the gasoline oxygenates methyl tert-butyl ether, ethyl tert-butyl ether, and tert-amyl methyl ether by propane-oxidizing bacteria. Appl Environ Microbiol 63(11):4216–4222
77. Pankow JF et al. (1997) The urban atmosphere as a non-point source for the transport of MTBE and other volatile organic compounds (VOCs) to shallow groundwater. Environ Sci Technol 31(10):2821–2828
78. Davidson JM, Creek DN (2000) Using the gasoline additive MTBE in forensic environmental investigations. Environ Forensics 1(1):31–36
79. Williams P (2001) MTBE in California drinking water: an analysis of patterns and trends. Environ Forensics 2(1):75–85
80. Rong Yue (2001) The MTBE paradox of groundwater pollution commentaries & perspectives. Environ Forensics 2(1):9–11
81. Stout SA et al (2015) Beyond 16 priority pollutant PAHs: a review of PACs used in environmental forensic chemistry. Polycyclic Aromat Compd 35(2-4):285–315
82. Douglas GS et al (1996) Environmental stability of selected petroleum hydrocarbon source and weathering ratios. Environ Sci Technol 30(7):2332–2339
83. Jewett SC et al (1999) 'Exxon Valdez' oil spill: impacts and recovery in the soft-bottom benthic community in and adjacent to eelgrass beds. Mar Ecol Prog Ser 185:59–83
84. Burns WA et al (1997) A principal-component and least-squares method for allocating polycyclic aromatic hydrocarbons in sediment to multiple sources. Environ Toxicol Chem Int J 16(6):1119–1131
85. Blumer Max (1976) Polycyclic aromatic compounds in nature. Sci Am 234(3):34–45
86. Youngblood WW, Blumer M (1975) Polycyclic aromatic hydrocarbons in the environment: homologous series in soils and recent marine sediments. Geochim Cosmochim Acta 39(9):1303–1314

87. Laflamme RE, Hites RA (1978) The global distribution of polycyclic aromatic hydrocarbons in recent sediments. Geochim Cosmochim Acta 42(3):289–303
88. Müller G, Grimmer G, Böhnke H (1977) Sedimentary record of heavy metals and polycyclic aromatic hydrocarbons in Lake Constance. Naturwissenschaften 64(8):427–431
89. Stout SA, Uhler AD (2003) Distinguishing, "background" hydrocarbons from contamination using chemical fingerprinting. Environ Claims J 15(2):241–259
90. Stout SA, Uhler AD, Emsbo-Mattingly SD (2004) Comparative evaluation of background anthropogenic hydrocarbons in surficial sediments from nine urban waterways. Environ Sci Technol 38(11):2987–2994
91. Pankow JF, Cherry JA (1996) Dense chlorinated solvents and other DNAPLs in groundwater: history, behavior, and remediation
92. Jackson RE, Dwarakanath V (1999) Chlorinated decreasing solvents: physical-chemical properties affecting aquifer contamination and remediation. Groundwater Monitor Remediation 19(4):102–110
93. Archer WL, Stevens VL (1977) Comparison of chlorinated, aliphatic, aromatic, and oxygenated hydrocarbons as solvents. Indus Eng Chem Prod Res Develop 16(4):319–325
94. Archer WL (1984) A laboratory evaluation of 1, 1, 1-trichloroethane–metal-inhibitor systems. Mater Corros 35(2):60–69
95. Morrison RD (2000) Critical review of environmental forensic techniques: Part I. Environ Forensics 1(4):157–173
96. Amter S, Ross B (2001) Was contamination of southern California groundwater by chlorinated solvents foreseen? Environ Forensics 2(3):179–184
97. Kannan K, Maruya KA, Tanabe S (1997) Distribution and characterization of polychlorinated biphenyl congeners in soil and sediments from a superfund site contaminated with Aroclor 1268. Environ Sci Technol 31(5):1483–1488
98. Wait AD (2000) Evolution of organic analytical methods in environmental forensic chemistry. Environ Forensics 1(1):37–46
99. Watanabe S et al (1996) Concentrations and composition of PCB congeners in the air around stored used capacitors containing PCB insulator oil in a suburb of Bangkok, Thailand. Environ Pollut 92(3):289–297
100. Jarman WM et al (1997) Levels and patterns of polychlorinated biphenyls in water collected from the San Francisco Bay and Estuary, (1993–95). Fresenius J Anal Chem 359(3):254–260
101. Johnson GW et al (2000) Resolving polychlorinated biphenyl source fingerprints in suspended particulate matter of San Francisco Bay. Environ Sci Technol 34(4):552–559
102. Rook JJ (1976) Haloforms in drinking water. J Am Water Works Assoc 68(3):168–172
103. Rook JJ (1977) Chlorination reactions of fulvic acids in natural waters. Environ Sci Technol 11(5):478–482
104. Kitis M et al (2010) Occurrence of trihalomethanes in chlorinated groundwaters with very low natural organic matter and bromide concentrations. Environ Forensics 11(3):264–274
105. Tidy CM (1879) XII—The processes for determining the organic purity of potable waters. J Chem Soc Trans 35:46–106
106. Douglas GS et al (1992) The use of hydrocarbon analyses for environmental assessment and remediation. Soil Sediment Contam 1(3):197–216
107. Pharr DY, Mckenzie JK, Hickman AB (1992) Fingerprinting petroleum contamination using synchronous scanning fluorescence spectroscopy. Groundwater 30(4):484–489
108. Bhoj Y et al. (2020) Chromatographic Techniques for Forensic Investigations. Technol Forensic Sci Sampling, Anal Data Regul 129–149
109. Dempster HS, Sherwood LB, Feenstra S (1997) Tracing organic contaminants in groundwater: a new methodology using compound-specific isotopic analysis. Environ Sci Technol 31(11):3193–3197
110. Kasprzyk-Hordern B, Dinsdale RM, Guwy AJ (2009) Illicit drugs and pharmaceuticals in the environment–forensic applications of environmental data. Part 1: estimation of the usage of drugs in local communities. Environ Pollut 157(6):1773–1777

111. Kasprzyk-H B, Dinsdale RM, Guwy AJ (2009) Illicit drugs and pharmaceuticals in the environment–Forensic applications of environmental data, Part 2: pharmaceuticals as chemical markers of faecal water contamination. Environ Pollut 157(6):1778–1786
112. Browne MA et al (2011) Accumulation of microplastic on shorelines worldwide: sources and sinks. Environ Sci Technol 45(21):9175–9179
113. Do Sul JAI, Costa MF (2014) The present and future of microplastic pollution in the marine environment. Environ Pollut 185:352–364
114. Woodall LC et al (2015) Using a forensic science approach to minimize environmental contamination and to identify microfibres in marine sediments. Mar Pollut Bull 95(1):40–46

Ilija Brčeski Prof. Dr. was born in 1962 in North Macedonia. He has graduated from the Faculty of Chemistry of the University of Belgrade on the Department of Biochemistry. He has worked as a researcher at the Chemistry, Technology and Metallurgy Institute in Belgrade. He has been working at the Faculty of Chemistry since 1988, currently with the title of Associate Professor. He teaches the subjects Inorganic Chemistry, Environmental Biotechnology, Natural Resources, as well as Inorganic Syntheses on the PhD chemistry studies. His scientific research activity is in the areas of inorganic, coordination and bioinorganic chemistry, as well as the Eco-chemical characterisation of the environment and natural resources. He is the author of the university textbook from the area of rare earth chemistry, numerous scientific works in international publications, lectures by invitation, etc. He is a co-editor of the publication "Journal of Environmental Protection and Ecology", published by Scientific Bulgarian Communications, as well as a member of the editorial board of the publication Archives of Public Health, published by the Institute for Public Health of the Republic of North Macedonia. He is a member of the European Academy of Sciences and Arts, Salzburg, Austria and the Serbian Chemical Society.

Ashok Vaseashta (M'79-SM'90) received Ph.D. in Materials Science and Engineering (minor in Electrical Engineering) from Virginia Polytechnic Institute and State University, Blacksburg, Virginia, USA. He is Executive Director of Research for International Clean Water Institute in VA, USA, Chaired Professor of Nanotechnology at the Academy of Sciences of Moldova and Professor, Nanotechnology and Biomedical Engineering at the Faculty of Mechanical Engineering, Transport and Aeronautics at the Riga Technical University. Prior to his current position, he served as Vice Provost for Research at the Molecular Science Research Center in Orangeburg, South Carolina. He served as visiting professor at the 3 Nano-SAE Research Centre, University of Bucharest, Romania and visiting scientist at the Helen and Martin Kimmel Center of Nanoscale Science at the Weizmann Institute of Science, Israel. He served the U.S. Department of State in two rotations, as strategic S&T advisor and U.S. diplomat. His research interests span nanotechnology, environmental/ecological science, and safety and security. His research on nanotechnology has been on improving the understanding, design, and performance of nanofibers and sensors/detectors, mainly for applications such as wearable electronics, target drug delivery, detection of biomarkers and toxicity of nano and xenobiotic materials. In the security arena, he has worked on counterterrorism, countering unconventional warfare and hybrid threats, critical-Infrastructure protection, biosecurity, dual-use research concerns, and mitigating hybrid threats. He has authored over 250 research publications, edited/authored eight books on nanotechnology, and presented many keynotes and invited lectures worldwide. He serves on the editorial board of several highly reputed international journals. He is an active member of several national and international professional organizations. He is a fellow of the American Physical Society, Institute of Nanotechnology, and the New York Academy of Sciences. He has earned several other fellowships and awards for his meritorious service including 2004/2005 Distinguished Artist and Scholar award.

Security Standards Applied to Drinking Water

Maria Popa and Ioana Glevitzky

Abstract This chapter presents security standards as applied to drinking water, based on the assumption that freshwater sources face huge threats for humankind both qualitatively and quantitatively due to uncertain global changes, namely population growth, rapid urbanization, and climate change. The study investigates whether the International Featured Standards (IFS—Food), which are complementary to HACCP (Hazard analysis and critical control point), can be used as a system of indicators to measure water security. The case study focuses on bottled water, by identifying structural and contingent risk factors for water as essential food and investigates the suitability of the IFS instrument as an indicator of food security. Given the variables of the food safety indicators, we consider research results which reflect the accessibility level of drinking water resources, the time in which they become unappealing and the risks that may arise in the bottled water production flow. According to the IFS principles, the security of the entire technical flow, for each danger (contaminants, acts of sabotage, vandalism, bioterrorism, and vulnerability to food fraud), and preventive controls associated at each stage of the production process needs to be ensured.

Keywords Security · Standards · Drinking water · Policy

1 Introduction: The Importance of Water for Humankind

According to the Drinking Water Directive [9], *"water is not a commercial good but patrimony to be protected, defended and treated as such"*. Water intended for human consumption means, all water either in its original state or after treatment, intended for drinking, cooking, food preparation or other domestic purposes, regardless of its

M. Popa (✉)
Faculty of Economic Sciences "1 Decembrie 1918", University of Alba Iulia, Gabriel Bethlen St., 5, 510009 Alba Iulia, Romania

I. Glevitzky
Doctoral School of the "Lucian Blaga", University of Sibiu, Victoriei Blvd, 10, 550024 Sibiu, Romania

origin [7]. Water is an important element for the existence of life and the development of human settlements. The first life forms ascended in the aquatic environment. The first human dwellings were established near rivers to ensure water for drinking and other household needs. Water is the environment in which all metabolic processes take place, all living creatures' tissues and organs contain water. All these prove the vital importance of water in the birth and preservation of life and the development of humankind over time [32].

The United Nations General Assembly declared the period 2005–2015 the International Decade "Water for Life" to raise knowledge which would lead to real actions [36]. World Water Day advocates "clean water for a healthy world." More than 2.5 billion people live without adequate sewer systems, 884 million people, mostly in Africa, have no access to drinking water, 1.5 million children under 5 die each year from water-borne diseases [2]. It is estimated that most of the diseases affecting humankind today find their source in poor water quality (P.S.E. Water policy in a new style [31]). According to a study within the United Nations Environment Program, 4 out of 5 common diseases in developing countries are caused by dirty water or by lack of sanitary facilities [16]. Therefore, drinking water is essential to ensure public health and quality of life. At present, drinking water sources—rivers, lakes and deep water are threatened by various contaminants [4].

2 The Status of Water Quantity and Quality at Global Level

Fresh water, representing only 2.70% out of the total water quantity on Earth, is found underground (22.40%), in ice and glaciers (77.24%), in surface fresh water sources (0.36%), such as rivers and lakes [24]. However, rivers and lakes are the main sources of water for daily use [34]. In an industrialized country, about 80% of the used water returns somewhat impurified into the natural circuit as wastewater. Daily, two million tons of wastewater is discharged in the world, and not all cities have purification stations consistent and in compliance with international or European requirements [36]. The average natural water reserve in many countries is estimated to be about 1700 m^3/inhabitant/year, compared to approx. 4200 m^3/inhabitant/year in developed countries [6].

Water has been considered an inexhaustible gift of nature, but nowadays water crisis emerges, especially at regional levels. Viktor Danilov-Danilyan, a professor at the Russian Academy of Sciences, believes that around 2020–2025, the water crisis will be felt globally [20]. Water is renewable within certain limits; and while water requirements and quality standards are increasing, the period between two uses is getting shorter [8]. Natural dilution was possible in the 60s–70s, but nowadays the dilution at river level is insufficient due to the large quantity of pollutants. The principle of "stopping pollution at source" is necessary to be applied to ensure the increase of water safety (P.S.E. Water policy in a new style 2001).

The tendency to deplete water resources is caused by extensive agricultural and industrial consumptions and by water quality depreciation. Its careless and wasteful use, coupled with the poisoning of the unique reservoirs puts this resource to risk, which nowadays is a global phenomenon [26]. Degradation of water quality in rivers, lakes and underground waters has a direct impact on ecosystems and human health [35]. It is an indescribable tragedy and a major obstacle to global development.

Water quality monitoring is a basic fundamental step in any water management program. Water resource management requires information on the following [27]:

- the surface and groundwater quality conditions at national level;
- the continuous monitoring of quality conditions;
- major issues related to water quality and the causes of their occurrence;
- programs that work effectively to prevent or remedy problems;
- compliance with quality standards and objectives.

Water is necessary for all life forms and for product manufacturing; it is one of the basic elements in varied domains. Water is a social requirement [5]; it is an essential element for maintaining and developing life on our planet. Water is the most important food that cannot be replaced. In extreme situations humans can manage without water in some uses, but not when it comes to drinking. Humans can live without food for a long time, but not without water and despite water is found in various foods, humans need water in liquid form [30].

3 Water Security Issues

According to the Food and Agriculture Organization (FAO) and World Health Organization (WHO)—Roma 1992 World Food Declaration and the FAO/WHO/1996 World Food Security Declaration, *"Food security exists when all people at all times have physical and economic access to safe and nutritious foods that meet food needs of the human body to lead a healthy and active life."* Recently climate change affected The Earth leading to great losses each year caused by the effects of global warming. Massive water drops cause extensive damage to households and agricultural land. In this context, authorities gave absolute priority to the management and security of water resources to protect their citizens. The World Resource Institute monitors and provides data on water scarcity [17]. We can identify the regions facing the so-called "water scarcity" (Fig. 1).

The concept of food security emerged in the 1970s, evolved from quantitative and economic considerations to a definition that includes food quality and human dimension [1]. The concept developed and its current established status encompasses aspects regarding population growth mainly in underdeveloped and developing countries, rapid urbanization in developed countries, increase in food commodity prices, and decline of product quality. "Basically, water security is the foundation for human and environmental security" (Fig. 2).

In the current ecological issues related to water scarcity, the research outlines the current situation regarding drinking water safety at global level. If public drinking water systems use only surface waters (lakes, rivers), groundwater (spring, wells,

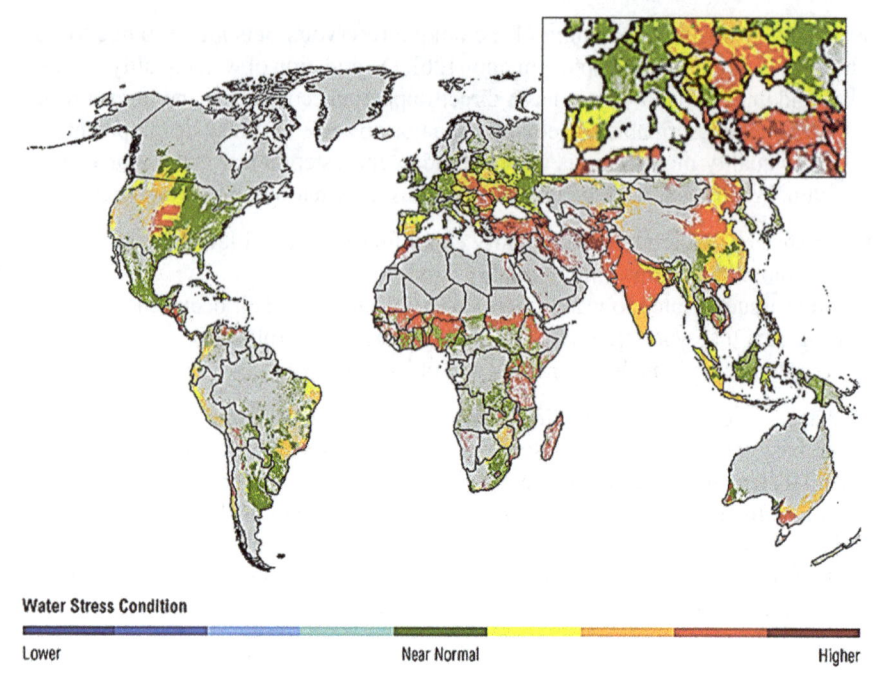

Fig. 1 Areas affected by water stress Condition (*source* World Resource Institute)

Fig. 2 The link between water, food; water and the environment [23]

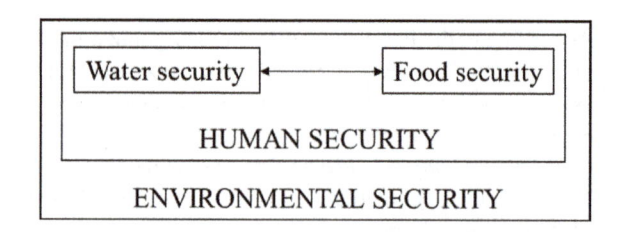

bore-whole) is an alternative little explored and exploited despite being a natural, untreated free source. Water security through the identification of possible frauds or even acts of terrorism is a concern for companies and for state control bodies regardless the source of water, its capture and supply to the population.

The current trend in quality, safety and security standards indicates the need for permanent monitoring of food safety and security to identify possible contaminants, including those induced by sabotage, vandalism, terrorism/bioterrorism, and fraud. The implementation and certification of an efficient water management system for water processing safety (such as International Featured Standards (IFS), British Retail Consortium (BRC), Food Safety System Certification (FSSC), etc.) ensures optimal monitoring and control conditions in water treatment and prevents water contamination risks. Studies assess the security vulnerability of organizations regarding location, staff, visitors, food fraud, deliberate evasion or contamination, and the ability to protect against acts of sabotage, vandalism, and terrorism/bioterrorism.

4 History and Main Types of Food Frauds

There are various methods for food fraud: substitution of products (the substitution of bottled water—table water declared as mineral or spring water), introduction of prohibited substances into food and animal feed (melanin crisis—2008 [15], dioxin crisis—1999 [21]), banned products of animal origin sold as food for human consumption (eggs from incubation), document forgery (horse meat crisis—2013 [3]), substitution and sale of horse meat containing phenylbutazone), dilution with liquids of lower value (olive oil—1981), change of geographical origin, replacement of organic products with conventional ones, false labelling or omission in labels, reporting the use of false research methods, use of prohibited substances in animal feed on farms (1970 EU growth hormones), changing the destination of a product animal by—products not intended for human consumption—chicken 2000 Rotterdam).

Prior to dealing with food fraud, food safety was considered a legislative priority. At international and European levels, there was no network for information exchange and management of fraud issues. In 2017, at EU level, the most common method of fraud was the omission of information on par with substitution, addition, replacement, followed by forgery and unapproved treatments mentioned on labels.

According to European legislation [10], in 2020, EU Member States elaborated and implement strategies to prevent and combat fraudulent practices in the food sector. In line with the newly adopted legislation, controls were established to identify misleading or fraudulent practices and food fraud related to food imported into the EU. Sanctions against food fraud were also established according to legislation. Requirements and guarantees were enforced regarding the issue of official certificates to prevent fraud. EU Centres were appointed to provide support to Member States in preventing and identifying food fraud and in securing administrative assistance and cooperation when crossing borders.

There is no precise definition of Food Fraud in the EU food safety legislation, except: "any suspected intentional action by businesses or individuals for the purpose of deceiving purchasers and gaining undue advantage therefrom, in violation of the rules referred to in Article 1(2) of Regulation (EU) 2017/625" [18], as shown in Fig. 3. The definitions differ from country to country depending on the internal organization. It is defined as food fraud if one or more or more of the following elements are proven:

- A breach of national and/or EU legislation on food safety
- Deliberate action of any person in the food chain
- An action meant to induce/mislead/deceive the consumer/beneficiary without any agreement.

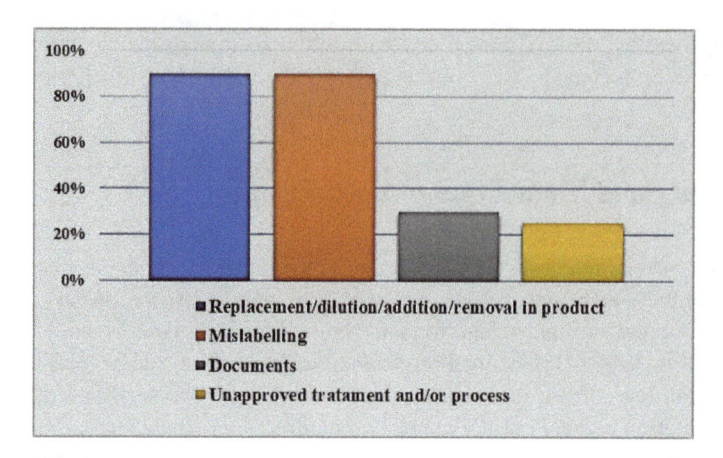

Fig. 3 Distribution of potential food fraud

- The pursuit of economic/financial profit that would not have been achieved without engaging in actions under (a), (b) and (c).

Food fraud is the combined action of three elements: opportunity, motivation, and control measures [33]. The most important element at the basis of food fraud is intention. An intentional deed is when the concerned operator/person foresees or expects a result benefits from the result of its deed. The intent is proven the hardest, because it requires a complex approach and must be accompanied by evidence. Negligence, as an act of unintentional action, can lead to food fraud by failing to comply with the imposed legal norms; the same holds true for persistence in identifying the same nonconformity over time and showing no intention of acting. Discussions are increasingly taking place on global level regarding the classification and differential approach of the two principles of food fraud and terrorism/food crime [22] (Fig. 4).

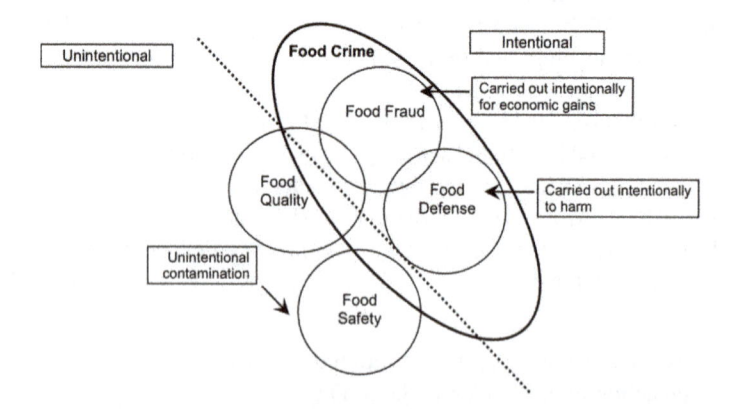

Fig. 4 Intentional vs. Unintentional Adulteration [11–13, 25]

To distinguish between the two aspects, one can ascertain that the scope of food fraud is direct economic profit. In the case of terrorism or "food crime", the aim is no longer to make economic profit; it is defined as the voluntary action of placing food on the market which contain toxic/harmful substances/compounds for consumer safety to induce panic, fear, or terrorism-like crises etc. [28].

5 Implementation of Management Systems (Standards) for Water Security

Application of modern food safety tools—Specific Standards—facilitates the investigation of the potential risks and the assessment of the contamination probability of water which was captured, treated and delivered to the population, within the drinking limit set by legislation. This standard can be applied to bottled water and for water from supply systems. The implementation and certification of an efficient water management system ensures optimal monitorization and control conditions for the treatment/process of bottling and transport of water while preventing contamination risks. The standards assess the companies' vulnerability in terms of location, staff and visitor security, food fraud, deliberate evasion or contamination, and the ability to protect against acts of sabotage, vandalism and terrorism/bioterrorism.

Numerous countries, including the EU sought to create instruments capable of implementing food safety and security objectives. At present, standards such as IFS (International Food Standard), FSSC 22000 (Food Safety System Certification), BRC (British Retail Consortium Food Standard), etc. are quality, safety especially, food safety standards which are designed to allow the assessment of food providers' systems in a unitary approach. These standards apply to all stages of water treatment starting at capture and they are recognized by the Global Food Safety Initiative (GFSI).

International Food Standard (IFS) is a standard applicable to food manufacturers, but also to businesses that transport, store, distribute/market food products. Its latest version of the IFS Food 6.1 standard was developed for food manufacturers and is used to audit food manufacturers regarding food safety and quality of processes and products. The list of requirements is organized in the following topics:

- Senior management responsibility
- Quality and food safety management system
- Resource management
- Planning and production process
- Measurements, analysis, improvements
- Food defence.

The latest version 2.2 of IFS Logistic standard was developed for businesses that are involved in transport, storage and distribution. Generally, these standards are addressed to food manufacturers that intend to produce under the private label of

European or International traders. Internationally, the IFS Food standard is a fully GFSI recognized standard that was developed to ensure that the food safety/security and quality requirements (product specifications, customer focus, etc.) are met and comply with the applicable regulatory requirements in the products' country of destination.

The IFS system was created to answer the following key question: "Is a food supplier capable of delivering safe food according to the agreed end product specification?" The standard is supported by retailers, food service companies and food manufacturers parties in the IFS technical committees and specifies the criteria that needs to be fulfilled at a given level.

Product fraud encompasses a wide range of deliberate fraudulent acts relating to food and food packaging, all of which are economically motivated and have serious ramifications to consumers and businesses [19]. The most serious of these fraudulent acts is the intentional and economically motivated adulteration (EMA) of food and packaging, where there is a high risk to consumer health. Food Industry faces challenges to the integrity and safety of its food supply chain, as the chain itself becomes more complex and global in nature. It is estimated that food fraud costs the global Food Industry US$20—$50 billion per year [19]. In addition to legislative requirements, which have been enacted to prevent product fraud and subsequent enforcement activity both nationally and internationally, industry bodies such as the Global Food Safety Initiative (GFSI) have driven for food safety schemes, such as IFS, to introduce and implement systems to mitigate the risk of food fraud. Factors that can contribute to food fraud are different. Food can be adulterated using a series of methods, all determined by the following factors:

- Historical evidence of substitution and falsification.
- The nature of raw material;
- The technical ease of doing the fraud;
- The possibility of illicit gain;
- Gaps in the technical and technological norms as well as in the product quality regulations;
- The technical difficulty of highlighting the counterfeit;
- Lack of organization and insufficient quality control;
- The degree of complexity of routine testing to identify substitutions or falsification of products.

The increasing complexity of food systems, the development of food processing technologies, changes in consumer needs linked to lifestyle and consumption, the introduction of new foods can lead to new or unforeseen (emerging) food safety concerns.

- Trade globalization, population migration, social inequalities, turbulent political and social situations can have a negative impact on human and animal safety, health, environment, and economy [14].
- Food has become a weapon in the hands of terrorists. These issues are of great general interest. Currently, responsible governments, private companies, cannot

ignore the possibility that terrorists, criminals and other anti-social groups can target the security of food supply [37].
- Food terrorism can be defined as an act of threat, deliberate contamination of food using chemical, biological or radiological agents to cause harm or even death and/or disrupt social, economic or political stability.
- Terrorists or other criminals exploit the lack of specific training and usually weaknesses are most vulnerable to deliberate activity, so implement ways to protect against a potential or actual threat is needed.

Proactive, risk-based methods, drafted according to best practice cases, proposed by experienced officials and experts and implemented by food manufacturers, seem to be the best solutions for new threats stemming from an unstable social, economic and political situation.

- Currently, the most recommended proactive methods for food safety, i.e., Food Defence, are CARVER + SHOCK, ALERT, Hazard Analysis Critical Control Defence Points (HACCDP) and Threat Analysis Critical Control Point (TACCP).
- Food Defence does not have an international and unique definition but there are two definitions offered by The United States authorities that describe their intent behind a Food Defence strategy. Food Defence is the collective term used by the US Food and Drug Administration (FDA), United States Department of Agriculture (USDA), Department of Homeland Security (DHS), etc. to include activities associated with protecting the nation's food supply from deliberate or intentional acts of contamination or tampering.

This term encompasses other similar verbiage (i.e., bioterrorism (BT), counterterrorism (CT), etc.)

- The United States Department of Agriculture (USDA) Food Safety and Inspection Service define Food Defence as "the protection of food products from intentional adulteration by biological, chemical, physical or radiological agents.
- A misconception of Food Defence is to consider it a synonym of Food Security.

Food Security is defined as the situation that exists when all people, at all times, have physical and economic access to enough food for an active, healthy life. Food security includes both physical and economic access to food that meet people's dietary needs and preferences. The purpose of a HACCP system is to identify unintentional physical, chemical, and biological hazards which are significant to food safety. The purpose of a Food Defence plan is to identify, mitigate, and monitor possible sources of intentional food contamination. While Food Safety and Food Defence programs stand independently, there are common elements (e.g., the sealing of transportation vessels). Food Defence is the implementation of measures that reduce the chances of deliberate contamination of food, using a variety of chemicals, biological agents, or other harmful substances to consumers. Food Defence refers to the prevention of intentional attacks on the production, supply, and consumption of food (Fig. 5), showing three pillars of food safety management system, viz. HACCP, TACCP and vulnerability ACCP.

Fig. 5 "Umbrella" of the GFSI food safety management (*Source* the global food Safety initiative)

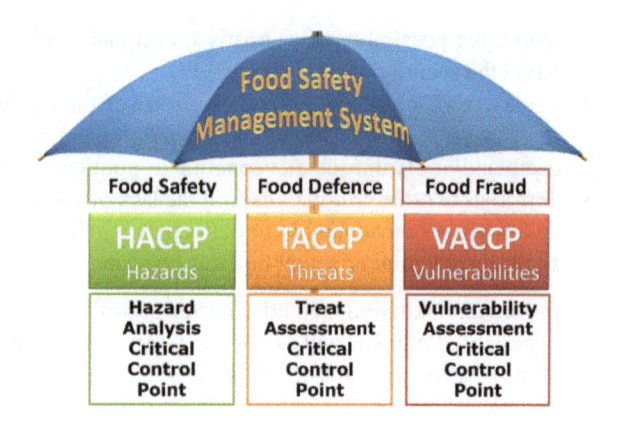

The team consists of representatives of all departments, positions or specialists who have responsibilities related to the company's defence plan. They draw up an inventory of existing processes to assess particularly the production flow analysis, internal controls that can lead to a safe working concept. The diagrams of these flows, with the existing key controls already presented and adopted, encompass the natural alignment of functions, for a common purpose, to encourage ownership of the Food Defence plan.

During this phase, the use of risk analysis studies and tools such as the ALERT initiative, the CARVER + SHOCK system, would be appropriate to help in identifying and addressing vulnerabilities.

The "ALERT" INITIATIVE. ALERT is an acronym for a food defence initiative developed in the U.S.A. to track the way preventive operations are carried out in a food processing unit.

ALERT identifies five key points that facilitate the use of raw material supply protection and reduces the risk of intentional food contamination.

A   Assure that the supplies and ingredients are from safe and secure sources
L   Look after the security of the products and ingredients obtained in facilities
E   Establish a system for identifying personnel and controlled access
R   Report about the security of the products obtained in the facilities
T   Announce authorities on any threats to food safety and security.

CARVER + SHOCK is the acronym of: **C**riticality = measure of public health and economic impacts of an attack; **A**ccessibility = ability to physically access and egress from target; **R**ecoverability = ability of system to recover from an attack; **V**ulnerability = ease of accomplishing contamination; **E**ffect = amount of direct loss from an attack as measured by loss in production; **R**ecognizability = ease of identifying target; **S**hock = psychological impacts.

Vulnerability is a weakness in handling, installation or manipulation or storage operation that would allow intentional contamination of a food product. A vulnerability assessment is carried out within the process of identifying and prioritizing weaknesses in a food related business. Specific points are identified in the food supply chain where intentional contamination has the greatest potential to harm public and economic health. The specific points represent the critical stages of the process.

Hazard Analysis Critical Control Defence Points (HACCDP) is designed as an extension of a HACCP plan or system. It covers vulnerability assessment issues where vulnerability can be defined as the ease of an attack at the point or location of food processing. The analysis provides an opportunity to build a successive approach against intentional contamination. However, there is a condition for success - the ability to differentiate Critical Control Points (CCP) from Critical Control Defence Points (CCDP) and to identify the consequences of a possible attack, which are more severe and more difficult to predict, than to assess the likelihood of accidental contamination. The goal of Threat Analysis Critical Control Point (TACCP) is to allow proportionate controls that are implemented to reduce the risk of a malicious attack and to reduce the consequences of an attack that may even be a food fraud.

The HACCDP system covers threats related to food terrorism, food sabotage, while TACCP distinguishes several types of economy related threats, espionage, counterfeiting, or cybercrime. HACCDP is suited for small and medium-sized organizations, while TACCP can be implemented by all food processors, regardless of size. As mentioned, HACCDP is an extension of the HACCP system and TACCP is rather a form of changed HACCP, including the specificity of food defence, towards food fraud. Food-Defence and Food-Safety share many concepts and techniques, merging into a wider approach of food protection, represented by a new concept called Food-Protection.

6 Hazard Analysis, Risk Assessment and Identification of Critical Safety Areas in Accordance with IFS Requirements

According to the IFS standard, hazards that may occur during all stages of production have been assessed for the Roua Apuseni product. In accordance with the IFS requirements, a risk analysis of physical chemical and biological contaminants was carried out in accordance with their maximum permissible levels in drinking water. To achieve this, the possible source of the hazard was investigated, and the risks were analysed according to both the Gravity (G) and the frequency (F). Depending on the severity and likelihood of occurrence, the risk class [29] may be determined using Table 1.

Table 2 presents the potential risks and risk analysis posed to the source compared to the acceptable levels required by the specific legislation on drinking water.

A risk analysis was performed to identify the main underlying causes of contamination of spring water, as shown in Table 3. The cause-effect diagram was used to determine the potential causes for a specific problem or effect.

A procedure setting the steps and methodology for risk assessment was established according to the IFS requirements at company level. This includes risks related to

Table 1 Establish the risk class (RC)

Gravity (G)	Probability of occurrence, frequency (F) (in the final product; to consume)		
High (H)	3	4	4
Medium (M)	2	3	4
Small (S)	1	2	3
	Small (S)	Medium (M)	High (H)

Source National good practice guide for food safety—HACCP food safety system, 2007

food safety and security, location security (for staff and visitors) vulnerability to food fraud, deliberate evasion and contamination resulting from acts of sabotage, vandalism and terrorism. Generally, a strict control of potential hazards is necessary to prevent products becoming contaminated by acts of sabotage, vandalism, bioterrorism or food fraud. The mechanisms of control are presented in Table 4.

A hazard analysis and risk assessment for food protection is carried out annually and after any changes that would affect the integrity of the food. The risk assessment aims to determine for each food-safety hazard the actions to eliminate or reduce contamination to the acceptable level, leading to the production of safe food. For the implementation of the IFS standard, a food safety assessment questionnaire may be used for areas where threat to food safety and security is present.

Thus, regarding the external side:

- Are doors, windows, and roof secure?
- Are the fence and delimitation walls in good condition?
- Are joints in the yard, perimeter, and access doors adequate?
- Is there a controlled access for people and vehicles?
- Are there alternative sources for critical utilities such as electricity, water, IT systems available in an emergency?
- Are there controlled and monitored parking areas?
- Are there ventilation systems adequately protected?
- Are there reception areas and secure storage areas?

Regarding the internal side:

- Are surveillance methods used—such as surveillance cameras, personnel, or security services?
- Are there effective systems to alert employees to security breaches?
- Is there controlled access?
- Are dangerous materials or regulated substances properly administered?
- Is the access of staff limited to their specific workplace/position/work area and working hours?

Regarding delivery and reception:

- Are the means of transport sealed?
- Are the drivers certified and are they trustworthy?

Security Standards Applied to Drinking Water

Table 2 Water contamination risk analysis

Type of contamination	Contaminant	Acceptable level	Source	Risk assessment			Control measures/PRPs[b]
				G	F	RC	
Biological	E. coli	0/250 ml	Water microbiota/everything that goes beyond is a risk	H	S	3	Water analysis report
	Enterococci	0/250 ml					
	P. aeruginosa	0/250 ml					
	Coliform bacteria	0/100 ml					
	Clostridium spp.	0/100 ml					
	TVC[a] at 22 °C	100/ml					
	TVC at 37 °C	20/ml					
Chemical	Aluminium	200 μg/l	Chemical properties of water/everything that goes beyond is a risk	H	S	3	Water analysis report
	Ammonium	0.50 mg/l					
	pH value	6.5–9.5					
	Conductivity	2500 μS cm^{-1}, 20 °C					
	Free residual chlorine	Approx. 0.1 mg/l					
	Iron	200 μg/l					
	Nitrite	0.50 mg/l					
	Oxides	5 mg O$_2$/l					
	Sulphides and H$_2$S	100 μg/l					
Physical	Colour	Acceptable	Physical properties of water/everything that goes beyond is a risk	M	S	2	Water analysis report
	Taste	Acceptable					
	Smell	Acceptable					

(continued)

Table 2 (continued)

Type of contamination	Contaminant	Acceptable level	Source	Risk assessment			Control measures/PRPs[b]
				G	F	RC	
	Turbidity	Normal					
	Conductivity	Normal					
Contamination with:	Parasites	Not allowed	Contaminants	M	S	2	Water analysis report
	Rust	Not allowed	Contaminants	M	S	2	
	Impurities	Not allowed	Contaminants	M	S	2	
	Other chemical substances	Not allowed	Contaminants	H	S	3	

[a]TVC-total viable count, [b]PRPs-prerequisite programmes

Table 3 Identify the hazards (HACCP analysis) that lead to contamination of spring water

Water contamination	C	D	O	Causes determination			Evaluation	
				Primary	Secondary	Tertiary	G	F
Microbiological - from the source, distribution system, piping, installation	x		x	Environmental contamination	Contamination from the water distribution system	Contamination from the internal installation	H	S
Nitrates -above the maximum admissible value	x			Environmental contamination	Contamination from the water distribution system	Contamination from the internal installation	H	S
Foreign bodies, organic and mineral impurities	x			Environmental contamination	Contamination from the water distribution system	Contamination from the internal installation	H	S

Where: C—contamination, D—development(growth), O—survival(outliving)

Table 4 Synthetic analysis of hazard control programs

Hazard	The effect (risk)	Control measures/preventive actions	The follow-up procedure	Corrections/corrective actions	Responsibility
Food defence	Foodborne illness	Access control, personal and products monitoring, transport conditions.	*method*: access cards, transport conditions, video surveillance system *frequency*: every day	Destroying of potentially contaminated products. Personal training and penalization	Quality control supervisor, quality assurance manager
Food Fraud	Foodborne illness	Supplier monitoring; documents control; Products analysis	*method*: supplier evaluation and approval, products inspection; documents control *frequency*: at each supply	Return or destruction of adulterated products	Quality control supervisor, quality assurance manager, materials supervisor

Security Standards Applied to Drinking Water 387

- Are deliveries and transports registered?
- Are the subcontractors approved for transport?
- Are the wrong or delayed deliveries investigated?
- Is the return of goods allowed? If so, are the situations being managed?
- Are the water and steam sources secured and monitored?
- Are there ways to verify the integrity of the chain of custody?
- Are all raw materials insured and monitored when not in use?
- Are packaging materials and product labels controlled?
- Is personal data from the employee's history checked on hiring?
- Are references to the latest job descriptions required?
- Are reasons for resigning employees analysed?
- Is the staff supervised/monitored? Is there a video monitoring system?
- Are the employees trained and conscious about food protection and identification/report of unusual events or suspicious behaviour of colleagues?
- Are the locks checked?
- Are personal items banned in the processing areas?
- Is there a written policy on the regime of legal or illegal arms and drugs?

Types of food fraud:

- Substitution of spring water or packaging with less valuable or other products. For example, use of tap/table water or the use of inferior quality PET bottles.
- Mislabelling of food (reclassifying a product, including falsification by imitation of a natural product).
- Different undisclosed issues such as the value of good production practice. For example, use of chlorination to increase spring water conservation or UV radiation treatment to reduce processing costs.

To prevent food fraud, strict control measures can be used at the factory level. For packaging, a declaration can be required from the supplier that the products are not tampered with. Additionally, analyses can be conducted on this issue once a year. Internally, according to a self-check program, products can be sent for specific analyses. Depending on the severity and likelihood of occurrence of these risks, we have established the HACCP risk class (Table 5). According to this methodology, controls should be placed in the specific areas or stages where they are able to be applied to prevent, eliminate any hazard to food safety or reduce it to an acceptable level. Classes 3 and 4 are considered significant hazards for food protection Class 2 hazards are considered to pose a medium risk and those of low risk belong in Class 1. Employees should be trained on the food protection plan. Visitors and inspectors must also be trained and accompanied throughout the visit. To identify vulnerabilities, an appropriate alert system can be set up, whose efficacy should be periodically tested. The control measures may be updated depending on the results.

Within the perimeter of the site, there are three large areas where acts may occur endangering the safety and security of food. These are: the catchment area, the transport to the production site, the bottling site, the packaging station, and the warehouses. For the classification of these areas, three levels were used: low, medium,

Table 5 Protection plan of spring water on process stages—level of risk indicators

Area/process step	Hazard	Acceptable level	G	F	RC	Risk indicator level
Spring source capture	Sabotage, discharges of toxic substances	Not allowed	H	S	3	critical
Pipeline transport	Sabotage—pipe piercing and introduction of toxic substances	Not allowed	H	S	3	critical
Water reception	Sabotage	Not allowed	M	S	2	medium
Packaging reception	Sabotage driver	Not allowed	S	S	1	small
	Personnel Sabotage	Not allowed	M	S	2	medium
Packaging storage	Fraudulent penetration into the warehouse	Not allowed	M	S	2	medium
	Sabotage	Not allowed	M	S	2	medium
Packaging intern transport	Sabotage	Not allowed	M	S	2	medium
Ozone water treatment	Sabotage by inappropriate dosing	Not allowed	H	S	3	critical
Water filing	Intentional bottling (with another label, other product, other weight, etc.)	Not allowed	S	S	1	small
Labelling	Product falsification	Not allowed	S	S	1	small
Shring-packaging	Damage to packaging	Not allowed	S	S	1	small
Final product storage	Fraudulent penetration into the warehouse	Not allowed	M	S	2	medium
	Sabotage	Not allowed	M	S	2	medium
Final product delivery	Intentional delivery of other products	Not allowed	S	S	1	small
Transport	Sabotage storage, sabotage transport	Not allowed	H	S	3	critical
	Sabotage by the staff that loads/unloads	Not allowed	M	S	2	medium
Internal and external communication	Violation of correspondence (recipes, specifications, etc.)	Not allowed	M	S	2	medium

Security Standards Applied to Drinking Water 389

and critical. Based on hazard analysis and the associated risk assessment, the critical security areas identified were the perimeter of the catchment area, the transport of water from the source to the bottling plant, the ozone treatment of the water and the transport of the finished product (Table 5). Table 6 describes the preventative actions, and the appropriate control measures which are designed to prevent, eliminate or reduce the risks to which the process is exposed.

Areas identified by the hazard analysis and associated risk assessment as critical to safety should be adequately protected.

Table 6 Preventive control measures associated with the dangers of the process steps

Area/process step	Preventive actions/control measures
Spring source capture	Check seals; install surveillance cameras; install motion sensors; provide protection perimeter of 50 m upstream from the source
Pipeline transport	Install sensors indicating the presence or absence of water on the pipe (pressure drop)
Water reception	Install an on-line sensor to measure conductivity and turbidity
Packaging reception	Check driver status
	Always check staff at the reception; organise staff training
Packaging storage	Secure access in the factory; check doors and windows; secure deposit areas; hire protection/guard services
	Ensure that packages are transported only by managers; organise staff training
Packaging intern transport	Organise staff training
Ozone water treatment	Verify batch product by residual ozone analysis
Water filing	Conduct unannounced control visits; offer practical training, supervise workers
Labelling	Verify through survey; organise staff training
Shrink-packaging	Organise staff training; restrict access only for people in charge
Final product storage	Secure access to factory and secure warehouses; hire protection/guard services
	Perform managers duties; organise staff training
Final product delivery	Organise staff training; restrict access only for people in charge
Transport	Check machinery/equipment before leaving the perimeter
	Supervise the personnel permanently; organise staff training; hire staff with proven experience and competence
Internal and external communication	Provide IT security using passwords; open envelopes only by recipients

7 Conclusions

In the context of the current problems that increase the risk of water scarcity for mankind, the chapter outlines the current situation regarding drinking water safety at global level, but also offers a case study on hazard analysis, risk assessment and identification of critical safety areas in water bottling process according to IFS requirements. If public drinking water systems use only surface waters (lakes, rivers), groundwater (spring, wells, drilling) is an alternative that will be less and less explored and exploited, even if it is a natural and untreated free source. Potential risks were investigated, and water contamination risks were assessed using modern food safety tools - International Featured Standards and IFS Food Standard. These were all correlated to the established limits set by current legislation for drinking water. Water contamination risk analysis is important, thus the evolution of the main drinking water indicators (pH, ammonium content, nitrite, nitrates, *E. coli*, Enterococci, *P. aeruginosa*, Coliform bacteria, *Clostridium* spp., TVC at 22 °C, TVC at 37 °C) must be investigated for a water that is harvested and bottled by an economic entity.

Additionally, an analysis was carried out to identify the main underlying causes of the contamination of source water, showing that regardless of the type of risk (biological, nitrite, impurities—in the case of spring water) danger can arise from contamination due to microorganisms. The use of the IFS Food tool instrument involves addressing food security in addition to product safety issues. In the survey, an inventory of the dangers that bottled water products are exposed to are identified: sabotage, vandalism, terrorism/bioterrorism, and fraud. These dangers were assessed at all stages of the product production process - from the source water catchment to customer delivery. The risk analysis was performed by establishing a risk class based on the level of the risk indicator, considering the severity and likelihood of occurrence for each hazard at each process area or stage. IFS Food analysis has shown that there are 4 critical levels of risk indicators, with 9 being classed as medium and 5 being classed as low. The areas and process steps that were shown to be critical were: the catchment area, transport of source water via pipe. In the latter the critical risks identified were sabotage by unauthorized persons entering the source area or breach the transport pipeline and discharge or introduce toxic substances in the water. Another critical area is the ozone treatment of the water with a risk of contamination due to inappropriate dosing or protection, and transport/distribution of the finished product. The primary risk is represented by sabotage that results in damage to the products' safety during transport, for example by storing the product with toxic substances. To ensure security for the entire technical flow covering each hazard, regardless of the severity level of the risk indicator there were established preventive controls associated with the stages of the production process according to the IFS principles.

Overall, this chapter is an investigation of the quality and safety of water bottled water and indicates the need for permanent monitoring on issues related to food safety and security to identify potential contaminants, through acts of sabotage, vandalism, bioterrorism, and vulnerability to food fraud. The implementation and certification of an efficient water resource management system (such as IFS, BRC, FSSC, etc.) ensures optimal monitoring and control conditions in water processing and mitigates the risk of contamination. The study assessed the vulnerability of a specific company, considering site security, staff and visitors, food fraud, deliberate evasion or contamination, and the ability to protect against acts of sabotage, vandalism, and terrorism/bioterrorism.

References

1. Barlow MC (2002) Blue gold. The battle against corporate theft of the world's water. Earthscan Publications Ltd. London, pp 57–64
2. Bokova I (2010) Nature and science. The 50th anniversary of the intergovernmental oceanographic commission, Paris
3. Brooks S, Elliott CT, Spence M (2017) Four years post-horsegate: an update of measures and actions put in place following the horsemeat incident of 2013. Sci Food 1:5
4. Cheese GM, Lupsa IR, Zora M, Negrea A, Iovi A, Bragea M (2008) Contribution to the knowledge of iron contains in water from West Zone of Romania. Chem. Bull. "POLITEHNICA" Univ. (Timişoara) 53:226–229
5. Chenoweth J (2008) Minimum water requirement for social and economic development. Desalination 229(1–3):245–256
6. Chiriac D, Humă C, Tudor C (2001) Socio-economic impact of water on the quality of life of the population in Romania. Qual Life 12:95–116
7. Council Directive 98/83/EC of 3 November 1998 on the quality of water intended for human consumption, OJ L 330, 5.12.1998
8. Diaconu M (2008) Water quality in the upper river basin of the Arges River. IWM Co-operatives
9. Directive 2000/60/EC of the European Parliament and of the Council of 23 October 2000 establishing a framework for Community action in the field of water policy, OJ L 327, 22.12.2000
10. EU Regulation 2017/625 of the European Parliament and EU Council of 15th March 2017 regarding official controls and other official activities performed to ensure the enforcement of legislation regarding food and animal feed, rules regarding animal health and welfare, plant health, and plant protection products
11. GFSI (2013) The global food safety initiative GFSI guidance document version 6.3. http://www.mygfsi.com/ Accessed 01 Sept 2013
12. GFSI (2014) GFSI position on mitigating the public health risk of food fraud. http://www.mygfsi.com/ Accessed 12 Oct 2015
13. GFSI (2014) Global food safety initiative—MyGFSI—food fraud mitigation. Available at: https://www.mygfsi.com/files/Information_Kit/GFSI_GMaP_FoodFraud.pdf, Accessed 02 Dec 2019
14. Goryakin Y, Lobstein T, James WP, Suhrcke M (2015) The impact of economic, political and social globalization on overweight and obesity in the 56 low- and middle-income countries. Soc Sci Med 133:67–76
15. Gossner CM, Schlundt J, Ben Embarek P, Hird S, Lo-Fo-Wong D, Beltran JJ, Teoh KN, Tritscher A (2009) The melamine incident: implications for international food and feed safety. Environ Health Perspect 117(12):1803–1808
16. Gore A (1995) The earth hanging in the balance. Ecology and Human Spirit, Series "Global Problems of Humankind", Technical, Bucharest, pp 42–76
17. https://www.wri.org, World Resource Institute Accessed 06 Dec 2018
18. https://ec.europa.eu, Food fraud: What does it mean? Accessed 04 Nov 2020

19. IFS Standards Product Fraud—Guidelines for Implementation, IFS Management GmbH, Germany, 2018
20. Jelev I (2008) Indicators for the assessment of climate change and some aspects specific to the mountain area, Romanian Academy, Terra National Conference and Life
21. Kennedy J, Delaney L, McGloin A, Wall PG (2009) Public perceptions of the dioxin crisis in Irish Pork. www.ucd.ie, UCD Geary Institute. Discussion Paper Series
22. Lotta F, Bogue J (2015) Defining food fraud in the modern supply chain. Eur Food Feed Law Rev 10(2):114–122
23. Lopez-Gunn E, De Stefano L, Llamas M (2012) The role of ethics in water and food security: balancing utilitarian and intangible values. Water Policy 14(S1):89–105
24. Lvovich MI, White GF (1990) Use and transformation of terrestrial water systems. Cambridge University Press, Cambridge, pp 47–78
25. Manning L, Mei Soon J (2016) Food science food safety, food fraud, and food defense: a fast-evolving literature. J Food Sci 81(4):823–834
26. Manoleli D, Platon V, Stănescu R, Prisecaru P, Georgescu L, Tilly J (2003) The impact of the implementation of environmental protection directives of the european union on several industrial sectors. European Institute of Romania Publishing, Bucharest, Bilingual, pp 109–110
27. Marinov AM, Dumitran GE, Diminescu MA (2007) Groundwater monitoring and aquifer remediation. POLITEHNICĂ Publishing House, Bucharest, pp 86–89
28. Morin JF, Lees M (2018) Food integrity handbook. A guide to food authenticity issues and analytical solutions Eurofins Analytics France, pp IX–XVII
29. National Good Practice Guide for Food Safety—HACCP Food Safety System (2007) Uranus, Bucharest, pp 101–102
30. Popkin BM, D'Anci KE, Rosenberg IH (2010) Water, hydration, and health. Nutr Rev 68(8):439–458
31. Public Services Employer (2001) Water policy in a new style. In: INFO Buletin. May, year XII
32. Rojanschi V (1985) The mysteries of a glass of water, Ceres, Bucharest, p 34 [publication in Romanian]
33. Saskia R, Wim H, Pieternel L (2017) Food fraud vulnerability and its key factors. Trends Food Sci Technol 67:70–75
34. Schneider SH (1996) Water resources. Encyclopaedia of climate and weather, vol. 2. Oxford University Press, New York, pp 817–823
35. Sharma S, Bhattacharya A (2017) Drinking water contamination and treatment techniques. Appl Water Sci 7:1043–1067
36. Stomff S (2009) World water day. Rev. Standard 4:20–23 [publication in Romanian]
37. World Health Organization (2002) Terrorist threats to food: guidance for establishing and strengthening prevention and response systems. Food safety issues. Available at: https://apps.who.int/iris/rest/bitstreams/50530/retrieve. Accessed 08 Jan 2018

Maria Popa was born in Cornereva Village, Caras Severin County, Romania, in 1966. She received the B.S. degree in chemical engineering in 1990 and the Ph.D. degree in environmental chemistry in 2004 both from Babes-Bolyai University of Cluj Napoca, Romania. In 2008, she obtained the M.S. degree in management of trade, tourism and services and the title of Internal Auditor for Quality Management Systems in accordance to Quality Management Systems: SR: EN ISO 9001, SR: EN ISO 190011. She also received the title Environmental Analyst with the following competences: environmental management, environmental legislation, lasting development, European norms in the field of environment, food safety and security. Between 2000 and 2006, she was lecturer at the Faculty of Science from "1 Decembrie 1918" University of Alba Iulia, and between 2006 and 2014 an associate professor. She has been Professor with the Business Administration and Marketing Department, Faculty of Economics, in the same university since 2014. Maria Popa was the dean of the Faculty of Economics from 2012 to 2020 (March). She has been the Ph.D. supervisor at the Doctoral School of Lucian Blaga University of Sibiu, in the field of engineering and management since 2017. Her interests in research include environmental management, food quality, safety and security control, waste management and improvement of the management process. She holds membership in: The Romanian Academic Management Society (SAMRO), The Balkan Environmental Association, The Society of Science and Engineering Environmental (SNSIM) (founding member), and The Editorial Board of the Journal of Environmental Protection and Ecology (JEPE). She has been the coordinator of UAB-BENA—International Centre of Environment, Sustainable Development and Food Quality Control since 2017. She is the author of 20 books, more than 150 articles (ISI and BDI), and more than 30 research projects.

Ioana Glevitzky was born in Beius, Bihor County, Romania in 1981. She received the B.S. (2005) and M.S. degrees (2006) in food engineering and organic chemistry from the Politehnica University of Timișoara, Romania. From 2018 follows the Ph.D. study in engineering and management at Lucian Blaga University of Sibiu. From 2006 to 2008, she was a Researcher in the Department of Research, Development and Quality Control Products, European Drinks Rieni, Romania. From 2008 to 2014 she has been a Quality manager at the Prefera Foods SA Company Oiejdea, Romania. From 2014 to 2016, she was an engineer at Alpin 57 Lux SRL Sebes, Romania. Since 2016 she works in Sanitary Veterinary and Food Safety Laboratory of Alba County, Romania. His research interests include water quality, groundwater pollution, detection of antibiotics in food, food safety, risk assessment, occupational health and safety. Ms. Glevitzky is a member of Romania's Chemical Engineering Association and Balkan Environmental Association. She was a recipient of the BENA International Conference on "Environmental Engineering and Sustainable Development", Award in 2017.

Water Security Safeguarded by Safe, Secure and Smart Water Management Solutions

Adrian Lucian Cococeanu and Teodor Eugen Man

Abstract These days water utility managers face the challenge of losing an unacceptable high level of revenue by compromised safety, security and loss of water. Aside from decreasing revenue, money for infrastructure improvement related investment, productivity, and service delivery, it also results in a consequential increase in the cost of the water production and delivery. The smart water solutions consist of numerous strategic methods and tools that could, under a consistent vision, contribute to build an efficient water management system targeting the overall objectives of the water security. This chapter outlines several of such initiatives to enhance water security.

Keywords Water security · ICS · SCADA · GIS · ERP · Asset management · Hydro informatics tools

1 Introduction to Water Security

The water security can be defined as "*the capacity of a population to safeguard sustainable access to adequate quantities of and acceptable quality water for sustaining livelihoods, human well-being, and socio-economic development, for ensuring protection against water-borne pollution and water-related disasters, and for preserving ecosystem in a climate of peace and political stability*", by UN-Water. To achieve and maintain water security, it is necessary to ensure access to safe and sufficient drinking water at an affordable cost in order to meet basic needs, including sanitation and hygiene, and to safeguard health and well-being. Furthermore, it is

---

A. L. Cococeanu
Department of Hydrotechnical Engineering, Civil Engineering Faculty, Politehnica University of Timişoara, Iuliu Podlipny Street, No. 23C, 300703 Timişoara, Romania
e-mail: adrian.cococeanu@student.upt.ro

T. E. Man (✉)
Department of Hydrotechnical Engineering, Civil Engineering Faculty, Politehnica University of Timişoara, Ulpia Traiana Street, No. 80/3, 300771 Timişoara, Romania
e-mail: eugen.man@upt.ro

also necessary for preservation and protection of ecosystems in water allocation and management systems to maintain their ability to deliver and sustain functioning of essential ecosystem of services.

According to the World Economic Forum, the "water crises" is believed to be among the most likely and the most impactful global risks of the next few decades. Similarly, the U.S. Intelligence Community Assessment of Global Water Security considers that water problems will contribute to destabilizing key states and, *"when combined with poverty, social tensions, environmental degradation, ineffectual leadership and weak political institutions, will contribute to social disruptions that can result in state failure"* (WEF, 2016). Global water crises are not limited solely to drought, but also include flooding, inadequate access to drinking water and sanitation, water-borne diseases and other challenges. Furthermore, the increasing frequency and severity of extreme weather events continues to place additional pressure on water supplies worldwide. Water resources management and water governance issues vary widely, however remain, by far one of the most common challenge faced by a majority of countries. This requires collaborative management to trans boundary water resources management within and between countries to promote freshwater sustainability and cooperation and evaluation of water supplies for socio-economic development and other activities, such as energy, transport, industry, tourism. It is critical to develop an ability to cope with uncertainties and risks of water-related hazards, such as floods, droughts and pollution and risks associated with water security to protect natural and built infrastructures and the cost-effective ecosystem services it provides.

The key issues that need to be addressed for proper planning, capacity development, operate and maintain infrastructure to ensure water security can be grouped, and described as below.

- Insufficient institutional skills and adaptation to climate change;
- Insufficient understanding of how policy instruments in water management can affect the economy and growth;
- Limited application of a holistic approach in water policymaking, multi-sector involvement and low coordination of roles and responsibilities;
- Hesitant trans-boundary cooperation in promoting the sustainable and equitable development of a shared water course;
- Unsatisfactory cooperation, joint research actions and knowledge sharing;
- Enabling factors which could support and enhance the water security and human security;
- Developing new technology, water harvesting, and desalination and water treatment;
- Raising the profile of water security on the political and developmental agendas of national governments;
- Including water in security policy planning;
- Involving local populations in the development process and running capacity-building programs at different levels;

- Encouraging investment in and increased collaboration on water management technologies;
- Generating better policies through dialogue, knowledge exchange and communication;
- Improving data quality to generate better policies [1, 2].

The issues are also critical for the governance and accountability, and the due consideration of the interests of all stakeholders through: appropriate and effective legal regimes for transparent, participatory and accountability.

2 Smart Water Solutions for Water Security

The subject of water domain, as it pertains to stakeholders, vary widely and covers a large number of business processes, especially if all of the domains and relevant activities are considered. This situation a mapping process and the prioritization of gaps that need to be bridged. Major areas are directly linked to the urban water use, where both expectations and possibilities are the highest. ICS (industrial control systems) are present in industrial processes such as water and wastewater, electric, oil and natural gas, chemical, pharmaceutical, food, and manufacturing. Industrial control system (ICS) are, in general, several types of control systems, including supervisory control and data acquisition (SCADA) systems, distributed control systems (DCS), and other control system configurations such as Programmable Logic Controllers (PLC) used in the industrial sectors and critical infrastructures.

ICS consists of combinations of control components (e.g., electrical, mechanical, hydraulic, pneumatic) that act together to achieve an industrial objective (e.g., manufacturing, transportation of matter or energy). The part of the system primarily concerned with producing the output is referred to as the process. The control part of the system includes the specification of the desired output or performance. Control can be fully automated or may include a human in the loop. Systems can be configured to operate open-loop, closed-loop, and manual mode. In open-loop control systems the output is controlled by established settings. In closed-loop control systems, the output has an effect on the input in such a way as to maintain the desired objective. In manual mode the system is controlled completely by humans. The part of the system primarily concerned with maintaining conformance with specifications is referred to as the controller. A typical ICS may contain numerous control loops, Human Machine Interfaces (HMIs), and remote diagnostics and maintenance tools built using an array of network protocols. ICS are critical to the operation of critical infrastructures that are often highly interconnected and mutually dependent systems [3, 4].

2.1 Evolution of Industrial Control Systems

Today's ICS evolved from the insertion of IT (Information Technology) capabilities into existing physical systems, often replacing or supplementing physical control mechanisms. For example, embedded digital controls replaced analog mechanical controls in rotating machines and engines. The improvements in cost-and performance have encouraged this evolution, resulting in many of today's "smart" technologies. This increases of the connectivity and criticality of these systems, it also creates needs for their adaptability, resilience, safety, and security. ICS engineering continues to evolve to provide new capabilities while maintaining the typical long lifecycles of these systems. The IT capabilities into physical systems presents emergent behavior that has security implications. Engineering models and analysis are evolving to address these emergent properties including safety, security, privacy, and environmental impact interdependencies. ICS are used to control geographically dispersed assets, often scattered over thousands of square kilometers, including distribution systems such as water distribution and wastewater collection systems, agricultural irrigation systems, oil and natural gas pipelines, electrical power grids, and railway transportation systems.

Water treatments plants are usually located within a confined factory or plant-centric area, when compared to geographically dispersed water or waste water distribution. Communications in plants industries are usually performed using local area network (LAN) technologies that are typically more reliable and high speed as compared to the long-distance communication wide-area networks (WAN) and wireless/RF (radio frequency) technologies used by distribution industries. The ICS used in distribution industries are designed to handle long-distance communication challenges such as delays, and data loss posed by the various communication media used. The security controls may differ among network types.

Critical infrastructures are highly interconnected and mutually dependent in complex ways, both physically and through a host of information and communications technologies. An incident in one infrastructure can directly and indirectly affect other infrastructures through cascading and escalating failures. SCADA systems monitor and control distribution by collecting data from and issuing commands to geographically remote field control stations from a centralized location. SCADA systems are used to monitor and control water distribution, including pipelines, as well as wastewater collection systems. SCADA systems and DCS are often networked together. The lack of monitoring and control capabilities could cause a large generating unit to be taken offline, an event that would lead to loss of power at a transmission substation. This loss could cause a major imbalance, triggering a cascading failure across the network. This could result in large area blackouts that could potentially affect, water treatment systems, wastewater collection systems, and pipeline transport systems that rely on the grid for electric power.

The basic operation of an ICS is shown in Fig. 1, critical processes may also include safety systems. Typical ICS contains numerous control loops, human interfaces, and remote diagnostics and maintenance tools built using an array of network

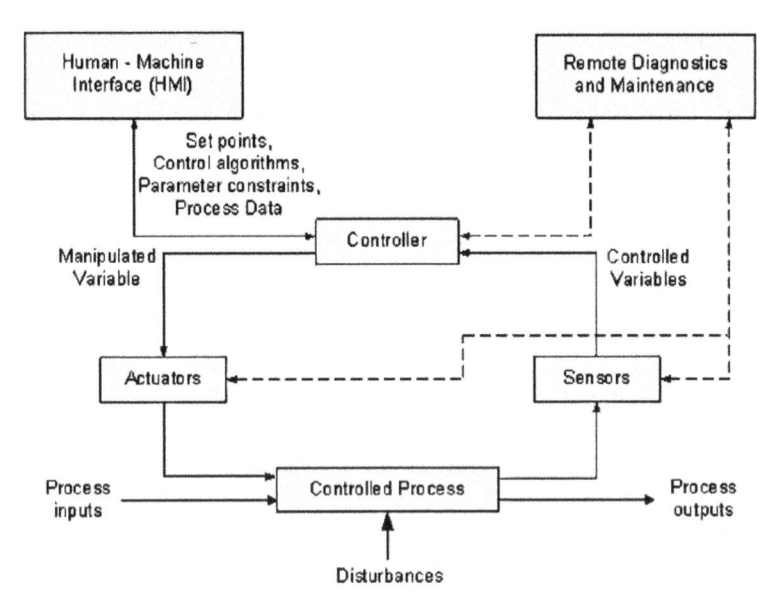

Fig. 1 ICS operation, *source* NIST 800-82

protocols on layered network architectures. A control loop utilizes sensors, actuators, and controllers (e.g., PLCs) to manipulate some controlled process. A sensor is a device that produces a measurement of some physical property and then sends this information as controlled variables to the controller. The controller interprets the signals and generates corresponding manipulated variables, based on a control algorithm and target set points, which it transmits to the actuators.

Actuators such as control valves, breakers, switches, and motors are used to directly manipulate the controlled process based on commands from the controller. Operators and engineers use human interfaces to monitor and configure set points, control algorithms, and to adjust and establish parameters in the controller. The human interface also displays process status information and historical information. Diagnostics and maintenance utilities are used to prevent, identify, and recover from abnormal operation or failures. Sometimes these control loops are nested and/or cascading—whereby the set point for one loop is based on the process variable determined by another loop. Supervisory-level loops and lower-level loops operate continuously over the duration of a process with cycle times ranging on the order of milliseconds to minutes [3, 4].

The design of an ICS, including a SCADA, DCS, or PLC-based topologies currently in use, depends on many factors, such as the design of ICS drive design decisions regarding the control, communication, reliability, and redundancy properties of the ICS. Because these factors heavily influence the design of the ICS, they will also help determine the security needs of the system. ICS processes have a wide range of time-related requirements, including very high speed, consistency, regularity, and synchronization. Humans may not be able to meet these requirements reliably and consistently; automated controllers may be necessary. Some systems may require the computation to be performed as close to the sensor and actuators as possible, to reduce communication latency and perform necessary control actions on time.

Systems have varying degrees of distribution, ranging from a small system (e.g., local PLC-controlled process) to large, distributed systems (e.g., water pipelines). Greater distribution typically implies a need for wide area and mobile communication. Supervisory control is used to provide a central location that can aggregate data from multiple locations to support control decisions based on the current state of the system. Often a hierarchical/centralized control is used to provide human operators with a comprehensive view of the entire system. Often control functions can be performed by simple controllers and preset algorithms. Complex systems require human operators to ensure that all control actions are appropriate to meet the larger objectives of the system. Availability system's requirements are also an important factor in design. Systems with strong availability/up-time requirements may require more redundancy or alternate implementations across all communication and control [3]. The failure of a control function could incur substantially different impacts across domains. Systems with greater impacts often require the ability to continue operations through redundant controls, or the ability to operate in a degraded state. The design needs to address these requirements. System's safety requirements area is also an important factor in design. Systems must be able to detect unsafe conditions and trigger actions to reduce unsafe conditions to safe ones. In most safety-critical operations, human oversight and control of a potentially dangerous process is an essential part of the safety system.

2.2 SCADA Systems in Critical Infrastructures—Water Sector

SCADA systems that is shown in Fig. 2, are used to control dispersed assets where centralized data acquisition is important as control. These systems are used in distribution systems such as water distribution and wastewater collection systems, water pipelines. SCADA systems integrate data acquisition systems with data transmission systems and HMI software to provide a centralized monitoring and control system for numerous process inputs and outputs.

SCADA systems are designed to collect field information, transfer it to a central computer facility, and display the information to the operator graphically or textually, thereby allowing the operator to monitor or control an entire system from a central location in near real time. Based on the sophistication and setup of the individual system, control of any individual system, operation, or task can be automatic, or it can be performed by operator commands. Typical hardware includes a control server placed at a control center, communications equipment (e.g., radio, telephone line, cable, or satellite), and one or more geographically distributed field sites consisting

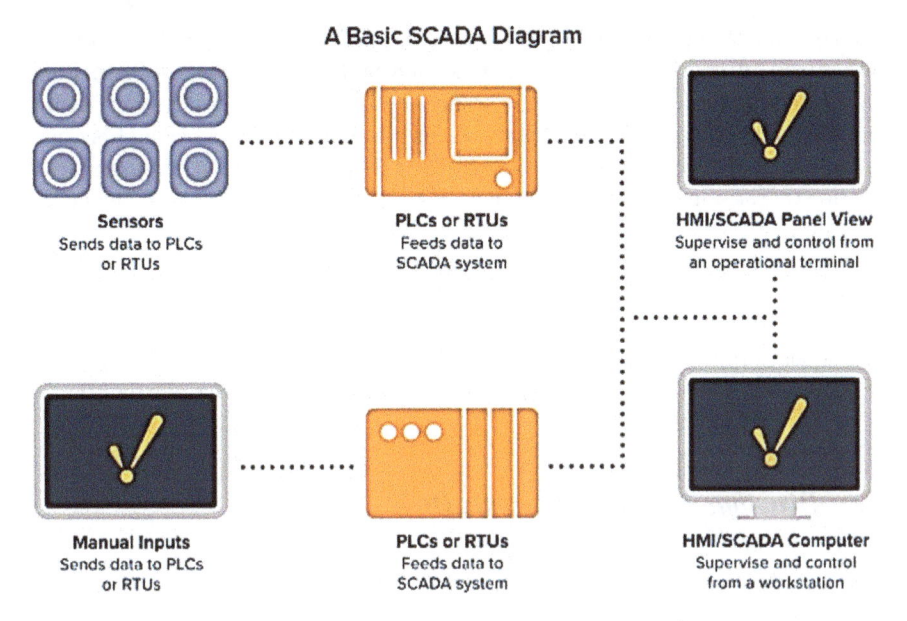

Fig. 2 Basic SCADA diagram, *source* https://inductiveautomation.com [5]

of Remote Terminal Units (RTUs) and/or PLCs, which controls actuators and/or monitors sensors. The control server stores and processes the information from RTU inputs and outputs, while the RTU or PLC controls the local process. The communications hardware allows the transfer of information and data back and forth between the control server and the RTUs or PLCs. The software is programmed to tell the system what and when to monitor, what parameter ranges are acceptable, and what response to initiate when parameters change outside acceptable values [6, 7, 8].

An Intelligent Electronic Device (IED), such as a protective relay, may communicate directly to the control server, or a local RTU may poll the IEDs to collect the data and pass it to the control server. IEDs provide a direct interface to control and monitor equipment and sensors. IEDs may be directly polled and controlled by the control server and in most cases have local programming that allows for the IED to act without direct instructions from the control center.

SCADA systems are usually designed to be fault-tolerant systems with significant redundancy built into the system. Redundancy may not be a sufficient countermeasure in the face of malicious attack. The control center houses a control server and the communications routers. Other control center components include the HMI, engineering workstations, and the data historian, which are all connected by a LAN. The control center collects, and logs information gathered by the field sites, displays information to the HMI, and may generate actions based upon detected events. The control center is also responsible for centralized alarming, trend analyses, and reporting.

The field site performs local control of actuators and monitors. Field sites are often equipped with a remote access capability to allow operators to perform remote diagnostics and repairs usually over a separate dial up modem or WAN connection. Standard and proprietary communication protocols running over serial and network communications are used to transport information between the control center and field sites using telemetry techniques such as telephone line, cable, fiber, and radio frequency such as broadcast, microwave and satellite. SCADA communication topologies vary among implementations. The various topologies used, including point-to-point, series, series-star, and multi-drop. Point-to-point is functionally the simplest type; however, it is expensive because of the individual channels needed for each connection. In a series configuration, the number of channels used is reduced; however, channel sharing has an impact on the efficiency and complexity of SCADA operations. Similarly, the series-star and multi-drop configurations use of one channel per device results in decreased efficiency and increased system complexity [3].

2.3 Distributed Control Systems DCS

Distributed Control Systems DCS are used to control production systems within the same geographic location for industries such water and wastewater treatment, electric power generation plants, chemical manufacturing plants, automotive production, and pharmaceutical processing facilities. These systems are usually process control or discrete part control systems.

DCS are integrated as a control architecture containing a supervisory level of control overseeing multiple, integrated sub-systems that are responsible for controlling the details of a localized process. DCS uses a centralized supervisory control loop to mediate a group of localized controllers that share the overall tasks of carrying out an entire production process. Product and process control are usually achieved by deploying feedback or feed forward control loops whereby key product and/or process conditions are automatically maintained around a desired set point. To accomplish the desired product and/or process tolerance around a specified set point, specific process controllers, or more capable PLCs, are employed in the field and are tuned to provide the desired tolerance as well as the rate of self-correction during process upsets. By modularizing the production system, a DCS reduces the impact of a single fault on the overall system. In many modern systems, the DCS is interfaced with the corporate network to give business operations a view of production. DCS encompasses an entire facility from the bottom-level production processes up to the corporate or enterprise layer. A supervisory controller (control server) communicates to its subordinates via a control network. The supervisor sends set points to and requests data from the distributed field controllers. The distributed controllers control their process actuators, based on control server commands and sensor feedback from process sensors.

Fieldbus networks eliminate the need for point-to-point wiring between a controller and individual field sensors and actuators. Additionally, a fieldbus allows greater functionality beyond control, including field device diagnostics, and can accomplish control algorithms within the fieldbus, thereby avoiding signal routing back to the PLC for every control operation. Standard industrial communication protocols designed by industry groups such as Modbus and Fieldbus are often used on control networks and fieldbus networks. In addition to the supervisory-level and field-level control loops, intermediate levels of control may also exist. For example, in the case of a DCS controlling a discrete part manufacturing facility, there could be an intermediate level supervisor for each cell within the plant. This supervisor would encompass a manufacturing cell containing a machine controller that processes a part and a robot controller that handles raw stock and final products. There could be several of these cells that manage field-level controllers under the main DCS supervisory control loop [3].

2.4 Programmable Logic Controller

Programmable Logic Controller that is shown in Fig. 3, based topologies PLCs are used in both SCADA and DCS systems as the control components of an overall hierarchical system to provide local management of processes through feedback control as described in the sections above. In the case of SCADA systems, they may provide the same functionality of RTUs. When used in DCS, PLCs are implemented

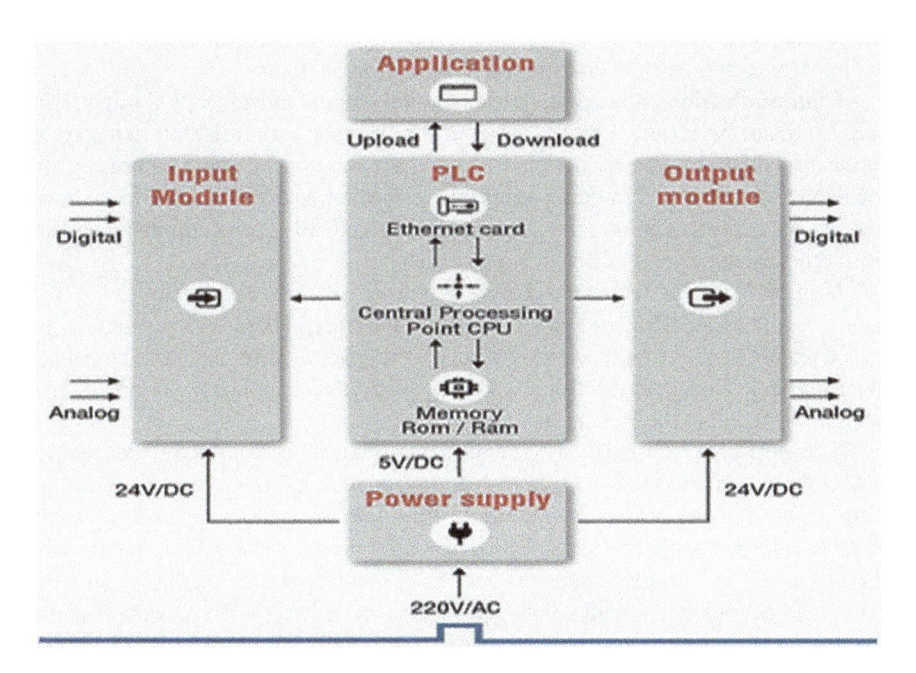

Fig. 3 Programmable logic controller—block diagram, *source* https://www.unitronicsplc.com

as local controllers within a supervisory control scheme. In addition to PLC usage in SCADA and DCS, PLCs are also implemented as the primary controller in smaller control system configurations to provide operational control of discrete processes such as automobile assembly lines and power plant soot blower controls.

These topologies differ from SCADA and DCS in that they generally lack a central control server and HMI and, therefore, primarily provide closed-loop control without direct human involvement. PLCs have a user-programmable memory for storing instructions for the purpose of implementing specific functions such as I/O control, logic, timing, counting, three mode proportional-integral-derivative (PID) control, communication, arithmetic, and data and file processing.

These controllers can automate a specific process, machine function, or even an entire production line. The PLC receives information from connected sensors or input devices, processes the data, and triggers outputs based on pre-programmed parameters. Depending on the inputs and outputs, a PLC can monitor and record run-time data such as machine productivity or operating temperature, automatically start and stop processes, generate alarms if a machine malfunction, PLC offer flexible and robust control solution, adaptable to almost any application. There are a few key features that set PLCs apart from industrial PCs, microcontrollers, and other industrial control solutions:

- **I/O**—The PLC's CPU stores and processes program data, but input and output modules connect the PLC to the rest of the machine; these I/O modules are what provide information to the CPU and trigger specific results. I/O can be either analog or digital; input devices might include sensors, switches, and meters, while outputs might include relays, lights, valves, and drives. Users can mix and match a PLC's I/O in order to get the right configuration for their application.
- **Communications**—In addition to input and output devices, a PLC might also need to connect with other kinds of systems; for example, users might want to export application data recorded by the PLC to a supervisory control and data acquisition (SCADA) system, which monitors multiple connected devices. PLCs offer a range of ports and communication protocols to ensure that the PLC can communicate with these other systems.
- **Human Machine Interface (HMI)**—In order to interact with the PLC in real time, users need an HMI, or Human Machine Interface. These operator interfaces can be simple displays, with a text-readout and keypad, or large touchscreen panels more similar to consumer electronics, but either way, they enable users to review and input information to the PLC in real time.

The use of programmable logic controller in water supply systems are used to ensure the automatic operation, implementing control loops to ensure protection when exceeding the maximum allowable flow rate, lowering the level of water or exceeding discharge pressures. Software-based programmable logic controllers (PLC) present numerous advantages over traditional hardware products. As with many applications in the industrial sector today, the addition of software systems enables greater productivity, detailed reporting for optimization and overall system flexibility [3, 9].

Advantages and disadvantages of programmable Logic Controller:

- Consumption efficiency improvement than the conventional automation panel;
- Implementation of changes and error correction;
- Testing possibilities: the program can be run and evaluated before it can be installed to perform the device management;
- Operating speed: the operating speed is dependent on the scan time of the inputs, while at present it is in the millisecond range;
- The programming mode: by introducing the diagrams, it was facilitated the access to the programming environment and for those who have no special knowledge in the field of programming;
- Documentation: very good documentation of the programs is possible by inserting comments in the spaces allocated to them, thus facilitating their continuation and debugging by other programmers;
- Increased security.

2.5 Frequency Converters

A frequency converter that is shown in Fig. 4, is an electronic component that adjust the speed and power of an electric motor by varying the frequency of the current and its voltage. Frequency converters, also called frequency variators or inverters, can automatically adjust the speed of a motor to the nominal speed required by a particular process, producing an energy saving of 10 up to 50% of the energy

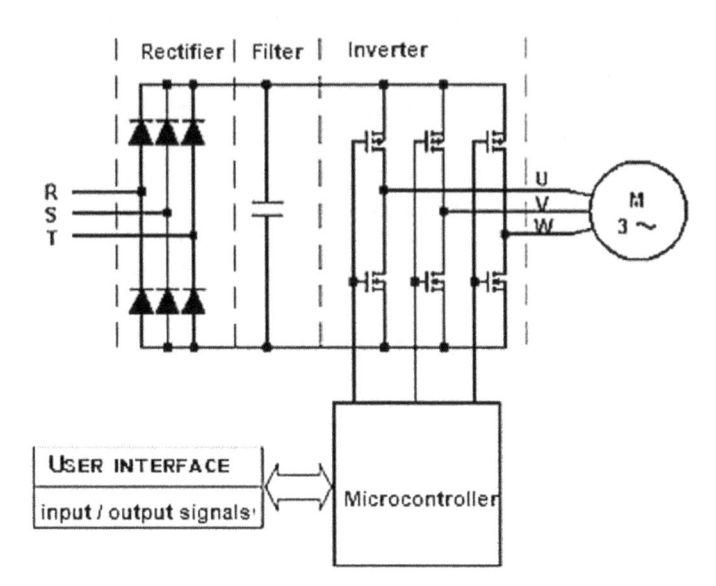

Fig. 4 Block diagram of a frequency converters

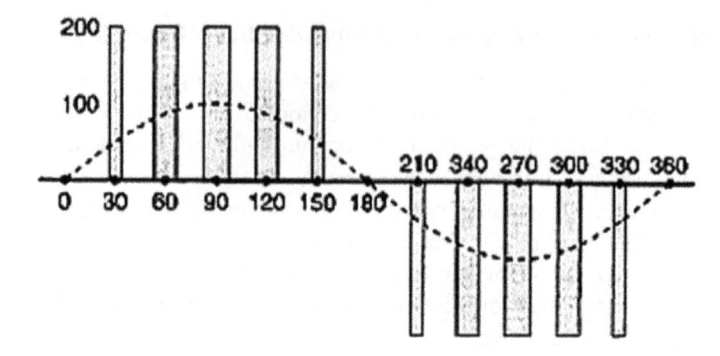

Fig. 5 Output voltage and current of frequency converters in pulse width modulation mode

consumed by the motor at maximum speed. The frequency converter adjusts and controls the frequency of the motor supply current. The frequency of the supply current is directly proportional to the rotation speed of the motor like is shown in Fig. 5. A frequency converter may be used to control the frequency and voltage of the supply current to control and to adjust the engine speed and as required by a specific application. With the change of the speed or power required by the engine, the frequency converter parameters are adjusted, so that the engine operates as per required specification for the application.

The frequency converters find their applicability in all type of electrical installations, from the appliances, to the power actuators used in the various equipment and operating the electric engine of the pumps or submersible pumps. The most important advantage of using frequency converters, both in new equipment and in refurbished ones, is the significant reduction in energy consumption. They operate groundwater capture, irrigation, sewerage, elevator, printing, fan, cutting, conveyor systems, to give just a few examples. About 65% of the electricity required for industrial consumers is consumed by electric engines, while only 10% of these engines used in the industry are equipped with frequency converters. In many cases, the engines are output controlled, using valves that regulate the flow of the actuated fluids, while the engine speed remains constant. These and other methods, such as using stepped engine or controlling the engines by stopping and starting them, are energy-inefficient methods.

There are estimates that the use of frequency converters in electric engine drives, leads to energy savings of about 120 million megawatts per hour, the equivalent of 15 nuclear reactors. In the last years, the technological advance of the electronic equipment has led to the significant reduction of the dimensions of the frequency converters. Also, modern converters use high-performance semiconductors, as well as advanced hardware and software technologies, which has led to an improvement in the performance of current frequency converters and to the widening range of benefits obtained from their use [10].

The essential benefit of using frequency converter is for reducing energy consumption and energy costs. Through the converter, both the torque and the power of the engine can be adjusted as required by your process. There is no other method of controlling the AC engines, which will allow you to achieve the same results. Systems that use actuators with electric engines consume about 65% of the electricity used in the industrial field. Optimizing the operation and control of electric engines, by using a frequency converter, will lead to increased productivity, overall, by lowering the costs of electricity used to drive motors by up to 50–60%. Thus, by using a frequency converter, depending on the type of industry in which it operates, the initial refurbishment investment can be amortized in less than 6 months. By operating the engines at the most efficient speed and power required in the application, the number of errors or failures decreases, thus increasing the efficiency level of the equipment.

The equipment used in the installations equipped with frequency converters will have a longer service life, as well as dead maintenance times. By optimizing the operation and control of the engines by the frequency converters, it reduces the time of the engines in overload, excessive heating, and overload protection. Also, a frequency converter will also act as a buffer at the start of the load of an engine, protecting the installation from the electric shock of starting the engine, regulating, instead, the load entry, by gradually increasing both speed and speed, engine power to rated values.

The advantages offered by the frequency converters:

- Dramatically reduce electrical energy consumption, implicating the cost of electricity.
- Are the most efficient means of control of engine power and rotation.
- Have the lowest current to start on all types of starters.
- Reduce over-loading and over-heating of engines and action belts.
- Are installed by simply connection to the current network.
- Can be used from small engines, used in the house, up to industrial engine groups, with high power.
- offer a high power factor, eliminating the need for other external power equipment.
- reduce the power current, protecting the electrical installation and other connected equipment [1, 4, 10, 9].

Submersible pumps used for water supply do not have a constant rotation speed. The rotation speed and water pressure must be kept constant so that the supply system is not destabilized. In order to consider keeping a linearity of the pressure values, as well as the pump rotation speed, the submersible pump speed variators are used. This variator will have automatic adjustment, without the need for human intervention at this level. Frequency converters for the submersible pump have great advantage of easy placement, in the sense that they are mounted on the pump discharge pipe, but their small size and shape allow them to be placed so that we have easy access to them. Horizontal as well as vertical mounting greatly facilitates occasional maintenance, as well as tracking variations, where necessary. The inverter protects the system even in the absence of water, stopping automatically so that the operation is not affected by this aspect. Water supply projects require safety in exploitation and energy efficiency, so these requirements can be achieved by equipping the motors with frequency converters [10].

2.6 Asset Management and Field Work Management

Major issue for water security improvement are specific measuring and control equipment that are used in water supply systems: electromagnetic flow meters, hydrostatic level transducers, pressure transducer [11]. Also, real-time quality management such disinfectant, turbidity, pH, temperature, conductivity, Redox, and installation of leak detectors in the networks.Optimized network operation and "just in time" repairs and investment programs.

2.6.1 Geographic Information System (GIS)

Applications that are covered include, development of hydraulic models, creation of thematic maps of the model output results, network simplification for hydraulic modeling, estimation of node demands, estimation of node elevations, water main isolation (i.e., identifying the valves to be closed for repairing or replacing a broken water main), and delineation of pressure zones. Here, GIS is used to organize the data for usage in water distribution networks design, and analysis.

In addition, GIS is used as a tool for number of created applications for network management; such as identifying valves to be closed in case of pipe break, service area for treatment plants, and network. GIS is used to provide graphical display of results obtained from both hydraulic simulation, and optimization models; linking tabular data with geographic locations, and graphical drawing. GIS application is used to create an integrated model for water distribution networks. Creation of digital vector maps; followed by geodatabase creation to store network data. Then, building geometric networks is needed to ensure accurate network drawing, followed by topology rules creation to ensure accurate spatial relationships. Finally, relationship classes are applied to link external model's data with GIS database like is shown in Fig. 6 [12].

In this GIS platform, water and sewage networks in Timisoara and peri-urban localities are modeled that is shown in Fig. 7. Almost 1200 km of water network and 950 km of sewerage networks [11]. The web portal that is integrated with GIS is a powerful instrument used to collect, store and represent the distribution networks and optimize network maintenance work, and provide a framework for continuous improvement represent in Fig. 8. A methodology for using GIS in water distribution networks to reduce the time needed to collect and store data in the distribution networks. Transferring data between the GIS system and hydraulic analysis models helped optimize the engineering design and analyses. Results of the hydraulic analyses (e.g. pressure, flow… etc.) and those of Optimization model (e.g. Pipe diameter) can be displayed in the GIS, and in combination with other layers such as the topographic layer of the city, greatly assist in the understanding of the network behavior and identifying the critical zones in network. Using GIS has enabled prompt action to identify problems (e.g., in case of pipe breakage, service areas) in the system

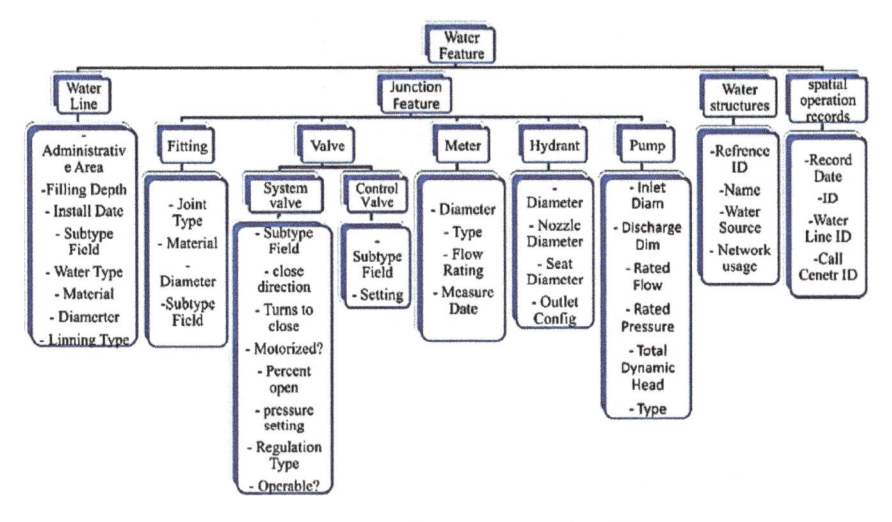

Fig. 6 Detailed GIS water distribution model data representation [12]

Fig. 7 GIS platform developed by ArcView GIS, *source* Aquatim S.A

followed by quick solutions to optimize network maintenance work and provide a framework for continuous improvement [11, 13].

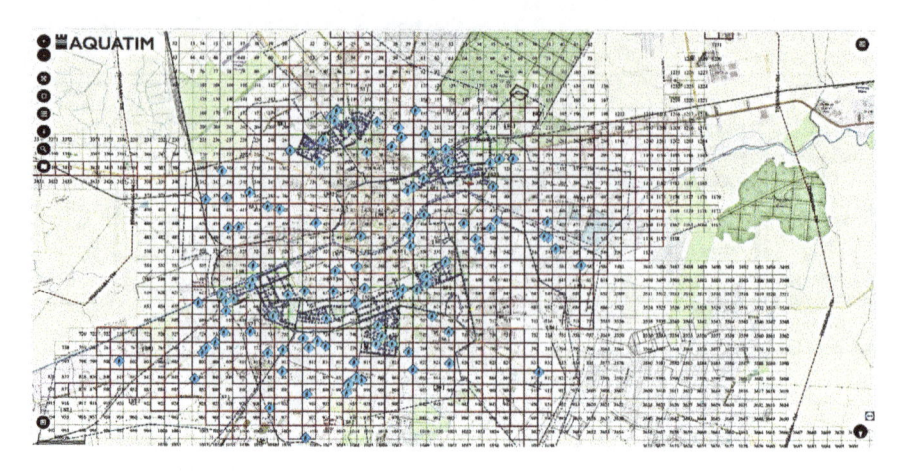

Fig. 8 Web portal platform based on GIS, *source* Aquatim S.A

2.6.2 Enterprise Resource Planning (ERP) and Asset Management

ERP software is used to manage the day-to-day business of a company. It helps manage things like accounting, human resources (HR), procurement, customer relationships (CRM), supply chain, manufacturing, engineering, maintenance, projects, service and more. An ERP solution integrates the above functions into one central piece of software, ensuring the business is operating as efficiently as possible and providing the organization with a single source of the truth for better decision-making. In short, ERP software is a suite of applications that encompasses the value stream of an organization in a single environment.

The benefits an ERP system can bring will depend on multiple factors, including the industry and business model. Initially, MRP systems were designed to speed production while reducing inventory. Among manufacturing companies, limiting cash tied up in inventory while speeding time to delivery and increasing order fill rates are frequently reported benefits of ERP. Inventory is just one resource ERP products touch though. As demand signals for human capital also travel from the sales forecast through to booked business, HR teams can hire and ensure enough people are available and adequately trained or certified to perform the work that is required. Many of today's water utility managers face the challenge of cutting their unacceptably high levels of nonrevenue water. Aside from decreasing revenue, money for investment, productivity, and service delivery, nonrevenue water also increases the cost of the delivered water [14].

Asset management involves achieving the least cost and least risk of owning and operating assets over their life cycle while meeting service standards for customers. Consequently, utility managers need to put in place policies, plans, and strategies. They must also develop and implement a suite of processes that cover asset acquisition, operation, maintenance, overhaul, and disposal. Asset management also means applying tools that help make these processes effective, such as setting service levels, computing life-cycle asset costs, maintaining an asset register, monitoring asset condition and performance, and carrying out risk analysis of possible asset failure. Asset management for water utilities is more complex than for most other sectors because of the number, variety, age, condition, and location of assets; the magnitude of asset investment; and the difficulty of inspecting and maintaining buried assets. This complexity is often compounded by lack of finance, information, and skills that can impede acquiring, commissioning, maintaining, overhauling, and replacing assets at the optimum time.

Asset management is a long-term commitment. Leading utilities that set benchmarks for others recognize that asset management is crucial to their results. Consequently, they have invested in improvements to systems and practices over many years and have fostered an asset management attitude in their workforce. Asset management is for all. Utilities of all sizes will have some form of asset management in place. Many utility managers will also be aware of inadequacies in their systems. However, they may be unaware of the scope and influence of sound asset management and how to improve their present policies, systems, practices, and attitudes.

This guide will have achieved its objective if utility managers start to improve asset management by checking the status of present practices. Utility managers should then develop and implement a remedial plan that recognizes that some solutions and targeted results may take years to achieve, while others can be implemented in much less time. Utility managers can seek help to improve asset management from system providers, consultants, leading utilities, utility associations. The ultimate beneficiaries of better asset management are the utilities' customers. Timely delivery and availability of quality water and wastewater services enables customers to enjoy a better quality of life and livelihood [14].

3 Discretization of Water Bodies Through Hydro Informatics Modeling Tools

A water body is a certain clearly distinguishable part of surface water, such as a lake, a stream, river or a part a stream or river. A groundwater body is a certain volume of water under the surface, a part of a so-called aquifer. The water body is an important entity in the Water Framework Directive (WFD) and enables us to relate water protection to natural hydrological units.

3.1 Modeling Underground Flow Using Hydro Informatics Tools

Groundwater is a natural heritage, which is critical for life and society. That is why it is necessary to use them rationally, as well as to protect against external factors that may affect the quality of the extracted water. Knowledge of the evolution over time of the piezometric load of aquifers in relation to the change of infiltration supply conditions, as well as the variation of chemical indicators of water used by catchments, are necessary to establish an efficient usage regime and maintain a good groundwater quality.

An intelligent information system is a set of people, equipment, software, processes and data designed to provide active information to the decision-making system. The informatics system is a part of the information system, which includes the automatic collection, processing and transmission of data and information within the information system. The informatics system is a system in which data and information about the discretized area through the prism of the available elements, can be reproduced automatically, in graphic format or in the form of documents specific to the discretized informatics system. However, this data and information must be spatially referenced so that all studies and analysis provided to the stakeholders which reflect reality and are precise located, the stakeholder's request, knowingly [15, 16].

3.2 Informatics Systems

The rapid and complex development of society has inevitably led to a significant increase in the volume of information, which tends to clutter and block information channels. On the other hand, the need for information for an operative knowledge and a conscious direction of the development of the society, is increasing. The purpose of informatics systems is precisely to resolve these contradictions between increasing the volume of information and the thirst for information. In order to make fast and optimal decisions, it is necessary to increase the efficiency, in the collection, updating, processing and presentation of information, as well as a superior capitalization of them. This requirement cannot be met in the face of an increasing volume of information other than by the use of means and techniques specific to informatics systems. The practical application of methods and techniques in the field of informatics systems also depends on the correct definition and acquisition of the basic concepts with which these systems operate. Informatics systems help us manage what we know, by organizing, storing, accessing, retrieving, manipulating, synthesizing and applying knowledge to solve problems. Basically, a system is a set of interconnected elements that act together to achieve a particular goal.

The main features of such a functional system are:

- **It has an objective**—any system has a purpose or an objective, which within the system, can be more difficult or easier to ascertain and define;
- **It is an ensemble**—any system consists of at least 2 distinct elements; each of these components has a definite role in achieving the objective of the system;
- **It has interconnections**—in order for the component elements (at least 2) to work together, they must be connected to each other; the connections between them are

called interconnections; the purpose of this link is transmitted to the results of its functions;
- **It allows processing**—in any system there is a transformation of a certain subject, subject to processing; any system receives something from the environment outside the system and transmits something else to the environment in which the system is located;
- **It has input / output**;
- **It allows hierarchy**—system of systems approach through subsystems;
- **It allows limitation**—any system is primarily limited in space (has a beginning and end) and has limits in time through life-cycle;
- **It allows homeostasis**—represents the property of a system to maintain its operating state within the limits of achieving its objectives (system capacity) and to change its operating parameters.

The Information System of an activity constitutes the ensemble: information, sources, consumer levels, traffic channels, procedures and the means of processing the information within the respective activity. Any specific activity that provides a specific information system, must provide complete information in sufficient quantity, accurate and at the level of efficiency required by consumer levels. The element that determined the qualitative leap of information systems was due to the development and improvement of data processing and automation procedures. This necessitates a system that represents the automated response with the help of an information system. Information Systems represent a technical and organizational set of: people, equipment, rules, methods, and aiming: collection, validation, storage, analysis (processing) [15, 16].

3.3 Advanced Hydroinformatics Tools

Groundwater flows are generally complex and difficult to delimit. In order to protect groundwater sources, it is essential to have detailed information on flow patterns and drainage to anticipate the potential spread of contamination. In many cases it is useful to know the behavior of the aquifer, especially about its flow velocity, accumulation capacity and permeability. By modeling and simulating groundwater flow underground, this information of interest can result.

Advanced computer modeling software allows us to determine the speed of water and the probability of pollution risks, to reach catchment wells, rivers and lakes. Based on the data collected from the field, numerical models can be created in several layers that represent the hydrogeological situation, including barriers, drainage and supply systems. By examining data and information such as precipitation, filtration and water infiltration, a detailed representation of groundwater runoff can be made.

Advanced hydro-computer modeling software allows us to simulate different patterns of water flow. By adding information about capture or drainage, we can anticipate the possible path of contamination migration and other risk assessment scenarios. The potential uses of groundwater models are limitless. The information may be useful in establishing a plan for the protection of aquifers, to assist in design work or to increase water supply or rational exploitation in terms of sustainability. Advanced informatics models for hydro-computer modeling are Mike 11—FeFlow module and PMWIN (Processing MODFLOW for Windows) respectively.

3.3.1 Informatics Program FEFLOW

FEFLOW software, shown in Fig. 9, is a professional software package for modeling underground flow and transporting dissolved elements and/or thermal transport processes underground. It offers pre- and post-processing functionality, an efficient simulation engine, developing a friendly graphical interface that allows easy access to advanced modeling options. Feflow is used by leading research institutes, universities, consulting firms and government organizations around the world. Its scope ranges from simple large-scale simulations to large-scale complexes [17].

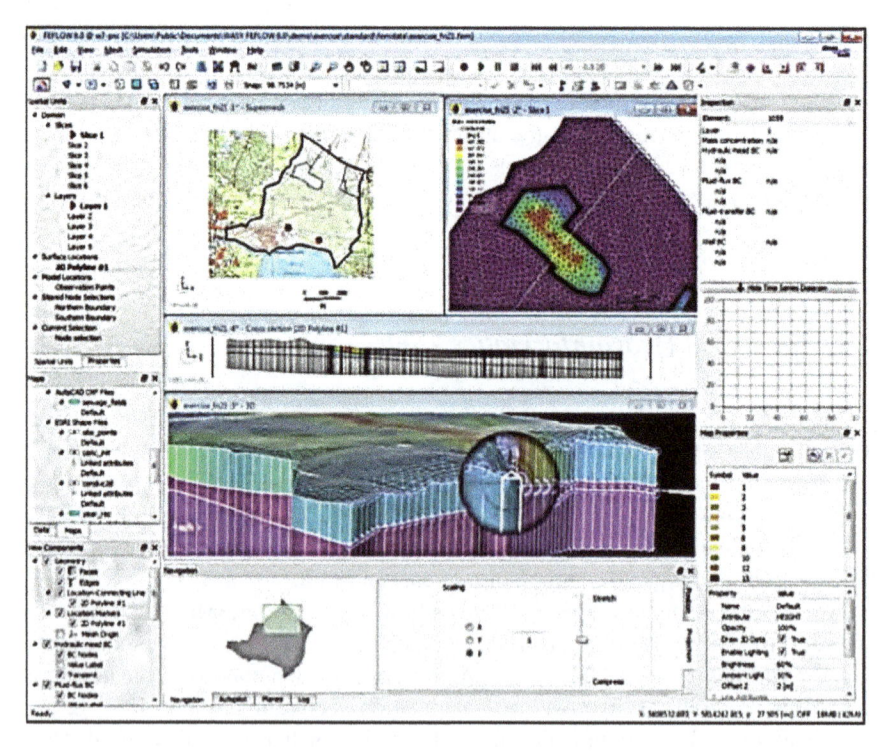

Fig. 9 Feflow software—interface

3.3.2 Informatics Software 3D-Groundwater Modeling—PMWIN

PMWIN comes with a professional graphical interface, supported models and programs, and several other useful modeling tools. The graphical interface allows you to create and simulate models. The program can import DXF and raster graphics and can manipulate models with up to 1,000 stress periods, 80 layers and 250,000 cells in each model layer. Modeling tools include a presentation tool, a results extractor, a field interpolator, a field generator, a water budget calculator, and a chart viewer.

Result Extractor allows the user to extract the simulation results from any period in a spreadsheet. The simulation results include hydraulic heads, extractions, cell–cell flow terms, compactions, subsidence, Darcy velocities, concentrations, and mass terms. The field interpolator takes the measurement data and interpolates the data to each model cell. The network pattern can be spaced irregularly. The water budget calculator not only calculates the budget of user-specified areas, but also the exchange of flows between such areas. This feature is very useful in many practical cases. It allows the user to determine the flow through a certain limit.

The field generator generates fields with heterogeneously distributed transmission values or hydraulic conductivity values. It allows the user to statistically simulate the effects and influences of small-scale unknown heterogeneities. The field generator is based on the algorithm of Mejia (1974). The graphical viewer temporarily displays development curves for simulation results, including hydraulic heads, pulls, decreases, compaction, and concentrations [18].

This is a 3D-Groundwater model that represent a groundwater borehole explorer that shows the concentration areas and catchment area of the drilling pump (blue paths). A block with low permeability is located in its center the pattern, that is shown in Fig. 10.

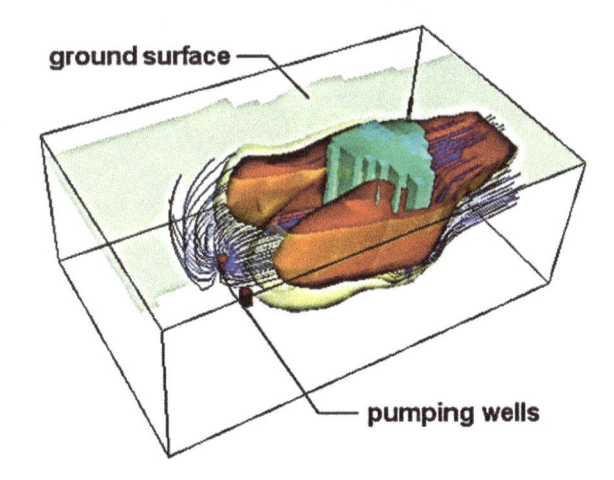

Fig. 10 PMWIN software—digital discretization project

3.3.3 Informatics Software Mike 11

Mike 11 is a 1D modeling system for rivers and canals, including built structures, MIKE 11 is synonymous with modeling rivers at the highest level covering several areas of application, that is shown in Fig. 11. It can be used for hydrodynamic modeling (rivers and estuaries, irrigation systems, dam/dam breakages, etc.), qualitative modeling (dissolved oxygen balance, eutrophication, heavy metals, swampy areas), advection/dispersion modeling (salt water intrusion, temperature) or for modeling sediment transport (cohesive sediments, morphological modeling). The connection of MIKE 11 with GIS and ArcGIS can be performed in order to extract cross-sections, contour river basins and flood maps. In order to achieve these goals or to be used as a forecasting tool, MIKE 11 can connect with ArcGIS, Google Earth or NASA Worldwide. MIKE 11 can be linked to other types of models, such as: Simulation of integrated captures (MIKE 11–WWTP); Mike Floods (MIKE 11–MIKE 21) [17].

MIKE URBAN—is a modeling program for urban water and is chosen when the important parameters for modeling are: degree of use, working mode, openness, flexibility, GIS integration as well as correctness, efficiency and stability of simulation engines. MIKE URBAN covers all types of water in cities, including:

- sewerage—separating, unitary or mixed systems;
- storm water drainage systems, including 2D flow;
- water distribution systems [17].

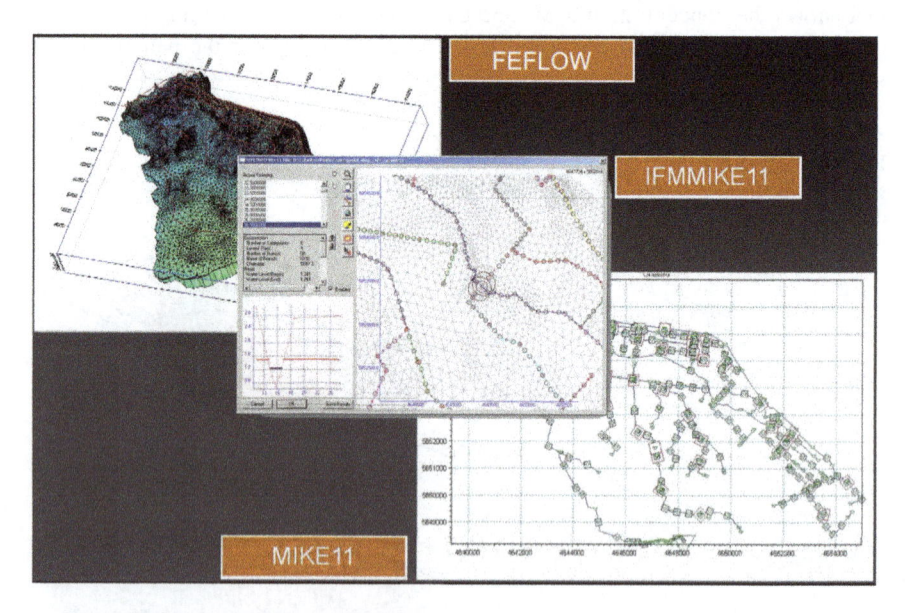

Fig. 11 MIKE 11 software digital schematic projects

MIKE URBAN is a complete integration of GIS and water modeling. WEST—is a powerful and easy-to-use tool for dynamic modeling and simulation of wastewater treatment plants (WWTP). The most interesting part of the model is that it allows modeling and evaluation of almost any modern treatment method (WWTP) [17].

WEST is designed for engineers or professionals who want to simulate the physical, biological or chemical processes that take place in water. With a mathematical model of a WWTP, it is possible to test any scenario of interest to track the efficiency of the treatment process, as well as the operational costs [17].

MIKE 21—is the most efficient coastal water modeling program. If it is necessary to simulate physical, chemical or biological processes in coastal and marine areas, MIKE 21 is accurate enough [17].

MIKE 3—provides simulation tools where 3D surface flow modeling and associated solid or water transport processes are needed. MIKE 3 is widely recognized as the gold standard for environmental and ecological studies.

ECO LAB—allows the transformation of any aquatic ecosystem into a reliable numerical model for accurate predictions [17].

MIKE FLOOD—is the most complete set of hydro informatics tools available for flood modeling. It includes a wide range of 1D and 2D flood simulation engines, allowing almost any type of flood to be modeled: rivers, floodplains—meadows, street floods, drainage networks, coastal areas, dams and discharges over dams, or any combination of the above [17].

MIKE SHE—is a software for modeling the major processes involved in the terrestrial component of the hydrological cycle. Thus, MIKE SHE models water flow, water quality and sediment transport, being successfully used in modeling small river basins. This software derives from System Hydrologique European (SHE), developed in 1977 by the British Institute of Hydrology, SOGREAH (France) and DHI (Denmark) [17].

MIKE 21C—is one of the most powerful hydro informatics tools used to simulate morphological changes of watercourses and canals due to changes in hydraulic regime [17].

4 Conclusion

To achieve and maintain water security is needed to ensure access to safe and sufficient drinking water at an affordable cost in order to meet basic needs, including sanitation and hygiene, and safeguard health and well-being. Also, preservation and protection of ecosystems in water allocation and management systems to maintain their ability to deliver and sustain functioning of essential ecosystem of services. Industrial control system (ICS) are in general several types of control systems, including supervisory control and data acquisition (SCADA) systems, distributed control systems (DCS), and other control system configurations such as Programmable Logic Controllers (PLC) use in the industrial sectors and critical infrastructures. SCADA systems are designed to collect field information, transfer it to a central computer facility, and display the information to the operator graphically or textually, thereby allowing the operator to monitor or control an entire system from a central location in near real time.

The use of programmable logic controller in water supply systems are used to ensure the automatic operation. As with many applications in the industrial sector today, the addition of software systems enables greater productivity, detailed reporting for optimization and overall system flexibility. The essential benefit of using frequency converter is reducing energy consumption and energy costs. Using GIS has enabled prompt action to identify problems in the system followed by quick solutions to optimize network maintenance work and provide a framework for continuous improvement.

ERP system can bring will depend on multiple factors, including the industry and business model. Initially, MRP systems were designed to speed production while reducing inventory. Advanced computer modeling software allows us to determine the speed of water and the probability of pollution risks, to reach catchment wells, rivers and lakes. Based on the data collected from the field, numerical models can be created in several layers that represent the hydrogeological situation, including barriers, drainage and supply systems. The smart water solutions are based on numerous methods and tools that could, under a consistent vision, contribute to build an efficiency water management targeting the objectives of the water security.

References

1. United States Department of Agriculture (2010) Natural resources conservation service. Engineering Technical No. MT-14
2. Dustin G, Hall JW (2014) Water security and society: risks, metrics, and pathways. Environ Res 39:611–639
3. NIST Special Publication 800–82—National Institute of Standards and Technology
4. Cococeanu A, Cretan I, Cojocinescu M, Man T, Pelea G (2017) Water wells monitoring using SCADA system for water supply network, case study: water treatment plant Urseni, Timis County, Romania, WMCAUS, Praque
5. https://inductiveautomation.com. Accessed 12 Mar 2021
6. Ritchie E (2011) The new face of SCADA. Water Efficiency 6(3):28–33
7. https://www.samanalyticsolutions.com/siemens-scada-software/. Accessed 12 Dec 2020
8. Eric L, Manou A, Annemarie Z, Eric L, Manou A (2011) Assessing and improving SCADA security in the Dutch drinking water sector. Int J Crit Infrastruct Prot 4(3–4):124–134
9. Ramazan B, Yucel C, Bayindir R, Cetinceviz Y (2011) A water pumping control system with a programmable logic controller (PLC) and industrial wireless modules for industrial plants—an experimental setup. ISA Trans 50(2):321–328
10. Alcock DN (1976) Understanding solid-state variable-frequency power-supplies for variable-speed pumping. Power Mag
11. Technical archive of Aquatim SA, Timisoara
12. Haythan A, Alaa Y (2012) Geografic information system in water distribution networks—conference
13. Bentley Systems Incorporated (2004) The fundamentals of SCADA—a white paper from Bentley systems

14. Dejan P, Mojca I, Andej K (2011) Enterprise resource planning (ERP) systems: use of reference models, perspectives in business informatics research. pp 178–189
15. Berry J (2005) Water quality sensor placement in water networks with budget constraints. In: Proceedings of the world water and environmental resources conference
16. Christie LD (1989) Understanding computer control options for water resources. In: Proceedings of the 16th annual conference of the ASCE water resources planning and management division. Sacramento, CA, pp 520–523
17. https://www.mikepoweredbydhi.com/products/mike-11. Accessed 10 Feb 2021
18. https://www.pmwin.net/index.htm. Accessed 10 Feb 2021

Adrian Cococeanu was born in Caransebes Village, Caras-Severin County, CS, Romania in 1986. He received the B.S. degrees in hydrotehnic engineering from the Polytechnic University of Timisoara, Romania, in 2011 and the M.S. degree in Optimization of hydrotehnic system, in 2013 and Sustainable Development postgraduate degree, in from Polytechnic University of Timisoara, Romania. He is currently pursuing the Ph.D. degree in civil engineering at Polytechnic University of Timisoara. From 2011 to 2013, he was a research engineer in Research and New Technologies Department of AQUATIM S.A. Since 2013, he has been an responsible process engineer in Surface Water Plant Timisoara. His research interest includes the development of groundwater resources (intended for water supply) through digital instruments and hydroinformatic systems. He is the author of 10 articles.

Teodor Eugen Man is affilitaed with the Polytechnic University of Timişoara. Currently he is Professor in the field of Civil Engineering and Installations, having completed 16 doctoral theses and in internship currently having 10 Ph.D. students. Leading positions held in UPT: Vice-Rector (1996–2000), Member of the Senate (1996–2000, 2000–2001), Dean of the Faculty of Hydrotechnics (2008–2011), Director of the Department of Hydrotechnics (2011–2012), Vice-Dean of the Faculty of Constructions and member of the Management Office of the Teachers' Council (2012–2014), Director of the Research Center in Hydrotechnics and Environmental Protection (2009–2018). He has developed a number of 23 volumes of courses, laboratory guides, monographs, textbooks, of which 6 as sole author and 10 as first author, over 300 published scientific papers of which 17 sole author, 145 first author, over 50 published abroad and over 30 articles listed ISI or ISI Proceedings, 3 patents, 8 innovations, 116 research and development contracts (grants), of which 70 was responsible/project manager, technical expertise: 70, verification projects: 242. He is a member of prestigious professional associations in the country and abroad, he was a visiting professor at Artois University, IUT Bethune in France, also for 14 years he taught specialized courses at the Faculty of Environmental Protection, University of Oradea. Prof. Man's CV was also presented in the "Dictionary of Romanian Personalities" (Contemporary Biographies—ed. 2012, 2017). *Member*

of the editorial board: the Scientific and Technical Bulletin of UPT, Department of Hydrotechnics; The scientific and technical bulletin of T.U. "Ghe. Asachi" Iași; Annals of the University of Oradea, Faculty of Environmental Protection and Faculty of Constructions and Architecture; Scientific Annals of Ovidius University Constanta; Revista Hidrotehnica, A.N. "Apele Romane", Bucharest. *Professional, cultural, educational associations*: Associate member of the Bucharest Academy of Agricultural and Forestry Sciences (ASAS); Member of the General Association of Romanian Engineers (AGIR); Founding Member of the Bucharest Association of Land Improvements and Agricultural Constructions (AIFCR); Member of the Romanian National Society for Soil Science Bucharest (SNRSS); Member of the National Committee for Irrigation and Drainage (CNRID) Bucharest, affiliated to the International Commission of Irrigation and Drainage, New Delhi, India (ICID).

Groundwater Safety by Monitoring Quality Parameters in Transylvania, Romania

Maria Popa and Ioana Glevitzky

Abstract This chapter presents the safety of groundwater in Transylvania, Romania by monitoring quality parameters. At first, presents the aspects regarding the importance of groundwater for humankind, the underground water being regarded as a natural, untreated and free source. Drinking water is used in human food that meets a number of physicochemical and microbiological conditions that allow consumption without endangering health. The case study is carried out in a Romanian region (Alba County—Transylvania), the main appreciation indicators of the underground water sources quality being monitored over time. Concentration/pollution levels were determined by physicochemical and microbiological analyzes as well as specific contaminants. The groundwater quality was investigated through an integrated approach to groundwater quality, in line with the Drinking Water Standards. The methodology used is in line with international requirements. The results obtained and presented in this chapter highlight areas with uncontaminated underground water as well as areas presenting risk of disease for the population. We highlight the main contaminants/pollutants present in the studied area, as well as the main solutions for the protection of the population.

Keywords Water quality · Groundwater · Spring water · Safety

1 Water as a Necessity for Humanity

Water is a fundamental social requirement and it is an essential "element" for maintaining and developing life on our planet. Theoretically, it seems that there would be enough water for mankind on Earth, both for the present and the future. Upon a more detailed examination, one can notice that "water is often in the wrong place, at

---

M. Popa (✉)
Faculty of Economic Sciences, "1 Decembrie 1918" University of Alba Iulia, Gabriel Bethlen St., 5, 510009 Alba Iulia, România

I. Glevitzky
Doctoral School of the "Lucian Blaga" University of Sibiu, Victoriei Blvd, 10, 550024 Sibiu, România

the wrong time, in the wrong form or the quantity is insufficient for current human needs". Over time, human necessities regarding water have often evolved, ignoring the fact that "water does not exist everywhere, everytime, and at alltime, no matter how much is the quantity on our planet" [2].

Water is the most important element that cannot be replaced. Although, water is found in various foods, human cannot live without liquid water. That is why the water has been the most important to us, still is and will be, for everyone [9]. Water is a biotope factor, as it determines different types of climate, secures the water cycle in nature and influences the distribution of living creatures on the surface of the Earth. The use of water sources is determined by their size and water availability in the geographical area. Studies are needed to ensure that the water flows will ensure consumption needs and that sources will not be polluted [12].

Clean drinking water is an essential element for ensuring public health and quality of life. Throughout human history, drinking water sources have been threatened by various types of contamination. Consumption of polluted water (with modified properties) generates various waterborne diseases, so the safety of water is of paramount importance. Water quality protection is a topical issue; therefore, it is essential to classify surface waters. By framing water into certain categories of quality, the developed benchmarks are achieved with aims to protect and purify water to produce drinking quality water, both quantitatively and qualitatively. Legislation, as well as health surveillance and inspection rules, regulate the quality of drinking water, but there are serious environmental problems across the world when it comes to environment due to pollution.

Economic and social development as well as the maintenance of human health are dependent on rapid access to adequate water supply [3]. In the world, the daily consumption of drinking water, relative to the number of inhabitants, is high, as it is not only used for drinking, but also for domestic activities, public services and in food industry. However, out of the total water consumption, drinking water has the smallest proportion, nevertheless it is of primary importance. Drinking water supplies and consumption have increased with the development of urbanization and increased industrialization of the economy.

The water issue is severely affected by two causes:

- complete lack of dams and reservoirs; insufficient facilities to make the water of rivers, lakes and groundwater usable. All this makes it impossible to bring water to certain areas, in the quantity and time required, making them impossible to use for social and economic purposes.
- the increasing pollution of inland waters, as well as of sea and ocean ones.

Drinking water must meet several requirements; these are drinking conditions: physico-chemical, bacteriological and biological (legally established). For water to fulfil its role in the body, it must be in balance and there should be no quantitative or qualitative deficiencies. When setting national drinking-water quality regulations and standards, many countries consider the World Health Organization Guidelines for drinking-water quality [28]. When setting national drinking water quality regulations and standards, many countries take into account the WHO Guidelines for drinking-water quality (GDWQ). The guidelines provides a basis for setting national regulations and standards for water safety in support of public health.

The Safe Drinking Water Act (SDWA) is the principal federal law in the United States and was established to protect the quality of drinking water for the public. This law focuses on all waters actually or potentially designed for drinking use, whether from above ground or underground sources [25].

The Drinking Water Directive [7] refers to the quality of water intended for human consumption. At European level, its aim is to protect human health from the negative effects of any contamination of water intended for human consumption by ensuring that it is wholesome and clean. Water meets certain drinking criteria also depending on the type of source (depth or surface), as per municipality regulations, the way water is distributed to consumers, etc. Wells water is generally good for direct consumption if the area, especially the soil, is not contaminated. But most of the time this water is contaminated due to anthropogenic factors and is not treated, often subject to chemical and microbiological contamination due to anthropogenic sources [14, 17].

Groundwater contamination caused by human activities is an important issue of the last century. Sewage systems, animal manure, fertilizers, pesticides, and irrigation salts get into groundwater and change its constitution and quality [18–20]. In rural areas that don't have distribution of potable water through public supply systems, the wells that represent sources of groundwater are very important. This water is not treated and is often subjected to chemical and microbiological pollution from anthropogenic sources [14]. The ground water in rural areas can be polluted as a result of farming activities. The quality of the drinking water drawn in wells can be affected especially for those wells that are placed in the valleys, and at insufficient distance from the stables [16, 17].

More than 1.2 billion people around the globe do not have access to a safe drinking water source. Compared to the situation in some parts of the globe, the status of Europe's water resources is relatively favorable: the continent does not face a general or total water shortfall, and extreme problems such as drought and floods are uncommon [21]. Poor access to sufficient quantities of water can also play a key factor in water-related disease and is closely related to ecosystem conditions. About one-third of the world's population lives in countries with moderate to high water stress, and problems of water scarcity are increasing, partly due to ecosystem depletion and contamination. Two out of every three persons on the globe may be living in water-stressed conditions by the year 2025, if present global consumption patterns continue [13].

Epidemics caused by water contamination occur explosively at any time of year and manifest in the territory fed by contaminated water source. They suddenly stop once the consumption of contaminated water is stopped [5, 6]. The main manifestations of the disease are the digestive illness (enteritis). Conditions to trigger such an epidemic are due to existence of a pollution source, germ's ability to survive in water for a sufficient time to cause the disease, existence of a population susceptible to illness. Hydric diseases classification can be based on the number of illness cases, their occurrence and development mode. We can find epidemics (a large number of cases in a short time); endemics (small number of cases found continuously in an area), and sporadic progress of diseases (isolated). Curious is that cases of epidemic and epizootic diseases transmitted via water, are encountered occasionally even after water decontamination [15].

Primary biological water pollution is the consequence of the presence of viruses, bacteria and parasitic agents in water. Some tests have shown that in some households' water remain polluted after decontamination process, although the values measured on total viable count and coliform bacteria were lower. These results suggest that groundwater was polluted, which causes water pollution even after several decontamination processes [26]. *Escherichia coli* can be transmitted through water consumption and gives gastro-enteritis and sepsis at children and various species of young animals [15]. The incidence of *E. coli* is about 17.5% in well water, surviving in drinkable water from wells for as long as six months, and in sewer water for four months. A severe epidemic of gastroenteritis transmitted by water appeared in England in 1980. Infection was possible due to a faulty sewage system, which allowed drinking water contamination [8, 22].

The problem of lack of water resources is affecting an increasing proportion of the world, primarily in arid and semi-arid regions. In these areas, the population's water need is considerable, and demand is constantly growing faster than at any time before [24]. The estimated global population growth therefore leads to an increase in water demand and therefore to an increase in the deficit of water sources. The water scarcity is an insufficient water to satisfy the normal requirements of pollution. The term is little used by policy makers because they fail to recognize the degrees of water deficit and how different societies adapt to this lack of water [3]. Worldwide, more than 1 billion people do not have acces to safe drinking water, while 2.6 bilion do not have adequate sanitation. The consequence is about 1.7 million deaths a year due to diseases related to unsafe water, sanitation and hygiene [29] (Fig. 1).

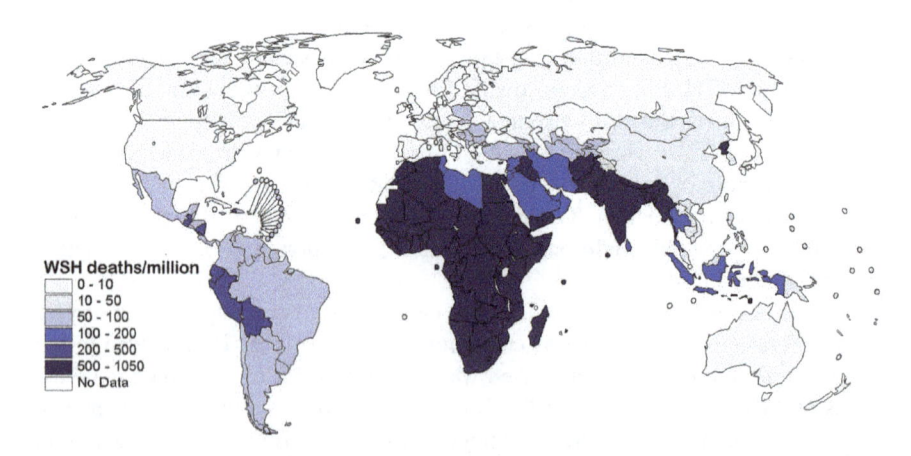

Fig. 1 Deaths for unsafe water, sanitation and hygiene. *Source* WHO World Health Report, 2002

However, after a closer examination, Europe's water is of no satisfactory quality. The freshwater deficit, predicted by specialists for the next century, will have disastrous consequences in some regions of the world, threatening the food industry, population health and, last but not least, international security. The degradation of highly polluted surface and underground waters, the reduction of the volume required for drinking water supply are alarming causes for the majority of the world's population. Another very important indicator pf significant concern is the quality of drinking water [27].

2 The Status of Aqua in Romania—Transylvania

In Romania, natural water sources are relatively poor and distributed unevenly over time and space. They generally consist of surface waters, including inland rivers, natural or artificial lakes and water from the Danube River, in addition to underground waters. Statistical data shows that urban population, but also much of the rural population, is already connected to centralized water supply systems. The flow, however, has proven to be insufficient and therefore the necessity of promoting new techniques and technologies for treating and increasing the capacity of water distribution installations has been considered and promoted by the European Union.

In Romania, drinking water is produced from surface water (60%) and groundwater (40%). Surface water, more exposed to anthropogenic pollution as groundwater contains, beside natural "pollutants", other organic/biological/microbial loads [4]. In order to cope with the ever increasing demand for potable water, the use of surface water (especially river water) as a source of drinking water has been considered as an optimal solution, as underground water was found to be insufficient to meet the requirements. The disadvantage is that surface water cannot be consumed as such but is subject to processing in order to correct physicochemical and biological properties to meet the drinking water quality requirements.

2.1 Study on the Hygienic Quality of Groundwater in Transylvania

In order to highlight the hygienic quality of the Transylvanian fountain waters, a number of 27 sources from different localities were studied for 5 quarters of the years 2018–2019, from which a sample of water was collected every three months. The figure shows the map of Europe and the location of Transylvania, Romania (Fig. 2).

The samples were taken with a quarterly frequency and the 405 analyzes were performed for conducting microbiological potability determination. The following parameters were determined for all water samples: total number of germs, *E. coli*

Fig. 2 The map of area under study—Transylvania, Romania

and enterococci. In this last part, the study summarizes the results of microbiological control performed on spring/well water samples, from different areas: plains, hills and mountains, in Transylvania, Romania. These were harvested in different seasons from springs or more or less properly maintained wells. In the Figs. 3, 4 and 5

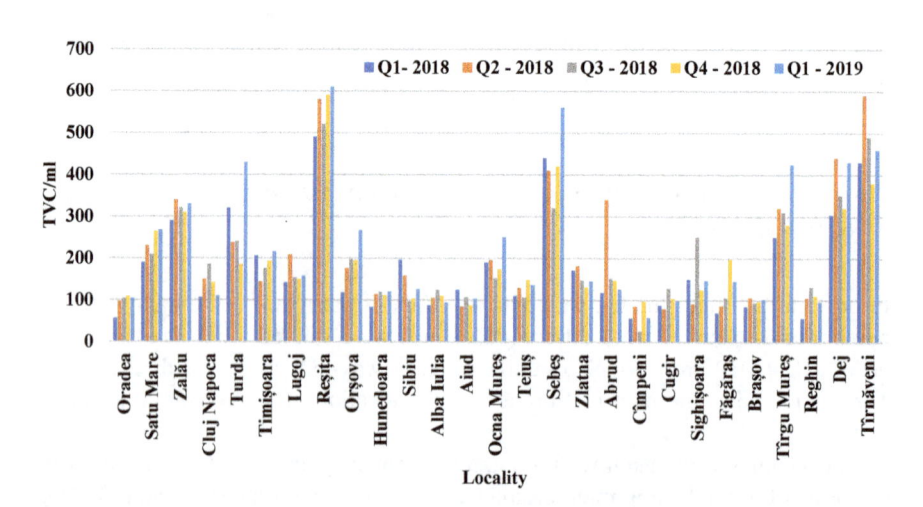

Fig. 3 Results for total viable count in water samples from Transylvania

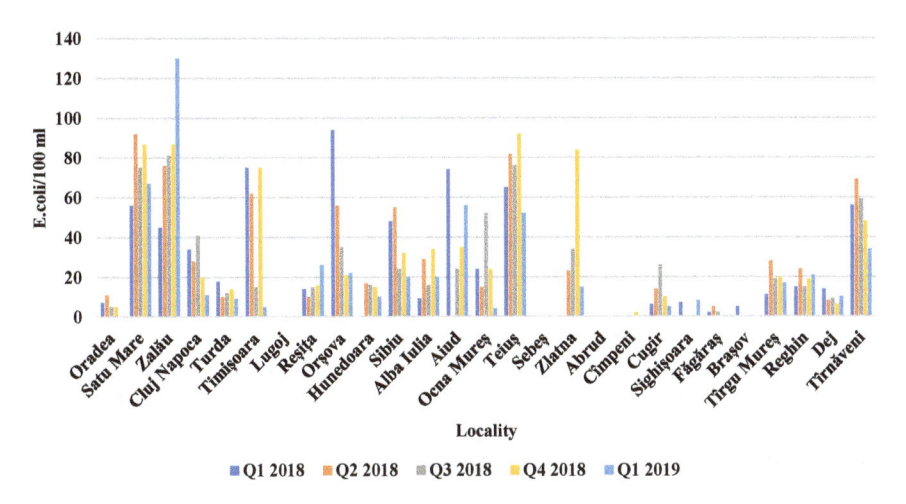

Fig. 4 Results for *E. coli* in water samples from Transylvania

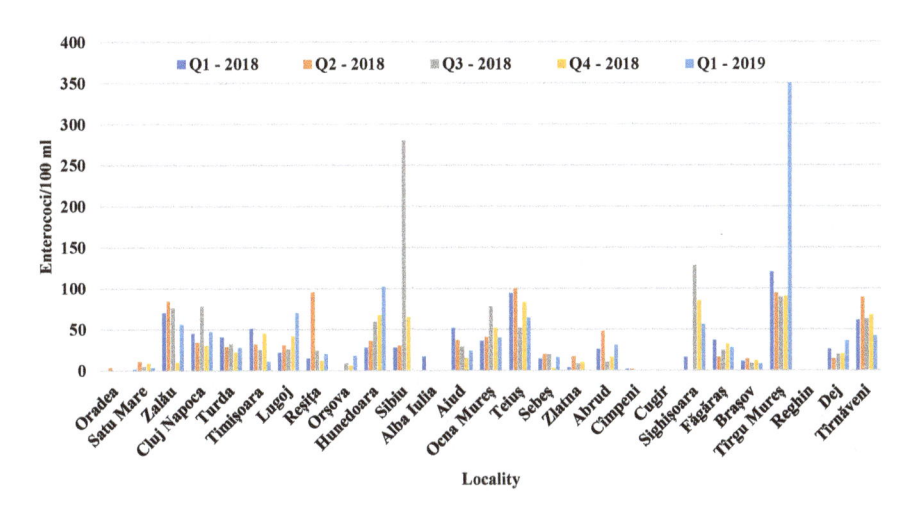

Fig. 5 Results for enterococi for the water samples from Transylvania

the microbiological parameters for the samples collected from 29 localities in the Transylvania area, Romania are presented.

In nearby areas, water samples taken from springs/wells located in different areas (plains, hills, mountains) have a different microbial load but close in value. There is therefore a diffuse microbiological pollution, due to either non-controversial fertilization of farmland, chaotic landfill, manure or defective sewage system. The importance of the total viable count (TVC) results from the ratio of microbial water load and the probability of pathogenic germs. If the number of colony-forming units it is high, also the risk of pathogens in water is higher. The data obtained from the ground water quality investigation aimed at identifying specific impurities, areas and sources of pollution. Thus, increased TVC/ml indicates the risk of the presence of other pathogens (bacteria, viruses, fungi, parasitic agents) in water.

As a result of exceeding the admissible values for coliform bacteria, including *E. coli*, water from the springs/wells can be considered to be polluted due to infiltrations from animal manure, defective sewage system or wastewater [1, 10]. *Escherichia coli* is mainly found in human and animal faeces. If the presence of *E. coli* bacteria is accompanied by high levels of nitrates and chlorides, this usually indicates contamination from wastewater or septic tank through infiltration or cracked pipes. Four or less colonies/100 ml coliforms in the absence of high nitrate and chlorine levels imply the presence of surface water infiltrations. At the same time, faeces may come into contact with natural waters, such as lakes, rivers, water reservoirs, under non-compliance with hygiene rules. By ingestion or aspiration, the body comes into contact with strains with pathogenic potential.

According to the experimental results, a severe contamination of groundwater is observed. Isolation of enterococci from the water sources of the area can cause a wide variety of infections. There is a variation of the bacteriological indicator—enterococci depending on the season (quarter), but also area. Faecal streptococci (enterococci) being specific for humans and animals, with higher resistance in the external environment compared to coliform bacteria and low variability, provides data on the source of pollution.

As a result of the microbiological analyzes carried out on the collected samples, it is observed that drinking water samples (from wells/water holes) are microbiologically polluted, irrespective of the source they come from and consequently the life of animals and humans is exposed to hydrological diseases. A large number of samples reacted positively to one or all of the determined parameters. The presented water samples are microbiologically polluted, irrespective of the way they work, the way the wells are fitted, the season or the area they come from; so the quality of water is no longer given by its depth, the type of construction and the mode of operation. The presence of *E. coli* in groundwater indicates a contamination due to wastewater, non-compliant sewage systems or manure from animals. These can easily lead to epidemics if water is contaminated in the water reserve of larger settlements and in the absence of efficient chlorination and disinfection systems.

There is also a diffuse microbiological pollution in some areas, which indicates that the groundwater is polluted and therefore the natural purification phenomena suffer transformations; nature is no longer able, through its own means, to fulfill its role as purifier. Depending on the season and place, the groundwater presents variations in flow, physical, chemical and biological properties, becoming unfit for consumption. The water supply from the fountains is of a relative safety, requiring minimal preventive measures: fitting works, wells protection (paved perimeter, covering, etc.); regular quality control; periodically performed decontamination and decontamination checks. Experimental determinations have validated the premise that pollution occurs as a result of human activities, as well as the fact that water is more contaminated than in the mountain area. Reducing groundwater pollution (which affects households having their own wells) can be achieved by removing absorbent wells that degrade the environment by quickly removing waste of any kind, but especially by connecting to the city's seweage system.

2.2 Study on Water Quality in Alba County

Taking into consideration global drinking water security issues as well as reports and studies on the water shortage on the Earth, the first step was to conduct an analysis of the data on spring water sources and public wells in Alba County—Transylvania. Thus, in 2017, a number of 132 public water sources in Alba County were monitored. For better data management, the area of Alba County was divided by the authors into 5 areas (Fig. 6) as follows: Zone I: Alba Iulia—Teiuş; Zone II: Sebeş—Cugir; Zone III: Cîmpeni—Zlatna zone IV: Blaj and Zone V: Aiud—Ocna Mureş.

The areas investigated had been established taking into account primarily the delimitations of the neighbouring urban and rural settlements, the relief, the demographic considerations, specifically the degree of development of the area. In the following sections results of the analyzes are presented, according to the requirements imposed by the Community legislation in the field of drinking water, at the same time highlighting areas with higher water potability compared to those unfit for

Fig. 6 Alba County with water sources investigated in 5 areas

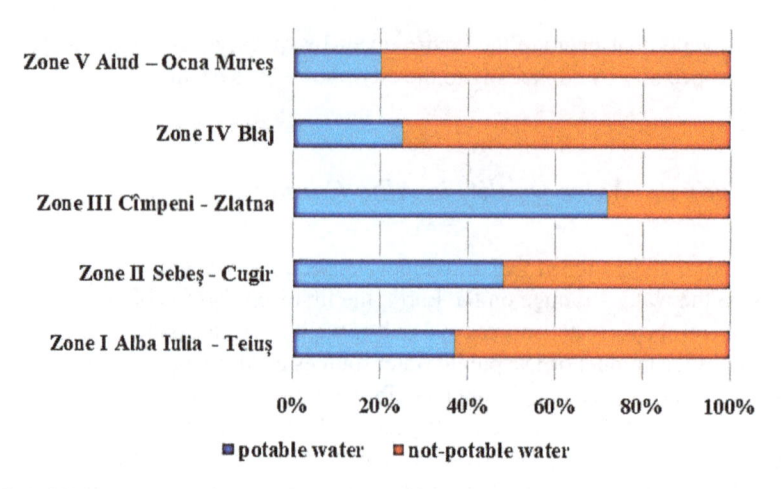

Fig. 7 Public water sources in Alba County in the 5 areas

human consumption. For the results of the analysis to be conclusive and relevant, the physico-chemical and microbiological parameters have been investigated according to the requirements agreed at European level. According to the analysis of water quality indicators the results are presented in Fig. 7.

The analysis of the obtained results showed that most of the investigated underground water sources (wells and springs) do not meet the standards required for drinking water. The furthest from the standard can be found in Zone V—Aiud-Ocna Mureş (80%), while the closest in zone III—Câmpeni-Zlatna (72%). We also observed that in the mountain region (zone III) water quality is superior when compared to the lower regions.

2.2.1 Study on Groundwater Quality in Zone I, Alba Iulia—Teius

Research on groundwater quality survey in Zone I, Alba Iulia—Teius, were directed on the simultaneous analysis of spring and well water samples collected from the Alba Iulia area and its neighborhoods. Thus, in March 2018, 10 samples of public spring water and 10 samples of fountain water from different areas were harvested. Sampling locations were randomly selected to have the widest possible coverage.

The results obtained from laboratory analysis—the physico-chemical parameters for samples of spring water and wells collected in zone I show a pH between 7.23 and 8.29 (the legal values in force are between 6.5 and 9.5). The total water hardness of samples is between 21.31 and 37.02° dH, acidity between 0.6 and 1.3 mg/l and alkalinity between 4 and 8.1 mg/l. The measured values of ammonium ions were in the maximum rated value of 0.50 mg/l, with one exception. Ammonia occurs after water pollution by organic substances that undergo decomposition, the first term of nitrogenous substance degradation. Its presence indicates recent pollution. The water samples from wells did not exceed the maximum rated value of nitrites

(NO_2^-). Nitrites are found due to the pollution of water with organic matter, by partial oxidation of the amino radical or by reducing nitrate. Their presence indicates an older water pollution, but together with higher concentrations of ammonia show that the pollution is continuous. The nitrates (NO_3^-) content of water from the analyzed samples did not exceed maximum legal value for drinking water of 50 mg/l, and varies between 0.2 and 28.9 mg/l. Table 1 presents the positive experimental results for the microbiological parameters of the analyzed waters—public water springs and fountain waters from zone I, Alba Iulia—Teius.

The most microbiological polluted waters are the fountains. In six of the ten samples, the values of the hygiene parameters were exceeded. The incidence of source water contamination is lower, only two of the ten sources investigated are not within the limits permitted by current legislation. The results for water analyzes according to the legislation are expressed only by reporting TVC/ml, *E. coli* and enterococci/100 ml. The total coliform bacteria indicator is no longer used, although, in the working method for *E. coli*, the steps of their determination must be followed. *E. coli* and coliform bacteria are present in most sources, and enterococci are present in half of the analyzed samples. The pollution may be caused by rain that has infiltrated into the ground. This indicates a contamination source due to poor planning of the fountains. *Pseudomonas aeruginosa* was not identified in analyzed sample.

Table 1 Positive laboratory results for microbiological parameters of Zone I waters

Source	Parameter					
	TVC/ml		Coliform bacteria/100 ml	*E. coli*/100 ml	Enterococi/100 ml	*Pseudomonas aeruginosa*/250 ml
	At 22 °C	At 37 °C				
Spring 3	1	32	Abs	Abs	Abs	Abs
Spring 7	37	76	42	34	17	Abs
Fountain 1	190	432	401	348	23	Abs
Fountain 3	3	80	22	15	Abs	Abs
Fountain 5	5	29	13	13	Abs	Abs
Fountain 6	20	95	31	23	49	Abs
Fountain 9	140	162	20	abs	130	Abs
Fountain 10	42	109	102	79	Abs	Abs
Normal values	100	20	Absent	Absent	Absent	Absent

2.2.2 Study on Groundwater Quality in Zone II, Sebeş—Cugir

For the study, 10 samples of fountain water from public sources from different localities of Zone II, Sebeş—Cugir were collected. From the point of view of the physical-chemical parameters, the pH of the analyzed waters varies between 6.9 and 7.8. Total water hardness is between 14.3 and 65° dH, acidity between 1 and 6 mg/l and alkalinity between 1 and 5.4 mg/l. For nitrites (NO_2^-) content, the obtained results are between 0.02 and 0.55 mg/l, respectively for nitrates (NO_3^-) between 10.36 and 50.02 mg/l. For the ammonium (NH_4^+) content, the experimentally results are between 0.15 and 2.17 mg/l. There are higher values tham the legally allowed limits, namely by 0.50 mg/l for nitrites, 50 mg/l for nitrates and 0.50 mg/l for ammonium ions. Nitrites (nitrates) occur as a result of pollution with organic substances, either by partial oxidation of the amine radical or by nitrate reduction. The presence of nitrates generally indicates the polution of water with faeces. Ammonium ions occur from water pollution with organic substances, which are decomposing, being the first term of nitrogenous substance degradation. Its presence indicates a recent pollution. Table 2 presents the values of the microbiological parameters that exceed the legal limits in the water sources analyzed in zone II, Sebeş—Cugir.

A major groundwater contamination has been identified. The values determined for coliform bacteria, *E. coli* and enterococci in the fountain waters taken from zone II are relatively high, making it unsuitable for human consumption. *E. coli* is one of the bacteria commonly found in water and is caused by coming into contact with the feces, or stool, of humans or animals [11]. The presence of the microorganism, along with high nitrate and chlorine values, show contamination from the sewage system: domestic water, septic tanks or cracked pipes. Overall, the results of laboratory analyzes show that anthropogenic pollution influences the quality of the Zone II Sebes—Cugir II fountains, where there is a major contamination of these, especially microbiological, all the activities that take place on the surface having a possible impact on the quality of aquifers.

2.2.3 Study on Groundwater Quality in Zone III, Zlatna—Cimpeni

For the study, 10 samples of spring water from Zone III, Cîmpeni—Zlatna a mountain area of Transylvania, were collected. It is determined that for all collected samples, the pH is within the limits, with values between 6.55 and 7.39. The hardness of spring water varies between 7.18 and 17.27° dH, acidity between 0.5 and 1.2 mg/l and alkalinity between 0.7 and 4.4 mg/l. The experimental results obtained for samples of spring water harvested in zone III, Zlatna-Cimpeni are shown in Table 3.

For NO_2^-, NO_3^- and NH_4^+ the measured values are within the legal limits, which indicates better hygienic water quality in the mountain area. In Table 4 the values of the microbiological parameters for the analyzed water sources in zone III, Zlatna-Cimpeni are presented.

All analyzed samples were negative for the presence of *Pseudomonas aeruginosa* and *E. coli*. At the same time no coliform bacteria have been identified. Enterococci

Table 2 Positive results for the microbiological parameters of Zone II waters

Source	Parameter					
	Number of colonies/ml		Coliform bacteria/100 ml	E. coli/100 ml	Enterococi/100 ml	Pseudomonas aeruginosa/250 ml
	At 22 °C	At 37 °C				
Fountain 1	162	150	141	120	2	Abs
Fountain 2	206	52	361	160	94	Abs
Fountain 3	148	198	204	145	56	Abs
Fountain 4	200	170	120	94	9	Abs
Fountain 5	120	17	Abs	Abs	16	Abs
Fountain 6	15	27	10	10	Abs	Abs
Fountain 7	47	89	72	53	Abs	Abs
Fountain 8	120	18	Abs	Abs	4	Abs
Fountain 9	13	74	29	18	abs	Abs
Fountain 10	110	4	Abs	Abs	Abs	Abs
Normal values	100/ml	20/ml	Abs	Abs	Abs	Abs

Table 3 Results of laboratory analyzes for physico-chemical parameters of harvested spring water

Source	Parameter		
	NO_2^- (mg/l)	NO_3^- (mg/l)	NH_4^+ (mg/l)
Spring 1	<0.02	22.0	0.15
Spring 2	<0.02	0.4	<0.05
Spring 3	0.05	21.0	0.37
Spring 4	<0.02	2.8	<0.05
Spring 5	<0.02	9.0	0.2
Spring 6	<0.02	1.5	<0.05
Spring 7	<0.02	2.04	<0.05

Table 4 Results for microbiological parameters of spring water harvested in area III, Zlatna-Cimpeni

Source	Parameter					
	Number of colonies/ml		Coliform bacteria/100 ml	E. coli/100 ml	Enterococi/100 ml	Pseudomonas aeruginosa/250 ml
	At 22 °C	At 37 °C				
Spring 1	160	25	Abs	Abs	Abs	Abs
Spring 2	190	22	Abs	Abs	Abs	Abs
Spring 3	120	8	Abs	Abs	Abs	Abs
Spring 4	36	4	Abs	Abs	2	Abs
Spring 5	1	Abs	Abs	Abs	Abs	Abs
Spring 6	87	23	Abs	Abs	4	Abs
Spring 7	4	Abs	Abs	Abs	Abs	Abs
Normal values	100/ml	20/ml	Abs	Abs	Abs	Abs

were identified in spring source collected from source 4 and source 6. There are high values for the total number of germs that develop at 22 and 37 °C in the spring waters harvested in zone III. In terms of physico-chemical and microbiological parameters, the study shows that the risk due to the consumption of spring water in zone III, Zlatna-Câmpeni is low but it is not entirely risk-free. There are water sources that exceed the limits allowed by standards for microbiological parameters: total viable count and enterococci.

2.2.4 Study on Groundwater Quality in Zone IV, Blaj

For the study, 25 samples of fountain water from different localities in Zone IV Blaj were collected. The results obtained show that the pH varies between 6.2 and 7.57 and doesn't fall in the "safe" pH range of 6.5–8.5. Non-compliance with normal values indicates pollution of water with inorganic or organic compounds. The pH has one of the basic properties of groundwater and surface water, namely buffering, thus ensuring a degree of natural supportability in the impact with acids or bases. The salts of Na^+, K^+, Ca^{2+} and Mg^{2+} play an important role in this respect. However, the pH of the water differs according to the geographical area: in the granite and forest regions, water is usually more acidic, and in the limestone soil it is alkaline.

Acid water does not mean it contains dangerous acids; in forest rivers and streams, water is loaded, for example, with acid organic materials, with humic acids, from the decomposition of plants (humus).

In almost all samples, values above 50 mg/l of nitrate content were recorded. Their presence indicates an organic contamination. The maximum determined value is 245 mg/l, except for two sources. In contrast, the nitrite and ammonium content falling within the acceptable range. Nitrates and nitrites are natural components of the soil and are part of the nitrogen cycle [23]. They occur with the mineralization of nitrogenous organic substances from plants and animals. Some nitrates are absorbed by plant roots and act as raw material for the synthesis of proteins and other nitrogen compounds. The remaining surplus contaminates the groundwater, found in rivers, lakes or underground waters.

Regarding wells water from Zone IV, Blaj, it is concluded that are high value of the nitrate content correlated with a low pH of the samples, this indicating that fountain water is not safe for drinking (non-potable). The microbiological results indicated that all parameters (mesophilic aerobic bacteria growing at 22 and 37 °C, *Escherichia coli* and most samples analyzed for faecal streptococci) of the fountain waters in Zone IV—Blaj was higher than the required values.

If we start from the hypothesis that human activity also influences the water quality of drinking water sources, we can say that at the level of zone IV—Blaj, the water from the individual sources (fountains) is not safe to drink unless it is treated to remove bacteria. The results obtained regarding the total number of mesophilic aerobic germs developing at 37 °C show that all analyzed samples do not fall within the limits accepted by the legislation because they exceed 20 CFU/ml. The minimum number obtained is 1.4×10^2, while the maximum number is 4.1×10^2 CFU/ml. Out of the 25 analyzed samples, all show a microbial load of the order of 10^2, i.e. over 100 CFU/ml.

2.2.5 Study on the Quality of Spring Water in Zone V, Aiud—Ocna Mures

In order to determine the quality of the spring waters in Zone V Aiud—Ocna Mureş, the physico-chemical and microbiological parameters were determined for the 7 public springs. The pH values in spring water range from 6.9 to 7.4, total hardness between 25.13 and 47.5° dH, total acidity between 5 and 14.5 mg/l, and total alkalinity between 1.8 and 7 mg/l. Table 5 presents the experimental results for the microbiological parameters of the analyzed public water springs in Zone V, Aiud—Ocna Mureş.

The analyzed spring water samples taken from the Ocna Mureş area were negative regarding the presence of microorganisms, except sources 5 and 7, where the microbiological growth does not fall within the free limits set out in the EU standards. The presence of both faecal streptococci and coliform bacteria is observed only in spring water taken from spring 5. Total viable count (TVC) at 22 and 37 °C is above the legal limits in the waters of the two springs. The results prove that the anthropic

Table 5 Results for microbiological parameters of spring water in Zone V, Aiud—Ocna Mures

Source	Parameter					
	Number of colonies/ml		Coliform bacteria/100 ml	E. coli/100 ml	Enterococi/100 ml	Pseudomonas aeruginosa/250 ml
	At 22 °C	At 37 °C				
Spring 1	4	Abs	Abs	Abs	Abs	Abs
Spring 2	22	Abs	Abs	Abs	Abs	Abs
Spring 3	1	Abs	Abs	Abs	Abs	Abs
Spring 4	20	Abs	Abs	Abs	Abs	Abs
Spring 5	3.5×10^2	1.5×10^2	12	Abs	4	Abs
Spring 6	25	1	Abs	Abs	Abs	Abs
Spring 7	1.20×10^2	23	Abs	Abs	2	Abs
Normal values	100/ml	20/ml	Abs	Abs	Abs	Abs

pollution influences the quality of the spring waters, being a major contamination especially at source 5 and 7. All activities that take place at ground surface have an impact on the quality of aquifers. But most investigated spring water were within the acceptable standard limits and are potable at its source.

As a result of the microbiological analyses carried out on the collected samples, the drinking water samples (water from the wells) are microbiologically polluted, regardless of the source and consequently the life of humans and animals is exposed to water-related diseases. In most localities of Alba County and Transylvania, groundwater is still used in the households sector because they are readily accessible, even if they exhibit variations in flow rate and if there are changes in terms of biological properties. As a result protection of water supply from contaminants from the outside is particularly important, with the exception of the possibility of internal pollution. The protection of the fountains is also a necessity by making a paved, cemented perimeter sloping towards the outside of the fountain to avoid undesirable contamination. The case studies on the quality of the ground water (spring and fountain) in different localities of Alba County show that their pollution is present in most of the analyzed sources. Their incidence is higher in the plain (lowland area) and lower in the mountains.

3 Conclusions

The results obtained evidence that anthropogenic pollution affect water quality of groundwater in Transylvania, Romania. There is a major water contamination with microorganisms: total number of germs, *E. coli* and enterococci, in urban area especially located in the plain area. It is clear that all activities taking place on the surface have an impact on groundwater quality.

Regarding public water sources in Alba County, the study shows that the risk due to consumption of water in the is reduced, but not completely. There are water sources that exceed the limits allowed by legislation for microbiological parameters—in particular. Some of the water samples are microbiologically contaminated and therefore the animals and humans are exposed to waterborne diseases. The presence of total viable count of mesophilic bacteria indicate the risk of water contamination with other dangerous pathogens to humans (bacteria, viruses, fungi, parasites). Obtained data of groundwater quality aimed to detect the microbiologically polluted soils and pollution sources in these areas. Water pollution in Alba County is more intense in urban and industrial areas. So, microbiological contamination incidence is higher in plains and lower at the mountains. An important role in ensuring water quality in rural areas it is introducing the water centralized system and sewer implicitly to avoid groundwater contamination.

References

1. Atalay A, Pao S, James M, Whitehead B, Allen A (2008) Drinking water assessment at underserved farms în Virginia's coastal plain. JEMREST 4:53–64
2. Bodor K (2005) Contributions to improving the quality indicators of drinking water with the help of new reagents and technologies. PhD thesis, "Politehnica" University of Timişoara, Timisoara, Romania
3. Chenoweth J (2008) A re-assessment of indicators of national water scarcity. Water Int 33(1):5–18
4. Cosma C, Nicolau M, Patroescu V, Stefanescu M, Ballo A, Florescu S (2009) The incidence of by-products (THMs) disinfection in drinking water. J Environ Prot Ecol 10(1):14–22
5. Decun M (1997) Igiena veterinară şi protecţia mediului. Ed. Helicon, Timişoara (in Romanian)
6. Decun M (2007) Igiena animalelor şi a mediului. Ed. Mirton, Timişoara (in Romanian)
7. Directive 98/83/EC (1998) On the quality of water intended for human consumption
8. Drăghici C (1996) Igiena animalelor şi protecţia mediului. Ed. Tipo Agronomia, Cluj-Napoca (in Romanian)
9. Ecoaqua (2002) Drinking water. Brochure, Bucharest, Romania
10. Entry JA, Farmer N (2001) Movement of coliform bacteria and nutrients în ground water flowing through basălt and sănd aquifers. J Environ Qual 30:1533–1539
11. Ercumen A, Pickering AJ, Kwong LH, Arnold BF, Parvez SM, Alam M, Sen D, Islam S, Kullmann C, Chase C, Ahmed R, Unicomb L, Luby SP, Colford JM (2017) Animal feces contribute to domestic fecal contamination: evidence from *E. coli* measured in water, hands, food, flies, and soil in Bangladesh. Environ Sci Technol 51(15):8725–8734
12. FAO—Food and Agriculture Organization of the United Nations (2017) Water for sustainable food and agriculture. A report produced for the G20 Presidency of Germany. http://www.fao.org/3/i7959e/i7959e.pdf

13. Global Environment Outlook. Nairobi, United Nations Environment Programme (1999) The state of the environment; freshwater. GEO-2000
14. Jamshidzadeh Z, Mirbagheri SA (2011) Evaluation of groundwater quantity and quality in the Kashan Basin, Central Iran. Desalination 270:23–30
15. Mănescu S (1989) Microbiologie sanitară. Ed. Medicală, București (in Romanian)
16. Muntean C, Negrea P, Ciopec M, Lupa L, Ursoiu I, Mosoarca G, Ghiga R (2006) Studies regarding the ground water pollution in a rural area. Chem Bull "POLITEHNICA" Univ (Timișoara) 51(65):75–78
17. Navarro A, Carbonell M (2007) Evaluation of groundwater contamination beneath an urban environment: the Besòs river basin (Barcelona, Spain). J Environ Manag 85(2):259–269
18. Papadakis N (2000) Public health correlation with water and sewage quality. J Environ Prot Ecol 1(1):63–71
19. Papadakis N, Nikolaou K, Ganidou M, Gregoriadou (2000) A temporal evolution of the surface water quality in Central Macedonia—Greece. J Environ Prot Ecol 1(3):336–340
20. Papadakis N, Tsoumbaris P, Doulgeris C (2007) Drinking water quality of the Thessaloniki prefecture, Greece. J Environ Prot Ecol 8(4):763–768
21. Prüss-Üstün A, Kay D, Fewtrell L, Bartram J (2004) Unsafe water, sanitation and hygiene. In: Majid E, Lopez AD, Rodgers A, Murray CJL (eds) Comparative quantification of health risks: global and regional burden of disease due to selected major risk factors editors. World Health Organization, Geneva, pp 1321–1352
22. Răpuntean G, Răpuntean S (2005) Bacteriologie veterinară specială. Academic Press, Cluj Napoca (in Romanian)
23. Richard AM, Diaz JH, Kaye AD (2014) Reexamining the risks of drinking-water nitrates on public health. Ochsner J Fall 14(3):392–398
24. Rodda JC (2001) Water under pressure. Hydrol Sci J Sci Hydrol 46(6):841–854
25. Safe Drinking Water Act Amendments (1986) United States. Pub.L. 99–359; 100 Stat. 642
26. Todoran A, Vică M, Glevitzky M, Dumitrel GA, Popa M (2010) Water environmental situation of wells in Galda de Jos Village, Romania: microbiological control. Chem Bull "POLITEHNICA" Univ (Timișoara) 55(69):95–98
27. Water: Quality, Treatment Processes. Methods of selecting control devices (2006) Specialty Magazine - Tehnica Instalațiilor (in Romanian)
28. WHO (2017) World Health Organization, Guidelines for drinking-water quality, 4th ed, incorporating the 1st addendum
29. WHO (2002) ed. The World Health Report 2002: Reducing Risks, Promoting Healthy Life. Geneva, World Health Organization

Maria Popa was born in Cornereva Village, Caras Severin County, Romania, in 1966. She received the B.S. degree in chemical engineering in 1990 and the Ph.D. degree in environmental chemistry in 2004 both from Babes—Bolyai University of Cluj Napoca, Romania. In 2008, she obtained the M.S. degree in management of trade, tourism and services and the title of Internal Auditor for Quality Management Systems in accordance to Quality Management Systems: SR: EN ISO 9001, SR: EN ISO 190011. She also received the title Environmental Analyst with the following competences: environmental management, environmental legislation, lasting development, European norms in the field of environment, food safety and security. Between 2000 and 2006, she was lecturer at the Faculty of Science from "1 Decembrie 1918" University of Alba Iulia, and between 2006 and 2014 an associate professor. She has been Professor with the Business Administration and

Marketing Department, Faculty of Economics, in the same university since 2014. Maria Popa was the dean of the Faculty of Economics from 2012 to 2020 (March). She has been the Ph.D. supervisor at the Doctoral School of Lucian Blaga University of Sibiu, in the field of engineering and management since 2017. Her interests in research include environmental management, food quality, safety and security control, waste management and improvement of the management process. She holds membership in: The Romanian Academic Management Society (SAMRO), The Balkan Environmental Association, The Society of Science and Engineering Environmental (SNSIM) (founding member), and The Editorial Board of the Journal of Environmental Protection and Ecology (JEPE). She has been the coordinator of UAB-BENA—International Centre of Environment, Sustainable Development and Food Quality Control since 2017. She is the author of 20 books, more than 150 articles (ISI and BDI), and more than 30 research projects.

Ioana Glevitzky was born in Beius, Bihor County, Romania in 1981. She received the B.S. (2005) and M.S. degrees (2006) in food engineering and organic chemistry from the Politehnica University of Timișoara, Romania. From 2018 follows the Ph.D. study in engineering and management at Lucian Blaga University of Sibiu. From 2006 to 2008, she was a Researcher in the Department of Research, Development and Quality Control Products, European Drinks Rieni, Romania. From 2008 to 2014 she has been a Quality manager at the Prefera Foods SA Company Oiejdea, Romania. From 2014 to 2016, she was an engineer at Alpin 57 Lux SRL Sebes, Romania. Since 2016 she works in Sanitary Veterinary and Food Safety Laboratory of Alba County, Romania. His research interests include water quality, groundwater pollution, detection of antibiotics in food, food safety, risk assessment, occupational health and safety. Ms. Glevitzky is a member of Romania's Chemical Engineering Association and Balkan Environmental Association. She was a recipient of the BENA International Conference on "Environmental Engineering and Sustainable Development", Award in 2017.

Sustainable Development, Management

Water Scarcity Management

Kaltoum Belhassan

Abstract Many regions and countries around the world are suffering from water scarcity due to the increasing demand for water, reduction of water resources, and increasing pollution of water driven by the continued growth in megacities, climate change, agriculture, and pollution. Actually, water scarcity is resulted from the lack of water resources to satisfy all humans' demand. This has two different aspects one is quantitative and other one is qualitative; therefore, water scarcity may be due to the water availability and/or to the bad quality of water. Since water is a scarce resource, it is critical to focus on managing water demand through the well-understanding of the impacts of climate change on freshwater demand, and thus adopt simultaneously mitigation and adaptation strategies that can tackle the causes of climate change and reduce the negative of its effects. In general, the main ways to save water resources are the water pollution prevention, water harvesting techniques, seawater desalination solutions, recharging aquifers/groundwater, and water reuse technologies. The governments through their policies and rules, organizations and groups should work all together to save water by increasing public awareness, and thus reducing the effects of water scarcity.

Keywords Water scarcity · Megacities · Climate change · Managing

1 Introduction

Water makes life possible since life without water is not sustainable. Water provides many beneficial functions, both for the earth and for humans—that help produce the abundance of life around us every day. Water covers about 70% of the planet, giving it the unique ability to foster and sustain life. Yet, <2.5% is freshwater, available for consumption, of course after processing. The three major sources of freshwater are rainwater, surface water and groundwater. Water is the most precious natural resource, as it is not only a drink, it is also used for growing crops, cooking meals,

---

K. Belhassan (✉)
Dewsbury, West Yorkshire WF13 4QP, UK

© Springer Nature Switzerland AG 2021
A. Vaseashta and C. Maftei (eds.), *Water Safety, Security and Sustainability*,
Advanced Sciences and Technologies for Security Applications,
https://doi.org/10.1007/978-3-030-76008-3_19

washing, and much more. Actually, almost 2 billion people around the world entirely rely on groundwater—underground stores of freshwater for their water-related needs [31]. In the 20th century, the population growth and industrialization, accompanied with the emergence of megacities and coupled with global increase of freshwater consumption, has rendered water a scarce resource. Megacities are a hotspot for higher greenhouse gas emissions that can have a devastating impact on air quality and on the climate at the regional to global scale. Megacities intensify the pressure on water resources and especially global groundwater storage and around 20% of global aquifers are being over-exploited [18]. Actually, the rise in freshwater demand and the groundwater depletion require implementation of quick actions to manage water scarcity. In fact, water scarcity can mean scarcity in availability due to physical shortage, or scarcity in access due to the failure of institutions to ensure a regular supply or due to a lack of adequate infrastructure. Water scarcity will be worsened as rapidly growing urban areas exercise heavy pressure on water resources. Climate change and bio-energy demands are also expected to increase already complex relationship between world development and water demand. This chapter aims to present the main causes of water scarcity and its aspects (quantitative and qualitative) and to provide perspectives on managing water scarcity through several actions to prevent and mitigate water scarcity. The ultimate goal is to move forward toward a water-efficient and water-saving economy.

2 Causes of Water Scarcity

Water (more precisely, freshwater), is as important to us, just as air. However, less than 1% of the planet's water is readily available for drinking or for agriculture. Freshwater is becoming increasingly scarce, which is rapidly becoming one of the world's biggest problems. The lack of freshwater is currently one of the main obstacles to the economic development of many countries, a problem that is widespread across the globe. On the socio-economic level, more than one billion people in the world do not have access to potable water, and 2.6 billion are without basic sanitation [10].

Water scarcity is the deficiency of available freshwater that human need for their routine activities around the globe. Water scarcity is not only the lack of enough water (quantity) but also the lack of access to safe water (quality). Worldwide, 40% of the total land area is arid, semi-arid and dry-sub-humid [2]. The Middle East and North Africa are considered as the most regions on earth suffering from water scarcity. Half of the European countries are facing water-stress, a term used to describe when demand for water is greater than the amount of water available at a certain period in time, and also when water is of poor quality and this restricts its usage. Water stress means deterioration in both the quantity of available water and the quality of available water due to factors affecting available water [7]. Generally, the main contributing factors to the water scarcity are megacities, climate change, agriculture, and increased pollution.

2.1 Megacities

The global population has increased significantly over the past century and this has included the proliferation of megacities. A megacity is a city with more than 10 million people [40]. A megacity has been described as the urban phenomenon of the 21st century. In 1950, only two cities in the world—New York and Tokyo—had populations of this size. In 2015, there were 33 listed megacities, and this is expected to increase to around 43 megacities by 2030, the UN projects that 60% of the global population will be urbanized; with Latin America and the Caribbean will be more than 80% urban [46]. The high rate of population growth influences the increase in the number of megacities. Most of the cities that have reached the 10 million population mark in recent years and are located in Asia and Africa. This large urban population is accompanied by the urban migrations and leading majority of people to live in overcrowded slums (lack of safe water and sanitation services). Megacities such as Mumbai and Mexico City are home to the largest slums in the world. Megacities conducted the emissions of greenhouse gases, such as carbon dioxide (CO_2), nitrogen oxides (NOx), ammonia (NH_3) and sulphur dioxide (SO_2). The increased amount of atmospheric carbon dioxide causes acidic rain, which deteriorates water availability and cause water scarcity through three different impacts: climate change, air pollution and soil contamination.

2.1.1 Megacities and Climate Change

The emissions of greenhouse gases and pollutants in megacities impact the composition of the atmosphere. Consequently, megacities contributed to global warming. The industries located in the megacities released large amounts of carbon dioxide and other greenhouse gases which settle into the atmosphere and cause the greenhouse effect (an increase in the air temperature and the rising of sea level lead to floods and droughts).

2.1.2 Megacities and Air Pollution

Humans are responsible for the excess emission of greenhouse gases by combustion of fossil fuels (produces the largest amount of the pollutant gases), deforestation and the destruction of natural ecosystems (forests are cleared to cater for these needs), industries, over use of single use-plastic, air conditioning, excess vehicles and other similar activities.

2.1.3 Megacities and Soil Contamination

The soil is contaminated mainly from acid rains and thus, affecting the quality and types of crops grown in these regions. Most of the soil is also contaminated with polluted water of factories, gutters. The contamination of soil affects water availability and leads to a reduction in the crop production efficiency. In addition, the high amounts of waste that are produced in these megacities cause more demand for water resources (for both industrial and private consumption) and have the added problem of over-using freshwater supplies in water-intensive industrial and agricultural practices.

2.2 Climate Change

As mentioned before that megacities impact on climate through the increase in greenhouse gas emissions, climate change has become a reality that humanity currently faces every day. The consequences of climate change include an increase in air temperature and rising sea-levels, which affect water availability (both as surface water and groundwater: quantity and quality). Climate change is likely to cause severe flooding in many regions in the world, which will contaminate water sources. Also, due to flooding, excess water will fall than vegetation and soil can absorb. The excess runoff travels to the nearest body of water and can pick up pollutants such as nitrogen and phosphorus on the way. Also, climate change has contributed to droughts (higher temperatures and increased variability in precipitation), which directly affect availability on freshwater and dependency on groundwater. People in water-scarce areas will increasingly depend on groundwater. Consequently, climate change will affect global water resources at varying levels. Reductions in surface water changes are accompanied with increasing demands on groundwater resources, as an expected consequence in the semi-arid regions of Mediterranean basin and in the semi-arid areas of the Americas, Australia, and southern Africa [38]. Africa is particularly vulnerable to the effects of climate change and global warming. Therefore, people living semi-arid climatic zones will suffer more from increased demand on water resources.

The Mikkes basin is located in the north center of Morocco (North Africa), it is part of the Mediterranean basin (semi-arid climate) whose climate has changed over a period of time and is affecting water resources at the present time. This basin is principally characterized by strong irregularity in rainfall, between year and another [6]. The yearly approach of the $P < 2T$ ratio reveals a very bad monthly distribution of precipitations and a seasonal drought which is apparently observed well after the Eighties era. The deficit of Mikkes River between the periods 1970–1979 and 1980–2009 is considerable with around 76%, this strong deficit shows the impacts of extent drought on flow Mikkes River [4]. The large decrease in the Mikkes groundwater recharge observed since the beginning of the 1980s is due to the combined effect of drought and overexploitation. This, in turn, has an influence on springs and River flow [5]. The overexploitation of groundwater reservoirs, sea level rise, and other factors contribute to the seawater intrusion into fresh groundwater. Once salt water has intruded into freshwater system it is difficult to reverse the process. Consequently, climate change has an indirect impact on the quality of groundwater.

2.3 Agriculture

The world's population is growing resulting in the increase of megacities around the globe. Increased human activities induce climate change and thus, we observe longer periods of drought in many areas around the world. All these factors lead to more water resources demand and more food production to support the population. Therefore, agricultural will be the greatest user of water. Since the nineteenth century, intensive agriculture produces were relying on huge amounts of inputs (such as pesticides and chemical fertilizers) and these inputs have increased dramatically in recent decades to provide enough food for several billion people. Nevertheless, the excess fertilization has a negative impact on ecosystems.

According to International Water Management Institute, agriculture which accounts for about 70% of global water withdrawals, plays a major role in water pollution and this due to inefficient agriculture methods and leaky irrigation systems [41]. Nitrate from agriculture is the most common chemical contaminant in the world's groundwater aquifers [42]. In the European Union, 38% of freshwater are under pressure from agricultural pollution [43]. In the United States of America, agriculture is the main source of surface water pollution [48]. This water pollution caused by agriculture has been amplified by the expansion of agricultural land which requiring land modifications such as clearing forests.

2.4 Pollution

Population growth, industries and rise of megacities are conducted to more water demand resources in agriculture and therefore increase the water pollution level. Water pollution is caused when a toxic substance gets into the water supply (such as surface water or groundwater) and contaminates it to the point that the supply becomes dangerous. Pollution is among the main causes of water scarcity. Industries and agriculture are considerate as the leading causes of water pollution. Acid rain is one major factor. In many places in the world, rivers, lakes, and groundwater are all polluted heavily. Thus, some water supply systems are too polluted to be suitable for human. The situation is even more serious in an area with a poor sewage system. Industry is the main responsible for dumping millions of tons of heavy metals, solvents, toxic sludge, and other wastes into freshwater each year. Around 80% of municipal wastewater is discharged into freshwater [44].

2.4.1 Major Water Pollutants

Polluted water can significantly damage life in aquatic ecosystem. Listed below are some of the main water pollutants:

Nutrients—is a form of water pollutant, where excessive inputs of nutrients (such as nitrogen or phosphorus), are added to freshwater bodies and causing excessive growth of algae that suppress other aquatic plants and animals.

Eutrophication—During rain events, the excess of nitrogen and phosphorus fertilizers can move via surface runoff and can be leached into the groundwater reservoirs and washed into waterways. With gradual increase in the concentration of nutrients in an aquatic ecosystem, the productivity or fertility of such an ecosystem naturally increases as the amount of organic material that can be broken down into nutrients increases. Water blooms, or great concentrations of algae and microscopic organisms, often develop on the surface, preventing the light penetration and oxygen absorption necessary for underwater life.

Pathogens—excreted in feces such as bacterial pathogens, enteric viruses, and enteric protozoa are strongly resistant in the water environment and can initiate waterborne infections [23]. Pathogens from livestock that are detrimental to public health include bacteria such as Campylobacter spp. and Clostridium botulinum and parasitic protozoa such as Giardia lamblia and Cryptosporidium parvum are all causing hundreds of thousands of infections every year [12].

Dam siltation—caused by the mobilization of sediment due to erosion; increased sediment deposition can eventually raise the level of the lake or riverbed, allowing land plants to colonize the edges, eventually converting the area to dry land [22].

Organic matter—from livestock farming is now significantly more widespread than organic pollution from urban areas, affecting a larger extent of water bodies [49]. And another booming sector, aquaculture is now releasing ever greater amounts of fish excreta (is high in ammonia, nitrate, and nitrite), excess feed, antibiotics, fungicides, and anti-fouling produce an effluent discharge to the water. Aquaculture can be a major contributor to the increasing level of organic waste and toxic compound in the aquaculture industry. In Scotland, the discharge of untreated organic waste from salmon production is equivalent to 75% of the pollution discharged by the human population. Shrimp aquaculture in Bangladesh generates 600 tons of waste per day [32].

Insecticides, herbicides and fungicides—are applied intensively in agriculture in many countries in the world [35]. When humans use these chemical substances improperly, they can pollute water bodies with carcinogens and other toxic substances. Water from excessive rainfall and irrigation cannot always be held within the soil structure. Therefore, pesticides and residues (also nitrates and phosphates) can be quickly transported to contaminate ground water and freshwater supplies over a large geographical area. The health effects of pesticides depend on the type of pesticide. Some, such as the organophosphates and carbamates, affect the nervous system. Others may irritate the skin or eyes. Some pesticides may be carcinogens (cancer causing). Others may affect the hormone or endocrine system in the body. Pesticide mixtures may be derived from common sources (such as point sources) or from multiple nonpoint sources and may include several different types of pesticide compounds with different mechanisms of toxicity.

Pesticides—are chemicals substances and they are used in the agriculture sector to control and/or eliminate pets and improving crop productivity and yields. However, when people use pesticides indiscriminately; they disrupt the balance of an ecosystem and also in many situations, they kill non-pest organisms and causing harm to wildlife. Moreover, pesticides, when sprayed on crop plants, settle on the ground, and when it rains, they mix with the rainwater and seep through the porous ground to reach the under-groundwater. Thereby, contaminating groundwater and making it unsuitable for both human and agricultural uses.

Salinization—of water resources is closely connected to irrigation and, in coastal areas, it is related to saltwater intrusions due to over-pumping of groundwater in coastal aquifers [16]. Salinization of freshwater is one of the most serious and widespread groundwater contamination issues throughout the world. The intrusion of saline seawater into aquifers—frequently the result of groundwater over-exploitation to maintain water demand for agriculture—is another important cause of salinization in coastal areas [26]. Intensive irrigation and overexploitation of groundwater—induced salinization is commonly reported in India and Pakistan. In general, due to increase in salinity, the biodiversity of micro-organisms, algae, plants, and animals decline [25].

Emerging pollutants—in agriculture encompass a diverse group of compounds, including antibiotics, vaccines, growth promoters and hormones. Currently, more than 700 emerging pollutants are listed as present in European aquatic environments [28].

In regions where the water in rivers or lakes is quite dirty, local people do not have much access to water and they often must rely on rainwater. Nevertheless, in drought periods, local people may have serious consequences especially where people are not connected to public water lines.

2.4.2 Water Pollution and Disease

Water pollution affects drinking water, rivers, lakes and oceans all over the world, which may cause many diseases referred to as water pollution diseases. Consequently, these could have serious health impacts. According to the World Health Organization, Africa is the most infected continent's health by water scarcity which encourages people to store water in their homes, which increases the risk of household water contamination and provide breeding grounds for mosquitoes. Actually, African people are forced to drink non-potable water, which has a direct impact in their health and cause waterborne diseases. The most common water pollution diseases include poisoning episodes which affect the digestive system. Here are some of the human infectious diseases [3]:

- Typhoid, usually transmitted by water or food in much the same way as cholera
- Intestinal parasites that populate the gastro-intestinal tract in humans and animals

- Enteric and diarrheal diseases caused by bacteria, parasites, and viruses
- Amoebiasis, caused by protozoan Entamoeba histolytica [36]. Amoebiasis is often asymptomatic but may cause dysentery and invasive extraintestinal disease
- Ascariasis, caused by the parasitic roundworm Ascaris lumbricoides
- Hookworm, a parasitic nematode worm that lives in the small intestine of its host. Two species of hookworms commonly infect humans: Ancylostoma duodenale and Necator americanus
- Gastroenteritis, an infection of the stomach and bowel (large intestine)
- Stomach cramps and aches
- Hepatitis, which is inflammation of the liver
- Encephalitis, an uncommon but serious inflammation of the brain
- Diarrhea
- Respiratory infections
- Vomiting
- Endocrine hormonal system
- Injury to kidneys
- Malaria is a serious disease, which is transmitted by mosquitoes. It is considerate as the most important parasitic infectious disease in the world
- Liver cancer (due to DNA damage)
- Damage to the nervous, kidney, blood forming, heart, and immune systems
- Less serious health effects like rashes, earaches, and pinkeye.

3 Water Scarcity Aspects

Water scarcity arises from multiple factors which are in strong combination such as megacities, population grow, climate change and growing pollution. Actually, water scarcity is a global issue of the environment that threatens human activities and also human health. Water scarcity has two aspects: one is quantitative and the other one is qualitative.

3.1 Water Scarcity Quantitative Aspects

This type of water scarcity aspect arises due to over utilization or over exploitation of water resources. Many countries suffer much from water scarcity as there is not enough water for their activities demands such as irrigation demands. Around 783 million people in the world do not have access to safe drinking water [39]. Sub-Saharan Africa has the largest number of water-stressed countries of any other region in the world. The Middle East and North Africa (MENA) region experiences an absolute water shortage [51]. The result of water demand outpacing the water availability will be significantly exacerbated by long period drought as a result of combined factors: decline in River flow and increase of water reservoir exploitation and thus depletion of recharge aquifer. Around 50% of the world's population will be face water shortages by 2030 [37]. One in three people are already facing water shortages [11]. Almost 1.6 billion face economic water shortage; lack of investment in water programs and water capacity [37]. And these are really extreme numbers. As population continues to grow, there will be more problems. Humans face drastic measures to make sure everyone has access to water [13]. By 2025, 1.8 billion people will be face absolute water shortages [37]. Actually, those people from areas which have lack the necessary infrastructure to take water from rivers and aquifers will suffer much from water shortages as they will not be able to maintain their water demands in agriculture and also to meet reasonable water needs for domestic, industrial, and environmental purposes.

3.2 Water Scarcity Qualitative Aspects

Water scarcity qualitative aspects, where water is sufficiently available (quantity) to meet the needs of the people, but, the people suffer from water scarcity. This scarcity may be due to bad quality of water, which is due to the pollution (domestic and industrial wastes, chemicals, pesticides and fertilizers used in agriculture), and thus, in this aspect water quality is degraded (hazardous for human use). Most of the geochemical and biochemical substances in water are acquired during the water cycle process and any changes in this water cycle process may lead to deterioration in water quality. Consequently, this poor quality of this water becomes unsafe for the required use [3].

The UN World Water Development Report from the World Water Assessment Program indicates that 40% of the world's population doesn't have access to safe water (sanitation and hygiene). Almost 2.5 billion do not have access to adequate sanitation [39]. Within twenty years, it is expected that an additional 2 billion will live in towns and cities, mainly in developing countries will suffer more from unsafe water: more than 90% of sewage is discharged untreated which lead to more water crisis [20]. Waterborne infections account for 80% of the world's infectious diseases [15]. Actually, long period of drought and waterborne diseases are behind the death of more than 2.2 million people in 2000 [45]. As direct consequences:

- 6 to 8 million people die annually from water-related diseases [39]
- 1.8 million people die every year from diarrhea diseases (including cholera) [50]
- More than 200 million people especially in rural areas are affected by Schistosomiasis, leading to the loss of 1.53 million disability-adjusted life years, according to the World Health Organization [19]
- Around 500 million people are at risk from Trachoma [50]
- Around 133 million people in the developing countries are suffering from high-intensity intestinal helminths infections which often leads to severe consequences

such as cognitive impairment, massive dysentery, or anemia. More than 9400 people die every year from intestinal helminths [50]
- Hepatitis A, there are around 1.5 million cases of clinical hepatitis A every year [50]
- More than 1.2 million from babies to older children and adults die annually from Malaria [9].

4 Managing Water Scarcity

Megacities exemplify some of the gravest problems the world faces as a global population accompanied with the growing pressure on global freshwater supplies as well as posing other environmental challenges such as climate change, agriculture, pollution and inefficient water management practices and these combined factors are most evident in the busiest cities. It is essential that water resources are managed effectively, and the infrastructure can cope with the demand. Managing water scarcity is the process of planning, developing, distributing, and making optimum use of surface water, groundwater and also marine water resources under defined water policies and regulations thus, managing water scarcity, in terms of both water quantity and quality, across all water uses. In addition, managing water scarcity includes consideration of several disciplines of hydrology such as water pollution, surface water, groundwater, and water chemistry. There are different approaches to decrease water shortage and water contamination. This chapter tackles some of these approaches: preventing water pollution, water harvesting, water desalination, recharging aquifers/groundwater, and water re-use—effective water treatment technologies.

4.1 Preventing Water Pollution

Human body needs clean water for healthy living, yet millions of people worldwide are deprived of it. Human can do many things to keep water cleaner.

> One person alone cannot save the planet's biodiversity, but each individual's effort to encourage nature's wealth must not be underestimated.
> —United Nations Environment Programme (UNEP)

Humans can decrease and prevent water pollution from water bodies either surface water or groundwater through some simple ways:

- Dispose of trash in the correct waste bin. Clean up any litter on beaches, at the riverside, and on and near water bodies
- Save water by turning off the tap water. This protects water resources from scarcity (quantity and quality)
- Never throw chemicals, oils, paints, or medicines down the sink drain or the toilet

- Take great care when using fertilizers and pesticides for gardens or farm and not to overuse them. This will reduce runoff. Consider composting and using organic manure instead
- Use environmentally responsible household products for laundry, household cleaning, and toiletries
- Living close to a water body needs more awareness such as growing more plants in yard so that rain drains fewer chemicals from the property into the water.

There are several directives legislation which many countries in the world adopted to prevent water pollution. These include:

- The Shellfish water directive aims to protect waters from pollution that are primarily used for fishing
- The bathing water directive aims to safeguard public health by protecting the quality of bathing water in freshwater and coastal water areas from pollution
- Monitoring of water and vector-borne diseases, particularly in the aftermath of disasters
- The drinking water directive aims to establish strict standards regarding the quality of drinking water by ensuring that it is wholesome and clean
- Capacity planning in health care and integration of health care and disaster risk management planning
- The Nitrate Directive aims to protect water quality by promoting the use of good farming practices either than using nitrates from agricultural sources which pollute water bodies. This will enable marine and freshwaters to be protected from eutrophication
- Preventive health maintenance programs promoting healthy lifestyles and improved nutrition and hygiene
- The urban wastewater directive aims to protect the environment from adverse effects of wastewater discharges by regulating the collection and treatment of urban wastewater
- The surface water regulations aim to protect all surface waters from pollution
- The groundwater regulations aim to prevent or limit the input of pollutants into the groundwater, by controlling the direct and indirect discharges of certain substances into the groundwater.

4.2 Water Harvesting

Many people in different areas of the world don't have access to clean drinking water. For potable water, huge investment costs and expenditure is needed. Most rainwater is naturally acidic, but it is not dangerous. Therefore, rainwater is fresh precipitation which is comparatively clean, and it can be used for several non-potable uses. In the early 1990s, several studies on traditional water harvesting infrastructures were published [30]. These techniques of water harvesting systems survive particularly in the arid and semi-arid areas (North and sub-Saharan Africa). There are three various types of water harvesting techniques: rainwater harvesting, groundwater harvesting and flood harvesting.

Rainwater harvesting is a great method for inducing, collecting, storing, and conserving local surface runoff for agriculture in arid and semi-arid regions [8]. It is used to provide drinking water, water for livestock, and so on during regional water restrictions (i.e., droughts) and allows groundwater levels to replenish. Actually, rainwater harvesting is one of the most effective methods of managing water scarcity. Because, rainwater harvesting in one side it relieves the pressure on sewers and the environment by mitigating floods, soil erosions and natural aquifers recharge. And on the other side it helps much in saving the high-quality of drinking water sources by reducing the consumption of potable water. There are three main types of rainwater harvesting system: direct pumped, indirect pumped and indirect gravity.

Direct pumped (submersible) is the most common type of rainwater harvesting system and it is generally the easiest to install. The pump is located within the underground tank and water collected from roof tops is simply pumped directly to the appliances and it is used for domestic purpose or garden crops.

Indirect pumped is a type of rainwater harvesting system in that it uses two separate tanks, one to store the water when harvested, this tank can be at any level in the building, as it does not rely on gravity to supply the outlets and another to supply the water when needed. The second tank can be located anywhere in the building.

Indirect gravity rainwater harvesting system in that the harvested water is first pumped to a high-level tank (header tank), stored, and then allowed to supply the outlets by gravity alone. With this arrangement, the pump only must work when the header tank needs filling.

4.2.1 Groundwater Harvesting

Groundwater harvesting is when the water flows across land (surface runoff), into the ground (infiltration and percolation), and through the ground (groundwater). The recharge areas of groundwater should be identified and monitored without any external disturbances and managed by a local community. Groundwater harvesting is a great method to save water for living things. The collected water is stored and pumped in a separate pipe distribution. It is considerate as artificial recharge to groundwater by which the groundwater reservoir is augmented at a rate exceeding that under natural conditions of replenishment. Actually, groundwater harvesting increases the availability of water and it is a very useful method in arid and semi-arid regions where people suffer much from water scarcity. This method is most commonly in India as it helps by reducing the cost and the demand for treated water.

4.2.2 Floodwater Harvesting

It is when the water that flows across the floodplains during a flood is collected and used later. Floodwater harvesting is a method of collecting and storing rainfall runoff that enters infiltration systems, watercourses, or sewers during or shortly after a rainfall event for irrigation use. Floodwater harvesting, also known as "large catchment water harvesting", and comprises two forms:

Flood water harvesting within stream bed, the water flow is dammed and as a result, inundates the valley bottom of the flood plain.

Flood water diversion, the River water is forced to leave its natural course and conveyed to nearby cropping fields.

Many people in the world continue to rely on water harvesting practices. There are many ways by which water harvesting can benefit a community:

- Decreasing the use of other valuable water sources like groundwater
- Groundwater replenishment
- Improvement in the quality of groundwater
- Increasing the water availability for irrigation
- Reduce stormwater pollutant discharges and flooding of roads
- Reduction of soil erosion as reducing the proportion of runoff
- Saving of energy to lift groundwater and maintenance costs.

4.3 Seawater Desalination

Megacities, climate change and environmental pollution are causing human beings to face a serious shortage of clean water. Seawater desalination is among the solutions to have enough freshwater to use in daily life. Therefore, some special processes are needed for desalination to overcome the water shortage [33]. This produces clean drinking water and is particularly useful in countries with coastlines but shortage in freshwater bodies. The three most popular methods of desalination to turn seawater into freshwater are distillation, reverse osmosis and electrodialysis.

Distillation is the method that most completely reduces the widest range of drinking water contaminants. It is the only process that replicates the hydrological cycle. Salt water is heated, producing water vapor. Then the water vapor is condensed into freshwater. It also removes more than 99.9% of the minerals dissolved in water. It is a simple evaporation-condensation-precipitation system. There are many types of distillation, but the most common ones are simple distillation, fractional distillation, vacuum distillation, and steam distillation. They differ mainly in the apparatus setup and their applications. Fractional distillation can guarantee high purity because it can provide separated chemicals due to a high number of theoretical plates, but simple distillation cannot guarantee high purity of the separated component. Thus, fractional distillation is more efficient than simple distillation and it is an important process in chemistry, industry, and food science.

Reverse osmosis uses a semi-permeable membrane to separate water and salts both in industrial processes and in producing potable water. This membrane allows only water molecules to pass through it and doesn't allow any other chemicals dissolved in the water from passing through it.

Electrodialysis (ED) is an electrochemical separation process that uses an electric potential to move selectively ions from one side to the other through a permeable membrane, leaving freshwater behind. Electrodialysis (ED) is driven by the development of ion exchange membranes produces high water recovery and does not require phase change, reaction, or chemicals. Therefore, ED has great environmental benefits without the use of fossil fuels and chemical detergents [1].

Desalination may be a viable option to overcome water shortage problems in many countries, such as Saudi Arabia, the United Arab Emirates, and Qatar, whose programs have been quite successful. Most countries of the Middle East, including Syria, have adequate energy to power desalination plants, which make them a vital medium-term solution [27]. Desalination still a costly water supply option compared to natural water resources (surface water or groundwater) and is outside the reach of poor nations.

4.4 Recharging Aquifers/Groundwater

Among the total water available on the earth surface, only 0.62% exists as groundwater [34]. Megacities and humans' activities demand much water and cause high groundwater depletion which has been tripled in the past five decades [47]. For this reason and others, it is necessary to protect groundwater resources through either natural recharge (NA) or artificial recharge (AR)—Managed aquifer recharge (MAR).

Natural recharge (NR) represents most groundwater systems receive, with both diffuse and localized recharge. Diffuse (direct) recharge is derived from precipitation and that occurs uniformly over large areas infiltrating and percolating through the unsaturated zone. As the aridity of a region increase, the importance of diffuse recharge decreases. Localized (focused) recharge refers to concentrated recharge from surface water bodies (such as streams, lakes) to the groundwater system.

Managed aquifer recharge (MAR) also called ***artificial recharge (AR)*** is the enhancement of natural groundwater supplies with excess surface water or reclaimed wastewater. It is a process by which excess runoff is directed into the ground accomplished by augmenting the natural infiltration to replenish an aquifer, using man-made conveyances such as infiltration basins, field flooding, infiltration galleries or injection wells. Often, artificial recharge is a great method to increase the groundwater level and is often incorporated into a broader water resource plan. Also, MAR is a general term that designates the set of methods used to recharge additional water to aquifers intentionally for its recovery and subsequent use or with the purpose to generate an environmental benefit [14]. MAR provides numerous benefits, including (1) storage to improve security of water supply, (2) treatment for all available water resources: surface water, groundwater, recycled water and desalinated seawater, (3) a low cost—low energy water supply option and (4) replenishing over-exploited aquifers. MAR in Africa is sourced from river water, treated wastewater, storm runoff, roof runoff and groundwater from the adjacent compartment. Surface water is the main water source for MAR in Africa, and the total annual recharge volume is about 158 Mm^3/year [17]. Treated wastewater is used in countries with limited water resources such as Tunisia for agricultural uses. Depending on the type of aquifer available for recharge, MAR projects will vary its nature. In confined aquifers, water is allowed to inject through a well. In unconfined aquifers, water infiltrates through permeable soils, where recharge can be augmented by basins and galleries [21].

4.5 Water Reuse and Effective Water Treatment Technologies

The growth of megacities, the economic development and climate change constitute the main factors which will put more pressure on water resources that lead to increased water polluted. Water reuse technology has become an accepted and reliable effective solution to cope with water scarcity conditions. Previously considered a disposal liability, reused wastewater can now become a valuable resource. In Mediterranean regions owing to water scarcity, water reuse becomes a common practice in agriculture while in some cities in Northern Europe, 70% of their water resource during the summer is obtained from indirect reuse [29]. A Zero-Liquid Discharge system (ZLD) is a treatment process designed to ensure that there will be no discharge of industrial wastewater into the environment. It is achieved by treating wastewater through recycling and then recovery and produce clean water that is suitable for reuse (e.g. irrigation, industrial purpose).

Wastewater reuse has many benefits such as:

- Lower energy costs compared to deep groundwater
- Preserve freshwater resources
- Decrease net water demand and adds value to water
- Increase the availability of potable water and reducing water cost
- Keeps water for drinking and reclaimed water for non- use
- Provide drought-proof water resource through reuse
- Improve visibility of long-term costs of water supply
- Enhanced water security and sanitation by minimize the use of surface water and groundwater resources
- Creation of wetlands and urban irrigation
- Attract new industry
- Manage the recharge of surface and groundwater to optimize quality and quantity
- Positively controls saline ingress and recharge aquifers to create sustainable water resource
- Reduces nutrient discharge to the environment and loss of freshwater.

5 Water Resources Policy and Recommendations

Water resource policy addresses both the quantity and the quality aspects of freshwater resources (surface water and groundwater) and also deals with the delivery of water services. Institutional and governmental water resource policy promotes a strong awareness of water scarcity and water use. Recycling of wastewater (reusing it after removing contaminants from domestic, commercial, and institutional applications) could be a big solution for local water shortages in many low-income countries [3]. Many organizations aim to promote water conservation, not only through raising awareness but also by conducting research into water resources and formulating effective water policies, such organizations include:

The Nature Conservancy is one of the largest environmental organizations in North America, founded in 1951 and with more than one million members. The Nature Conservancy is actively protecting rivers, lakes, and natural lands in 30 countries and all 50 states.

Water Aid is an international non-governmental organization established in 1981, focused on water, sanitation, and hygiene in developing countries. It operates in 34 countries.

The International Water Association (IWA), being the largest membership association in the global water sector. The Association builds on a 70 years heritage of connecting water professionals (scientists, researchers, technology companies, and water and wastewater utilities) from over 140 countries. The IWA is an international reference for the water and sanitation industry.

Water 1st International is based in Seattle, Washington (United States) and it is a non-profit organization, focuses on building a worldwide community of concerned individuals in poor countries and mobilizing them to act for better water access.

Water Charity is a non-profit organization, is doing to provide clean, safe drinking water to people in developing countries. And focuses on educating people about the need and benefits of proper hydration. These actions involve the implementing projects geared toward water safety and effective sanitation.

The International Water Management Institute (IWMI) is a non-profit research, IWMI undertakes projects intending to use effective water management to underpinning food security and reducing poverty while safeguarding vital environmental ecological processes.

In developing world, the proportion of people with access to safe water varies between 30% in 1970 and 84% in 2004 [24]. Demand for freshwater is rising daily with world population growth, megacities and climate change. All these factors lead directly to increasing the problem of water scarcity. These are some recommendations which the world needs as good vision for managing water scarcity:

- Develop comprehensive policies and strategies for prioritizing sharing and managing available resources, taking into account multiples take holder demands
- Designing climate-resilient infrastructure—to protect against future flood risks, cities can look to redesign themselves to better absorb stormwater, e.g. by moving towards the "sponge city" model

- Keep the multiple water use sectors engaged in analyses, choices and decisions related to managing water scarcity
- Moving from a linear water use model (groundwater and freshwater resources) to a circular model of water (reusing wastewater, harvesting rainwater…), thereby, reduce the water scarcity of the world's megacities
- All working together (water companies, regulators, academics…) to create a long-term planning framework for managing water scarcity
- Manage water efficiently by maintaining adequate water supplies for customers, while balancing their needs against those of the environment—this can be done by local authorities (municipal water management) or it can be done at household, commercial and governmental level and involves making use of the newest smart technology (e.g. smart sensors, smart water-leak detectors, smart meters and data sharing equipment) to reduce waste and use water supplies in the most efficient ways possible.

6 Conclusion

Humans use daily freshwater, and this will put much pressure on water demand and conducting to water scarcity. Actually, the direct cause of water scarcity is the insatiable need for freshwater in modern industry, agriculture and extractive technologies. Megacities, climate change, agriculture and also pollution act as powerful catalysts on water scarcity. Megacities impact on atmospheric composition and climate and likely lead to increases in temperatures, changes in precipitation patterns, thereby, resulting long period of droughts and flooding in many regions in the world such as arid and semi-arid areas. Agriculture is the greatest user of water. Polluted drinking water causes problems for health and leads to waterborne diseases. All these factors combine to increase the water scarcity or stress on water supply. In fact, water scarcity has two different aspects: One is quantitative aspects, which is associated with the availability of water resources. Other one is qualitative aspects; water scarcity may be due to the bad quality of water. Therefore, water has to be treated as a scarce resource, with a far stronger focus on managing demand. Managing water scarcity—in megacities—is to measure the impacts of climate change on freshwater demand, while providing a quantitative framework to address both adaptation and mitigation strategies simultaneously that can help current and future freshwater challenges. There are many ways to save water and prevent water scarcity such as preventing water pollution, water harvesting, seawater desalination, recharging aquifers/groundwater and also water re-use. Also, the proper policies making by the governments and strict the rules are a guide to reduce water scarcity. Essentially, many organizations and groups play an important role in increasing public awareness and to know the importance of water and the negative effects of water pollution. Henceforth, managing water scarcity should be proactive rather than being reactive.

References

1. Al-Amshawee S, Bin Mohd Yunusab MY, Mohd Azoddeina AA, Hassell DG, Dakhil IH, Abu Hasane H (2020) Electrodialysis desalination for water and wastewater: a review. Chem Eng J 380 (Elsevier)
2. Becerra-Castro C, Lopes AR, Vaz-Moreira I, Silva EF, Manaia CM, Nunes OC (2015) Wastewater reuse in irrigation: a microbiological perspective on implications in soil fertility and human and environmental health. Environ Int 75:117–135
3. Belhassan K (2014) Water and life. CreateSpace Independent Publishing Platform. ISBN-10:1494718995 (182 pp)
4. Belhassan K (2011) Relationship between hydrology and climate in the stream Mikkes (Morocco). Res J Earth Sci 3(1):27–38
5. Belhassan K, Hessane MA, Essahlaoui A (2010) Interactions eaux de surface – eaux souterraines: bassin versant de l'Oued Mikkès (Maroc). Hydrol Sci J 55(8):1371–1384
6. Belhassan K (2009) Climate change and degradation of surface water: Mikkes basin (Morocco). In: Thirteenth international water technology conference, IWTC 13 2009, Hurghada, Egypt, pp 33–46
7. Bixio D, Thoeye C, De Koning J, Joksimovic D, Savic D, Wintgens T, Melin T (2006) Wastewater reuse in Europe. Desalination 187:89–101
8. Boers TM, Ben-Asher J (1982) A review of rainwater harvesting. Agric Water Manag 5:145–158
9. Boseley S (2012) Malaria kills twice as many as previously thought. http://www.theguardian.com/society/2012/feb/03/malaria-deaths-research
10. Bouzane B (2010) UN declares water, sanitation as human rights. Postmedia News
11. CAWMA (2007) Comprehensive assessment of water management in agriculture. In: Water for food, water for life, a comprehensive assessment of water management in agriculture. Earthscan, and Colombo: International Water Management Institute, London
12. Christou L (2011) The global burden of bacterial and viral zoonotic infections. Clin Microbiol Infect 17(3):326–330
13. DeCapua J (2013) Will there be enough water for everyone? Voice of America News
14. Dillon PJ (2005) Future management of aquifer recharge. Hydrogeol J 13(1):313–316
15. Donya CA (2007) Pollution causes 40% of all deaths. Nation's Health 37(8):11
16. Fohrer N, Chícharo L (2011) Interaction of river basins and coastal waters—an integrated ecohydrological view. Treat Estuar Coast Sci 10:109–150 (Elsevier)
17. Girma E, Jonathan FL, Karen V (2020) Managed aquifer recharge in Africa: taking stock and looking forward. Water 12(7):1844
18. Gleeson T, Wada Y, Bierkens M, Van Beek L (2012) Water balance of global aquifers revealed by groundwater footprint. Nature 488(7410):197–200. https://doi.org/10.1038/nature11295
19. Gryseels B, Polman K, Clerinx J, Kestens L (2006) Human schistosomiasis. Lancet 368(9541):1106–1118
20. Günter L, Elke M (2005) Ecological sanitation—a way to solve global sanitation problems? Environ Int 31(3):433–444
21. Jomol TJ, Rema KP (2016) Managed aquifer recharge: a review. Int J Modern Trends Eng Res (IJMTER) 03(09), ISSN (Online):2349–9745; ISSN (Print):2393–8161
22. Lawrence E, Jackson ARW, Jackson JM (1998) "Eutrophication". In: Longman dictionary of environmental science. Addison Wesley Longman Limited, London
23. Leclerc H, Schwartzbrod L, Dei-Cas E (2002) Microbial agents associated with waterborne diseases. Crit Rev Microbiol 28(4):371–409
24. Lomborg B (2001) The skeptical environmentalist: measuring the real state of the world. Cambridge University Press, Cambridge, UK and New York
25. Lorenz JJ (2014) A review of the effects of altered hydrology and salinity on vertebrate fauna and their habitats in north-eastern Florida Bay. Wetlands 34:189–200

26. Mateo-Sagasta J, Burke J (2010) Agriculture and water quality interactions: a global overview. SOLAW background thematic report-TR08. Rome, Food and Agriculture Organization of the United Nations (FAO)
27. Morvan HP, Wardeh S, Wright NG (2005) Desalination for Syria. In: Hamdy A, Monti R (eds) Food security under water scarcity in the Middle East: problems and solutions. Options Méditerranéennes: Série A. Séminaires Méditerranéens 65. CIHEAM, Bari, pp 325–336
28. NORMAN (2016) List of emerging substances. Network of Reference Laboratories, Research Centers and related Organizations for Monitoring of Emerging Environmental Substances (NORMAN)
29. Oyegoke SO, Adeyemi AO, Sojobi AO (2012) The challenges of water supply for a megacity: a case study of Lagos Metropolis. Int J Scient Eng Res 3(2):1–10. ISSN 2229-5518
30. Prinz D, Wolfer (1999) Traditional techniques of water management to cover future irrigation water demand. Zeitschrift für Bewässerungswirtschaft 34(1):41–60
31. Puri S (2009) Transboundary aquifer resources publishing by ScienceDirect in Encyclopedia of Inland Waters
32. SACEP (2014) Nutrient loading and eutrophication of coastal waters of the South Asian Seas—a scoping study. South Asian Co-Operative Environmental Programme (SACEP)
33. Sadrzadeh M, Mohammadi T (2008) Sea water desalination using electrodialysis. Desalination 221:440–447. https://doi.org/10.1016/j.desal.2007.01.103
34. Satyendra K, Kamra SK, Yadav RK, Bhaskar N (2014) Effectiveness of horizontal filter for artificial ground water recharge structure. J Agric Eng 51(3):24–33
35. Schreinemachers P, Tipraqsa P (2012) Agricultural pesticides and land use intensification in high, middle and low income countries. Food Policy 37:616–626
36. Stanley SL (2003) Amoebiasis. Lancet 361(9362):1025–1034
37. UN (2007) United Nations "International Decade for Action: 'Water for Life' 2005–2015.", UN-Water
38. UN (2010) Climate change adaptation: the pivotal role of water, UN-Water
39. UN (2013) With world water day being celebrated on 22 March 2013, United Nations (UN), New York
40. UNCHS (1996) United Nations Centre for Human Settlements An Urbanizing World: global report on human settlements. Oxford University Press, Oxford. World Commission on Environment and Development (1987) Our Common Future, Oxford University Press, p 8
41. UN FAO (2017) United Nations Food and Agriculture Organization: water pollution from agriculture: a global review, Executive summary
42. UN WWAP (2013) The United Nations World Water Development Report. United Nations World Water Assessment Programme (UN WWAP). Paris, United Nations Educational, Scientific and Cultural Organization
43. UN WWAP (2015) The United Nations World Water Development Report: Water for a sustainable world. United Nations World Water Assessment Programme (UN WWAP). Paris, United Nations Educational, Scientific and Cultural Organization
44. UN WWAP (2017) The United Nations World Water Development Report 2017: Wastewater, the untapped resource. United Nations World Water Assessment Programme (UN WWAP). Paris, United Nations Educational, Scientific and Cultural Organization
45. UN WWDR (2003) United Nations World Water Development Report: United Nations, New York
46. UN WWDR (2006) United Nations World Water Development Report 2nd World Water Development Report
47. UN WWDR (2012) United Nations World Water Development Report 4: managing water under uncertainty and risk. UNESCO Publishing, Paris
48. US EPA (2016) United States Environmental Protection Agency: water quality assessment and TMDL information. Washington, DC
49. Wen Y, Schoups G, van de Giesen N (2017) Organic pollution of rivers: combined threats of urbanization, livestock farming and global climate change. Scient Rep 7:43289

50. WHO (2004) World Health Organization: Water Sanitation and Health (WSH). Facts and figures updated November 2004. https://www.who.int/water_sanitation_health/publications/facts2004/en/
51. World Bank (2018) Beyond scarcity: water security in the Middle East and North Africa. MENA Development Report. Washington, DC: World Bank. ©World Bank. https://openknowledge.worldbank.org/handle/10986/27659 License: CC BY 3.0 IGO

Kaltoum Belhassan was born in Fez, Morocco in 1979. She received the master's degree in Hydrogeology, Department of Geology, Faculty of Science and Technology, Fez, Morocco, in 2003, postmaster degree in Hydrogeology, Department of Geology, Dhar El Mehraz Sciences University, Fez, Morocco, in 2005 and she got her Ph.D. in Hydrogeology, Department of Geology, Dhar El Mehraz Sciences University, Fez, Morocco in 2011. She is the author of one book Water and Life (CreateSpace Independent Publishing Platform, 2014) and 12 articles such as: Relationship between River Flow, Rainfall and Groundwater pumpage in Mikkes Basin, Morocco (Iranian Journal of Earth Sciences, 2011), Interactions eaux de surface–eaux souterraines: bassin versant de l' Oued Mikkes, Maroc (Taylor & Francis online "Hydrological Sciences Journal", 2010). She is a reviewer of some Scientific Journals such as Iranian Journal of Earth Sciences. Her researches interest water and climate change. Dr. Belhassan was a supervisor of American Association of International Management for one year (2009–2010). She is a member of British Hydraulic Society. Award for Excellence in Ph.D. in 2011 from Sidi Mohammed Ben Abd Allah University Fez Morocco.

Sustainable Approaches for the Treatment of Industrial Wastewater Using Metal-Organic Frame Works

Madhavi Konni, Saratchandra Babu Mukkamala, R. S. S. Srikanth Vemuri, and Manoj Kumar Karnena

Abstract In the urban aquatic ecosystem, environmental pollution due to waste disposal and treatment is becoming a significant concern worldwide. Pollution levels are increasing alarmingly in the marine ecosystem, according to the available scientific literature. Hazardous components, such as recalcitrant chemicals that are already present in the environment, will enhance the strength of upcoming contaminants. New technologies have been advanced for water treatment by utilizing material science engineering in the past few decades. However, water treatment technologies have suffered various limitations like more power consumption, high operation cost, and difficulty in operation. Thus, there is a need to develop new water treatment methods, such as composites and components based on Nanomaterials which have shown to be technologically advanced and some are already employed for treating wastewater currently. These resources are operative, cost-effective, and eco-friendly, highly recommended for conserving natural resources and remediating the environment. This chapter mainly focuses on novel materials, such as MOFs and Carbon porous materials, to treat wastewater. The production, synthesis, adequacy of these materials in applications have been comprehensively recognized and the several special benefits of these materials in water treatment are presented.

Keywords Sustainability · Metal-organic frameworks · Water treatment · Adsorption · Adsorbent · Adsorbate

---

M. Konni
Department of Basic Science and Humanities, Dadi Institute of Engineering, Visakhapatnam, India

S. B. Mukkamala
Department of Chemistry, GITAM Institute of Science, GITAM (Deemed to be) University, Visakhapatnam, India

R. S. S. Srikanth Vemuri
Department of Basic Science and Humanities, Vignan's Institute of Engineering for Women, Visakhapatnam, India

M. K. Karnena (✉)
Department of Environmental Science, GITAM Institute of Science, GITAM (Deemed to be) University, Visakhapatnam, India

© Springer Nature Switzerland AG 2021
A. Vaseashta and C. Maftei (eds.), *Water Safety, Security and Sustainability*, Advanced Sciences and Technologies for Security Applications, https://doi.org/10.1007/978-3-030-76008-3_20

1 Introduction: Water Pollution Due to Industrial Waste

Worldwide demand for quality water is increasing due to the rise in population and industrial growth. For achieving these water demands, sustainable and efficient water treatment technologies are essential for reusing the water, as the availability of natural water resources is inadequate. Contaminants, such as, heavy metals and organics engendered by industrial activities need treatment before releasing them into the nearest water bodies, and to treat these polluted waters, cost-effective and efficient treatment technologies are required in the current scenario [24, 96]. Most of the industries produce a large amount of wastewater that has numerous pollutants. Textile industries are significant trades that release various organic waste types, such as, dyes into the streams and contaminate them. Use of unscientific practices and discharge of partially treated dyes into the natural waters can cause severe pollution. Releasing moderately treated dye into the streams will cause colorations. Thus, it results in a decrease in the sunlight and the aquatic ecosystem is affected adversely. These substances are even harmful and deplete the dissolved oxygen levels in the marine system. Industrial laundries, softeners, and surfactants also release such dyes into the aqueous environment, however the impact is less than in the textile industries. Industries like pharmaceuticals and chemical manufacturing release organic complexes and solvents that also found in wastewater. Various anthropogenic activities and industrial activates like mining, steel, battery, tannery, and thermal release heavy metals, as recalcitrant, are also found in wastewater and these substances pose a serious threat to nature [73, 108]. Numerous examples of contaminants can be listed and found in wastewater streams. There is also a need to treat such wastewaters before releasing them into the environment to meet sustainability goals.

1.1 Various Techniques Used for the Treatment of Wastewater in the Current Scenario

Due to the complexity of pollutants in wastewater, the treatment options are usually not adequate in removing impurities to reuse the wastewater. The physicochemical and biological methods are the main methods in eliminating the contaminants in wastewater. Microbial treatment is the most promising and cost-effective method in removing organic pollutants in sewage [107]. These approaches have limitations as they are not efficient in removing recalcitrant contaminants [111]. Thus physicochemical methods are employed in treating these industrial effluents. The physicochemical methods include coagulation, flocculation, filtration, ozone/UV treatment, Fenton reactions, etc. However, as these techniques are technically complicated and require more capital for the operation, these are limited to practical functions. For instance, filtration technologies are inefficient in meeting environmental standards, precisely discharge standards upon releasing the same into natural water bodies.

Additionally, methods like coagulation followed by flocculation release a massive amount of sludge; their disposal and handling of this sludge results in a severe problem. To apply chemical oxidation techniques, a considerable volume of hazardous and chemical substances is compulsory and requires more operation [134]. As mentioned above, the foremost drawbacks are low competence, need more capital for the process, and they hinder other constituents in wastewater. The organic components in the wastewater eliminated by biological treatment; But, these treatment options don't provide acceptable results in removing the pollutants in the sewer as some of these pollutants remain resilient towards biological activities [45, 82, 106, 110, 130].

Consequently, innovative methods are still required for treating wastewater, which is reliable, efficient, and low cost and can function suitably. Expensive technologies like catalytic degradation, membrane process, ion exchange, etc. were used to remove heavy metals but suffer from drawbacks requiring frequent maintenance and enhancing the costs [42, 135]. So, there is a need to develop technologies that can install and operate sustainably in any environment. The adsorbent physisorption depends on the capacity of adsorption, endurance, and renewability of the adsorbents [75], and these techniques are considered to be more efficient, since they have a wide variety of application owing to its simplicity in design, and operation [75, 148]. Employing adsorbents with appropriate physiognomies will enhance the efficiency of this technique. Exploring adsorbents in terms of their structural, texture, and Physico-chemical properties are exceptionally significant to deliver the information for choosing suitable adsorbents for future operations. This technology is feasible and economical for eliminating harmful substances from wastewater [6]. Due to the adaptable design and smooth process, adsorption techniques have more significant advantages than other technologies. Furthermore, these techniques are rescindable and recyclable by appropriate methods [42, 91].

The present chapter mainly focuses on novel materials, such as MOFs and Carbon porous materials, to treat water. In recent years these two materials are extensively synthesized and used in treating industrial wastewater. Considerable research attention is being drawn towards the MOFs as these materials can be tailored according to our needs, both texturally and chemically. According to the literature available, MOFs are promising materials, especially for wastewater treatment, as the pore size and volume can be customized. This mechanism helps enhance the adsorption capacities. Nano-porous carbons also gained particular research interest in the present scenario. These materials are derived from the MOFs and have unique characteristics like good porosity, higher hydrophobicity, and aqueous constancy. In keeping a view of the present book's scope, which mainly focuses on sustainability, we have discussed low-cost nanoporous materials and biological precursors that can be enhanced and readily used for treating wastewater.

2 Treatment of Wastewater by Adsorption Process

2.1 Adsorption Theory

Adsorption is a procedure in which compound accumulates on the alliance between the phases, i.e., either with liquid-solid or gas-solid. Adsorbate is a compound where a partnership builds up, and adsorbent is the compound on which adsorption occurs [39]. In wastewater treatment, the adsorption phenomenon is confined to the liquid-solid interface, which is why our discussion is limited. The adsorption phenomenon can be either a physical or chemical process based on the interactions and nature amongst the adsorbents and adsorbates. The chemical reactions between the adsorbent and adsorbate molecules form a primary bond, i.e., covalent, ionic, or dative, during the chemisorption process. Chemisorption is an equilibrium process and irreversible, and these properties enhance the adsorption phenomenon. Calcination is generally performed for regeneration [75, 148]. Adsorbent pores adhered to adsorbates by secondary bonds during the adsorption process. The solvent exchange method will help recover the adsorbent in most cases, and the physical adsorption mechanism is reversible [149].

Adsorption isotherms and kinetic models will define the adsorption efficacy of a compound. The adsorption isotherms depict the quantity of material adsorbed per mass of adsorbent at adsorbate concentrations in equilibrium conditions. In these experiments, equilibrium phases are achieved by an adsorbent at a specific mass and volume at precise adsorbate concentrations. The Eq. 1 below, introduces mass-balance [128].

$$AE = \frac{(EC0 - ACE) \times Vo}{AM} \quad (1)$$

where, EC0 = Equilibrium Initial adsorbate concentration, ACE = Final adsorbate concentrations, V0 = Volume, AM = Adsorbent mass and AE = Adsorption equilibrium.

From Eq. 1, constants of adsorption isotherms will be obtained, and values signify adsorbent properties and efficiency. The adsorption isotherms help understand the relationship between the technical and equilibrium data [95] for the adsorption process, numerous isotherms used. Still, Langmuir and Freundlich's model considered being more prominent as it has a wide range of concentrations to describe experimental settings [81]. Adsorption mechanisms and adsorption rates can be understood/determined by applying kinetics and applying viable applications. However, sluggish rates of adsorption have limitations for use. The adsorption phenomenon has three stages: diffusion by the film, distribution by pore, and solute molecule adhesion to the adsorbent's surface. Delivery by film means dispersal of adsorbate molecules on to the surface in films; diffusion of pore includes distributing particles on to the adsorption site. In contrast, adhesion occurs when molecules of the adsorbate diffuse into the surface's pores [109].

2.2 Factors Influencing the Adsorption Phenomenon

Numerous parameters will affect the capacity of adsorptions of various adsorbents. Adsorbentsinclude functionalized groups and solubility and the size of the molecules of an adsorbate, besides adsorbent, encompasses the area, size of the pore, and volume of the pore [136, 157]. In the adsorption phenomenon, conditions like the strength of a solution, the solution's temperature, and the solution's pH play a vital role [3, 19]. Surface interactions depend on the size of the adsorbent molecules [102]. The unit's size is directly proportional to the adsorption process; as particles' size increases, the polarizability will be enhanced, resulting in the formation of strong forces of dispersion on the surface. Contrarily, the molecules' scope is limited as the steric effects will slow down the absorption rate, blockage the pores, and reduce the surface area. For instance, a study conducted by Lachheb et al. [80] stated that due to aromatic (naphthyl group) bonds, a sizeable steric hindrance arose, and this results from low adsorption of Congo dye analogy to methylene blue [80]. Valix et al. [142], stated that the acid dyes' molecule size was hindered by steric effects and reduced the diffusion onto adsorbent carbon materials [142]. Adsorbate functionalized groups will regulate the properties like chare of ions, hydrogen bonding, acid and base interaction, and solubility. The pKa value of adsorbate describes the basicity and acidity. In the process design of adsorbates, the adsorption properties of parent molecules show distinct features compared to the protonated/deprotonated forms, and thus we have to contemplate with pKa values.

Functionalized groups are responsible for healthy bond formation, and thus molecules will get adsorbed onto the adsorbents by interactions between positive-negative charges and acid-basic behaviors. Pereira et al. [114] conducted a study on AC (activated carbon) for the adsorption of dyes and deliberated the chemical groups and their effects [114]. He also stated that dye removal in the solution is influenced by the chemical vicissitudes between activated carbon's surface-interface. Adsorption of dyes is mainly due to the surface's basicity, and anionic dyes are reactive and acidic. Lewis's primary sites with oxygen-free molecular interactions will enhance the process of adsorption. The molecules comprehending oxygen sites are generally acidic and have limitations for adsorption to anionic dyes. Whereas cationic dyes have a higher affinity in adsorption shows a positive(+ve) effect. Radovic et al. [116] stated that the aromatic solute molecule adsorption on carbon compounds is mainly dominant due to the interaction between pie-pie dispersion [116] that equilibrium could be affected by interactions of dispersive and electrostatic effects. Ghaffar and Younis [48], showed that the adsorption process is affected instantaneously by the interactions between covalent-electrostatic, bonding of hydrogen, and Vander walls forces; High internal volumes are required for the material, which is acting as an adsorbent, and these materials should be accessible to the complexes in solvents that are needed to eliminate. The adsorption phenomenon is immensely affected by the pore size/volume and surface area.

For the adsorbates to move across the membrane, the requirements include their distribution in the solution and pore size of the membrane [37, 36]. Transport of some of the substances across the membrane is in the order of pore size as described: for phenolic compounds, the appropriate pore size is less than 2 nm (micropores), for proteins 2–50 nm (mesopores), and bacteria 50 nm [36]. The possessions of adsorbents are affected by the solutions' pH; adsorbate chemical properties are altered by these conditions and result in sorbate particles' formation. Acid-base interactions with major groups will change the interface of the adsorbents. In an aqueous solution, the solid surface's charge is determined by the Zeta potential (Z), and it is a parameter related to pH. The pH of the solution will influence the adsorbent's surface, the less contaminated surfaces obtained in case of lower pH, and the more negative surface obtained in case of pH. These conditions will affect adsorption, resulting in a change of the electrostatic interactions adsorbent zeta potential. For instance, to remove the blue-106 dye, the pH of the solution should be reduced, if we increase the pH, it will reduce the molecular adsorption as the repulsion occurs due to the electrostatic effect [7]—the electro-kinetic properties of the adsorbents demarcated by *Pzc* (point of zero charge). *Pzc* is a condition where the pH of the adsorbent has zero surfaces. The solution's temperature will affect the water molecular hydration degree and adsorption [40]. The adsorption equation given by the van't Hoff reveals untenanted sites throughout the adsorption of Gas-phase is exothermic. In the case of wastewater treatment, untenanted sites are not available as these phases are always liquid and involve dampened surfaces during the adsorption process. In a liquid phase, solvent desorption is always associated with solute adsorption as these molecules occupy the sites. Thus, the thermodynamics process is more complicated. Adsorption of dyes in wastewater is generally endothermic [59]. The solution temperatures determine the diffusion rates and viscosity; results, only at higher temperatures, the adsorption equilibrium is attained.

3 Adsorbent Materials

3.1 *Adsorbents in the Current Scenario*

As we discussed in the earlier sections that eliminating absorptive impurities from the waste stream and its porosity plays an essential role. Based on the sizes of solid materials, pores are categorized. Molecular pores of a compound that are less than 2 nm termed as micropore, range between 2–50 nm called mesopores, and higher than 50 nm considered as macropores. Figure 1 depicts the categorization of pores of porous materials and demonstrates how these pore sizes exude particles with a larger scale.

The materials like silica, resins, activated carbon, and alumina gained more importance as a novel adsorbent material than conventional adsorbents (Fig. 2). The main criteria for selecting these materials over conventional adsorbents are their low cost and high efficiency in removing pollutant parameters as untreated zeolites and silica can remove the organic solvents in wastewater as they are hydrophilic

Fig. 1 Illustrates the schematic representation of pores' types and their connectivity in porous materials (Reproduced with permission from [90].)

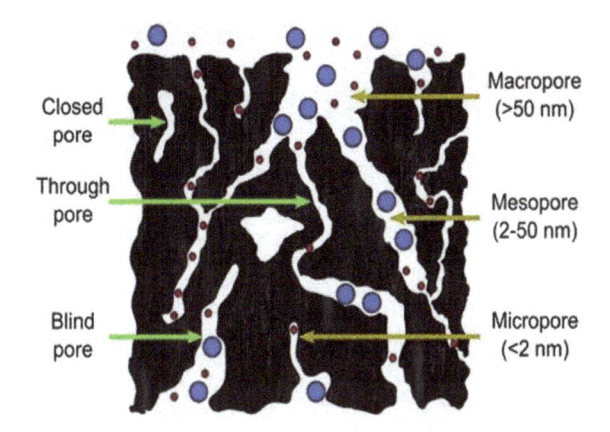

Closed pore
Through pore
Blind pore
Macropore (>50 nm)
Mesopore (2-50 nm)
Micropore (<2 nm)

[33]. Contrarily, pollutants can be removed efficiently with the help of activated carbon materials as they are hydrophobic. Thus, for the treatment of wastewater, activated carbon is used as an adsorbent commonly. Functionalized groups like N_2 (Nitrogen), O_2 (oxygen), H_2 (hydrogen), S (sulfur) are present in activated carbon. The adsorption can remove contaminants in the wastewater; carbonyl, lactones, carboxyl, quinone's, and phenol functionalized groups are responsible for adsorption. The hydroxyl groups of phenols and carboxylic acids formed during the adsorption process give acidic nature to activated carbons [87]. Starting the functionalized groups by thermal treatment or chemical treatment will significantly affect the adsorption process [13]. The adsorption capacities can be adsorbates enhanced using diverse O_3 (ozone) treatment, plasma, microwave, and acid-base treatment. Pi electrons on sheets of condensed polyaromatic compounds are the basic physiognomies of activated carbons [11, 88]. During the heat treatment, the oxygen molecules are removed from the activated carbon surface in the inert atmosphere, resulting in Lewis base sites' formation with abundant electrons. Due to this treatment, activated carbon basicity will amplify as functional groups containing oxygen and acidic groups are eliminated. Removal of an oxygen molecule from the activated carbon surface will favor the adsorption of pollutants in the wastewater as the surface becomes less polar. Monser and Adhoum [101], stated that activated carbons are not efficient in removing organic pollutants and heavy metals than organic compounds [101].

Yao et al. [153] reported that the adsorption of organic compounds in aqueous solutions is reduced by binding oxygen to the sites of activated carbon and functional groups, which are acidic [153]. Boehm [15], in his study, showed that hydrogen bonds present in the molecules of water adsorb functional groups containing oxygen atoms, and molecules of water that are additional will cluster at sites. Cluster formation at the adsorption sites will decrease the interactions between pollutants and activated carbon. It will allow the contaminant to attach at hydrophobic ends, which will block the pollutant molecules, and micropores' access will be inadequate [29]. Organic pollutants in wastewater will be removed with the help of adsorbents, having an even hydrophobic surface. Nevertheless, the adsorbents used for water treatment should

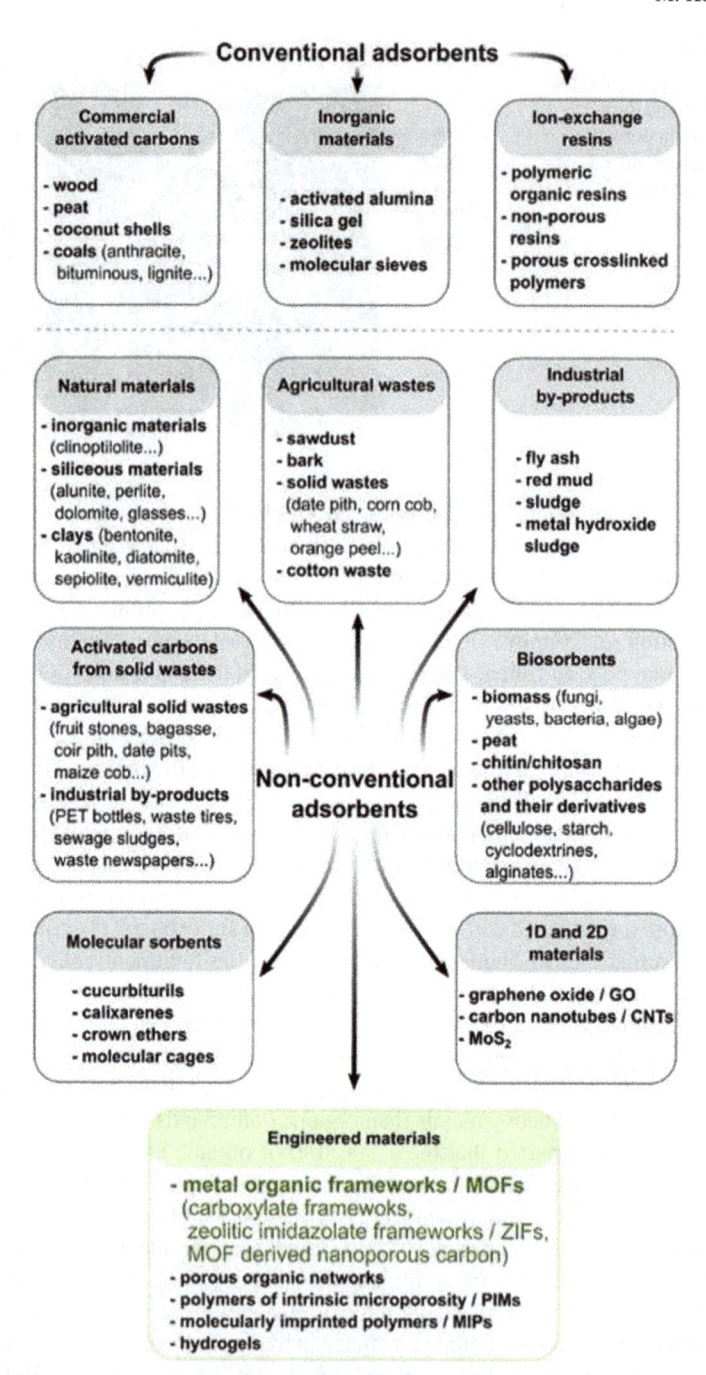

Fig. 2 Crini et al. [31] listed and categorized the materials used as an adsorbent for wastewater treatment (Reproduced with permission from [31].)

be hydrophobic [34]. Recent research about the adsorption materials mainly emphasizes manufacturing the specific adsorbents for optimizing them for application. Manufacturing of low-cost adsorbents is highly indeed for the treatment of various effluents from the different industries. Contrarily, there is a need for the development of an adsorbent with excellent capacities of adsorption. One-dimensional and two-dimensional adsorbents like CNTs and Graphene oxides (GO) have gained much importance in recent years as they possess ideal surface area and unique interface. These adsorbents are still under emerging state due to the inconsistency in batch operation and production cost. For instance, molybdenum disulfide sheets (MoS2) were used to absorb the dye; later, it was observed that it is unfeasible for dye removal due to its partial adsorption in aqueous solutions [121]. Molecular sorbents can adsorb the pollutants into their cavity as they are macrocyclic compounds. The comparable size of the macrocyclic compounds enables multiple points of interaction and helps absorb impurities efficiently [22, 23, 127, 132]. After proper provisions and immobilization, for the treatment of wastewater, these compounds can be used. Non-conventional adsorbents are considered engineered material as they can be assembled. Absorbents will have control of their structure, interface, and porosity. The properties of the adsorbents are altered by building block selection and post-synthetic modification. Compared to traditional adsorbents, these materials gained much more importance due to their salient features and gained more interest among the researchers.

3.2 MOF (Metal-Organic-Framework) for Water Treatment

3.2.1 Carboxylates-Frameworks

Due to metallic locations, porosity, and functionality on pores, these materials are considered novel. In analogous to zeolites, which are microporous and inorganic, MOFs (Metal-organic frameworks) structure is flexible and modified by pore functionalization (Fig. 3). Thus, these materials gain more attention as nanomaterials due to their flexibility [20]; Examples of these materials are shown in Fig. 3.

Molecular structures of MOFs comprise metal ions through organic linkers, and MOFs' porosity is one of the main features. Even though zeolites are inorganic compounds, both zeolites and MOFs have similar structures and uniform pore sizes. But, they are different in some crucial aspects. For instance, the synthesis of MOFs is facile compared to zeolites. A wide variety of these substances can be manufactured by changing their ions and linkers that are organic towards adjacent areas [84]. Nearly 75,000 MOF structures are recently deposited in "Cambridge's Crystallographic Datacenter" [99, 160]. Among these deposited structures, most of the systems are carboxylates. The carboxylate group of MOFs form bonds that are coordinating with metal ions and creates a robust framework. These frameworks can endure the alkyl, sulfonic, nitro functionalized groups on the linker of carboxylate. Thus, using carboxylic acid for synthesis can help in producing the MOFs quickly.

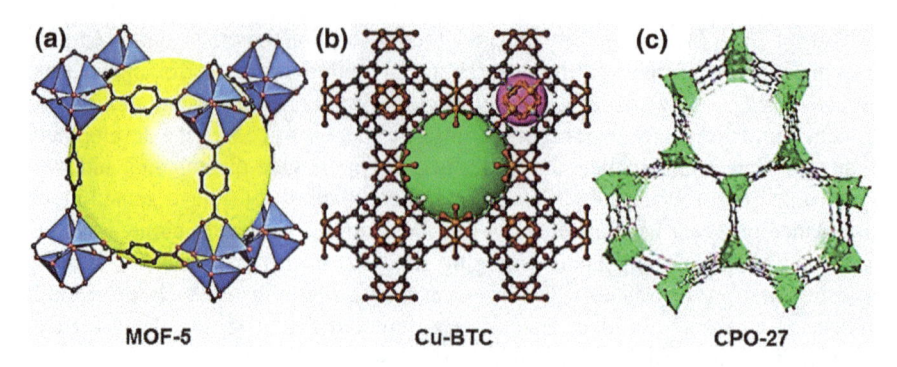

Fig. 3 Illustrates examples of MOFs' structure at a molecular level (Reproduced with permission from [131].)

Due to their salient features of MOFs, they have a wide variety of applications like gas-liquid adsorption [97, 103, 129], catalysis [54, 143], and separations [18, 21, 62, 64]. Nonetheless, most of the MOFs in water is uneven, and this problem has to be overcome for the application of MOFs in industries. For instance, the structure of MOF-5 is altered when they are exposed to molecules of water. Interactions of MOF-5 with water molecules will decrease the surface areas and transform into MOF-69c because the carboxylic groups are replaced, coordinating with zinc atoms [146]. Thus, there a need for selective MOFs like MILs-100/UiOs-66 for water treatment. MOF materials' functionalization will control the wettability and improve stability [122].

3.2.2 Zeolite-Imidazole-Frameworks (ZIFs)

ZIFs are the subsection of MOFs with good porosity and enormous potentiality for adjustment and development, as these compounds have high chemical and thermal stability [66, 152]. The ZIFs consist of metal ions like Co, Zn, Cu, etc., and nitrogen atoms interlink them with linkers of imidazole. Zeolites and ZIFs consist of analogous frameworks where Z-Im-Z replaces X-O-X bridges; Here X represents Si, Al, or P, where Z represents Cu, Co, or Zn having 145 degrees bond angle [57]. Different structures of ZIFs can be manufactured by altering the organic imidazole molecules and metal atoms [10]. ZIF-8, 11 are proven to be a suitable adsorbent even in water and alkaline aqueous medium as they possess high chemical and thermal stability [69, 67, 112, 158]. Park et al. [112], revealed in their study that the structures of ZIF-8, 11 are not altered for seven days even they are subject to water at 50 °C. However, by increasing the water temperature to the boiling points, the Structure of ZIF-11 is altered after three days, but ZIF-8 exhibited stability [112]; this has proven ZIF-8 is a more stable compound among the MOFs. The reason for strength in ZIF-8 is that imidazole linkers form strong bonds between ligands and metals compared to carboxylate linkers. Strong bonding abilities in ZIF-8 led to more adsorption [46, 70].

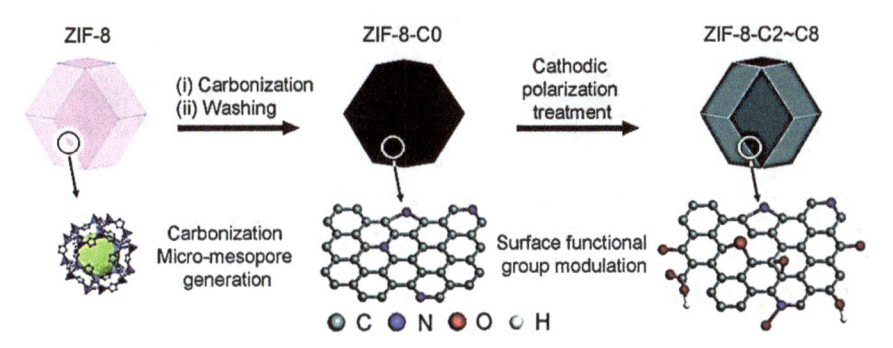

Fig. 4 Carbonization of ZIF-8 results in forming a highly porous template (Reproduced with permission from [86].)

3.2.3 Nano-structured Porous Carbon Materials (NSPCMs)

NSPCMs are the derivatives of MOFs and act as an adsorbent, catalyst, fuel cell, etc., as they have exceptional chemical and physical properties [17, 68, 69]; NSPCMs have a wide range of environmental applications because it helps in removing organic contaminants and hazardous substances from water [8, 73, 139]. Methods like chemical vapor depositions (CVDs) or carbonization fabricate the NSPCMs [20]. Some NSPCMs are resilient to alterations like carbonization with the hydrothermal process; the generated materials are non-porous and exhibit low surface area. The materials produced by hydrothermal carbonization don't have microporosity, which is necessary for adsorption. Conventional methods are used to activate these materials and thus results in enhancing the surface areas. For instance, they are performing hydrothermal carbonization cannot crystalline cellulose materials as resistant [141]. MOFs are contemplated as a substitute for the production of NSPCMs as they have larger surface areas. Examples of NSPCMs obtained from MOFs are MOF-8, AL-PCP, and ZIF-8 [26]. Direct carbonization of MOFs will result in the production of porous materials. Hu et al. [61] showed in their study that decarbonization is a single-step process and can be done at 800 °C, and he used this process to convert Al-PCP to nanocarbon. He concluded that temperatures are crucial to reaching the large pore size and high surface area [61]. Figure 4 depicts the production of NSPCMs with ZIF-8. Lim et al. [92] stated that if MOFs that are non-porous containing zinc metal ions will give highly porous NSPCMs by direct carbonization treatment [92].

3.2.4 Adsorption Mechanisms of MOFs

Compared with conventional adsorbents, MOFs perform similar mechanisms; however, the MOFs with open metal sites show enhanced adsorption. In the present subdivision, the adsorption mechanisms that are specific to MOFs and in ordinary are discussed. Figure 5 illustrates the pollutant removal with the help of MOFs by

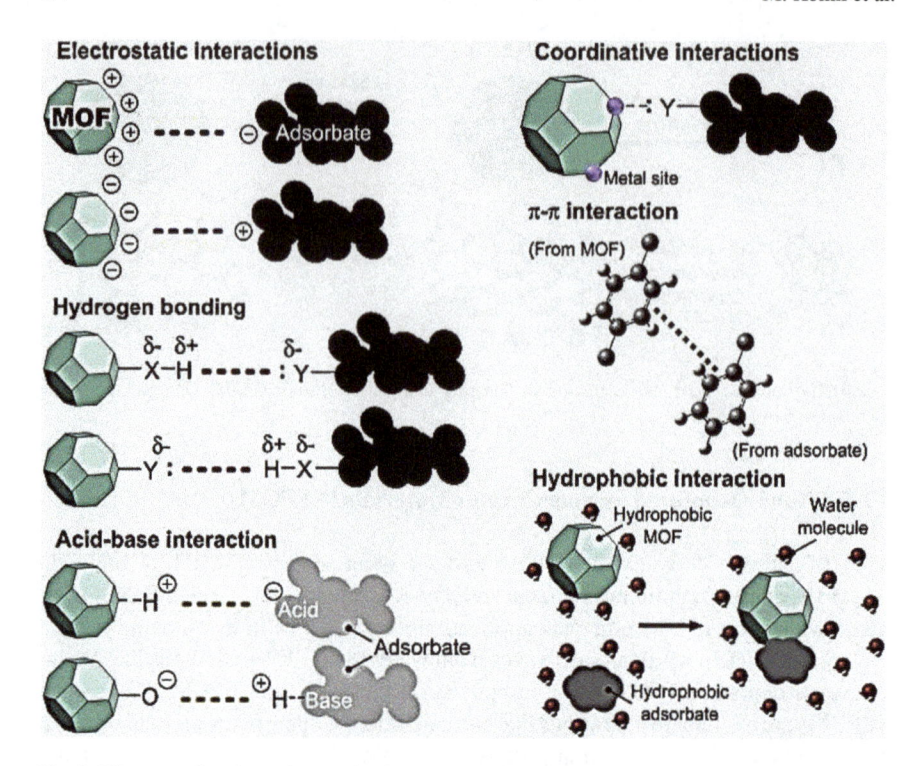

Fig. 5 Illustrates the adsorption mechanism using MOFs (Reproduced with permission from [56].)

various means in adsorption. The effective way to remove hazardous substances in water is electrostatic interactions, as shown in Fig. 5.

Haque et al. [53] conducted a study on MIL-101Cr and MIL-53Cr (MOFs) to remove anionic methyl orange dye. In comparison with MIL-53Cr, MIL-101Cr showed better adsorption due to the large pore size and surface area. However, doping ethylene diamine (ED) to MIL-101Cr results increase in adsorption capacity and an insignificant decrease in the size of the pore observed electrostatic interactions between MIL101, PEDMIL-101, and EDMIL-01 (Fig. 6) [100].

Lin et al. conducted a study to adsorb Methylene Blue (MB) on Cu-BTC and revealed that interactions (electrostatic) between adsorbent and Molecules of MB are strongly pretentious by the pH of the solution. An increase in pH results in a decrease of negative surface area and therefore created favorable conditions for exchanging the molecules between MB (positive charge) and Cu-BTC (negative control) and enhances the adsorption. H_2 bonds are usually formed between adsorbent and adsorbate, and it may lead to adsorption. However, molecules of water vie for bonding sites of hydrogen in aqueous solutions. Thus, for efficient adsorption, multiple interactions between these systems are essential.

Liu et al. [96] stated that MOFs showed low adsorption in removing the phenol groups like PNP (p-nitrophenols), phenol during interaction with 100-Cr, 100-Fe

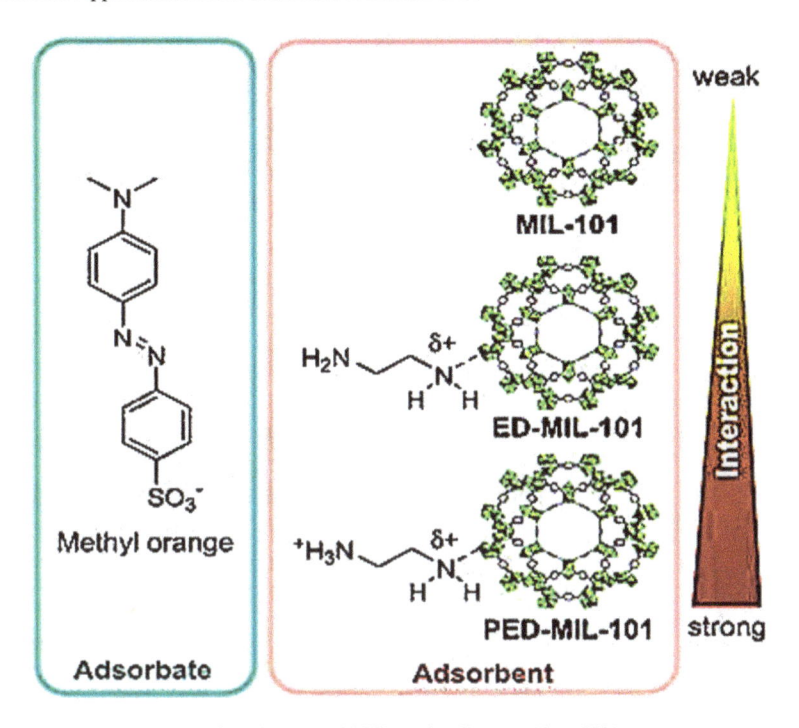

Fig. 6 Electrostatic interactions between MOFs and pollutants (dye) [53]

NH2MIL-101 (Al) type MOFs. Comparing, NH2MIL-101 (Al) exhibited more effective mechanisms of adsorption of Phenols than other MOFs. This removal of PNP through NH2MIL-101 (Al) type MOFs is due to their hydrogen bonding as shown in Fig. 7; no interaction of acid-base is observed in removing contaminants in the adsorption process.

For instance, in aqueous solutions, Khan et al. [76] stated that thallic acid adsorption mechanisms on various MOFs revealed that ZIF-8 showed higher adsorption without amino groups at neutral pH. In aqueous solutions, MOFs consist of metal ions, and these ions stimulate the adsorption of dye, and several researchers reported the same. For example, Tong et al. [138] stated that MOFs like MIL-100-Fe and MIL-100-Cr showed different adsorption behaviors in removing dyes like methyl blue (MB) methyl orange (MO) even though they have analogous surface areas. Low adsorption of methyl orange by MIL-100Cr is due to water molecules. However, MB adsorption is more due to electrostatic interactions.

In adsorbing organic molecules in aqueous solutions, hydrophobic interactions are frequent. To demulsify and clear the oil spills in the aqueous medium, MOFs generally used, and Yang et al. [150] utilized FMOF-1 to demulsify the oil droplets in water. The presence of perfluorinated surfaces inside the MOFs; these substances act as hydrophobic. Due to this character, FMOF-1 adsorb additional oil droplets because of interactions between nonpolar adsorbates and enhance adsorption. Lin

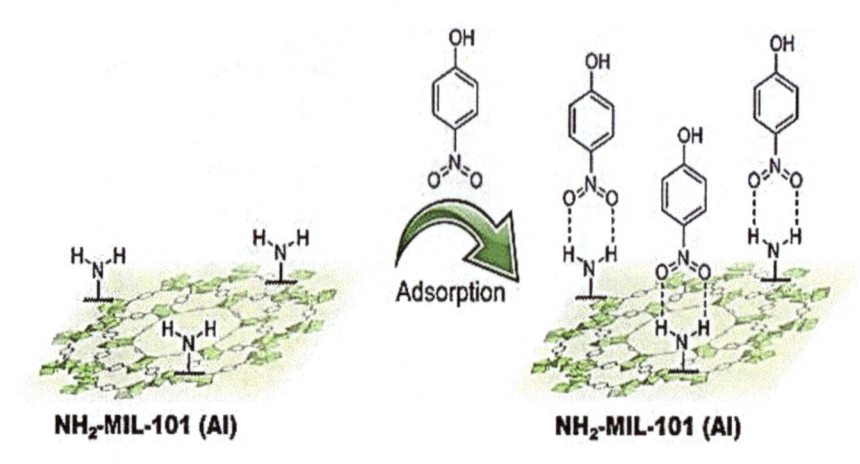

Fig. 7 Adsorption mechanism exhibited by NH$_2$MIL-101AL to eliminate PNP (Reproduced with permission from [96].)

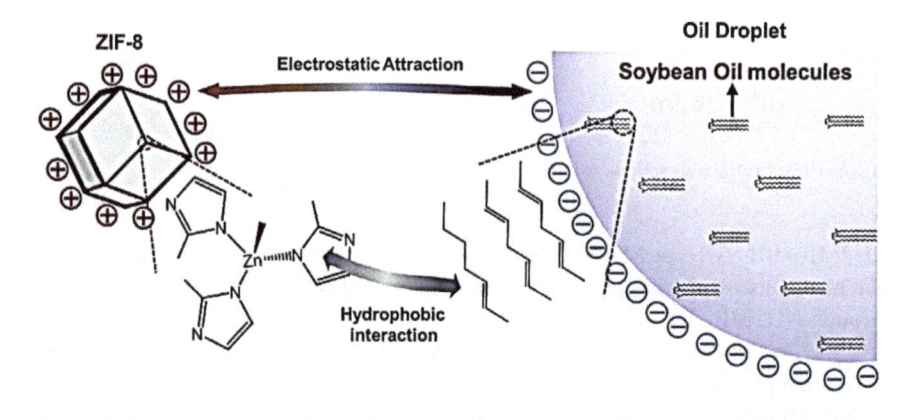

Fig. 8 Illustrates the utilization of ZIF-8 to remove oil (Reproduced with permission from [35].)

et al. [94], adsorption and demulsification of oil in water are due to electrostatic and hydrophobic interactions. The oil used in his study is soybean, and this mechanism was showed in Fig. 8.

4 Sustainable Applications of MOFs as Adsorbents

Over a few decades, the metal-organic frameworks and their derivatives gained prominence for several applications. Research on these materials has provided more feasibility for viable applications. Furthermore, sustainable applications of MOFs have gained much importance among the researchers for their development. Treatment of wastewater and the removal of organic pollutants with adsorptive mechanisms are some of its sustainable applications. MOFs reduce stress on the environment and helps in recovering the valuable materials in the wastewater. Nevertheless, adsorbent sustainability depends on the performance and depends on their production, replicability, and preparation throughout their application. Many researchers reviewed the technologies of MOFs and summarized them for progress and sustainable utilization. Reinsch [117] summarized the synthesis of MOFs in the view of sustainable chemistry [117]. Julien et al. [71] also reviewed various MOFs, their production, synthesis, and utilization in sustainability [71].

In the above analysis, they evaluated the criteria for MOFs for sustainable application and recommended the following, along with several areas where these substances can manufacture:

- Should utilize the biocompatible and safe elementary unit.
- The utilization of energy should be reduced.
- Reaction media should be H_2O.
- Solvent utilization should be minimal or prohibited.
- Replication and production of the materials should be continuous.

Connolly et al. [27], in their minireview, listed out various procedure for the sustainable synthesis of MOFs and explained in detail their applications. Mon et al. [100], in their review, provided multiple types of MOF materials currently using for remediating the water. Even though several MOFs exist and are presently utilized in the science field, researchers' principal focus is on the materials, which are long-lasting and low cost. Li et al. [89], drafted various MOF materials to remove the water's pollutants through membrane separation techniques. In the present chapter, MOFs' stability is considered to be a significant field for producing sustainable compounds. Furthermore, to authenticate investigation in the science field, the real-life applications of these substances are accentuated.

4.1 Production of MOFs

Conventionally MOFs are manufactured by using polar aprotic solvents with a solvothermal process. Polar Aprotic solvents are generally toxic; the solvothermal process occurs under internal pressure for 24–72 h below 200 °C. Transition metals (nitrate or chloride salts) are added to this process, resulting in HNO_3 and H_2SO_4. As the strong acids produced along with mother liquor after the filtration process during the conventional MOFs production, they will pollute the water stream by form complex materials. Thus, for the sustainable operation of adsorption, the output of MOF is considered a crucial process. The choice of metal is essential for the greener production of MOFs in the research. Researchers proposed salts like sulfate as the alternative to chloride and sulfate salts as they are less corrosive, but the major drawback with these salts is low solubility.

Later, the nitrate salts are replaced by the acetate salts as slight alternatives as these salts fasten the rate of nucleation and require less in quantity compared with nitrate salts [14]. For MOF production, utilization of OH^-, HCO^{3-} and metal oxides produces by-products, i.e., less carbon dioxide and nontoxic water. Nevertheless, due to these salts' reduced solubility, different approaches are required then the solvothermal process to overcome these disadvantages. Direct utilization of metal in electrochemical conditions is considered a new approach because they don't generate any by-products during MOFs synthesis. Mueller et al. [104], produced highly porous materials by using copper cathode with the same method. For the integration of MOFs along with the selection of salts, metals are even to be considered. Notably, leaching of transition metals from the MOF adsorbents will cause severe consequences during water treatment. Amongst other transition metals, manganese, zirconium, titanium, and iron will exhibit less toxicity [71]. Many researchers are developing and investigating MOFs based alkaline earth metals to minimize the toxicity during the production process. The organic linker used in MOFs production will not cause any toxicity, but their manufacturing causes stress on relic resources. Though some investigators found a replacement for these metals, there is a lack of literature, and these studies are limited. For instance, MOFs that are interlinked by the acid of terephthalate are formed with xylene's help. Ren et al. [119], synthesized MOFs using similar settings used in the production of Polyethylenterephthalate hydrolysis. Ren et al. [119] used waste polyethylenterephthalat bottles for MOF-Cr production, and in this experiment, waste bottles acted as a source for terephthalic linkers. Many researchers showed interest in developing a bio-based organic linker for carboxylate complexes.

Dreischarf et al. [38] and Wang et al. [145], conducted a study and stated that amino acid, asparagine acid, and 2.5 furan dicarboxylic acid are analog terephthalic acid to produce MOFs. Fumaric acid's low cost and renewable nature enabled them to utilize as a linker in MOFs production. Conventional methods like the solvothermal process are still used to produce MOFs in the bench-scale experiments as they are reliable and the earliest approach. This process's popularity is associated with one main feature, and this process doesn't require sophisticated instruments for the generation of high crystallinity MOFs. Nevertheless, this process requires a high amount of energy and generates more waste and safety and health concerns. As mentioned earlier, using polar aprotic solvents causes toxicity. Thus, researchers are replacing them with alternatives for greener synthesis-Cseri et al. [32] listed out the options for harsh solvents. For instance, Zhang et al. [159] proposed Cyrene instead of conventional solvent for greener MOF production, and the results are adequate. At synthetic conditions, its application for the synthesis of MOF is challenging. Ionic liquids are also investigated in MOF synthesis as they have low volatility and can use at high temperatures [113]. However, the impregnation of ionic liquids into the crystal structure and recycling is very difficult. MOFs synthesis with water as solvent gained more interest among the researchers. Gaab et al., used the hydrothermal process to synthesize MOF by using a solvent like water. ZIF 8 was synthesized at room temperatures with an aqueous solution that has also been confirmed [151]. Approaches for the synthesis of many MOFs in aqueous solutions are limited due to low solubility and instability. Synthesis of MOFs near the critical waters is a challenging method, and it should be overwhelmed; Ibarra et al. [65] used near-critical water to produce MOFs.

Nevertheless, this process needs a high amount of energy, and this will lead to safety-related problems. These drawbacks conduct the research to synthesize MOF materials with high efficiency and minimal waste production. Many researchers summarized their reviews to integrate MOFs at the lab-scale and later utilize them for various industrial applications [118, 123]. For the continuous synthesis of MOFs, flow chemistry is used as a useful tool. Polyzoidis et al., in their study, used flow chemistry in a microreactor for synthesizing MOFs by controlling the particle size. During the MOF synthesis, to reduce solvent utilization, the mechanochemical process was used, and this process involves ball milling and crushing under ionic-liquid conditions. For the Greener synthesis of MOFs, mechanochemistry enables to use of metal hydroxides or oxides. Even though this procedure is solvent-free, but low crystallinity and conversion were detected. Tanaka et al. [134], used zinc acetate in mechanosynthesis to produce ZIF-8 to overcome these problems. The rate of reaction and crystallinity in mechanochemical synthesis is enhanced by adding additives that are ionic or liquid [72]. For a greener and lab-scale synthesis of MOFs, accelerated aging [25] and aerosol synthesis [98] are the other approaches.

The utilization of MOFs as preliminary materials to synthesize nanoporous carbon materials has an advantage towards greener synthesis. Materials during carbonization and heat treatment will not have enough porosity and surface area; the conventional process activates these materials to enhance the pore size and surface area. Though chemical or physical activation is not required for the MOFs that have been synthesized from nanoporous carbon [61]. The subsequent substances will have suitable porosity and larger surface area, which helps treat wastewater (Fig. 9).

4.2 MOFs Preparation and Fabrication

For practical applications like adsorption, catalysis, and separation, the MOF powder should be converted into macroscopic substances [4, 43, 140]. The synthesized MOF after the conversion should have functional porosity and surface area like the original [104]. Thus, it is essential to choose suitable methods to develop unique MOFs with excellent characteristics. Adsorbent preparation and fabrication should be taken into consideration for assessing these materials for sustainability. To turn out the MOFs into pellets or granules or to prepare them as adsorbents, techniques like extrusion, spray drying, and granulation are commonly used [2, 125, 155]. The major challenge is, the fabricated MOFs should be analogous in properties like chemical and mechanical stability to the original MOFs powder. Porous materials fabrication was generally done by mixing them with inorganic/organic additives and thermal treatment to obtain a substantial mechanical framework [85].

However, under high-pressure conditions, the fabricated MOFs show weak mechanical strengths, and the same was reported by many researchers [85]. For instance, ZIFs made up of Zinc show weak mechanical strength with shear stability

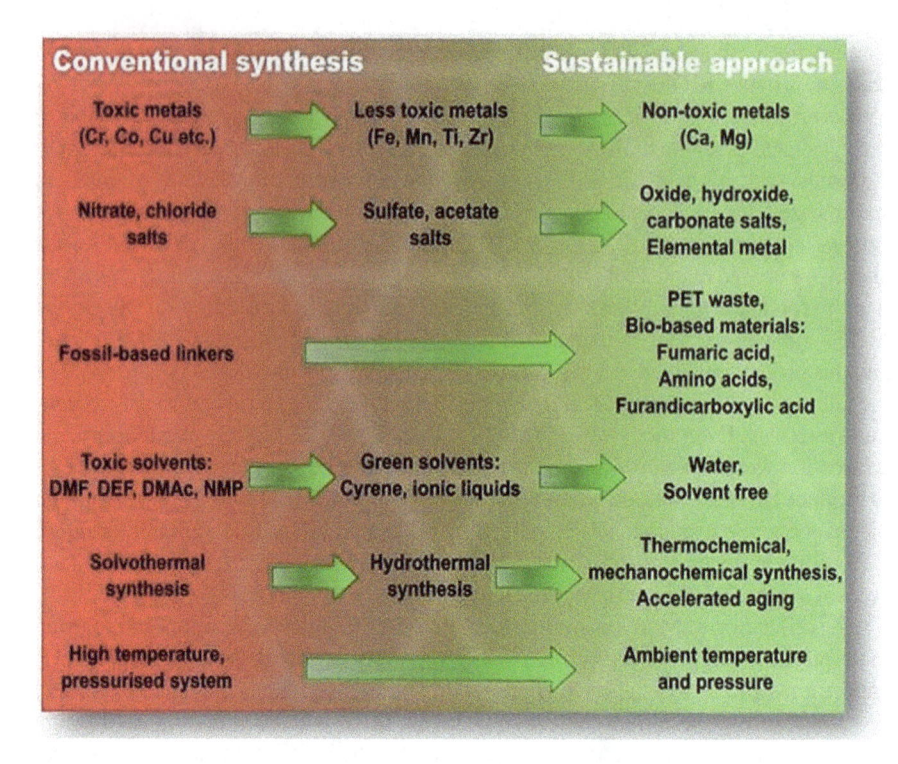

Fig. 9 Sustainable procedures for the synthesis of MOFs

between 0.2–0.11 GPa. Amorphization of MOFs even occurs when they are subjected to external pressures. Furthermore, unlike zeolites and ceramics, MOFs at high temperatures are not thermally stable, and their dopped additives can't be removed by subjecting to high temperatures [154]. The granulation shaping methods are used to prepare Spherical granule Cu-BTC, and the obtained granules are in the range of 1–3 mm [16]. Additionally, Ren et al. [120], used sucrose as a binder in the Zr-MOF preparation and obtained 0.5–0.15 mm (diameter) spheres. Nevertheless, due to the Zr-MOF pore blockage by binding molecules, the BET surface areas and pore size of these Zr-MOFs spheres are reduced by fifty percent. Thus, it is challenging to obtain an equal size sphere using granulation methods; to enhance the adsorption, the MOFs' particle size should be uniform. This formulation can be obtained by regulating many parameters. The granulation method's main limitation is that the fracturing of MOF materials occurs due to collision stress during synthesis and leads to agglomeration [85]. MOFs for an industrial application like adsorbent/catalysts can also be synthesized by the spray drying method. These methods can be used for the synthesis of materials from the range of millimeters to large particles. Nevertheless, large particles' presence will not allow the millimeter-sized particles to free-flowing [85]. The suspension's viscosity is altered as the additives/binders are used in the spray drying process and cause problems. Additionally, the general design rules do not apply to this process because the interaction between additives and MOFs are very intricate. The fabrication of MOF materials is complicated compared to other generally used materials in the spray drying method [85]. However, Avci-Camur et al. [9], produced Zr-MOFs (spherical) through the spray-drying process, and this research gained attention by many researchers as a green solvent was used for the synthesis of MOFs.

The fabrication of MOFs can be done even by the extrusion technique. Crawford et al. [30] used this technique to fabricate the size of the monolith of Cu-BTC to millimeter Cu-BTC. The major drawback of this technique is uniform bodies are not formed and not resilient to abrasion. The complex mixture (MOF + additive + solvent) should have unique properties like malleability to cast out, and it should also exhibit cohesion to avoid the defects in creating surface area. Küsgens et al. [79], developed monoliths of Cu-BTC with high stability in centimeter-scale. In comparison to zeolite, these MOF showed more efficiency and adsorption capacities.

The pellet and powder preparation involves a pressing process, and these materials have excellent mechanical strength. The MOFs mechanical properties will determine the quality of pellets during the MOFs materials processing. The critical process will lead to the deformation of MOF properties and have significant effects. The pressure is one of the vital parameters for pellet preparation to regulate the various properties [5, 105, 115]. These pressing methods' significant drawbacks are crystallinity, and the MOFs' amorphous properties are altered due to pressure. Tagliabue et al. [133] developed Ni-doped COP-27 to investigate the adsorption mechanism without suing any binder; the study revealed that due to the pressure, the crystalline structures of the MOFs are altered. Bazer-Bachi et al. [12] used direct methods of compression and prepared pellets. They showed that the substance's deformation depends on the amount of pressure applied; thus, this method is not suitable, affecting its performance. The compression method for MOFs [28, 126] is not appropriate or feasible for amorphization, but NH_2MIL 53 pressure does not affect the structure; however, it does not apply to the other MOFs preparation. In situ methods (seeding growth/layer-layer deposition) develop MOFs with silica/aerogel substrates' help. Still, theses process is tedious as it involves stepwise procedures [60]. Methods like phase inversion can also be used for the synthesis of spherical shape MOFs used for adsorption. Abbasi et al. [1], used the phase inversion method for removing oil in water by preparing spherical shape ZIF 8, and these composites showed high efficiency of adsorption. Gu et al. [49] and Zeng et al. [156] converted the prepared composites into carbon and utilized them for dye removal. These studies showed that phase inversion methods are useful for organizing the MOFs as these methods will not alter the mechanical properties or crystallinity of the substances. Many researchers are using 3D printing methods for the fabrication of MOFs, as these methods show versatility. These methods will not alter the materials' properties [137] and provide a possibility for optimizing the adsorbent printed by this method [78].

4.3 Applications of Adsorbents for Water Treatment

4.3.1 Application of MOFs for the Removal of Organic Pollutants in Water

Many researchers have studied a wide variety of MOFs to remove pollutants in water, chiefly, the pore size/volume and charge interaction between adsorbent-adsorbate affect the adsorption mechanism [75]. MIL-100-Fe/Cr adsorbed the organic pollutants like clofibric and naproxen from the water; this study revealed that the adsorption is more significant than activated carbon [55]. Jiang et al. [70], developed a ZIF-8, revealed that ZIF-8 adsorbed the benzotriazole organic pollutants in water quickly and efficiently. Khan et al. [76], in their study, used ZIF-8 to remove diethyl phthalate and phthalic acid from the water. They achieved the right amount of adsorption in comparison to the other adsorbate due to electrostatic interactions. Huo et al., reported that the interaction between Lewis base and acid results in enhanced adsorption of the malachite green on MIL 100-Fe due to open metal ends and pi-pi electrons benzene rings between them. Huang and co-workers [63], synthesized MIL 100-Cr and revealed that these composites adsorbed Malachite green rapidly. Lin and Chang [93], synthesized ZIF-67 and stated that ultra-high adsorption of ZIFs was found, and they adsorbed malachite green in the water; this is due to pi-pi interactions. The oil spills can even be cleared with hydrophobic MOFs; Yang et al. [150], synthesized FMOF-1, which is hydrophobic, and removed the oils from water. Another scientific study showed that Cu-BTC-MOF cleared the soybean oil from water, and this high adsorption is due to benzene rings of MOF [144]. Lin et al. [94] removed the oil droplets with the help of ZIF-8 from the water and concluded that ZIFs have oil adsorption capacities.

4.3.2 Application of Nanoporous Carbons for the Removal of Organic Pollutant in Water

Many researchers investigate nanoporous materials derived from different sources to extract organic contaminants and colored dyes in the aqueous media (water). Han et al. [52] compared the nanoporous materials developed by silica to activated carbon and reported that nanoporous materials adsorbed the molecules of humic acid fastly and efficiently. Sandeman et al. [124], compared these materials for adsorption of charged dyes to AC (activated carbons) and revealed that two factors influenced the ionic and anionic colored dyes' interactions with activated carbon. First, the interaction depends on the charge of the dyes and the symmetry of molecules. Second, on the textural characteristics of nanoporous materials and their nature. The hydrogen bonding and dispersion of adsorption between heterogenous complex and nanoporous materials are provided by the atoms of N_2(Nitrogen), H_2(Hydrogen), and O_2(Oxygen); properties of the solvents also influence these mechanisms [50, 77, 83]. Ghaedi et al. [47], reported that a high amount of MB (methylene blue) from the aqueous medium is adsorbed by AC (Activated Carbon) doped with $AgNO_3$(silver nanoparticles). Faria et al. [41], studied the adsorption of colored dyes on AC materials and reported that the acid-colored-dye adsorption phenomenon on the surface of activated carbon is due to its surface's basicity. AC (Activated Carbon) is the standard material used by many researchers to remove organic-dyes in the water [44, 58, 74]. For instance, Hameed and Daud [51] used the activated carbon resultant of agricultural waste to absorb dyes' adsorption. Activated carbon materials are used to absorb heavy metals; however, these materials' utilization is limited due to difficulty replicating and high costs. Nevertheless, there is a lack of literary works related to the adsorption of complex substances and organic contaminants on carbon nanoporous materials; Torad et al. [139] synthesized nanoporous material with the help of MOFs for the elimination of MB from aqueous medium(water) by using one-step carbonization and achieved excellent results without using any extra precursor. [147] did similar works, produced nanoporous carbons using MIL 100-Fe with ion thermal methods, and removed the water's methylene blue.

5 Conclusions

In summary, the treatment of water by adsorption has more advantages than other treatment methods as its operation is straightforward and eco-friendly. This process is an efficient treatment option for removing noxious substances from the wastewater. In the adsorption and water treatment, the porous material shows salient features like good porosity and larger surface areas. The current review mainly focused on research advancements to eliminate pollutants through the water's adsorption process with ZIFs and MOFs' help. However, challenges are there for the application of these materials for adsorption. The prime challenge about MOFs and ZIFs is that these are prepared in powdered form, which may not be suitable for some practical applications like catalysis and adsorption. Thus, modification of these materials is required for management and reprocessing. Approaches for the improvement of MOFs in powder form are available. Nevertheless, these modifications are issues as the modified substance's mechanical strengths and crystallinity is highly unstable. Thus, there is a need to develop new cost-efficient methods that should maintain high surface area without losing its stability and crystallinity. So, various strategies for the modification of MOF powders are discussed. Adsorption of dye with the help of ZIF-8 having larger molecules with smaller pores is discussed as a second limitation. However, this problem can be rectified by the carbonization (heat treatment) method. The unique properties, i.e., having a "high surface area" and "pore size/volume" of MOFs and ZIFs, allowed the materials for utilization as a precursor. The MOFs are not stable in an aqueous solution as the metal ions are leached; this problem must be resolved for the industrial application.

The utilization and production of MOFs/Porous carbon-based materials are regarded as a third major challenge. As mentioned in the text, these substances are not naturally produced and prepared synthetically using organic ligands. This process of production requires analytical grade organic solvents. For the production of analytical grade, organic solvents require relic fuels, and utilization of these relic fuels will liberate also generates waste as by-products. In the meantime, porous materials based on carbon can be produced by natural resources; however, these materials don't process greater surface area than MOFs. The regeneration of these materials is difficult for cyclic use. For Overcoming these challenges, research requires new methods and methodology which can be reproducible, effective, ecofriendly, and frugally viable is needed. The examination of these substances' performance change as adsorbents under these novel methods will help us recognize and examine this material behavior for the discretion of porous adsorbents and different procedures for impending real-time requests in handling water.

References

1. Abbasi Z, Shamsaei E, Fang XY, Ladewig B, Wang H (2017) Simple fabrication of zeolitic imidazolate framework ZIF-8/polymer composite beads by phase inversion method for efficient oil sorption. J Colloid Interface Sci 493:150–161
2. Akhtar F, Andersson L, Ogunwumi S, Hedin N, Bergström L (2014) Structuring adsorbents and catalysts by processing of porous powders. J Eur Ceram Soc 34(7):1643–1666
3. Al-Degs YS, El-Barghouthi MI, El-Sheikh AH, Walker GM (2008) Effect of solution pH, ionic strength, and temperature on adsorption behavior of reactive dyes on activated carbon. Dyes Pigm 77(1):16–23
4. Alaerts L, Kirschhock CE, Maes M, Van Der Veen MA, Finsy V, Depla A, De Vos DE (2007) Selective adsorption and separation of xylene isomers and ethylbenzene with the microporous vanadium (IV) terephthalate MIL-47. Angew Chem Int Ed 46(23):4293–4297
5. Alcaniz-Monge J, Trautwein G, Pérez-Cadenas M, Roman-Martinez MC (2009) Effects of compression on the textural properties of porous solids. Microporous Mesoporous Mater 126(3):291–301
6. Allen SJ, Brown PA (1995) Isotherm analyses for single component and multi-component metal sorption onto lignite. J Chem Technol Biotechnol: Int Res Process Environ Clean Technol 62(1):17–24
7. Amin NK (2009) Removal of direct blue-106 dye from aqueous solution using new activated carbons developed from pomegranate peel: adsorption equilibrium and kinetics. J Hazard Mater 165(1–3):52–62
8. Angamuthu M, Satishkumar G, Landau MV (2017) Precisely controlled encapsulation of Fe_3O_4 nanoparticles in mesoporous carbon nanodisk using iron based MOF precursor for effective dye removal. Microporous Mesoporous Mater 251:58–68
9. Avci-Camur C, Troyano J, Pérez-Carvajal J, Legrand A, Farrusseng D, Imaz I, Maspoch D (2018) Aqueous production of spherical Zr-MOF beads via continuous-flow spray-drying. Green Chem 20(4):873–878
10. Banerjee R, Phan A, Wang B, Knobler C, Furukawa H, O'Keeffe M, Yaghi OM (2008) High-throughput synthesis of zeolitic imidazolate frameworks and application to CO_2 capture. Science 319(5865):939–943
11. Barton SS, Evans MJB, Halliop E, MacDonald JAF (1997) Acidic and basic sites on the surface of porous carbon. Carbon 35(9):1361–1366
12. Bazer-Bachi D, Assié L, Lecocq V, Harbuzaru B, Falk V (2014) Towards industrial use of metal-organic framework: impact of shaping on the MOF properties. Powder Technol 255:52–59

13. Bhatnagar A, Hogland W, Marques M, Sillanpää M (2013) An overview of the modification methods of activated carbon for its water treatment applications. Chem Eng J 219:499–511
14. Biemmi E, Christian S, Stock N, Bein T (2009) High-throughput screening of synthesis parameters in the formation of the metal-organic frameworks MOF-5 and HKUST-1. Microporous Mesoporous Mater 117(1–2):111–117
15. Boehm HP (1994) Some aspects of the surface chemistry of carbon blacks and other carbons. Carbon 32(5):759–769
16. Cameron IT, Wang FY, Immanuel CD, Stepanek F (2005) Process systems modelling and applications in granulation: a review. Chem Eng Sci 60(14):3723–3750
17. Carlsson JM, Scheffler M (2006) Structural, electronic, and chemical properties of nanoporous carbon. Phys Rev Lett 96(4):
18. Castarlenas S, Téllez C, Coronas J (2017) Gas separation with mixed matrix membranes obtained from MOF UiO-66-graphite oxide hybrids. J Membr Sci 526:205–211
19. Cañizares P, Carmona M, Baraza O, Delgado A, Rodrigo MA (2006) Adsorption equilibrium of phenol onto chemically modified activated carbon F400. J Hazard Mater 131(1–3):243–248
20. Chaikittisilp W, Ariga K, Yamauchi Y (2013) A new family of carbon materials: synthesis of MOF-derived nanoporous carbons and their promising applications. J Mater Chem A 1(1):14–19
21. Chen Y, Lv D, Wu J, Xiao J, Xi H, Xia Q, Li Z (2017) A new MOF-505@ GO composite with high selectivity for CO_2/CH_4 and CO_2/N_2 separation. Chem Eng J 308:1065–1072
22. Chen B, Ma G, Kong D, Zhu Y, Xia Y (2015) Atomically homogeneous dispersed ZnO/N-doped nanoporous carbon composites with enhanced CO_2 uptake capacities and high efficient organic pollutants removal from water. Carbon 95:113–124
23. Chen RH, Qiao HT, Liu Y, Dong YH, Wang P, Zhang Z, Jin T (2015) Adsorption of methylene blue from an aqueous solution using a cucurbituril polymer. Environ Prog Sustain Energy 34(2):512–519
24. Chong M, Jin B, Chow CWK, Saint C (2010) Recent developments in photocatalytic water treatment technology: a review. Water Res 44:2997–3027
25. Cliffe MJ, Mottillo C, Stein RS, Bučar DK, Friščić T (2012) Accelerated aging: a low energy, solvent-free alternative to solvothermal and mechanochemical synthesis of metal–organic materials. Chem Sci 3(8):2495–2500
26. Comotti A, Bracco S, Sozzani P, Horike S, Matsuda R, Chen J, Kitagawa S (2008) Nanochannels of two distinct cross-sections in a porous Al-based coordination polymer. J Am Chem Soc 130(41):13664–13672
27. Connolly BM, Mehta JP, Moghadam PZ, Wheatley AE, Fairen-Jimenez D (2018) From synthesis to applications: metal–organic frameworks for an environmentally sustainable future. Curr Opin Green Sustain Chem 12:47–56
28. Couck S, Gobechiya E, Kirschhock CE, Serra-Crespo P, Juan-Alcañiz J, Martinez Joaristi A, Denayer JF (2012) Adsorption and Separation of Light Gases on an Amino-Functionalized Metal-Organic Framework: an Adsorption and In Situ XRD Study. Chemsuschem 5(4):740–750
29. Coughlin RW, Ezra FS (1968) Role of surface acidity in the adsorption of organic pollutants on the surface of carbon. Environ Sci Technol 2(4):291–297
30. Crawford D, Casaban J, Haydon R, Giri N, McNally T, James SL (2015) Synthesis by extrusion: continuous, large-scale preparation of MOFs using little or no solvent. Chem Sci 6(3):1645–1649
31. Crini G, Lichtfouse E, Wilson LD, Morin-Crini N (2019) Conventional and non-conventional adsorbents for wastewater treatment. Environ Chem Lett 17(1):195–213
32. Cseri L, Razali M, Pogany P, Szekely G (2018) Organic solvents in sustainable synthesis and engineering. In: Green chemistry. Elsevier, pp 513–553
33. Davis ME (2002) Ordered porous materials for emerging applications. Nature 417(6891):813–821
34. Davis SW, Powers SE (2000) Alternative sorbents for removing MTBE from gasoline-contaminated ground water. J Environ Eng 126(4):354–360

35. Dhaka S, Kumar R, Deep A, Kurade MB, Ji SW, Jeon BH (2019) Metal–organic frameworks (MOFs) for the removal of emerging contaminants from aquatic environments. Coord Chem Rev 380:330–352
36. Dias JM, Alvim-Ferraz MC, Almeida MF, Rivera-Utrilla J, Sánchez-Polo M (2007) Waste materials for activated carbon preparation and its use in aqueous- phase treatment: a review. J Environ Manage 85(4):833–846
37. Dias EM, Petit C (2015) Towards the use of metal–organic frameworks for water reuse: a review of the recent advances in the field of organic pollutants removal and degradation and the next steps in the field. J Mater Chem A 3(45):22484–22506
38. Dreischarf AC, Lammert M, Stock N, Reinsch H (2017) Green synthesis of Zr-CAU-28: structure and properties of the first Zr-MOF based on 2, 5- furandicarboxylic acid. Inorg Chem 56(4):2270–2277
39. Dąbrowski A (2001) Adsorption—from theory to practice. Adv Coll Interface Sci 93(1–3):135–224
40. Dąbrowski A, Podkościelny P, Hubicki Z, Barczak M (2005) Adsorption of phenolic compounds by activated carbon—a critical review. Chemosphere 58(8):1049–1070
41. Faria PCC, Orfao JJM, Pereira MFR (2004) adsorption of anionic and cationic dyes on activated carbons with different surface chemistries. Water Res 38(8):2043–2052
42. Fu F, Wang Q (2011) Removal of heavy metal ions from wastewaters. J Environ Manag 92(3):407–418
43. Férey G (2008) Hybrid porous solids: past, present, future. Chem Soc Rev 37(1):191–214
44. Gao Y, Xu S, Yue Q, Wu Y, Gao B (2016) Chemical preparation of crab shell-based activated carbon with superior adsorption performance for dye removal from wastewater. J Taiwan Inst Chem Eng 61:327–335
45. Garcıa MT, Ribosa I, Guindulain T, Sanchez-Leal J, Vives-Rego J (2001) Fate and effect of monoalkyl quaternary ammonium surfactants in the aquatic environment. Environ Pollut 111(1):169–175
46. Ge D, Lee HK (2011) Water stability of zeolite imidazolate framework 8 and application to porous membrane-protected micro-solid-phase extraction of polycyclic aromatic hydrocarbons from environmental water samples. J Chromatogr A 1218(47):8490–8495
47. Ghaedi M, Heidarpour S, Kokhdan SN, Sahraie R, Daneshfar A, Brazesh B (2012) Comparison of silver and palladium nanoparticles loaded on activated carbon for efficient removal of methylene blue: kinetic and isotherm study of removal process. Powder Technol 228:18–25
48. Ghaffar A, Younis MN (2015) Interaction and thermodynamics of methylene blue adsorption on oxidized multi-walled carbon nanotubes. Green Process Synth 4(3):209–217
49. Gu X, Zhou K, Li Y, Yao J (2016) Millimeter-sized carbon/TiO_2 beads fabricated by phase inversion method for oil and dye adsorption. RSC Adv 6(20):16314–16318
50. Gun'ko VM, Turov VV, Skubiszewska-Zięba J, Leboda R, Tsapko MD, Palijczuk D (2003) Structural characteristics of a carbon adsorbent and influence of organic solvents on interfacial water. Appl Surf Sci 214(1–4):178–189
51. Hameed BH, Daud FBM (2008) Adsorption studies of basic dye on activated carbon derived from agricultural waste: Hevea brasiliensis seed coat. Chem Eng J 139(1):48–55
52. Han S, Kim S, Lim H, Choi W, Park H, Yoon J, Hyeon T (2003) New nanoporous carbon materials with high adsorption capacity and rapid adsorption kinetics for removing humic acids. Microporous Mesoporous Mater 58(2):131–135
53. Haque E, Lee JE, Jang IT, Hwang YK, Chang JS, Jegal J, Jhung SH (2010) Adsorptive removal of methyl orange from aqueous solution with metal-organic frameworks, porous chromium-benzenedicarboxylates. J Hazard Mater 181(1–3):535–542
54. Harding JL, Metz JM, Reynolds MM (2014) A tunable, stable, and bioactive MOF catalyst for generating a localized therapeutic from endogenous sources. Adv Func Mater 24(47):7503–7509
55. Hasan Z, Jeon J, Jhung SH (2012) Adsorptive removal of naproxen and clofibric acid from water using metal-organic frameworks. J Hazard Mater 209:151–157

56. Hasan Z, Jhung SH (2015) Removal of hazardous organics from water using metal-organic frameworks (MOFs): plausible mechanisms for selective adsorptions. J Hazard Mater 283:329–339
57. Hayashi H, Cote AP, Furukawa H, O'Keeffe M, Yaghi OM (2007) Zeolite A imidazolate frameworks. Nat Mater 6(7):501–506
58. Hoda N, Bayram E, Ayranci E (2006) Kinetic and equilibrium studies on the removal of acid dyes from aqueous solutions by adsorption onto activated carbon cloth. J Hazard Mater 137(1):344–351
59. Hong S, Wen C, He J, Gan F, Ho YS (2009) Adsorption thermodynamics of methylene blue onto bentonite. J Hazard Mater 167(1–3):630–633
60. Hu Y, Lian H, Zhou L, Li G (2015) In situ solvothermal growth of metal–organic framework-5 supported on porous copper foam for noninvasive sampling of plant volatile sulfides. Anal Chem 87(1):406–412
61. Hu M, Reboul J, Furukawa S, Torad NL, Ji Q, Srinivasu P, Yamauchi Y (2012) Direct carbonization of Al-based porous coordination polymer for synthesis of nanoporous carbon. J Am Chem Soc 134(6):2864–2867
62. Huang K, Liu S, Li Q, Jin W (2013) Preparation of novel metal-carboxylate system MOF membrane for gas separation. Sep Purif Technol 119:94–101
63. Huang XX, Qiu LG, Zhang W, Yuan YP, Jiang X, Xie AJ, Zhu JF (2012) Hierarchically mesostructured MIL-101 metal–organic frameworks: supramolecular template-directed synthesis and accelerated adsorption kinetics for dye removal. CrystEngComm 14(5):1613–1617
64. Huang C, Sun R, Lu H, Yang Q, Hu J, Wang H, Liu H (2017) A pilot trial for fast deep desulfurization on MOF-199 by simultaneous adsorption-separation via hydrocyclones. Sep Purif Technol 182:110–117
65. Ibarra IA, Bayliss PA, Pérez E, Yang S, Blake AJ, Nowell H, Schröder M (2012) Near-critical water, a cleaner solvent for the synthesis of a metal–organic framework. Green Chem 14(1):117–122
66. James JB, Lin YS (2017) Thermal stability of ZIF-8 membranes for gas separations. J Membr Sci 532:9–19
67. Jiang M, Cao X, Zhu D, Duan Y, Zhang J (2016) Hierarchically porous N-doped carbon derived from ZIF-8 nanocomposites for electrochemical applications. Electrochim Acta 196:699–707
68. Jiang HL, Liu B, Lan YQ, Kuratani K, Akita T, Shioyama H, Xu Q (2011) From metal–organic framework to nanoporous carbon: toward a very high surface area and hydrogen uptake. J Am Chem Soc 133(31):11854–11857
69. Jiang H, Yan Q, Chen R, Xing W (2016) Synthesis of Pd@ ZIF-8 via an assembly method: Influence of the molar ratios of Pd/Zn2+ and 2-methylimidazole/Zn2+. Microporous Mesoporous Mater 225:33–40
70. Jiang JQ, Yang CX, Yan XP (2013) Zeolitic imidazolate framework-8 for fast adsorption and removal of benzotriazoles from aqueous solution. ACS Appl Mater Interfaces 5(19):9837–9842
71. Julien PA, Mottillo C, Friščić T (2017) Metal–organic frameworks meet scalable and sustainable synthesis. Green Chem 19(12):2729–2747
72. Karadeniz B, Howarth AJ, Stolar T, Islamoglu T, Dejanović I, Tireli M, Užarević K (2018) Benign by design: green and scalable synthesis of zirconium UiO-metal–organic frameworks by water-assisted mechanochemistry. ACS Sustain Chem Eng 6(11):15841–15849
73. Karnena MK, Konni M, Saritha V (2020) Nano-catalysis process for treatment of industrial wastewater. In: Handbook of research on emerging developments and environmental impacts of ecological chemistry. IGI Global, pp 229–251
74. Khaled A, El Nemr A, El-Sikaily A, Abdelwahab O (2009) Removal of Direct N Blue-106 from artificial textile dye effluent using activated carbon from orange peel: adsorption isotherm and kinetic studies. J Hazard Mater 165(1–3):100–110
75. Khan NA, Hasan Z, Jhung SH (2013) Adsorptive removal of hazardous materials using metal-organic frameworks (MOFs): a review. J Hazard Mater 244:444–456

76. Khan NA, Jung BK, Hasan Z, Jhung SH (2015) Adsorption and removal of phthalic acid and diethyl phthalate from water with zeolitic imidazolate and metal–organic frameworks. J Hazard Mater 282:194–200
77. Kowalczyk P, Terzyk AP, Gauden PA, Gun'ko VM, Solarz L (2002) Evaluation of the structural and energetic heterogeneity of microporous carbons by means of novel numerical methods and genetic algorithms. J Colloid Interface Sci 256(2):378–395
78. Kreider MC, Sefa M, Fedchak JA, Scherschligt J, Bible M, Natarajan B, Hartings MR (2018) Toward 3D printed hydrogen storage materials made with ABS-MOF composites. Polym Adv Technol 29(2):867–873
79. Küsgens P, Zgaverdea A, Fritz HG, Siegle S, Kaskel S (2010) Metal-organic frameworks in monolithic structures. J Am Ceram Soc 93(9):2476–2479
80. Lachheb H, Puzenat E, Houas A, Ksibi M, Elaloui E, Guillard C, Herrmann JM (2002) Photocatalytic degradation of various types of dyes (Alizarin S, Crocein Orange G, Methyl Red, Congo Red, Methylene Blue) in water by UV-irradiated titania. Appl Catal B 39(1):75–90
81. Langmuir I (1918) The adsorption of gases on plane surfaces of glass, mica and platinum. J Am Chem Soc 40(9):1361–1403
82. Lapertot M, Pulgarín C, Fernández-Ibáñez P, Maldonado MI, Pérez-Estrada L, Oller I, Malato S (2006) Enhancing biodegradability of priority substances (pesticides) by solar photo-Fenton. Water Res 40(5):1086–1094
83. Leboda R, Turov VV, Tomaszewski W, Gun'ko VM, Skubiszewska-Zięba J (2002) Effect of adsorption of nitroaromatic compounds on the characteristics of bound water layers in aqueous suspensions of activated carbons. Carbon 40(3):389–396
84. Lee J, Farha OK, Roberts J, Scheidt KA, Nguyen ST, Hupp JT (2009) Metal–organic framework materials as catalysts. Chem Soc Rev 38(5):1450–1459
85. Lee UH, Valekar AH, Hwang YK, Chang JS (2016) The chemistry of metal–organic frameworks
86. Lei Y, Wei L, Zhai S, Wang Y, Karahan HE, Chen X, Chen Y (2018) Metal-free bifunctional carbon electrocatalysts derived from zeolitic imidazolate frameworks for efficient water splitting. Mater Chem Front 2(1):102–111
87. Leon CLY, Radovic LR (1994) Interfacial chemistry and electrochemistry of carbon surfaces. Chem Phys Carbon 24(24):213–310
88. Leon CLY, Solar JM, Calemma V, Radovic LR (1992) Evidence for the protonation of basal plane sites on carbon. Carbon 30(5):797–811
89. Li X, Wang B, Cao Y, Zhao S, Wang H, Feng X, Ma X (2019) Water contaminant elimination based on metal–organic frameworks and perspective on their industrial applications. ACS Sustain Chem Eng 7(5):4548–4563
90. Li R, Zhang L, Wang P (2015) Rational design of nanomaterials for water treatment. Nanoscale 7(41):17167–17194
91. Liao X, Wang F, Wang Y, Cai Y, Liu H, Wang X, Hao Q (2020) Constructing Fe-based bi-MOFs for photo-catalytic ozonation of organic pollutants in Fischer-Tropsch waste water. Appl Surf Sci 145378
92. Lim S, Suh K, Kim Y, Yoon M, Park H, Dybtsev DN, Kim K (2012) Porous carbon materials with a controllable surface area synthesized from metal–organic frameworks. Chem Commun 48(60):7447–7449
93. Lin KYA, Chang HA (2015) Ultra-high adsorption capacity of zeolitic imidazole framework-67 (ZIF-67) for removal of malachite green from water. Chemosphere 139:624–631
94. Lin KYA, Chen YC, Phattarapattamawong S (2016) Efficient demulsification of oil-in-water emulsions using a zeolitic imidazolate framework: adsorptive removal of oil droplets from water. J Colloid Interface Sci 478:97–106
95. Liu X, Wang M, Zhang S, Pan B (2013) Application potential of carbon nanotubes in water treatment: a review. J Environ Sci 25(7):1263–1280
96. Liu B, Yang F, Zou Y, Peng Y (2014) Adsorption of phenol and p-nitrophenol from aqueous solutions on metal–organic frameworks: effect of hydrogen bonding. J Chem Eng Data 59(5):1476–1482

97. Ma J, Guo X, Ying Y, Liu D, Zhong C (2017) Composite ultrafiltration membrane tailored by MOF@ GO with highly improved water purification performance. Chem Eng J 313:890–898
98. Marquez AG, Horcajada P, Grosso D, Ferey G, Serre C, Sanchez C, Boissiere C (2013) Green scalable aerosol synthesis of porous metal–organic frameworks. Chem Commun 49(37):3848–3850
99. Moghadam PZ, Li A, Wiggin SB, Tao A, Maloney AG, Wood PA, Fairen-Jimenez D (2017) Development of a Cambridge Structural Database subset: a collection of metal–organic frameworks for past, present, and future. Chem Mater 29(7):2618–2625
100. Mon M, Bruno R, Ferrando-Soria J, Armentano D, Pardo E (2018) Metal–organic framework technologies for water remediation: towards a sustainable ecosystem. J Mater Chem A 6(12):4912–4947
101. Monser L, Adhoum N (2002) Modified activated carbon for the removal of CU, Zn, Cr and CN from wastewater. Sep Purif Technol 26:00155–1
102. Moreno-Castilla C (2004) Adsorption of organic molecules from aqueous solutions on carbon materials. Carbon 42(1):83–94
103. Mounfield WP III, Tumuluri U, Jiao Y, Li M, Dai S, Wu Z, Walton KS (2016) Role of defects and metal coordination on adsorption of acid gases in MOFs and metal oxides: an in situ IR spectroscopic study. Microporous Mesoporous Mater 227:65–75
104. Mueller U, Schubert M, Teich F, Puetter H, Schierle-Arndt K, Pastre J (2006) Metal–organic frameworks—prospective industrial applications. J Mater Chem 16(7):626–636
105. Mueller U, Lobree L, Hesse M, Yaghi O, Eddaoudi M (2005) U.S. Patent No. 6,893,564. U.S. Patent and Trademark Office, Washington, DC
106. Munoz R, Guieysse B (2006) Algal–bacterial processes for the treatment of hazardous contaminants: a review. Water Res 40(15):2799–2815
107. Neoh CH, Noor ZZ, Mutamim NSA, Lim CK (2016) Green technology in wastewater treatment technologies: integration of membrane bioreactor with various wastewater treatment systems. Chem Eng J 283:582–594
108. Ngah WW, Hanafiah MM (2008) Removal of heavy metal ions from wastewater by chemically modified plant wastes as adsorbents: a review. Biores Technol 99(10):3935–3948
109. Okolo B, Park C, Keane MA (2000) Interaction of phenol and chlorophenols with activated carbon and synthetic zeolites in aqueous media. J Colloid Interface Sci 226(2):308–317
110. Oller I, Malato S, Sanchez-Perez JA (2011) Combination of advanced oxidation processes and biological treatments for wastewater decontamination review. Sci Total Environ 409:414
111. Panizza M, Cerisola G (2009) Direct and mediated anodic oxidation of organic pollutants. Chem Rev 109(12):6541–6569
112. Park KS, Ni Z, Côté AP, Choi JY, Huang R, Uribe-Romo FJ (2006) HK 27 Chae, M. O'Keeffe, OM Yaghi, Exceptional chemical and thermal stability of 28 zeolitic imidazolate frameworks. Proc Natl Acad Sci 103:10186–10191
113. Parnham ER, Morris RE (2007) Ionothermal synthesis of zeolites, metal–organic frameworks, and inorganic–organic hybrids. Acc Chem Res 40(10):1005–1013
114. Pereira MFR, Soares SF, Órfão JJ, Figueiredo JL (2003) Adsorption of dyes on activated carbons: influence of surface chemical groups. Carbon 41(4):811–821
115. Plaza MG, Ferreira AFP, Santos JC, Ribeiro AM, Müller U, Trukhan N, Rodrigues AE (2012) Propane/propylene separation by adsorption using shaped copper trimesate MOF. Microporous Mesoporous Mater 157:101–111
116. Radovic LR, Silva IF, Ume JI, Menendez JA, Leon CLY, Scaroni AW (1997) An experimental and theoretical study of the adsorption of aromatics possessing electron-withdrawing and electron-donating functional groups by chemically modified activated carbons. Carbon 35(9):1339–1348
117. Reinsch H (2016) "Green" synthesis of metal-organic frameworks. Eur J Inorg Chem 2016(27):4290–4299
118. Ren J, Dyosiba X, Musyoka NM, Langmi HW, Mathe M, Liao S (2017) Review on the current practices and efforts towards pilot-scale production of metal-organic frameworks (MOFs). Coord Chem Rev 352:187–219

119. Ren J, Dyosiba X, Musyoka NM, Langmi HW, North BC, Mathe M, Onyango MS (2016) Green synthesis of chromium-based metal-organic framework (Cr-MOF) from waste polyethylene terephthalate (PET) bottles for hydrogen storage applications. Int J Hydrogen Energy 41(40):18141–18146
120. Ren J, Musyoka NM, Langmi HW, Swartbooi A, North BC, Mathe M (2015) A more efficient way to shape metal-organic framework (MOF) powder materials for hydrogen storage applications. Int J Hydrogen Energy 40(13):4617–4622
121. Roobakhsh S, Rostami Z, Azizian S (2018) Can MoS2 nanosheets be used as adsorbent for water treatment? Sep Purif Technol 200:23–28
122. Rubin HN, Reynolds MM (2017) Functionalization of metal–organic frameworks to achieve controllable wettability. Inorg Chem 56(9):5266–5274
123. Rubio-Martinez M, Avci-Camur C, Thornton AW, Imaz I, Maspoch D, Hill MR (2017) New synthetic routes towards MOF production at scale. Chem Soc Rev 46(11):3453–3480
124. Sandeman SR, Gun'ko VM, Bakalinska OM, Howell CA, Zheng Y, Kartel MT, Mikhalovsky SV (2011) Adsorption of anionic and cationic dyes by activated carbons, PVA hydrogels, and PVA/AC composite. J Colloid Interface Sci 358(2):582–592
125. Schwab MG, Senkovska I, Rose M, Koch M, Pahnke J, Jonschker G, Kaskel S (2008) MOF@polyhipes. Adv Eng Mater 10(12):1151–1155
126. Serra-Crespo P, Stavitski E, Kapteijn F, Gascon J (2012) High compressibility of a flexible metal–organic framework. RSC Adv 2(12):5051–5053
127. Shetty D, Jahovic I, Raya J, Asfari Z, Olsen JC, Trabolsi A (2018) Porous polycalix [4] arenes for fast and efficient removal of organic micropollutants from water. ACS Appl Mater Interfaces 10(3):2976–2981
128. Silva TL, Ronix A, Pezoti O, Souza LS, Leandro PK, Bedin KC, Almeida VC (2016) Mesoporous activated carbon from industrial laundry sewage sludge: adsorption studies of reactive dye Remazol Brilliant Blue R. Chem Eng J 303:467–476
129. Singha S, Maity SK, Biswas S, Saha R, Kumar S (2016) A magnesium-based bifunctional MOF: studies on proton conductivity, gas and water adsorption. Inorg Chim Acta 453:321–329
130. Sule R, Mishra AK (2020) MOFs-carbon hybrid nanocomposites in environmental protection applications. Environ Sci Pollut Res 1–15
131. Sumida K, Rogow DL, Mason JA, McDonald TM, Bloch ED, Herm ZR, Long JR (2012) Carbon dioxide capture in metal–organic frameworks. Chem Rev 112(2):724–781
132. Sun X, Li B, Wan D, Wang N (2016) Using a novel adsorbent macrocyclic compound cucurbit [8] uril for Pb2 + removal from aqueous solution. J Environ Sci 50:3–12
133. Tagliabue M, Rizzo C, Millini R, Dietzel PD, Blom R, Zanardi S (2011) Methane storage on CPO-27-Ni pellets. J Porous Mater 18(3):289–296
134. Tanaka S, Nagaoka T, Yasuyoshi A, Hasegawa Y, Denayer JF (2018) Hierarchical pore development of ZIF-8 MOF by simple salt-assisted mechanosynthesis. Cryst Growth Des 18(1):274–279
135. Tchobanoglus G, Burton F, Stensel HD (2003) Wastewater engineering: Treatment and reuse. Am Water Works Assoc Journal 95(5):201
136. Terzyk AP (2004) Molecular properties and intermolecular forces—factors balancing the effect of carbon surface chemistry in adsorption of organics from dilute aqueous solutions. J Colloid Interface Sci 275(1):9–29
137. Thakkar H, Eastman S, Al-Naddaf Q, Rownaghi AA, Rezaei F (2017) 3D-printed metal–organic framework monoliths for gas adsorption processes. ACS Appl Mater Interfaces 9(41):35908–35916
138. Tong M, Liu D, Yang Q, Devautour-Vinot S, Maurin G, Zhong C (2013) Influence of framework metal ions on the dye capture behavior of MIL-100 (Fe, Cr) MOF type solids. J Mater Chem A 1(30):8534–8537
139. Torad NL, Hu M, Ishihara S, Sukegawa H, Belik AA, Imura M, Yamauchi Y (2014) Direct synthesis of MOF-derived nanoporous carbon with magnetic Co nanoparticles toward efficient water treatment. Small 10(10):2096–2107

140. Tranchemontagne DJ, Mendoza-Cortés JL, O'Keeffe M, Yaghi OM (2009) Secondary building units, nets and bonding in the chemistry of metal–organic frameworks. Chem Soc Rev 38(5):1257–1283
141. Unur E (2013) Functional nanoporous carbons from hydrothermally treated biomass for environmental purification. Microporous Mesoporous Mater 168:92–101
142. Valix M, Cheung WH, McKay G (2004) Preparation of activated carbon using low temperature carbonisation and physical activation of high ash raw bagasse for acid dye adsorption. Chemosphere 56(5):493–501
143. Wang D, Li Z (2016) Coupling MOF-based photocatalysis with Pd catalysis over Pd@ MIL-100 (Fe) for efficient N-alkylation of amines with alcohols under visible light. J Catal 342:151–157
144. Wang HN, Liu FH, Wang XL, Shao KZ, Su ZM (2013) Three neutral metal–organic frameworks with micro-and meso-pores for adsorption and separation of dyes. J Mater Chem A 1(42):13060–13063
145. Wang S, Wahiduzzaman M, Davis L, Tissot A, Shepard W, Marrot J, Serre C (2018) A robust zirconium amino acid metal-organic framework for proton conduction. Nat Commun 9(1):1–8
146. Wu T, Shen L, Luebbers M, Hu C, Chen Q, Ni Z, Masel RI (2010) Enhancing the stability of metal–organic frameworks in humid air by incorporating water repellent functional groups. Chem Commun 46(33):6120–6122
147. Xiao JD, Qiu LG, Jiang X, Zhu YJ, Ye S, Jiang X (2013) Magnetic porous carbons with high adsorption capacity synthesized by a microwave-enhanced high temperature ionothermal method from a Fe-based metal-organic framework. Carbon 59:372–382
148. Yagub MT, Sen TK, Afroze S, Ang HM (2014) Dye and its removal from aqueous solution by adsorption: a review. Adv Colloid Interfac 209:172–184
149. Yamagiwa K, Katoh M, Yoshida M, Ohkawa A, Ichijo H (1997) Potentiality of temperature-swing adsorption for removal of hydrophobic contaminants in water. Water Sci Technol 35(7):213–218
150. Yang C, Kaipa U, Mather QZ, Wang X, Nesterov V, Venero AF, Omary MA (2011) Fluorous metal–organic frameworks with superior adsorption and hydrophobic properties toward oil spill cleanup and hydrocarbon storage. J Am Chem Soc 133(45):18094–18097
151. Yao J, Li L, Wong WHB, Tan C, Dong D, Wang H (2013) Formation of ZIF-8 membranes and crystals in a diluted aqueous solution. Mater Chem Phys 139(2–3):1003–1008
152. Yao J, Wang H (2014) Zeolitic imidazolate framework composite membranes and thin films: synthesis and applications. Chem Soc Rev 43(13):4470–4493
153. Yao Y, Xu F, Chen M, Xu Z, Zhu Z (2010) Adsorption behavior of methylene blue on carbon nanotubes. Biores Technol 101(9):3040–3046
154. Yin H, Kim H, Choi J, Yip AC (2015) Thermal stability of ZIF-8 under oxidative and inert environments: a practical perspective on using ZIF-8 as a catalyst support. Chem Eng J 278:293–300
155. Zacher D, Shekhah O, Wöll C, Fischer RA (2009) Thin films of metal–organic frameworks. Chem Soc Rev 38(5):1418–1429
156. Zeng Y, Wang K, Yao J, Wang H (2014) Hollow carbon beads fabricated by phase inversion method for efficient oil sorption. Carbon 69:25–31
157. Zhang X, Li A, Jiang Z, Zhang Q (2006) Adsorption of dyes and phenol from water on resin adsorbents: effect of adsorbate size and pore size distribution. J Hazard Mater 137(2):1115–1122
158. Zhang H, Liu D, Yao Y, Zhang B, Lin YS (2015) Stability of ZIF-8 membranes and crystalline powders in water at room temperature. J Membr Sci 485:103–111
159. Zhang J, White GB, Ryan MD, Hunt AJ, Katz MJ (2016) Dihydrolevoglucosenone (Cyrene) as a green alternative to N, N-dimethylformamide (DMF) in MOF synthesis. ACS Sustain Chem Eng 4(12):7186–7192
160. Øien-Ødegaard S, Shearer GC, Wragg DS, Lillerud KP (2017) Pitfalls in metal–organic framework crystallography: towards more accurate crystal structures. Chem Soc Rev 46(16):4867–4876

Madhavi Konni holds a doctorate in Nanoscience and Nanotechnology in the Department of Chemistry from GITAM (Deemed to be University), Visakhapatnam, India. Currently, she is working as Associate Professor at Dadi Institute of Engineering and Technology, Anakapalle. Her research interests include synthesizing Metal-Organic Frameworks, Material Chemistry, Carbon Nano Materials, Hydrogen Storage Technologies, and Nano Catalysis. She published many research articles and book chapters in reputed international journals and holds two patents. Dr. Madhavi is a fellow and received HEAM Scholar Award (2018), the Young Scientist Award (December 2017 and March 2017), and few other awards from different societies.

Saratchandra Babu Mukkamala has been working in GITAM since 1991 and currently heading the GITAM Institute of Science since January 2019. He received his master's degree in Chemistry in 1988 and Ph.D. in 1992 from Andhra University, Visakhapatnam. He has been a Visiting Scientist at the University of Durban-Westville, South Africa (2001), a DFG Postdoctoral Fellow at Institute of Inorganic Chemistry, University of Karlsruhe, Germany (2001–2004), and an NSF Postdoctoral Fellow at National Tsing Hua University, Taiwan (2005–2007). He visited several universities in the USA, Canada, Japan, Australia, and Singapore on an academic assignment. He is a recipient of the Best Researcher award twice from GITAM in 2010 and 2012. His research interests include the Synthesis of Metal-Organic Frameworks and Carbon nanotubes for gas storage applications, Luminescent nanomaterials for Technological and biomedical applications.

R. S. S. Srikanth Vemuri received the B.S. degree (2005) in Chemistry, and M.S. degree (2007) in Applied Environmental Chemistry from Andhra University, India awarded his Doctoral degree in Kinetic studies on some micellar catalyzed electron transfer reactions from same institution in 2014. He was awarded the of UGC N-SAP fellowship during the period 2008–2014. He had also achieved the Indian Academy of Sciences Summer Research Fellowship in 2019. He is currently working as an Associate Professor in Chemistry at Vignan's Institute of Engineering for Women, Visakhapatnam, India, since 2014. His research interests are Environmental Chemistry, Computational Chemistry, and Physical Chemistry. He is currently working on homogenous and heterogeneous catalysis, multivariate statistical analysis of soil samples, and theoretical studies on tuning of molecules for semiconductors.

Manoj Kumar Karnena received the B.S. degree in Environmental Science and the M.S. degree in Environmental science from GITAM (Deemed to be) University, India. Later, he pursued his Ph.D. in Water and Wastewater treatment at the same institution. He was the university topper and gold medalist in bachelors and post-graduation. He holds an Environmental Law degree from National Law University, Delhi, and Industrial Safety Degree from Annamalai University, Tamilnadu. Since 2019, he has been working as a teaching assistant with the Department of Environmental Science, GITAM Institute of Science, GITAM (Deemed to be) University, India. His greatest strengths are his research and communication skills. He started teaching and working in technical work in 2019 in various backgrounds. Manoj Kumar Karnena has written more than 50 papers since he started writing in 2017. He has written several book chapters and books on the topic within his research interest and holds two patents. His research interest is in Water and Wastewater Treatment, Natural Coagulants, Nanocatalysis, Storage Technologies, and membrane technology. Manoj Kumar Karnena had won several engineering designs and environmental protection awards in water and wastewater treatment.

Household Water Treatment and Safe Storage

Stefan-Ionut Spiridon, Eusebiu Ilarian Ionete, and Roxana Elena Ionete

Abstract Decentralized treatment of drinking water and its safe storage in specially designated storage vessels, prior to be used for household daily needs, with the purpose of preventing illnesses and improving health, is defined as Household Water Treatment and Safe Storage philosophy. Moreover, primary water collection and storage in different types of containers, from leather bellows, wood vessels, and ceramics, to glass and metal, has been a common practice for years, demonstrating that the concept of safe storage for domestic drinking water and the eventual treatment, is not new. Society and technological development, nowadays, transformed the concept of drinking water as a local need to a global need. Presently, more than half of population worldwide is supplied with tap water in their homes by using individual household connections. Access to safe, adequate, sufficient and physically accessible and cost-effective tap water is a fundamental right, essential for our daily comfort. Seen as a long-term goal, water availability determines the quality of drinking water supplied to households. Since water is not always considered safe for use in terms of quality, in most of homes the use of treatment devices is sometimes recommended. This safety measure typically provides a final "extra" step following a treatment process granted by a centralized water treatment system, administered by a municipal authority or private entity. The benefits of Household Water Treatment and Safe Storage (HWTS) technologies is used to improve water aesthetics and/or to remove certain harmful contaminants that are presented in this chapter, including possible organic, inorganic or microbiological substances present in the household's drinking water supply. Considering the location of the device, either at the point where the main water supply is defined or at the point where the drinking water is withdrawn and enters the house, they can be identified as "point-of- entry"

S.-I. Spiridon · E. I. Ionete · R. E. Ionete (✉)
National Research and Development Institute for Cryogenics and Isotopic Technologies, Uzinei Street; no. 4; O.P Raureni; CP 7, 240050 Ramnicu Valcea, Romania
e-mail: roxana.ionete@icsi.ro

S.-I. Spiridon
e-mail: ionut.spiridon@icsi.ro

E. I. Ionete
e-mail: eusebiu.ionete@icsi.ro

or "point-of-use". In the industrialized countries and regions, household drinking water treatment is foreseen as an additional barrier of safety to a water source which has already been treated upstream or it is of a certain known quality. For households, that are more likely to bear the burden of water-related diseases due to the lack of a drinking water supply through a connection or other "improved" water supply such as a public pipe, a borehole, a protected well/spring water or rainwater, the use of drinking water treatment devices is imperative. Due to various factors, namely minimal maintenance designs, use of local materials, applications under demanding special conditions, public acceptability and economic sustainability, the study and development of household drinking water treatment devices has been a process of adaptation to the concrete geographical conditions of that area and public health practices. Nowadays, on the international market, there are a plethora of options for home drinking water treatment and safe water storage that are feasible and cover the safety needs of a local small community. Optimizing the potential of HWTS technologies and the scale-up of these systems will make available the use of this technology globally for millions of users to meet the enormous need for safe water for the decades to come.

Keywords Treatment · Storage · Drinking water · Feasibility · Water quality

1 Introduction to Household Water Treatment and Safe Storage

A major contributor to risk of disease is acute water diarrhea (AWD)—the most common cause of death among children under the age of 5, worldwide, predominantly in developing countries), is the contaminated water, and is linked with the use of unsafe drinking water sources, gaps in sanitation and hygiene education, but also the lack of knowledge regarding common and simple techniques to purify and store the drinking-water in households (World Health Organization [24] 2013). This is due a multitude of factors that contribute to the human behavior in terms of the water treatment necessity, from individual beliefs and values, to the social regulations on public health, and demographic, economic and environmental factors [6, 15].

Ensuring each household with affordable, safe and piped-in water is an imperative goal, that may yield optimal health gains and a sustainable environment for the population. Thus, improvement of water treatment operations must take into consideration relevant elements operating in each specific context of human life, in order to obtain a proper design and to develop a viable and efficient household water treatment and safe storage (HWTS) system.

Simple, effective and low-cost technologies for treating drinking water that will serve a large community of people, improving dramatically the microbial water quality and significantly reducing the chemical contamination and the incidence of water diseases, are hard to achieve. Even so, a challenge that promoters of household water treatment systems have reported is the compliance of their utilization by the end-users, in terms of lack of *correct, consistent* and *sustained* use [6].

HWTS is a batch process based on a series of steps: (i) protection of the water source—water extraction from a deep clean source; (ii) sedimentation—use of coagulants; (iii) filtration—separating solid particles; (iv) disinfection—using chlorine and ozone to decrease the risk of recontamination during distribution; and (v) safe storage. If correctly applied, the HWTS must be applied by the end users at a well determined frequency to ensure a consistent protection against waterborne pathogens, can be used as an effective tool for an emergency situation response to contaminated water treatment.

1.1 Water—Source of Life

Water dominates the Earth's surface: 70.8% of it is covered by the oceans, and about 3.16%, by ice on land. Lakes cover approximatively 0.39%, while rivers and wetlands about 0.53%, the total water-covered surface being above 75% [14]. In the troposphere water exists in various forms (liquid, gaseous and solid), and also very small amounts of water from the troposphere can eventually reach the stratosphere due to turbulences.

Water is the key element responsible for all processes associated with life on Earth, outside as well as inside organisms. The high-water content between 50 to 80% in life forms as bacteria, plants and animals including humans, indicates the relevance of this element to all organisms. Water is also the transport medium of different nutrients of the living organisms. Thus, water has many functions within an organism decisively involved in a great number of biochemical processes and serving as the solvent for most metabolic reactions.

Water is not responsible only for shaping the landscape and enabling an abundant flora and fauna, but also constitutes the base of human communities, and the development of cities. It is well known that communities in early civilizations that had direct access to a drinkable and clean water resource, developed and thrived. It is a common fact that big cities/communities around the world are generally located in areas adjacent to water bodies that can be also used as transport routes and/or irrigation systems for agriculture (to grow crops).

Since the amount of drinking water recommended for a good health varies (around 1–2 L per day) and relates with the age, level of physical activity, health-related issues, and environmental conditions, ensuring a good quality for the water is a must. This means that proper quality standards should be respected, meaning a water free of toxic compounds (e.g. pesticides, heavy metals, industrial/pharmaceutical residues) and pathogenic germs.

As a matter of fact, a growing household's consumption is reported nowadays, and it is evaluated to be over 6% of the global water consumption. An estimated global share of 9% is used in the industry (e.g. for production of cement, paper, and various chemical products), also dependable on each country economical and industrial power. In order to prevent the degradation of the water sources (surface and groundwater), treatment of domestic and industrial wastewater must be performed prior to its release into the environment [20–21].

Implementing water and sewage treatment systems and also adopting a sustainable strategy for water resources is a great first step in water quality security. A minor element but with a major impact in the sustainable strategy for the drinkable water resources safety is the implementation of HTWS knowledge and technologies to small and big communities.

1.2 What Is HWTS?

Household water treatment and safe storage (HWTS) system refer to quantitative and qualitative water treatment technologies used in homes, which can provide and improve the drinking-water quality and reduce water diseases, particularly among those who rely on water from unimproved sources, and in some cases, unsafe or unreliable piped water supplies.

HTWS integrates different single treatment solutions, including chemical disinfectants, coagulants, filters (ceramic, biological, sand), solar or ultraviolet disinfection processes [7, 23], and combined ones (e.g. both coagulant and disinfectant). These technologies reported a success in improving the microbiological quality of drinking water and in preventing and reducing water diseases [8, 10, 23].

Water quality degradation is considered to be one of the key impact factors generated by the climate change against drinkable water resources. Increases in flooding, drought, decreasing water availability, algal blooms, coastal inundation, and sea level rise seem to have direct and indirect effects on drinking water quality [12]; WHO 2009). Direct negative effects occur through transport of fecal and other wastes into water supplies, growth of harmful algal blooms, etc. Indirect negative effects on drinking water quality occur when users are forced to use lower quality drinking water supplies as for example contaminated surface water. HWTS allows an increase of resilience to water quality degradation by giving the users the option for water of better quality at the point of use and improves public safety against both intended and unintended water supply contamination.

2 Water Contamination and HWTS Options

Water is essential for the metabolic processes taking place in living organisms since is acting as both solvent [18], lubricant, transport and reaction medium [3]. The quality of water varies with the region/place, along with the seasons, and geological (rocks, soil, etc.) pathways through which it flows. It is influenced by anthropic factors such as open defecation, inappropriate waste management, indigent agricultural practices, negative industrial effect, and by naturally occurring contaminants.

When the water quality assessment is foreseen, three major aspects have to be considered:

(a) *the biological quality aspect*: presence of bacteria, viruses, protozoa and worms in water;
(b) *the chemical quality aspect*: presence of minerals, metals and other chemicals in water; and
(c) *the physical quality aspect*: temperature, color, smell, taste and turbidity of water.

2.1 Safe-Drinking Water Definition

Water is now considered as a fundamental right and for sustainable development, it is essential to have accessible sources of safe drinking water. Even though, no universally accepted definition of "safe drinking water" exists, as concept, this term is defined as the water quality that does not pose any significant risk to human health over a lifetime of its consumption, including different sensitivities that may occur between life stages (World Health Organization [26] 2017).

2.1.1 Biological Quality

Water is a natural medium for living organisms. Most of them are harmless or even beneficial, but others can cause health problems for humans and animals. The living things that cause disease are identified as microorganisms, microbes or germs, classified as pathogens. The main classes of biological pathogens (bacteria, viruses and protozoa) are commonly identified as waterborne and considered to pose significant threats to human health. Therefore, in addition to poor water quality, water inaccessibility and lack of hygiene amenities, are commonly viewed as a gateway for a variety of infectious diseases caused by helminths, protozoa, bacteria and viruses.

The major threat from biological pathogens is diarrhea. As reported by World Health Organization, diarrheal diseases are responsible for the death of 1.9 million people each year, and it considered among infectious diseases, as the third-leading cause of both mortality and morbidity, after respiratory infections and HIV/AIDS, positioning it above tuberculosis and malaria as health safety threat. It is a symptom of gastrointestinal infection, usually caused by a diversity of bacterial, viral and parasitic organisms (World Health Organization [25], UNICEF 2014).

Even if, water is microbiologically safe when deposited into containers, it can be quickly contaminated during storage and use, primarily by contact with human hands or contaminated utensils (e.g. bucket). Also, if the water storage vessel is inadequately covered, water contamination can be associated with other factors like dust, animals, birds and insects. Therefore, subsequent contamination of the water can often be so great as to cancel the initial disinfection stage.

The minimum number of pathogenic agents that can cause illness to a person, is referred as the infective dose. If the existence of pathogen in water do not promote illness in a human being it means that the infective dose level was not achieved. The infective dose is distinctive based on the type of pathogen. Normally, bacteria present a higher infective dose than most of the viruses and protozoa. Thus, for some bacteria to cause illness, a larger number needs to be ingested, as compared to other pathogens. This level of infective dose is directly connected to the age of the individual and also to its health status.

As a conclusion, investigations conducted to detect and determine each pathogen in water are a time-consuming, complicated and expensive step in determining the water quality and safety. To simplify this complex methodology, a technique based on establishing the presence or absence of certain bacterial indicator organisms is used, since there are no well-established testing techniques feasible for viruses and protozoa. Bacterial indicator tests are cheaper, easier to perform and provide faster results compared to direct pathogen testing. There is no universal indicator to establish that water is pathogen free, but there are certain types of indicators, each with special characteristics for pathogen detection and identification. Coliform bacteria are one of the frequently used indicators due to their high ratios' existence into the pathogens, making them detectable in a water sample.

2.1.2 Chemical Quality

Chemicals in drinking water can present a threat for human health and life. Chemicals take part in the composition of the drinking-water in different ways. Some of them, such as arsenic, calcium, fluoride, magnesium and sulphur are naturally occurring in groundwater, while others are the consequence of anthropogenic activities. Most of the countries, especially during their industrialization phase, have a limited compliance to environmental rules and regulations compared with their intensive development in agricultural and industrial activity. Therefore, water sources are progressively becoming contaminated with chemical residues (e.g. pharmaceuticals, personal care products, pesticides, hormones, microplastics, food additives).

Regardless of the certain household water treatment (HWT) methods used, chemicals in drinking-water can have an adverse impact. As example, excess of organics and other chemicals in water conducts to decreased lifetime of absorption media, such as activated carbon filter, or can increase the chlorine demand. Nonetheless there are many chemicals that may appear in drinking-water, only some of them, as arsenic and also excess fluoride, can be the major issue for negative health effects on a large scale. Other chemicals, such as nitrates and nitrites, may also be a health safety problem in certain situations (World Health Organization [26] 2017). The most regulated and monitored chemicals that are critical in the drinking water content are identified as arsenic, fluoride, nitrate and nitrite, and inorganic salts classified as total dissolved solids (TDS).

Arsenic

Arsenic is poisonous for human health, so if it is ingested by drinking water or eating food for several years, chronic health problems known as arsenicosis develops. Arsenic can naturally occur in water. Worldwide, arsenic in high levels has been identified as naturally occurred in water from deep wells in more than 30 countries, including Bangladesh, Brazil, Cambodia, El Salvador, India, Indonesia, Mexico, Nepal and Nicaragua. Is reported that in the South Asia region, over 60 million people are affected and have health problems caused by the unsafe levels of arsenic in their drinking-water. The most severely affected country, with a population of 35–60 million of its 130 million people, confirmed to be directly exposed to arsenic-contaminated water is Bangladesh [9]. There is currently no recognized method to effectively treat and cure arsenic poisoning. Despite of this fact, the health issue produced by the arsenic poisoning may be reversed in the early stages by eliminating the direct exposure to arsenic. Drinking and using water that has low arsenic levels, that is considered to be within the safe limit is the only prevention method to avoid long term health problems. There are few HWT technologies that are capable to remove arsenic from drinking-water to safe levels.

Fluoride

Fluoride may also be identified naturally in groundwater and some surface water and also in the human and animals body structure. As presented in World Health Organization's (WHO) publications, high levels of fluoride are found naturally in regions like Africa, the Eastern Mediterranean and Southern Asia. Naturally occurring increased levels of fluoride have been identified to extend from Turkey through Iraq, Iran, Afghanistan, India, northern Thailand and China (World Health Organization [26] 2017). In a small amount, fluoride is generally beneficial for people's teeth and bone structure, but in a higher dose over time, it inflicts damage that can be observed in teeth color changing and pitting effect. Fluoride is also a key element in bones structure builds up and excess can cause crippling skeletal damage. High amounts of fluoride are a major health risk for infants and young children because of body growth and development processes. There is currently no specific method to cure or treat fluoride poisoning. The only measure of safety is to drink and use water that has safe levels of fluoride. Emerging household water treatment technologies have been developed with the capacity to remove fluoride from drinking-water.

Nitrate and Nitrite

Nitrate and nitrite are two different types of chemical compounds that are also found naturally in the environment, especially in vegetables, animals and humans in non-harmful quantities. Nitrate is commonly used in agriculture as fertilizers compound, and nitrite is used especially as a food preservative, in the sector of processed meat and canned foods. Nitrate presence in groundwater and surface water normally is low but is influenced directly by the agricultural and industrial activities and as result can reach high levels. For developing countries that practice intensive livestock farming (cattle, pig and poultry sectors), high nitrate levels detected are often associated with higher levels of microbiological contamination from the nitrates content in feces.

This poisoning with nitrate and nitrite can cause methemoglobinemia, commonly called "blue baby syndrome". This disease occurs in newborns that are bottle fed with food prepared with the high-level nitrate and nitrite drinking-water. The effect observed produce by this poisoning are difficulty of breathing and skin color turning blue from a lack of oxygen, untreated leading to death.

Inorganic Salts Classified as Total Dissolved Solids

Total dissolved solids (TDS) in a water sample are represented by inorganic salts such as sodium chloride, magnesium, calcium, potassium and also small amounts of organic matter that are dissolved in water sample. TDS in drinking-water is present due to the natural sources, sewage, urban runoff and industrial wastewater. In some areas of the world, high levels of TDS naturally occurring have been identified in their drinking-water.

Depending of the very high or low levels of TDS presence, the water is identified as "hard" or "soft" water, respectively. To understand the difference between "hard" and "soft" water, it is established that the hard water is classified as such because it requires more soap to get a good lather and it is "hard" to work with [13]. Soap is less effective with hard water due to its reaction with the magnesium and calcium high content, this fact leading to an increase of the quantity of used soap or foam detergents for laundry and bathing. In fact, hard water leaves a residue and causes scale to build up on cooking pots and water pipes. Generally, the taste of hard water is preferred in consumption due to the dissolved minerals, but very high concentrations of TDS cause a bitter or salty taste to the drinking water. Soft water is mainly preferred for laundry, bathing and cooking activities because of low deposition of precipitate. However, water with extremely low TDS concentrations (e.g. rainwater) is considered unacceptable because of its taste characteristics.

2.1.3 Physical Quality

The measured and identified physical characteristics of drinking-water are turbidity, color, taste, smell and temperature. According to World Health Organization, drinking-water is considered to have good physical qualities if it is clear, tastes good, has no smell and is cool (World Health Organization [26] 2017). Physical properties of water are also the chemical content that can negatively impact certain HWT methods. For example, high turbidity can cause malfunction due to clogging of mechanical filtration technologies, such as ceramics, or interfere and can also negatively impact the performance of chlorine, solar or ultraviolet disinfection. As an extension for the example above mentioned, low or high pH (acidity/alkalinity) and very cold temperatures are responsible for a negative impact on the performance of chlorine disinfection.

Turbidity

Turbidity influences how the water looks (e.g. cloudy, dirty or muddy). Turbidity is caused by suspended sediments like silt and clay suspensions, by sand, and even by chemical precipitates that are considered a form of suspended solids, such as mostly iron precipitates. Drinking water with high turbid content will not really influence the human health. Due to the fact that viruses, parasites and some bacteria can sometimes stick to the suspended particles in water, it can be considered as a health safety issue. Taking this into consideration, since the turbid water usually contains more pathogens, drinking it can significantly increase the probability of health issues and diseases. High turbidity levels minimize the efficiency of household water treatment technologies, such as chlorination, and solar disinfection (SODIS). Sand in water can also damage pipes, valves and pumps shortening their lifetime. Furthermore, it can also be noted that clear water does not necessarily mean that it is free of pathogens and safe to drink.

Color

Colored water is not necessarily dangerous for human health. However, it can be a factor indicating a potentially contaminated water source. The different colors of water are determined by the content of:

- plant residues (e.g. leaves, bark and peat) gives water a dark brown or yellow color.
- high content of clay, sand and silt in water usually cause brown or red color.
- iron suspensions change watercolor in orange or brown color.
- high manganese content turn water black.
- algae content gives water a bright- or blue-green tinge and can produce toxins; and
- bacterial growth turn water black.

As a result, color characteristic is a simple water quality indicator and also sometimes effective.

Taste and Smell

For water quality primary determination, taste and smell are indicators of some sort of contamination, especially when a rapid change occurs. It is very likely that an unpleasant smell or test will not affect health, but most probably will not be accepted for drinking. The following are some reasons that distort the olfactory and gustatory properties of water:

- algae and certain bacteria modify the taste and smell of water to an unpleasant one.
- sulphate (SO_4) gives the water a bitter taste.
- sulphate (SO_4), under the action of certain bacteria, turns into hydrogen sulphide (H_2S) giving the water a smell of rotten eggs.
- excess iron in water for tea or coffee causes a harsh taste to the drink.
- chlorinated water has a distinct taste; and
- salinity can be due to saline intrusion.

Temperature

When refer to drinking water of good quality, we think to cold and not warm water. The recommended temperature range is from 4 °C to 10 °C, since in warm water some bacteria can grow and evolve, thus further degrading the water characteristics in terms of taste, smell and appearance.

2.2 *Treatment Methods of Household Water*

Before establishing the reliable solution for water treatment at household level, identifying and securing the best possible water source is essential. Drinking water quality can be affected by pollution sources both at the source and point of collection, due to different reasons as:

- inappropriate choice of location.
- aquifer or water source poorly protected against pollution.
- inadequate structural design deteriorated or damaged water collecting and distribution elements.
- lack of hygiene and sanitation knowledge and practice in the community.

Ensuring the protection of the water source minimize the safety risks and improve water quality and health. In this regard, various actions can be considered at community level to ensure the safety of the water source:

- periodic cleaning of the area adjacent to the water source,
- location of latrines and sewers downstream the water sources,
- delimitation perimeter with fences of the open water sources in order to prevent the entry of animals.
- covering wells to prevent groundwater contamination with rainwater/surface water.
- development of proper drainage systems for wastewater,

Household Water Treatment and Safe Storage

- stabilization of springs against erosion and ensuring protection from surface run-off contamination, and
- non-polluting use of watershed.

2.2.1 Water Source Protection and Safety

Use of appropriate water safety plans allows the supply of drinking water, under controlled conditions, from source to consumer, but also assess the adequacy of the HWTS. Water safety planning is considered one of the most reliable approaches to provide consistent supplies of safe drinking water at end users through identification of water clean source, regular monitoring of control measures, prioritization and management of risks prior to problems arise [4]. Regular monitoring is required as a control measure and also periodic review of the water safety plan ensures the effectiveness of water safety plan. This planning is based on the principles of sanitary inspections and on hazard assessment and critical control points. Water safety plans have the flexibility to adapt to the multitude of water supply sources, from point sources (e.g. wells, drilling), to centralized pipeline systems, regardless of their generation, old or new.

By definition, water treatment is the result of any process that improves water quality, giving it the qualities needed for end use, regardless of its specificity (e.g. drinking, industrial water supply, irrigation, river flow maintenance, including water safely returned to the environment). By treatment, contaminants and unwanted components in the water are removed or reduced so that the water becomes suitable for the desired end user. This is critical to human health and allows users to benefit from both drinking and industrial or agricultural use.

The water treatment methods may vary slightly at different locations, depending on the technology and water safety for the water source plan, but in all the cases the basic treatment principles are largely the same. This section describes simple and standard water treatment processes, following treatment methods described used individually in the decontamination process, or mixed according to the end user needs. They are classified depending on the contaminant and the quality aspect factors as:

(a) Microbiological water treatment processes.
(b) Chemical water treatment processes; and
(c) Physical water treatment processes.

2.2.2 Microbiological Water Treatment Processes

Large quantities of wastewater sewage are generated from many different sources. Bioremediation of wastewater for large communities is accomplished by wastewater strategies enclosing various technological elements that refer to oxidation ponds, aeration/anaerobic lakes, aerobic/anaerobic bioreactors, sludge, percolating/trickling filters, and removal of nutrients by biological processes. These types of methods are not well suited for a HTWS system, where the end user is the one involved in controlling the wastewater treatment process [11].

Wastewater treatment with the help of aerobic and anaerobic bioreactors and active sludge methods for a HTWS system is conceived to initiate and maintain the natural process of decomposition of pollution under controlled conditions by the end user. By simplifying the description of method, we can say that the use of certain microorganisms to convert the organic nutrients and matter from the water into materials, that are beneficial for the environment, is a wastewater treatment technology. Sewage contains nutrients of various types, from elements such us nitrogen (N), calcium (Ca), sodium (Na), potassium (K) or iron (Fe), to sugars, proteins and fats known as energy-generating vectors for microorganisms. Depending on the contaminant (s) of water, the organisms that prevail are the ones that are best equipped to the "environment" or conditions in the system. The used microorganisms are settled as a solid material for an easy removal of the organic wastes. In this stage of wastewater treatment, is critical to preserve the adequate conditions in relation to the type of microorganisms. If the optimum conditions are not present, the unwanted microorganisms will dominate and their removal from the system will be very difficult. For the HTWS's microbiological water treatment processes design, different types of microorganisms can be used as bacteria, protozoa, metazoan separately or a combination thereof in different concentration.

The mixture of microorganisms that have been in contact with the organic contaminants in the wastewater as final product is known as activated sludge. When most of the material is removed from wastewater, the microorganisms form floc and settle down as sludge. Even, some microorganism will anyway evolve in the system, the one that dominate, will be the ones that is best suited with the medium.

Bacteria are characterized as single celled microorganisms, come in three basic forms: bacillus, coccus, spirillum. There are classified in terms of their oxygen response. To live, aerobic bacteria need oxygen, but can survive for a time period without it, while anaerobic bacteria cannot live in the presence of oxygen. When activated sludge system is envisaged, the main microorganisms are aerobic bacteria. They have the tendency to consume the biodegradable material found in wastewater. They are used in the microbiological water treatment processes to consume soluble organic material in the wastewater. For solid particles of certain undesired contaminants that can't be removed with the help of bacteria, a two-step process is needed that involves adsorption and absorption phases. In the adsorption process the bacteria secrete enzymes, which dissolve the contaminant that will pass through the bacteria's cell wall. In the absorption process, dissolved units of organic water contaminant pass into the cell membrane of the bacteria and are decomposed. In a typical system configuration with activated sludge, wastewater flowing in or wastewater leaving first stage of purification enters the aeration basin. Microorganisms captured in the settled sludge, in view of future usage, are returned to the aeration basin as shown in Fig. 1.

Protozoa represents approximatively 4% of the microorganisms used in the wastewater microbiological treatment processes. Although its contribution to the removal of organic nutrients is insignificant, the protozoa strongly improve the clarity

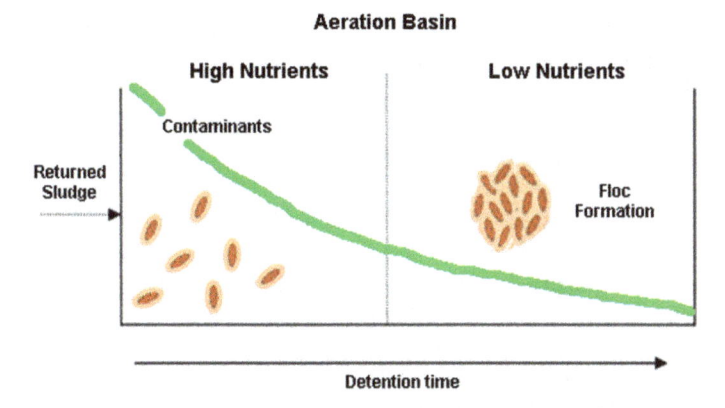

Fig. 1 Microorganisms floc formation and settle out as sludge in a aeration basin

of water. Their impact on the removal of organic substances is small, but other factors that provide strong indications of treatment system performance, such as their number and overall behavior, are significant. Protozoa, as group of one-celled microorganisms, come in various sizes and shapes, from 1 Âμm to millimeters. Their role in the treatment process is to eliminate non-flocculent bacteria and also the very small floc, that normally would not settle [17]. They can be classified based on the way they nourish. Thus, based on their behavior in activated sludge, there are categories, such as: Amoebae, Flagellates and Ciliates.

Amoebae are identified as the most primitive single celled organisms. The food source of amoebae vary, from bacteria to dead organic material present in water. They slowly extend lobe-like projections called pseudopodia until it has completely surrounded its food, and secrets enzymes that will break the food particle into smaller unit that will absorb into the cell.

Flagellates absorb nutrients, just like bacteria so they represent the completion of bacteria for the dissolved nutrients. Ciliates have a lower contribution to the removal of organic material from the wastewater. Their main purpose is to feed on bacteria, and not on the dissolved organics. If bacteria and flagellates compete for dissolved nutrients, ciliates are in competition with other ciliates in devouring bacteria. The presence of ciliates is usually an indication of good treatment of the wastewater. They have the tendency to dominate after the formation of floc and when most of the organic nutrients have been removed. Their function is to remove excess algae and bacteria from the fluid and clarifying the effluent.

Metazoan (multi-cellular microorganisms) feed with bacteria, algae and protozoa. Different types of metazoan can be used in a HTWS system based on microbiological water treatment processes and they are identified as: rotifers, nematodes, tartigrades, annelids, ostracods, copepods [1].

Each of the metazoan, as described above, have a key role in eliminating and feeding on the other microorganism. The efficiency of this type of treatment system is dependent on a number of factors, from temperature, pH, to oxygen availability and microbial competition and therefore needs to be very well controlled and also combined with other water treatment solutions.

2.2.3 Chemical Water Treatment Processes

Several chemical processes are applied for wastewater treatment, namely precipitation, ion exchange, neutralization, adsorption and disinfection (e.g. chlorination/dechlorination, ozone, ultraviolet radiation). For the HTWS only the chemical precipitation and disinfection are suited because of the economical factor and also complexity. Important drawbacks of this treatment process are the high costs of used chemical additives and the necessity of disposing large amounts of chemical sludge.

The dissolved inorganic components in a polluted water source can be removed by different procedures: adding an acid or alkali, changing the temperature, or by precipitation as a solid. The resulting precipitate can be removed by various solid removal process such as sedimentation of flotation. Precipitation processes such as coagulation and flocculation are still implemented, but nowadays it is desirable that any chemical process be replaced by environmentally sustainable process, in order to reduce any negative environmental impact produced by the use of chemicals. Coagulation process is essential in water treatment, since through chemical reaction the particles between coagulant and colloids in the wastewater are destabilized, as shown in Fig. 2.

By definition, as a water treatment method, coagulation involves the addition of polymers that have the ability to clump the small, destabilized particles together into larger aggregates so that they can be more easily separated from the water. Coagulation is classified as a chemical process that involves neutralization of charge. Iron and aluminum salts are the most widely used coagulants but salts of other metals such as titanium and zirconium have been found to be highly effective, as well. In a colloidal suspension, particles have the tendency to settle very slowly or not at all because the colloidal particles carry surface electrical charges that mutually repel each other. A coagulant, like a metallic salt with the opposite charge, added into the

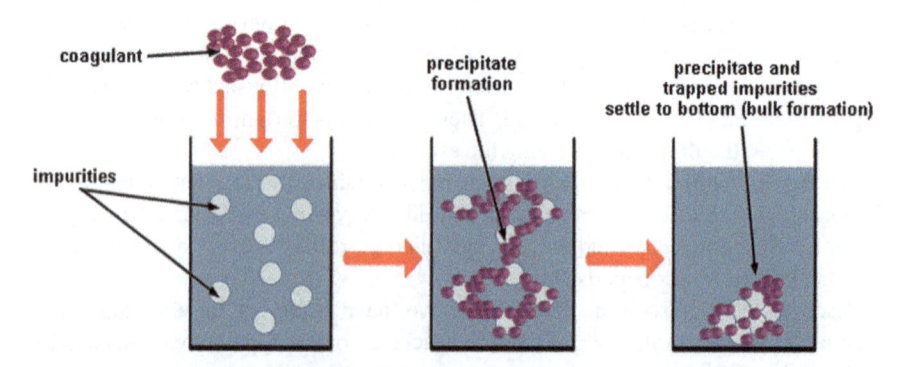

Fig. 2 Coagulation water treatment process

water will overcome the repulsive charge and will destabilize the suspension, van der Waals forces causing the particles to agglomerate together and to form micro floc.

Another approach at HTWS level for treating water in house is to eliminate or inactivate pathogens through disinfection. The disinfection of wastewater is identified as a chemical treatment process implemented by treating the effluent stream with proper disinfectant solution to eliminate or at least disable the pathogens. It is used to exterminate or inactivate pathogens (e.g. microbes, viruses, protozoan) and meet the wastewater discharge standards. The ultimate goal of disinfection is to protect human health, by reducing or even eliminating microbial risk. The desired ideal disinfectant should be toxic to bacteria, is inexpensive, safe to free handling and have reliable means of detecting the presence of a residual. Agents used for chemical disinfection include chlorine, ozone, ultraviolet radiation, chlorine dioxide, and bromine.

Chlorine is, by far the best and most known and time-tested disinfection agent, for safety and reliability. Its remarkable properties make it an ideal disinfectant. The effectiveness of disinfection using chlorine is dependent upon its chemical form, and of factors like pH, temperature, and organic content in the wastewater. To complete this treatment process in the water safety plan, the dechlorinating method is recommended. Dechlorination is the complementary process of chlorination, where reducing agents such as activated carbon, sulfur compounds, hydrogen sulfide and ammonia can be injected to reduce, close to total removal, residual chlorine present in the disinfected effluent prior to discharge.

2.2.4 Physical Water Treatment Processes

Sedimentation, filtration, and distillation are the main physical processes used in wastewater treatment that proven its effectiveness and simplicity. Taking into account that the end user of a water source can apply these methods easily compared with the other types of water treatment processes with a lower negative impact to the environment, for a HTWS system, they are essential in its design.

The most widely used physical treatment process used to reduce wastewater turbidity is sedimentation. It consists of leaving the water in a container to settle and be in an equilibrium, separating thus the water from the sedimentable particles. The process can be accelerated by adding to the water special chemicals or plants, known as coagulants, which helps particles (sand, silt and clay) join together and form larger clumps that deposited at the bottom of the container. As chemical coagulants, aluminum sulphate, poly aluminum chloride, alum potash and iron salts, are used. Depending on their availability, plants like prickly pear cactus, moringa seeds, or corn cobs are also used for sedimentation.

In a sedimentation basin, also called a clarifier or settling basin tank low water velocities are used for allowing floc to settle to the bottom. Their physical form may be of rectangular shape, allowing water to flow from end to end, or circular shape where water flow is from the center outward. The design of sedimentation basin outflow is made to allow only the thin top layer of water to exit, since that layer is the most free of mud. It has been experimentally proven that the efficiency of a

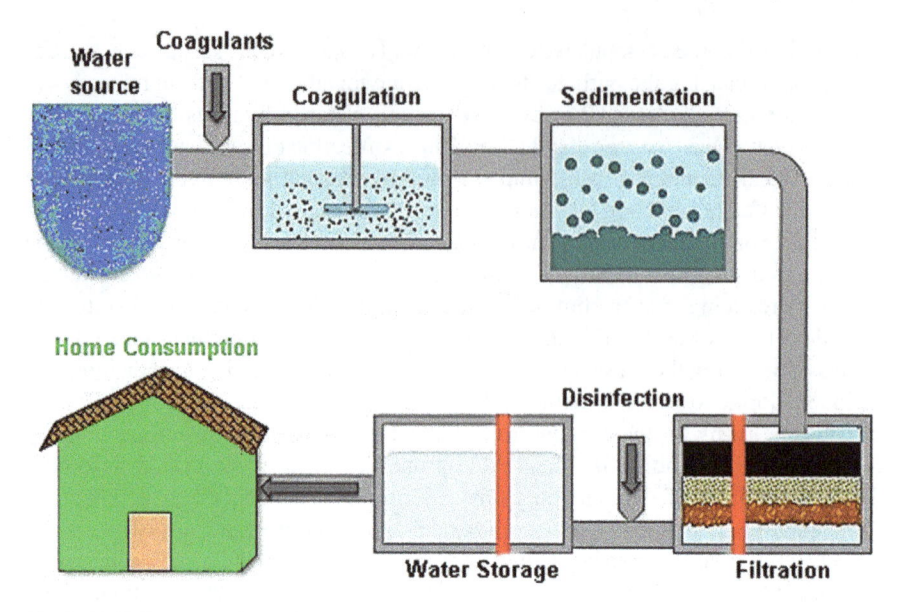

Fig. 3 HTWS system with sedimentation and filtration

sedimentation basin do not depend on the holding time nor on the depth of the basin. For a HTWS set-up, the end-user can design his sedimentation basin tacking into consideration the ground sedimentation area as in Fig. 3.

Therefore, the depth of the basin must be large enough so that the currents caused by water jets do not disturb the deposited mud (sludge) and to facilitate interactions between particles already deposited and established. The presence of sloping plates and tubes drastically increases the area available for particle removal in accordance with Hazen's original theory. The footprint of a sedimentation basin with sloping plates or tubes can be much smaller than that of a conventional sedimentation basin. As the process evolves and the particles settle to the bottom of the sedimentation basin, the floor is covered with formed sludge which must be further discharged and treated.

A sedimentation subcategory is the particles removal by trapping them in a layer of a suspended coagulated floc by pumping the water stream upward. The major advantage of such floc blanket filters is that they occupy a smaller surface area than conventional sedimentation [22]. The disadvantages are related to the fact that the efficiency of particle removal is not constant, it can show very large variations depending on the changes in the quality of the water introduced and also by the flow of extracted water. If the particles to be eliminated are not easily removed from the solution, dissolved air flotation (DAF) can be a way, as illustrated in Fig. 4.

By installing air diffusers on the bottom of tank, mechanisms for attachment (collision, trapping, and absorption) of air bubbles to floc (suspended solids) are created, resulting thus a floating mass of concentrated floc. By this method of sedimentation, the bubbles-solid particles formed layer rises to the surface of the flotation tank, and

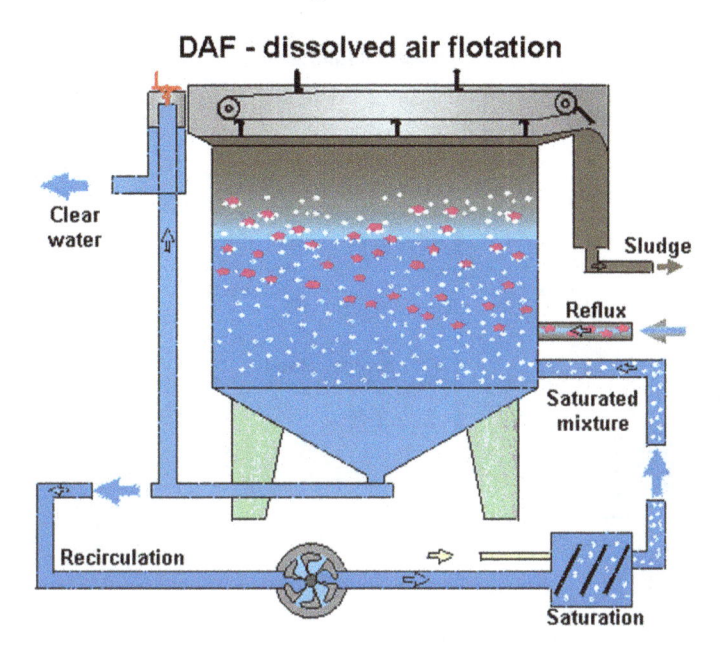

Fig. 4 HTWS system with dissolved air flotation

easily removed, while extracting the clarified water through the tank lower area. In the specific case when water sources are vulnerable to unicellular algae blooms and have low turbidity and high color, then DAF method is recommended [16]. Flocculation is identified as the process in which colloids come out of suspension in the form of floc or flake, either spontaneously or due to the addition of a clarifying agent. The action differs from precipitation process in that, prior to flocculation, colloids are merely suspended in a liquid and not actually dissolved in a solution.

Filtration is a method commonly used to ensure a reduced turbidity and removal of pathogens from water sources. It is the physical process that eliminates turbidity by passing wastewater through a simple filter media. Sand and ceramic are commonly used filtration media, despite the fact that special membranes and other media can also be used, the difference between them being the cost. Some filters are designed to grow a biological layer to eliminate or inactivate pathogens for a better removal process. Others, such as activated carbon, adsorb and hold contaminants like a sponge rather than mechanically remove them like a sieve. Their filtering capacity is consumed once the absorption is complete, regeneration or exchange of the filter being critical.

Distillation is the process by which water is boiled to produce vapors, vapors that are condensed on a cold surface to become liquid again. As the dissolved substances are not normally vaporized, they remain trapped in the solution to be boiled. It was observed that the distillation process cannot completely purify the water because the contaminants have different boiling points, in some situations similar with water, and there is a possibility that drops of unvaporised impure liquid to be transported

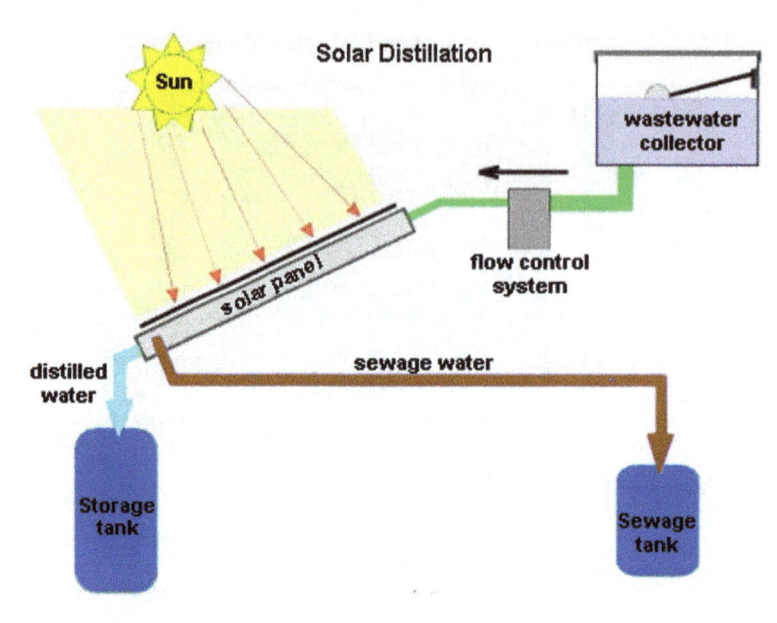

Fig. 5 Distillation methods used for HTWS design

by the stream. However, distillation process can provide 99.9% pure water, but it is considered demineralized water.

To heat water by solar energy, a system of mirrors that automatically tracks the sun is used, thus focusing sunlight on thin layers of water. During the process of heating by solar rays, the water molecules are transformed into steam, leaving behind the contaminants in the pool. In a particular design, the steam is condensed at the bottom of the glass roof and slides along the walls of the glass panel being collected in the collection gutters. The collection gutters located at the bottom of glass panel, thus constructed, must allow water to flow freely to the pipe carrying the distillate to the freshwater tank as shown in Fig. 5 [2].

Distillation process do not have the ability to eliminate all the chemicals but clears away soluble minerals (e.g., calcium, magnesium, phosphorous) and toxic heavy metals (e.g. lead, arsenic, mercury). There are some chemicals that can produce dangerous compounds during the heating process. The vaporization process strips salt, metals, and biological threats. Stripping of minerals is considered not to be harmful to humans due to the fact nowadays the vast majority of minerals is obtained from food or supplements, not from drinking water.

2.2.5 Water Storage Safety

Households that implement HTS systems are designed to collect, transport, and treat their drinking water. Even after treatment, the water should be handled and stored

Household Water Treatment and Safe Storage

properly to keep it safe. If not, the quality of treated water could become worse than the original water source and may cause diseases. Recontamination of drinking water considered safe at a given time, is not a negligible problem. One of the most common diseases recorded today is diarrhea, caused by the water contamination during storage in households, first noted in the 1960s [27]. Distribution and safe storage are achieved by eliminating potential interactions of treated water with contamination sources and also by keeping the container clean and covered. The container should prevent any external contaminant factors to reach and touch the water.

There are many designs for water containers around the world. A safe container for water storage must meet some minimum characteristics:

- have a durable and seal-tight cover,
- have a tap or narrow water outlet opening,
- have a compact base for stability,
- be resistant to external factors, and
- be easy to maintain.

Safe water handling practices for a working HTWS system demands:

- collecting and storing untreated water in a separate container;
- storing treated water in a separate container.
- providing cleaning routine of the storage container; and
- using water in short term after treated, preferably on the same day.

2.3 HTWS Selection Criteria

There are specific criteria that must considered when deciding the solution to be implemented for the household water treatment. Some of these include:

1. Effectiveness—refer to the performance of the applied technology
2. Appropriateness—characterise the flexibility of the selected solution
3. Acceptability—the applied technology implementation acceptance
4. Cost—the applied technology costs for the household
5. Implementation—the applied technology impact.

Other criteria can also be added. For example, to extent the wastewater treatment technology for a household incorporating in the HTWS also a product such as chlorine or a ceramic filter, will involve exhaustible components, sustainability in this case is considered a criterion of selection.

2.3.1 Effectiveness

Effectiveness refers to the ability of the chosen technological solution to ensure an appropriate quantity and quality of water. Standards that indicate the effectiveness of the HTWS include:

- the quality of water—based on which microbiological, physical and chemical contaminants can be removed by the technology and how much.
- the quantity of water—based on the quantity of treated water that can be provided every day.
- the local water source—(i) capacity of the technology to be able to treat the specific microbiological, physical and chemical contaminants of the local water source; (ii) capacity to treat water from different sources to the same level.

2.3.2 Appropriateness

Some technologies for wastewater treatment in HTWS are more affordable depending on the needs and conditions of the community. The following criteria are used to help to match a technology with a particular community:

- Local availability—(i) the technology availability to be locally manufactured in or near the community, using local materials and labor; (ii) technical support and service proven by the need for imported spare parts or consumables and the reliability of the spare parts chain.
- Efficiency—effective time for a household to treat enough water to meet their daily needs.
- Operational costs—(i) operation and maintenance periodicity; (ii) technology lifespan till maintenance or replacement.

2.3.3 Acceptability

Consumer point of view about the implemented HTWS technology will directly influence its widespread adoption and persistent use. It is difficult for a lot of people to accept and implement a new technology until they personally experience the benefits of it. People's approach regarding the use of new HTWS technology is influenced by following criteria of taste, smell and color of water as:

- improving the aspect, taste and smell of the treated water.
- motivation for implementing the technology.
- benefits from the technology use.
- granting comfort, improving health and standard of living, saving time or money.

2.3.4 Costs

Most HWTS options that offer effectiveness of the technology must be purchased. Thus, several costs and aspects need to be considered:

- Capital costs—representing (i) the basic purchase cost of a reliable product; (ii) transportation fee.

- Ongoing costs—representing (i) the procurement of consumable products; (ii) equipment operating and maintenance; (iii) replacement parts and potential reparation interventions.
- Affordability—(i) households economically affordance of the technology cost; (ii) households payment willingness of capital costs; (iii) households payment willingness for on-going operation and maintenance procedures; (iv) technology impact over the household income fluctuations.

Costs reimbursement plays an important role for HTWS sustainability and economical effectiveness.

2.3.5 Implementation

The HTWS technology can be implemented taking in consideration several factors as:

- manufacture quality of the technology for HTWS,
- local manufacturing and repair skills and spare parts availability,
- time needed to implement the technology in the household,
- knowledge needed for properly use of the technology adopted,
- proper monitoring of the wastewater treatment technology in household, and
- integration level into current government programs for HTWS.

3 Advantages and Limitations of the HWTS Technologies

Laboratory studies and also field trials with simple HWTS have proven a significant improvement of the drinking-water quality, technologies implemented in the development strategies of many countries [8]. The tested HWTS methods were based on filtration, chemical disinfection, disinfection by heating (boiling or distillation) and flocculation. Furthermore, a mix of these methods is considered to increase the efficacy of water treatment. A simple example of microbial removal performance, presenting the advantages and limitations of each method is provided in Table 1, in order to demonstrate the principles that influence the technology of HTWS system.

Although, HWTS options categorized and split into consumable and durable products, each of them need different approach for implementation stage to satisfy the criteria of affordability and availability, as presented in Table 2. Using the multi-barrier approach is the best way to reduce the risk of drinking unsafe water. To use a proper HTWS system means to use more than one single technology to improve water quality. Both community and household water treatment systems follow the same water treatment process. The only difference is the scale of the systems that are used by communities and households.

Table 1 Microbial removal performance and the advantages and limitations example

Method	Removal Performance	Advantages	Limitations
Filtration (ceramic)	• Bacteria: 2–6 • Protozoa: 4–6 • Viruses: 1–4	• Easy to use • Visual improvement in treated water • One-time expense • Small or medium local production that can benefit the economy	• Low/absent residual protection could cause recontamination (despite the fact that the treated products are in safe storage containers) • Variation of water quality of locally manufactured filters • Filter deterioration requires a reliable supply chain • Operation of filters clean must have cyclicality • Minimal or low flow rate of 1–3 L per hour (slower in turbid waters) • Potential user taste objections
Filtration (slow sand filtration – bio-sand)	• Bacteria: 1 ± 3 • Protozoa: 2 ± 4 • Viruses: 0.5 ± 2	• Flow rate ($\approx$20 L/h) • Easy to use • Simple to implement and operate • Visual improvement of the obtained treated water • Production from available materials • Longer life • One-time expense	• Low/absent residual protection with recontamination potential • Difficulty in manufacturing and transporting heavy concrete and plastic (45–160 kg) filter housing and sand • Need of periodic cleaning and difficulty in estimating when cleaning is required
• Micro-Filtration (MF) • Ultrafiltration (UF) • Nanofiltration (NF) • Reverse osmosis (RO)	• Bacteria: 2 MF; 3 UF, NF or RO—4 MF; 6 UF, NF or RO • Protozoa: 2 MF; 3 UF, NF or RO—6 MF; UF, NF or RO • Viruses: 0 MF; 3 UF, NF or RO—4 MF; 6 UF, NF or RO	• Visual improvement of the obtained treated water • Longer lifespan of spare parts if accessibility is ensured • One-time expense	• Low/absent residual protection with recontamination potential • multi step system to obtain a product of high quality and requires additional end user technical support • critical to have a reliable supply chain for spare parts

(continued)

Household Water Treatment and Safe Storage

Table 1 (continued)

Method	Removal Performance	Advantages	Limitations
Combined flocculants/disinfectant powders	• Bacteria: 7 ± 9 • Protozoa: 3 ± 5 • Viruses: 4.5 ± 6	• Minimise of heavy metals (e.g. arsenic) and some pesticides • Residual protection counter to recontamination • Visual improvement of the obtained treated water • Easy to classify as non-hazardous, long shelf life	• Multi step system to obtain a product of high quality and requires • Critical to have a reliable supply chain for spare parts • Designed for implementation in locations with high turbidity • Higher relative cost per treated water liter
Thermal (boiling and pasteurization)	• Bacteria: 6–9+ • Protozoa: 6–9+ • Viruses: 6–9+	• Possibility in many households to distillate or boil the water • Acceptability at socio-cultural level water boiling procedure	• Low/absent residual protection could cause recontamination • Potential increased risk of injuries from burning and respiratory infections caused by exposure to indoors stoves or fires • high cost due to use of carbon-based fuel source • Potential user taste objections
Solar disinfection (solar disinfection + thermal effect)	• Bacteria: 3–5+ • Protozoa: 2–4+ • Viruses: 2–4+	• Easy to use • Small cost gathers the plastic bottles • Possible change in taste of the treated water • Low/absent residual recontamination	• Additional treatment method (filtration or flocculation) of waters of that have higher turbidity • Regarding the availability of the clean plastic bottles, is a determined volume of treated • Low visual improvement in treated water • Need a relative longer time to obtain qualitative treated water
Chlorination	• Bacteria: 3 ± 6 • Protozoa: 3 ± 5 (non-*Cryptosporidium*) • Protozoa: 0.5 ± 11 (*Cryptosporidium*) • Viruses: 3 ± 6	• Low/absent residual recontamination • Easy to use • Small or medium local production that can benefit the economy • Cheap to implement	• Very low/absent efficiency for turbid waters • Strange taste and smell of the treated water • critical to have a reliable supply chain • Need of water quality control check • Obtain of chlorination by-products

Table 2 Comparative assessment of consumable and durable HWTS products

Consumable products	Durable products
• Must be constantly refilled • Present little to no expenses, but depending on the usage it inflicts additional costs • Can be self-sustainable • Has distribution and marketing strategy similar to the commercial products • Entrusted to be implemented by private-sector organizations	• One-time or occasional purchase • Are viewed as an expensive investment but present lower maintenance and operating costs • The technology can be supported financially • Implementation of the technology is related to the community infrastructure development • Entrusted to be implemented by non-governmental organizations

In Tables 1 and 2, a case study conducted by the World Health Organization (WHO) and United Nation Children Fund (UNICEF) is shown, emphasizing harmonized statistical indicators at global scale for better monitoring and evaluating of the HWTS implementation and operation. These indicators have been created based upon available HWTS technologies and are organized in accordance with the following themes: reported and observed water consumption; water use and storage in appropriate manner and consistent conditions; knowledge and behavior of end-user; environmental health impact and water quality improvement (World Health Organization [25], UNICEF 2014).

4 Cost-Effectiveness and Cost–Benefit of the HWTS Technologies

Household water treatment technology at first evaluation is very profitable from an economical point of view as compared to the common water treatment methods. Furthermore, if an economic analysis is performed, and the cost to provide safety measurement for drinkable water by HWTS is compared to the cost curing the water-borne illness responsible for health issues in a community, the result indicates that HWTS direct costs are lower both for individuals and governments. Correct Implementation of HWTS can actually result in net savings to the public sector and also a lower negative impact to the environment. As shown in this chapter, HWTS can be effective in preventing a lot of factors that have adverse effects on public health and on the environment. However, the efficiency in preventing diseases of the HWTS is directly influenced by the implementation ratio into a vulnerable community. The implementation ratio of the HWTS will be also dependable of cost effectiveness and also economically compared to alternative interventions and common practice costs for water treatment.

A study reported by Blanton et al. [5], regarding the impact of HWTS at community level in Kenya demonstrate that HWTS implementation in school raised the attendance with minimum 26%, by simply granting access to clean water and promoting hand-washing [5]. Based on this school-project, cost savings were registered as both direct reductions from health-care and treatment costs and additionally from the children respecting the proper hygiene at school. At a large scale it was concluded that overall expenditures for health-care to treat diarrheal disease can be reduced by investing in HWTS and by implementing the technology in the country infrastructure [7].

To understand and achieve the full potential of HWTS, it is necessary for this system to perform well, to be affordable and also cost-effective. Identification and successful implementation of HWTS is received as a beneficial promoting way to increase the adoption of HWTS on a large scale in a country infrastructure. HWTS implementation as a strategy has the goal to reach the most vulnerable populations at scale (coverage), and also to cover the population needs on the long term (adoption) in an efficient and consistent way (World Health Organization [25], UNICEF 2014).

5 Summary of Key Messages, Conclusions and Path Forward

The HWTS is nowadays a key component of the global strategy to ensure safe water to a large population in economically emerging nations that currently have limited access or sources of safe clean water, and also for the countries in which the population is exposed to contaminated water sources or faces a high risk of potable water contamination.

Different studies regarding sustainability of HTWS and implementation experience confirm that proper design of HWTS systems can be rapidly implemented and accessed by vulnerable populations that do not have access to a proper water source, thus dramatically improving the microbiological water quality and significantly reducing diseases if used correctly. Accomplishing a safe, reliable, piped-in water system to every household is the main goal of a healthy infrastructure for water resource direct access and also a plus of comfort, but the time and resources needed to construct, the operation and maintenance costs of a piped distribution water supply system are not always affordable or available. HWTS systems ensure the health benefits of safe drinking-water for end-users and can also eliminate the lack of water supplies by unwanted scenarios caused by natural disastrous events, as flooding or drought. HWTS systems could be considered as complementary to the existing water supply and drainage systems having the main objective to provide safe water, easy access of the end user and health improvement.

World Health Organization (WHO) confirms that safe water and improved water don't have the same characteristics. They consider that he improved water source is represented by source that has been secured from outside contamination through active intervention and construction design. It is considered that some improved water sources are protected as compared to others, but not all can provide safe drinking-water. Distinctive HTWS technologies have different levels of performance in removing or minimizing the main classes of pathogens and contaminants. The multi-barrier approach is the best water safety plan to adequately reduce water contamination and protect human health. This approach is a safe principle that can be utilized at both community and household levels, considering the water treatment process as a complex one. Operational continuity of HWTS relies on a number of factors and has the advantage to allow end users a variety of options, based on the desired performance. This gives households and end users the freedom of choice and the opportunity to use a reliable technology designed for improving their life and health.

References

1. Abebe E, Andrássy I, Traunspurger W (2006) Freshwater nematodes—ecology and taxonomy. CABI Publishing
2. Aybar HŞ, Egelioğlu F, Atikol U (2005) An experimental study on an inclined solar water distillation system. Desalination 180(1–3):285–289
3. Ball P (2008) Water—an enduring mystery. Nature 452(7185):291–292
4. Bartram J, Corrales L, Davison A, Deere DA (2009) Water safety plan manual—step by step risk management for drinking-water suppliers. World Health Organization (WHO)
5. Blanton E, Ombeki S, Oluoch GO, Mwaki A, Wannemuehler K, Quick R (2010) Evaluation of the role of school children in the promotion of point-of-use water treatment and handwashing in schools and households—Nyanza Province, Western Kenya, 2007. Am J Trop Med Hyg 82(4):664–671
6. Brown J, Clasen T (2012) High adherence is necessary to realize health gains from water quality interventions. PLoS ONE 7(5): https://doi.org/10.1371/journal.pone.0036735
7. Clasen TF (2008) Scaling up household water treatment: looking back. Seeing Forward World Health organization, Geneva
8. Clasen T, Schmidt WP, Rabie T, Roberts I, Cairncross S (2007) Interventions to improve water quality for preventing diarrhoea: systematic review and meta-analysis. Br Med J https://doi.org/10.1136/bmj.39118.489931.be
9. Feroze MA (2001) Technologies for Arsenic removal from drinking water. Bangladesh University of Engineering & Technology and United Nations University: 267. Retrieved 20 October 2012
10. Fewtrell L et al (2005) Water, sanitation, and hygiene interventions to reduce diarrhea in less developed countries: a systematic review and meta-analysis. Lancet Infect Dis 5:42. https://doi.org/10.1016/S1473-3099(04)01253-8
11. Gray NF (2005) Water technology: an introduction for environmental scientists and engineers (2nd edn). Elsevier Science & Technology Books, ISBN 0750666331. The Netherlands, Amsterdam
12. Kundzewicz ZW et al (2007) Freshwater resources and their management. In: Parry ML, Canziani OF, Palutikof JP, van der Linden PJ, Hanson CE (eds) Climate change 2007: working group II: impacts, adaptation and vulnerability. Cambridge University Press, Cambridge, pp 173–210
13. Leurs LJ, Schouten LJ, Mons MN, Goldbohm RA, van den Brandt PA (2010) Relationship between tap water hardness, magnesium, and calcium concentration and mortality due to ischemic heart disease or stroke in The Netherlands. Environ Health Perspect 118(3):414–420. https://doi.org/10.1289/ehp.0900782
14. Marcinek J, Rosenkranz E (1996) Das Wasser der Erde. Gotha. 3. Auflage, p 328

15. Mäusezahl D, Christen A, Pacheco GD, Tellez FA, Iriarte M, Zapata ME et al (2009) Solar drinking water disinfection (SODIS) to reduce childhood diarrhoea in rural bolivia: a cluster-randomized, controlled trial. PLoS Med 6(8):
16. Palaniandy P, Adlan H, Aziz HA, Murshed MF, Hung Y (2017) Waste treatment in the service and utility industries, 5 dissolved air flotation (DAF) for wastewater treatment
17. Pauli W, Jax K, Berger S (2005) Protozoa in wastewater treatment: function and importance, biodegradation and persistance
18. Reichardt K, Timm LC (2020) Water, the universal solvent for life. In: Soil, plant and atmosphere. Springer, Cham, pp 7–13
19. Sandru C, Iordache M, Iordache AM, Ionete RE (2018) Assessment of heavy metal sludge quality in a municipal wastewater treatment plant. Progress Cryogen Isotope Sep 21(1)
20. Sandru C, Iordache M, Iordache AM, Zgavarogea R, Ionete RE (2019) Distribution of heavy metals in water and sediments from lakes of the Olt watershed. Progress Cryogen Isotope Sep 22(1)
21. Shiklomanov WI (1997) Assessment of water resources and availability in the world. In: Comprehensive assessment of the freshwater resources of the world. SEI, Stockholm
22. Shrikrishna DA, Kiran MT (2014) S. Chand's basics of civil engineering
23. Sobsey MD, Handzel T, Venczel L (2003) Chlorination and safe storage of household drinking water in developing countries to reduce waterborne disease. Water Sci Technol 47(3):221–228
24. World Health Organization (WHO) (2013) Household water treatment and safe storage. ISBN-13 978 92 9061 615 3
25. World Health Organization (WHO), UNICEF (2014) Progress on drinking water and sanitation: 2014 update world health organization and UNICEF. Switzerland, Geneva
26. World Health Organization (WHO) (2017) Guidelines for drinking-water quality (report) (4 ed.), p. 631. ISBN 978-92-4-154995-0
27. Wright J, Gundry S, Conroy R (2004) Household drinking water in developing countries: a systematic review of microbiological contaminationbetween source and point-of-use. Trop Med Int Health 9(1):106–117

Stefan-Ionut Spiridon was born in Râmnicu Vâlcea city, Vâlcea county, Romania in 1984. He received his B.S. degree in Engineering and Environmental Protection in Industry from University of Craiova, Craiova city, Romania in 2008, his M.S. degree in Radiation protection and Nuclear Safety from University POLITEHNICA of Bucharest, Bucharest city, Romania in 2011 and the Ph.D. degree in Electrical Engineering from University of Craiova, Craiova city, Romania in 2014. From 2008 to present, he is a Senior researcher (technology development engineer TDE II) at PESTD (Experimental Pilot Plant for tritium and deuterium separation) unit, INCD-TCI ICSI Râmnicu Vâlcea. He is the author of more than 14 articles in ISI journals, 15 articles in conference proceedings, 1 book chapter, 4 patents. His research interests include CAD (computer aided design) engineering in nuclear technology; theoretical and experimental research in the field of sensors/chemosensors/biosensors and sensor design/development; study of materials in nuclear technology for hydrogen isotopes storage and development of getter bed materials, LPCE process (Liquid Phase Catalytic Exchange) and CD process (Cryogenic Distillation) for hydrogen isotopes; Heat exchanger development and design; Arduino programmable and electronics. Dr. Stefan Ionut Spiridon is member of General Association of Engineers in Romania (A.G.I.R.) from 2008.

Eusebiu Ilarian Ionete was born in Drăgășani city, Vâlcea county, Romania in 1969. He received the B.S. and MD degree in electrical engineering, electronics, from University Politehnica of Bucharest, Romania, in 1993 and the Ph.D. degree in electrical engineering from the Gheorghe Asachi Technical University of Iasi, Romania, in 2014. He is a Senior Researcher with the ICSI Nuclear Department, National Research and Development Institute for Cryogenics and Isotopic Technologies (ICSI), Râmnicu Vâlcea, Romania. He has published ten articles in ISI journals, more than ten articles in conference proceedings, one book and one book chapter, and holds five patents. His research interests include theoretical and experimental studies in the fields of sensors, chemosensors and biosensors; sensor design and development; carbon nanotubes and sensors based on carbon nanotubes; engineering in nuclear technology; materials in nuclear technology for hydrogen isotopes storage and the development of getter bed materials.

Roxana Elena Ionete was born in Râmnicu Vâlcea city, Vâlcea county, Romania in 1971. She received the B.S. and MD degree in mechanical engineering, from Transilvania University of Brașov, Romania, in 1994 and the Ph.D. degree in engineering sciences from the Technical University of Construction of Bucharest, Romania, in 2003. She is a Senior Researcher, head of department, with the ICSI Analytics, National Research and Development Institute for Cryogenics and Isotopic Technologies—ICSI, Râmnicu Vâlcea, Romania. She has published over 50 articles in ISI journals, more than ten books and books chapters, and holds five patents. Her research activities have been rewarded with more than 10 prices and medals at national and international conferences. Her research interests include theoretical and experimental studies in the fields of stable isotope applications for environmental/hydrology and health. She has dedicated a lot of interest to the study of isotopic exchange processes in the interrelationships between abiotic factors and living systems, with direct applications in environmental science, health and agriculture/food safety and security.

Hydrocarbons Removal from Contaminated Water by Using Expanded Graphite Sorbents

Anatoly Kodryk, Alexander Nikulin, Alexander Titenko, Fedor Kirchu, Yurii Sementsov, Kateryna Ivanenko, Yuliia Grebel'na, Alex Pokropivny, and Ashok Vaseashta

Abstract Every year more than ~0.6 MT of oil is discharged in ocean waters. Consequently, considerable amount of oil contaminants is found in and near water surface, which have a great impact on the world ocean ecology. This chapter describes ways to clean oil and oil containing contaminants at and near the surface of water and in events, when there is an underwater leak caused by natural fractures, leaks or mining accidents. An effective sorbent consisting of expanded graphite (EG) is proposed,

---

A. Kodryk · A. Nikulin · A. Titenko
The Ukrainian Civil Protection Research Institute, 18 Rybalska str., Kiev 01011, Ukraine
e-mail: kodrik@ukr.net

F. Kirchu
Department of Aviation Engines, National Aviation University, Ave., Kosmonavta Komarova, 1, Kiev 03058, Ukraine
e-mail: fkirchu@yandex.ru

Y. Sementsov · Y. Grebel'na
Chuiko Institute of Surface Chemistry of NAS Ukraine, 17 General Naumov Street, Kiev 03164, Ukraine
e-mail: ysementsov@nas.gov.ua

Y. Grebel'na
e-mail: liza50@ukr.net

K. Ivanenko
Institute of High-Molecules Compounds Chemistry, National Academy of Science of Ukraine, Kharkiv highway, 48, Kyiv 02160, Ukraine
e-mail: k_ivanenko@i.ua

A. Pokropivny (✉)
Frantsevich Institute for Problems of Materials Science of the NAS of Ukraine, 3, Krzhyzhanovskyi str., Kiev 03142, Ukraine
e-mail: apokr@ukr.net

A. Vaseashta (✉)
International Clean Water Institute, Manassas, VA, USA
e-mail: prof.vaseashta@ieee.org

Institute of Electronic Engineering and Nanotechnologies "D. Ghitu", ASM, Chisinau, Moldova

Institute of Biomedical and Nanotechnologies, Riga Technical University, Riga, Latvia

© Springer Nature Switzerland AG 2021
A. Vaseashta and C. Maftei (eds.), *Water Safety, Security and Sustainability*,
Advanced Sciences and Technologies for Security Applications,
https://doi.org/10.1007/978-3-030-76008-3_22

which can be used as good sorbent, not only for cleaning oil pollution, but also as filter, since it has bulk density as low as 4 kg/m^3, noting that and 1 g EG has specific surface ~80 m^2, which binds ~80 g of oil or petrol. Physical and chemical properties of EG, as a sorbent, are analyzed in comparison with activated carbon. Furthermore, since drinking water supplied through corroded pipes is of poor quality, conventional filters clog quickly. Therefore, filters using EG can be used efficiently in the first stage of cleaning. Mobile equipment for producing EG onsite of accidents is described. The equipment consists of several structural units: gas supply unit, burner unit, graphite feed unit, reaction chamber, separation and output units. The reaction chamber was designed, such that a standing vortex is formed, by which there is a thermal shock and additional expansion of unopened graphite particles. Automation and simulation of the combustion process in the chamber has shown to improve the quality of EG, which reduces equipment dimensions and enhance efficiency.

Keywords Expanded graphite · Oil contamination · Water · Sorbent · Ecology

1 Introduction

The problem of preservation and rehabilitation of the environment, ensuring the reduction of anthropogenic contaminants have been and remain a challenging task due to its large scale nature. With new and emerging contaminants, megacities development, oil explorations worldwide and global industrialization, the relevance of contamination and its impact on health is growing every year. Oil and its products are highly toxic substances and remains one of the most common contaminants on water surface and in wastewater today. Despite of the fact that a variety of existing and available methods to capture and mitigate contaminants, a satisfactory level of treatment of natural and industrial wastewater from hydrocarbons from petroleum in nature at minimal cost, has not been achieved and hence, there is an urgent need for the improvement of such processes, and is grand environmental technical challenge of the twenty-first century [1]. Many processes and technologies have been explored, however, of particular importance is the use of adsorption technologies [2, 3]. An investigation of great interest consists of search for and study of materials of organic and inorganic nature with high sorption characteristics [4]. It is important to study the origin of the sorbent, because when solving the issues of their utilization and (or) regeneration, in the case of their organic basis, there are a number of additional issues associated, primarily with the low re-generatability to their technological cycle.

Due to the fact that oil production in the world is on the rise for ubiquitously desired oil-independence, there is a growing need to protect the environment from resulting pollution is increasing constantly. There is an increasing need for the development of economically sound production of appropriate sorbents and the development of effective mobile devices for their onsite application and solution to mitigate such environmental problems. The need to solve these problems is also confirmed by the fact that up to 300–400 g of sorbents per person per year are produced and used for environmental measures in developed countries, and, in addition, reserves are created in case of environmental disasters.

Companies, such as "Chemviron Carbon", "Norit", "Sutcliffe Speakman Carbon" and other similar manufacturers, produce a significant large quantity of high-quality, which are also extremely expensive sorbents. The use of synthetic sorbents has increased in the last few years in the form of sheets or rolls of sorbent. Synthetic sorbents can often be reused by squeezing the oil out of them, although extracting small amounts of oil from sorbents is sometimes more expensive than using new sorbent. Additionally, oil-soaked sorbent is problematic to handle and causes minor releases of oil between the regeneration area and the area where the sorbent is used. These sorbents are normally produced at sites which may be far away from the site of an environmental disaster, where they may need to be used for collecting oil and oil products from the water surface. An option to produce these sorbents near an area of use, along with high hydrophobicity and buoyancy in the saturated absorbed product state, are few very important considerations. Furthermore, the quality of sorbents is determined by several factors, such as their capacity to absorb the ingredients, desorption, regeneration of the sorbent and its overall utilization. It is well known that carbon-based sorbents are promising materials [5]. One such material is expanded graphite (EG), which not only has a significantly higher adsorption capacity compared to widely used sorbents, but also has additional advantages, such as the ability to desorb sorbate and recover the sorbent, or its disposal after contamination. In the next section, we describe selection criteria of choosing sorbents, methods to produce and properties of EG and also discuss several practical applications for the use of EG.

2 Chemical Sorbents: COREXIT® and EG

One of the most harmful chemical pollution of the marine environment, arises from oil and petroleum products. Over 6 million tons of oil enters the global marine environment annually. The most common reasons for pollution of the marine environment, are offshore oil production, tanker accidents, and shipping and maritime activities. On an annual basis, normal shipping, tanker accidents and illegal dumping into the oceans, alone creates over 0.6 million tons of oil [6]. This has detrimental impact on the environment, harmful to marine birds and mammals as well as fish and shellfish, halts the germination and growth of algae, seaweed, kelp, and other plant life and has economic impact on coastal communities, including their livelihood.

When eliminating oil pollution from the water surface, typically containment of spilled oil or oil products is conducted with booms, which is a routine for the most cleaning technologies. Since, Oil Spill Response Data (OSRD), is generally recorded in terms of the rate and weight of product harvested from the surface, it is recommended to localize an oil spill without significant damage to the environment, depending on weather conditions, and the effort not to exceed ~24–72 h, from the beginning moment of the accident. Beyond that period, the oil products begin to emulsify and begin to sink, thus forming "oil clouds" density in water, or settle to the bottom surface, thus damaging plants and coral reef. Hence, it is necessary to speed up the process of destroying the oil film by using special dispersant preparations that accelerate the rate of natural dispersion, reduce surface tension between oil molecules, promote the formation of emulsions and externally clean the surface of oil products [7–9]. Additional steps include applying a sorbent to the contaminated surfaces by any manual and/or mechanized mechanisms, until oil is completely absorbed, and the formation of a floating conglomerate has, in one way or another, removed waste sorbent from the water surface [10], as much as possible. The primary criteria for oil-absorbing materials are being harmlessness to the environment, weight absorbed oil product per unit weight of the sorbent, buoyancy in the initial and saturated state, hydrophobicity, capability of regeneration and re-use, manufacturability, ease of application to the surface and removal, and affordability. It would seem that the use of powder sorbents that maintain buoyancy for a long period of time, can significantly increase time reserved for preparing and oil gathering. Chemical sorbents are actively used for mitigating oil spills, as research on some of the desirable options continues to be carried out.

The largest oil spill in recent decades was caused by an industrial accident that began on April 20, 2010, in the Gulf of Mexico on the BP-operated Macondo Prospect. Named as Deepwater Horizon, it was considered to be the largest marine oil spill in the history of the petroleum industry and estimated to be 8–31% larger in volume than the previous largest, oil spill, also in the Gulf of Mexico. The U.S. federal government estimated that the total discharge more than 4.9 million barrels (210 million US gal; 780,000 m^3), leaked came from the bottom of the rig, from a poorly plugged well. The oil spill was regarded as one of the largest environmental disasters in American history. BP sprayed more than 40 million gallons of toxic COREXIT® EC9500A dispersant for oil sorption. Although it was approved by EPA at the time of its use, there were no known toxicology studies at that point. In 2011, a toxicology study was conducted, and it reported that there were over 57 chemicals in the ingredients, including chemicals associated with skin irritation from rashes to burns, eye irritation, cancer, respiratory toxins or irritants; and kidney toxins.

As per report by Vorobiev et al., the data shows that the main part of oil does not reach surface and should remain between the bottom and the surface [10]. This was later verified by observation from under the surface of the water in the Gulf of Mexico, where "oil clouds" [6], with the maximum dimensions of 16 km long, 5 km wide and 100 m high were observed. BP has not tracked and reported these spots via its satellites and continued to spray the surface of the bay water with a toxic dispersant, preventing most of the oil rise to the surface, which in turn would simultaneously contaminate all life forms in water and air. Thus, despite being the worst oil spill and environmental disaster in American history, it seemed that consent by authorities for BP to use COREXIT® has caused long-term human and ecological consequences that may be perceived as worse than the original spill [11]. Officially, about 20 percent of nearly 5,000 Coast Guard personnel who responded to the BP spill were exposed to the toxin and reported persistent coughing. Others experienced wheezing and trouble breathing, according to a 2018 study commissioned by the

National Institutes of Health [12]. Under pressure, Nalco eventually produced a list of COREXIT® ingredients—one of them is 2-butoxy-ethanol, a carcinogen toxic chemical that can cause liver and kidney damage and other related health problems [13]. (Nalco is a company which manufactures trademark COREXIT® EC9527A and EC9500A for BP [14]).

"*COREXIT®-9500 is one of the most toxic oil spill containment reagents ever invented*," said Anita George Ares and James R. Clark of the University of Louisiana, who presented a report for Exxon Biomedical Sciences, titled "Acute Aquatic Toxicity of Three COREXIT® Products—An Overview" [15]. COREXIT's® core is 2-butoxy-ethanol, a toxic substance that strikes the circulatory and central nervous system, kidneys and liver. Because of butoxyethanol, COREXIT® is 4 times more toxic than crude oil. When mixed with crude oil in warm water, the toxicity of the dispersant increases by almost 11 times, according to the said report. It seems that this is exactly what happened at a 100-m depth of the bay, where environmentalists discovered a 10-m poisonous layer that killed all sea-life that came in contact with the dispersant. It was once quoted, "*Now, under the influence of an underwater loop current, a part of the toxic spot goes into the World Ocean. The other part goes to the coast and gradually rises to the surface*" [16]. The consequences of the accidental oil spill still linger on and affects the environment [17], despite BP extended tremendous resources to rehabilitate the environment and all similar consequences of the accident, that will last for years or decades to come. In their justification, BP still maintained that there is no effective alternative to this dispersant and also that each solvent is toxic in its own way.

Analyzing the above situation, a relevant research question that has become relevant is about the process of mitigation of petroleum products not only on the water surface, but also at the time of their outflows from faults, breakthroughs of oil pipelines, "oil clouds", as well as natural and artificial accumulations of oil under the water surface. Unlike toxic COREXIT®, EG binds oil and does not have a negative impact on the environment. Moreover, it can then be assembled mechanically, and collected oil can be reused. All of the above indicates that there is a strong need to use new, modern, environmentally friendly methods for water purification. These methods, obviously, include use of EG to capture oil and petroleum from water. This compound is non-toxic, environmentally friendly and has an extremely high developed surface and adsorption properties. Hence, technical problems associated with using COREXIT® could be resolved, using this new and improved approach. This also necessitates additional studies of its processing and production.

3 Expanded Graphite as a Sorbent

A useful form of graphite is expanded graphite, which has been known for years. The first patent related to this topic appeared as early as in 1910 [18, 19]. Since then, numerous patents related to the fabrication methods and resulting applications of expanded graphite have been issued. For example, many patents have been issued related to the expansion process [20, 21], expanded graphite-polymer composites [22–27], flexible graphite sheet and its fabrication process by compressing expanded graphite [28–36], and flexible graphite sheet for fuel cell elements [37, 38]. Also, there are patents relating to grinding/pulverization methods for expanded graphite to produce fine graphite flakes [39–41]. All of these patents use heat treatments, typically in the range of 600–1200 °C., as the expansion method for graphite. The direct application of heat generally requires a significant amount of energy, especially in the case of large-scale production [42].

Use of expanded graphite as perspective sorbent, however, is relatively new theme for this compound and is gaining attention of researchers. In our recently reported works [43, 44], we have systematically presented data on the efficiency of using various types of sorbents, carried out a comparative analysis, as well as, articulated boundaries of the use of this compound. Particular attention is paid to the study of the sorption properties of expanded graphite (EG) in relation to oil and oil products. As per literature, bulk mass of EG is orders of magnitude less $\sim$(2–5 kg/m^3) than the bulk mass of activated carbon matrices, which raises the question of the difficulties of its transportation due to large volume and the possibility of obtaining the required amount directly in the disaster area. A mobile production facility, which can generate vast amounts of EG, on demand and on-site would be a preferable method to mitigate such events and their consequence management.

It is well known that the wastewater, nowadays is contaminated with oil and/or its products. Once the contamination is introduced, accidently, inadvertently or intentionally, it pollutes water bodies and soil, thereby violating the ecological state of the region. Despite of the fact that the extraordinary measures on production, processing and transportation of oil are mandated to avoid accidental disasters, their number have not decreased around the world. This is attributed due, primarily, to the fact that most of the main oil and gas pipelines, facilities for the extraction and transportation of oil and oil products have already been operating to their full current capacity, under standardized terms and new and innovative projects, such as Keystone XL-pipeline project was revoked, negatively impacting ground breaking commitments. Emergency situations arising from these circumstances, not only violate the ecological situation in the region, but also create an increased regional fire hazard. To remove the thick layers of oil from the water surface, a number of effective tools (so-called "skimmers") have been developed and successfully applied these days, which are placed on the special floating resources. Removing of the thin layers of oil and oil products from the water surface, especially in storm conditions, and from the soil surface, is only possible by using sorbents. Quality of absorbent materials is determined primarily by their relative capacity of the ingredients absorbed, the possibility of regenerating adsorbents and desorption and recycling conditions. For the collection of oil and petroleum products from the water surface in the utilization zone, its hydrophobicity and continuous buoyancy in its saturated absorption state is critically important. In addition, the ability to release the sorbed oil products and multiple use of sorbents and their disposal, while maintaining environmental situations are significant as well. Inorganic sorbents (clay, sand, tuffs, zeolites, pumice, etc.) do not meet these specified requirements. Furthermore, organic adsorbents such as activated charcoal (AC), viscose, hemp flour and others have very low efficiency.

At present, the most effective sorbents used are foamed synthetic materials (polypropylene, polyethylene, polystyrene, and other related polymers). The effectiveness of these sorbents is quite high, but it is expensive from cost effectiveness standpoint. In addition, there is no way to produce and use these sorbents directly at the place of ecological catastrophe. The results of these studies [45–48], suggest that EG may be one of the leading candidates to be used as a sorbent for collecting oil and petroleum products.

A patented method for producing EG [49], and a mobile installation for its production, as well as the model of a light craft for water way, have been developed for preliminary investigations. At the experimental site (the Novaya Devina field, Ukraine—https://zvgraphit.com.ua/ru/analitic/5557-zavalevskiy-grafit--unikalnoe-mestorojdenie-gordost-strany.html, accessed on 03/24/2021), the oil density was 0.8 g/cm^3, viscosity (η) ~23 mPa s, and the temperature was set at about 20 °C [50]. The density of the EG samples oil was 0.6 g/cm^3, and EG dispersion with water 0.67–0.7 g/cm^3. The value of the absorptive capacity of the EG (Fig. 1) was determined by the formula:

$$\delta(Q) = (m_2 - m_1)/m_1 \qquad (1)$$

where m_1 and m_2 are the masses of adsorbent before and after sorption [50].

From Fig. 1 it is seen that the sorption capacity is a linear function of a decrease in bulk density, that is, increasing the surface area of the EG samples.

Table 1 shows the values of absorption capacity and the uptake of the oil from the water surface (τ) for EG in comparison with the sorbents that are most commonly used for cleaning oil contaminated surfaces.

As can be seen from Table 1, EG has greater absorption capacity in comparison with the other commonly used sorbents. Additional advantages are the ability to obtain the necessary amount directly in the disaster area, the possibility of desorption of sorbate and sorbent recovery, recycling or elimination after contamination.

Fig. 1 Dependence of the EG absorption capacity on the bulk density

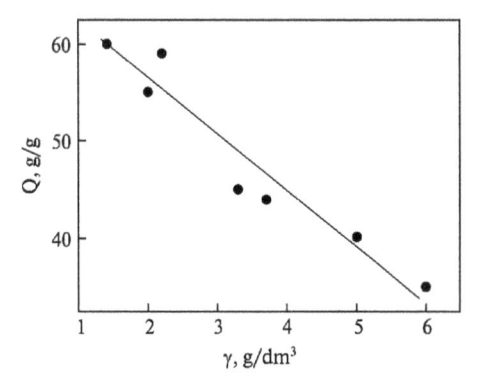

Table 1 Comparative characteristics of EG and adsorbents used for cleaning oil-contaminated surfaces

Oil product viscosity, M Pa s	Sorbent type					
	EG		Perlite hydrophobized		Basalt fibre	
	τ, s	m_a/m_s^*	τ, s	m_a/m_s^*	τ, s	m_a/m_s^*
Decan, 0.8	3	55	3	4.4	4	7.7
Oil, 915	960	65	2100	5.0	1100	7.0
Oil, 245	150	66	180	4.4	600	8.2
Crude oil, 16	55	60	60	4.4	60	8.0

Note $*m_a$—absorbed mass of oil products, m_s—mass of sorbent

Economic factor includes low cost of sorbent. The only disadvantage of EG as an oil sorbent is its low bulk density, and therefore even a small mass of material takes a considerable amount. This is a significant disadvantage for transportation by air due to limited volume, especially in case of emergency. Therefore, it is logical to increase the density of EG due to its modification by the oleophilic liquids and further granulation. A paraffin solution "white spirit" (a mixture of gasoline solvents) and oil of the brand MS-8P were used as modifiers. Table 2 shows an oil absorption of EG modified samples designated, for example, P-0.08 (0.08 STA paraffin 1 g of EG), M-0.51 (0.51 g of oil for 1 g of EG) relative to MS-8P oil.

After the oil saturation an increase in EG granules up to 5–10% is observed. As can be seen from Table 2, oil absorption has a maximum, depending on the content of the modifier in both per unit weight and per unit volume. The value of the optimum content of modifier is for paraffin 0.16 g/g and for oil 0.88 g/g. The results for the oil absorption depending on the density of the granulated samples with optimal modifier content are listed in Table 3.

Table 2 Oil absorption for the EG samples modified by paraffin and oil (density of the granular samples is about 0.03 g/cm^3), with respect to the MS-8P oil

Sample	EG	P-0.08	P-0.16	P-0.33	M-0.33	M-0.51	M-0.8S	M-1.37
		Paraffin			Oil			
Modificator content, g/g	0	0.08	0.16	0.33	0.33	0.51	0.88	1.37
Bulk density, g/dm^3	3.4	6.5	9.5	14.3	7.1	8.3	10.0	14.3
Density of granules, g/cm^3	0.030	0.031	0.027	0.030	0.027	0.027	0.028	0.030
Specific volume, cm^3/g	33.3	32.2	37.0	33.3	37.0	37.0	35.7	33.3
Oil adsorption, g/g	7.8	12.9	18.6	16.7	15.2	15.4	17.2	16.2
Oil adsorption, g/cm^3	0.23	0.40	0.50	0.50	0.41	0.42	0.48	0.49

Table 3 Dependence of the oil absorption by EG on its density

Material	P-0.16					P-0.33		
Density, g/cm^3	0.013	0.017	0.021	0.027	0.040	0.030	0.042	0.069
Oil adsorption, g/g	42.1	32.2	24.2	18.6	9.5	16.7	10.4	5.3
Oil adsorption, g/cm^3	0.54	0.55	0.50	0.50	0.38	0.50	0.44	0.37

It is seen that with an increase in specific density of samples, the oil absorption per unit of mass of the modified EG is decreased, however, it is possible to achieve a slight increase in oil absorption per unit volume of the sorbent due to the modification. Let us compare of sorption properties for EG and activated carbon during purification of water from soluble and insoluble organic pollutants. In [51], the comparison of the sorption properties of EG, its granular samples and the activated carbon used to purify water from soluble and insoluble organic contaminants has been made along with analysis of these properties with the structural characteristics of the sorbents. For the study four EG samples (EG1-EG4) were selected, as well as a granulated sample of the EG shredded foil (EG-F); activated carbon (AC), granulated (SCN) and powder (lightening) carbon (LC); carbon black (CB) (Table 4).

The effective specific surface area of the sorbents was determined by thermal desorption of nitrogen; the total volume of micro- and mesopores—by condensation of benzene vapor. To determine the oil absorption the weighing bottle of sorbent was filled until the excess solvent, which was then removed. Rhodamine C adsorption isotherms were measured under static conditions. For this purpose, the sample of sorbent was maintained in a dye solution in water for 14 days, with occasional stirring until equilibrium. The concentration of dye stuff was measured by absorbance of solution at 550 nm. Water purification of rhodamine C was carried out under dynamic conditions by passing the dye solution through the sorbent cartridge upward (once or several times). This cartridge filled with the EG had the same low flow resistance as SCN filled with granular coal. The water samples contaminated with oil were mixed with the samples of sorbent using bubbling air. In order to determine the

Table 4 Sorbent characteristics: specific surface area (S, m^2/g), pore volume (V, cm^3/g), bulk density (m, g/dm^3), oil absorption (M, cm^3/g)

Samples	S, m^2/g	V, cm^3/g	m, g/dm^3	M, cm^3/g
EG1	100	0.14	7	60
EG2	61	0.17	7.5	56
EG3	34	–	7.9	51
EG4	20	–	7.8	52
EG-F	30	–	–	–
CB	8	0	–	–
SCN	820	0.45	470	1.4
LC	1500	0.52	590	2.6

Fig. 2 Comparison of the wastewater treatment processes by bubbling using 1: EG; 2: activated carbon (500 ml of water, 2 g of sorbent)

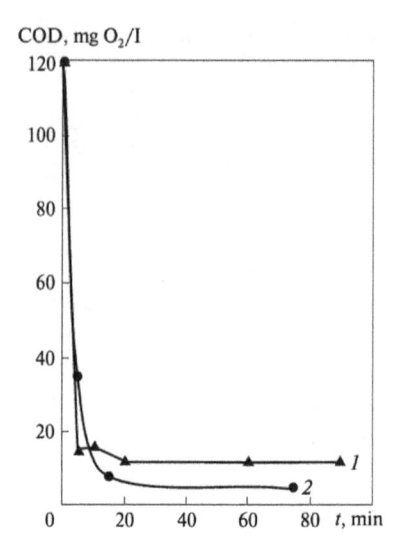

concentration of oil in water, the chemical oxygen demand (COD) in milligrams of oxygen was measured.

From the characteristics of the sorbents listed in Table 4 it is seen that the bulk density of the EG is several times less than the mass of bulk activated charcoal and foil. At this value, the oil absorption of EG1 (60 cm^3/g) is 25–40 times higher than that of activated carbon. These data indicate that the use of EG is extremely effective for water purification from the insoluble liquid oils. Figure 2 shows data on the oil absorption while bubbling of 500 ml water with 2 g of sorbent having 120 mg O_2/l. It can be seen that the absorption rate of contamination in the first 5 min of EG experience is higher than the powder (lightening) carbon (LC).

Unique properties of the EG described above were used for the sorption purification of water contaminated with oil at the primary treatment stage, and after preliminary water purification by filtration through sand. To investigate the possibility of using EG on a primary sewage purification step, water heavily polluted with the oil products was used. Besides, two kinds of impurities in the form of organic substances (insoluble floating in the form of droplets on the surface and dissolved in water) were used. The total concentration of organic substances by COD was 2.0 mg O_2/l, the concentration of dissolved was 400 mg O_2/l (for potable water, it should not exceed 5 mg O_2/l).

Figure 3 shows a picture of scanning microscopy for the spherical nitrogen-containing SCN activated carbon (Fig. 3a, b) and EG samples (Fig. 3c) and structural transformation of graphite to EG (Fig. 3d). As can be seen, EG surface has a porous structure with a pore size of about 10 μm, while the outputs of the surface porous SCN granules have dimensions of below 1 μm.

The best kinetic characteristics of EG samples are due to the fact that EG has a structure that is open to the diffusion of the dissolved molecules. A large part of the EG surface comprises an outer surface of particles of very small size, obviously,

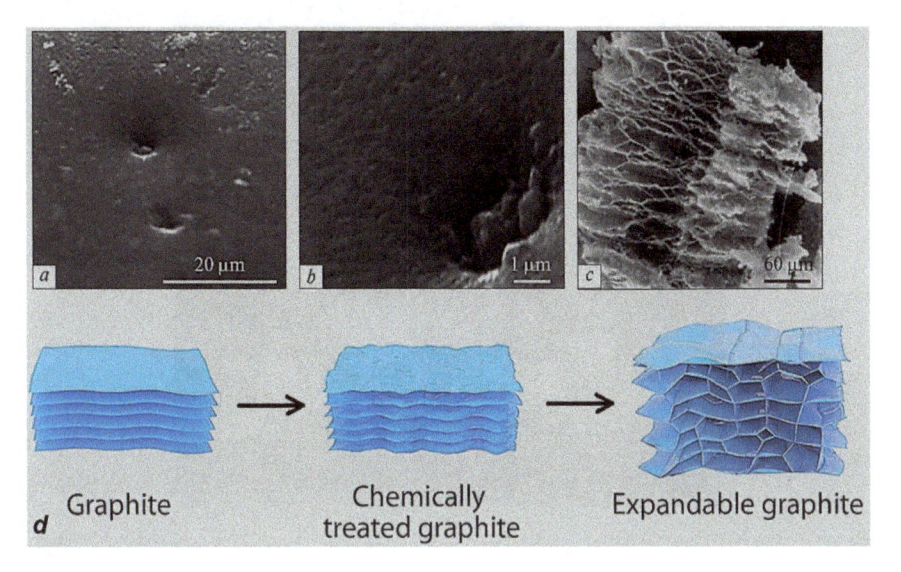

Fig. 3 The surface structure of the spherical sorbent SCN (**a, b**) and EG (**c**) according to SEM, and representative model of structural transformation of graphite to EG (**d**)

hundreds of nanometers. The surface of AC is almost all included in relatively large pores in the particle size of a few millimeters. Therefore, a small particle size of the EG (much smaller than that for AC) led to its high adsorption activity in dynamic conditions.

4 Mobile Equipment and Simulation of Operating Parameters for EG Production

At the Ukrainian Research Institute of Civil Protection, a number of research activities were carried out aimed at studying the process of thermal expansion of graphite and determining the factors influencing its sorption capacity, which made it possible to create relevant technology and mobile equipment [1], protected by the Ukrainian patent [52], for producing a carbon sorbent for anthropogenic liquids based on EG, to be applied directly at the oil spill site. Diesel fuel, propane-butane mixture and electric power can be used as fuel.

When using the technology of combustion of diesel fuel and propane-butane mixture, a two-stage combustion of fuel in a reactor is proposed using two chambers with different excess air ratios. A photo of the mobile equipment and the main unites is shown in Fig. 4. The equipment consists of several structural units: gas supply unit, burner unit, intercalated graphite (IG) feed unit, reaction chamber, separation and output units. The schematic diagram of the reactor itself is shown in Fig. 5. The

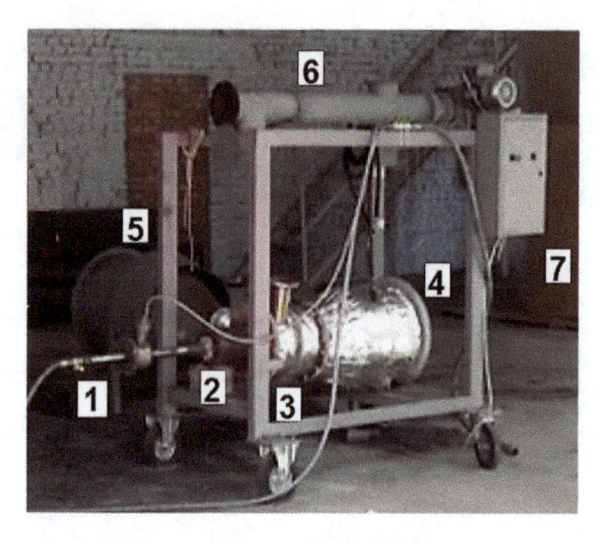

Fig. 4 General scheme of the equipment: (1): gas supply unit, (2): burner unit, (3): intercalated graphite feed unit, (4): reaction chamber, (5): separation and output units, (6): auger, and (7): control systems

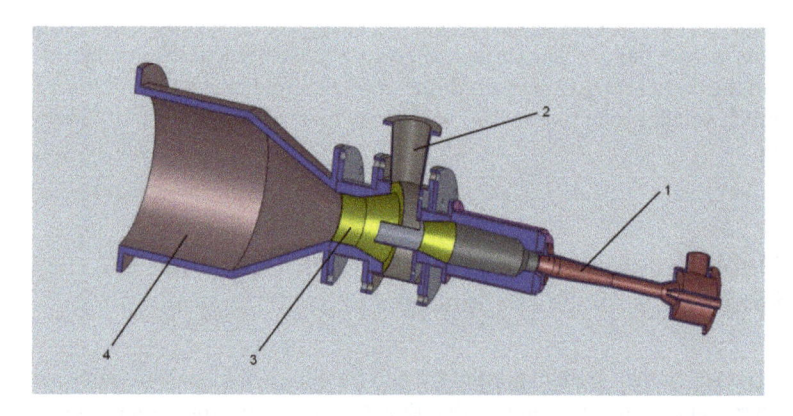

Fig. 5 Schematic diagram of a reactor for two-stage fuel combustion: (1): preliminary combustion chamber, (2): device for graphite supplying, (3): main combustion chamber with an outlet of combustion products and graphite into the reaction chamber, and (4): reaction chamber

proposed technology ensures stable maintenance of the required temperature and reduces energy costs.

To control the residence time of graphite of different fractionation in the reaction chamber, which is necessary for the complete expansion of each individual particle of intercalated graphite (IG), a new technology for producing EG, based on the organization of annular gas-dispersed vortices in the reactor, has been proposed and devices for its implementation have been developed. At the same time, both the creation of special reactor designs and the modification of existing reaction chambers,

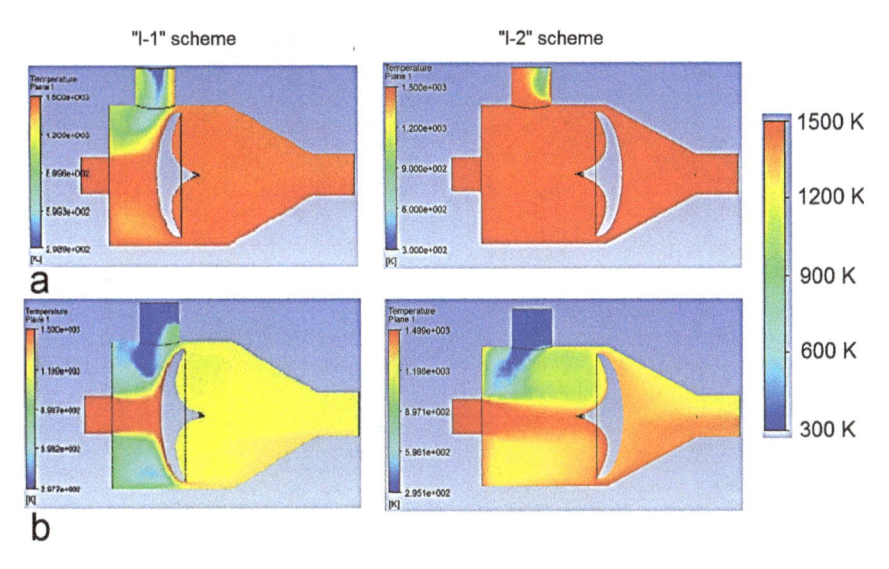

Fig. 6 Fields of parameters of temperatures of the gas flow in the reaction chamber at different locations of the splitter—a screen to change the pressure and temperature in the reaction chamber to produce split graphite. Boundary conditions on the surface: "IG entrance"—free entrance, while **a** without the presence of the IG, **b** with the supply of the IG to the chamber

for example, the installation of flow-forming elements, is considered. Figure 6 shows the temperature distribution fields in the reactor for two variants of the arrangement of the divider: "I-1" and "I-2".

Figure 7 shows the trajectories of graphite particles with color indication of the temperature field of the gas flow in the reaction chamber.

Obviously, the I-2 scheme has the most favorable arrangement of the splitter. This is due to the fact that a more uniform distribution of dispersed IG in the reaction chamber is provided. The "I-2" scheme also provides more favorable conditions at the graphite feed inlet. This is facilitated by the dilution at the inlet of the IG into the chamber of the inlet pipe. With this arrangement, the diffuser at the outlet from the inlet pipe does not create an increased back pressure, which is observed in the "I-1" scheme. The "I-2" scheme allows to provide a larger volume, in comparison with the "I-1" scheme, the reaction chamber in which the IG is heated. With an optimal choice of the gap between the splitter and the position of the splitter in the chamber, we achieve high quality EG. Changing the position of the splitter allows you to control the thermodynamic characteristics of the reactor over a wide range.

In the process of conducting research, a comparative analysis of the efficiency of convection and radiation heating methods was carried out and the decisive role of thermal radiation for the process of thermal expansion under conditions of thermal shock was shown. On the basis of the created general mathematical model, it was concluded that for the flow of particles under the influence of thermal shock, the mutual shading of particles plays a significant role, as the flow of one fraction overlaps the other, due to direct thermal (infrared) radiation. This is critical during heating

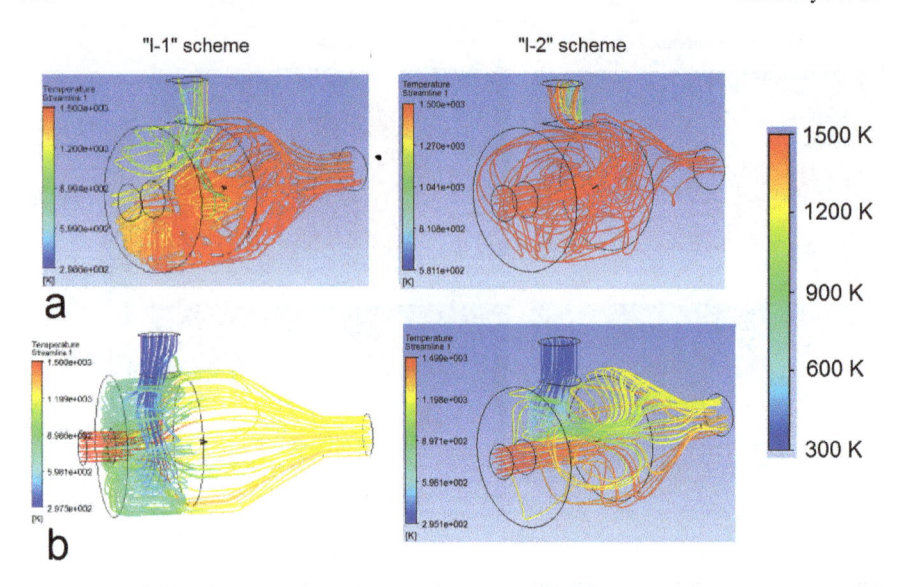

Fig. 7 Trajectories of graphite particles with color indication of the temperature field of the gas flow in the reaction chamber with different placement of the splitter. The boundary condition on the surface "entrance of the IG" is free entrance, while: **a** without the presence of the IG, **b** with the supply of the IG into the chamber

and splitting of graphite since the shading of particles depend both the productivity of the installation and the sorption capacity of the final product. Additionally, a mathematical model for heating by radiant heat exchange, i.e. heat transfer by radiation from more heated particle to less heated particles in intercalated graphite, during the process of its structural transformation in thermo-exfoliated graphite is developed. A new procedure is proposed for building the mathematical model, based on approximated numerical iterations with the first-order finite differences scheme [53]. A set of physical processes that take place during the structural transformation of graphite into thermally expanded graphite under thermal shock conditions is created and experimentally confirmed on the operating equipment. Analyzing the mathematical model presented here, it is possible to make a comparative analysis of the efficiency of convection (Fig. 8a), versus radiation (Fig. 8b) methods of heat transfer, which shows the decisive role of thermal irradiation.

An important result of the mathematical model developed here was found to be dependent of the time until the completion of the thermal expansion process on the brightness temperature of the reaction chamber of the setup, Fig. 9a. The general mathematical model that was created [6] confirmed that the mutual shading of particles plays a significant role for the flow of particles under the influence of thermal shock (Fig. 9b), in which both the production capacity and the sorption capacity of the final EG product depends. The simulation results were confirmed by the experimental results.

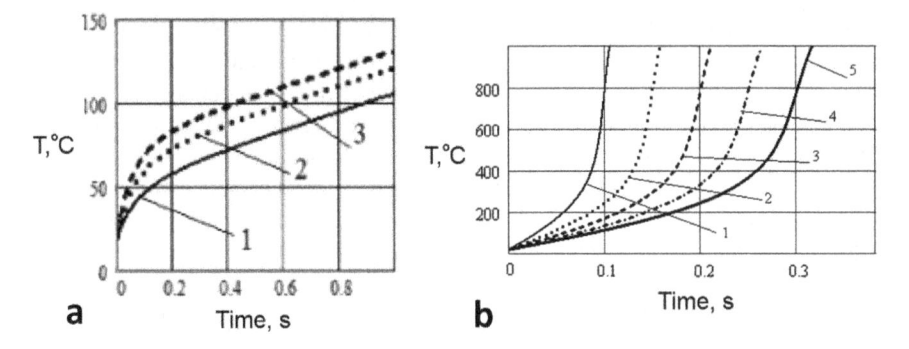

Fig. 8 a Dependence of the temperature of a graphite particle on the **a**: residence time in the flow of a gaseous medium heated to a temperature of 1200 °C, depending on its speed: 1–10, 2–20, 3–30 m/s; and **b**: time of its stay in conditions of thermal irradiation at a temperature of 1200 °C for particle diameters in mm: 1–0.2; 2–0.3; 3–0.4; 4–0.5; 5–0.6

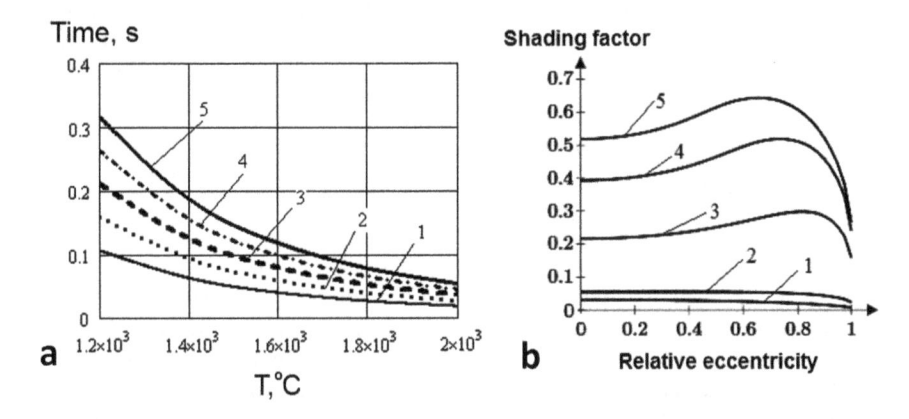

Fig. 9 a Dependence of the heating time of a particle to full thermal expansion on the brightness temperature of radiation for various particle diameters (in mm): 1–0.2, 2–0.3, 3–0.4, 4–0.5, 5–0.6; **b** Dependence of the shading coefficient on the relative eccentricity of the point in the process dynamics of particle temperature rise. Curves: 1–20, 2–250, 3–500, 4–800, 5–1000 °C

The simulation results also allowed us to achieve sustainable work of an ejector, by means of which intercalated graphite is introduced into the reaction chamber. Work of an ejector is the work which can be performed taking into account the parameters (pressure, temperature) directly at the exit from the nozzle, or at the entrance to the reaction chamber. Reprofiling of the reaction chamber is needed for,

(1) providing a more stable vortex flow inside it,
(2) ensuring the separation of split and un-cleaved graphite,
(3) increasing the time of the expansion, and
(4) creation of uniform thermal field in the reactor [54].

These simulations allowed us to reduce density and increase the sorption capacity of the expanded graphite. Taking into account the foregoing, a model range of mobile units with a capacity of 5–20 kg/h has been created, which can operate using liquid fuel, propane-butane mixture or electricity using the principles of thermal irradiation. The sorption property of the obtained EG in relation to oil and oil products is at least 80 g of oil per gram of sorbent. The installations have good technical and economic indicators. Although, some new methods using nanomaterials have been reported to mitigate petroleum products [55], EG, still has advantages for large area spills.

Notwithstanding the availability of technologies and equipment for the production of high-quality sorbents, their use is inhibited due to the lack of an effective technology for their application to a large contaminated surface and collection. In this regard, we are developing methods, technologies and equipment for the possible effective use of sorbents both by existing oil skimmers and using new technologies and equipment for its application and collection. The methods developed here will allow for the extraction of oil products with simultaneous regeneration. Furthermore, considering the peculiarities of existing sorbents, it is effective for this technique to use sorbents that meet the requirements of the oil collector in terms of the absorption rate of 1–2 s.

5 Conclusions

Carrying out work on the use of EG as an effective sorbent, with such a combination of physical, chemical and commercial qualities, is relevant and makes this material truly unique and applicable for the protection of critical infrastructure objects in the elimination of the consequences of oil products and toxic hydrocarbon-based liquids entering the ecosystem. Thus, EG sorbent is intended for cleaning waters and surfaces from oil products contained in them, wherever oil, oil products and other types of fatty materials are extracted, stored, transported and processed (ports, oil storages, oil terminals, car washes, gas stations, etc.). In practical terms, a typical medium-sized gas carrier "Q-Flex", has the capacity of production of about 200,000 m^3. Suppose it is fully equipped with EG with a density of 5 kg/m^3, hence 1000 tons of EG can be taken to the accident site, which may adsorb more than 80,000 tons of oil. This value is near the capacity of a single middle "Panamax" oil tanker. Mobile equipment can be used for additional producing EG, by which a 1 ton of EG may be produced for two days at the site of a local accident. A separate and important issue is finding a way of practical means of application of EG at an accident site. It can be a mixture with foam, which has already been implemented in practice, or use in the composition with a polymer or other matrix. This issue requires a fundamental study.

Although, recent studies have indicated that there are additional environmentally friendly methods and materials, such as use of magnetic nano silica synthesized using barley husk to remove petroleum in polluted water [55], the use of EG is still preferred for large scale collection of oil. It is promising to use it as a filler for filters to remove anthropogenic components from aqueous solutions in a flow-through mode, which is confirmed by the calculations and experiments. The use of granular adsorbents and liquids with magnetic properties, which are easily removed by a magnetic field after adsorption of oil, is promising. An important use of the sorbent consists in the purification of wastewater contaminated with oil products and other chemical compounds, both for each individual pollutant and for their mixtures. In these cases, the absorption capacity of EG for oil is more than 80 kg per 1 kg of sorbent. For widespread use of the proposed sorbent, it is necessary to develop effective methods and means for applying it to a contaminated surface, collecting it from the surface and subsequent processing and disposal. Finally, it is important to note that the produced EG sorbent acts as a volumetric absorber not only of hydrocarbons (including oil products), but, in addition, for a wide range of other pollutants, which include, for example, nitric, sulfuric, phosphoric acids, toxic substances such as sarin, soman, mustard gas, lewisite, etc., which significantly expands the range of its use both in the water purification and, in common, in the field of civil protection.

References

1. Nikulin AF, Kodrik AI, Titenko AN (2016) Use sorbent based on exfoliated graphite to protect critical infrastructure. In: Proceedings of the VII international scientific and practical conference "emergency situations: warning and elimination" dedicated to the 60th anniversary of the first scientific unit in the Republic of Belarus in the field of prevention and elimination of emergencies and fires. Kolorgrad, Part 2, pp 240–250
2. Packs GE, Petryashin LF, Lysyany GN (1986) Environmental protection in the oil and gas industry. M. Nedra, 244 pp
3. Kotov SG, Lupei AYu (2001) Sorption materials for liquidation of emergency spills of oil and oil products. Sci Supp Fire Safe 10:92–98
4. Sementsov YuI, Pyatkovskiy ML (2008) Expanded graphite. Inorganic materials science. In: Gnesin GG, Skorokhod VV, Encyclopedic edition in 2 volumes, Kiev, vol 2, Book 2, pp 410–425
5. Pakhovchishin SV (2000) The role of the nature of the surface of dispersed particles in the processes of wetting and creation of structure. Abstract of the dissertation of the doctor of chemistry sciences. K. IBKH NAS of Ukraine, 36 pp
6. Kodryk A, Nikulin S, Nikulin M (2013) Oil pollution utilization problems on water surface and under water. Naukovy visnik UkrNDIPB 2(28):35–38
7. Mochalova OS, Gurvich LM, Antonova NM (2014) Institute of Oceanology named after P.P. Shirshov, RAS. Oil accidental spills and the role of dispersants in their elimination. "NefteGazoPromy-slovy Engineering"
8. Patin SA (2001) Oil and ecology of the continental shelf. M, VNIRO, p 247
9. US-Russia Energy Working Group (2003) Prevention workshop materials oil spills and liquidation of their consequences. Moscow, December 4–5, 2003
10. Vorobiev YuL, Akimov VA, Sokolov FI (2005) Prevention and elimination of emergency spills of oil and oil products, Moscow, p 347
11. https://whistleblower.org/gulftruth/
12. Fears DI (2019) Chemical that EPA allows to help clean up oil spills sickens people and fish, lawsuit claims. The Washington Post, March 26, 2019
13. Sheppard K, BP's bad breakup: how toxic is Corexit? https://www.motherjones.com/environment/2010/08/bp-ocean-dispersant-corexit/

14. Wise JP (2011) A review of the toxicity of chemical dispersants. Rev Environ Health 26(4):281–300
15. George-Ares A, Clark JR (1997) (1997) Acute aquatic toxicity of three corexit products: an overview. Int Oil Spill Conf Proc 1:1007–1008. https://doi.org/10.7901/2169-3358-1997-1-1007
16. Tikhomirov V, Shpilko A (2010) Toxic defense. Magazine "Ogonyok" no 30 dated 02.08.2010, p 38
17. Hamdan LJ, Salerno JL, Reed A, Joye SB, Damour M (2018) The impact of the deepwater horizon blowout on historic shipwreck-associated sediment microbiomes in the northern Gulf of Mexico. Sci Rep 8:9057. https://doi.org/10.1038/s41598-018-27350-z
18. U.S. patent 1,137,373
19. U.S. patent 1,191,383
20. U.S. patent 4,915,925
21. U.S. patent 6,149,972
22. U.S. patent 4,530,949
23. U.S. patent 4,704,231
24. U.S. patent 4,946,892
25. U.S. patent 5,582,781
26. U.S. patent 4,091,083
27. U.S. patent 5,846,459
28. U.S. patent 3,404,061
29. U.S. patent 4,244,934
30. U.S. patent 4,888,242
31. U.S. patent 4,961,988
32. U.S. patent 5,149,518
33. U.S. patent 5,294,300
34. U.S. patent 5,582,811
35. U.S. patent 5,981,072
36. U.S. patent 6,143,218
37. U.S. patent 5,885,728
38. U.S. patent 6,060,189
39. U.S. patent 6,287,694
40. U.S. patent 5,330,680
41. U.S. patent 5,186,919
42. Drzal L, Fukushima H, Expanded graphite and products produced there from US patent # 20060148965A1. https://patents.google.com/patent/US20060148965A1/en
43. Sementsov YuI, Yatsyuk OP, Nikulin MO (2014) Adsorption properties of thermo-expanded graphite in relation to petroleum products. Technogenic Environ Safety Protect 7:129–136
44. Sementsov YuI, Revo SL, Ivanenko KO, Hamamda S (2019) Expanded graphite and its composites. Kyiv: "Akademperiodyka", 226 pp
45. Patent 2128624 (1999) Russia, MPK6 S 01, V 31/04, S 25 V 1/00, Byul no 2
46. Patent 2134657 (1999) Russia, MPK6 S 01, V 31/04, Byul no 23
47. A.S. 1761667 A (1992) SSSR, MPK5 S 01, V 31/04, Byul no 34
48. Kotov SG, Lupey AYu (2001) Sorbtsionnye materialy dlya likvidatsii avariynykh razlivov nefti i nefteproduktov. Nauchnoe obespechenie pozharnoy bezopasnosti 10:92–98
49. Patent 45084 A (Ukraine). S04V35/536, S01V31/04. Publ Byul no 3 (2002)
50. Yanchenko VV, Revo SL, Sementsov YuI, Piatkovskyi ML, Yatsiuk OP (2003) Termorozshyrenyi grafit – sorbtsiinyi materIal dlia zbirannia nafty ta naftoproduktiv z poverkhni vody ta zemli. Naukovyi visnyk UNDI pozhezhnoi bezpeky 1(7):139–144
51. Vlasenko EV, Godunov IA, Lanin SN, Nikitin YuS, Khokhlova TD, Shoniya NK (2005) Cravnitelnyy analiz strukturnykh i sorbtsionnykh kharakteristik termorasshirennykh grafitov i aktivnykh ugley v ochistke vodyot organicheskikh veshchestv. Vestn Mosk universiteta Ser 2. Khimiya 46(4):10–18

52. Nikulin OF, Nedbaev NY, Titenko OM (2015) The method of obtaining thermally extended graphite: 102725 Pat. on the utility model; CO1B 31/04 (2006.1), HO5B 6/00; declared 24/06/2015 (u201506313)
53. Nikulin AF, Titenko AN (2015) Mathematical model of particle heating during the formation of thermally expanded graphite under conditions of thermal irradiation. Tech Mech 2:118–127
54. Kirchu FI, Nikulin OF, Kodrik AI, Titenko OM, Moroz OI (2019) Application of ANSYS CFX software for simulation of gas flow movement in the reaction chamber at preparation of sorbent on the basis of thermally splinted graphite. Sci Bull Civil Protect Fire Safe 2(8):74–80
55. Akhayere E, Vaseashta A, Kavaz D (2020) Novel magnetic nano silica synthesis using barley husk waste for removing petroleum from polluted water for environmental sustainability. Sustainability 12:10646. https://doi.org/10.3390/su122410646

Anatoly Kodryk graduated in 1972 from the Kramatorsk Industrial Institute with the title of mechanical engineer in the field of mechanical engineering. From 1975 to 1978 he studied in graduate school at the Institute of Superhard Materials of the National Academy of Sciences of Ukraine, in 1982 he defended his Ph.D. thesis in the field of the theory of cutting materials. He worked in scientific and managerial positions at the Institute of Superhard Materials of the National Academy of Sciences of Ukraine, the Research Institute of Electro-Mechanical Devices, the Kiev association "MAYAK", the Ukrainian Research Institute of Oil and Oil Refining "MASMA", the Ukrainian Research Institute of Civil Protection DSNS Ukraine. Currently, he is a leading researcher at the Institute of Public Administration and Scientific Research in the Field of Civil Protection of the State Social Security Service of Ukraine. Research interests include the development of technologies and methods for biological and ecological protection of the environment, technologies for the prevention and elimination of the consequences of technogenic emergencies. Author of near 100 scientific articles.

Alexander Nikulin was born in 1955 in Melitopol, Zaporizhzhya region. In 1989 he defended Ph.D. thesis on the topic Theoretical and experimental justification of the flat process superfinishing, Kiev Polytechnic Institute, Kiev. In 2010 he defended Doctoral thesis on the theme Methods for building information technology for change engineering environments based on evolutionary principles, Lviv National Press Academy, Lviv. He is Academician of the Academy of Energy of Ukraine, Chief Researcher, Fire Protection Research Center, Institute of Public Administration and Research on Civil Protection of the State Emergency Service of Ukraine. Scientific awards: Gold Medal named after V.G. Shukhov 2008, Russian Academy of Engineering, Moscow, for exceptional achievements in the field of energy saving and energy efficiency; Gold Medal named after V.P. Glushkov 2009, International Academy of Computer Systems, Kiev, for real successes in the field of non-traditional energy. He have developed innovative technology for decontamination of hazardous biological wastes due to oxidation in supercritical water medium and design of a supercritical reactor for decontamination of hazardous biological wastes and innovative technology for the production of a highly efficient sorbent based on thermally dispersed graphite for the elimination of emergency spills of oil and oil products, and the design of stationary and mobile plants for the production of thermally separated graphite in the city for the elimination of emergency spills of oil.

Alexander Titenko was born in Kyiv, Ukraine in 1954. He received a diploma in mechanical engineering from the National Technical University of Ukraine "Igor Sikorsky Kyiv Polytechnic Institute" in 1977 and the Ph.D. degree in engineering science from State Institution "Institute of Environmental Geochemistry of National Academy of Science of Ukraine" in 2016. From 1995 to 2013, he was a Chief Designer of an Industrial Enterprise. From 2013 to 2021, he was a Research Assistant with Institute of Public Administration and Research in Civil Protection. He is the author of more than 40 scientific articles and holds five patents. His research interests include the theory of foam formation and foam flow, the creation of mathematical models in the design of fire extinguishing devices based on CAFS technology and new fire-resistant structural materials.

Fedor Kirchu works as Associate Professor, Department of Aircraft Engines, National Aviation University, Kiev. He is the author and co-author of more than 30 research articles. He is a specialist on engines and power plants, in particular in the numerical investigation of hub losses of pushing air propellers, distribution of dispersed particles in a two-phase gas-laser flow, energy support for the permanent mission to the planets of Solar system.

Yurii Sementsov was born in Budapest, Hungary, in 1951. He received the B.S. and M.S. degrees in general physics, X-ray metal physics from the T.G. Shevchenko Kyiv State University, Kyiv, in 1973 and the Ph.D. degree in Physical and Mathematical Sciences (Solid State Physics) from Ivan Franko State University, Lviv, in 1985, and Dr. Sc. degree in Physical and Mathematical Sciences (Physics and Chemistry of Surfaces) from Chuiko Institute of Surface Chemistry of National Academy of Sciences of Ukraine, Kyiv, in 2020. From 1973 to 1987, he was a Research with the Physical Faculty of the T.G. Shevchenko Kyiv State University, Kyiv, Ukraine. Since 1987, he has been a Senior Researcher of Department of Physical Chemistry of NanoCarbon Materials, Chuiko Institute of Surface Chemistry, National Academy of Sciences of Ukraine, Kyiv. Since 2000 Science consultant (part-time)

of "TMSpetsmash" ltd., Kyiv, Ukraine. He is the author of five books, more than 120 articles, and more than 40 patents. His research interests include Physical Materials Science and Carbon Materials Science, in particular physics, physical chemistry and engineering of carbonaceous nanostructural materials, their technical, medical and ecological applications. Since 2019 an academic of all-Ukrainian public organization "Academy of Technological Sciences of Ukraine", specialized in high technologies.

Kateryna Ivanenko was born in Rzhyshchiv town, Kyiv oblast (province), Ukraine, in 1968. She received the B.S. and M.S. degrees in physics from the T.G. Shevchenko Kyiv State University, Kyiv, in 1991 and the Ph.D. degree in Candidate of Physical and Mathematical Sciences (Solid State Physics) from Taras Shevchenko National University of Kyiv, Kyiv, Ukraine, in 2004. From 1986 to 2019, she was an Engineer and Senior Researcher with the Physical Faculty of the Taras Shevchenko National University of Kyiv (T.G. Shevchenko Kyiv State University), Kyiv, Ukraine. Since 2020, she has been a Senior Researcher with the Institute of Macromolecular Chemistry of National Academy of Sciences of Ukraine. She is the author of three books, more than 70 articles, and 2 patents. Her research interests include Metal Physics and Ceramics, Polymer Composites, in particular physics, applying carbon nanomaterials in metal physics, ceramics and polymer composites and their technical, medical and ecological applications.

Yuliia Grebel'na was born in Snovsk, Chernihiv region, Ukraine in 1979. She received the M.S. degree in Technical Electrochemistry from the National Technical University "Kyiv Polytechnic Institute", Ukraine. Since 2010 she has been working as the technical director of the plant for the production of sealing materials from thermally-expanded graphite "TMSpetsmash" ltd., Kyiv, Ukraine. Since 2018, she is a Ph.D. student at the. She is the author of near 10 articles, and 2 patents. Her research interests include the synthesis and study of carbon-carbon nanocomposite materials on the basis of thermally-expanded graphite and carbon nanotubes.

Alex Pokropivny was born in Bila Tserkva, Kyiv region, Ukraine, in 1972. He received the M.S. degree in applied physics and mathematics from the Moscow Institute for Physics and Technology (MIPT) in 1996 and the Ph.D. degree in solid state physics from Institute for Problems of Materials Science NAS of Ukraine, in 2003. From 2000 up to now he works with some interruptions as senior researcher in Institute for Problems of Materials Science NAS of Ukraine. In 2001 he was a recipient of the Award of Young Scientists titled as Investigation of contact interactions in a scanning probe microscope. In 2006–2007 he received Max-Planck postdoctoral fellowship with MPI of Polymer Research, Germany. In 2011–2013 he held postdoc position in Laboratory of Energetics and Combustion (E.M2.C, CNRS), France. He is the author and co-author of two books and more than 100 research articles. His research interests include molecular dynamics, quantum-chemistry hybrid DFT/MD simulations of tip tops for atomic force microscopy, different types of polymerized structures, namely graphenes, carbon chains, nanotubes, fullerenes, zeolites and other carbon and inorganic phases and their experimental synthesis with supercritical fluid and CVD methods.

Ashok Vaseashta (M'79-SM'90) received Ph.D. in Materials Science and Engineering (minor in Electrical Engineering) from Virginia Polytechnic Institute and State University, Blacksburg, Virginia, USA. He is Executive Director of Research for International Clean Water Institute in VA, USA, Chaired Professor of Nanotechnology at the Academy of Sciences of Moldova and Professor, Nanotechnology and Biomedical Engineering at the Faculty of Mechanical Engineering, Transport and Aeronautics at the Riga Technical University. Prior to his current position, he served as Vice Provost for Research at the Molecular Science Research Center in Orangeburg, South Carolina. He served as visiting professor at the 3 Nano-SAE Research Centre, University of Bucharest, Romania and visiting scientist at the Helen and Martin Kimmel Center of Nanoscale Science at the Weizmann Institute of Science, Israel. He served the U.S. Department of State in two rotations, as strategic S&T advisor and U.S. diplomat. His research interests span nanotechnology, environmental/ecological science, and safety and security. His research on nanotechnology has been on improving the understanding, design, and performance of nanofibers and sensors/detectors, mainly for applications such as wearable electronics, target drug delivery, detection of biomarkers and toxicity of nano and xenobiotic materials. In the security arena, he has worked on counterterrorism, countering unconventional warfare and hybrid threats, critical-Infrastructure protection, biosecurity, dual-use research concerns, and mitigating hybrid threats. He has authored over 250 research publications, edited/authored eight books on nanotechnology, and presented many keynotes and invited lectures worldwide. He serves on the editorial board of several highly reputed international journals. He is an active member of several national and international professional organizations. He is a fellow of the American Physical Society, Institute of Nanotechnology, and the New York Academy of Sciences. He has earned several other fellowships and awards for his meritorious service including 2004/2005 Distinguished Artist and Scholar award.

Sustainable Approach to Water-Energy Nexus: Sea-Water Reverse Osmosis Desalination and its Future Directions

Anirban Roy and Asim K. Ghosh

Abstract It is well established that to sustain the growth in human population commensurate growth in potable water resources is a necessity. Thus, to produce "extra" water in potable form, technologies like sea-water or brackish water desalination is resorted which include thermally driven technologies like multi-stage flash (MSF) or multi effect distillation (MED) or pressure driven like reverse osmosis (RO). It is well established that operation of "State of Art" RO plants are energetically closest (3–6 kWh/m^3) to thermodynamic limits (1.56 kWh/m^3 for 50% recovery of 35,000 ppm feed) of desalination whereas MSF or MED are more energy intensive (20 kWh/m^3 and higher). Hence, RO has gained widespread popularity over the last few decades. However, as fresh water is extracted from seawater, what remains is concentrated brine, which has almost twice the salt concentration of sea-water and is(increasingly becoming) an environmental concern. Therefore, Zero Liquid Discharge (ZLD) desalination processes for brine or hypersaline streams management from land locked desalination is an absolute necessity. As RO itself is a mature technology, there is little scope in improvement of the same as far as membrane configuration and chemistry is concerned. Hence, in this book chapter, the authors discuss sustainability of desalination from two perspectives. One is to explore the potential to integrate solar/wind energies to desalinate water through RO. This would lead future efforts into solar and wind related developments as well as energy storage devices influencing RO. Then the chapter delves into handling the RO reject. The reject brine of RO (or any hypersaline stream) can be subjected to three approaches: (i) Water for Energy (WFE), (ii) Energy for Water (EFW) and (iii) Brine to Chemicals (BTC). The selection of WFE and EFW depends on the degree of salinity (DOS). If the DOS is large, indicating high osmotic pressures, then it is advisable to recover the osmotic energy of the stream to generate power (WFE) through technologies like

A. Roy
Department of Chemical Engineering Goa, BITS Pilani, Goa 403726, India
e-mail: anirbanr@goa.bits-pilani.ac.in

A. K. Ghosh (✉)
Desalination & Membrane Technology Division, Bhabha Atomic Research Centre, Trombay, Mumbai 400085, India
e-mail: akghosh@barc.gov.in

Pressure Retarded Osmosis or Reverse Electrodialysis. However, EFW approach can also be undertaken utilizing brine concentrator (BC) to generate water and recover salts from brine. Zero Liquid Discharge approach utilizing BC and aided by membrane-based forward osmosis (FO)/electrodialysis reversal (EDR)/membrane distillation (MD) to achieve 10% concentration of brine followed by BC (to attain 22% concentration) is also discussed. To achieve ZLD, Forced Circulation Crystallizer (FCC) is utilized downstream of BC. This process will recover salts as valuable byproduct and thereby improve the overall techno-economics. Challenges involved in design and development of ZLD desalination processes for wider scale application of these technologies in desalination and other industries is discussed. A new avenue worth exploring is BTC, which involves integration of Chlor-Alkali processes to recover Chlorine and Hydrogen from brine which are extremely important to the chemical industry to produce water treatment chemicals (using Chlorine) as well as for producing urea (using hydrogen through Haber process) for agriculture. Thus, the book chapter focuses not only on making RO sustainable but also provides futuristic avenues to make the Water-Energy and Water-Energy-Food Nexus more sustainable.

Keywords Desalination · Reverse osmosis · Water-energy nexus · Brine handling · Sustainability

1 Introduction

Sustainability is quite often, a misunderstood term. The fact that the success or failure of a particular technology is not dependent only on the novelty of the technology, but also on the associated paraphernalia. In order to develop a technology that is sustainable in the "true sense" of the term, it should satisfy the constraints of technical feasibility, economic affordability, emotional acceptability as well as guarantee future security. This has to be the case for the technology as well as the product that it is delivering. Often technologists and economists' debate in two different planes, where technological achievement and financial feasibility seldom coincide. This gap has been bridging very rapidly in recent years, as policy makers have been realizing the need to address the challenges related to limited resources being available to an ever growing population.

In 2005, the Late Nobel Laureate (Professor) Richard Smalley came up with the list of 10 challenges [45] which pose a serious threat to humanity for the upcoming decades. That list is widely cited and interesting to observe, energy and water capture the top two ranks, respectively. Also, the rest of the challenges are related in some way or the other to the energy and water constraints. The world recently witnessed the Syrian crisis, which stemmed from the "low standard of living" faced by the population [43]. The root of the problem lies in the access to clean drinking water, transportation and lowering of prices (can be related to energy) [43]. The visionary Prof. Smalley was not wrong after all! The approach, which has been popularized over the years, therefore, is "Water—Energy Nexus" (WEN) [12]. This signifies a holistic perspective towards developing any technological solutions related to either Energy or Water. WEN can be easily explained on the basis of the fact that for energy generation there is always a water footprint and for producing clean water, there is a need for energy. Thus, it is indeed myopic to develop a water solution without considering energy impact and vice versa [42]. Over the years, the concept has been encouraged, advocated and researched by various groups worldwide [8, 20, 23, 25, 35]. Although tracing the origin of this approach is debatable, however, a report by Peter Gleick in 1994 can be considered as a reference point, where the interdependency of "Water and Energy" has been discussed very comprehensively [15]. In fact, the report begins with the line that "Energy and freshwater resources are intricately connected" [15]. This concept has gained tremendous visibility and acceptance (hence sustainability of the approach is unquestionable!) and therefore "Water—Energy—Food Nexus" [26], "Water—Energy—Food—Health Nexus" [36] have also been coined, researched and techno-socio-economic viability is being evaluated for policy making [11].

This chapter focuses on WEN and the authors believe that water security (which is also related to food security, as 70% of the underground freshwater is used for agriculture [47]) can be addressed in a more "sustainable" manner if we start looking ahead and start addressing the challenges posed by current technological solutions. In this regard, desalination, i.e. removing salts from water, to yield water of a more acceptable quality, is the only way to support the increasing food, energy and water stress experienced worldwide [6]. The challenge in desalination lies in recovery of a generally more acceptable quality of water with a downside. To satisfy overall mass balance, the constituents of the feed water (various salts) will be found in the "reject" or non-usable stream, but as water has been extracted, this concentration will be higher than the feed itself. This downside may be neglected at smaller scales, but when desalination is considered for a city or an entire nation, then the million gallons per day of reject stream poses an environmental challenge along with multiple other risks [41]. If the feed is seawater, then typical desalination plants produce reject streams with almost twice the salinity, which is known as "brine". Brine handling (BH) poses great challenges but presents multiple opportunities too, if understood properly. Thus, this chapter is based on a viewpoint with a futuristic perspective. This chapter first explains the fundamentals related to desalination and poses the problem of BH. Then the authors provide an analysis on the possible ways to handle the brine through approaches like (i) Water for Energy (WFE), (ii) Energy for Water (EFW) and (iii) Brine to Chemicals (BTC). The fundamental aspect in this chapter is the focus on BH as if this issue is addressed and the full potential of brine is utilized, desalination indeed is well poised to serve water in a sustainable way eon into the future.

2 Fundamentals of Water Energy Nexus

Thermodynamically, mixing is a spontaneous process. For a layman, it is easy to understand through the example of mixing of gases. Everyday events are related to these mixing phenomena, and knowingly or unknowingly we utilize this to our advantage. Cooking, washing, assimilation of nutrients in the body, breathing, chemical manufacturing, all these commonplace events involve mixing of multiple components and at times in multiple phases. The advantage of mixing is that it is favored by nature, i.e., it is spontaneous, which can happen on its own, given enough time. The challenge is "demixing", i.e., separation of these mixed components. Demixing does not occur spontaneously, in fact, the Second law of Thermodynamics prohibits it [21]. Thus, if demixing of two gases is to be carried out or if one needs to precipitate out excess sugar from a coffee cup, it requires energy. The age old and beautiful story of Moses parting the red sea too required divine intervention in form of strong east winds! [16]. Thus, separation requires energy, and this is the beginning of "Water—Energy Nexus" understanding.

Thermodynamically, separation requires a form of energy transfer and in nature the freely available form is heat. Men, through technological innovations, have managed to devise a more efficient form of energy transfer and that is work. However, work is ultimately derived from heat, and when this exercise is performed in cycle (like in an engine), it leaves a portion of heat unutilized [4]. This gives rise to efficiency of conversion between heat and work and for a heat engine; the famous Carnot efficiency dictates the achievable limit of such efficiencies [40]. This is the inherent concept in desalination.

The evolution of planet earth and formation of water started taking place billions of years ago [39]. Oceans started forming and the water bodies, as we witness today, are the result of billions of years of contact of water with geological formations. Essentially, it can easily be assumed that the ocean and sea are in chemical equilibrium with the mineral formations and hence the salinity which is encountered is uniform throughout, between 35,000 ppm to 38,000 ppm [17]. The challenge in desalination lies in recovering water from this feed to quench the thirst of a developing society as well as for other purposes required for sustainability. Taking a cue from the previous discussion, this requires separation of salt and water and is "non-spontaneous" and requires energy transfer. This energy transfer can be in form of heat and work only [30]. Thus, it is easy to understand that desalination can be carried out either using thermal route (multi stage flash, multi effect desalination) or using (pump) work (like in reverse osmosis). Invariably one question arises, and that is which is preferable? Needless to mention (and stemming from previous discussions), heat is a lower grade of energy transfer than work. Again, for a layman, anything that is random and directionless (heat) will be possessing less potential to carry out a specific task, than with respect to a form of energy which has direction (work). In this regard, it is important to understand the specifics of energy requirements required for a given desalination operation.

Sustainable Approach to Water-Energy ...

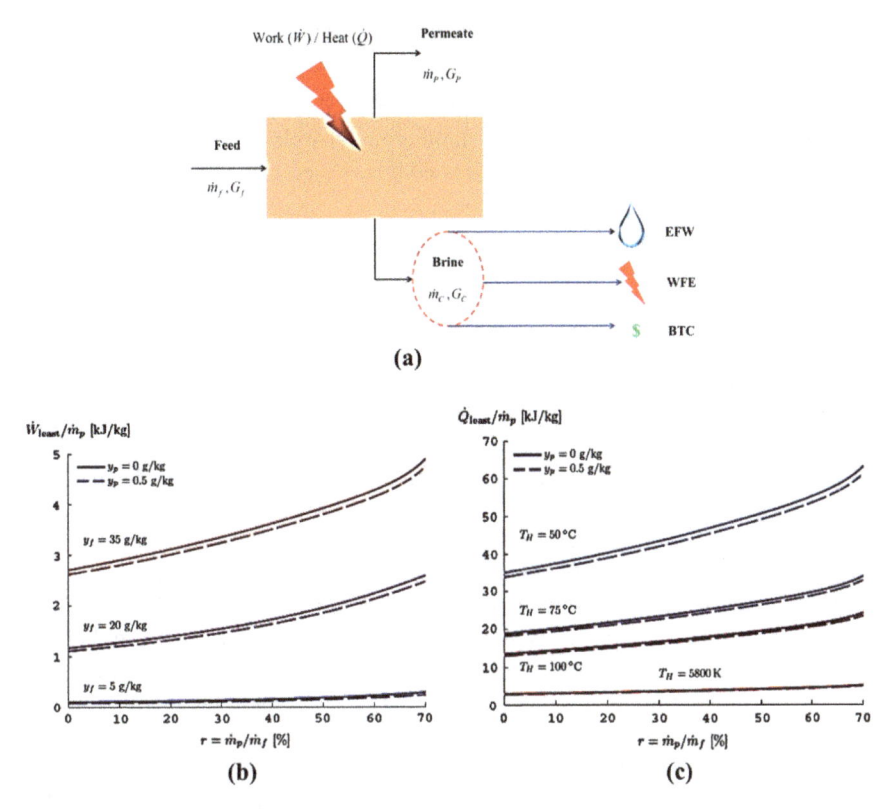

Fig. 1 a General desalination process carried out with either heat or work, b least work of separation (adapted from [29]) at 25 °C, c least heat of separation (adapted from [29]) at 25 °C (with feed being 35 g/kg ~ 35,000 ppm)

As depicted in Fig. 1a, the general desalination process can be represented as a separating box which facilitates the feed being split into two streams, the permeate (little or no salinity) and a brine (more saline than feed due to stream concentration). Each of the stream, have their own Gibbs energies and hence the minimum work of separation can be denoted as [29]:

$$\frac{\dot{W}_{least}}{\dot{m}_p} = (G_P - G_C) - \frac{1}{r}(G_f - G_C) \quad (1)$$

$$r = \frac{\dot{m}_P}{\dot{m}_f} \quad (2)$$

where, $\dot{W}_{least}$ is the least rate of work of separation (kW), G_f is the Gibbs free energy (kJ/kg) of feed, G_P for Gibbs free energy (kJ/kg) of permeate and G_C is the Gibbs free energy (kJ/kg) of the brine, r is the recovery of the permeate, $\dot{m}_f$ is the mass flow rate (kg/s) of feed, $\dot{m}_P$ is the mass flow rate (kg/s) of permeate and $\dot{m}_c$ is the mass flow (kg/s) rate of brine.

Figure 1b depicts the least work of separation obtained from the approach and is agnostic to the process and represents the thermodynamic limits. The desalination can be carried out using thermal energies and it is related to work as [29].

$$\dot{W}_{least} = \left(1 - \frac{T_0}{T}\right) \dot{Q}_{least} \qquad (3)$$

where, $\dot{Q}_{least}$ is the least rate of heat (kW) of separation, T is the temperature (K) and T_0 is the ambient temperature (25 °C).

These equations show that as the T tends to get higher (say the temperature of sun), the $\left(1 - \frac{T_0}{T}\right)$ tends to 1 and hence the work rate tends to equal to the heat rate [29]. Keeping these calculations in perspective, it is interesting to understand the current energetics for desalination and how those compare with these limits.

The energy required for desalination is expressed in terms of kWh/m^3, i.e., the energy required in kWh for producing 1 m^3 of desalinated water. The thermodynamic approach discussed above yields a value of minimum work to be 1.06 kWh/m^3 (for 50% recovery of from a seawater feed of 35,000 mg/L) [14]. Figure 2 shows the comparison of various technologies [28] employed for desalination against the thermodynamic minimum. It is seen that while RO is closest to the thermodynamic limit, the thermal technologies require twice equivalent work as that of RO. This is due to the involvement of phase change and factors discussed above. Thermal route inherently requires evaporation and there are two issues due to which the energy penalty is higher. First is the latent heat of vaporization which is multiple times greater than the Gibbs energy of mixing. Second, due to boiling point elevation (more salt presence results in higher boiling point), the energy requirement increases as more water is recovered. In fact, in one of the recent works, a probable reason for the salinity levels (35,000–38,000 ppm) as encountered in sea and oceans has

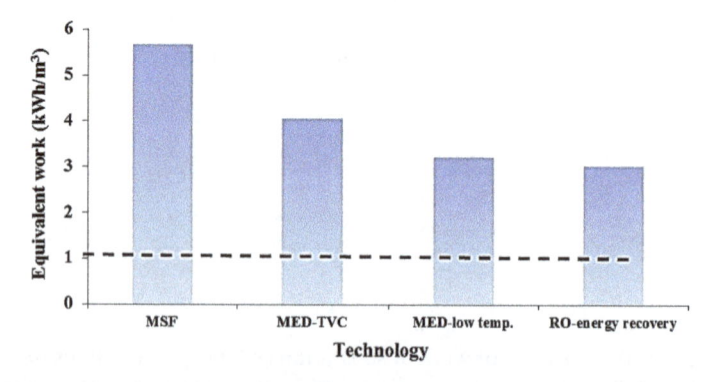

Fig. 2 Equivalent work comparison between various technologies [28] and the thermodynamic limit of desalination for 50% recovery (in dotted line)

been explained [37]. In fact, it has to be appreciated that any higher salinity would actually challenge the RO and thermal desalination energetics very abruptly. This chapter explores the avenues of brine handling (having higher salinities than the feed water) and how it can be viewed as a potential source of energy and value added chemicals for sustaining desalination.

Thus, production of water through energy transfer is termed in this chapter as "Energy for Water" (EFW). This is an easy concept to understand. However, realizing the opposite is difficult and will be attempted by the authors. The fact that while energy is required for demixing or separation, it is easy to realize that the converse is true, that energy is released during mixing process. Essentially all our regular "observable" mixing processes are constant pressure and temperature (depending on the initial and final states), simplest one being mixing a spoonful of table salt in water. The heat of mixing is negligible while salt mixes with water, hence it is not easy to imagine extracting work! Human tendency is always to extract work from heat and hence this energy of mixing has gone unnoticed for many years, till Loeb discussed harvesting salinity gradient power. The phenomenon of two different salinity streams mixing together yields tremendous amount of free energy (Gibbs free energy), which can be harvested. These include simple phenomena like rivers mixing with oceans and sea! This is what is dealt in the chapter as "Water for Energy" (WFE). The last aspect is utilizing the brine generated from EFW approach or even some from WFE and generating value added chemicals, which is being discussed by "Brine to Chemicals" (BTC). Thus, the chapter deals with brine handling (EFW) towards extracting maximum water through technologies like Zero Liquid Discharge or utilize brine by mixing with other sources to get energy (WFE) and lastly, utilize brine to get value added chemicals (BTC).

3 Energy for Water (EFW) for Brine

Brine handling has been traditionally carried out through ocean outfall, sewer discharge, evaporation ponds or deep well injection [44]. However, for handling brine at high discharge rates (million gallons per day capacity), it is important to resort to simpler and scalable technologies. The approach for EFW is that energy penalty being the cost, maximum water recovery is the primary objective. This approach has been popularized recently through the Zero Liquid Discharge (ZLD) strategy. The advent of ZLD can be traced to 1970s in the USA when ZLD was adopted as a regulatory mandate by the power plants near Colorado river [48]. ZLD is a very ambitious approach where the energy penalties increase as the recovery increases, however, due to current situation, the global investment on ZLD processes is projected at around 100–200 million USD [48]. Initially, ZLD used to be thermal processes with brine crystallizers, but due to its size and other constraints, requires huge capital and costs. Later reverse osmosis (RO) started to be integrated into the process, but the limitation of RO lies in its limited ability to handle salinity beyond a threshold. Thus, it has become imperative to look for better technologies which can be modular, less expensive and can be operated with lesser energy penalties. USA, China and India are three nations who are focused on ZLD implementations for the various sectors catering to manufacturing [48]. It will be interesting to understand the various technologies (thermal or membrane based) which has promise in achieving ZLD.

3.1 Mechanical Vapor Compression (MVC):

MVC uses mechanical energy to desalinate water and it consists of an evaporator-condenser and a compressor. In MVC process, the entire efficiency of the process hinges on heat transfer occurring in falling film inside tube bundles. The brine feed falls inside the heat transfer tubes where evaporation occurs and water vapor thus formed is taken up by a vapor compressor (Fig. 3a) [34]. The vapor compressor delivers the superheated vapor to the outside of the tube bundles, and thus, the latent heat of condensation is transferred to evaporate the brine feed inside. The condensate is the distillate water recovered from the process and the heat rejected is used to preheat the incoming brine to the tube bundle. MVC has been used very extensively particularly for ZLD approach and sets the "gold standards" for energy comparisons. MVC can achieve 98% recovery of water, with 250,000 mg/L of salinity with an

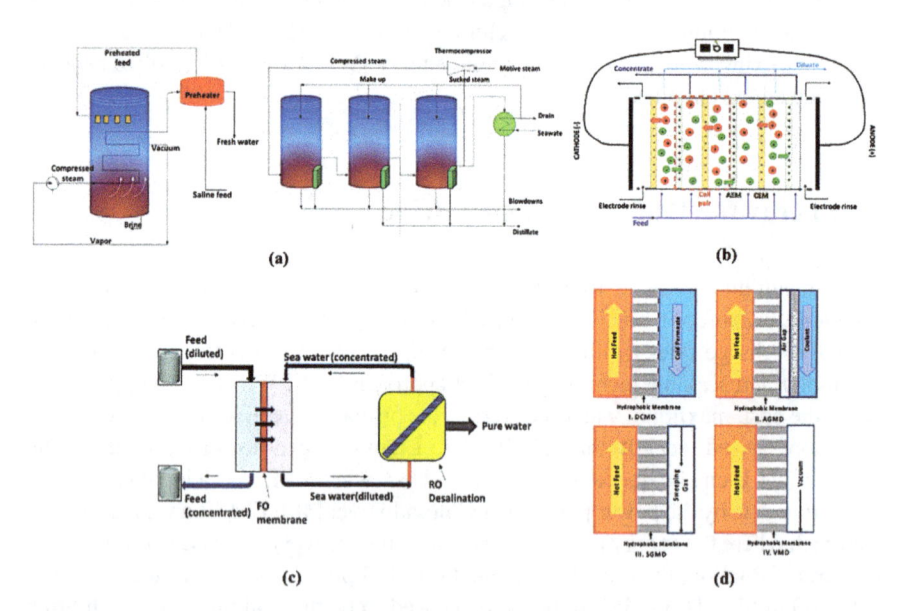

Fig. 3 EFW technologies **a** MVC (adapted from [34] with permission, **b** ED / EDR (adapted from [9] with permission), **c** FO (adapted from [18] with permission) and **d** MD (adapted from [2] with permission)

energy footprint of 40 kWh/m³ of product water [48]. Generally; MVC is coupled with any of the thermal desalination technologies.

3.2 Electrodialysis (ED)/Electrodialysis Reversal (EDR)

The driving force for ED/EDR is electrical voltage applied across ion selective membranes. The membranes allow passage of counterions but prevent transport of co-ions. The cations move towards negatively charged cathode and anions move in the opposite direction. The cation exchange and anion exchange membranes are employed to facilitate this transport (Fig. 3b) [9]. This conventional technology is known as ED. EDR is a modified form of ED, where the polarity of the electrodes is reversed frequently to minimize fouling. ED and EDR can concentrate feed to around 100,000 mg/L with an energy penalty of 7–15 kWh/m³ [48].

3.3 Forward Osmosis (FO)

FO has gained a lot of popularity in the last decade with implementation of pilot plants. This technology does not use hydraulic pressure (pump work) against osmotic pressure, rather, it uses natural osmosis to draw water from a feed water to a higher concentration draw solution (Fig. 3c) [18]. The concentrated feed is sent to the evaporation pond or crystallizer. FO can concentrate typical RO rejects (60,000 mg/L TDS) to around 220,000 mg/L, consuming ~21 kWh/m³ of feedwater [48].

3.4 Membrane Distillation (MD)

The driving force in this process is thermal energy which induces partial vapor pressure difference between the feed and permeate side of a hydrophobic microporous membrane (Fig. 3d) [2]. The feed side water is heated and the differences in temperature between heated feed water and relatively cold permeate side results in a vapor pressure difference which governs the transport.

There can be four modes of MD: (i) direct contact MD (DCMD), (ii) air gap MD (AGMD), (iii) sweeping gas MD (SGMD) and (iv) Vacuum membrane distillation (VMD)[31]. MD requires phase transition; hence thermodynamically the process inherently is more energy intensive than RO or ED / ERD, consuming 22–67 kWh/m³ of product water [48]. MD is relatively new and less matured than the previous mentioned technologies; however, its effectiveness has definitely been demonstrated in a lot of installations (bench scale as well as pilot). It has capability to concentrate RO brine to around 175,000 mg/L with > 98% water recovery [48].

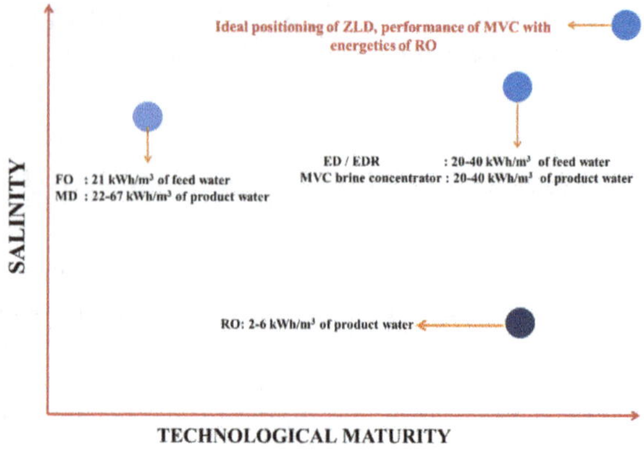
Ideal positioning of ZLD, performance of MVC with energetics of RO
FO : 21 kWh/m³ of feed water
MD : 22-67 kWh/m³ of product water
ED / EDR : 20-40 kWh/m³ of feed water
MVC brine concentrator : 20-40 kWh/m³ of product water
RO: 2-6 kWh/m³ of product water
SALINITY
TECHNOLOGICAL MATURITY

Fig. 4 Technological maturity versus salinity

A conceptual diagram can be constructed with technological maturity and ability to handle salinity levels as reference for comparison. This is depicted in Fig. 4. It can be understood that the ideal positioning of any ZLD must have the recovery of MVC coupled with energetics of RO. But that is thermodynamically improbable, but the technologies which are closer to such limits can be sustainable. RO is technologically mature but cannot handle higher salinities. MVC is matured too and can handle high salinities but comes with an energy penalty. ED/EDR and FO and MD are still to be implemented at scales of RO or MVC hence their maturity is not high, but their capabilities and promise to handle high salinity brine is between RO and MVC.

4 Water for Energy (WFE) for Brine

Reference [32] first described this non-conventional form of energy and was later endorsed and more explored by Loeb in 1976 [27]. There was a period of silence from the engineering and scientific community since then in this field, primarily due to the matured energy sector heavily dependent on the conventional sources [1]. However, renewed interest was shown from 1990s onwards [1]. As discussed previously, this process uses the concept of mixing to harvest Gibbs energies in form of shaft work. Interestingly, the WFE approach is yet to be scaled up and is still in its nascent stage of development. Statkraft commissioned and studied the first plant in 2009 and operated before decommissioning it due to unavailability of suitable membranes. There are various types of salinity gradient power technologies which are discussed below:

4.1 Reverse Electrodialysis (RED)

As the name suggests, the function is opposite to the ED process (EFW) (Fig. 5a) [19]. The ion exchange membranes are arranged in stack with alternate cation and anion exchange membranes. Two streams of differing salinities are fed in the stack in alternating channels and the concentration gradient acts as driving force for ion transport [24]. The resistance of membranes towards passage of ions determines the

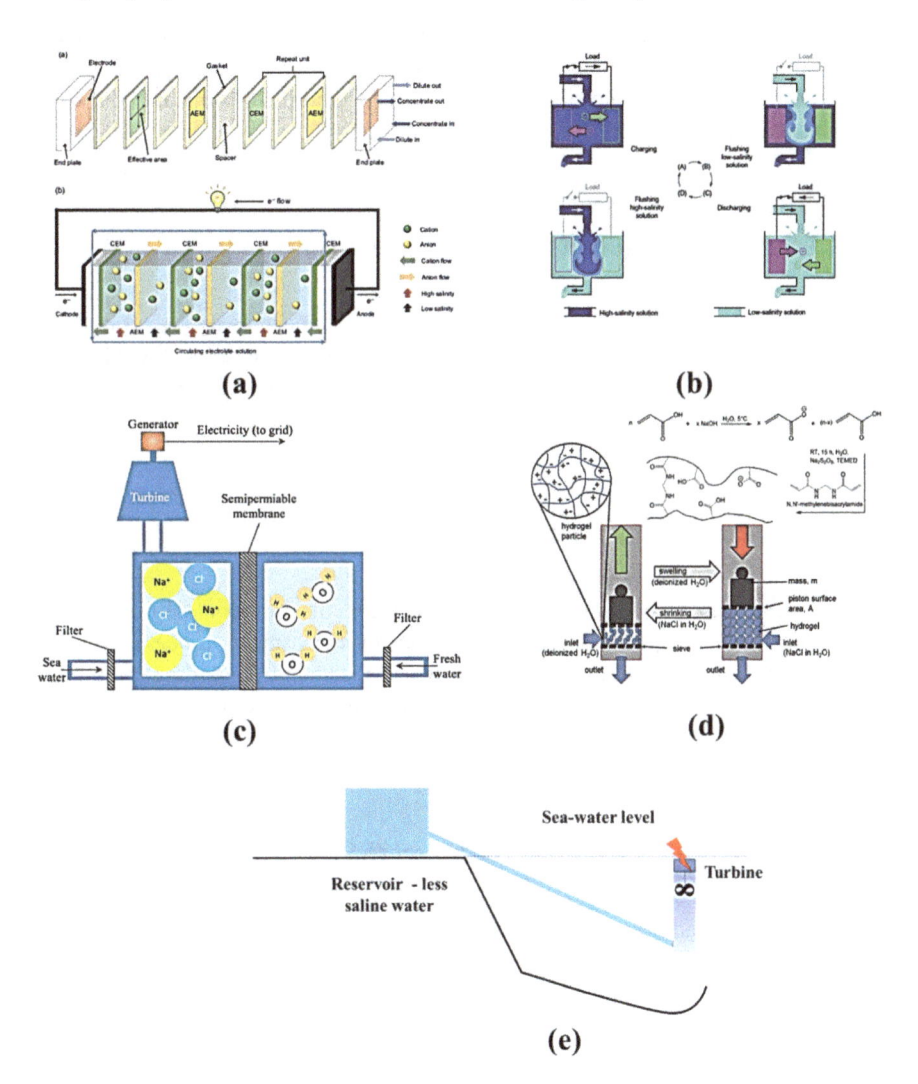

Fig. 5 **a** RED process (adapted from [19] with permission, **b** Accmix adapted from [22] with permission, **c** PRO (adapted from [46] with permission), **d** SSH (adapted from [7] with permission, **e** HG

rates and more selective the membranes are, more perfection is achieved in separation of ions and, hence generating current.

4.2 Microbial RED

Microbial RED is a hybrid process of placing a RED stack in a microbial fuel cell (MFC). A MFC is a bio-electro-chemical system resulting in electrical power production from organic wastes present in industrial or domestic wastewater. The hybrid combination results in reduced energy losses due to thermodynamically favorable reactions at both electrodes [49].

4.3 Accumulator Mediated Mixing (Accmix)

This process involves the ion exchange transport mechanism as described in RED, but with help of special electrodes and does not require membranes (Fig. 5b) [22]. The electrodes accumulate charge in presence of high salinity stream and discharge it while in contact with a low salinity stream.

4.4 Pressure Retarded Osmosis (PRO)

This process utilizes the mixing of two streams of different salinities using membranes and is opposite in function to the reverse osmosis process (Fig. 5c) [5]. The higher salinity stream is partially pressurized hereby retarding the mixing process and hence the terminology [38]. The pressurized draw side solution is then used to rotate a turbine.

4.5 Swelling and Shrinking Hydrogel (SSH)

SSH is the newest addition to the genre of salinity gradient power technology. The hydrogels absorb water from low salinity stream/freshwater and swell (Fig. 5d) [7]. This swelling action is utilized by weights placed on top of it being raised thus increasing the potential energy. The hydrogel shrinks once it is brought in contact with a high salinity stream and therefore reversing back to its original state.

4.6 Hydrocratic Generator (HG)

It is a form of direct mixing driven salinity gradient energy generation process [10]. Like PRO, this allows mixing of two different salinity streams but without membranes. HG consist of vertical tubes with lateral openings and submerged in sea water (Fig. 5e). The low saline stream is injected from the bottom and sea water mixes with the stream causing an upwelling phenomenon thereby generating power.

4.7 Reverse Vapor Compression

As the name suggests, it functions opposite to the vapor compression mechanism for desalination [10]. The two different salinity streams generate different vapor pressures when exposed to vacuum conditions, and the vapor in the low salinity chamber will be higher in pressure. This vapor flows to the chamber having low pressure and thus possessing the capability of rotating a turbine in the process.

Regarding these above technologies, it can be said that none have actually matured enough to be implemented as EFW alternatives. These technologies will take more time to mature and then can be used either as standalone or hybridized with other processes. The most promising in terms of developmental stage is PRO with recent projects of SEAHERO (South Korea), Megaton project (Japan) and MEDINA (Europe), PRO and advanced process intensification strategies are being employed to have sustainable water and energy [33].

5 Brine to Chemicals (BTC)

The brine handling process is perhaps best utilized in the form of extraction chemicals from the stream. Undoubtedly, chlorine and caustic soda are two most popular chemicals finding applications in various industrial applications. Chlorine is used in manufacture of vinyl chloride, methyl chloride, ethyl chloride as well as applications in paper and pulp industry, water treatment sector and various other sectors too. For this purpose, the Chlor-Alkali process is an extremely important process producing both these along with evolution of hydrogen. There have been very few reports on the integration of Chlor-Alkali process being utilized for brine stream handling [3, 31], however, very few reports are there which comprehensively models the process. Recently, researchers from MIT [13] have proposed an integrated strategy to produce NaOH from desalinated brine for consumption during the process itself and generate Chlorine and Hydrogen. Such strategies can be the "Holy Grail" for WEN issues and future direction of technology development and propagation (Fig. 6).

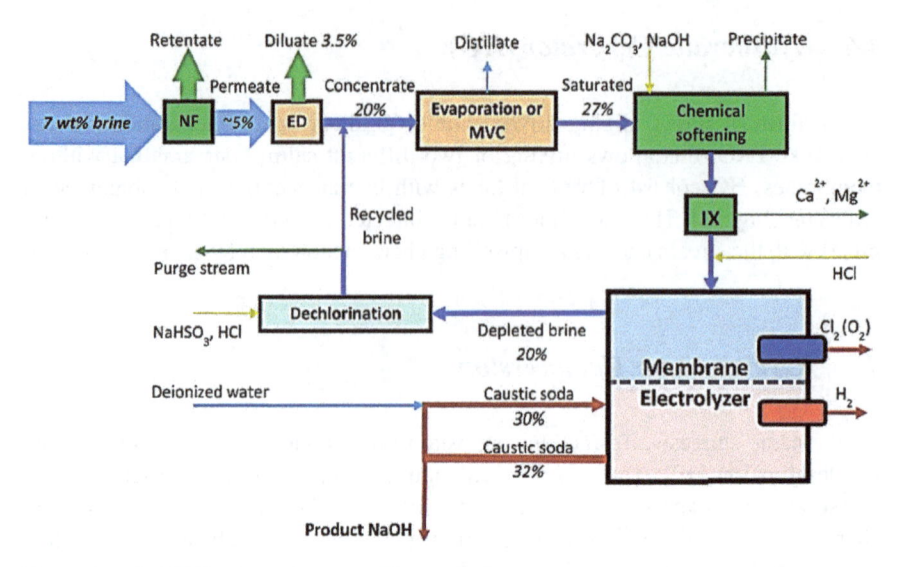

Fig. 6 Integrated Chlor-Alkali process for NaOH, Hydrogen and Chlorine production from desalination brine (adapted from [45])

6 Conclusions and Future Trends

It is important to appreciate the fact that each of the technologies discussed above are individually capable of performing a certain objective of either producing water or producing energy or producing chemicals. It is important to appreciate that such individual treatment cannot be considered if a holistic perspective is to be adopted and sustainability is to be ensured. Sustainability has been defined by the authors in the beginning of the chapter and for satisfying all the four constraints, it is important to devise a process addressing WFE, EFW and BTC and extracting every ounce of resource from a stream with minimum energy and water footprint. The authors believe that hypersaline streams should not be looked at as waste streams, but as a resource from which both energy and value-added chemicals can be obtained. A conceptual process is presented in Fig. 7, where a hypersaline stream acts as a motive fluid driving an engine, which absorbs water in the dilution step and rejects pure water at the concentrator step and the Gibbs energy of mixing is harvested as power in the turbine. This functions like a heat engine with heat absorption and heat rejection steps and production of power. The Gibbs energy harvester can be used as standalone skids to treat, harvest water as well as generate osmotic power and the concentrated stream can be used to generate value added chemicals and can be the future solution of WEN related issues.

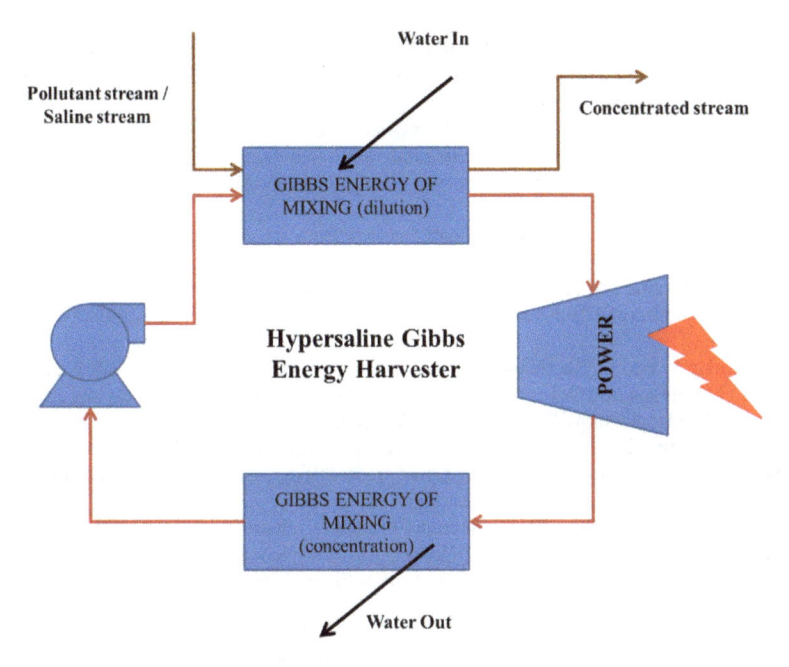

Fig. 7 Gibbs energy harvesters for producing water and power

Acknowledgements AR would like to acknowledge the Department of Science and Technology grant CRG/2018/001538 dt 28/01/2020, for supporting the work.

References

1. Achilli A, Childress AE (2010) Pressure retarded osmosis: from the vision of Sidney Loeb to the first prototype installation—review. Desalination. https://doi.org/10.1016/j.desal.2010.06.017
2. Ahmed FE, Lalia BS, Hashaikeh R, Hilal N (2020) Alternative heating techniques in membrane distillation: a review. Desalination. https://doi.org/10.1016/j.desal.2020.114713
3. Al-mutaz IS, Wagialla KM (1988) Techno-economic feasibility of extracting minerals from desalination brines. Desalination. https://doi.org/10.1016/0011-9164(88)80031-6
4. Allahverdyan AE, Hovhannisyan KV, Melkikh AV, Gevorkian SG (2013) Carnot cycle at finite power: attainability of maximal efficiency. Phys Rev Lett https://doi.org/10.1103/PhysRevLett.111.050601
5. Altaee A, Zaragoza G, Sharif A (2014) Pressure retarded osmosis for power generation and seawater desalination: performance analysis. Desalination. https://doi.org/10.1016/j.desal.2014.03.022
6. Amy G, Ghaffour N, Li Z, Francis L, Linares RV, Missimer T, Lattemann S (2017) Membrane-based seawater desalination: present and future prospects. Desalination. https://doi.org/10.1016/j.desal.2016.10.002
7. Arens L, Weißenfeld F, Klein, CO, Schlag K, Wilhelm M (2017) Osmotic engine: translating osmotic pressure into macroscopic mechanical force via poly(acrylic acid) based hydrogels. Adv Sci https://doi.org/10.1002/advs.201700112

8. Barik B, Ghosh S, Saheer Sahana A, Pathak A, Sekhar M (2017) Water-food-energy nexus with changing agricultural scenarios in India during recent decades. Hydrol Earth Syst Sci https://doi.org/10.5194/hess-21-3041-2017
9. Campione A, Cipollina A, Calise F, Tamburini A, Galluzzo M, Micale G (2020) Coupling electrodialysis desalination with photovoltaic and wind energy systems for energy storage: dynamic simulations and control strategy. Energy Convers Manag https://doi.org/10.1016/j.enconman.2020.112940
10. Cipollina A, Micale G (2016) Sustainable energy from salinity gradients. https://doi.org/10.1016/C2014-0-03709-4
11. Daher BT, Mohtar RH (2015) Water–energy–food (WEF) nexus tool 2.0: guiding integrative resource planning and decision-making. Water Int https://doi.org/10.1080/02508060.2015.1074148
12. Deshmukh A, Boo C, Karanikola V, Lin S, Straub AP, Tong T, Warsinger DM, Elimelech M (2018) Membrane distillation at the water-energy nexus: limits, opportunities, and challenges. Energy Environ Sci https://doi.org/10.1039/c8ee00291f
13. Du F, Warsinger DM, Urmi TI, Thiel GP, Kumar A, Lienhard JH (2018) Sodium hydroxide production from seawater desalination brine: process design and energy efficiency. Environ Sci Technol https://doi.org/10.1021/acs.est.8b01195
14. Elimelech M, Phillip WA (2011) The future of seawater desalination: energy, technology, and the environment. Science (80). https://doi.org/10.1126/science.1200488
15. Gleick PH (1994) Water and energy. Annu Rev Energy Environ https://doi.org/10.1146/annurev.eg.19.110194.001411
16. Harris MJ (2012) How did Moses part the Red Sea? science as salvation in the Exodus tradition. In: Moses in biblical and extra-biblical traditions.https://doi.org/10.1515/9783110901368.5
17. Helm KP, Bindoff NL, Church JA (2010) Changes in the global hydrological-cycle inferred from ocean salinity. Geophys Res Lett https://doi.org/10.1029/2010GL044222
18. Ibrahim GPS, Isloor AM, Yuliwati E (2018) A review: desalination by forward osmosis. In: Current trends and future developments on (Bio-) membranes: membrane desalination systems: the next generation. https://doi.org/10.1016/B978-0-12-813551-8.00008-5
19. Jang J, Kang Y, Han JH, Jang K, Kim CM, Kim IS (2020) Developments and future prospects of reverse electrodialysis for salinity gradient power generation: influence of ion exchange membranes and electrodes. Desalination. https://doi.org/10.1016/j.desal.2020.114540
20. Kahrl F, Roland-Holst D (2008) China's water-energy nexus. Water Policy. https://doi.org/10.2166/wp.2008.052
21. Kolesnikov I (2001) Thermodynamics of spontaneous and non-spontaneous processes. Choice Rev https://doi.org/10.5860/choice.39-2250
22. La Mantia F, Brogioli D, Pasta M (2016) Capacitive mixing and mixing entropy battery. In: Sustainable energy from salinity gradients https://doi.org/10.1016/B978-0-08-100312-1.00006-7
23. Lanjewar S, Mukherjee A, Khandewal P, Ghosh AK, Mullick A, Moulik S, Roy A (2020a) Thermodynamics of synthesis and separation performance of Interfacially polymerized "loose" reverse osmosis membrane: benchmarking for greywater treatment. Chem Eng J https://doi.org/10.1016/j.cej.2020.127929
24. Lanjewar S, Mukherjee A, Muzamil Rehman L, Roy A (2020b) Blue energy and its potential: the membrane based energy harvesting. In: Advances in membrane technologies https://doi.org/10.5772/intechopen.86953
25. Larsen MAD, Drews M (2019) Water use in electricity generation for water-energy nexus analyses: the European case. Sci Total Environ https://doi.org/10.1016/j.scitotenv.2018.10.045
26. Leck H, Conway D, Bradshaw M, Rees J (2015) Tracing the water-energy-food nexus: description, theory and practice. Geogr Compass https://doi.org/10.1111/gec3.12222
27. Loeb S (1975) Osmotic power plants. Science (80). https://doi.org/10.1126/science.189.4203.654
28. McGinnis RL, Elimelech M (2007) Energy requirements of ammonia-carbon dioxide forward osmosis desalination. Desalination. https://doi.org/10.1016/j.desal.2006.08.012

29. Mistry KH, Lienhard JH (2013) Generalized least energy of separation for desalination and other chemical separation processes. Entropy. https://doi.org/10.3390/e15062046
30. Mukherjee A, Lanjewar S, Kumar R, Chakraborty A, Abdelrasoul A, Roy A (2020) Role of thermodynamics and membrane separations in water-energy nexus. In: Modeling in membranes and membrane-based processes. https://doi.org/10.1002/9781119536260.ch4
31. O'Brien TF (1990) Dechlorination of brines for membrane cell operation. In: Modern chlor-alkali technology. https://doi.org/10.1007/978-94-009-1137-6_21
32. Pattle RE (1955) Properties, function and origin of the alveolar lining layer. Nature. https://doi.org/10.1038/1751125b0
33. Quist-Jensen CA, Macedonio F, Drioli E (2015) Membrane technology for water production in agriculture: desalination and wastewater reuse. Desalination. https://doi.org/10.1016/j.desal.2015.03.001
34. Rabiee H, Khalilpour KR, Betts JM, Tapper N (2018) Energy-water nexus: renewable-integrated hybridized desalination systems. In: Polygeneration with polystorage: for chemical and energy hubs. https://doi.org/10.1016/B978-0-12-813306-4.00013-6
35. Radcliffe JC (2018) The water energy nexus in Australia. Water-Energy Nexus. https://doi.org/10.1016/j.wen.2018.07.003
36. Rasul G, Sharma B (2016) The nexus approach to water–energy–food security: an option for adaptation to climate change. Clim Policy https://doi.org/10.1080/14693062.2015.1029865
37. Rehman LM, Dey R, Lai Z, Ghosh AK, Roy A (2020) Reliable and novel approach based on thermodynamic property estimation of low to high salinity aqueous sodium chloride solutions for water-energy nexus applications. Ind Eng Chem Res https://doi.org/10.1021/acs.iecr.0c02575
38. Rehman LMAMZLAR (2020) Membrane technology: transport models and application in desalination process. Model Membr Membr Process 327–373
39. Robert F (2001) The origin of water on earth. Science (80). https://doi.org/10.1126/science.1064051
40. Şahin B, Kodal A (1995) Steady-state thermodynamic analysis of a combined Carnot cycle with internal irreversibility. Energy. https://doi.org/10.1016/0360-5442(95)00076-S
41. Sauvet-Goichon B (2007) Ashkelon desalination plant—a successful challenge. Desalination. https://doi.org/10.1016/j.desal.2006.03.525
42. Scott CA, Pierce SA, Pasqualetti MJ, Jones AL, Montz BE, Hoover JH (2011) Policy and institutional dimensions of the water-energy nexus. Energy Policy. https://doi.org/10.1016/j.enpol.2011.08.013
43. Selby J (2019) Climate change and the Syrian civil war, part II: the Jazira's agrarian crisis. Geoforum https://doi.org/10.1016/j.geoforum.2018.06.010
44. Semblante GU, Lee JZ, Lee LY, Ong SL, Ng HY (2018) Brine pre-treatment technologies for zero liquid discharge systems. Desalination. https://doi.org/10.1016/j.desal.2018.04.006
45. Smalley RE (2005) Future global energy prosperity: the terawatt challenge. In: MRS bulletin. https://doi.org/10.1557/mrs2005.124
46. Tawalbeh M, Al-Othman A, Abdelwahab N, Alami AH, Olabi AG (2021) Recent developments in pressure retarded osmosis for desalination and power generation. Renew Sustain Energy Rev 138. https://doi.org/10.1016/j.rser.2020.110492
47. Todd Jarvis W (2013) Groundwater around the world: a geographic synopsis. Groundwater. https://doi.org/10.1111/gwat.12072
48. Tong T, Elimelech M (2016) The global rise of zero liquid discharge for wastewater management: drivers, technologies, and future directions. Environ Sci Technol https://doi.org/https://doi.org/10.1021/acs.est.6b01000
49. Zhu X, Hatzell MC, Cusick RD, Logan BE (2013) Microbial reverse-electrodialysis chemical-production cell for acid and alkali production. Electrochem Commun. https://doi.org/10.1016/j.elecom.2013.03.010

Anirban Roy received his BTech in Chemical Engineering from Heritage Institute of Technology, Kolkata in 2007 and Masters in Chemical Engineering from Jadavpur University, Kolkata in 2009. Thereafter he had a short stint at the University of Rhode Island and returned to India in 2011. He worked as an Engineer and Consultant for M.N. Dastur and Co. Pvt. Ltd. designing plants for Iron and Steel sector. Thereafter he pursued his PhD at Indian Institute of Technology Kharagpur from 2012-2016. His focus was on developing indigenous hemodialysis membranes and developing business and financial model for making it affordable. He worked on biocompatibility tests, developing extrusion technology, optimizing the same as well as develops polymer blends to yield the correct porosity and permeability. He developed and patented three technologies. After his experience, he joined BITS Pilani Goa as Assistant Professor in 2017 and has involved himself in research along the lines of "Water—Energy Nexus". He works on thermodynamics and membrane technology for water treatment and energy generation (through salinity gradient). He also works on processes including advanced oxidation and adsorption and integration of such technologies for sustainable water treatment. He has published 20 journal articles of international repute, filed 8 patents (4 US patents), co-authored 1 book and co-edited another 1 and has published 5 book chapters.

Asim K. Ghosh received M.Sc degeree in Chemistry from Indian Institute of Technology Kharagpur, India in 1994 and the Ph.D degree in Chemistry from Mumbai University, India in 2003. From 2005 to 2007, he was a post-doctoral research fellow at Department of Civil & Environmental Engineering, University of California-Los Angeles, USA. Since 1995, he has been working as Scientist at Bhabha Atomic Research Centre, Mumbai. His research interests include development of advanced membranes having significant technological advancements in the field of separation science and technology for societal, nuclear & industrial applications which can be adapted at domestic, community and industrial scales. The thin film composite based sea water reverse osmosis (SWRO) membrane, he developed enables production of high-performance SWRO membranes for upcoming desalination plants in India. He has published 90 articles in international journals &conference proceedings, filed 3 patents and co-authored 4 book chapters. He is an Editorial Board Member of the 'Journal of Polymer Materials'. He is a recipient of the Scientific & Technical Excellence Award by the Department of Atomic Energy, the 'Make-in-India' award by Indian Desalination Association, and Dr. Pathak Memorial Award by Association of Separation Scientists & Technologists.

River Bank Filtration System: Cost Effective Water Supply Alternative

Sachin Saxena, Aparna Satsangi, and Vuppulury Soamidas

Abstract Prominent indicators of a major looming water crisis are low ground water tables, dry wells and depletion of drinking water resources at exponential rates; therefore, there is an urgent need for the development and adoption of alternative innovations that could reduce the huge burden of scarce water resources. Riverbank filtration (RBF) system is one such cost effective sustainable approach. The RBF technology has been in widespread use in USA and many European countries. In the past few years this has been applied effectively in India. This paper presents the effectiveness and application of RBF on the banks of rivers in India and other countries adopting the technique for consumption of naturally filtered water. It will also highlight the important parameters to consider and methodology, while selecting and installing the RBF technology onsite.

Keywords River bank filtration · Treatment · Maintenance

1 Introduction

The use of surface water to supplement the essential needs of increasing population emerges from the annual water budget and exchanges that take place on earth surface. Out of 114,000 km^3 of water that precipitates on land, 40,000 km^3 is fed into the oceans each year as flow of river water [4]. This water is the main source of fresh water for the survival of human beings and animals in several areas. Ground water flows into the oceans at a rate of around 1000 km^3 per year [50]. However, the level of contamination and range of acceptance of the acquired water may reach beyond the permissible limits. and therefore, certainly requires another step of filtration. Understanding the indispensable value of potable water, and the fact, that the ground water sources are highly vulnerable, despite many urban cities are exponentially exploiting the ground water sources for all kinds of use (bathing, cleaning, drinking

---

S. Saxena (✉) · A. Satsangi · V. Soamidas
Department of Chemistry, Faculty of Science, Dayalbagh Educational Institute, Dayalbagh, Agra, India
e-mail: sachinusic@gmail.com

© Springer Nature Switzerland AG 2021
A. Vaseashta and C. Maftei (eds.), *Water Safety, Security and Sustainability*,
Advanced Sciences and Technologies for Security Applications,
https://doi.org/10.1007/978-3-030-76008-3_24

and flushing). Large populated centers and high density areas in India are facing the adversity of not getting access to clean water, in spite of the fact; the rate of misuse is very little governed by the concerned authorities. Although Jal Jeevan Mission by Ministry Of Jalshakti, India (established in 2014 [29]) is now governing the rules for proper usage and management of water resources with the aim of providing Functional Household Tap Connection (FHTC) to every rural household i.e., *Har Ghar Nal Se Jal* (HGNSJ) by 2024 as one of the prime objectives.

Reverse Osmosis (RO) has become the major mode of filtration for making water fit to drink. RO utilizes only around 50% of water drawn. The remaining 50% water drains out and is highly contaminated making it unsuitable for even agriculture use. Further, the occurrence of demineralization in the RO process is another cause of concern. Essential minerals from the water get filtered out, which is detrimental to human health [52]. The other alternatives of filtration exist but cost of getting purified water is very high. Here comes the role of RBF system which uses surface water and natural filtration process. If the system is constructed with proper design, RBF system is the most cost-effective water treatment system with low maintenance and operational cost. Sharma et al. [43] reported that if the conventional water systems of the cities like Blantyre (Malawi) are replaced by the RBF, it would save almost 80% of the actual annual cost [43]. As per Saph Pani, the average operating cost of drinking water production from RBF ranges from 4.45 INR/m^3 to 18 INR/m^3 (INR—Indian Rupees) [41]. In Republic of Serbia, operational and construction cost of RBF wells is around 0.15 €/m^3. It is also observed to be much superior treatment system for removal of organic micro pollutants when compared with adsorption and membrane filtration phenomenon [15, 46, 47, 53].

RBF develops a hydraulic gradient that leads to movement of surface water from rivers (through bed) to feed the neighboring aquifer, thus potentially utilizing the percentage of water in rivers that would ultimately flow into the oceans. In light of the existing demands and severity of water shortage, river bank filtration opens up a cost effective and eco-friendly solution at locations where the river bed and soil strata on its banks can be utilized for filtration of river water. This natural phenomenon travels horizontally to feed nearby aquifers which are at a gravitationally lower level. In the absence of any abstraction, the aquifer may achieve the same water level as the surface water source. Abstraction of water from these aquifers causes a drop in the level of water in the region immediately surrounding that point, thus causing water to flow towards itself. Depending upon the distance of the point of abstraction from the surface water source and other conditions, it is possible that a major portion of the water flow is from this source. As the water travels through the porous media many undesirable substances (bacteria, protozoa and viruses, biodegradable and toxic substances) are removed in a natural process of filtration. This reduces the cost of any further treatment that may be required before supply [22]. It provides safe drinking water without any investment in coagulation and filtration. The RBF process is illustrated schematically in Fig. 1.

For a successful implementation of RBF some conditions are required to be satisfied. Some of the important ones are:

Fig. 1 RBF Schematic representation

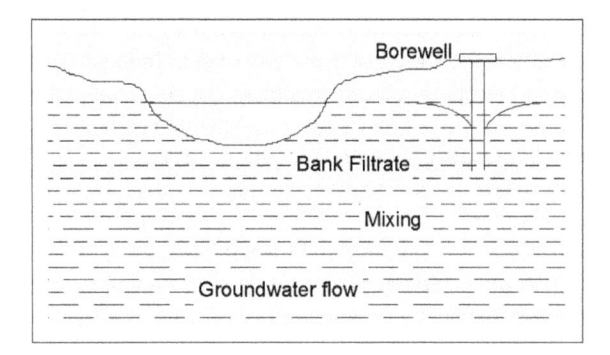

i. The soil on the riverbank must be permeable
ii. There should not be any natural or artificial barrier to the flow of water from the river to the abstraction site
iii. The distance from the river bank to the abstraction site should neither be very large (else bank filtrate may not be available in sufficient quantity) nor should it be too small (else filtration will not be effective). The value of these distances will depend on the type of soil on the riverbank.
iv. The depth of the bore should be slightly (5 to 10 ft.) more than the maximum depth of the river.

The aquifer often serves as a natural filter and also biochemically attenuates potential contaminants present in the surface water [16]. The direct abstraction of surface water for drinking is problematic due to the high seasonal discharge variations, resulting in disturbance of the daily production. This in turn causes consumer dissatisfaction and creates management challenges. Large capital investments are required to achieve sufficient quality and quantity of drinking water from direct surface water abstraction. In addition, heavy metal toxicity, pesticide contamination and water borne diseases are major challenges [8, 17]. It can be concluded that a permanent and long-term remedy to provide sufficient quantity of good quality of water round the year does not exist at present.

This chapter presents an overview of the site selection procedure and effectiveness of techniques and methodology used for selection of RBF site. It also discusses about the RBF system installed at Dayalbagh, Agra, India which is utilizing the system for feeding the clean drinking water (with minimal post treatment) to villages and D.E.I community.

2 RBF Across Countries

The simple and effective natural water filtration system has been exploited from ancient times across the world, due to significant improvement in the physico-chemical properties found in bank filtrate. Reduction in turbidity has been observed in Columbia River in Washington State [28] while Weiss [54] reports reduction in turbidity in surface water in Ohio river at Indiana. The benefits of this system are not only in terms of natural treatment but if sustainability cost and time is considered the removal of organic micro pollutants has been found to be superior over much prevailing adsorption and membrane processes [47]. The decrease in concentrations and values of total organic carbon, dissolved organic matter, BOD, COD has been observed in Missouri river in Missouri, South Platte River at Colorado, Elbe River, Germany, Nakdong River in South Korea and many other RBF sites [19, 20, 23, 25]. Further, pathogens also get filtered out when seepage and percolation take place through the bank of the water body [35].

RBF process is reliable for removing total coliform and fecal coliform and therefore, has significantly removed microbial counts at almost every RBF site in different countries [11, 44, 45]. However, water interaction occurring through limestone (Karst aquifer), stony soil and coarse gravel water formations show poor microbial filtrations [31]. Study by factor analysis shows that RBF water is much chemically stable as compared to river water [26]. The microbiological examination is also dependent on the temperature and climatic variations as the rate of biological (degradation process) and physical processes increases with increase in temperature [37]. The case study of Lower Rhine valley in Germany found that there was an increase in chemical pollutants in bank filtrate, during low rainfall and seasons with less lake or river water [9]. Isotopic tests such as ^{222}Rn and isotopes of oxygen determination in river water and the abstraction well is helpful in finding the surface water seepage in the abstraction well [36]. The water signatures of river water and bank filtrate are of immense help in understanding the groundwater and surface water mix in the borewell and also provide the insights of the hydraulic conductivity, velocity of water seepage and travel time [18]. The RBF benefits have not only benefited Europe and USA but recently developing countries are actively building the technology to get consumable water for its population. Egyptian cities like Assiut, Aswan etc. are using Nile River as a valuable source for getting quality water through RBF unit [2]. Thus, RBF system is an essential and integral part of water treatment technology (Fig. 2), this not only reduces the cost of pretreatment, maintenance but can be directly used to meet high quality water demand in different countries. The filtration technique is already prevalent in developed nations from decades and now it is efficiently spreading in the developing countries.

RBF technology not only provides natural treatment of raw water but also minimizes color, odor, suspended particles and algae that are present in the water reservoir. No use of chemicals is involved when dealing with bank filtration system. Large floating debris is also not observed in RBF. Though seasonal variations, effect of temperature, distance of the borewell from the water reservoirs are the factors which governs the quality and quantity of water supply from RBF wells.

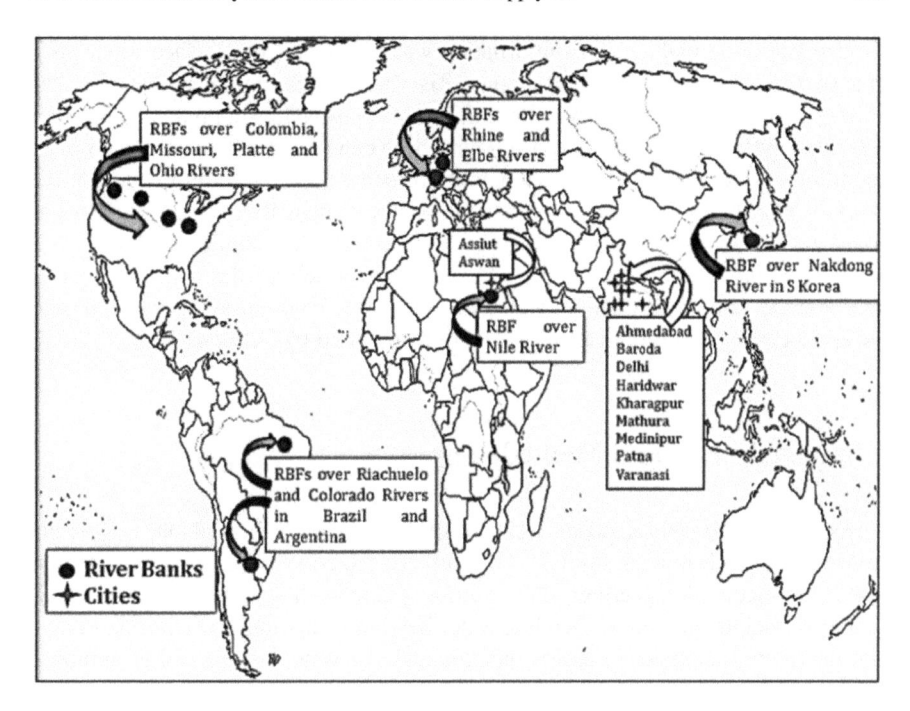

Fig. 2 General overview of riverbank filtration system used around the world including India

3 RBF in India

European countries and USA have been using RBF for more than 100 years [14]. In India, RBF has been introduced within the past couple of decades. When applied to rivers that have passed through major cities and towns, RBF can significantly reduce the pollutants present. Thus, RBF has the potential to provide drinking water to many cities currently using surface water as a source for their public water supply.

According to SaphPani in Mathura, subsurface water is collected from a well centered in the river bed. The water bearing stratum at this RBF site at Mathura was as much as 26 m below the bed of the river. Here, bank filtration significantly reduced color, UV absorbents, and dissolved organic contaminants [41]. Further, the cities of Ahmedabad, Baroda, Kharagpur, Delhi, and Medinipur partially meet their drinking-water demand by extracting bank filtrate through the use of horizontal collector wells [44, 45].

The Uttarkhand Jal Sansthan (UJS) has been one of the pioneering agencies using RBF in India. With the support of DST and also independently they have set up RBF facilities in many locations. These have proven the potential for application of RBF in supply of safe drinking water to the residents of the hill regions of Uttarakhand [42].

In most cities in India where RBF is used, no significant additional treatment is provided to the filtrate and RBF is considered as the crucial treatment process to remove microbial pathogens from drinking water derived from surface water [40]. In a Two-Day International Workshop on Riverbank Filtration the results of a fact-finding study on the use of bank filtration for drinking-water supplies in India were presented [34]. Scientific investigations have been conducted on the hydrogeological conditions, water quality, and sustainability of using bank-filtration systems in the cities of Haridwar, Patna and Varanasi along the Ganga River in Muzaffarnagar along the Kali River in Uttar Pradesh; in Nainital by Lake Nainital [6, 33, 48, 49]; in Delhi along the Yamuna River [27, 32]. An overview of the hydrogeological, hydrological, and water-quality parameters relevant to bank filtration in Europe and the applicability of this data to India has been compiled by Grischek [39].

4 Site Selection and Methodology

Selection of RBF sites depends on the tools and techniques available to find the strata of soil. Hydrological studies with computerized geo-electric monitors are used to evaluate electrical signatures of the aquifer system leading to the determination of saturated thickness of subsurface like water bearing formations, which is an important hydraulic parameter to assess quantification of water withdrawal at optimum level. Electrical resistivity techniques are another kind of geophysical investigations. Electrical resistivity tomography (ERT) and vertical electrical sounding (VES) are very helpful in finding out the freshwater zone, saline range and soil structure in the selected area [14, 55]. These methods are most reliable, fast and economic in delineating changes in lithology of sub surface formations. The technique can effectively identify presence of water, clayey, sandy zones and hard rocks. An electric current is passed through the ground by connecting with the battery, the terminals of a pair of electrodes, driven in the ground; called as current electrode pair. The voltage developed due to current flow across another pair of electrodes, known as potential electrode pair and driven in the ground is measured. The number of Schlumberger depth sounding measurements is carried out with maximum current electrode spacing in the study area. The field VES curves are converted to geo-electric parameters, which are true resistivity and used for defining the thickness of individual sub surface layers. The true resistivity value of the layer ultimately represents lithology of the layer including the expected ground water quality. Figure 3 represents the VES 3D representation, where site selection was on the basis of VES data (with different water zones with depth). Figure 4 shows the cross-section of the profiles (geophysical points at the river bank area from VES 1 to VES 5). Therefore, this data becomes essential before the installation of RBF plant.

Other techniques used for site characterizations are geographical information system (GIS), pumping test and borehole logging studies [12]. These techniques further help in selecting the type of aquifer design needed to be setup at an area near river. Figure 5 shows a clear methodology for installing RBF technology near rivers or lakes. Important outlines and work needed for proposing RBF unit can be summarized as:

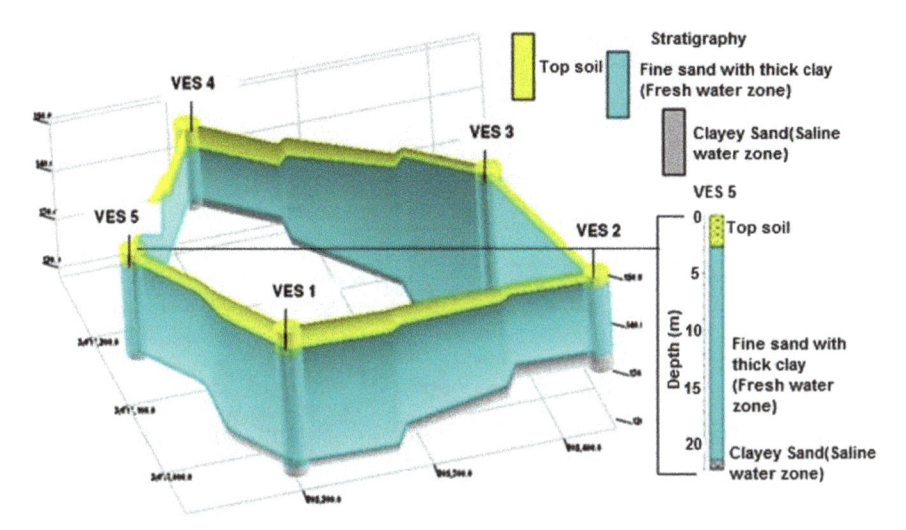

Fig. 3 VES (3D representation) of subsurface strata on a RBF site with 5 different geophysical points (VES 1 to VES 5) selected in an area (Inset shows 2D strata of VES 5 with different water zones with depth)

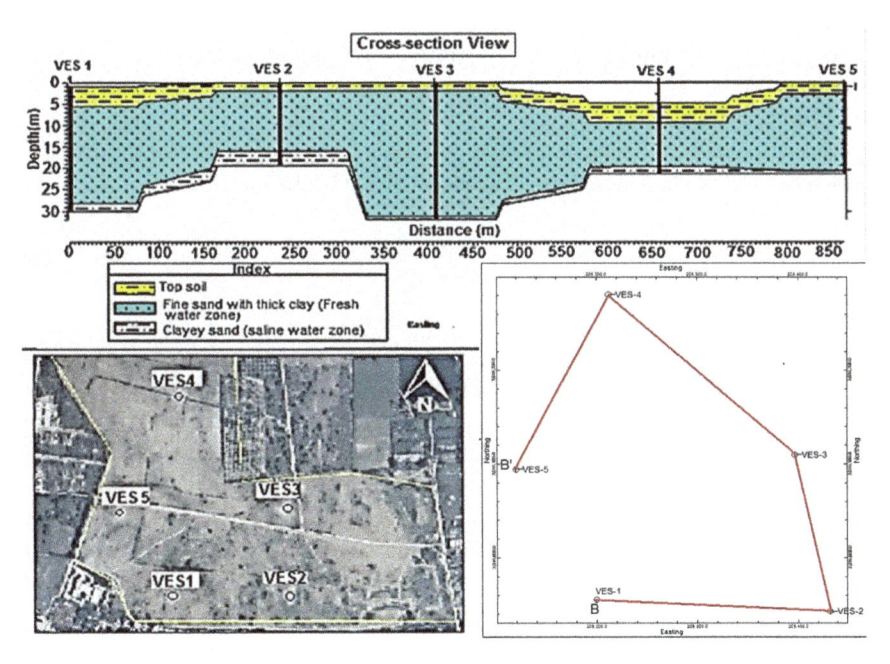

Fig. 4 Resistivity cross-section view of the profile (VES 1 to 5—geophysical survey points selected near riverbank area)

Methodology

Geodetic survey of the site for measurement of
- Site level with respect to Rivers/lakes
- Surface water level
- Ground water level
- Depth of proposed well

Drilling of wells

Field Investigations
- Pumping and yield test
- Onsite measurement of water quality parameters

Laboratory measurements
- Grain size distribution of borehole soil material
- Quality parameters of RBF percolated water (IS: 10500-2012) including toxic metals and pesticides
- Bacteriological investigations

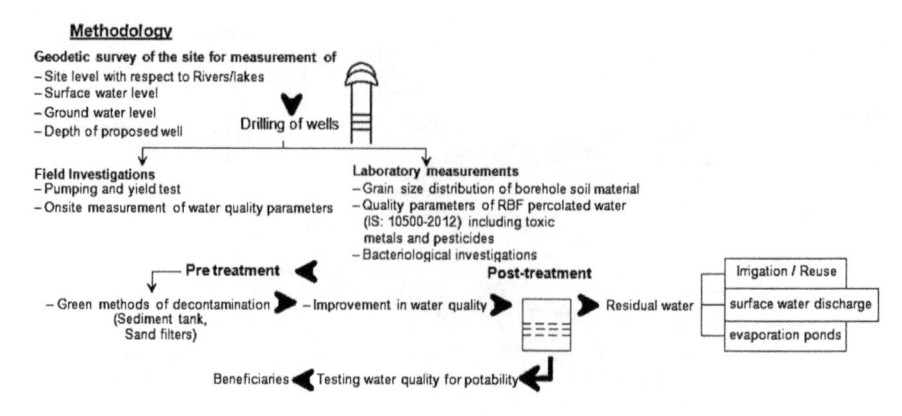

Pre treatment
- Green methods of decontamination (Sediment tank, Sand filters)

Post-treatment
- Improvement in water quality

Residual water → Irrigation / Reuse, surface water discharge, evaporation ponds

Beneficiaries ← Testing water quality for potability

Fig. 5 Outline of general methodology for installing and operating RBF system near rivers or lakes

- Preliminary work which involves soil analysis at the proposed site, river water quality (pH, EC, TDS, turbidity total alkalinity, total hardness, chloride, calcium, magnesium, sodium, potassium, fluoride, toxic metals and pesticides), river slope analysis.
- Hydrogeological analysis covering sub-surface layers and their resistivity, mapping of lithological sections, determination of granulometric properties, baseline data on hydro-chemical properties, geological surroundings, pumping tests, infiltration studies.
- Measuring and tabulating data on the surface and sub-surface geological environments, aquifer characteristics; surface and groundwater quality.

5 Design of RBF System

The major goal of using RBF system is to feed essential clean water to people and further, sustainable use of this resource system in long run by utilizing the beds and banks of lakes and rivers as naturally filtered medium. This technology not only reduces the cost of the water treatment but also balances the fluctuations occurring in the ionic concentration of water resources. The design of the technology is dependent upon water body movement on land (rivers or lakes), the hydrogeology of the selected site and the aim of abstracting water. The functioning of any RBF well relies on the connectivity occurring between the river and adjacent aquifer. The distance of the recharging aquifer, therefore, must be appropriate which may not be too large, that the hydraulic connectivity breaks off, nor too small that the filtration concept gets hindered [1]. It has been observed that the working of filtration unit can be examined using the flow path length (distance between river and aquifer), infiltration area and thickness of aquifer (made of sand or alluvial etc.). If the demand for daily consumption of water is large, flow rate should not be too large, though the effectiveness in filtration is observed when water take more time to interact with the bank of the river. Studies have reported horizontal aquifers (Radial collector wells) to be more efficient than vertical wells, due to low operational cost and less cleaning operations required. It has lower drawdown and seepage take place at lower velocity. The water seeps horizontally; thereby interaction of the surface water with the riverbank takes place improving the quality of water [13].

6 Challenges in RBF System

Lot of research and analysis is needed before making borewell or abstracting well on the bank of river which should focus on the quality and quantity of filtered water needed daily. The optimization of the distance of the aquifer from the river govern the hydraulic connectivity and flow rate of the pretreated water. In spite of this tedious work, further problems of contamination persist where heavy rainfall and floods directly contaminate the treated water that is collected in the abstraction aquifer.

Thus, the variation in climatic conditions and seasonal rainfalls do affect the percolated treated water. Recently, flood proof RBF wells have been constructed in Uttarakhand, India to rectify such problems [3]. Further, low infiltration rates towards aquifer have been reported to show decline in dissolved oxygen (DO) of water. It is evident from the fact, that DO parameter is an indicator of quality water. Dynamics of DO at the interface of surface water and particles (sand or alluvial etc.) has been studied to find the reason behind the depletion taking place [7, 30].

7 Case Study: RBF in Dayalbagh

The Dayalbagh Educational Institute (D.E.I.) with the aid of Department of Science and Technology under Water Technology Initiatives (DST-WTI) India; installed a RBF plant. The site at Dayalbagh, Agra where RBF is being implemented is shown in Fig. 6. This location is in Chandmari-ka-tila, close to River Yamuna. In order to find groundwater potential zones at Chandmari-ka-tila in Dayalbagh, VES studies were implemented over the area. The D.E.I. RBF site was selected on the basis of strata of the freshwater available with depth and were evaluated to determine the saturated thickness of the subsurface, which is an important hydraulic parameter to assess quantification of water withdrawal at optimum level [10, 51].

The RBF system along the river Yamuna at Dayalbagh can be observed in the Fig. 7, where the filtered solar operated RBF unit and the location of the site with respect to D.E.I can be observed in the map. It is represented as engulfed by the dark narrow lane of river Yamuna. The RBF setup shown here consists of 3 large tanks connected one after the other, two control rooms and solar panel room for clean and sustainable working of unit.

The bore well installed at some distance from the river gets the surface water filtered after seeping through the sand, gravel and silt layers of the banks of Yamuna.

Fig. 6 RBF Site and its Map—DEI

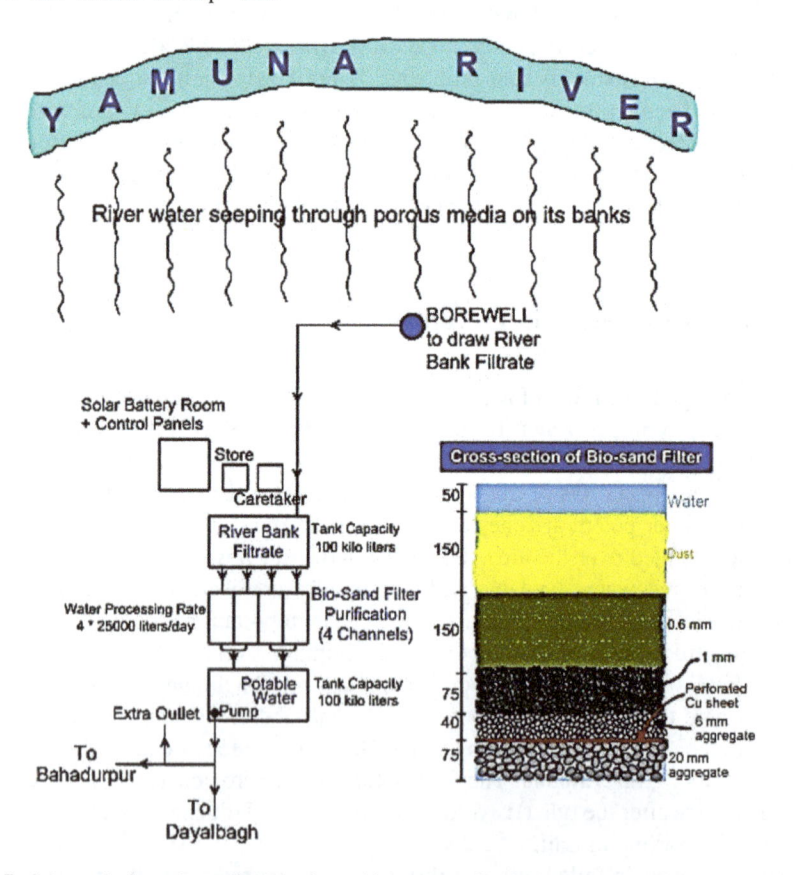

Fig. 7 Schematic flow chart of RBF unit working at Dayalbagh, Agra

Figure 7 clearly shows water coming in the bore well through river in an horizontal manner. Water is collected from the bore well to Tank 1 as naturally filtered riverbank filtrate (100 kl); it is followed by post filtration process using bios and purification channel. The percolated water from bios and filter gets collected in tank 2 which consists of 4 channels with a capacity of 25 kl of water each. The post treated water is finally transferred to tank 3, from where it is distributed to villages (one of the villages Bahadurpur) and DEI (after periodic check of the need of the residual chlorine while covering distance to the villages).

Bios and filter constructed allow water to percolate through the aggregates layer with gradation (bottom to top), from 20 mm to 0.6 mm aggregates followed by layer of dust sieving particles to even less than 5 microns. This helps in filtering out the pathogens and suspended solids from the riverbank filtrate if present. At present, DEI is supplying RBF water to Bahadurpur village and soon it will be supplied to the DEI community.

A drastic decrease in the turbidity levels was observed in the borewell water as compared to river water (decrease from 5.09 NTU to 0.8 NTU). This decrease in the turbidity is due to the natural soil filters which trap the suspended particles responsible for unclear water of the river. Total suspended solids show a remarkable fall of 85% and more in all samples. This again is due to the capacity of the natural soil to block the movement of suspended solids. According to BIS 10500 [21], the acceptable limit for Total Dissolved Solids is 500 mg/L. Another important parameter is the pathogen and coliform content in the water. These tests were performed only on RBF water samples. There was high level of coliform count in the river water which reduced in the borewell sample to much extent. On the other hand, E. coli was found absent per 100 ml in the samples. The bank filtration analysis work along Yamuna river has also been reported by Krishan et al. [24] in Agra [24].

8 Conclusion

RBF treatment systems are simple and involve low-cost pretreatment steps in drinking-water production. In some cases, serve as a final step in providing potable water. This technology is useful in developing new sustainable water sources and further, can be effectively applied to the existing surface water bodies like rivers, lakes, streams etc. It has been observed that resistivity test like VES signatures help in finding the soil structure and water zones w.r.t depth for digging well; while advance Isotopic analysis techniques (presence of ^{222}Rn and ^{18}O in borewell) help in studying the heterogeneity of water in well. The mixing of the groundwater and percolated water is well interpreted by these techniques. The location of abstraction well from surface water, rate at which volume of water gets filtered daily to meet population demand; aquifer thickness, hydraulic gradient dropdown cone formation etc. are some of the studies which are needed for installation and further consistent operation of a RBF unit. Different RBF designs and modifications according to the need have been observed, though no specific guidelines and standards are there for bank filtration system. In Dayalbagh, DEI is implementing RBF on a small scale for meeting drinking water needs of the small population in villages near the RBF site including DEI and Dayalbagh.

Acknowledgements The work presented in this paper has a part of the RBF Project sanctioned to DEI and funded by the Department of Science and Technology, New Delhi, under its Water Technology Initiative. The authors are also thankful to Dr. P. C. Kimothi, Uttarakhand Jal Sansthan, Dehradun, and Dr. N. C. Ghosh, National Institute of Hydrology, Roorkee, for their critical insight and helpful suggestions.

References

1. Abba UD, Firuz M, Zaharin A, Azmin WN, Umar N, Ibrahim A (2017) An overview assessment of the effectiveness and global popularity of some methods used in measuring riverbank filtration. Hydrogeol J550:497–515
2. Abdalla FA, Shamrukh M (2011) Riverbank filtration as an alternative treatment technology: AbuTieg case study, Egypt. Riverbank filtration for water security in desert countries. Springer, Dordrecht, pp 255–268
3. Ahmed AKA, Marhaba TF (2016) Review on river bank filtration as an in-situ water treatment process. Clean Tech Environ. https://doi.org/10.1007/s10098-016-1266-0
4. Dai K, Trenberth KE (2002) Estimates of freshwater discharge from continents: latitudinal & seasonal variations. J Hydrometeorol 3:660–687
5. Dash RR, Mehrotra I, Kumar P, Grischek T (2008) Lake bank filtration at Nainital, India: water-quality evaluation. Hydrogeol J 16(6):1089–1099
6. Dash RR, Prakash EVPB, Kumar P, Mehrotra I, Sandhu C, Grischek T (2010) River bank filtration in Haridwar, India: removal of turbidity, organics and bacteria. Hydrogeol J. https://doi.org/10.1007/s10040-010-0574-4
7. Diem S, Cirpka OA, Schirmer M (2013) Modeling the dynamics of oxygen consumption upon riverbank filtration by a stochastic– convective approach. J Hydrol 505:352–363
8. Doussan C, Poitevin G, Ledoux E, Detay M (1997) Riverbank filtration: modeling of the changes in water chemistry with emphasis on nitrogen species. J Contam Hydrol 25:129–156
9. Eckert P, Lamberts R, Wagner C (2008) The impact of climate change on drinking water supply by riverbank filtration. Water Sci Technol 8(3):319–324
10. Fritz B, Sievers J, Eichhorn S, Pekdeger A (2002) Geochemical and hydraulic investigations of river sediments in a bank filtration system. In: Dillon P (ed) Management of aquifer recharge for sustainability. Swets and Zeitlinger, Lisse, pp 95–100
11. Ghodeif K, Grischek T, Bartak R, Wahaab R, Herlitzius J (2016) Potential of riverbank filtration (RBF) in Egypt. Environ Earth Sci 75:1–13
12. Gomez E, Broman V, Dahlin T, Barmen G, Rosberg JE (2019) Quantitative estimations of aquifer properties from resistivity in the Bolivian highlands. H2 Open J 2(1):113–124
13. Grischek T, Schoenheinz D, Ray C (2002) Siting and design issues for riverbank filtration schemes Water Science and Technology Library (WSTL) 43:291–302
14. Grischek T, Schoenheinz D, Sandhu C, Hiscock K (2005) River bank filtration—its worth in Europe and India. J Indian Water Resour Soc 25(2):25–30
15. Grooters S (2007) Role of riverbank filtration in reducing the costs of impaired water desalination. Desalination and water purification research and development. Program report no. 122
16. Gutierrez JP, Halem DV, Reitveld L (2017) Riverbank filtration for the treatment of highly turbid Colombian rivers. Drink Water Eng Sci 10:13–26

17. Hiscock KM, Grischek T (2002) Attenuation of groundwater pollution by bank filtration. J Hydrol 266(3–4):139–144
18. Hoehn E, Von Gunten H (1989) Radon in groundwater: a tool to assess infiltration from surface waters to aquifers. Water Resour Res 25:1795–1803
19. Hoppe-Jones C, Oldham G, Drewes JE (2010) Attenuation of total organic carbon and unregulated trace organic chemicals in US riverbank filtration systems. Water Res 44:4643–4659
20. Ibrahim N, Abdul AH, Yusoff MS (2015) River bank filtration: study of Langat river water and borehole water quality. Applied mechanics and materials. Trans Tech Publication, Stafa Zurich, pp 1194–1198
21. Indian standard 10500 (2012). Drinking Water, pp 1–2
22. Jaramillo M (2012) Riverbank filtration: an efficient and economical drinking-water treatment technology. Dyna 79(171):148–157
23. Kim M, Oh S, Choi N, Park C, Ko C (2013) The assessment of water-quality and well yield for operation of riverbank filtration in field scale. In: EGU general assembly conference abstracts, p 3672
24. Krishan G, Singh S, Sharma A, Sandhu C, Grischek T, Ghosh NC, Gurjar S, Kumar S, Singh RP, Glorian H, Börnick H (2016) Assessment of water quality for river bank filtration along Yamuna River in Agra and Mathura. Int J Environ Sci 7:56–67
25. Kwon DY (2015) Study on water quality improvement by bank filtration. Desalin Water Treat 54:1385–1392. https://doi.org/10.1080/19443994.2014.899522
26. Lee JH, Hamm SY, Cheong JY, Kim HS, Ko EJ, Lee KS, Lee SI (2009) Characterizing riverbank-filtered water and river water qualities at a site in the lower Nakdong River basin. Repub Korea J Hydrol 376:209–220
27. Lorenzen G, Sprenger C, Taute T, Pekdeger A, Mittal A, Massmann G (2010) Assessment of the potential for bank filtration in a water-stressed megacity (Delhi, India). Environ Earth Sci. doi: 10.1007/s12665-010-0458-x
28. Mikels MS (1992) Characterizing the influence of surface water on water produced by collector wells (PDF). J Am Water Works Assoc 84:77–84
29. Ministry of Jalshakti, India https://jaljeevanmission.gov.in/. Accessed Oct 2020
30. Musche F, Sandhu C, Grischek T, Patwal PS, Kimothi PC, Heisler AA (2018) Field study on the construction of a flood—proof riverbank filtration well in India—Challenges and opportunities. Int J Disaster Risk Reduct 31:489–497
31. Pang L (2009) Microbial removal rates in subsurface media estimated from published studies of field experiments and large intact soil cores. J Environ Qual 38:1531–1559
32. Pekdeger A, Lorenzen G, Sprenger C (2008) Preliminary report on data of all inorganic substances and physicochemical parameters listed in the Indian and German Drinking Water Standards from surface water and groundwater at the 3 field sites. TECHNEAU integrated project deliverable www.techneau.org
33. Prasad T, Varma AK, Singh BP (2009) Water supply to Patna through river bank filtration: problems and prospects. Proceedings of the international conference on water, environment, energy and society WEES, New Delhi, 12–16, 3, pp 348–1357
34. Ray C, Ojha CSP (eds) (2005) Riverbank filtration—theory, practice and potential for India. In: Proceedings of international workshop riverbank filtration, 1–2 March (2004), Indian Institute of Technology Roorkee, India. Water Resources Research Center, University of Hawaii, Manoa, Cooperative Report CR-2005-01
35. Ray C (2011) Riverbank filtration concepts and applicability to desert environments. Riverbank filtration for water security in desert countries. Springer, Dordrecht, pp 1–4
36. Regli C, Rauber M, Huggenberger P (2003) Analysis of aquifer heterogeneity within a well capture zone, comparison of model data with field experiments: a case study from the river Wiese. Switz Aquatic sci 65:111–128
37. Rudolf von Rohr M, Hering JG, Kohler H-PE, von Gunten U (2014) Column studies to assess the effects of climate variables on redox processes during riverbank filtration. Water Res 61C:263–275

38. Sandhu C, Bornick H, Feller J, Grischek T, Jacob T, Kimothi PC, Patwal PS (2014) Report on bank filtration economics & cost-estimates. Saph Pani Project Deliverable D1.4. Available: http://www.saphpani.en/downloads. Accessed 21 Feb 2021
39. Sandhu C, Grischek T, Ray C, Schoenheinz D, Thakur AK (2009) Case studies from India: RBF potential for cities. Abstracts NATO advanced research workshop—riverbank filtration for water security in desert countries, Luxor (Egypt), 24.–27.10.2009, pp 49–50
40. Sandhu C, Grischek T, Schoenheinz D, Ojha CSP, Irmscher R, Uniyal HP, Thakur AK, Ray C (2006) Drinking water production in India—bank filtration as an alternative. Water Digest 1(3):62–65
41. Sandhu C, Grischek T, Kumar P, Ray C (2011) Potention for riverbank filtration in India. Clean Tech Environ Policy 13:295–316
42. Sharma B, Uniyal DP, Dobhal R, Kimothi PC, Grischek T (2014) A sustainable solution for safe drinking water through bank filtration technology in Uttarakhand. India 107:1118–1124
43. Sharma SK, Chaweza D, Bosuben N, Holzbecher E, Amy G (2012) Framework for feasibility assessment & performance analysis of riverbank filtration systems for water treatment. J Water Supply 61:73–81
44. Singh P, Kumar P, Mehrotra I, Grischek T (2010) Impact of riverbank filtration on treatment of polluted river water. J Environ Manage 91(5):1055–1062
45. Singh P, Kumar P, Mehrotra I, Grischek T (2010) Impact of riverbank filtration on treatment of polluted river water. J Environ Manage 91:1055–1062
46. Stauder S, Stevanovic Z, Richter C, Milanovic S, Tucovic A, Petrovic B (2012) Evaluating bank filtration as an alternative to the current water supply from deeper aquifer: a case study from the Pannonian Basin. Serb Water Resour Manag 26:581–594
47. Sudhakaran S, Lattemann S, Amy GL (2013) Appropriate drinking water treatment processes for organic micropollutants removal based on experimental and model studies—a multi-criteria analysis study. Sci Total Environ 442:478–488
48. Thakur AK, Ojha CSP, Grischek T, Ray C, Jha R, (2009a) River bank filtration in extreme environment conditions. In: Proceedings of international conference on water, environment, energy and society (WEES-2009), New Delhi, 12–16 Jan, 3, pp 1340–1347
49. Thakur AK, Ojha CSP, Grischek T, Sandhu C, Jha R (2009b) Assessment of water quality at RBF site Haridwar. In: Proceedings of international conference on water, environment, energy and society (WEES-2009), New Delhi, 12–16 Jan, 3, pp 1235–1241
50. Trenberth KE, Fasullo J, Mackaro J (2011) Atmospheric moisture transports from ocean to land and global energy flows in reanalyses. J Clim 24:4907–4924. https://doi.org/10.1175/2011JCLI4171.1
51. Umar AB, Ladan B, Gado AA (2017) Groundwater evaluation study using electrical resistivity measurements in Bunza area of Kebbi State, Nigeria. Int J Environ Bioenergy 12:100–114
52. Verma KC, Kushwaha AS (2014) Demineralization of drinking water: is it prudent? Med J Armed Forces India 70:377–379
53. Wang JZ (2005) American experience with riverbank filtration system—a detailed case study at Louisville Kentucky. J HarbinInst Technol (New Series) 12:75–81
54. Weiss WJ (2005) Water quality improvements during riverbank filtration fate of disinfection by-product precursors, pathogens, and potential surrogates. Dissertation, The Johns Hopkins University
55. Zarroca M, Bach J, Linares R, Pellicer X (2011) Electrical methods (VES and ERT) for identifying, mapping and monitoring different saline domains in a coastal plain region (Alt Emporda, Northern Spain). Hydrogeol J 401(1–2):407–422

Sachin Saxena was born in Aligarh, U.P, India in 1987. He received his Ph.D.(GATE) in Chemistry (Electrochemistry) from Dayalbagh Educational Institute, Dayalbagh, Agra, India in 2016. He was awarded Young system scientist Award in 2010 and Best Paper Award in IEEE sponsored 41st conference National Systems Conference 2017. From 2016 to 2017, he worked as a Research Associate in Riverbank filtration project on Yamuna river (Department of Science and Technology, India). He has published number of quality research papers in journals of International repute. Further, He is working as a Faculty in Dayalbagh Educational Institute, Dayalbagh, Agra, India from October 2017 onwards.

Aparna Satsangi is Assistant Professor in Department of Chemistry, Dayalbagh Educational Institute, Dayalbagh, Agra, Uttar Pradesh. She completed her M.Sc. and Ph.D. in *Atmospheric Sciences* from Department of Chemistry, Dayalbagh Educational Institute, Dayalbagh, Agra, Uttar Pradesh. She has qualified CSIR (NET) and GATE. She also had a Women Scientist Scheme Project (WOS A, DST Project). Her areas of interest are *Atmospheric Chemistry* and *Air Quality Monitoring*. She has published several research articles in various national and international journals. She is also a member of several national and international organizations.

Vuppulury Soamidas obtained his B.Tech. from the Regional Engineering College, Warangal and his M.Tech. and Ph.D. from the Indian Institute of Technology Madras. He taught at CBIT, Hyderabad, for 4 years before joining Dayalbagh Educational Institute, Dayalbagh, Agra, India in 1994. He teaches Mechanics of Solids, Mechanics of Machines, Mechanical Engineering Design and Mechanics of Composite Materials. His research interests are Composites, Solid Mechanics, Mechanical Design, Finite Element Methods, Biomechanics, Severe Plastic Deformations and Eco-friendly design. He is currently working on a project on hybrid vehicles. He is a life member of various Technical Societies.

Case Studies

Drought Land Degradation and Desertification—Case Study of Nuntasi-Tuzla Lake in Romania

Carmen Maftei, Gabriel Dobrica, Constantin Cerneaga, and Nicusor Buzgaru

Abstract The aim of this chapter is to present the status of a case study of Nuntasi-Tuzla Lake. Situated on the Black-Sea Coast (Dobrogea region-Romania), Nuntasi-Tuzla Lake is a part of Razim-Sinoe lagoon complex and a component of Danube Delta Biosphere Reserve. Due to its position, this area has a great importance for natural, scientist and cultural tourism. The human interventions on the Razim-Sinoe lagoon complex started over 100 years ago and the first interventions were intended to improve the navigation and water circulation between the lakes and Danube arms for fishery purposes by excavation of backwaters (small canals which link the lakes with each other). The second bigger intervention is represented by irrigation system which was built in 1970. The canals of the old network have been dredged out to provide the necessary volume for irrigation from Danube. In this way, Razim-Golovita Lake, has become the main sources for the irrigation system. The human intervention has led to a negative evolution of the Nuntasi-Tuzla ecosystem and in August 2020, the lake almost completely disappeared. In following September, due to extensive measures implemented by the authorities a fragile equilibrium was re-established. However, a management strategic plan is proposed in order to ensure proper management both protected natural area and to implement concrete measures which minimize the effects of natural hazards.

Keywords Drought · Desertification · Management · Dobrogea

C. Maftei (✉)
Transilvania University of Brasov, Brasov, Romania
e-mail: cemaftei@gmail.com

G. Dobrica · C. Cerneaga · N. Buzgaru
Doctoral School of Applied Sciences, Ovidius University of Constanta, Constanta, Romania
e-mail: nicusor.buzgaru@dadl.rowater.ro

1 Introduction: Water Policy Context

According with ONU (United Nations Organization) water security refers to "*the capacity of a population to safeguard sustainable access to adequate quantities of acceptable quality water for sustaining livelihoods, human well-being, and socio-economic development, for ensuring protection against water-borne pollution and water-related disasters, and for preserving ecosystems in a climate of peace and political stability.*" To achieve this goal, it is necessary to have an adequate policy regime related to: transboundary cooperation, financing schemes for a sustainable development, political stability, education, health, and integrated management of water resources.

Since 1975 a massive effort has been made by Europe in order to implement the water security policy. The document adopted in 2000—Water Framework Directive—WFD (2000/60/EC) [1] represents the current legal instrument in the field of water protection in Europe. Among the key objectives of WFD is a unitary management system based on watersheds (river basins/catchments) which represent an innovative system to investigate the water budget. In this respect each member country has the obligation to establish a basin management plan (BMPs). This plan, updated every six years, contains the principal objectives related to the protection of water quality applicable to both surface and groundwater bodies. WFD forms together with Urban Wastewater Directive (1991), Nitrates Directive (1991), Drinking Water Directive (1975 and revised in 1998), Groundwater Directive (2006), Environmental Quality Standards Directive (2008), Floods Directive (2007), Bathing Water Directive (2006) and Marine Strategy Framework Directive (2008) the principal EU specific law ensuring the water security in river basin. In the context of climate change, Europe is facing increased risk of damage by flood and drought. According with EEA (European Environment Agency) flooding and drought are the main natural hazards in Europe [2]. Concerning the drought, 17% of the European territory is affected by water scarcity. The cost of drought in Europe is at least 100 billion EU [2]. Recognizing the impact of drought hazard and the prediction for the next period, European Commission adopted in 2007 the Communication "*Addressing the challenge of water scarcity and droughts*" [3]. The Environmental Council asked the European Commission to review and develop the "*European policy concerning the water scarcity and drought*" which was completed in November 2012 (4) being part of *Water Blueprint* adopted also in November 2012 [4]. The EU Commission recognized the limited progress achieved in implementing of the instrument proposed in 2007 and decided that the revised RBMPs were to include the drought risk management and aspects concerning climate change [3]. The measures envisaged to reduce the drought risk are green infrastructure, water re-use for irrigation or industrial purposes, developing of the European Drought Observatory. To support the efforts of all stakeholders involved in drought management GWP (Global Water Partnership) and WMO (World Meteorological Organization) launched the Integrated Drought Management Programme (IDMP) in 2013. Simultaneously, GWP for Central and Eastern Europe (GWPCEE) launched an IDMP for this region. Analysis of the drought in the CEE region and integration of this phenomenon in the first RBMPs showed that the Drought Management Plan (DMP) development in the region investigated is extremely weak [5, 6]. In this context, a *Guidelines for the preparation of DMP* was prepared by GWPCEE [7]. The principal objective was to support the responsible authorities in national/regional drought development planning to prepare and integrate DMP within BMPs. Additional documents were prepared: Handbook of Drought Indicators and Indices [8], Practical Guidelines on Planning Small Water Retention in River Basins.

Starting with 2007 Romania became a member of the UE. In the pre-adherence period Romania started implementing the WFD by transposition in national legislation (Article 3) and proposed a schedule to implement Article 4 of WFD (2000/60/EC). According with the proposed schedule, Romania achieved the objectives set out in Article 4(1a, b) of WFD in 2015 [9]. In the 2004–2009 period Romania established the monitoring program (Article 8) and has accomplished the Romanian Basin Management Plan—RBMPs (Article 13). In the period 2021–2027 the objectives of 4(1c)—environmental objectives for protected areas—must be completed and also the revision of RBMPs (Article 4 and 13). The second edition of RBMPs is in preparation for 2021.

Concerning the drought, the *National Strategy for Drought Mitigation and Prevention and Combating Land Degradation and Desertification* was elaborated in 2008 but is not yet adopted. The Strategy mentioned some reactive measures enabling emergency management.

The objective of this chapter is to present a case study concerning the status of Nuntasi-Tuzla Lake. In august 2020 a breaking news article was spreading through the Doborgea region (Fig. 1) and Romania: "a lake of 850 hectares dried!" The "alarm bell" has been sounded by Istria city hall which announced the responsible authorities and public body. An extensive rehabilitation campaign of Nuntasi Lake has started in September. This ecological disaster is worth being investigated from the WFD perspective and the forthcoming second edition of Dobrogea Littoral RBMPs. The second section of this paper offers information about the study area and the importance of this site in the region. In the discussion section we describe the human intervention which seriously disrupted the lake ecosystem and water budget.

2 Study Area

Situated on the Black Sea Coast, approximately 35 km North of Constanta, Nuntasi-Tuzla Lake is a part of Razim-Sinoe lagoon complex, being a component of Danube Delta Biosphere Reserve (Fig. 1). Located north of Grindul Lupilor, southeast of Histria acropolis-Grindul Saele, south of Grindul Chituc and west of the localities Corbu, Nuntași and Istria (Fig. 1) the areal (ROSPA0031) is a natural-mixed reserve (botanical and fauna) with an important archeological value (the ancient Greek Histria acropolis).

From the geomorphological point of view the region has a coastal lowland aspect (Prispa Hamangia) developed between the green schist Casimcea plateau and Black

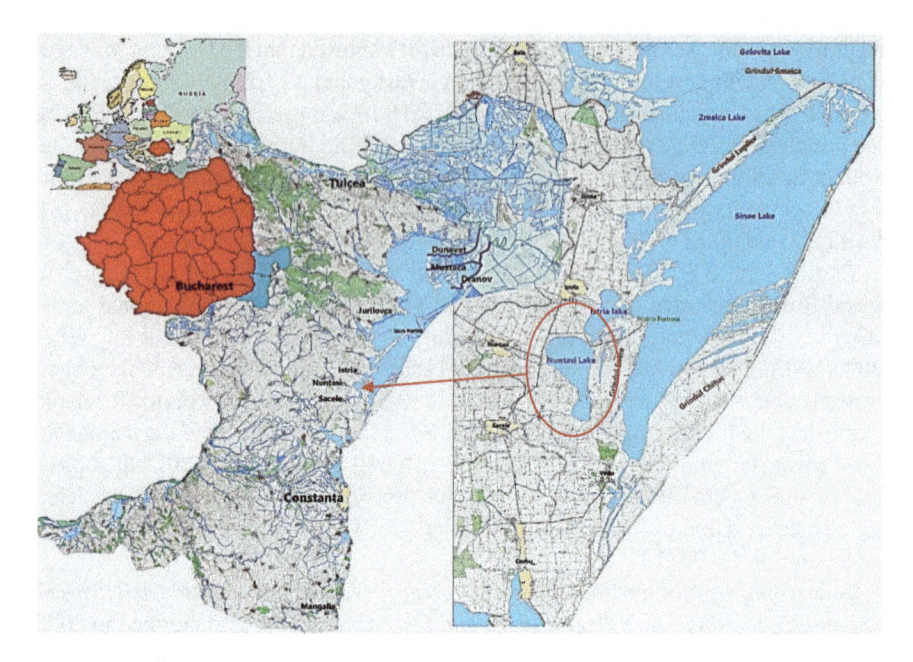

Fig. 1 Location of Nuntasi-Tuzla lake

Sea being a combination of beach ridge plains (Sacele and Chituc), sandy barriers and shallow lakes (Sinoe, Histria and Nuntasi) [10]. The relief here has a height of no more than 25 m and the slope decreases from East to South east. From a geological point of view the region fundamentally consists of green schist covered with thick layers of loess (2–15 m). The climate is temperate-continental influenced by Black Sea. The mean annual temperature in the region is about 11 °C and annual rainfall amount of about 400 mm [11–13].

From the hydrological point of view, Dobrogea basin is divided into two main sub-basins: one tributary to the Danube (Danube basin) and other tributary to Black Sea basin- Dobrogea Littoral basin. The rivers from Dobrogea-Littoral basin are more developed in the north part of Dobrogea region. The Nuntasi-Tuzla Lake (ROLN05–RORW15-1-7_B1) is situated at the confluence of two rain water systems of small dimensions Nuntași (22 km long) and Săcele (10 km long) Fig. 2. The catchment area is 22,334 ha. According with Dobrogea RBMPs [14] the Nuntasi valley multiannual average discharge is about of 0.356 mc/s and Sacele valley 0.092 mc/s. A particular importance is held by the connection between the lakes of the Razim-Sinoe lagoon complex same as the connection between the lakes, Black Sea and Danube (sf. Gheorghe arm). This connection will be explained in the context of anthropic activity.

The groundwater is present in the proluvial deposits (west part of region), sand or gravel deposit (littoral belt) and above an impermeable or relatively impermeable layer (perched water).The groundwater level varies between 2 m (around the lakes)

Fig. 2 Nutasi-Tuzla lake basin

and 15 m in the interior of the region [15]. The aquifer flow rate is around 0.5–2 l/s. In the region with sand or gravel layer, the flow could be 10 l/s [15].

According with the public administration [16] the region land use is predominantly agricultural (arable land 36.69%, grassland 9.3%, vineyards and orchards 0.1%). The rest of territory is occupied by water body (50%) and forest (0.01%). Region's natural vegetation is characteristic of the steppe area. In this perimeter there are 30 types of ecosystems that shelter 5,137 species of which 1,689 species of flowers, most of them protected by national and international regulations.

The region's economic activity is based on agrarian, animal husbandry and fishing. To sustain the economic development of Dobrogea, during 1968–1974 period important hydraulic works were made in the region aimed at supporting the agriculture. In the study area the Razim-Sinoe hydraulic system was built during the 1971–1975 period. This system will be described in the following section. The region is known for ancient Greek Histria acropolis and the existence of sapropelic mud in Nuntasi-Tuzla Lake with therapeutic properties. In the last decades, there has been a significant improvement of tourism activity thanks to the existence of archeological sites and the Biosphere Danube Delta Reserve.

3 Lake Genesis and Its Characteristics

The first scenario concerning the Nutasi Lake genesis is based on the marine transgression and coastal dynamic process [17]. Geomorphological and archeological arguments point to the neo-tectonic activity as the main factor that controlled the reconfiguration of the coastal region around Histria, before, during and after the decline of the ancient city [18]. Archeological arguments that point to the later formation of the Istria and Nuntasi lakes are based on the reconstruction of the paths of ancient roads identifiable on aerial photographs as having a clear orientation towards the ancient city of Histria. For the Roman period, the aqueducts that remained suspended on the western bank of the Istria Lake as well as the submerged artefacts found in these lakes point towards the conclusion that in that period, the Nuntasi-Tuzla, Istria and Sinoe lakes did not exist. As proven by the new absolute ages obtained for these units, in the interval comprised between 1200 and 600 years ago the quasi-concomitant formation of the contact ridge in the Old Sacele, New Sacele and the Sacele North—Chituc marine fields occurred which led to the appearance of the Histria, Nuntasi and Sinoe lakes. To conclude the lake was formed 1200–600 years ago, as a result of tectonic activities.

In 1976 Braier [19] published a hydro geographical study about the Dobrogea Lakes, where she mentioned that the Nuntasi –Tuzla Lake have the following characteristics: surface of 1,050 hectares, volume of 9.28×10^6 mc, 6.2 km long, width varies between a minimum of 1.7 and maximum of 3 km, coefficient of maximum elongation of 2.00, depth varies between 2.15 and 6.15 m. All the documents consulted (e.g. Dobrogea RBMPs, Emergency Situation Inspectorate's emergency preparedness plan) give no other information about the lake's evolution which would help the anticipation of the August 2020 event. The lake hydrochemistry parameters investigation is based on the data provided by different sources. Braier (1976) showed that the Nuntasi-Tuzla Lake hydrochemistry investigation started in 1935 but the measurement was insufficient [19]. Some investigations was made by the author during 1967–1971 period. Systematic studies started in 2000. In 1976 Braier has demonstrate that Dobrogea brackish lakes are classified according to their salinity into 3 categories that indicate the influence of the sea [19]: oligohaline—0.5–2.5 g/l (e.g. Razim, Golovita, Babadag, Tasaul,); mesohaline—3–10 g/l (e.g. Zmeica, Sinoie) and polyhaline—10–17 g/l (e.g. Istria, Nuntasi, Tuzla). Due to the saline water, climatic conditions and water lakes fauna and flora, the sapropelic mud is developing in the Nuntasi-Tuzla Lake area. The therapeutic properties of this mud are well-known due to Techirghiol Balneotherapy Center. Braier (1976) pointed out two major problems related to Nuntasi Lake: (i) the water salinity decrease and (ii) the mud production of lakes as Nuntasi decrease due to fresh water input [19]. The environmental report for 2001–2003 highlighted the water lake was being eutrophic. Starting with 2005 the lake water was hypertrophic.

4 Ecosystem Lake Degradation Under Human Intervention

In order to describe the ecosystem degradation, a description of anthropic activity should be given. Before human intervention the Razim-Sinoe lagoon complex was connected with the Black Sea through a kind of natural gateway as Gura Portiței (Razim Lake) and Periboina (Sinoe Lake). Through these straits an equilibrium between fresh and salty water was maintained in the lagoon complex. Human intervention in the study area could be divided into three categories: canalization phases, polder construction and irrigation phase.

The first stage of human intervention started in 1856 when European Commission for Danube proposed to transform the reed areas into polders for agriculture in order to increase the grain production. This concept has failed but during 1938–1940 and 1955–1965 some polders were embanked for the same purposes. In the absence of drainage and irrigation works, the agricultural production has been weak [20]. The canalization phases starting during 1903–1912 period when the main woks consisted of channels' bed regulation and deepening of Dunavăț (ancient Danube Arm) and Dranov and their connection to Sfântul Gheorghe Arm and Razim Lake [21] for the purpose of creating fisheries. In the second stage (1930–1949) some canals were excavated in order to improve the circulation between the small villages, mostly located between Sulina and Sf. Gheorghe Arms. In the last stage of canalization phase (1955–1965) many more canals were constructed which connect the Danube with the lakes and the lakes with each other in order to improve the water circulation, the oxygen content and to raise the natural fisheries production.

During 1971–1975 the irrigations system around the Razim Lake were built. The water sources for irrigation system is the Razim-Golovita-Zmeica lagoons (Fig. 1). For this purpose, the lagoons were isolated from the Black Sea as was Sinoe Lake with different hydraulic works (Gura Portiței, the main connection with the sea of Razim Lake, was closed in 1976), to which a strengthening of the littoral belt was added. The alimentation of lagoons with fresh water is provided by several canals (Dunavat, Dranov, Mustaca re-dimensioned) from Sf Gheorghe arm. The total discharge of three canals is 260 mc/s (for a probability of occurrence of 1%) with an average of 80 mc/s. From analyses conducted in the study area by the former IEELIF (society of execution and exploitation of land improvement works) it is clear that starting with 1973 the lagoons water lost salinity and became fresh [15]. The irrigation system was fully working during the 1976–1989 period. After 1990 water demand decreased to 20% of its capacity.

During the experiments performed in the 1976–1982 period the researchers of the former Design and Research Institute of Water Management demonstrated that the Nuntasi-Tuzla Lake ecological balance is disturbed due to fresh water intrusion and pollution [22]: (i) the water mineralization decrease from 58.7 g/l (1934) to 1.6 g/l in 1982, (ii) bacterial pollution of water and sapropelic mud and (iii) in 1980 the production of sapropelic mud was zero.

During the experiments carried out in the 2002–2003 period, Drăgan-Bularda (2004) has demonstrated through Enzymatic Indicator of Mud Quality (EIMQ) that the mud of the Nuntași lake is categorized as poor (EIMQ = 0.3–0.45) from the therapeutic point of view in relation to the values obtained for Techirghiol Lake and the process of mud formation is very low [23]. The studies show a degradation of the Nuntasi-Tuzla Lake due to the freshwater input but there is no information on the decrease of the lake surface or lake water sources input. In order to investigate the situation that occurred in August 2020, a short analysis was performed concerning the water budget using remote sensing data.

The water supply of the lake is given by precipitation and Nuntasi and Sacele Rivers flows (Fig. 2). The evolution of Nuntasi and Sacele River discharge data during 1965–2010 show a constant decreasing trend for both data time series (Fig. 3). Generally, the average ratio of discharge value is 4.6:1 for Nuntasi River. Starting with 2001 this ratio decreased to 3.1:1, the discharge of Nuntasi River become smaller (Fig. 3).

The analysis of precipitation and evapotranspiration is based on the anomaly chart. This anomaly chart shows the difference between annual rainfall and evapotranspiration respectively and the multiannual values during a baseline period. The precipitation and temperature are achieved at Jurilovca meteorological station situated 30 km north of Nuntasi (Fig. 1). The evapotranspiration is calculated with the well-known Thornthwaite formula based on the monthly temperature. The evolution of the precipitation anomalies (1965–2015) is presented in Fig. 4 (the baseline—0–0 on the graphic—represents the multiannual precipitation—394 mm).

Starting with 1974 and until 2011 the annual precipitation was less than the baseline average (with some exceptions in 1975, 1978, 1991 and the 1995–1998 period. Starting with 2012 a positive anomaly is noticed. The analysis of evapotranspiration is based on the anomaly chart (Fig. 5). The period 1965–2019 contains more years

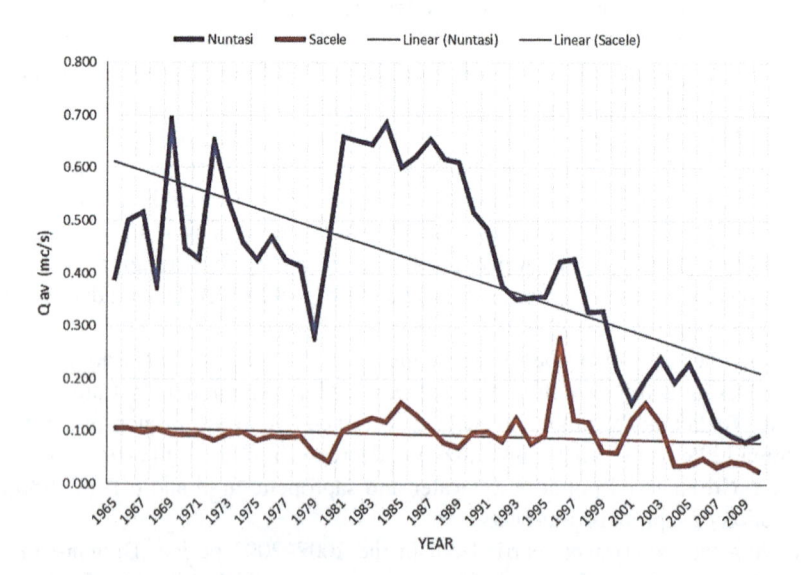

Fig. 3 Evolution of discharge rate for Nuntasi and Sacele rivers

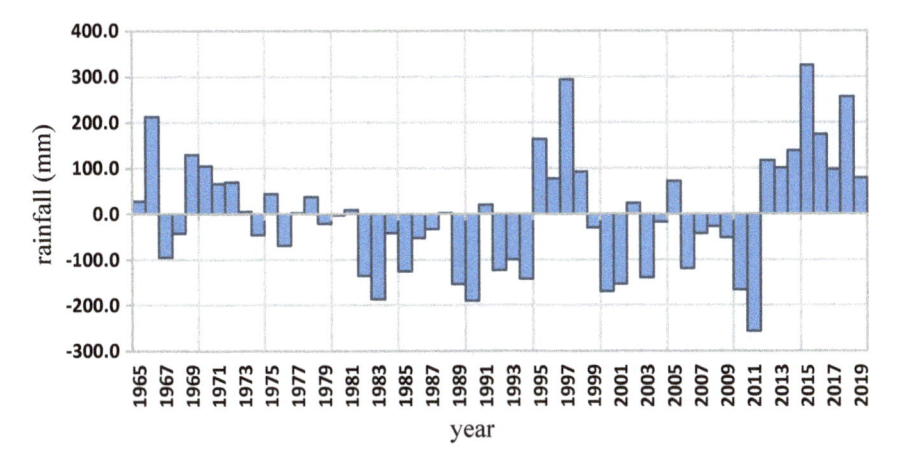

Fig. 4 Anomaly chart precipitation (Jurilovca meteorological station)

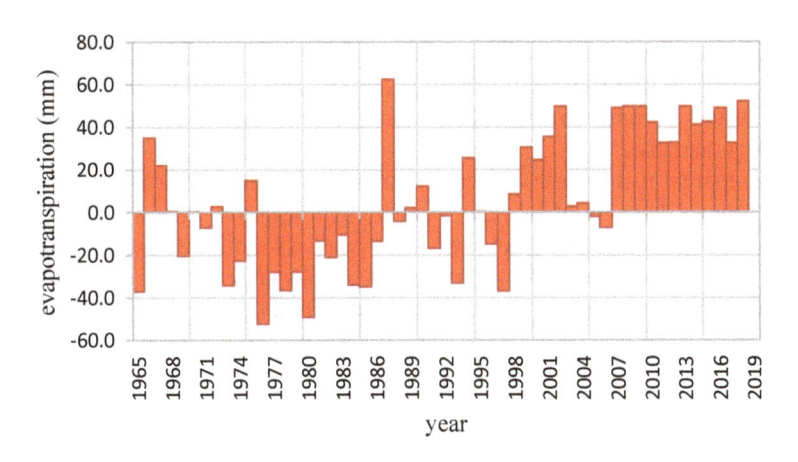

Fig. 5 Anomaly evapotranspiration chart

with a negative anomaly (especially during 1967–1997 period) but starting with 1998 only positive anomaly has been registered.

The annual water budget is shown in Fig. 6. It is clear that the annual budget is negative and starting with 1992 the quantity of water brought by the two tributary rivers (Nuntasi and Sacele) has dropped significantly. The volume of precipitation increased starting with 2012 but the evapotranspiration losses are much larger than the precipitation increase. There is no information related to the groundwater regime and its proportion in the water budget.

Research about the water losses through the irrigation canals [15] point to the following conclusion: during the 1976–1989 irrigation period, the rate of losses through the waterproofed canals were around 200–300 l/m^2 while for those unwaterproofed the water losses were 700–900 l/m^2. It is, therefore, should be mentioned

■ annual rainfall　■ annual flow　■ annual evapotranspiration

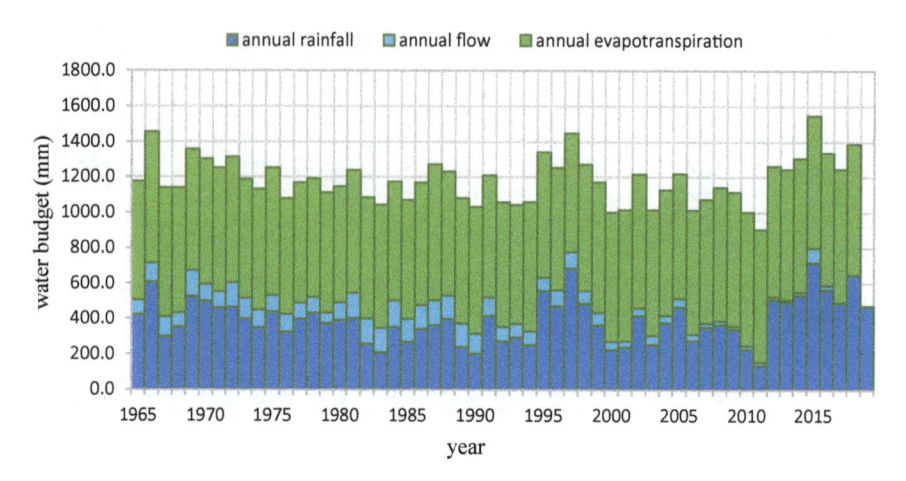

Fig. 6 Water budget for Nuntasi-Tuzla Lake

that the canal which links the Nuntasi-Tuzla Lake with the Istria Lake was silted before 1976 [19]. It means the water budget in that period was somewhat equilibrate but the water of Nuntasi-Tuzla Lake became fresh, a fact which contributed to the ecosystem degradation.

It is apparent that there has been a fundamental change after 2012 in the lake morphology: a decrease of its water budget could lead to a decrease of the lake's surface. Remote sensing images have been consulted for different period in order to observe the changes in the lake's surface. The results are presented in Fig. 7. It can be see that, the south part of Nuntasi-Tuzla Lake dried in 2013 for the first time.

5 Conclusion

Situated in the South East part of Romania, Dobrogea represents an area where due to its topography and location, between the Black Sea and lower Danube, it presents a very diverse and unique character. Here, in the North Eastern part of Dobrogea the oldest mountain range in Romania, the Macin Mountains, represent remnants of the Hercynian Mountains. The territory has been inhabited by humans since Middle and Upper Paleolithic but the area is best known for its Greek and Roman citadels (Histria, Tomis—Constanta, Callatis—Mangalia, etc.). The river Danube ends its journey to the Black Sea where it has developed a beautiful delta which comprises the most diverse ecosystem of Europe. Today this area is the largest nature reserve of Europe and over 50% of its area was placed on the World Heritage List. Nuntasi-Tuzla Lake is part of the Danube Delta Biosphere Reserve, located in its south part.

In August 2020 the lake almost completely disappeared due to various complicating factors that are more about anthropic factors than climate change or natural factors (Fig. 8). In September 2020 the Nutasi Lake was reconnecting with Istria and

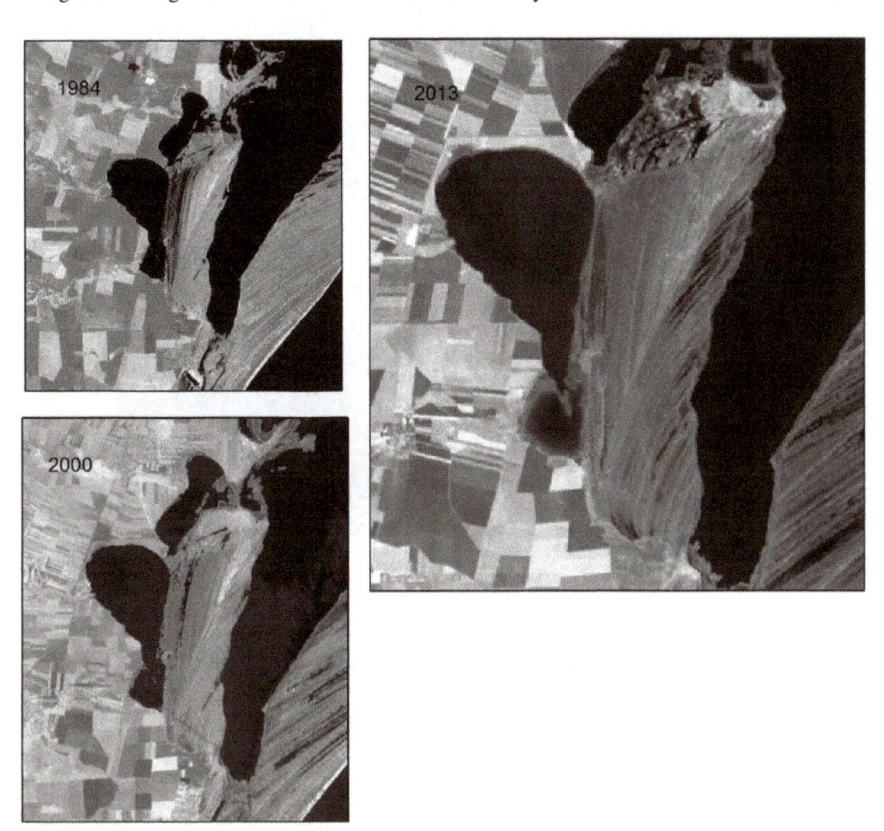

Drought Land Degradation and Desertification—Case Study ... 593

Fig. 7 Remote sensing images for study area (all images Landsat TM, are collected for August)

Fig. 8 Nuntasi Tuzla lake (August 2020— *source* ABADL)

Fig. 9 Lake restoration (September 2020—*source* ABADL)

Sinoe Lake (Fig. 9). A fragile equilibrium was re-established and there's now hope that the ecosystem will recover.

Human interventions on the Razim-Sinoe lagoon complex have started over 100 years ago. Obviously the human intervention has led to a negative evolution of the Nuntasi-Tuzla ecosystem discussed in the previously sections. In this context, a strategic management plan is needed. The overall objective of such a plan should ensure proper management both protected natural area and to implement concrete measures which minimize the effects of natural hazards (droughts, floods, etc.). All the stakeholders (public and administrative bodies, research institutes, universities) must to work jointly and put together their resources (human, time, equipment, etc.) to develop a common strategy leading to sharing data, information and competencies concerning the entire monitoring procedure to fully respond to the WFD, Romanian and European water policy and to help improve the operational efficiency of the administration. Joint decisions and parallel actions combined with educational activities will improve the capacity to preserve the valuable environmental heritage of this area.

Given the specific conditions (from ecological point of view Nuntasi-Tuzla Lake is monitored by the Danube Delta Biosphere Reserve Romanian Agency—ARBDD; from the hydrological point of view, Nuntasi-Tuzla catchment is a part of the Dobrogea Littoral Basin) a future management plan should contain several principal directions:

- Current Status Assessment. Define the specific problems, limitations, constrains, needs.
- Protect and Preserve the Ecological and Cultural values.

- Enhance Sustainable Management having as objective the harmonization of activities towards a sustainable development of the area with respect to the environment.
- Strengthen Institutional Capacity and Collaboration (enhance collaboration for research, planning and policy development).
- Monitoring and Evaluation of environmental parameters and indicators (including a geo database containing the factors which influenced such ecosystems, which should be estimated using harmonized methodologies).
- Raising Public Awareness. Involving people in Environmental Protection activities.
- Increasing Resources and ensuring financing.
- Without a doubt, the second revision of RMBP's plan ABADL should include the drought plan management and should pay particular attention to Nuntasi area.

Furthermore, we should never forget George Vâlsan (Romanian Geographer, 1885–1935): "Dobrogea has many hidden beauties, you only need to come close and search and it will reveal the unique traits it possesses, with love."

References

1. Water Framework Directive—WFD (2000/60/EC)
2. EEA—report available at: https://www.eea.europa.eu/data-and-maps/indicators/direct-losses-from-weather-disasters-3/assessment-2
3. COM (2007) 414 Addressing the challenge of water scarcity and droughts
4. COM (2012) 0673—European policy concerning the water scarcity and drought
5. GWP CEE, Fatulova E (2014) Report on review of the current status of implementation of the drought management plans and measures
6. Kindler J, Thalmeinerova D (2012) Inception report for the GWP CEE part of the WMO/GWP integrated drought management programme
7. GWPCEE (2015) Guidelines for the preparation of DMP
8. World Meteorological Organization (WMO) and Global Water Partnership (GWP) (2016) Handbook of drought indicators and indices. In: Svoboda M, Fuchs BA (eds) Integrated Drought Management Programme (IDMP), integrated drought management tools and guidelines series 2. Geneva
9. www.mmediu.ro
10. Preoteasa L, Vespremeanu-Stroe A, Hanganu D, Katona O, Timar-Gabor A (2013) Coastal changes from open coast to present lagoon system in Histria region (Danube delta). J Coast Res (Special Issue No. 65):564–569
11. Maftei C, Bărbulescu A (2008) Statistical analysis of climate evolution in Dobrudja region. Lecture notes in engineering and computer sciences, vol II. pp. 1082–1087
12. Maftei C, Barbulescu A (2009) Frequential models for the precipitation evolution in Romania, Analele Universitatii Ovidius, Seria Construcții 1(11):73–81 ISSN: 12223-7221
13. Maftei C, Barbulescu A, Hubert P, Dobrica G (2012) Statistical analysis of the precipitation from Constanța (Romania) Meteorological Station, in recent researches in applied computers and computational science. In: The 11th WSEAS international conference on applied computer and applied computational science (ACACOS'12), Rovaniemi Finlanda, pp 52–57. ISBN: 978-1-61804-084

14. ABADL—Dobrogea RBMPs Romanian Basin Management Plan (2015) available at: http://arhiva.rowater.ro/SCAR/Planul%20de%20management.aspx
15. Grumezea N, Kleps C, Tusa C (1990) Evolutia nivelului si chimismului apei freatice din amenjarile de irigatii in interrelatie cu mediul inconjurator. Bucuresti, Ed. Intrepreinderea poligrafica Oltenia
16. http://www.primaria-istria.ro/
17. Ștefan AS (1987) Évolution de la côte dans la zone des bouches du Danube durant l'Antiquité. în: Déplacements des lignes de rivage en Méditerranée d'après les données de l'archéologie, Editions du CNRS, Colloques Internationaux, Paris, pp 191–209
18. https://histria.geo.unibuc.ro/?page_id=2
19. Breier A (1976) Lacurile de pe litoralul românesc al Mării Negre. Studiu hidrogeografic, Edit. Academiei Române, București
20. Pons JL (1992) Natural resources in environmental status report, vol 4 (Conservation status of Danube Delta), IUCN Publications Unit
21. Gastescu P, Bretcan P, Teodorescu DC (2016) The lakes of the Romanian black sea coast man-induced changes, water regime, present state. Rev. Roum. Géogr./Rom. J Geogr 60(1):27–42, București
22. Design and Research Institute of Water Management project
23. Drăgan-Bularda M, Grigore CE, Țura D (2004) Utilizarea indicatorului enzimatic al calității nămolului în scopul valorificării și protecției lacurilor saline (Utilisation of the enzymatic indicator of mud quality for exploitation and protection of salt lakes) Studia Universitatis Babeș-Bolyai (Biologia series), vol XLIX, pp 129–140

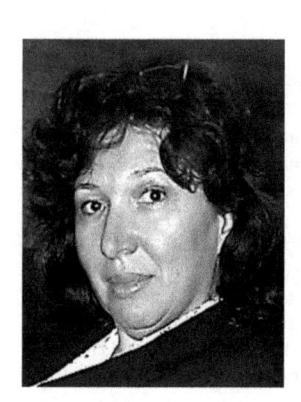

Carmen Maftei is Professor at the Building Services Department of Transylvania University of Brasov, Romania. She attended Polytechnic Institute of Iasi where she majored in Land Reclamation in 1988 and from 1996–1998 she attends a specialization in Water resources and protection of water resources, equivalent of Master of Science at Technical University of Construction Bucharest. In 2002 she finished the Ph.D. studies in Water Science in Continental Environment in joint supervision at Montpellier University and Ovidius University. She's research activity is focused on: (i) hydrology and hydraulic modelling; (ii) Geographic Information Systems applications in environmental sciences; (iii) remote Sensing applications in environmental sciences. As a result of research, she holds 11 books and books chapter, over 100 scientific papers published in different journals or conference proceedings, 1 patent, 3 international projects, 18 national grants, 10 research contracts with economic partners. She is members in many editorial boards of scientific journals. Her 27 years teaching experience includes hydrology, hydraulics, Geographic Information Systems & Applications at "Ovidius" University of Constanta and Transylvania University of Brasov.

Gabriel Dobrică was born in Medgidia City, Constanta, Dobrogea, Romania in 1988. He received the B.S. degree in civil engineering from "Ovidius" University of Constanta, Romania, in 2012, also degree in geodetic engineering from University of Agronomic Sciences and Veterinary Medicine of Bucharest, Romania, in 2014 and the M.S. degree in civil engineering from "Ovidius" University of Constanta, Romania, in 2014. He is currently pursuing the Ph.D. degree in Civil engineering and installations at "Ovidius" University of Constanta, Doctoral School of Applied Sciences, Romania. From 2016 up to present, he is working at the National Administration of "Romanian Waters". His research interest includes the evolution of coastal erosion on the Romanian coast, this is also the title of the research topic of the doctoral thesis. Mr. Gabriel Dobrică was member of the target group of the project "Academic excellence and entrepreneurial values scholarship system to ensure opportunities for training and development of entrepreneurial skills of doctoral and postdoctoral students—*ANTREPRENORDOC*", in 2019.

Constantin Cerneagă was born in Babadag City, Tulcea, Dobrogea, Romania in 1984. He received the B.S. degree in civil engineering from "Ovidius" University of Constanta, Romania, in 2014 and the M.S. degree in civil engineering from "Ovidius" University of Constanta, Romania, in 2016. He is currently pursuing the Ph.D. degree in Civil engineering and installations at "Ovidius" University of Constanta, Doctoral School of Applied Sciences, Romania. From 2016 up to present, he is working at the "Romanian Waters" National Administration—Dobrogea Water Branch. His research interest includes flood hazard mapping and modelling using GIS application.

Nicusor Buzgaru born in Constanta, Romania in 1983, received his B.S. degree in environmental engineering at the University of Agronomic Sciences and Veterinary Medicine of Bucharest, Romania in 2007 and his M.S. degree in Integrated Environmental Management at the "Ovidius" University of Constanta, Romania in 2008. In 2020 he started Ph.D. studies in engineering and installations at "Ovidius" University of Constanta, Doctoral School of Applied Sciences, Romania. He is currently deputy director at the Dobrogea-Littoral Water Basin Administration. In 2013 he was appointed Chief of the Danube Hydrotechnical System for the Constanta County where he started working in the flood defence, drought measures and accidental pollutions strategy for the Constanta County. In 2016 he was part of the collective that elaborated the first Flood Risk Management Plan for the Constanta and Tulcea counties, as well as the Danube basin.

Statistical Assessment of the Water Quality Using Water Quality Indicators—Case Study from India

Alina Bărbulescu, Lucica Barbeș, and Cristian-Ștefan Dumitriu

Abstract Surface water quality is permanently subject to major changes due to anthropogenic and natural factors. Therefore, its permanent monitoring needs to be realized by different methods, one of them being the multiparametric statistical analyzes that could contribute to a possible medium and long term forecast. In this study, we use different water quality indexes (WQIs) to assess the water quality of Sutlej, one of the most important rivers in India. To achieve our goal, the most significant water parameters were utilized: water temperature (T), pH, dissolved oxygen (DO), electrical conductivity (C), biochemical oxygen demand (BOD), nitrites and nitrates (NO_2^- and NO_3^-), fecal coliform (FC), total coliform (TC). The results show that the water quality can be classified as *borderline* (British Columbia WQI), *fair* (CCME WQI), or *very poor* (Weighted WQI). The water quality is better at the first ten observation stations and worse when the WQI is computed using the fecal coliform (or total coliform). The chapter demonstrates that the main source of water pollution for the river under investigation is the total coliform.

Keywords Statistical assessment · Water quality · Total coliform · Water quality index

---

A. Bărbulescu
Department of Civil Engineering, Transilvania University of Brașov, 5, Turnului Str., Brașov, Romania
e-mail: alina.barbulescu@unitbv.ro

L. Barbeș (✉)
Department of Chemistry and Chemical Engineering, Ovidius University of Constanta, 124, Mamaia Av., Constanța, Romania
e-mail: lbarbes@univ-ovidius.ro

C.-Ș. Dumitriu
S.C. Utilnavorep S.A., 55 Aurel Vlaicu Av., 900055 Constanța, Romania

1 Introduction

Assessing the water quality conditions in most countries has become a critical issue in recent years, especially due to the concerns that freshwater will be a scarce resource in the future. Water quality monitoring is a useful tool not only to assess the impact of pollution sources but also to ensure efficient management of water resources and protection of aquatic life [1]. Available sources of drinking water are surface waters (rivers, lakes, streams), groundwater (deep and shallow aquifers), and glaciers. The high content of chemicals and microorganisms (natural or anthropogenic origin) characterizes the quality of water (contamination or pollution). Although 70% of the earth's surface is covered by water, 2.5% of this water is fresh and only 0.3% of this water is available for human use, according to the United Nations World Water Assessment Program [2]. Also, the pressure on these resources is constantly increasing. Surface water undergoes continuous changes that influence water quality parameters due to erosion and sedimentation processes [3, 4].

The quality of surface water is increasingly affected by pollution with various chemical compounds. Sources of surface water pollution are classified as diffuse and point. The diffuse contaminants are represented by pollutants from agricultural and industrial activities, while the rest come from the chemical industry, mining, or households [5]. Their transport can be described and predicted by mathematical models. Descriptive statistics, graphical representations, and multivariate statistics are used to study the hydrogeochemical characteristics of water. Graphic methods include the Durov diagram, the Stiff diagram, the Piper diagram, and the Gibbs diagram. Common methods of multivariate statistical analysis include factorial, Principal Component Analysis, and cluster analysis [6–11].

Drinking water quality is an issue of primary interest for the population's survival. Water is essential to sustain life, and human beings must assure the availability of this resource for them and the future generation. Therefore, continuous surveillance and management of this limited resource are necessary. This preoccupation is reflected, as a hot topic, in the researches all over the world [12–16]. Failure to treat efficient water sources and unsafe distribution of treated drinking water can expose the community to the risk of disease or other adverse health effects [17, 18].

Surface water is much more vulnerable than groundwater because it can be contaminated by various sources. Therefore, when choosing a water source from which a system will be supplied, the geological and hydrological conditions should be taken into account, together with the climatic influences in the respective area. Cleansing waters are also important for their use and reuse for drinking and industrial activity. Therefore, modern technologies have been developed mainly based on ecological—friendly materials and technologies [19–21].

Water quality is evaluated according to the following categories of indicators: (1) hydromorphological (water depth, flow, width, level); (2) physical (water temperature, pH, electrical conductivity, transparency, turbidity, temporary, permanent, and total hardness); (3) chemical (chemical oxygen demand—COD and biochemical oxygen demand BOD); (4) nutrients (nitrogen and phosphorus compounds) and different metals (Cd, Hg, Zn, Cr, Cu, Ni, As, Ag, Mo, Se, Co); (5) organic and inorganic micropollutants (detergents, pesticides, phenols, cyanides, hydrocarbons); (6) biological (plankton, benthic algae, macrozoobenthos) and (7) microbiological (coliforms, streptococci) [19, 22].

The knowledge about the water resources is important for the sustainable development of any country, the sources conservation, and for initiating and implementing strategic programs that monitor physicochemical, microbiological, and ecological parameters. The Report of the Ministry of Environment & Forest from India presents the situation of the different water bodies, showing that more than 70% of the freshwater can't be used for consumption [15, 23]. The Ganges and the Yamuna are among the most polluted rivers in the world, the last one being covered with toxic froth, provoked by industrial waste deposition [24–27]. The interest in monitoring and solving the pollution issues in India is emphasized by numerous articles that estimate the river water quality, especially by using water quality indicators [15, 23, 28–31]. It is mentioned that the official water quality indicator for India is the National Sanitation Foundation Water Quality Index (NSFWQI) [29, 30].

According to international criteria, the water quality is assessed by using several parameters, as water temperature (T), pH, electrical conductivity (EC), dissolved oxygen (DO), biological oxygen demand for 5 days at 20 °C (BOD5), chemical oxygen demand (COD), total dissolved solids (TDS), total suspended solids (TSS), total phosphate, nitrite, nitrate, ammonia ions, heavy metals, total salt concentration, fecal coliform (FC) and total coliform (TC). The water quality management in India is performed under the provision of the Water (Prevention and Control of Pollution) Act, 1974 [32]. The Central Pollution Control Board introduced the concept of "designated best use", for the classification of water bodies, based on the criteria: pH, DO, EC, BOD5, total coliform, and fecal coliform, establishing critical limits for these criteria. The pH value represents the degree of acidity or alkalinity of a water sample. Dissolved oxygen represents the amount of total oxygen dissolved in a water body. Biochemical oxygen demand represents the quantity of oxygen required by aerobic microorganisms for complete degradation of organic waste from a water body, being an indicator of organic pollution. All these parameters are essential for water quality [23, 29]. In this context, to assess the water quality of Sutlej, one of the most important rivers from India, the authors employed different water quality indexes (WQIs), using the above parameters and compared the values obtained.

2 Materials and Methods

2.1 Water Quality Indicators

Numerous water quality indicators have been formulated all over the world for prompt and efficient estimating the overall water quality. Some of these parameters are, for example, the U.S. National Sanitation Foundation Water Quality Index (NSFWQI),

British Columbia Water Quality Index (BCWQI), Canadian Council of Ministers of the Environment Water Quality Index (CCMEWQI), Oregon Water Quality Index (OWQI), etc. These indexes are based on the comparison of the water quality parameters to regulatory standards and give a single value to the water quality of a source. In the following, we present three WQI, used in the present study. They are BCWQI, CCME WQI, and a weighted arithmetic index.

2.1.1 British Columbia Water Quality Index (BCWQI) [33]

BCWQI incorporates three elements:

- F_1—the percentage of variables that do not meet their objectives at least once during the study period, relative to the total number of measured variables;
- F_2—the percentage of individual tests that do not meet the objectives;
- F_3—the amount by which objectives are not met as the maximum deviation for any objective.

This factor is computed as

$$\frac{measured\ value - objective}{measured\ value} * 100,$$
$$\text{or} \quad (1)$$
$$\frac{objective - measured\ value}{measured\ value} * 100,$$

for objectives defined as a maximum value or as a minimum value, respectively.

BCWQI is calculated as:

$$\text{BCWQI} = 100 - \sqrt{F_1^2 + F_2^2 + (F_3^2/9)}/1.453 \quad (2)$$

The factor 1.453 assumes that the index maximum value is 100. BCWQI takes values between 0 and 100, and is utilized for categorizing the water quality as *Excellent* (0–3), *Good* (4–17), *Fair* (18–43), *Borderline* (44–59), or *Poor* (60–100).

2.1.2 Canadian Council of Ministers of the Environment Water Quality Index (CCMEWQI) [34]

CCME WQI is used for emphasizing the water quality, by collecting information about different physicochemical parameters during a certain period and synthesizing the data into a single number, easy to communicate to a large audience. The CCME WQI incorporates three elements:

- *Scope*—F_1—computed as for BCWQ,
- *Frequency*—F_2—computed as for BCWQ, and

- *Amplitude*—F_3—the amount by which failed test values do not meet their objectives, computed in the following stages:

a. Compute the "excursions", by:

$$excursion_i = \frac{Failed\ test\ value_i}{Objective_j} - 1, \qquad (3)$$

when the test value must not exceed the objective, or by:

$$excursion_i = \frac{Objective_j}{Failed\ test\ value_i} - 1, \qquad (4)$$

when the test value must not fall below the objective;

b. Compute the normalized sum of excursions, *nse*, as:

$$nse = \frac{\sum_{i=1}^{n} excursion_i}{total\ number\ of\ tests} \qquad (5)$$

c. Finally, determine F_3 by:

$$F_3 = \frac{nse}{0.01\ nse + 0.01}. \qquad (6)$$

The CCME Water Quality Index (CCME WQI) is calculated by:

$$CCME\ WQI = 100 - \sqrt{F_1^2 + F_2^2 + F_3^2}/1.732,$$

and takes values between 0 and 100, that are used for ranking the water quality as *Excellent* (95–100), *Good* (80–94), *Fair* (65–79), *Marginal* (45–64) or *Poor* (0–44). BCWQI and CCME WQI were designed to be applied using the physicochemical parameters.

The third index employed here is a weighted index, computed based on the formula [30]:

$$WWQI = (\sum_{i=1}^{n} w_i Q_i) / \left(\sum_{i=1}^{n} w_i\right), \qquad (7)$$

where Q_i and w_i are the quality index and the weights allocated to the ith parameter,

$$Q_i = 100 \times (V_i - V_0)/(S_i - V_0), \qquad (8)$$

V_i is the estimated concentration of the ith parameter in the analyzed water, V_0 is the ideal value of this parameter in pure water. $V_0 = 0$, for all parameters but pH for which $V_0 = 7.0$, and DO (for which, $V_0 = 14.6$ mg/l). S_i is the recommended standard value of the ith parameter.

The unit weight (w_i) for each water quality parameter is calculated by using the formula:

$$w_i = \frac{1/S_i}{\sum_{i=1}^{n}(1/S_i)} \tag{9}$$

Taking into account the index values, the water quality is classified as follows: *Excellent* (0–25), *Good* (26–50), *Poor* (51–75), *Very poor* (76–100), or *Unsuitable for drinking* (above 100).

Here, we present the indexes computed in different scenarios:

1. using six physico-chemical parameters;
2. utilizing the physicochemical parameters and the fecal coliform;
3. using the same parameters as in 1. and the total coliform.

The WQIs have been determined for the series containing the recorded data at all the hydrological stations from Sutlej River and at each station, during the study period (2007–2017). Comparisons of the values of the WQIs are provided in Results and Discussion.

2.2 Study Area

The Satluj River (Fig. 1) has the origin in the Tibetan Plateau (the southern slopes of Mount Kailash) at an elevation of more than 4,500 m above the see level. It flows from west to southwest, entering India in Himachal Pradesh.

The River's total length is 1448 km, being the largest among the rivers from Himachal Pradesh. After leaving this province, it enters the plains of Punjab at Bhakra, where the Bhakra Reservoir has been built. The total drainage area up to Bhakra Reservoir is about 56,500 km². The Sutlej finally drains into the Indus valley in Pakistan [35]. To assess the water quality of Sutlej River, the mean annual data of the following water parameters were used: temperature (T), pH, dissolved oxygen (DO), electrical conductivity (EC), biological oxygen demand (BOD), fecal coliform. They were collected for the period 2007–2017 at 22 hydrological stations (denoted by S1–S22) situated along the river and are published on the official site ENVIS Centre on Control of Pollution Water, Air, and Noise, Govt of India [36].

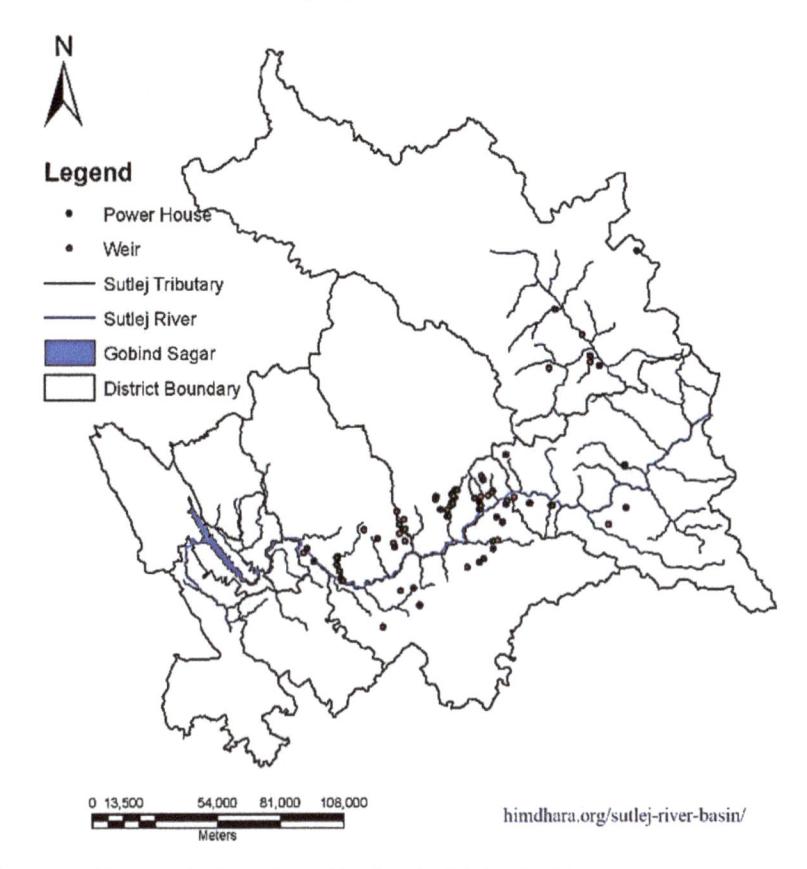

Fig. 1 Sutlej River basin (https://www.himdhara/sutlej-river-basin)

3 Results and Discussions

Figure 2 shows the data series of the 22 hydrological stations for DO and EC, whereas Table 1 contains the basic statistics of the study series, that are: mean, standard deviation (sd), variance (var), minimum (min), maximum (max), range, skewness (skew), kurtosis (kurt), coefficient of variation (cv).

In terms of coefficient of variation, the smallest variations of the parameters' values are recorded for pH, DO and temperature, the highest ones corresponding to the fecal and total coliform. The highest kurtosis is that of the FC series. The standard deviations of TC, FC, and EC series are very high, showing a high dispersion about the average. These variations are due to the uneven distribution of pollutants (especially TC and FC) along the river. A large amount of fecal and total coliform are noticed at stations S11, S14–S17 during the study period.

The boxplots (Fig. 3) emphasize the existence of outliers for the majority of the series, especially for TC, FC, BOD, Nitrate + Nitrites.

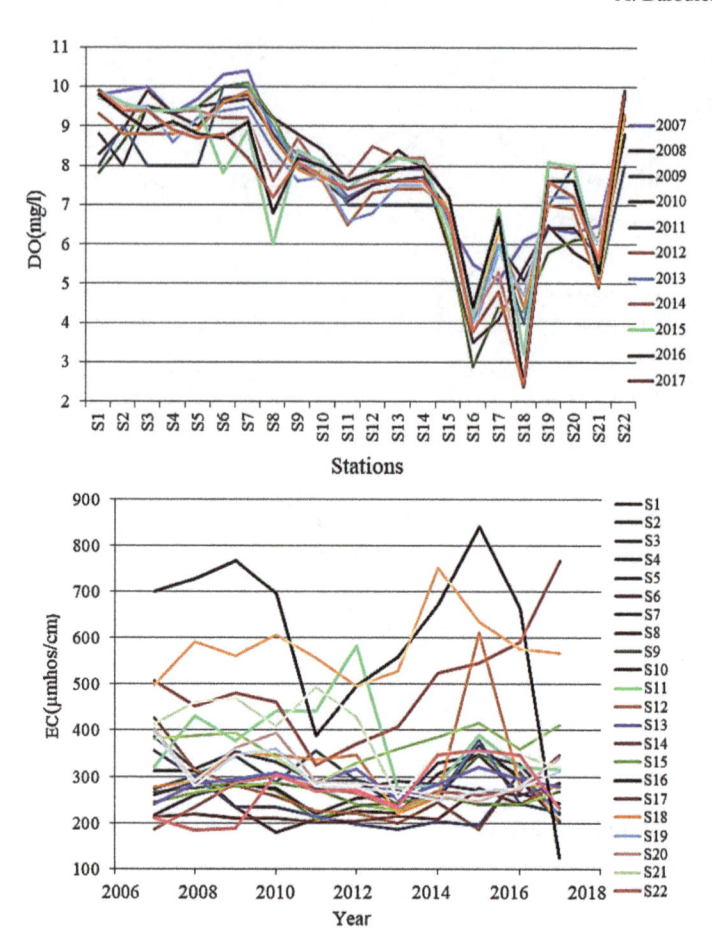

Fig. 2 DO (mg/l) and EC (μmhoms/cm) series recorded at the hydrological stations (S1–S22) during the study period

The results of the computations of BCWQI and CCME WQI for each station and the entire river are summarized in Table 2. The data in the second and fifth columns are obtained using T, DO, pH, EC, BOD, nitrate + nitrite, those from the third and sixth columns (fourth and seventh) using the previous parameters, and the fecal coliform (total coliform).

According to CCME WQI, the water quality is *fair*, and according to BCWQI, it is *borderline* (see the last line of Table 2). The water quality is *excellent* at the stations S1–S10, and S14. Based on BCWQI, the water quality is *poor* at S11–S13, and S19–S22, *borderline* at S17 and S18, and *fair* at S16. According to CCME WQI, the water quality is *poor* (or *marginal*) at S16, when using FC or TC in the calculation (or when using only the physicochemical) water parameters. The water quality is classified as *marginal* at S18 and S17 (when using FC and TC) and belongs to the category *fair* at S11, S22, S15 (only when using FC or TC in computation), and S12

Table 1 Recommended values of the water parameters and basic statistics of the study series

	Temp. (°C)	DO (mg/L)	pH	EC (μmhos/cm)	BOD (mg/l)	Nitrate + nitrite (mg/l)	FC (MPN/100 ml)	TC (MPN/100 ml)
	>22	>5	6.5–8.5	<1000	<3	<45	<2500	<5000
Mean	17.160	7.666	7.686	327.30	3.031	1.8853	2514.828	6448.89
sd	5.035	1.734	0.371	126.48	7.621	1.9415	11098.992	21124.24
var	25.354	3.007	0.138	15997.77	58.077	3.7694	123187613.715	446233601.59
min	4	2.4	6.4	2.18	0	0	0	3
max	28.70	10.4	8.5	840	88	12	140,000	210,000
Range	24.70	8	2.1	837.82	88	12	140,000	209,997
skew	−0.510	−0.839	−0.402	1.53	7.090	2.1864	8.971	5.71
kurt	−0.444	0.288	−0.326	2.96	67.891	6.9000	100.860	41.98
cv (%)	29.28	22.62	4.83	38.64	251.47	102.98	441.34	327.56

(only when using TC in computation). At stations S13, S19, S20, S22, S12 (when using the physicochemical parameters and the physicochemical parameters + FC) the water quality is *good*. Remark that there are discrepancies in the classifications of the water quality when using BCWQI and CCME WQI.

The charts of the BCWQI and CCME WQI variations for the series S11–S22 are presented in Figs. 4 and 5.

The WWQI computed for the entire series has the values 77.91, 79.54, and 79.55, respectively, so the water can be classified as very poor. The water quality at the stations can be classified as good at S1–S9, S14, S22; poor at S10, S12, S13, S15, S19, S20; very poor at S21 and unsuitable for drinking purposes at S11, S16–S18. Remark that are based on different computational methods, the category to which the water belongs is many times different due to the absence of a unified classification scale. Another drawback is that the methodology doesn't always provide standardized values of the index (between 0 and 100), a fact that raises questions related to the 'best' WQI to be employed.

Most of the approaches for determining WQIs that are using the pH series don't take into account both boundaries (6.5 and 8.5, in our case), a fact that is questionable, and, in our opinion, could introduce bias in the water quality classification provided. This is the case of the first and third indexes used here. The class to which a certain type of water belongs could be different when using the upper bound instead of the lower one for the WQI calculation. For example, if a WQI is computed for each year and each station, utilizing formula (7) for Sutlej River, in 35 cases the class changed when using the upper limit (8.5) instead of the lower one (6.5) for pH. The corresponding percentage (14.45% of the individual WQIs) is high, so it is critical that the issue should be addressed accordingly in the future.

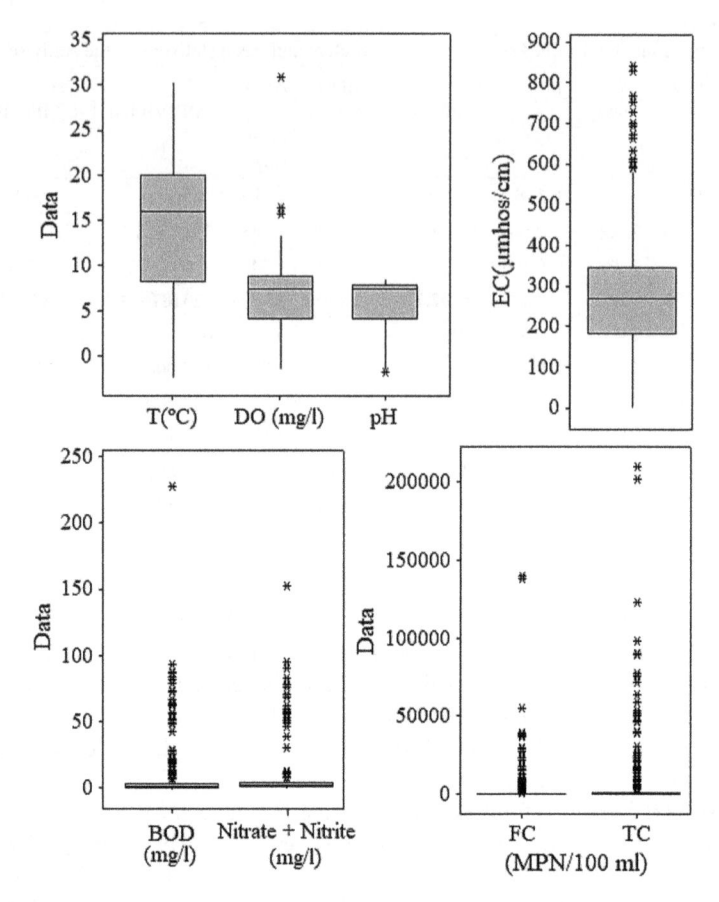

Fig. 3 Boxplots of data series

4 Conclusions

In this study, we demonstrated that the water quality of the Sutlej River, India, necessitates the authorities' attention due to its quality, which is not the best one, especially due to the high fecal (and total coliform) concentrations noticed at three stations during the entire study period. Despite the limitations related to the BCWQI and CCME WQIs' computation (some of them already mentioned), the use of these indicators is an important tool that gives a comprehensive view of the water quality. Employing different indexes in the same analysis could reduce the risk of water quality misclassification.

Table 2 BCWQI and CCME WQI

Index	BCWQI			CCME WQI		
Station	Phys-ch	FC	TC	Phys-ch	FC	TC
S1	0	0	0	100.00	100.00	100.00
S2	0	0	0	100.00	100.00	100.00
S3	0	0	0	100.00	100.00	100.00
S4	0	0	0	100.00	100.00	100.00
S5	0	0	0	100.00	100.00	100.00
S6	0	0	0	100.00	100.00	100.00
S7	0	0	0	100.00	100.00	100.00
S8	0	0	0	100.00	100.00	100.00
S9	0	0	0	100.00	100.00	100.00
S10	0	0	0	100.00	100.00	100.00
S11	70.18	64.89	64.89	78.62	73.77	73.79
S12	76.87	80.15	68.68	80.67	83.44	75.14
S13	88.44	90.08	90.08	90.34	91.72	91.72
S14	0	0	0	100.00	100.00	100.00
S15	73.83	64.61	64.84	80.22	74.31	74.36
S16	42.77	37.53	37.12	50.27	39.10	38.62
S17	57.39	51.08	49.40	65.38	57.89	52.96
S18	55.65	49.75	48.57	63.89	57.00	52.84
S19	75.98	79.26	79.26	80.26	83.08	83.08
S20	74.74	77.78	77.78	80.42	83.22	83.22
S21	62.81	67.41	67.41	70.76	74.94	74.94
S22	88.39	90.02	90.02	90.34	91.72	91.72
Total	**58.84**	**54.43**	**54.46**	**70.41**	**65.39**	**65.00**

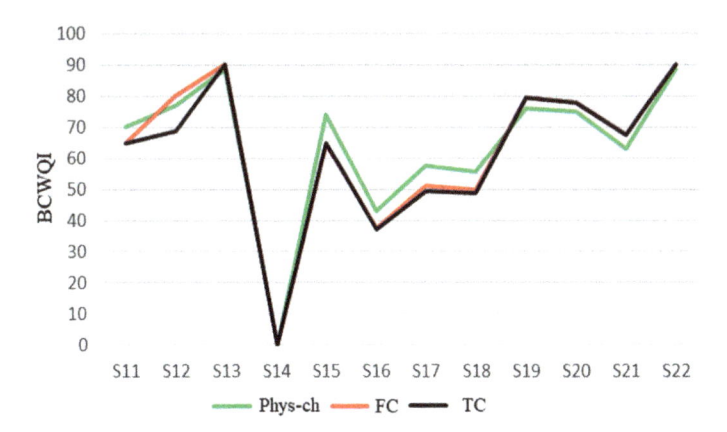

Fig. 4 BC WQI for the series S11–S22

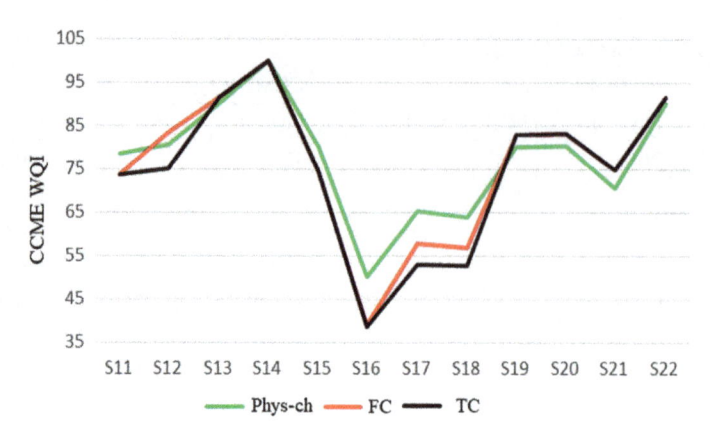

Fig. 5 CCME WQI for the series S11–S22

References

1. Water Research Center (2020) Monitoring the quality of surface waters. https://water-research.net/index.php/water-treatment/water-monitoring/monitoring-the-quality-of-surfacewaters. Accessed 13 Nov 2020
2. UNESCO WWAP (2020) Water and climate change. The UN World Water Development Report 2020, Paris, UNESCO. https://unesdoc.unesco.org/ark:/48223/pf0000372985.locale=en. Accessed 13 Nov 2020
3. Aminiyan MM, Aminiyan FM, Heydariyan A (2016) Study on hydrochemical characterization and annual changes of surface water quality for agricultural and drinking purposes in semi-arid area. Sustain Water Resour Manag 2:473–487
4. Issa IE, Al-Ansari N, Knutsson S, Sherwany G (2015) Monitoring and evaluating the sedimentation process in Mosul Dam reservoir using trap efficiency approaches. Engineering—London 7:190–202
5. Ustaoğlu F, Tepe Y (2018) Water quality and sediment contamination assessment of Pazarsuyu Stream, Turkey using multivariate statistical methods and pollution indicators. Int Soil Water Conserv Res 7(1):47–56
6. Bărbulescu A (2016) Studies on time series. Applications in environmental sciences. Springer
7. Bărbulescu A, Barbeş L (2020) Assessing the water quality of the Danube River (at Chiciu, Romania) by statistical methods. Environ Earth Sci 79(6):122. https://link.springer.com/article/10.1007/s12665-020-8872-1
8. Bărbulescu A, Nazzal Y, Howari F (2020) Assessing the groundwater quality in the Liwa area, the United Arab Emirates. Water 12(10):2816. https://doi.org/10.3390/w12102816
9. Chounlamany V, Tanchuling MA, Inoue T (2017) Spatial and temporal variation of water quality of a segment of Marikina River using multivariate statistical methods. WA Sci Technol 76(6):1510–1522
10. Mishra A (2010) Assessment of water quality using principal component analysis: a case study of the river Ganges. J Water Chem Technol 32:227–234. https://doi.org/10.3103/S1063455X10040077
11. Popa P, Murariu G, Timofti M, Georgescu LP (2018) Multivariate statistical analyses of water quality of Danube River at Galati, Romania. Environ Eng Manag J 17:491–509
12. Bărbulescu A (2020) Assessing the groundwater vulnerability: DRASTIC method and its versions. A review. Water 12(5):1356. https://doi.org/10.3390/w12051356
13. Bărbulescu A, Barbeş L (2013) Assessment of surface water quality Techirghiol Lake using statistical analysis. Rev Chim-Bucharest 64(8):868–874

14. Bărbulescu A, Barbeş L, Dani A (2020) Statistical analysis of the quality indicators of the Danube River (in Romania). In: Naddeo V, Balakrishnan M, Choo KH (eds) Frontiers in water-energy nexus, nature-based solutions, advanced technologies and best practices for environmental sustainability, advances in science, technology & innovation (IEREK interdisciplinary series for sustainable development), Springer, Cham, pp 277–279 (ISBN: 978-3-030-13067-1). https://link.springer.com/chapter/10.1007/978-3-030-13068-8_69
15. Dani A, Bărbulescu A (2020) Statistical analysis of the water quality of the major rivers in India. In: Naddeo V, Balakrishnan M, Choo KH (eds) Frontiers in water-energy-nexus-nature-based solutions, advanced technologies and best practices for environmental sustainability, advances in science, technology & innovation (IEREK interdisciplinary series for sustainable development). Springer, Cham, pp 281–283 (ISBN: 978-3-030-13067-1). https://doi.org/10.1007/978-3-030-13068-8_70
16. Jiang C et al (2020) Characteristics and causes of long-term water quality variation in Lixiahe abdominal area, China. Water 12(6):1694. https://doi.org/10.3390/w12061694
17. Le Bradford et al (2016) Drinking water quality in Indigenous communities in Canada and health outcomes: a scoping review. Int J Circumpol Heal 75(1):32336. https://doi.org/10.3402/ijch.v75.32336
18. Chirila E, Bari T, Barbeş L (2010) Drinking water quality assessment in Constanta town. Ovidius Univ Ann Chem 21(1):87–90
19. Bărbulescu A, Barbeş L (2021) Statistical methods for assessing water quality after treatment on a sequencing batch reactor. Sci Total Environ 752:141991. https://doi.org/10.1016/j.scitotenv.2020.141991
20. Bărbulescu A, Duteanu N, Negrea A, Ghangrekar MM (2018) New trends in monitoring and removing the pollutants from water. J Chem U A R 2018:8394086. https://doi.org/10.1155/2018/8394086
21. Gabor A et al (2017) Optimizing the lanthanum adsorption process onto chemically modified biomaterials using factorial and response surface design. J Environ Manage 204(3):839–844. https://doi.org/10.1016/j.jenvman.2017.01.046
22. Teodosiu C, Robu B, Cojocariu C, Barjoveanu G (2015) Environmental impact and risk quantification based on selected water quality indicators. Nat Hazards 75(1):89–105
23. Bărbulescu A, Dani A (2019) Statistical analysis of the water quality of the major rivers in India. Rom Rep Phys 71(4):716. http://www.rrp.infim.ro/2019/AN71716.pdf
24. Chalasani R (2018) Toxic foam pollutes India's sacred Yamuna River. https://abcnews.go.com/International/toxic-foam-pollutes-indias-sacred-yamuna-river/story?id=57995346
25. GhaggarReport (2010) Report on pollution problem of River Ghaggar. www.indiaenvironmentportal.org.in/files/file/GhaggarReport.pdf
26. Kaushik A, Sharma HR, Jain S, Dawra J, Kaushik CP (2010) Pesticide pollution of river Ghaggar in Haryana, India. Environ Monitor Assess 160(1–4):61–69
27. CPCB Polluted river stretches in India. Criteria and status. Central Pollution Control Board. https://www.cpcb.nic.in/wqm/RS-criteria-status.pdf. Accessed 13 Nov 2020
28. Banerjee T, Srivastava RK (2009) Application of water quality index for assessment of surface water quality surrounding integrated industrial estate-Pantnagar. Water Sci Technol 60(8):2041–2053. https://doi.org/10.2166/wst.2009.537
29. Bora M, Goswami DC (2017) Water quality assessment in terms of water quality index (WQI): case study of the Kolong River, Assam, India. Appl Water Sci 7(6):3125–3135. https://doi.org/10.1007/s13201-016-0451-y
30. Bhutiani R et al (2016) Assessment of Ganga river ecosystem at Haridwar, Uttarakhand, India with reference to water quality indices. Appl Water Sci 6:107–113. https://doi.org/10.1007/s13201-014-0206-6
31. Ponsadailakshmi S et al (2018) Evaluation of water quality suitability for drinking using drinking water quality index in Nagapattinam district, Tamil Nadu in Southern India. Groundwater Sustain Develop 6:43–49. https://doi.org/10.1016/j.gsd.2017.10.005
32. Water Act (1974). https://pcbassam.org/rules/WaterAct.pdf

33. Zandbergen PA, Hall KJ (1998) Analysis of British Columbia Water Quality Index for warweshed managers: a case study of two small watersheds. Water Qual Res J Canada 33(4):519–549
34. Canadian Council of Ministers of the Environment (2001) CCME Water Quality Index 1.0 User's Manual 2017 Update. www.ceqg-rcqe.ccme.ca/download/en/138. Accessed 10 Oct 2020
35. Pal I et al (2013) Predictability of Western Himalayan river flow: melt seasonal inflow into Bhakra Reservoir in northern India. Hydrol Earth Syst Sci 17:2131–2146. https://doi.org/10.5194/hess-17-2131-2013
36. ENVIS Centre on Control of Pollution Water, Air and Noise, Water Quality Database. http://www.cpcbenvis.nic.in/water_quality_data.html#. Accessed 10 Sept 2020

Alina Bărbulescu graduated from the University of Craiova (Romania), Faculty of Mathematics, and from the Petre Andrei University of Iași (Romania), Faculty of Law. After a Master in Mathematics at Bucharest University (Romania), she got a Ph.D. in Mathematics from Al. I. Cuza University of Iasi (Romania), a Ph.D. in Cybernetics and Economic Statistics from the Academy of Economic Studies Bucharest (Romania), and a Ph.D. in Civil Engineering, with Magna cum Laude, at the Technical University of Civil Engineering, Bucharest (Romania). She got the habilitation in Civil Engineering in 2014, with the thesis *Modeling the spatially distributed precipitations*, and in Cybernetics and Economics Statistics in 2019, with the thesis *Contribution to statistical analysis and modelling in economics and finance*. She is a Professor at Transylvania University of Brașov, Department of Civil Engineering. Before her current position, she taught at the Ovidius University of Constanța, Romania, Higher Colleges of Technology, U.A.E and served as a visiting professor at Abu Dhabi University. Her research interests span applied statistics, econometrics, hydrological modelling, water and air pollution. She has authored over 170 articles, 30 books and book chapters, was co-editor of 17 special issues of scientific journals and conference proceedings and presented invited lectures at international conferences. She is a member of IAHS, IAENG, SSMR, and topic editor at different journals among which *Water, Atmosphere* (*Clarivate Analytics*).

Lucica Barbeş got the B.Sc. in Organic Substances of Technology in 1995 and Ph.D. in Chemical Engineering from the Politehnica University of Bucharest, Romania, in 2003. She is currently Associate Professor of Chemistry and Chemical Engineering Department, Ovidius University of Constanţa, Romania. She is the single author or co-author of 4 books, holds one patent (2003), published 50 research articles, of which 30 are included in WOS, Scopus, or ACS database. She has earned the Marie Curie grant from the French National Centre for Scientific Research (2003–2004), the individual mobility training staff Lifelong Learning Programme—Erasmus at universities from Italy (2011), or Spain (2012). She was invited speaker to the Interdisciplinary Nanoscience Center, University of Aarhus, Denmark (2006), a participant of the NATO Advanced Study Institute on Chemicals as Intentional and Accidental Global Environmental Threats in Borovetz, Bulgaria (2005), and she has more than 32 participations to national and international Conferences. She serves on the Reviewer Board of Water and other highly reputed international journals. She is vice-president of the Chemistry Society of Romania—Constanta Branch, member of the Romanian Society of Chemical Engineering, the Romanian Association of Food Industry Specialists, and the Internationale Gesellschaft fur Warenwissenschaft und Technologie, Austria. Her research interests include indicators and integrated environmental monitoring, emerging pollutants in the air, water, and soil, water and wastewater treatment, marine fouling; VOCs monitoring and analysis, spectral analyses (FTIR, IRRAS), techniques of enzymes immobilization on different metallic surfaces, conventional and alternative fuels combustion.

Cristian-Ştefan Dumitriu got the B.Sc. and M.Sc. in Naval Engineering from Dunărea de Jos University of Galaţi, Romania. He got the Ph.D. in Engineering and Materials Science from Dunărea de Jos University of Galaţi, in 2006. He got the Project Management certificate and that of Project Management Process Framework in 2007. He got the Certificat Profesional în Management in 2008, from CODECS Romania (and The OPEN University, U.K). He is Head of Technical Office for Design, Technologies and Research at UTILNAVOREP S.A., Constanţa, Romania. Before his current position, he was design engineer in the Basic Design Department at Daewoo Shipyard Mangalia, Romania. His research interest includes the materials corrosion in different media, strength of materials, and environmental sciences.

Developing Village-Level Water Management Plans Against Extreme Climatic Events in Maharashtra (India)—A Case Study Approach

Aman Srivastava and Pennan Chinnasamy

Abstract Remote villages of the Sahyadri region in the Maharashtra state of India are witnessing water scarcity issues regarding drinking and irrigation. Issues are deepened by the limited surface water storage capacities against high rainfall, mismanagement of available water resources, and inadequate groundwater potential to support primary livelihood prospects in the villages. To combat this, the study developed village-level water management plans against extreme climatic events with a case study approach in semi-arid villages of Nashik district in Maharashtra state of India. Hydrological components were scientifically investigated over the five land-use classifications viz., agriculture (having 78% land area), hills (16%), water bodies (3%), built-up (1.5%), and forest (1.5%). Results indicated a sharp decline in water availability across check dams from the South-West monsoon season (storage volume attaining 567.2 thousand cubic meters (TCM) between June and September) to the following summer (0.2 TCM between March and May). Despite receiving high rainfall, (~2000 mm annually), water scarcity issues persisted accounting for outflows (in terms of evaporation, evapotranspiration, and surface runoff) as high as 94% during South-West monsoon to 64% during post-monsoon. Consequently, events of flash floods remained rampant during the South-West monsoon followed by hydrological drought (from January to May). The study concluded that preparing seasonal water budget projections could better examine the temporal variations of the various hydrological components, surface water storage capacities, and water management provisions. Such strategies may increase awareness among land and water resource managers, policy-makers, and politicians to formulate region-specific water management and security plans.

A. Srivastava (✉) · P. Chinnasamy
Centre for Technology Alternatives for Rural Areas, Indian Institute of Technology Bombay, 400076 Mumbai, India
e-mail: amansrivastava1397@gmail.com

P. Chinnasamy
e-mail: p.chinnasamy@iitb.ac.in

P. Chinnasamy
Rural Data Research and Analysis (RuDRA) Lab, Indian Institute of Technology Bombay, 400076 Mumbai, India

© Springer Nature Switzerland AG 2021
A. Vaseashta and C. Maftei (eds.), *Water Safety, Security and Sustainability*,
Advanced Sciences and Technologies for Security Applications,
https://doi.org/10.1007/978-3-030-76008-3_27

Keywords Water resources management · Extreme climatic events · Floods · Droughts · Groundwater · Sustainable development · Hydrology

1 Introduction

India, with the world's 18% human population and 11.5% livestock population on 2.4% of the landholding, uses the world's 4% of the water resources [21, 22]. Additionally, India is the largest user of groundwater resources, extracting 251 billion cubic meters (bcm) annually, and the second-largest agricultural producer in the world [8]. Being having an extensive network of river systems coupled with irrigation, there exist unique complexities in the spatial and temporal components of water availability across the country. Of the total 4,000 bcm of rainfall received annually, more than 3,000 bcm is received during the South-West (S-W) monsoon (between June and September). Approximately, 1,869 bcm is the average annual surface runoff flows in rivers and aquifers, while the remaining (~53% of the rainfall) is lost due to evaporation and evapotranspiration [18–20]. The disparity in the Indian monsoon systems has been demarcated by its magnitude. For instance, wet states like Meghalaya (North-East India) receive 12,000 millimeters (mm) of annual rainfall, while dry states like Rajasthan (West-India) receive a mere 100 mm rainfall. Out of the total utilizable water resources in the country, 690 bcm is available as surface water and 433 bcm as groundwater [21, 34]. As per the census data of 2011, the per capita water availability in India is 1,545 cubic meters/year (m^3/yr.) which is predicted to decline to 1,340 m^3/yr. by 2025 and further to 1,140 m^3/yr. by 2050. As per the international standards, the water availability per capita in the range of 1,000–1,700 m^3/yr. categorizes a nation as "water-stress" and in the case, if it falls below 1,000 m^3/yr., the nation is categorized as "water-scarce" [22]. Currently, India is witnessing water stress conditions with high vulnerability towards the water-scarcity. A rise in water demand due to the increasing cases of water scarcity simultaneously with a demographic explosion in different parts of the country is further underscoring the water scenarios [9, 15, 30]. One of the major attributes contributing to the declining water resources is the challenge associated to minimize mismanagement issues owing to irrigation apart from domestic and industrial water requirements [30].

The threat to water management and security in developing countries like India, especially the villages located in arid and semi-arid zones where rainfed agriculture is in prevalence, have been exacerbated due to climatic fluctuations, the imbalanced spatial distribution of water resources, urbanization induced land-use rapid alterations, and increased vulnerability to extreme events such as floods and droughts [2, 9, 13, 26, 33]. According to the United Nations, water security is the need of the time and it is defined as *"the capacity of a population to safeguard sustainable access to adequate quantities... of water for sustaining livelihoods, human well-being, and socio-economic development...."* [36]. But the persistent issues of spatial imbalance of water resources across rural-India have undermined the water management and security provisions, thus pushing the rural population (in view of migration) to sub-optimal livelihood alternatives [27]. As agriculture is the predominant profession in the rural areas, reduced water availability decreases agricultural productivity, and hence the income of those who are directly (farm holders) or indirectly (farm laborers) dependent on the agriculture-based income sources. The force on the rural communities to make the livelihood choice is due to the vulnerability of the rural regions towards water scarcity in terms of physical (lack of surface water and groundwater management and security for irrigation) and economical facets (lack of financial capital of an individual to extract or use available surface water and groundwater resources) [25, 31]. Moreover, the drinking water supply and distribution infrastructure is observed to be limited due to funding constraints. It is observed that the villages not having access to the water infrastructure are dependent on the nearby villages/water infrastructure. This subsequently increases the efforts and time in fetching out water from distant sources. As a result, the villages having water scarcity issues are not affected in the mere isolation, instead simultaneously affect the clusters of villages located in its vicinity, thereby causing consequential water stress [3]. As the efficacy of the water management policies and implementing strategies in the villages, requiring sustainable water management, lacks comprehensiveness, there is an urgency to devise and implement the holistic water management strategies. This will necessitate an all-inclusive participatory planning approaches by defining the roles and responsibilities of the decentralized institutions actively involved in serving to address issues of water safety, security, and sustainability [10, 32, 35].

In an attempt to address the water scarcity challenges, various schemes ensuring water supply to all households such as the National Rural Drinking Water Programme (NRDWP) in 2009 which is restructured and subsumed into Jal Jeevan Mission (JJM) in 2020 have been launched by the Government of India [18]. Besides this, research attempts have taken rigorously to understand and resolve the issues associated with water imbalance and management issues on the village-scale. Chintalapudi et al. [10] developed a groundwater security plan in water scarcity villages in the Madhya Pradesh state of India with a participatory-based approach of data collection, thereby analyzing results using GIS at a micro-watershed scale. Singh et al. [30] used the modified version of Sledger's framework of risk perception to explore and demonstrate the effectiveness of the socio-cognitive factors to highlight the adaptation behavior of the villagers on water scarcity as a case study in the Rajasthan state of India. Similar studies on a regional and global scale have been conducted for water management against surface water exploitation and groundwater depletion [1, 6, 9, 11, 12, 16, 17, 24, 29, 32]. Under these contexts, the current paper presents evidence-based research on mismanagement of water resources, issues associated with water scarcity, and challenges in addressing the problems of unreliable seasonal water availability using the knowledge of hydrological science as a case study approach in Shivaji Nagar and Holdar Nagar villages of Maharashtra state in the West-India. The overall approach is to propose location-specific sustainable conservation measures and establishing local mechanisms to secure water resources for basic needs.

1.1 Study Objectives

The primary objectives of this research are to explore the spatio-temporal variation of the various hydrological components and their impact on the available water resources within the village watershed boundary across winter, summer, S-W monsoon, and post-monsoon seasons. The secondary objectives of this study are first to develop land use land cover maps showing different land use classifications, thereby establishing the relation between land use and hydrological processes, and second to prepare watershed development strategies for achieving sustainable water management amidst recurrent events of floods and droughts.

The above-mentioned objectives require good quality and quantity of observation data, however, obtaining such data is challenging. In such instances, when observation data is limited, the use of Remote Sensing (RS) data, especially satellite imagery (using Google Earth Pro), has been widely used to address data gaps [9, 32, 33]. This paper presents an integrated study using primary data obtained through field surveys and experimentations, secondary data, and RS data, with a case study approach under rural contexts.

2 Study Area

Trimbakeshwar block (an administrative unit) of Nashik district in the Maharashtra state of India has 125 villages where despite heavy rainfall (>2,000 mm), the pre-monsoon (summer) groundwater level trend shows a fall in water level in the range of 0–0.2 m [7]. The Shivaji Nagar village and its adjacent Holdar Nagar village are the study site and situated in the South of the Trimbakeshwar block (Fig. 1). It is geographically located between longitudes 20° 6′ 49.86″ N and 20° 5′ 57.09″ N and latitude of 73° 34′ 6.93″ E and 73° 33′ 20.27″ E at an average mean sea level of 740 meters (m). The study site is a remote area located at a distance of 27 kilometers (km) from the Trimbakeshwar block and about 42 km from the Nashik district. The Shivaji Nagar village consists of three hamlets viz., Kavlyachapada (KCP), Shivaji Nagar (SN), and Satya Nagar (STN), and its Gram Panchayat (GP; village-level self-governance body in India) is located in SN. While Holdar Nagar (HN) village doesn't have any hamlets and its GP is located within the village. The population of the villages in 2019 was 1,415 and 948 for Shivaji Nagar village and Holdar Nagar villages, respectively (Table 1). Physiographically, the villages are situated in the Western region of Nashik district at the edge of Sahyadri hill (the main system of hills in the Western Ghat) having Deccan trap basaltic formations. The region is underlain by the basaltic lava flows of Upper Cretaceous to Lower Eocene age that occupies about 90% of the district area and gives rise to table type topography called a plateau. The soil, being the weathered product of basalt was observed with shades from black, brown to red with widespread outcrops of basalt rocks [7].

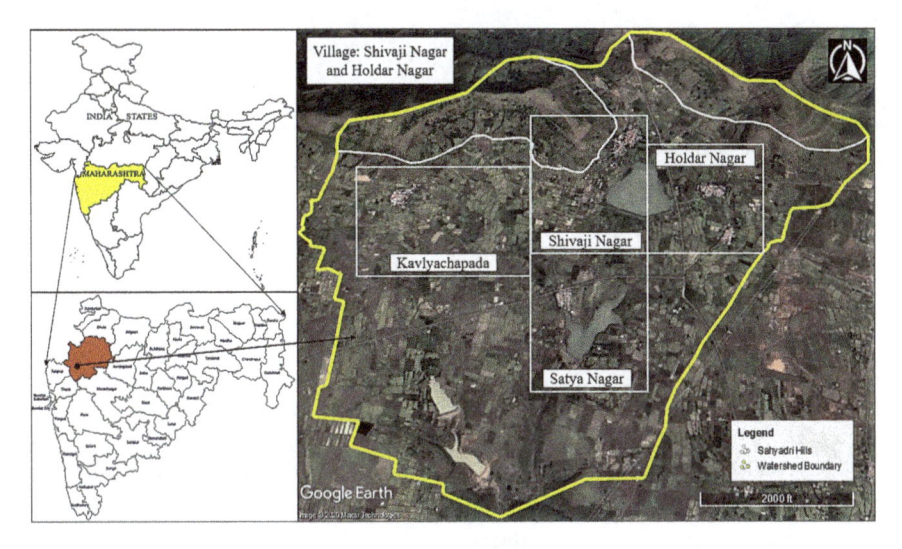

Fig. 1 Location of study site, Shivaji Nagar village (and its hamlets) and Holdar Nagar village of Nashik district (insets showing the location in Maharashtra and India)

Hydro-meteorologically, the region in and around the study site is hilly having semi-arid climatic conditions, thereby experiencing heavy annual rainfall of more than 2,000 mm, temperature averagely fluctuating between 5 and 40 °C, and relative humidity ranging from 43 to 62%. January is the driest month (averagely 0.88 mm rainfall) with the lowest temperature (averages 16 °C), while May is the warmest month (averages 28 °C) and July is the wettest month (averagely 320 mm rainfall) [4, 28]. Agriculture is the predominant occupation of the village and more than 90% of the population are engaged in the cultivation of paddy and wheat. Paddy is the major Kharif crop (cultivated in 77% of the culturable command area) cultivating extensively between July and October (during the S-W monsoon period), while wheat is the major Rabi crop (cultivated in 80% of the culturable command area) that is cultivated by few progressive farmers in the villages who are having access to water from October to February (non-monsoon period).

3 Methodology

The primary data on sectoral-specific information at large (Table 1) and water resources in specific (Table 2) of the Holdar Nagar village and the three hamlets of Shivaji Nagar village was obtained by visiting the offices of and by conducting Participatory Rural Appraisal (PRA) activities. The purpose behind conducting PRA along with training and orientation programs was to develop a common understanding of the physical and social dynamics of water resources and their management systems in the villages. Sarpanch (head of the village), Gram Sevak (village-level Government

Table 1 Basic information about study site viz., Shivaji Nagar and Holdar Nagar villages of Nashik district in Maharashtra state

S. No.	Particulars	Details	Details
1	Name of village	Shivaji Nagar	Holdar Nagar
2	Number of hamlets	03	00
3	Name of taluka/block	Trimbakeshwar	Trimbakeshwar
4	Name of the district	Nashik	Nashik
5	Population		
	2011	1,169 (M—589; F—580)	611 (M—314; F—297)
	2019	1,415 (M—703; F—712)	948 (M—488; F—460)
6	Total no. of households	150	95
7	Livestock population	320	178
8	Available water resources	Check dams and dug wells	Check dams and dug wells
9	Toilet facility in households	Partial	Partial
10	Water and sanitation committee	Absent	Absent
11	Educational facility		
	Anganwadi	03	01
	Primary school	03	01
	Middle school	00	00
	Secondary school	00	00
12	Institutions		
	Primary Healthcare Centre (PHC)	00	00
	Self-Help Groups (SHGs)	00	00
	Credit cooperative society	00	00
	General shop	03	01
	Water User Associations (WUAs)	00	00
13	Village electrification	Yes	Yes
	Power supply duration	<10 h a day	<10 h a day
14	Connectivity by all-weather road	Yes	Yes

M: Male; F: Female

secretary to GP), members of the GP, teachers from the Zilla Parishad (village level administrative body) schools, Anganwadi (rural child care center in India) workers, and local villagers participated in PRA for the preparation of the water budget and management plans.

Table 2 Existing status of water resources in the study area viz., Shivaji Nagar and Holdar Nagar villages of Nashik district in Maharashtra state

S. No.	Particulars	Shivaji Nagar village	Holdar Nagar village
1	Drinking water sources	Dug wells and springs	Dug wells and springs
2	Agricultural water sources	Rainfed, check dams, reservoirs, and private dug wells	Rainfed and check dams
3	Potential of groundwater	Poor (no tubewells/borewells found)	Poor (no tubewells/borewells found)
4	Availability of piped water supply scheme	Absent	Absent
5	Availability of rainwater harvesting scheme	Absent	Absent
6	Alternate water sources during drought/water scarcity	Tanker water supply (from private agencies, Government bodies, and NGOs)	Tanker water supply (from private agencies, Government bodies, and NGOs)
7	Sustainability of water sources across seasons	Poor water availability during winter (Jan–Feb) and summer (Mar–May)	Poor water availability during winter (Jan–Feb) and summer (Mar–May)
8	Key issues identified during field investigation	• Flash floods during S-W monsoon and hydrological drought during summer • Unmaintained check dams (silted) with reduced storage potential • Limited cropping seasons (mostly one) with unreliable rainfed based irrigation source	• High dependency on drinking water sources from Shivaji Nagar village • Unmaintained check dams (silted) with reduced storage potential • Limited cropping seasons (mostly one) with unreliable rainfed based irrigation source
9	Possible solutions identified during field investigation	Construction of additional check dams; rejuvenation of existing water storage structures; forestation; Strengthening local institutions	Construction of additional check dams; rejuvenation of existing water storage structures; forestation; mechanisms for groundwater extraction

The entire study period was classified into four seasons viz. winter (January and February), summer (March to May), S-W monsoon (June to September), and post-monsoon (October to December). The hydrological components were normalized in the units of thousand cubic meters (TCM) for comparative analysis. A reconnaissance survey using the Global Positioning System (GPS essential tool) followed by a detailed field survey and experimentations were conducted during the interface of summer and S-W monsoon season in 2019. Following this, Remote Sensing (RS) data, obtained using Google Earth Pro and substantiated with ground truth verifications, were used for demarcating watershed boundary and developing a thematic LULC map through a manual supervised classification method [32, 33] for the study site (Fig. 2). The disparity in spatial and temporal components of the monthly rainfall was analyzed using statistical methods. Soil Conservation Services—Curve Number (SCS-CN) method [34] was then employed to determine seasonal surface runoff across diverse land uses (Table 3). The data for the rates of evaporation from surface water bodies and evapotranspiration from agricultural land use were obtained from secondary sources [21, 34] (Table 4). The existing surface water storage structures (check dams) and dug wells were studied to ascertain their present status concerning quantity and functionality (Table 5). Net groundwater flow across watersheds was estimated using the water budget (hydrological) equation [34].

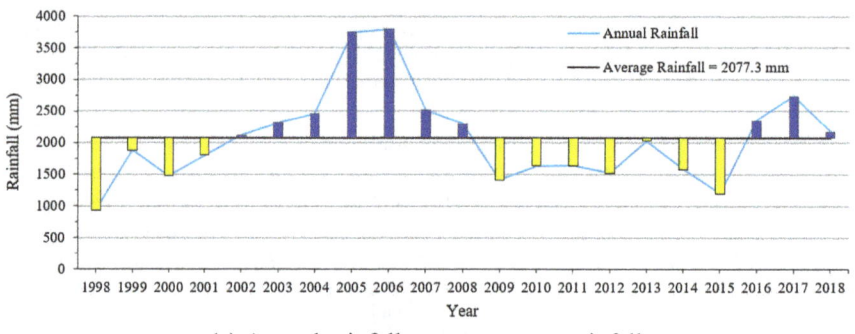

(a) Annual rainfall versus average rainfall

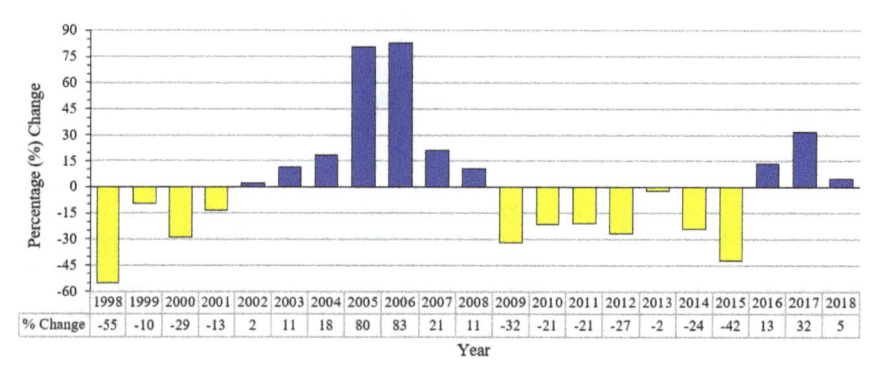

(b) Percentage departure of annual rainfall from average rainfall

Year	1998	1999	2000	2001	2002	2003	2004	2005	2006	2007	2008	2009	2010	2011	2012	2013	2014	2015	2016	2017	2018
% Change	-55	-10	-29	-13	2	11	18	80	83	21	11	-32	-21	-21	-27	-2	-24	-42	13	32	5

Fig. 2 Temporal analysis of annual rainfall between 1998 and 2018 received in Trimbakeshwar block of Nashik district in Maharashtra state

Table 3 Land use land cover assessment using Soil Conservation Services—Curve Number (SCS-CN) technique of the study area [34] viz., Shivaji Nagar and Holdar Nagar villages of Nashik district in Maharashtra state

S. No.	Land use	CN(II)	Area (m^2)	Remarks
1	Study site	79	4,353,260	Mostly observed mixed red and black soils; soil group—C (moderately high surface potential) was identified; Antecedent Moisture Conditions (AMC) were average (AMC–II)
2	Agriculture	76	3,402,671	Cultivation was based on bunded practice with good hydrologic conditions
3	Hills	91	699,128	Hard surface areas with outcrops of basalt rocks over Sahyadri range inside the study area
4	Water bodies	100	123,178	Check dams and reservoirs were mostly silted allowing higher infiltration against percolation
5	Built-up	88	65,209	Mostly a mix of the paved and unpaved roads and houses of both cement and non-cement type were observed
6	Forest	60	63,074	Open forest (moderately dense) with a diminutive scrub cover was observed in patches over Sahyadri hills

Table 4 Seasonal rates/magnitude for evaporation, evapotranspiration, and rainfall of Trimbakeshwar block in Nashik district between 1998 and 2018 [21, 28, 34]

Season	Months	Days	Evaporation (mm/d)	Evapotranspiration (mm/d)	Rainfall (mm)	% Rainfall
Winter	Jan, Feb	59	5.0	2.0	4.15	0.20
Summer	Mar, Apr, May	92	11.0	5.0	25.34	1.22
S-W Monsoon	Jun, Jul, Aug, Sept	122	6.0	1.0	1,913.40	92.11
Post-Monsoon	Oct, Nov, Dec	92	4.5	2.0	134.40	6.47
Total	Year	365			2,077.30	100.00

4 Results

4.1 Rainfall Analysis

Trimbakeshwar block, with an average annual rainfall of 2,077 mm, was the second-highest rain receiving block (among 15 blocks) between 1998 and 2018 across the Nashik district. This was ~198% more than the average rainfall received in the Nashik district (~1,051 mm) with a coefficient of variability estimating 35%, in the rainfall

Table 5 Water storage capacity assessment in the study area viz., Shivaji Nagar and Holdar Nagar villages of Nashik district in Maharashtra state

Water sources	Location	Area (m^2)	Depth (m)	Total capacity (m^3)
Check Dam—1	SN-HN	64,684	3.75	242,565.00
Check Dam—2	STN	36,772	5.38	197,833.36
Check Dam—3	STN	9,007	6.12	55,122.84
Check Dam—4	STN	7,079	7.65	54,154.35
Reservoir—1	SN	2,807	2.87	8,056.09
Reservoir—2	KCP	2,454	2.36	5,791.44
Dug Well—1	SN	33.28	16.30	542.55
Dug Well—2	SN	24.02	10.73	257.72
Dug Well—3	SN	40.83	7.60	310.29
Dug Well—4	HN	36.96	5.48	202.54
Dug Well—5	HN	40.26	10.64	428.41
Dug Well—6	STN	35.05	8.08	283.20
Dug Well—7	STN	29.03	8.28	240.40
Dug Well—8	STN	33.18	6.40	212.37
Dug Well—9	KCP	35.68	13.51	482.02
Dug Well—10	KCP	35.05	8.68	304.20
Dug Well—11	KCP	31.57	13.20	416.72
Total	**Study site**	**123,177.91**		**567,203.50**

SN: Shivaji Nagar; HN: Holdar Nagar; STN: Satya Nagar; KCP: Kavlyachapada

[5, 14, 28]. In the past two decades, the maximum actual rainfall (>2,500 mm) was recorded between 2002 and 2008 with a percentage departure ranging from 2 to 83%, while the least average rainfall was recorded (~1,600 mm) between 2009 and 2015 with a percentage departure ranging from −42 to −2% (Fig. 2).

The year 2015 was declared a drought year in the Nashik district where all the blocks witnessed one of their least rainfall since 1998 [28]. Seasonally, more than 92% of the rainfall was received during the S-W monsoon followed by 6.5% during the post-monsoon (Table 4). In general, the rainfall trend in the study site was observed highly erratic and fluctuating. Field investigations and focused group discussions with local communities further confirmed the sufferings witnessed by them from extreme climatic events such as flash flood events during S-W monsoon and drought-like situations during the non-monsoon period.

4.2 Land Use and Land Cover Analysis

The surface area of the watershed (inclusive of both Shivaji Nagar and Holdar Nagar villages), as obtained by analyzing satellite imagery, was 4.35 km^2 (Fig. 3). About 80% of the watershed area was under green cover (agriculture and forest) while merely 1.5% of the land was under human and livestock settlement. The forest cover (1.5%) inside the study area was observed limited to four small patches located over Sahyadri hills. The hills were found mostly barren with widespread outcrops of basalt rocks with forest cover embodying merely 9% of the hills. Household surveys

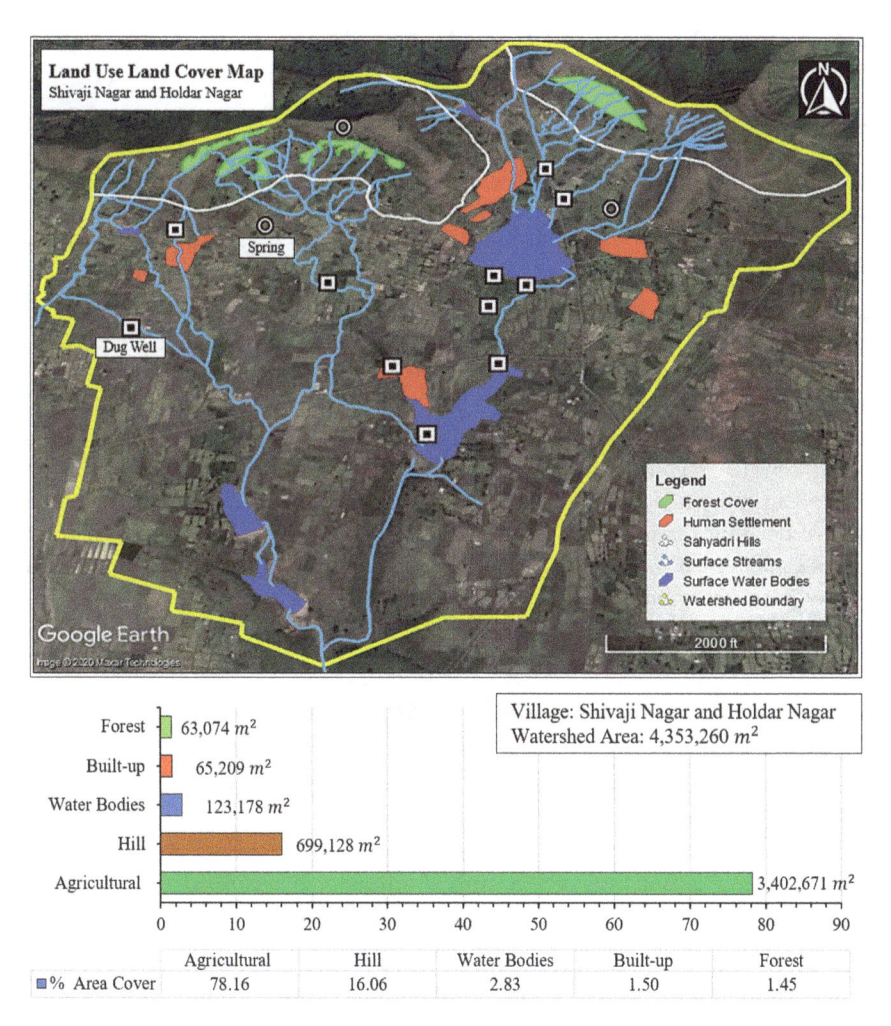

Fig. 3 Land Use Land Cover (LULC) map (top panel) of Shivaji Nagar and Holdar Nagar villages of Nashik district with percentage-wise analysis of the area under different land use classifications (lower panel)

revealed mass deforestation activities prevailing since the 1980s for construction and cooking purposes. Due to declined green cover, secondary issues, such as soil erosion from the hilly region and deposition of the silt inside water bodies during monsoon seasons was observed periodic and widespread. Additionally, a high imbalance in the proportion of irrigation water (from check dams, reservoirs, and dug wells covering ~3% of the land-use) to the agricultural land was observed. Against every ~30 m^2 area of agricultural land required to be irrigated, merely ~1 m^2 area of water bodies were available.

4.3 Seasonal Analysis of Hydrological Components

The entire hydrological analysis of the study site categorized climatic conditions into dry (winter and summer seasons) and wet (S-W monsoon and post-monsoon seasons). Surface runoff (R) was observed as a dominating hydrological component during monsoon season ranging from 33 to 89% followed by surface storage (S) ranging from 6 to 36% (Fig. 4). Both R and S showed a steep declining trend from the S-W monsoon until the summer of the following calendar year. Despite moderate groundwater flow (G) throughout seasons (ranging between 4 and 82%), there were no active tubewells or borewells identified in the villages. Though three attempts were successively taken in 2012, 2016, and 2019 for establishing tubewells at a depth of 45 m, 92 m, and 107 m, respectively, the bores didn't yield water, as a result of which, agriculture remained mostly rainfed and limited to one season. Evaporation (E) was highly correlated with the water availability across storage structures due to which it showed a similar trend as S. While the evapotranspiration (ET) was observed more or less consistent across every season. The reason was attributed to the large land-use under agriculture. Overall, both E and ET remained comparatively under prevalent hydrological behavior in the village catchments against R and S.

4.4 Seasonal Analysis of Water Balance

The water budget of the study site was estimated based on the difference between total input (from rainfall) and total output (through evaporation, evapotranspiration, surface runoff, and groundwater flow) of the hydrological cycle. The hydrological difference provided total storage (in terms of water stored in check dams, reservoirs, and dug wells). Due to high temporal variability in the magnitude of the rainfall, all other hydrological parameters witnessed considerable fluctuations across seasons (Fig. 5). The summer season recorded the highest outflows which remained >99% from the watershed followed by the S-W monsoon with 94% outflows. Post monsoon season recorded 64% outflows while the winter season recorded the least outflows (~16%). Extremely high outflows during the S-W monsoon resulted in flash flood events. Furthermore, the situation with regards to water availability in the villages

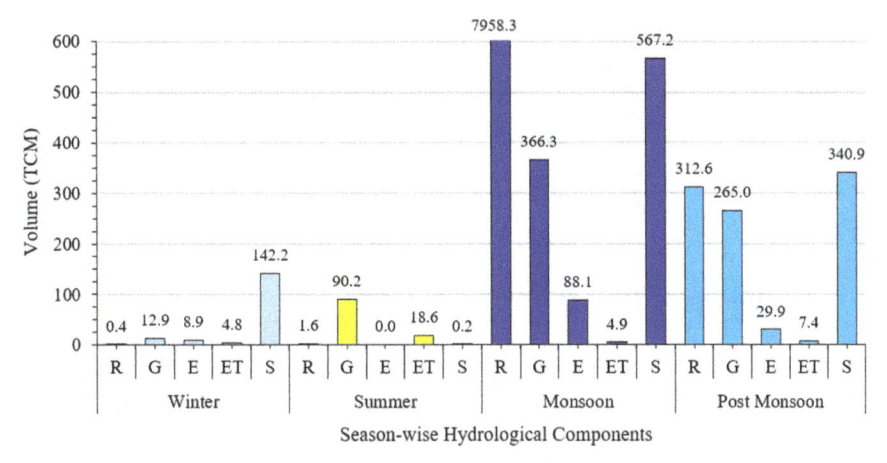

(a) Magnitude-wise analysis of hydrological parameters

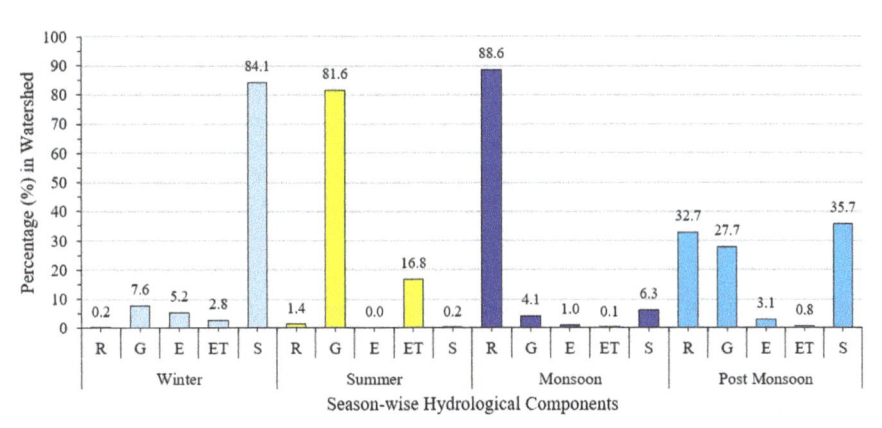

(b) Percentage-wise contribution of hydrological parameters

Fig. 4 Analysis of seasonal variation in the hydrological components across the study site viz., Shivaji Nagar and Holdar Nagar villages of Nashik district in Maharashtra state [*Note* R: Surface runoff; G Groundwater flow; E: Evaporation; ET: Evapotranspiration; S: Surface storage; the magnitude of R (Fig. 4a) in monsoon is 7,958 thousand cubic meters (TCM), which has not been accommodated within the scale of the graph considering high variation in the overall magnitudes of the R across seasons.]

remained pathetic and one of the major reasons identified during field investigation was the severe mismanagement of water resources. Lower water storage during summer (mere 0.2 TCM) resulted in agricultural, hydrological, and socio-economic droughts. In addition, the study site was located on moderately steeper slopes of the Sahyadri range, natural gravity and poor infiltration due to rocky substrata caused greater surface runoff during S-W monsoon season. This anomaly was evident, as the water balance analysis indicated ~81% surface runoff. Also, the leftover water

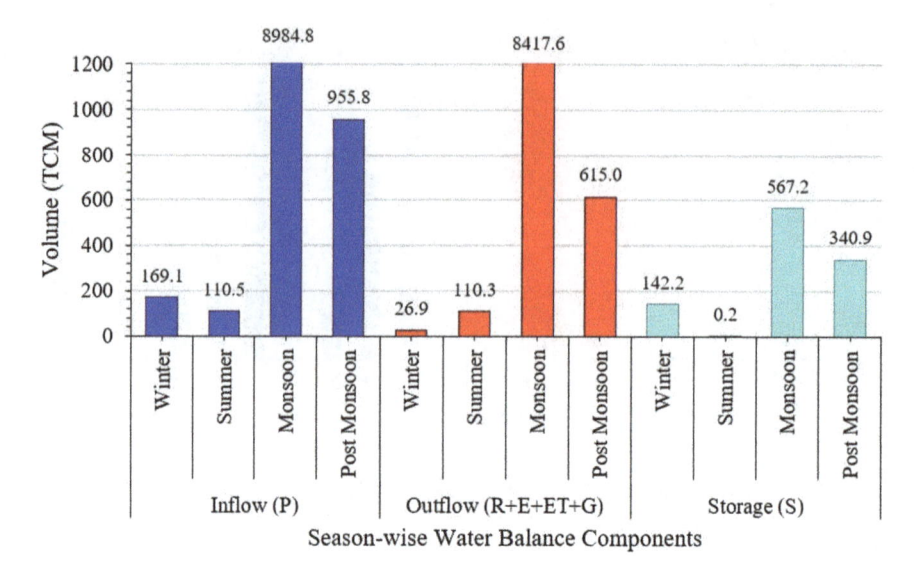

Fig. 5 Analysis of the variation in water balance (budget) components in the units of thousand cubic meters (TCM) across the study site viz., Shivaji Nagar and Holdar Nagar villages of Nashik district in Maharashtra state [*Note* P: Rainfall; R: Surface runoff; E: Evaporation; ET: Evapotranspiration; G Groundwater flow; S: Surface storage; the magnitude of R in inflow and outflow is ~8985 and ~8418 thousand cubic meters (TCM), respectively, which has not been accommodated within the scale of the graph considering high variation in the overall magnitudes of other components across seasons.]

in check dams and reservoirs during winter was observed with high silt content as a result of which, it couldn't be utilized for irrigation and drinking purposes. Measures in the direction of arresting excess surface runoff by increasing storage potential emerged as one of the major recommendations for the study site in view of achieving sustainability in watershed management.

5 Discussions

5.1 Watershed-Specific Management Measures

The water management plans in the study site were prepared considering *water availability*, *water use*, and *water entitlement* with equal priority. Significance to *water availability* was configured through water budgeting studies indicating mismanagement in supply-demand coupled with hydrological imbalance. *Water use* was ascertained in order to make sustainable efforts towards the preliminary phase of executing plans involving participatory-based monitoring and conservation. As the check dams and reservoirs in the study site were contemplated as common pool

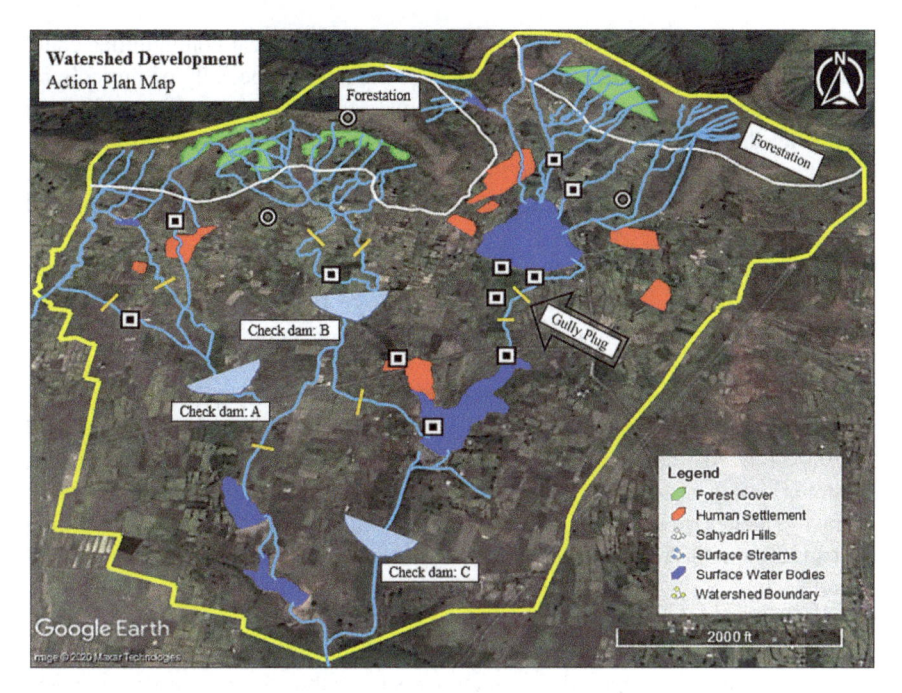

Fig. 6 An action plan for watershed development for the water resources management in Shivaji Nagar and Holdar Nagar villages of Nashik district in Maharashtra state

resources, *water entitlements* to weaker and vulnerable sections given achieving equitable water distribution were considered. An action plan showcasing structural interventions, such as establishing low head check dams, gully plugs, etc., at appropriate locations was developed (Fig. 6). Additionally, through participatory planning approaches and geophysical interpretations, recommendations for addressing diverse water scarcity issues in the study site were proposed, which are described hereunder:

- Adequate water storage potential was available in the study site, however, scarcity problems persisted due to a lack of initiatives aiming towards water conservation ensuring water security in a long run. Dry weather conditions for a prolonged period followed by a short spell of heavy rainfall obligated reconsideration towards increasing storage structures. In this regard, three low head check dams were proposed at three different locations on existing channels based on field investigation (Fig. 6).
- Lack of proper interconnectedness between upstream and downstream storage structures caused poor conveyance for flood routing. Creating a cascading network among existing and forthcoming storage structures by channelizing surplus runoff through gravity flow from the upstream to the downstream structure was observed impactful given buffering the negative consequences of the flash flood events [9, 32].

- The operation and maintenance of these structures by local communities could minimize excessive siltation and impacts of the flood, especially on agriculture. Furthermore, green initiatives in terms of mass forestation over barren hills and creating thick vegetation along channels were a couple of the pressing recommendations that were anticipated to address issues of soil erosion, high discharge surface runoff, high siltation rates, and structural deformations to check dams and reservoirs [33].
- The dependency of agricultural practices on rainfed-based irrigation given partially explored groundwater resources caused high unreliability in production, thereby enforcing farming communities to discontinue agriculture for a season or year, especially during monsoon failure or droughts. Curtailing agricultural troubles required holistic and community-based participatory approaches. Forming and enabling Water User Associations (WUAs) and other dedicated groups towards making collective decisions on the cropping pattern and maintaining storage structures could ensure long-term sustainability. This may require clear guidance on institutional responsibilities and roles of GP and strengthening of local capabilities so as to ensure that the required water resources remained protected.
- Training programs on practicing better irrigation techniques, crop diversification, measures to control evaporation, and evapotranspiration were a few other approaches proposed for attaining sustainable water conservation and management.

5.2 Future Directions

The study urges the following future studies on account of the limitations and scopes of the present work:

- *Determining water supply-demand potential*: The present study though aimed at estimating diverse water balance components, determining population water demand and assessing supply potential of existing water resources can help estimate unmet domestic water demand and unmet agricultural water draft across critical water-scarce seasons. Subsequently, future studies can focus on developing supply-side and demand-side water management plans given issues such as water scarcity, floods, and droughts.
- *Advancing LULC studies*: Conducting an advanced analysis in LULC to differentiate between vegetation cover, barren land, and the forest (cover) leaves would allow further precision to the findings on percentage cover under diverse land-use. Additionally, studies on LULC trend analysis, for instance between 2000 and 2020, could help identify the correlation between water demand and changing land-use patterns. Overall, this can aid in developing a priority index that can prioritize diverse watershed management and water conservation measures considering the hydrological conditions prevailing at the micro-watershed level.

- *Provisioning Rainwater Harvesting (RWH)*: Although less than 5% of the houses in the study site were having a terrace (most houses were constructed using locally available material such as mud and wood), construction of new houses under the housing scheme by the Government of India would be providing pucca houses (houses constructed primarily from cement-brick). As the field investigation identified the RWH option as an appropriate alternative to minimize water scarcity issues, inclusion of the appropriate designs for roof-top-based RWH in housing scheme could help provide alternatives to villagers for storing water during S-W monsoon. This approach, in general, can additionally provide strong policy recommendations for converging housing schemes with the RWH model.
- *Groundwater exploration*: Limited attempts were taken for groundwater extraction from the farming communities at the study site, as establishing tube well for the same was not only expensive but resulted in poor or no yield. There is a need to estimate actual groundwater recharge from the study site, as the present study estimated groundwater flow using an indirect method (water budget equation). This may be attained by conducting evidence-based field experiments regarding recharge and discharge rates inside the water spread area or with the existing dug wells. This may provide a basis for conducting the desiltation program of existing check dams and reservoirs in view of maximizing groundwater percolation.
- *Developing open-source web-based applications*: The development of open-source web-application-based water budgeting tools can be an effective way to arrive at the water balance of the given catchment or village-level watershed through minimal inputs. The development of such user-friendly applications provides scope to the local administration and communities to remain aware of the water balance scenarios of their catchment. This technology has the potential to direct a decentralized water governance model among stakeholders as an early decision support system.

6 Conclusions

The present study was grounded on an evidence-based case study in Shivaji Nagar and Holdar Nagar village of Maharashtra state in the West-India. The study highlighted the experience of the mismanagement associated with the utilization of water resources, issues on the scarceness of the water in supporting primary livelihood activities such as agriculture, and challenges in addressing the problems of the disproportionate availability of water resources across winter, summer, South-West monsoon, and post-monsoon seasons. The study demonstrated that preparing seasonal water budget projections could better examine the temporal variations of the various hydrological components, surface water storage capacities, and management provisions for humans, institutions, livestock, and agriculture water needs. For instance, water budgeting, in the current study site, provided better quantification and analysis about the fundamental problem that a short spell of monsoon causes flash floods while a prolonged dry period causes hydrological and socioeconomic droughts. The study found public-participatory-based holistic approaches for preparing watershed management strategies and associated water security and conservation plans as an imperative proposition towards addressing water scarcity issues. These methodologies were having the potential to sustain a greater quantity of water resources across critical seasons like summer. Moreover, the study emphasized that the engagement activities with the village communities can empower them to assess their own viable options for ground and surface water management including the measures to conserve water through their own local solutions. The study specified that the systematic scientific investigation of the spatio-temporal components of water resources could comprehend better water management policies. Such approaches may increase awareness among land and water resource managers, policy-makers, and politicians to formulate appropriate and region-specific water management and monitoring plans. This would require the strengthening of local capabilities, social institutions, decentralized water governance, and ownerships towards common pool resources.

Acknowledgements The authors thank the Indian Institute of Technology (IIT) Bombay and Hindustan Aeronautics Limited—Nashik Division from India for providing a fellowship platform for this work to be conducted. The authors would like to thank the Center for Technology Alternatives for Rural Areas (CTARA) of the IIT Bombay and the Rural Data Research and Analysis (RuDRA) Lab for providing infrastructure to conduct this research. Thanks are due to the active participation of the villagers, farmers, teachers, and students from the Shivaji Nagar and Holdar Nagar villages and officials from Gram Panchayat, Zilla Parishad, and District Headquarter of the Nashik district.

Funding The work was supported by the Center for Technology Alternatives for Rural Areas of the Indian Institute of Technology Bombay, India (Grant No. 183350021-2019) and Hindustan Aeronautics Limited—Nashik Division of the Maharashtra state, India (Grant No. RD/0114-HAL00E0-002).

Conflict of Interest. The authors declare that the research was conducted in the absence of any commercial or financial relationships that could be construed as a potential conflict of interest.

References

1. Aeschbach-Hertig W, Gleeson T (2012) Regional strategies for the accelerating global problem of groundwater depletion. Nat Geosci 5(12):853–861. https://doi.org/10.1038/ngeo1617
2. Aziz F, El Achaby M, Ouazzani N, El-Kharraz J, Mandi L (2020) Rainwater harvesting: a challenging strategy to relieve water scarcity in rural areas. In: Patnaik S, Sen S, Mahmoud M (eds) Smart village technology. Springer, Cham, pp 267–290. https://doi.org/10.1007/978-3-030-37794-6_13
3. Balaei B, Wilkinson S, Potangaroa R, Adamson C, Alavi-Shoshtari M (2019) Social factors affecting water supply resilience to disasters. Int J Disaster Risk Reduct 37:101187(1–15). https://doi.org/10.1016/j.ijdrr.2019.101187
4. Biswas B, Jadhav RS, Tikone N (2019) Rainfall distribution and trend analysis for upper Godavari basin, India, from 100 years record (1911–2010). J Indian Soc Remote Sens 47(10):1781–1792. https://doi.org/10.1007/s12524-019-01011-8

5. Borse K, Agnihotri PG (2020) Trends of rainfall, temperature and rice yield of Nashik region of Maharashtra. Innov Technol Explor Eng 9(5):395–398. https://doi.org/10.35940/ijitee.e2276. 039520
6. Burnham M, Ma Z, Zhang B (2016) Making sense of climate change: hybrid epistemologies, socio-natural assemblages and smallholder knowledge. Area 48(1):18–26. https://doi.org/10.1111/area.12150
7. CGWB—Central Ground Water Board (2014) Ground water information: Nashik district Maharashtra, Ministry of Water Resources, Government of India. http://cgwb.gov.in/district_profile/maharashtra/nashik.pdf. Accessed 23 Oct 2020
8. Chinnasamy P, Agoramoorthy G (2015) Groundwater storage and depletion trends in Tamil Nadu State, India. Water Resour Manag 29(7):2139–2152. https://doi.org/10.1007/s11269-015-0932-z
9. Chinnasamy P, Srivastava A (2021) Revival of traditional cascade tanks for achieving climate resilience in drylands of South India. Front Water 3:639637. https://doi.org/10.3389/frwa.2021.639637
10. Chintalapudi P, Khadse G, Pujari P, Sanam R, Labhasetwar P (2017) Integrated water security plan for water scarcity villages in central India. Environ Dev Sustain 19(6):2547–2564. https://doi.org/10.1007/s10668-016-9850-3
11. Coe R, Stern RD (2011) Assessing and addressing climate-induced risk in sub-Saharan rainfed agriculture: lessons learned. Exp Agric 47(2):395–410. https://doi.org/10.1017/S00144797110 0010X
12. Fukuda S, Noda K, Oki T (2019) How global targets on drinking water were developed and achieved. Nat Sustain 2(5):429–434. https://doi.org/10.1038/s41893-019-0269-3
13. Khadke L, Pattnaik S (2021) Impact of initial conditions and cloud parameterization on the heavy rainfall event of Kerala (2018). Model Earth Syst Environ. https://doi.org/10.1007/s40 808-020-01073-5
14. Kulkarni PP, Pardeshi SD (2019) Evaluation and interpolation of rainfall trends over the semi-arid upper Godavari basin. Meteorol Atmos Phys 131(5):1565–1576. https://doi.org/10.1007/s00703-018-0652-z
15. Kumar MD (2018) Water management in India: the multiplicity of views and solutions. Int J Water Resour Dev 34(1):1–5. https://doi.org/10.1080/07900627.2017.1351333
16. Liu J, Hull V, Godfray HC, Tilman D, Gleick P, Hoff H, Pahl-Wostl C, Xu Z, Chung MG, Sun J, Li S (2018) Nexus approaches to global sustainable development. Nat Sustain 1(9):466–476. https://doi.org/10.1038/s41893-018-0135-8
17. Maheshwari B, Varua M, Ward J, Packham R, Chinnasamy P, Dashora Y, Dave S, Soni P, Dillon P, Purohit R, Shah T (2014) The role of transdisciplinary approach and community participation in village scale groundwater management: insights from Gujarat and Rajasthan, India. Water 6(11):3386–3408. https://doi.org/10.3390/w6113386
18. MoJS—Ministry of Jal Shakti (2020) Jal Jeevan Mission, Ministry of Jal Shakti, Government of India, New Delhi. https://ejalshakti.gov.in/jjmreport/JJMIndia.aspx
19. MoWR—Ministry of Water Resources (2017) Impact of climate change on water resources, Government of India. http://www.indiaenvironmentportal.org.in/files/file/Impact%20of%20C limate%20Change%20on%20Water%20Resources.pdf. Accessed 23 Oct 2020
20. NCIWRD—National Commission on Integrated Water Resources Development (1999) Integrated water resources development—a plan for action. Ministry of Jal Shakti, Department of Water Resources, River Development and Ganga Rejuvenation, Government of India
21. NIH—National Institute of Hydrology (2020) Hydrology and water resources information system for India, Water Resources Systems Division, National Institute of Hydrology. http://117.252.14.242/rbis/India_Information/Water%C2%A0Budget.htm. Accessed 23 Oct 2020
22. NWM—National Water Mission (2018) State-water budgeting report. Ministry of Jal Shakti, Department of Water Resources, Government of India. http://nwm.gov.in/sites/default/files/A%20note%20on%20State%20Water%20Budgeting%2019.3.2018.pdf. Accessed 23 Oct 2020

23. Narasimhan TN (2008) A note on India's water budget and evapotranspiration. J Earth Syst Sci 117(3):237–240. https://doi.org/10.1007/s12040-008-0028-8
24. Nguyen TP, Seddaiu G, Virdis SG, Tidore C, Pasqui M, Roggero PP (2016) Perceiving to learn or learning to perceive? Understanding farmers' perceptions and adaptation to climate uncertainties. Agric Syst 143:205–216. https://doi.org/10.1016/j.agsy.2016.01.001
25. Opiyo F, Wasonga O, Nyangito M, Schilling J, Munang R (2015) Drought adaptation and coping strategies among the Turkana pastoralists of northern Kenya. Int J Disaster Risk Sci 6(3):295–309. https://doi.org/10.1007/s13753-015-0063-4
26. Poonia A, Punia M (2019) Associates and determinants of drinking water supply: a case study along urbanrural continuum of semi-arid cities in India. Urban Water J 16(10):749–755. https://doi.org/10.1080/1573062X.2020.1729387
27. Ranganathan T, Ranjan R, Pradhan D (2018) Water scarcity and livelihoods in Bihar and West Bengal, India. Oxford Dev Stud 46(4):497–518. https://doi.org/10.1080/13600818.2018.1447097
28. Sanjay C, Bhasker P, Damodare SL, Abhale AP (2018) Statistical analysis of seasonal rainfall variability in Nasik district by using GIS interpolation. J Pharmacogn Phytochem 7(4):2072–2077. https://www.phytojournal.com/archives/2018/vol7issue4/PartAI/7-4-115-107.pdf. Accessed 23 Oct 2020
29. Singh VP (2017) Challenges in meeting water security and resilience. Water Int 42(4):349–359. https://doi.org/10.1080/02508060.2017.1327234
30. Singh C, Osbahr H, Dorward P (2018) The implications of rural perceptions of water scarcity on differential adaptation behaviour in Rajasthan, India. Reg Environ Change 18(8):2417–2432. https://doi.org/10.1007/s10113-018-1358-y
31. Smiley SL (2016) Water availability and reliability in Dar es Salaam, Tanzania. J Dev Stud 52(9):1320–1334. https://doi.org/10.1080/00220388.2016.1146699
32. Srivastava A, Chinnasamy P (2021) Water management using traditional tank cascade systems: a case study of semi-arid region of Southern India. SN Appl Sci 3:281. https://doi.org/10.1007/s42452-021-04232-0
33. Srivastava A, Chinnasamy P (2021) Investigating impact of land-use and land cover changes on hydro-ecological balance using GIS: insights from IIT Bombay, India. SN Appl Sci 3:343. https://doi.org/10.1007/s42452-021-04328-7
34. Subramanya K (2013) Engineering hydrology, 4th edn. Tata McGraw-Hill Education, New Delhi, pp 73–232
35. UNICEF—United Nations International Children's Education Fund (2012) Manual for the preparation of water safety and security plan. UNICEF, New York
36. Water UN (2013) Water security and the global water agenda—a UN-water analytical brief. United Nation University, Institute for Water, Environment and Health

Aman Srivastava obtained his Master's degree in Technology and Development from the Centre for Technology Alternatives for Rural Areas (CTARA) of the Indian Institute of Technology (IIT) Bombay. After his Research in Development Fellow position with Hindustan Aeronautics Limited (HAL) fellowship in association with CTARA-IIT Bombay, he joined the Ministry of Rural Development, Government of India at New Delhi as Research Associate in 'Mahatama Gandhi National Rural Employment Guarantee Scheme' (MGNREGS), which is the largest running rural employment scheme across the world. He currently works here and primarily involved in developing sustainable solutions in convergence with MGNREGS works to address the water scarcity issues, prevailing in semi-arid regions across rural-India. His research interests include water accounting and budgeting, water resources system analysis, rainwater harvesting, and watershed management with a primary focus on achieving climate change resilience across the drylands of India. Mr. Srivastava is two times recipient of the Institute Silver Medal viz, one for Bachelor's degree in Civil Engineering and second for Master's degree in Technology and Development.

Pennam Chinnasamy obtained his Master's degree in Physics from Wesleyan University, Connecticut—US, followed by a doctoral degree, with a focus on hydrology, from Missouri University, US. After his Research Fellow position with Ashoka Trust for Research in Ecology and the Environment, he joined the International Water Management Institute as a Researcher and was stationed in Nepal and Indian offices, where he focused on climate change impacts on underdeveloped and developing nations. He then joined Nanyang Technological University, Singapore, as a Senior Researcher developing real-time flood predicting models for Singapore. He is currently an Assistant Professor with the Indian Institute of Technology, Bombay—India, under the Centre for Technology Alternatives for Rural Areas (CTARA), where his work primarily focuses on natural resources assessment, monitoring, and management in rural regions. He is the founding Director of the Rural Data Research and Analysis (RuDRA) lab, which is the first big data lab for rural regions, housed in an academic institution in India. Over the past decade, Dr. Pennan has experience working in NGOs, National and regional Government agencies, and academic institutions, focusing on sustainable surface and groundwater management plans, climate change impacts, large data analysis, and hydrological simulation models.

Web Application Tool for Assessing Groundwater Sustainability—A Case Study in Rural-Maharashtra, India

Aman Srivastava, Leena Khadke, and Pennan Chinnasamy

Abstract Assessment of aquifer properties is important for the effective management of groundwater resources, especially in India—the world's largest extractor of groundwater. Spatio-temporal variations in aquifer properties exist, therefore, estimations are expensive and time-consuming. While equations to estimate properties are available, computational difficulties and the availability of secondary data to estimate aquifer properties are challenging. Under such circumstances, this study developed a free open-source web application-based Groundwater Calculator (G-Cal) tool to generate groundwater properties, estimate flow between two wells, and evaluate components of well hydraulics in steady-state flow under both confined and unconfined aquifer conditions. The field surveys and data, collected from the Kavlyachapada hamlet in Shivaji Nagar village, India (as a case study), were used to run the G-Cal tool for estimating the aforementioned properties. The G-Cal outputs were validated against manual estimations, and results indicate a high accuracy, across parameters. Field experimentations indicated that the discharge into the well-located away from the hills (*W2*) was estimated to be 500% greater than the well-located closer to the hills (*W1*). Furthermore, *W2* recuperated approximately 1.5 times faster than *W1*. As a result, agricultural practices were observed prevalent in and around the region of *W2*, especially during South-West monsoon (June to September). The study concluded that the development of such user-friendly web applications followed by their validation through grounded case studies can be significant for groundwater

---

A. Srivastava (✉) · P. Chinnasamy
Centre for Technology Alternatives for Rural Areas, Indian Institute of Technology Bombay, Mumbai 400076, India

P. Chinnasamy
e-mail: p.chinnasamy@iitb.ac.in

L. Khadke
Department of Civil Engineering, Indian Institute of Technology Bombay, Mumbai 400076, India

P. Chinnasamy
Rural Data Research and Analysis (RuDRA) Lab, Indian Institute of Technology Bombay, Mumbai 400076, India

© Springer Nature Switzerland AG 2021
A. Vaseashta and C. Maftei (eds.), *Water Safety, Security and Sustainability*, Advanced Sciences and Technologies for Security Applications, https://doi.org/10.1007/978-3-030-76008-3_28

budgeting and developing water security plans. Such approaches are much-needed in a rural context where agricultural practices remain the predominant occupation.

Keywords Groundwater · Water management · Water security · Sustainable development · Free open-source · Agriculture

1 Introduction

India is the highest groundwater extractor in the world [1, 13, 20]. Almost 85% population resides in rural areas [29] where 60% of the irrigation land and 85% of domestic activities are dependent on groundwater [35]. More than 80% of India's total freshwater is used in agriculture, of which 63% is from groundwater resources, mostly using the wells to meet irrigation demand [24]. This excessive groundwater extraction has caused the falling of groundwater to unsustainable levels and, at the same time, resulted in degraded groundwater quality at various locations in India [2, 5, 8]. In a pan India study, Khetwani and Singh [23] witnessed a four-meter decline in the groundwater level in the last 20 years, across 286 districts in 18 states [23]. The impacts of the declining groundwater to unsustainable levels have been reflected in over 70% of the rural households where the primary livelihood, such as agricultural water demand, mostly relied on groundwater sources. The major reason for the groundwater issues has been attributed to the mismanagement, over-exploitation, and wastage of the available water resources [10, 25, 26, 34]. In addition, challenges in calculating and estimating the groundwater and aquifer parameters partly due to data-related problems besides problems to access observation data from government agencies, are key reasons for groundwater mismanagement [6].

The estimation of groundwater availability and the study of the influence of hydrogeological and meteorological parameters on groundwater occurrence, circulation, and distribution provides an understanding of groundwater dealing with exploration, development, and careful utilization [33]. Groundwater information is time-reliant and dependent on the geologic settings of the location [18]. Effective groundwater management requires an understanding of the spatio-temporal variations [28] in aquifer properties, especially specific yield and hydraulic conductivity [15]. Chinnasamy et al. [7] stressed the fact that accurate estimations of groundwater properties are necessary for establishing an effective dialogue with policymakers, land managers, and farm stakeholders [7, 8]. The handbook by the Groundwater Estimation Committee [16] provides details on estimating groundwater aquifer properties, however, such estimations require knowledge of physical drivers that influence aquifer properties. In addition, information on climate change extreme events like floods and droughts, impacts of rapid urbanization on land use and land cover changes, and strategies to combat them becomes of paramount importance [14, 32] while proposing sustainable groundwater management and security plans [30]. Howe et al. [19] indicate that the Indian monsoon rainfall season has become more erratic, characterized by frequent floods [21] and droughts along with a noticeable shift in the peak monsoon precipitation [19, 27, 31]. However, the major concern is in identifying methodologies to estimate the usage of groundwater through advanced computing and modeling systems. Although the understanding of the complex relationship between groundwater parameters has improved, due to advances in modeling perspectives in the current decade, these approaches are expensive [9, 22]. This is attributed to the high cost associated with proprietary modeling software, costly secondary data, even though the bulk of empirical formulae, and underlying equations, had been suggested by researchers to estimate the groundwater parameters, and are freely available. Thus, there is a need for an economically viable, socially acceptable, and technically appropriate solution that can aid in estimating groundwater parameters and creating an associated database. Under this context, the present research focuses on the accurate and open-source model-based estimation of groundwater parameters (e.g., specific yield, discharge, and groundwater flow into and in between the wells). This study will provide fundamental information to enhance the study of groundwater hydrology of a catchment or a watershed systematically, especially at the village-scale where capacity is limited and groundwater needed security against over-exploitation.

The primary objective of this study is to develop a Groundwater Calculator (G-Cal) based on an empirical and textbook approach for estimation of the groundwater parameters at a micro-watershed or a village level using JavaScript as a front-end technology backed up by Laravel, an open-source PHP framework, and MySQL to store and analyze data. The secondary objectives include the analysis of various characteristics of an aquifer and its correlation with the irrigation practices, the estimation of groundwater flow between two wells, and the analysis of steady-state groundwater flow problems associated with wells under both confining and unconfining aquifer conditions. In order to validate the efficiency of G-Cal, the tool will be used for the Kavlyachapada hamlet in the Shivaji Nagar village, India, as a case study.

2 Materials and Methods

2.1 Development of Groundwater Calculator (G-Cal)

The G-Cal is designed using HTML (HyperText Markup Language) and CSS (Cascading Style Sheet) with client-side logic coded with JavaScript and jQuery (an open-source framework of JavaScript). It is connected to a web server that performs the function of serving static files to the client or user (Fig. 1). It can access an Application Programming Interface (API) which is connected to the MySQL database. The API, which is developed using Lumen, a PHP (Hypertext Preprocessor) micro-framework, serves the purpose of storing the results (as backup) obtained during G-Cal calculations. The communication between the client and the API server is established using Representation State Transfer Web Services (RESTful Web Services). The API server responds to all HTTP (HyperText Transfer Protocol) requests of the

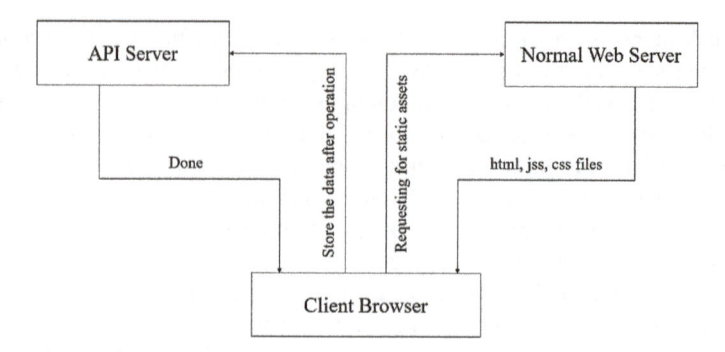

Fig. 1 Path flow network for developing a web application for Groundwater Calculator (G-Cal)

client, such as Create, Read, Update, or Delete (CRUD) for the particular operation under progress that subsequently reflects over the stored database.

2.2 Operation Design in G-Cal

The G-Cal web tool has been designed to perform three operations viz., (1) analysis of aquifer characteristics, (2) analysis of groundwater flow between two wells, and (3) analysis of well hydraulics (Fig. 2). Being a Multi-Page Application (MPA), the G-Cal can perform the aforementioned operations in both, online and offline mode.

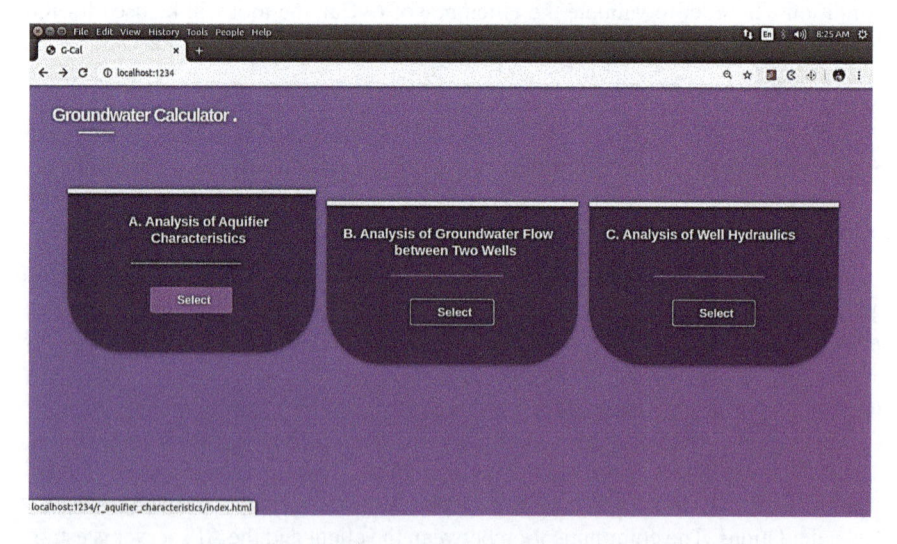

Fig. 2 Cover page of Groundwater Calculator (G-Cal) web tool showing the basic operations

Table 1 Percentage (%) range of porosity and specific yield for the selected aquifer materials [15, 33]

Aquifer material	Range of % porosity (n)	Range of % specific yield (S_y)	% Average specific yield (S_y)
Clay	45–55	1–10	5
Sand	35–40	10–30	20
Gravel	30–40	15–30	22.5
Sandstone	10–20	5–15	10
Basalt	15–50	1–7	4
Shale	1–10	0.5–5	2.75
Limestone	1–10	0.5–5	2.75

Table 2 Permeability coefficient (K) for granular materials in the units of meters per day (m/d) [15, 33]

Granular material	K (m/d)
Clean gravel	43,200
Clean coarse gravel	436
Mixed sand	6.48
Fine sand	22
Silty sand	0.91
Silt	0.22
Clay	0.0008

However, in the offline mode, the G-Cal performs all the operations without storing (backup) the generated data (results) in the database.

The first aspect of designing the G-Cal considered estimating the parameters associated with aquifer properties. From the knowledge of engineering hydrology, the composition of the aquifer decides the two essential properties of aquifer viz., (1) its capacity to release the water from the pore spaces, and (2) the ability with which the water transmits with ease [15]. The former was expressed in G-Cal by percentage porosity (*n*) of the sample of aquifer material (Earth formations) obtained from the field. The data required to calculate *n* depends upon the volume of void in aquifer material (V_v) and the volume of the porous medium (V_o), as shown in Eq. 1. While the latter was expressed in the form of percentage average specific yield (S_y) and specific retention (S_r), which depends on the type of the aquifer material and its porosity [15], as shown in Eq. 2. The G-Cal estimates both the properties of the aquifer for the different Earth formations such as clay, sand, gravel, sandstone, shale, and limestone (refer to Tables 1, 2 and 3). In the case of no data on aquifer material to be input to G-Cal, the data from Table 1 is displayed to the user against the selected type of formation.

$$n = \frac{V_v}{V_o} \times 100 \qquad (1)$$

$$n = S_y + S_r \tag{2}$$

The second aspect of designing the G-Cal considered estimating the discharge (Q) for the groundwater flow conditions between two well for both confining and unconfining aquifer conditions (Fig. 3). In the case of confined aquifers, Darcy's law [11] was considered for estimating Q. The user is required to enter the values of piezometric heads of both upstream well (h_o) and downstream well (h_1) spaced at a distance (L) along with the depth of the aquifer (B). The discharge for confined aquifers between two wells is shown in Eq. 3. While in the case of an unconfined aquifer, Dupits's assumption [12] was considered for estimating Q. The user is required to enter the values of the water table of both upstream well (h_o) and downstream well (h_1) spaced at a distance (L), as shown in Eq. 4. In both types of aquifers, the user is supposed to enter the value of the permeability coefficient (K). If the value of K is not available, the G-Cal uses the representative values of K, as given in Tables 2 and 3. The user is directed to choose the preferred type of the Earth material viz., granular material, or consolidated material, based on which the value of K is input

Table 3 Permeability coefficient (K) for consolidated materials in the unit of meters per day (m/d) [15, 33]

Consolidated material	K (m/d)
Sandstone	0.43
Carbonate rocks	0.44
Shale	8.64×10^{-8}
Fractured/weathered rock	0.43
Permeable Basalt	15.00

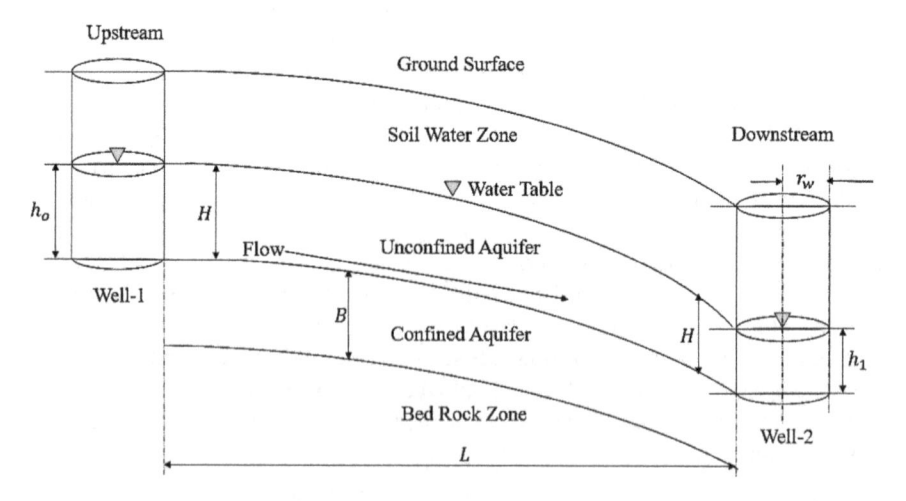

Fig. 3 Conceptual model of groundwater flow into the wells and in between the wells under confining and unconfining aquifer conditions for Kavlyachapada hamlet in Shivaji Nagar village of Nashik district in Maharashtra, India

to the equation of discharge.

$$Q = \frac{h_o - h_1}{L} \times K \times B \tag{3}$$

$$Q = \frac{h_o^2 - h_1^2}{2 \times L} \times K \tag{4}$$

The third and final aspect of designing the G-Cal considered the wells extracting water from both confined and unconfined aquifers under steady-state groundwater flow conditions (Fig. 3). The G-Cal considered estimating well discharge (q), the specific capacity (K_o), and recuperation time (T_r) of the well using a set of equations (Eqs. 5–7). In the case of confined aquifers, Thiem's equation [34] was considered to estimate q. The user is required to enter the values of the thickness of the confined aquifer (B), permeability coefficient (K), drawdown at the pumping well (s_w), the radius of the pumping well (r_w), and radius of influence (R) (Eq. 5). Apart from the discharge, Eq. 5 also provided results on transmissibility (T) of the well, as shown in (Eq. 6). While in the case of an unconfined aquifer, Dupit's assumption [12] was considered on a similar approach to groundwater flow conditions to estimate q. The user is required to enter the values of saturated thickness of the unconfined aquifer (H) and depth of water (h_w) in pumping well apart from the values of K, r_w, and R, as shown in Eq. 7.

$$q = \frac{2 \times \pi \times (K \times B) \times s_w}{\ln \frac{R}{r_w}} \tag{5}$$

$$T = K \times B \tag{6}$$

$$q = \frac{\pi \times K \times (H^2 - h_w^2)}{\ln \frac{R}{r_w}} \tag{7}$$

The specific capacity of the well (K_o) was calculated based on the values of Q, s_w, and r_w, as shown in Eq. 8. The s_w was calculated as the difference between initial drawdown (H_1) and drawdown (H_2) after time t (Eq. 8). This exercise also provided results on the cross-section area of the pumping well (A) (Eq. 9) and specific capacity per unit well area of the aquifer (K_s) (Eq. 10). Using K_s, H_1, and H_2, the recuperation time (T_r) was calculated using Eq. 11.

$$K_o = \frac{q}{s_w} = \frac{q}{H_1 - H_2} \tag{8}$$

$$A = \frac{\pi}{4} \times (2 \times r_w)^2 \tag{9}$$

$$K_s = \frac{K_o}{A} \tag{10}$$

$$T_r = \frac{1}{K_s} \times \ln \frac{H_1}{H_2} \qquad (11)$$

2.3 Study Area

The study site, Kavlyachapada (KCP) hamlet, is located in the Shivaji Nagar village at latitude 20°6′25.68″ N and longitude 73°33′27.39″ E, has a mean sea level of about 732 meters (m) and lies under the administration of Trimbakeshwar taluka (block) of Nashik district in the Maharashtra state of India (Fig. 4). Shivaji Nagar village consists of three hamlets viz., KCP, Shivaji Nagar, and Satya Nagar. The study site is bounded by Shivaji Nagar hamlet and Holdar Nagar village in the East, Somnath Nagar village in the West, Vele village at the South-West, Satyanagar hamlet and Devargaon village at South, and Peth taluka in the North. The Gram Panchayat (GP; defined as the village level governance system in India) is located in the Shivaji Nagar hamlet at latitude 20°6′37.07″ N and longitude 73°34′7.13″ E. The location of the study site lies adjacent to the Sahyadri range of Maharashtra (in the North), having mostly black cotton soil (clayey) along with outcrops of basalt rock [17]. The KCP hamlet experiences a semi-arid climate with temperature ranges from 5 to 40 °C [4]. The study site receives heavy rainfall, having annual precipitation

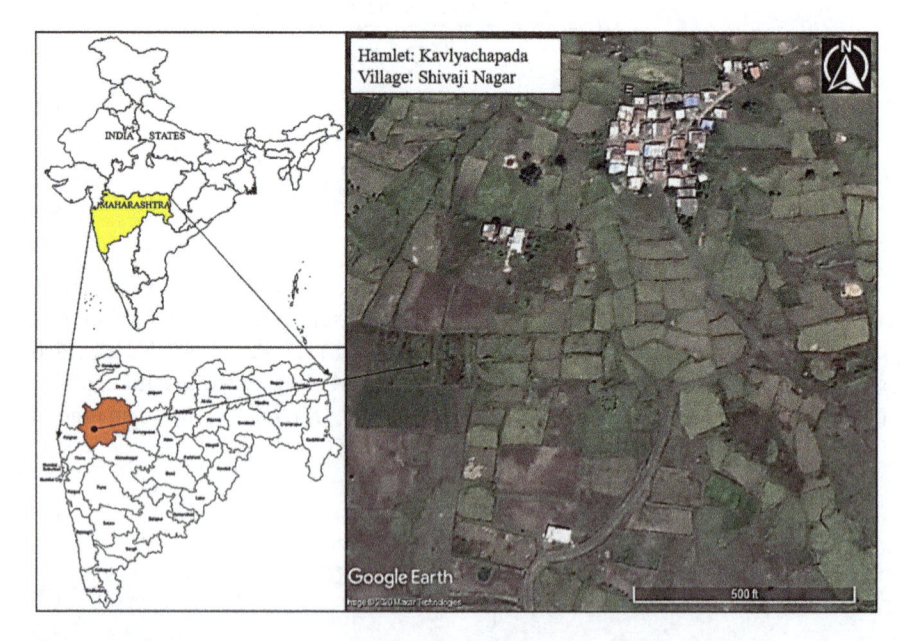

Fig. 4 Location of the study site, Kavlyachapada (KCP) hamlet in Shivaji Nagar village of Nashik district (insets showing the location in Maharashtra state and India)

of about 2,000 mm of which, more than 90% falls between June and September (South-West monsoon in India) [3, 27]. The primary source of water to the village is from the South-West monsoon, however, during the non-monsoon period (especially summer), other sources such as check dams, small water storage reservoirs, and dug wells become the primary source for drinking water and irrigation. More than 90% of the population in the village practice rainfed-based agriculture (mostly Kharif cropping) as their primary source of livelihood. The agriculture activities are limited to one season due to scarce water availability. On the contrary, for the last couple of decades, water-guzzling crop such as paddy continues to be the major crop cultivated. There are three wells in the study site of which one is a GP well (i.e., owned by the community), while the remaining two are private wells. Although private wells are mostly employed for irrigation, the GP well is typically used to meet drinking water demands in KCP hamlet.

2.4 Data Collection and Processing

The field surveys, experiments, and qualitative interviews were conducted with the residents of the Kavlyachapada hamlet between May and July 2019. The purpose was to study and collect the information regarding the background of the field conditions concerning the functionalities of the surface water storage structures such as dug wells, and the availability of water across the study period. The study comprehensively focused on the data collection for the various groundwater parameters such as n, V_v, V_o, S_y, S_r, h_o, h_1, L, K, q, H_o, H_1, h_w, R, and r_w.

The soil type in and around the study site was mostly black cotton soil (clayey) along with brick-red laterites. Since the location of the study site was over and adjacent to the hills, the soil layers were shallow with widespread outcrops of basalt rocks. They are derived from the underlying basalts. The basalt aquifers were found with perched water tables contributing to the wells in the study site. The samples of soil from three different regions of the study site were collected and tested for its porosity. The V_v and V_o of the tested sample were 2.33 cubic meters (m^3) and 5.21 m^3, respectively. The survey identified soil type as clay in the upper layer of the soil profile (between the surface and averagely 1 m below ground level (bgl)) and aquifer material type as basalt rocks (approximately at a depth of 2 to 5 m bgl). Therefore, the corresponding values of S_y for basalt rock (from Table 1) were taken as the input in G-Cal. In order to determine the discharge between wells, two wells from the study site, namely Well-1 and Well-2, separated by a distance of 425 m (L) was selected (Fig. 5), and required data on each well was obtained during field experiments and surveys. Field survey identified unconfined basalt aquifer as the major source of groundwater in the village. Hence, this study analyzed the groundwater parameters under unconfined aquifer conditions. Based on the layering of the rocks and seepages into the wells (Fig. 5), the aquifer material was classified as consolidated and sub-classified as a fractured rock (permeable basalt). Therefore, the value of K for the study site was considered as about 15 m per day (m/d) (Table 3).

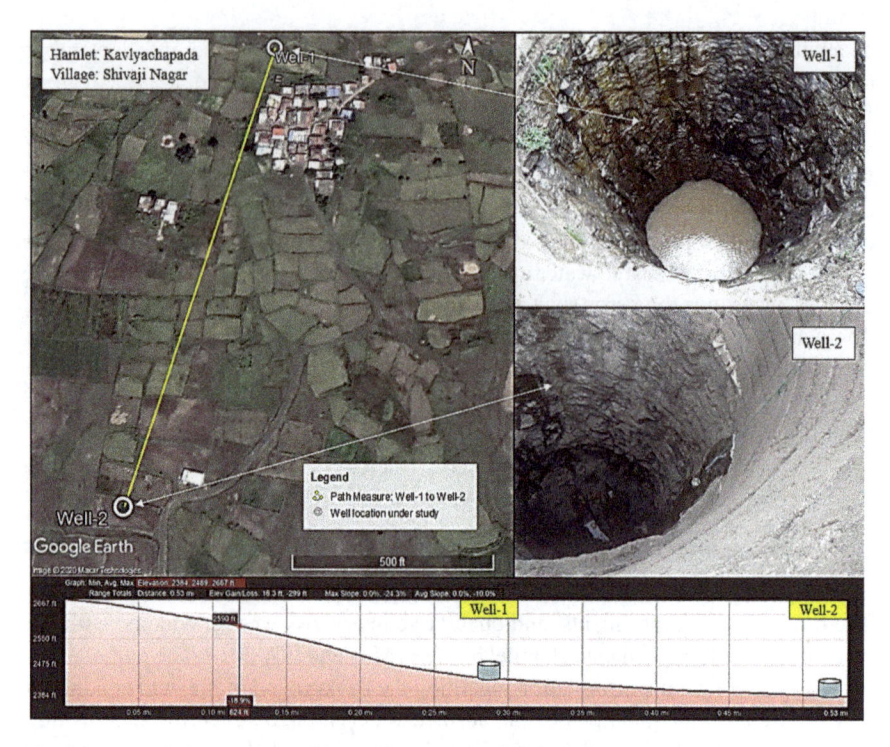

Fig. 5 Study site for groundwater flow between Well-1 and Well-2 at Kavlyachapada (KCP) hamlet in Shivaji Nagar village (insets showing wells and elevation profile)

The depth to the water level of each well was measured 4-times (every week) in a month across May to July 2019. The Well-1 having a radius (r_w) of 3.31 m is a GP well located close to the hills at latitude 20°6′28.19″ N and longitude 73°33′26.01″ E and majorly used to meet the domestic and drinking water demand of the entire KCP hamlet. The depth of the Well-1, when it was completely dry, was measured as 13.51 m bgl, which was considered as the thickness of the unconfined aquifer (H) in and around the region of Well-1. However, the average depth to the water level (h_o) in May, June, and July was measured as 12.07 m bgl, 13.01 m bgl, and 1.52 m bgl, respectively. While the Well-2, having a radius (r_w) of 3.25 m, is a private well located away from the hills at latitude 20°6′14.65″ N and longitude 73°33′23.38″ E and was used mostly for meeting irrigation demand. The depth of this well, when it was completely dry, was measured as 6.40 m bgl, which was considered as the thickness of the unconfined aquifer (H) in and around the region of Well-2. While, the average depth to water level (h_1) in May, June, and July was measured as 4.45 m bgl, 5.90 m bgl, and 0.68 m bgl, respectively.

3 Results and Discussions

The efficiency of the G-Cal web tool was analyzed in two ways viz., (1) by solving the groundwater problems manually with the data collected during the field survey, and (2) by comparing the output of G-Cal against (1). The analysis of later provided a better understanding of the accuracy of the G-Cal web tool.

3.1 Key Findings from the Recuperation Test

The recuperation test was conducted for both Well-1 and Well-2 in the first week of July. The water from the wells was pumped at a constant rate, and a drawdown was obtained. In the case of Well-1, the drawdown at the start of the recuperation test was recorded at 6.48 m (H_1) while the drawdown after 11.53 h (h) (t) was recorded at 1.52 m (H_2). Whereas, in the case of Well-2, the drawdown at the start and after 9.2 h (t) was recorded 4.95 m (H_1) and 0.42 m (H_2), respectively. The values of H_1, H_2, K_o, and K_s were used to estimate the T_r individually for Well-1 and Well-2 for the July period. However, to estimate q, parameters such as R were assumed 300 m, and the average depth of water of each well (h_w) for May, June, and July was considered as input to G-Cal. The value of K for the study site was considered between 15 m/d and 150 m/d based on the number of potential fractures (recharging the wells) observed in the well.

3.2 Unconfined Aquifer Analysis

The data obtained on the type of soil (upper strata) and aquifer material (lower strata) during the field survey were manually solved using Eqs. 1 and 2 to obtain n, S_y, and S_r. The average porosity of the collected sample was found at 45% (Eq. 12). It is evident from Table 1 that the porosity obtained by manual calculation is identifiable as clay whose corresponding values of n is between 45 and 55% and S_y between 1 and 10%. This result was found consistent with field observations (Eq. 13). Based on the average value of S_y (calculated as 5% for clay), the S_r was calculated close to 40% (Eq. 14). However, the lower strata of the soil profile were of basalt rock whose n, S_y, and S_r would be different from upper strata, these parameters were thus estimated averagely as 34.5% (average of 15 and 50%, Table 1), 4%, and 30.5%, respectively (Eqs. 15 and 16).

$$n_{sample} = \frac{2.33}{5.21} \times 100 = 44.72\% \tag{12}$$

$$n_{clay} = n_{sample} = S_y + S_r \tag{13}$$

$$44.72\% = 5\% + S_r \; i.e., \; S_r = 39.7 \tag{14}$$

$$n_{basalt} = S_y + S_r \tag{15}$$

$$34.50\% = 4\% + S_r \; i.e., \; S_r = 30.50\% \tag{16}$$

3.3 Discharge Analysis for the Groundwater Flow Conditions

The steady-state groundwater flow case of the study site was manually solved for unconfining aquifer conditions between Well-1 and Well-2 using Eq. 4. The Well-1 was geographically located (at upstream) closer to the hills such that during the South-West monsoon, the surface runoff under gravity was observed in the direction from Well-1 to Well-2 (evident from the elevation profile, as shown in Fig. 5). The discharge, estimated manually, for May was 2.22 m³/d (cubic meters per day), and increased to 2.37 m³/d in June, and reduced significantly to 0.03 m³/d in July (Eqs. 17–19). This anomaly was attributable to the shift from dry summer (May to mid-June) to South-West monsoon (mid-June onwards). The depth to the water level in wells was observed rapidly declining between May and mid-June (peak of the summer). In fact, for the first two weeks in June, the wells dried up completely, however, during South-West monsoon (initial phase), the depth to the water level in the wells improved. At the same time, the aquifer medium and upper soil layers gradually became saturated, thereby allowing a significant portion of rainwater to run off rather than infiltrating into the ground.

$$Q_{May} = \frac{12.07^2 - 4.45^2}{2 \times 425} \times 15 = 2.22 \; m^3/d \tag{17}$$

$$Q_{June} = \frac{13.01^2 - 5.90^2}{2 \times 425} \times 15 = 2.37 \; m^3/d \tag{18}$$

$$Q_{July} = \frac{1.52^2 - 0.68^2}{2 \times 425} \times 15 = 0.03 \; m^3/d \tag{19}$$

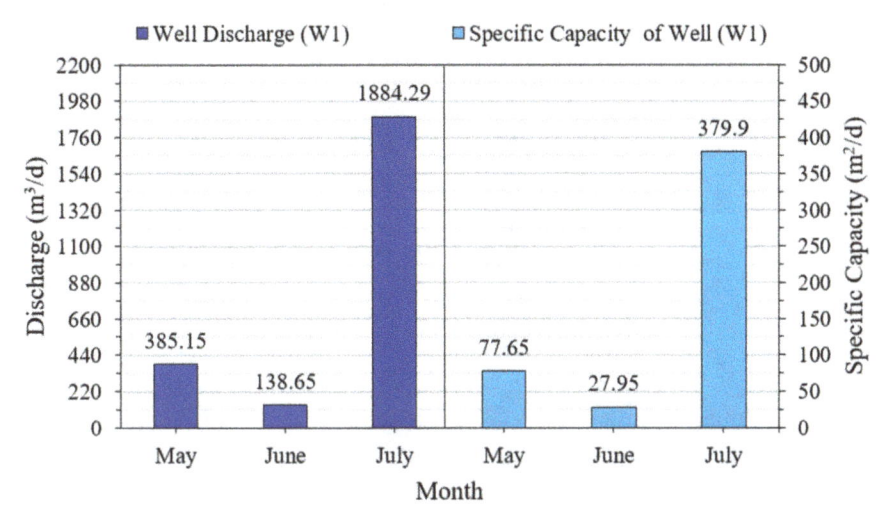

Fig. 6 Monthly well discharge and specific capacity of well (W1) at Kavlyachapada hamlet of Shivaji Nagar village in Nashik district of Maharashtra state

3.4 Analysis of Wells Extracting Water from Unconfined Aquifers

The q, K_o, A, and K_s were estimated manually for both Well-1 and Well-2 under unconfining aquifer conditions using Eqs. 7–11. In the case of Well-1, q and K_o for May (denoted as 1), June (as 2), and July (as 3) were calculated (Fig. 6 and Table 4). The q was recorded maximum during July with 1,884 m³/d, while June recorded the least value of q with 139 m³/d. A similar trend was followed by K_o. This was attributed to the enormous availability of the water during South-West monsoon as compared to dry summer. At the same time, K_s (K_{s13} for July) was calculated as 11/d and the T_r (T_{r13} for July) of the well was calculated as about 3 h (Eqs. 20 and 21), which implies that the total time taken by Well-1 to recuperate to its original level was 14.7 h (as t was observed 11.53 h).

$$K_{s13} = \frac{379.90}{34.42} = 11.04/d \qquad (20)$$

$$T_{r13} = \frac{1}{11.04} \times \ln\frac{6.48}{1.52} = 0.13\,d = 3.15\,hr \qquad (21)$$

Similarly, in the case of Well-2, q and K were obtained (Fig. 7 and Table 4). The Well-2 showed a similar trend as Well-1 for q and K, however, the corresponding values of the same were found averagely 456% and 500% higher. This was attributed to the geographical location of the wells. The Well-1, located closer to the hills, was exposed to comparatively higher surface runoff. Also, the depth of the soil layer at the hillside was averagely found between 0.2 and 1 m with widespread outcrops of

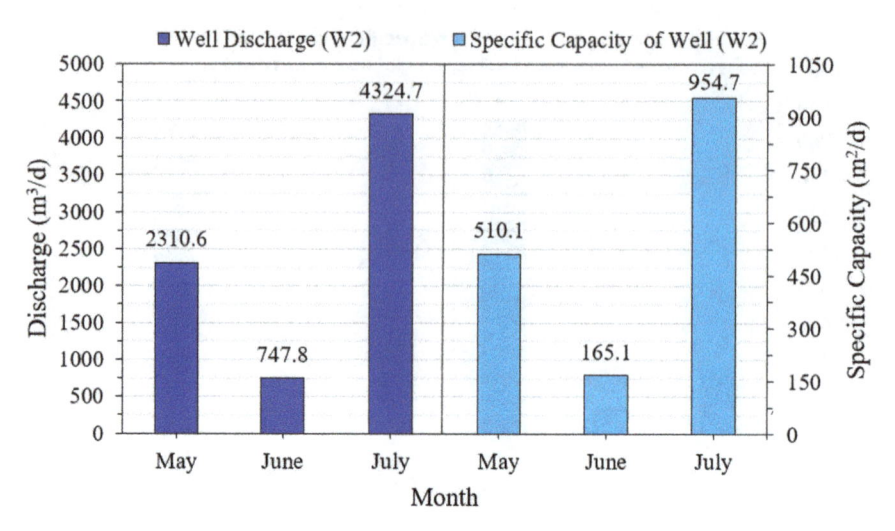

Fig. 7 Monthly well discharge and specific capacity of well (W2) at Kavlyachapada hamlet of Shivaji Nagar village in Nashik district of Maharahstra state

basalt rock limiting the natural infiltration. On the contrary, the slope was gentler when traversed from Well-1 to Well-2 resulting in less surface runoff. Also, the region in between wells was mostly paddy fields having an average depth of soil 0.5–3 m. Therefore, the conditions for the infiltration around the region of Well-2 were favorable for providing the possibility of higher discharge to the well with a greater probability of groundwater flow. The q was recorded maximum during July with 4,325 m^3/d, while June recorded the least value of q with 748 m^3/d. A similar trend was followed by K_o. The K_s (K_{s23} for July) was calculated as about 29/d and the T_r (T_{r23} for July) of the well was calculated as about 2 h (Eqs. 22 and 23). The total time for Well-2 to recuperate to its original level was found to be 11.2 h (as t was 9.2 h). In general, the field experimentations indicated that the discharge into the well-located away from the hills (W2) was estimated to be 500% greater than the well-located closer to the hills (W1). Furthermore, the well W2 was observed recuperating approximately 1.5 times faster than the well W1, as a result of which, agricultural practices were observed prevalent in and around the region of

W2, especially during monsoon (refer to Fig. 5).

$$K_{s23} = \frac{954.7}{33.18} = 28.77/d \qquad (22)$$

$$T_{r23} = \frac{1}{28.77} \times \ln\frac{4.95}{0.42} = 0.086\,d = 2.06\,h \qquad (23)$$

3.5 Comparative Analysis Between Manual Estimation and G-Cal Outputs

The outputs obtained from the G-Cal web tool for the entire case study in the KCP hamlet were validated against manually estimated outputs for the aforementioned groundwater components. For this, the data for (1) analysis of aquifer properties (n, V_v, V_o, S_y, and S_r), (2) analysis of groundwater flow between two wells (h_o, h_1, L, and K), and (3) analysis of well hydraulics of the two wells (q, H_o, H_1, h_w, R, and r_w) were input to the G-Cal web tool. For example, the output obtained from G-Cal for A, K_o, and K_s under the unconfining aquifer condition in July for Well-2 was obtained as about 33.18 square meters (m²), 954.7 m²/d, and 28.8/d, respectively which was identical and consistent with the outputs obtained manually (Fig. 8). The comparative analysis for the rest of the parameters also indicated that the outputs from the G-Cal operation and manual calculations were identical and consistent (listed in Table 4).

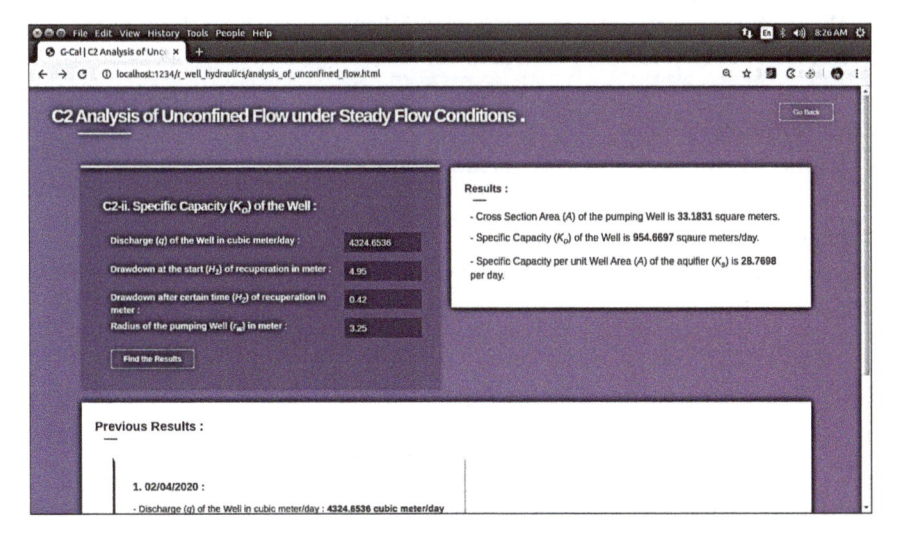

Fig. 8 Web application for Groundwater Calculator (G-Cal) showing an estimation of specific capacity (K_o) in July for Well-1 under unconfined aquifer conditions

Table 4 Comparative analysis of the outputs between the results obtained manually and the results obtained using the Groundwater Calculator (G-Cal)

Groundwater parameters	Units	Manually calculated	G-Cal calculated
Porosity (n)	%	44.72	44.7217
Specific yield (S_y) of clay	%	5.00	5.0000
Specific retention (S_r) of clay	%	39.72	39.7217
Discharge from well-1 to well-2 (Q)			
1. May (Q_{May})	m³/d	2.22	2.2214
2. June (Q_{June})	m³/d	2.37	2.3726
3. July (Q_{July})	m³/d	0.03	0.0326
Case of well-1: cross section area	m²	34.42	34.4196
Discharge into the well (q)			
1. May (q_{11})	m³/d	385.15	385.1524
2. June (q_{12})	m³/d	138.65	138.6478
3. July (q_{13})	m³/d	1884.29	1884.2898
Specific capacity (K_o)			
1. May ($K_{o,11}$)	m²/d	77.65	77.6517
2. June ($K_{o,12}$)	m²/d	27.95	27.9532
3. July ($K_{o,13}$)	m²/d	379.90	379.8971
Specific capacity/Area in July (K_{s13})	/day	11.04	11.0372
Recuperation time in July (T_{r13})	h	3.15	3.1530
Case of well-2: cross section area	m²	33.18	33.1831
Discharge into the well (q)			
1. May (q_{21})	m³/d	2310.61	2310.6093
2. June (q_{22})	m³/d	747.75	747.7544
3. July (q_{23})	m³/d	4324.65	4324.6536
Specific capacity (K_o)			
1. May ($K_{o,21}$)	m²/d	510.07	510.0683
2. June ($K_{o,22}$)	m²/d	165.07	165.0672
3. July ($K_{o,23}$)	m²/d	954.67	954.6697
Specific capacity/Area in July (K_{s23})	/day	28.77	28.7698
Recuperation Time in July (T_{r23})	h	2.06	2.0579

3.6 Opportunities for Groundwater Management and Security Using G-Cal

Along with the findings from G-Cal, the study from the qualitative interviews on the study site also validated the high rates of surface runoff from the village during South-West monsoon due to its location over the steeper slopes of the Sahyadri hills. At the same time, the observation regarding depth to the water level in dug wells

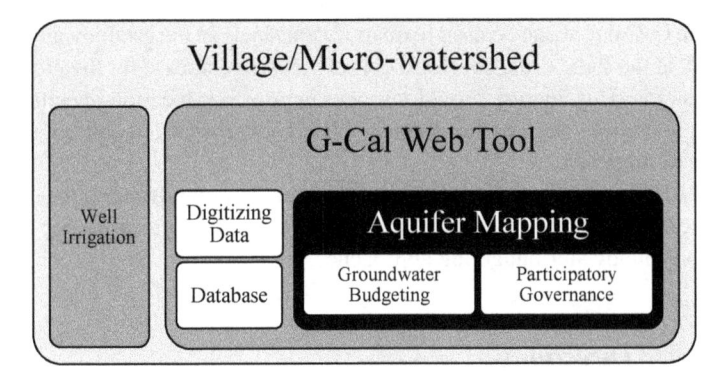

Fig. 9 Village-level facets for groundwater (well irrigation) management, governance, and security plan using Groundwater Calculator (G-Cal)

showed the reduced potential of the groundwater resources in that region. This could be attributed to the fact that the wells and the surface storage structures usually dry up post-mid-winter (till the end of the following summer) restricting the agricultural activities to one season. Thus, from consideration of vulnerability to groundwater storage and security at a village scale, there is a clear need to develop a robust mechanism on participatory forms of groundwater management and governance coupled with digitizing data collection using open-source web tools such as G-Cal (summarized in Fig. 9). The following series of steps could form the building block for village-level groundwater governance strategies. This can help attaining groundwater management and security with a considerable orientation towards G-Cal based web tools:

- It is recommended that the user should first develop a conceptual model for the given micro-watershed by specifically identifying the patterns of access to groundwater and the associated social processes that regulates its usage including agrarian practices.
- Status of groundwater resources can be obtained through conducting Participatory Rural Appraisals (PRA) such as developing resource maps, social maps, Venn diagram, and by organizing focused group discussions, problem ranking exercises, qualitative interviews, transect walk, and other relevant community-based interactions involving villagers, farming and non-farming communities, representatives from educational institutions, and village-level administrative bodies.
- Once acquired the primary information on the prevailing trends of groundwater usage and security issues, scientific investigation through the aquifer mapping approach can be adopted. It should aim to demarcate location-specific groundwater typologies to develop disaggregated hydrogeological regimes by mapping micro-level aquifers and by defining watersheds boundaries.
- To achieving participatory groundwater governance, a pilot program on community mobilization on digital methods of collecting hydrologic and hydrogeologic data and aquifer mapping can be organized at this stage.

- Then the G-Cal tool can be used to quantify and analyze the groundwater parameters and, at the same time, develop a comprehensive database for the groundwater flow systems. This, in turn, provides scopes to prioritize location-specific groundwater applications such as for domestic, livestock, irrigation, and other primary livelihood purposes.
- Further, the key findings from the location-specific groundwater attributes could be directed for effective decentralized management of groundwater resources through groundwater budgeting approaches.

3.7 Future Directions

In general, the present research is an attempt to transmute general empirical equations into a calculating tool to minimize laborious and tedious manual estimations inherent to groundwater analysis. More specifically, the scope was limited to estimating the well parameters in the hard rock region by developing a G-Cal tool whose operation was grounded on the textbook methods on groundwater (well) hydrology. In view of key findings and discussion made in this study, future studies should consider developing tools and methodologies for achieving a better understanding of regional groundwater characteristics, as follows:

- The G-Cal web tool developed in this study is not merely limited to the case study of KCP hamlet, but it is also technically sound for analyzing the groundwater parameters in diverse locations. However, not all the parameters, for which the G-Cal was designed, were explored and covered in the present case study. Future studies can employ such tools for varying hydrogeologic and climatic conditions as a local solution to precisely locating the groundwater access.
- This research primarily focused on the development and functioning of a software G-Cal. While such attempts may be useful in the application of known methods in characterizing open-well parameters, the availability of inputs required for such estimations shall remain a constraint. This study urges future investigation to devise methodologies to overcome issues of the primary data acquisition during field investigation.
- This study compared G-Cal outputs with manual estimations for validation purposes given the same data inputs, which point towards the limited data collection for analyzing various aspects of groundwater typologies. However, this step can also be performed using diverse data inputs given the sample size of the field experimentations. The latter can allow a further account for robustness and inclusivity of the G-Cal tool in real-life applications.
- Although general empirical equations were used while developing G-Cal, enciphering location-specific empirical equations may enhance the precision of diverse aquifer properties. Additionally, this can provide scope to develop the groundwater model at a village level so as to generate past, current, and future well irrigation usage scenarios. This would require methodological interventions

towards linking availability, accessibility, and demand in a more scientific and participatory manner.

4 Conclusions

Groundwater extraction is unsustainable in many regions of India, while proper management of groundwater aquifers also faces challenges, especially in estimating aquifer properties. While models exist, most are expensive and need a steep learning curve, which could be difficult with limited capacity. Under such a scenario, the primary objective of this study was to develop a free and open-source web application-based Groundwater Calculator (G-Cal) tool, that can estimate groundwater properties, and also host these properties for known geological formations as a database. The G-Cal web tool was designed to estimate the groundwater parameters at the micro-watershed or the village level that included calculations on porosity and specific yield of aquifer material, discharge from upstream well to the downstream well, well-discharge under steady-state groundwater flow conditions, and specific capacity and recuperation time of the well under both confining and unconfining aquifer conditions.

The outputs obtained for the aforementioned groundwater parameters using G-Cal were validated with the grounded field observations at the Kavlyachapada hamlet located in the Shivaji Nagar village of the Maharashtra state in India. The comparative analysis of the outputs obtained using the G-Cal web tool, against that of manually solved outputs indicated the high accuracy. Although the present study used the same input parameters for both manual estimation and G-Cal operation, future studies can conduct such studies at other study areas with diverse input parameters to achieving robustness in the usage of the web-tool. The user-friendly web applications, such as G-Cal, provide scopes to comprehend groundwater phenomena holistically, develops a groundwater database, and supports in drafting groundwater budget at the village-level. This, in turn, provides a check over the sectoral specific water deficiency or surpluses, thereby keeping the scope for careful groundwater utilization and adequate security towards its sustainable usage. Furthermore, the application of tools like G-Cal via mobiles/androids in the arid and semi-arid regions can influence the primary decision of crop selection, and redefining of distribution of water resources. Additionally, it can promote involving local community participation for common-pool resources management under the participatory governance approach.

Acknowledgements The authors thank the Centre for Technology Alternatives for Rural Areas (CTARA) of the Indian Institute of Technology Bombay and Hindustan Aeronautics Limited (HAL), Nashik Division, India for providing a fellowship platform for this work to be conducted. The authors would like to thank Mr. Sandeep R. Mahajan's Lead India Jalgaon Group for providing the infrastructure to conduct this research. Technical assistance provided by Mr. Ankush Patil in the development of the web application is highly acknowledged. Sincere gratitude to all the participants, farmers, village people, and Gram Panchayat from Shivaji Nagar village, and local Government officials from Nashik district for their kind support in the field survey and data acquisition.

Conflict of Interest The authors declare that the research was conducted in the absence of any commercial or financial relationships that could be construed as a potential conflict of interest.

References

1. Bhanja SN, Mukherjee A, Rodell M (2020) Groundwater storage change detection from in situ and GRACE-based estimates in major river basins across India. Hydrol Sci J 65(4):650–659. https://doi.org/10.1080/02626667.2020.1716238
2. Bhatnagar I, Jain K (2020) Simple methodology for estimating the groundwater recharge potential of rural ponds and lakes using remote sensing and GIS techniques: a spatiotemporal case study of Roorkee tehsil, India. Water Resour 47(2):200–210. https://doi.org/10.1134/S0097807820020025
3. Biswas B, Jadhav RS, Tikone N (2019) Rainfall distribution and trend analysis for upper Godavari basin, India, from 100 years record (1911–2010). J Indian Soc Remote Sens 47(10):1781–1792. https://doi.org/10.1007/s12524-019-01011-8
4. CGWB (2014) Central groundwater board report, Ministry of Water Resources, Government of India, Delhi. https://cgwb.gov.in/district_profile/maharashtra/nashik.pdf. Accessed 21 Sept 2020
5. Chinnasamy P (2017) Depleting groundwater–an opportunity for flood storage? A case study from part of the Ganges River basin, India. Hydrol Res 48(2):431–441. https://doi.org/10.2166/nh.2016.261
6. Chinnasamy P, Agoramoorthy G (2015) Groundwater storage and depletion trends in Tamil Nadu State, India. Water Resour Manag 29(7):2139–2152. https://doi.org/10.1007/s11269-015-0932-z
7. Chinnasamy P, Maheshwari B, Dillon P, Purohit R, Dashora Y, Soni P, Dashora R (2018) Estimation of specific yield using water table fluctuations and cropped area in a hardrock aquifer system of Rajasthan, India. Agric Water Manag 202:146–155. https://doi.org/10.1016/j.agwat.2018.02.016
8. Chinnasamy P, Srivastava A (2021) Revival of traditional cascade tanks for achieving climate resilience in drylands of South India. Front Water 3:639637. https://doi.org/10.3389/frwa.2021.639637
9. Conant B Jr, Robinson CE, Hinton MJ, Russell HA (2019) A framework for conceptualizing groundwater-surface water interactions and identifying potential impacts on water quality, water quantity, and ecosystems. J Hydrol 574:609–627. https://doi.org/10.1016/j.jhydrol.2019.04.050
10. Coyte RM, Jain RC, Srivastava SK, Sharma KC, Khalil A, Ma L, Vengosh A (2018) Large-scale uranium contamination of groundwater resources in India. Environ Sci Technol Lett 5(6):341–347. https://doi.org/10.1021/acs.estlett.8b00215
11. Darcy HPG (1856) The public fountains of the city of Dijon. Exhibition and application of principles to follow and forms to be used in matters of water supply, etc. V. Dalamont
12. Dupit JÉJ (1863) Theoretical and practical studies on the movement of water in open canals and through permeable soils: with considerations relating to the regime of large waters, the outlet to give them, and the alluvial walk in moving bottom rivers. Dunod
13. Dutta S (2018) Cause-effect analysis between irrigation and agricultural expansion on subsurface water resources: a case study of Kanksa Block in Ajay-Damodar Interfluve of Barddhaman District, West Bengal, India. Sustain Water Resour Manag 4(3):469–487. https://doi.org/10.1007/s40899-017-0128-1
14. Earman S, Dettinger M (2011) Potential impacts of climate change on groundwater resources–a global review. J Water Clim Change 2(4):213–229. https://doi.org/10.2166/wcc.2011.034
15. Freeze RA, Cherry JA (1979) Groundwater. Prentice-Hall, Englewood, Cliffs, New Jersey

16. GEC (2017) Groundwater Resource Estimation Committee, Ground Water Resource Estimation Methodology, Ministry of Water Resources Report, Government of India, Delhi. https://cgwb.gov.in/Documents/GEC2015_Report_Final%2030.10.2017.pdf. Accessed 21 Sept 2020
17. Geological Survey of India (2012) District resource Map. GSI, Nashik district. https://www.gsi.gov.in/webcenter/ShowProperty;jsessionid=Sg6Fv1RMR4d2qhULGNbDgUWsuqwTxLiT2ps1LIWEPXdLECOWjB7S!-316375636!871958704?nodeId=/UCM/DCPORT1GSIGOVI063304//idcPrimaryFile&revision=latestreleased. Accessed 21 Sept 2020
18. Gyeltshen S, Tran TV, Teja Gunda GK, Kannaujiya S, Chatterjee RS, Champatiray PK (2020) Groundwater potential zones using a combination of geospatial technology and geophysical approach: case study in Dehradun, India. Hydrol Sci J 65(2):169–182. https://doi.org/10.1080/02626667.2019.1688334
19. Howe PD, Thaker J, Leiserowitz A (2014) Public perceptions of rainfall change in India. Clim change 127(2):211–225. https://doi.org/10.1007/s10584-014-1245-6
20. Joshi D, Kulkarni H, Aslekar U (2019) Bringing aquifers and communities together: Decentralised groundwater governance in rural India. In: Singh A, Saha D, Tyagi A (eds) Water governance: challenges and prospects. Springer Water. Springer, Singapore, pp 157–185
21. Khadke L, Pattnaik S (2021) Impact of initial conditions and cloud parameterization on the heavy rainfall event of Kerala (2018). Model Earth Syst Environ 1–14. https://doi.org/10.1007/s40808-020-01073-5
22. Khadri SF, Pande C (2016) Ground water flow modeling for calibrating steady state using MODFLOW software: a case study of Mahesh River basin, India. Model Earth Syst Environ 2(1):39(1–17). https://doi.org/10.1007/s40808-015-0049-7
23. Khetwani S, Singh RB (2018) Groundwater dynamics in Marathwada region: a spatio-temporal analysis for sustainable groundwater resource management. Conserv Sci 9(3):537–548. https://ijcs.ro/public/IJCS-18-47_Ketwani.pdf
24. Li Q, Lian B, Wang Y, Taylor RA, Dong M, Lloyd T, Liu X, Tan J, Ashraf MM, Waghela D, Leslie G (2018) Development of a mobile groundwater desalination system for communities in rural India. Water Resour 144:642–655. https://doi.org/10.1016/j.watres.2018.08.001
25. Rao NS, Rao PS, Reddy GV, Nagamani M, Vidyasagar G, Satyanarayana NLVV (2012) Chemical characteristics of groundwater and assessment of groundwater quality in Varaha River Basin, Visakhapatnam District, Andhra Pradesh, India. Environ Monit Assess 184(8):5189–5214. https://doi.org/10.1007/s10661-011-2333-y
26. Rossetto R, De Filippis G, Borsi I, Foglia L, Cannata M, Criollo R, Vázquez-Suñé E (2018) Integrating free and open source tools and distributed modelling codes in GIS environment for data-based groundwater management. Environ Model Softw 107:210–230. https://doi.org/10.1016/j.envsoft.2018.06.007
27. Sanjay C, Bhasker P, Damodare SL, Abhale AP (2018) Statistical analysis of seasonal rainfall variability in Nasik district by using GIS interpolation. Pharmacogn Phytochem 7(4):2072–2077. https://www.phytojournal.com/archives/2018/vol7issue4/PartAI/7-4-115-107.pdf
28. Sarah S, Ahmed S, Boisson A, Violette S, De Marsily G (2014) Projected groundwater balance as a state indicator for addressing sustainability and management challenges of overexploited crystalline aquifers. J Hydrol 519:1405–1419. https://doi.org/10.1016/j.jhydrol.2014.09.016
29. Selvakumar S, Ramkumar K, Chandrasekar N, Magesh NS, Kaliraj S (2017) Groundwater quality and its suitability for drinking and irrigational use in the Southern Tiruchirappalli district, Tamil Nadu, India. Appl Water Sci 7(1):411–420. https://doi.org/10.1007/s13201-014-0256-9
30. Sikka AK, Alam MF, Pavelic P (2020) Managing groundwater for building resilience for sustainable agriculture in South Asia. Irrig Drain. https://doi.org/10.1002/ird.2558
31. Srivastava A, Chinnasamy P (2021) Water management using traditional tank cascade systems: a case study of semi-arid region of Southern India. SN Appl Sci 3:281. https://doi.org/10.1007/s42452-021-04232-0
32. Srivastava A, Chinnasamy P (2021) Investigating impact of land-use and land cover changes on hydro-ecological balance using GIS: insights from IIT Bombay, India. SN Appl Sci 3:343. https://doi.org/10.1007/s42452-021-04328-7

33. Subramanya K (2013) Engineering hydrology. McGraw Hill Education, New Delhi
34. Thiem G (1906) Hydrologische methoden. Gebhardt, JM Leipzig, Germany
35. Wagh VM, Panaskar DB, Muley AA, Mukate SV, Lolage YP, Aamalawar ML (2016) Prediction of groundwater suitability for irrigation using artificial neural network model: a case study of Nanded tehsil, Maharashtra, India. Model Earth Syst Environ 2(4):1–10. https://doi.org/10.1007/s40808-016-0250-3

Aman Srivastava obtained his Master's degree in Technology and Development from the Centre for Technology Alternatives for Rural Areas (CTARA) of the Indian Institute of Technology (IIT) Bombay. After his Research in Development Fellow position under Hindustan Aeronautics Limited (HAL) fellowship in Association with CTARA-IIT Bombay, he joined the Ministry of Rural Development, Government of India at New Delhi as Research Associate in 'Mahatama Gandhi National Rural Employment Guarantee Scheme' (MGNREGS), which is the largest running rural employment scheme across the world. He currently works here and primarily involved in developing sustainable solutions in convergence with MGNREGS works to address the water scarcity issues, prevailing in semi-arid regions across rural-India. His research interests include water accounting and budgeting, water resources system analysis, rainwater harvesting, and watershed management with a primary focus on achieving climate change resilience across the drylands of India. Mr. Srivastava is two times recipient of the Institute Silver Medal viz, one for Bachelor's degree in Civil Engineering and second for Master's degree in Technology and Development.

Leena Khadke pursued her Masters in Climate Science and Technology from the Indian Institute of Technology (IIT) Bhubaneswar. Currently, she is a Ph.D. scholar at the Department of Civil Engineering IIT Bombay. She is working in the field of hydrology, hydro-climatology, ecology, numerical modeling, complex networks, and climate and weather extremes for the past 3 years. She has an excellent grip on Weather Research and Forecasting (WRF) model and extensively used analysis tools like MATLAB, Python, GrADS, ArcGIS. Ms. Leena is a recipient of the Institute Gold Medal for her Bachelor's degree in Civil Engineering and Institute Silver Medal for her Master's degree in Climate Science and Technology. She received the Ministry of Human Resource Development (MHRD) based Teaching Assistant Project (TAP) fellowship for the Ph.D. Program at IIT Bombay.

Pennam Chinnasamy obtained his Master's degree in Physics from Wesleyan University, Connecticut - US, followed by a doctoral degree, with a focus on hydrology, from Missouri University, US. After his Research Fellow position with Ashoka Trust for Research in Ecology and the Environment, he joined the International Water Management Institute as a Researcher and was stationed in Nepal and Indian offices, where he focused on climate change impacts on underdeveloped and developing nations. He then joined Nanyang Technological University, Singapore, as a Senior Researcher developing real-time flood predicting models for Singapore. He is currently an Assistant Professor with the Indian Institute of Technology, Bombay - India, under the Centre for Technology Alternatives for Rural Areas (CTARA), where his work primarily focuses on natural resources assessment, monitoring, and management in rural regions. He is the founding director of the Rural Data Research and Analysis (RuDRA) lab, which is the first Big data lab for rural regions, housed in an academic institution in India. Over the past decade, Dr. Pennan has experience working in NGOs, national and regional government agencies, and academic institutions, focusing on sustainable surface and groundwater management plans, climate change impacts, large data analysis, and hydrological simulation models.

Modeling

On the Semiconductor Spectroscopy for Identification of Emergent Contaminants in Transparent Mediums

Surik Khudaverdyan, Ashok Vaseashta, Gagik Ayvazyan, Mane Khachatryan, Aigars Atvars, Mihail Lapkis, and Sergey Rudenko

Abstract In this chapter, we present a theoretical study of photoelectronic processes in experimental silicon n^+-p-n^+ structures with applications in identifying emergent contaminants in aqueous medium. Contribution due to various mechanisms of photon absorption to the total photocurrent is calculated. Various mechanisms, such as the influence of tunneling on the spectral characteristic and selective spectral photosensitivity of samples under investigation were investigated. The nature of the relationship between energy parameters of the absorbed waves and the structural parameters is revealed. Expressions are obtained for photocurrent with and without external diffusion current due to potential silicide barriers arising from p-n junction

S. Khudaverdyan (✉) · G. Ayvazyan · M. Khachatryan
National Polytechnic University of Armenia, Yerevan, Armenia
e-mail: xudaver13@mail.ru

G. Ayvazyan
e-mail: agagarm@gmail.com

M. Khachatryan
e-mail: kmane@mail.ru

A. Vaseashta
International Clean Water Institute, Manassas, VA, USA

Institute of Electronic Engineering and Nanotechnologies "D. Ghitu", ASM, Chisinau, Moldova

Institute of Biomedical and Nanotechnologies, Riga Technical University, Riga, Latvia

A. Vaseashta
e-mail: prof.vaseashta@ieee.org

A. Atvars
Photonics Laboratory, Institute of Astronomy, University of Latvia, Riga, Latvia
e-mail: Aigars.Atvars@lu.lv

M. Lapkis · S. Rudenko
RD Alfa Microelectronics, Riga, Latvia
e-mail: mihails.lapkis@rdalfa.lv

S. Rudenko
e-mail: rudenko@rdalfa.lv

or n^+-p-n^+ structures. We also calculate expressions for the absorption coefficient. In some samples, the injection of electrons through the direct-mesh n-p junction and the enhancement of the spectral photocurrent was observed. In samples without injection amplification of the photocurrent, an inversion of the sign of the spectral photocurrent took place. The inversion point is linear with the offset voltage, which can be used to find the unknown wavelength. Mutually compensating transitions in silicon structures provide shift of the maximum of the spectral photosensitivity from an intrinsic (~850 nm) to the short-wavelength region ~590 nm and 530 nm. The study suggests that, with the choice of structural parameters, it is possible to obtain different short-wavelength spectral maxima more accurately for a specific application such as for new and emergent contaminants in aqueous medium.

Keywords Absorbed waves · Photocurrent · Spectral photosensitivity · Injection amplification · Water contamination

1 Introduction—State-of-the-Art Semiconductor Spectroscopy

Currently, the importance of the applied science has grown significantly, primarily due to the necessity for the humanity to accept the new challenges and to solve the grand challenges associated with human health and environmental safety, climate change, contamination in water and food monitoring, creation of powerful army, and in general overall improvement of quality of life on the Earth. In this chapter, we describe how the spectral analysis of the optical information signal, where a semiconductor photodetector acts as a primary sensor, can play an important role in sensing/detection of emergent contaminants in aqueous medium due to spectral enhancement of the signal. In view of health and safety risk associated with such contaminants, observed improvement of the parameters of the photodetector with the help of new spectrophotometric functions is an urgent task.

Generally, the spectral analysis is carried out by means of a device that contains either a monochromator, a diffraction grating and a prism, or optical filters. Their use makes the device less efficient and very expensive. In order to eliminate these disadvantages, it is necessary to develop a semiconductor photodetector with selective spectral sensitivity and to carry out the relevant spectral analysis. The use of photodetector in spectrophotometry will eliminate the use of diffraction gratings, prisms and high-precision mechanical systems and will ensure high resolution, reliability, the fast registration of the spectrum, notwithstanding reduced cost and size. So far, the work carried out in this direction has remained at the research level, since it requires special conditions and sophisticated technology [1–9]. The research has high commercial value by industry [10, 11] since the identification of the composition of the environment in the field is critical for safety. In addition, use of such photodetectors in recent multipurpose monitoring systems is of paramount importance [12, 13]. The authors of this work have developed and studied two-barrier structures [14–20].

The structures are based on silicon and have a vertically oriented silicon junction and oppositely directed p-n or n-p junctions. The new physical principle applied in the photodetector ensures the selective spectral sensitivity. Such structures, as well as the structures with two oppositely placed p-n or n-p junctions, have a number of new features, and are described below using additional research.

2 Innovative Method of the Spectroscopy

An innovation method is employed by using two vertically placed junctions, in which the photocurrents created by longitudinal illumination, partially or completely compensate each other. The widths of the depleted regions of the junctions are changed by external voltage, which leads to the change in the fraction of the absorbed quanta of each wave and, consequently, result in change in the photocurrents, one at the expense of the other one. This process is also connected with the absorption depth of the wave. Taking these factors into account, the expressions for the resultant photocurrent were received based on the energy parameters of the radiation and the structural parameters. This makes it possible to create an algorithm for the spectral distribution of the intensity.

2.1 Photoelectronic Processes in Experimental Samples. Selective Spectral Sensitivity

To understand the mechanism of the photoelectronic processes occurring in longitudinally illuminated silicon two-barrier structures, it is necessary to estimate the contribution of various photon absorption mechanisms on the photocurrent. From the Einstein relation, the diffusion coefficients of electrons and holes in silicon are, respectively, $D_n = \sim 59 \, \text{cm}^2/\text{s}$, $D_p = \sim 16 \, \text{cm}^2/\text{s}$, and the diffusion lengths are $L_n = \sim 6 * 10^{-3}$ cm, $L_p = \sim 4 * 10^{-3}$ cm at the diffusion time of 10^{-6} s. In real samples, the diffusion time may be ten times as much, which makes the reading of diffusion lengths several times.

In the structures under consideration, the base region has the thickness of 2–6 μm. It is covered by the space charges of both barriers which can be regarded as thin n-p junctions since their width is much smaller than the diffusion length of electrons and holes [21]. From this perspective, the charge carriers fly through the space-charge layer without recombination, i.e. the recombination in the regions of the n-p junction is absent. In p^+ or n^+ regions, when the ionized impurity density is 10^{18} cm^{-3}, the Fermi level approaches the conduction-band bottom or the valence band top closely. When the density is higher, the Fermi level enters the allowed band, thus increasing the energy of the absorbed photons. Such degree of degeneracy is absent in the structures under consideration, and the interband absorption occurs. When the radiation is absorbed, it is likely that excited an electron does not pass from the valence band to the conduction band, instead it forms a bound system with the hole, i.e. an exciton. Thus, the energy of the exciton formation is less than the energy gap width. The formation of a stable system of an electron and a hole is possible, only at sufficiently low temperatures [21]. In the cases when the exciton binding energy becomes commensurable with the energy of lattice thermal vibrations, the exciton breaks down and the corresponding lines in the absorption spectrum disappear. It is worth noting that the energy of thermal vibrations is 25.8 meV at 300 K and the exciton binding energy in silicon is 28 meV.

In bulk semiconductors, the exciton state manifests itself only upon deep freezing of the samples, which makes this process unsuitable for use. In thin-film nanoscale semiconductor structures, the exciton states are well expressed at room temperature. In such samples, it is possible to change the binding energy and other parameters of excitons by changing the size of nanostructures. Thus, it is possible to control excitons in reduced dimensional structures [22] and to create devices based on the physical processes involving excitons. The absorption spectrum of excitons in the thin layers of the semiconductor, at the transverse electric field, moves into the red region in the system with quantum restrictions. The radiation detection takes place because of the decay of the excitons formed during the resonant excitation by the radiation. The structures under consideration are much thicker than the nanoscale structures, hence the exciton absorption at room temperature is practically absent. The free carrier absorption, and the impurity and lattice absorptions display infrared spectrum (at low temperatures) and low values of absorption coefficient. Hence, relatively thick samples are needed for tangible absorption.

2.2 The Determination of the Influence of the Tunneling on the Spectral Response

According to the literature [23], the numerical dependence of the absorption coefficient on the wavelength in silicon has been extensively characterized. It contains various contributions, including due to the long-wave part of the spectrum after the absorption edge except for the bandgap absorption. This contribution is significantly small, and when it is present, the main spectral distribution of the intensity may slightly expand towards longer waves. In view of the above-stated interpretation, we used dependence of the absorption coefficient on the wavelength for silicon, which is widely used in the development of silicon-based solar cells.

2.3 Tunneling Issues

We discuss separately the influence of tunneling on the photoelectronic processes occurring in the samples under consideration. It is known that the electric field result due to the slopes of the energy bands. Here, electrons can tunnel through the triangular barrier (Fig. 1a). The barrier height is E_g and the thickness d is

$$d = E_g/qE, \qquad (1)$$

where E is the electric field strength. With the increase in E, the slope increases and d decreases, and the tunneling probability increases. Upon the absorption of a quantum with the energy hv, the barrier thickness decreases to the value d' (Fig. 1b)

$$d' = (E_g - hv)/qE, \qquad (2)$$

As a result, the tunneling junction becomes more probable. In the strong electric field, the decrease in the barrier thickness is equivalent to the decrease in the energy gap width. This shifts the absorption edge towards lower energies (long waves).

It is known that the tunneling of electrons through the potential barrier in silicon begins at the thickness less than $d \sim 8E^{-7}$ cm. When the voltage incident on the base $V = 1$ V and the base thickness is equal to 2 µm, the field strength $E = 1/(2E^{-4}) = 5000$ V/cm. Thus, from the formula (1), the barrier thickness $d' = (E_g - hv)/q E$, and $(E_g - hv) = 8E^{-7} * 5000 = 0.004$ eV. It is by this small amount that the energy gap width decreases. With the voltage incident on the base equal to 2 V, the decrease will make 0.008 eV, etc. At high voltages and narrow bases, the absorption

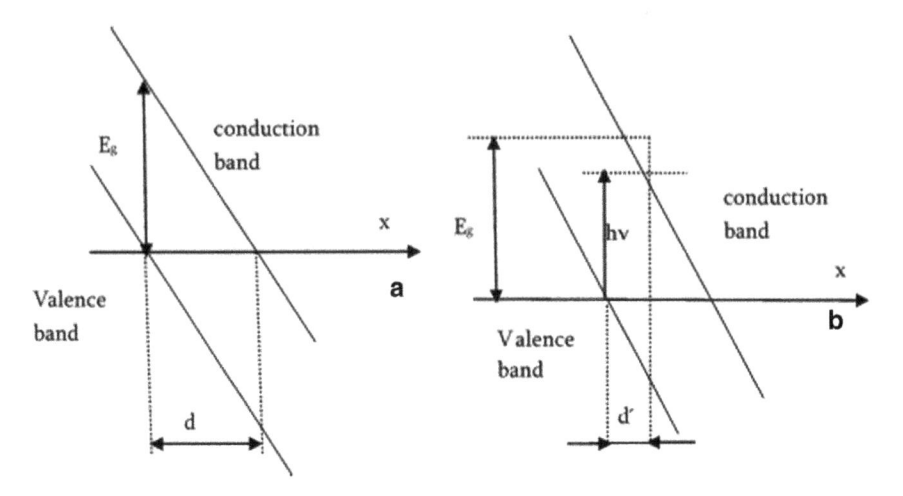

Fig. 1 a The electrons can tunnel through the triangular barrier, **b** absorption of a quantum with the energy hv, decreases barrier thickness to the value d'

edge shifts towards lower energies due to tunneling. This shift is however, small and hence, we can say that in the structures under consideration, the basic mechanism is the bandgap absorption.

3 Interrelation of the Structural Parameters

To solve the given problem, it is necessary to obtain an expression connecting the structure, energy and technological parameters of the photodetector. The p^+-n-p^+ and n^+-p-n^+ photodetectors, in which the base is covered with $x_m - x_0$ and $d - x_m$ depleted layers of the oppositely directed potential barriers (Fig. 1a, n^+-p-n^+ structures, x_m—the connection point of the barriers) are studied. As opposed to the previously discussed structures, here the width of the near-surface layer with the thickness x_0 is considered. The potential distribution is determined by the solution of the Poisson equation which connects the field potential $V(x)$ with the volume density of the charges N_a creating this field (Fig. 2) [22]. Let us solve the Poisson equation which has the following form,

$$\frac{d^2 V(x)}{dx^2} = -\frac{\rho}{\varepsilon \varepsilon_0} \qquad (3)$$

We proceed with the potential $V(x)$ to the potential energy of electrons, $\varphi(x)$, $\varphi(x) = -qV(x)$ in the Poisson equation. Since $\rho = qN_a$, we receive the following,

$$\frac{d^2 \varphi}{dx^2} = -\frac{q^2 N_a}{\varepsilon \varepsilon_0}, \qquad (4)$$

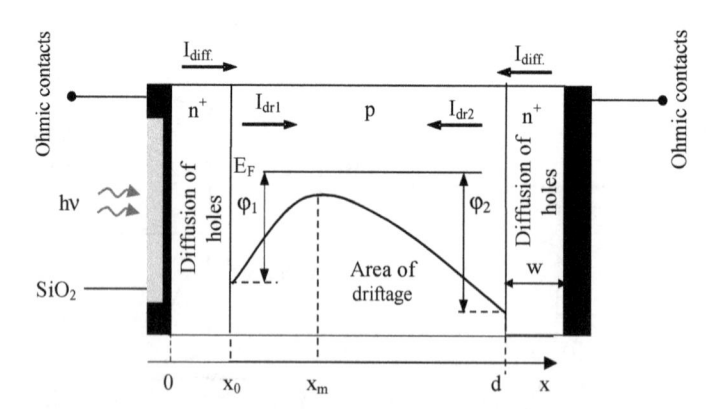

Fig. 2 n^+-p-n^+ structure. The distribution of the potential energy of holes in the valence band, the direction of the photocurrents

Fig. 3 The energy change of the valence band under the influence of the external voltage

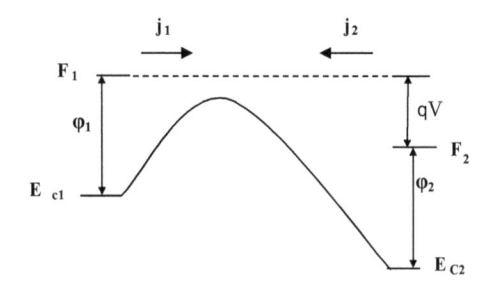

where N_a is the p-type impurity density, ε is the relative permeability of the substance, ε_0 is the permittivity of free space and q is the electron charge.

The boundary conditions for this equation are $\left(\left|\frac{d\varphi}{dx}\right| = 0\right)$ at $x = x_m$ (x_m is the maximum of the potential energy of holes) and $\varphi(x) = \varphi_{b1}$ at $x = x_0$ (Fig. 2). Taking into account the integrated Eq. (4) and that at $x = d$, when the external bias voltage is present, $\varphi(d) = \varphi_{b2} + qV$ (Fig. 3), we obtain an expression for x_m and $d - x_m$ depending on the external bias voltage,

$$x_m = \frac{d + x_0}{2} - \frac{\varepsilon\varepsilon_0(\Delta\varphi - qV)}{N_a q^2(d - x_0)}, d - x_m = \frac{d - x_0}{2} + \frac{\varepsilon\varepsilon_0(\Delta\varphi - qV)}{N_a q^2(d - x_0)} \quad (5)$$

If the near-surface n⁺ layer is thin, then $x_0 = 0$ and the expression (5) will have the form as obtained previously [17, 18],

$$d - x_m = \frac{d}{2} + \frac{\varepsilon\varepsilon_0(\Delta\varphi + q(-V))}{q^2 N_a d} \text{ or, } x_m = \frac{d}{2} - \frac{\varepsilon\varepsilon_0(\Delta\varphi + q(-V))}{q^2 N_a d} \quad (6)$$

By Eqs. (5) and (6), we can find the limiting values of the voltage at $x_m = d$ and $x = x_0$.

4 On the Development of the Optimal Design of Photodetector

Taking into account the volt-ampere characteristic (Fig. 4), at the radiation absorption of the blue LED, for instance, an abrupt change in the photocurrent takes place within the voltage range of 0–0.24 V, when the difference in the heights of the potential barriers is overcome (the near-surface barrier becomes equal to the rear barrier). With the expansion of the near-surface barrier, the further increase in the photocurrent slows down, which is probably due to the depth decrease in the number of the absorbed quanta. Table 1 shows the absorption depths of the waves corresponding to the maxima of blue, green and red LEDs.

Fig. 4 Light volt-ampere characteristic at the radiation absorption of the blue LED

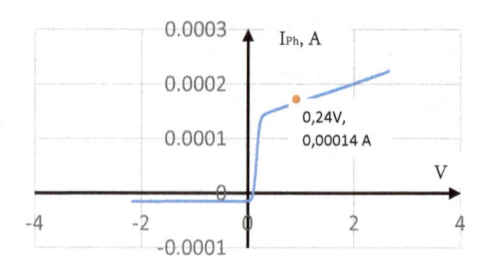

Table 1 Presents the parameters of the waves of blue, green and red radiations

Wavelength, nm	Absorption depth, nm	Absorption coefficient, cm^{-1}
460	476	$2.1 * 10^4$
560	1730	$5.78 * 10^3$
660	3870	$5.58 * 10^3$

It is obvious that at the base width of 2 μm, the blue and green radiations can be effectively absorbed. For the red radiation, the base width must be ~4 μm.

5 Photocurrent with and Without the External Diffusion Current: Case Studies of Silicide Potential Barriers—p–n Junction and n^+-p-n^+ Structures

Taking into account the exponential law of the absorption of the given wave, the expressions for the drift currents (Fig. 1a) formed in the base of the structures under consideration have the form:

$$I_{dr1} = qF_0 S(1 - e^{-\alpha x_m}), \tag{7}$$

$$I_{dr2} = qF_0 S(e^{-\alpha x_m} - e^{-\alpha d}), \tag{8}$$

where α is the absorption coefficient of the electromagnetic radiation, S is the photosensitive area and $F_0 = P_{opt}(1 - R)/Sh\nu$ is the total flux of the incident photons per unit area, P_{opt} is the radiation power, R is the reflection coefficient, h is Planck's constant, ν is the frequency of the electromagnetic radiation and q is the electron charge. The presence of an out-of-the-base region in the structure creates a diffusion component in the public photocurrent.

By neglecting thermal generation current, we derive an expression for the total photocurrent, taking into consideration the drift and diffusion components. To determine the diffusion photocurrent in the structure, it is necessary to find the minority carrier density p_n in the n-semiconductor by the one-dimensional diffusion equation,

$$\frac{\partial^2 p_n}{\partial x^2} - \frac{p_n}{L_p^2} = -\frac{p_{n0}}{L_p^2} - \frac{G(x)}{D_p}, \qquad (9)$$

where $L_p = \sqrt{D_p \tau_p}$ is the hole diffusion length in the n-region, $G(X) = F_0 \alpha e^{-\alpha x}$ is the hole-electron generation rate, D_p is the hole diffusion constant in the n-region, τ_p is the lifetime of the excess-carriers (holes), p_{n0} is the equilibrium concentration of the holes in the n-region.

The solution of the Eq. (3) under the boundary conditions $p_n = p_{n0}$ at $x = \infty$ and $p_n = 0$ at $x = d$ (p_n is the equilibrium concentration of the minority charge carriers in the n^+—region of Fig. 1a) has the form,

$$I_{diff} = S\left(qp_{n0}\frac{D_p}{L_p} + qF_0\frac{\alpha L_p}{1+\alpha L_p}e^{-\alpha d}\right) \qquad (10)$$

Taking into account (7), (8) and (10), the expression for the total current through the structure can take the form,

$$I_{tot} = I_{dr1} - I_{dr2} - I_{diff} = SqF_0\left(1 - 2e^{-\alpha x_m} + \frac{e^{-\alpha d}}{1+\alpha L_p}\right) - Sqp_{n0}\frac{D_p}{L_p} \qquad (11)$$

Since, in case of the normal operation in (11), the term containing p_{n0} is considerably smaller than the second term, it can be neglected and the expression (11) will take the form,

$$I_{tot} = SqF_0\left(1 - 2e^{-\alpha x_m} + \frac{e^{-\alpha d}}{1+\alpha L_p}\right) \qquad (12)$$

When irradiated by the integral flux (e.g. of the Sun), the expression for the photocurrent can be presented as,

$$\sum_{i,j} I_{Phi,j} = \sum_{i,j} I_{dr1i,j} - \sum_{i,j} I_{dr2i,j} - \sum_{i,j} I_{difi,j}$$

$$= Sq\sum_i\sum_j F(\lambda_i)\left(1 - 2e^{\alpha_i x_{mj}} + \frac{e^{-\alpha_i d}}{1+\alpha_i w}\right), \qquad (13)$$

where i = 1, 2, 3, ... changes in the integral flux with the change in the emission wavelength and j = 1, 2, 3, ... changes with the change in the bias voltage, $F(\lambda_i)$ is the total flux of the incident photons with the wavelength λ_i. If, in the formula (12) the width w of the n—region is taken to be less than L_p, then the value of L_p can be replaced by w, which is presented in (13).

The analysis of the spectral characteristics can be carried out with the help of the expression (12). Here, the external bias voltage V is determined with the help of the dependence of x_m on the voltage. From Fig. 2 it is obvious that in the two-barrier structure under consideration, as opposed to the single p-n junction, the dependence of the width of the depleted regions on the bias voltage has a linear character. Thus, it is possible to uniformly change x_m from x_0 up to d by the external voltage.

6 The Solution of the Expression for the Absorption Coefficient

Having the experimental values of the photocurrents, it is possible to determine the absorption coefficient of the wave, α depending on the change in x_m. If the depth of x_m corresponds to the most deeply penetrated wave from the integral flux, then a small voltage change will correspond to a change in x_m, within which only that part of the wave will be absorbed. Since the most deeply penetrated wave reaches x_m, the three values of the photocurrent I_1, I_2, I_3 corresponding to the three values of the voltage with the difference of 1 mV, are conditioned only by the absorption of that wave. Thus, when the diffusion component of the photocurrent is absent, with the help of the expression (12), we obtain the following equation:

$$\frac{I_1 - I_2}{I_2 - I_3} = \frac{e^{-\alpha x_{m2}} - e^{-\alpha x_{m1}}}{e^{-\alpha x_{m3}} - e^{-\alpha x_{m2}}}. \tag{14}$$

By denoting $A = \frac{e^{-\alpha x_{m2}} - e^{-\alpha x_{m1}}}{e^{-\alpha x_{m3}} - e^{-\alpha x_{m2}}}$ or $A = \frac{I_1 - I_2}{I_2 - I_3}$ and taking into account that the photocurrents have small values, it makes sense to transform the equation into the Maclaurin series around the point "x_m". Then, $e^{-\alpha x_m}$ for x_1, x_2, x_3 will have the form,

$$e^{-\alpha x_{m1}} = 1 - \alpha x_{m1} + \frac{\alpha^2 x_{m1}^2}{2}, \quad e^{-\alpha x_{m2}} = 1 - \alpha x_{m2} + \frac{\alpha^2 x_{m2}^2}{2}, \quad e^{-\alpha x_{m3}} = 1 - \alpha x_{m3} + \frac{\alpha^2 x_{m3}^2}{2}$$

With the help of that transformation, we will receive,

$$\frac{e^{-\alpha x_{m2}} - e^{-\alpha x_{m1}}}{e^{-\alpha x_{m3}} - e^{-\alpha x_{m2}}} = A = \frac{x_{m1} - x_{m2} - \alpha\left(\frac{x_{m1}^2 - x_{m2}^2}{2}\right)}{x_{m2} - x_{m3} - \alpha\left(\frac{x_{m2}^2 - x_{m3}^2}{2}\right)}.$$

Based on the latter and the experimental data of the photocurrent, we will determine the value of α,

$$\alpha = \frac{2A(x_{m2} - x_{m3}) - x_{m1} + x_{m2}}{A(x_{m2}^2 - x_{m3}^2) - x_{m1}^2 + x_{m2}^2} \tag{15}$$

When calculated, the diffusion component of the photocurrent gets cancelled, and we again obtain the expression (15). Using α, and the experimental data of the photocurrent, we determine the intensity of the absorbed wave with the help of the expression (12).

$$F_{0i} = \frac{I_i}{\left(1 - 2e^{-\alpha_i x_{mj}}\right) + \frac{e^{-\alpha_i d}}{1+\alpha_i w}} \qquad (16)$$

The next step is to obtain the dependence of the photocurrent of a given wave on x_m and subtract it from the total photocurrent. Further, a new small voltage change will move x_m towards the surface, and the registration region will involve in the next wave based on the penetration depth. It will help to determine the length and the intensity of that wave. The step-by-step repetition of the cycle will help to obtain the spectral dependence of the absorbed wave.

7 Case Studies of Three Typical Samples

7.1 8_Silicide-n-p

The positive photocurrents are due to the reverse-biased surface barrier. In the wavelength range of 350–600 nm, the photocurrent of the reverse-biased surface barrier, down to the zero voltage, is much higher than the photocurrent of the reverse-biased rear barrier (Fig. 5a). The short-wave maximum is in the region of 530 nm.

At positive voltages, the number of the absorbed short-wave quanta near the surface barrier is greater than that near the rear barrier. When the voltage sign is

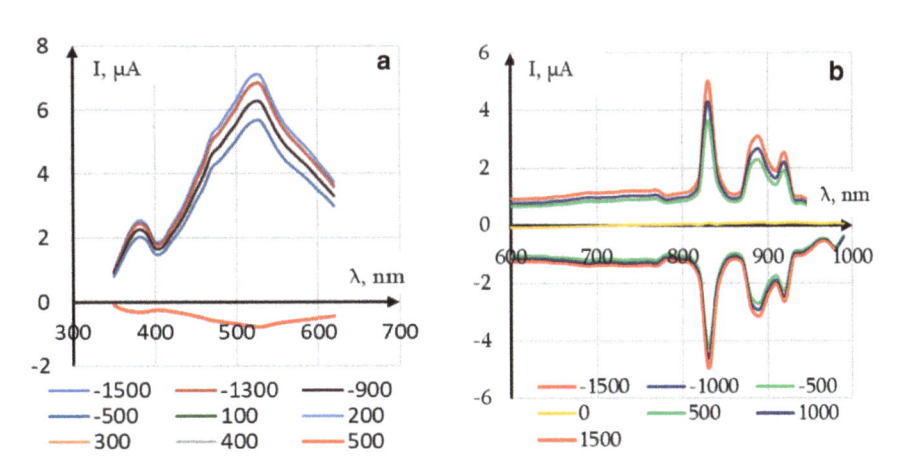

Fig. 5 The shortwave **a** and the longwave **b** spectral dependence of the photocurrent for the sample 8_Silicide-n-p at different bias voltages V in millivolts

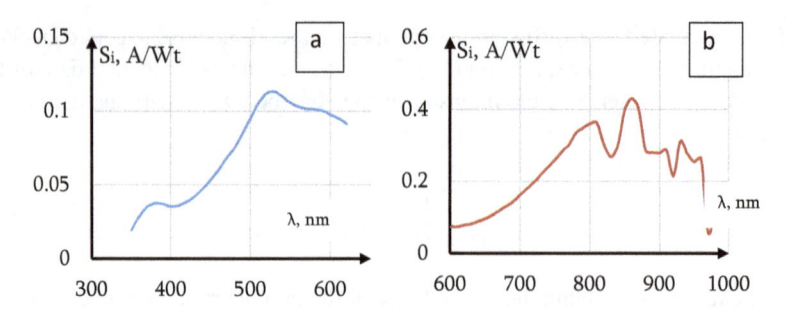

Fig. 6 The shortwave **a** and the longwave **b** current spectral photosensitivity for the sample 8_Silicid-n-p

changed, the behavior of the spectral photocurrent stays the same, due to the smaller number of the quanta that reach the rear junction region, the value of the photocurrent is lower for all of values of the voltage (Fig. 5a). When the wavelength range is 600–1000 nm, the quanta go deeper, and the photocurrent of the rear junction becomes comparable with the photocurrent of the surface barrier (Fig. 5b). The main maximum is located at the wavelength of 830 nm, which is close to the bandgap absorption of silicon.

The spectral distribution of the current photosensitivity (Fig. 6a, b) shows that the maximum is in the region of the bandgap absorption (860 nm) and has the value of 0.43 A/W. This is comparable with the sensitivity of the commercial silicon photodiodes. The photosensitivity at 530 nm is at 0.11 A/W.

7.2   37_n-p-n

As compared to 8_Silicide-n-p structures, here both in the shortwave and longwave regions, the negative photocurrents are comparable with the positive ones (Fig. 7a, b).

The longwave maximum is located at the wavelength of 830 nm (Fig. 7b), which is close to the bandgap absorption of silicon. This can also be seen on the spectrum of the current photosensitivity (Fig. 8). The spectral distributions of the current photosensitivity (Fig. 8a, b) show that the maximum in the short-wave region at 560 nm is 4.1 A/W, and in the long-wave region at 830 nm is 1.4 A/W. These values are high, if we consider that the best samples of photodiodes, e.g. HAMAMATSU, have the photosensitivity up to 0.7 A/W.

An identical behavior was observed with the sample 45_n-p-n.

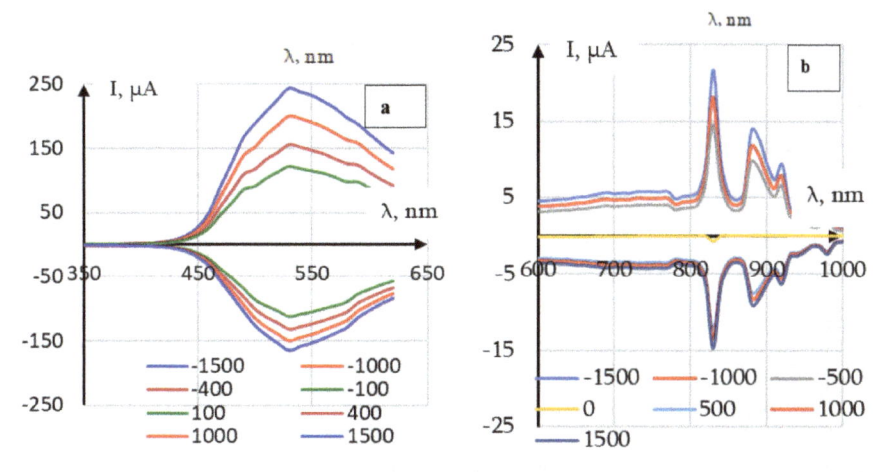

Fig. 7 The shortwave **a** and the longwave **b** spectral dependence of the photocurrent for the sample 37_n-p-n at different bias voltages V in millivolts

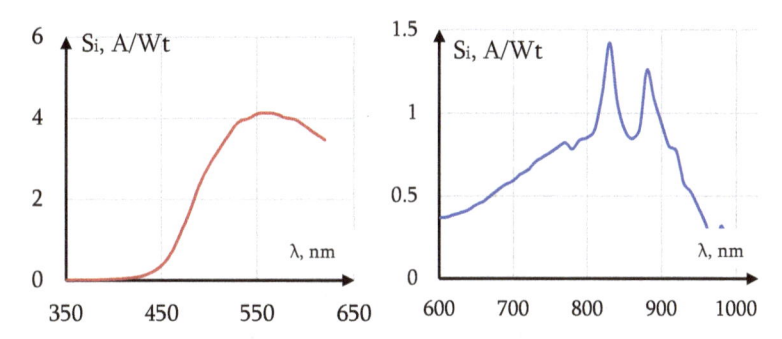

Fig. 8 The shortwave **a** and the longwave **b** current spectral photosensitivity for the sample 37_n-p-n

7.3 253_n-p-n

According to its indicators, the structure 253_n-p-n has some special features. The negative photocurrent, as opposed to the previous cases, belongs to the reverse-biased surface barrier (Fig. 9). The change in the sign of the long-wave photocurrent in the absence of the bias points to that (Fig. 9b, yellow line).

The short waves are absorbed near the surface barrier, creating a negative photocurrent, while the long waves are absorbed near the rear barrier, creating a positive photocurrent. The spectral distributions of the current photosensitivity (Fig. 10) show that the maximum in the short-wave region at 480 nm is 0.03 A/W, and in the long-wave region at 830 nm is 0.045 A/W.

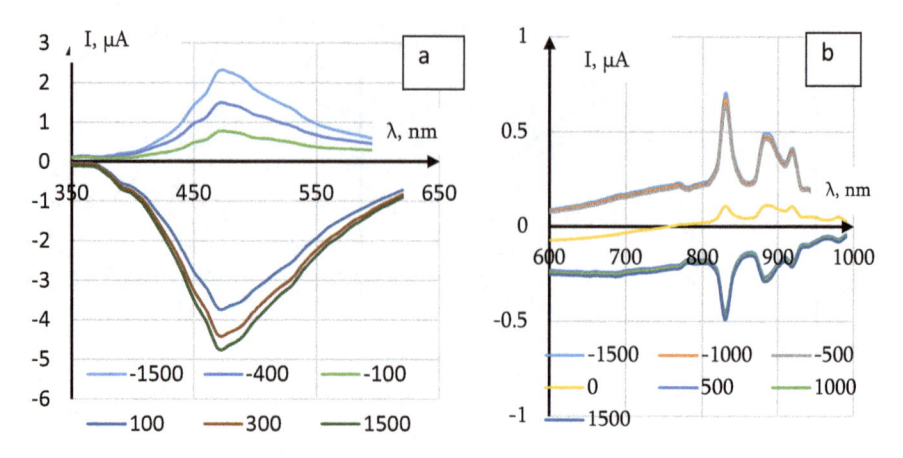

Fig. 9 The shortwave **a** and the longwave **b** current spectral photosensitivity for the sample 253_ n-p-n at different bias voltages V in millivolts

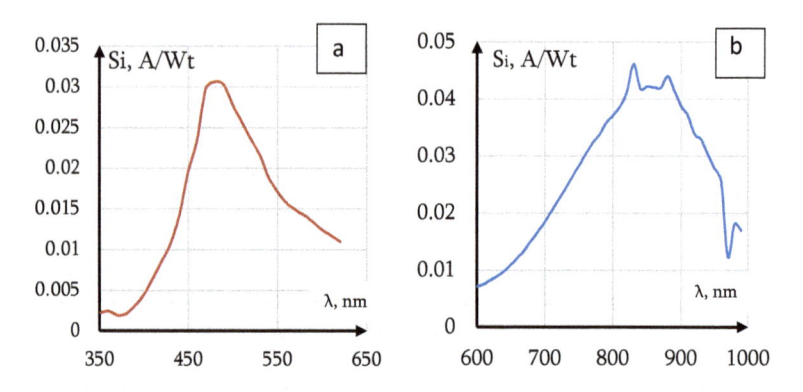

Fig. 10 **a** The shortwave, and **b** at longwave. Spectral photosensitivity for the sample 253_ n-p-n

Thus, there is a low photosensitivity, apparently due to the great compensation of the photocurrents of both barriers. In conclusion, we can say that Structures 37 and 45_n-p-n have abnormally high photo sensitivities. It is probably due to the presence of the mechanism of the internal amplification of the photocurrent. The speed of the photodetectors 37 and 45_n-p-n at the voltage of 20 mV is about 0.1–0.15 MHz upon the pulse rise (Fig. 11a), and about 0.04 MHz upon the pulse drop (Fig. 11b) of the photocurrent. With the increase in the voltage, the speed can increase. The device can be recommended for recording of the changes of not very transient optical signals.

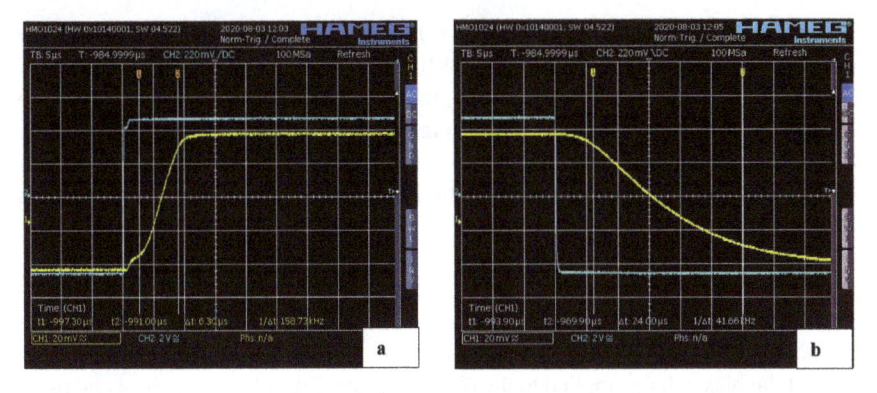

Fig. 11 **a** The rise and **b** the drop of the photo-signal pulse for the sample 45_n-p-n

7.4 On the Photocurrent Amplification Mechanism

Initially, it was assumed that the radiation distribution in the sample, in terms of the theory, obeys the exponential law. That is, by changing the voltage and the position of the junction point of the spectral photocurrent, it is possible to redistribute the radiation absorptions between the regions of the barriers. At that point, it is necessary to implement the mechanism of the bandgap absorption, and the n-p-n regions must be uniform so that no other fields could interfere.

The above stated structures are somewhat different from the ordinary n-p structures. For certain wavelengths, the following information was observed: (a): The absorption depths of the waves of 900 nm and 830 nm are 32.68 and 15.46 μm, respectively, which is larger than the base width (5.8 μm). Consequently, with the change in x_m, the change in the number of the quanta is insignificant (the slope of the absorption curve is small) and the response of the photocurrent change to the wavelength may be erroneous. (b): When the wavelengths are 700 nm and 600 nm, the absorption depths are 5.26 μm and 2.42 μm, respectively, which is comparable with the base width. With the change in x_m, the intensity modulation efficiency is much better, and the response to the wavelength is more accurate (Fig. 12 a, b). At

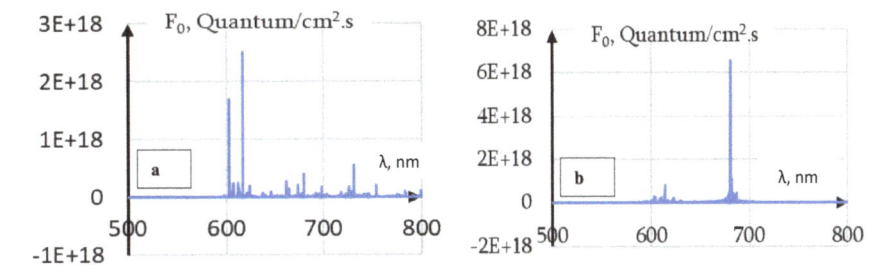

Fig. 12 The spectral distribution of the intensity of the wave with the length of **a** 600 nm, **b** 700 nm

600 nm, the spectral line of 617 nm and 603 nm is obtained at the radiation power of 1.34 µW, and at 700 nm, the spectral line is 681 nm at the radiation power of 7.5 µW. Thus, to obtain accurate results, the absorption depth of the wave must be comparable with or less than the base depth.

The spectral distribution of the radiation intensity is obtained based on the experimental data and the power of the radiation incident on the sample 45_n-p-n. It is obvious that starting from the wavelength of 460 nm (Fig. 13, curve 1), the photocurrent and the corresponding intensity are greater than those obtained from the radiation absorption at the external quantum efficiency equal to 1 (Fig. 13, curve 2). It is possible only when the photocurrent is amplified.

The spectral distributions of the current photosensitivity are identical at both signs of the bias voltage applied to the photodetector (Fig. 14—curves 1–4 and 6–8). Such behavior can be explained as follows. The applied positive voltage mainly falls on the reverse-biased near-surface n-p junction, and the public current depends on the reverse current of that junction. When the sample is irradiated, the radiation is absorbed in both junctions. At that, the part of the radiation that reaches the rear junction is absorbed in it, and the photogenerated carriers decrease the height of the

Fig. 13 The spectral distribution of the radiation intensity for the sample 45_n-p-n

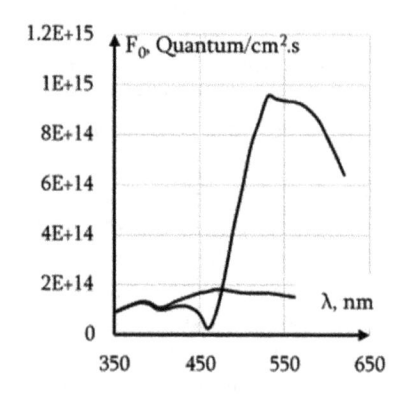

Fig. 14 The spectral distribution of the current photosensitivity for the sample 45_n-p-n

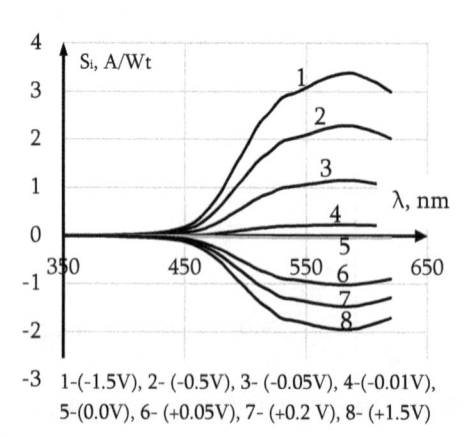

1-(-1.5V), 2- (-0.5V), 3- (-0.05V), 4-(-0.01V), 5-(0.0V), 6- (+0.05V), 7- (+0.2 V), 8- (+1.5V)

barrier of the forward-biased rear junction by compensating the volume charges. In accordance with this, the rear barrier, when irradiated, opens at lower voltages than in the absence of the irradiation, and the electrons passing through it, get injected into the base and get pulled into the n^+—region towards the positive electrode. This increases the photocurrent near-surface barrier. The higher the blocking voltage on it and the more absorbed quanta in the region of the rear junction, the more the photocurrent. This happens when the wavelength is extended from 350 to 590 nm, i.e. with the increase in the absorption depth or in the number of the quanta that reach the rear junction (Fig. 14, curves 6–8). The maximum of the spectral current photosensitivity is in the wavelength range of 590 nm. With the further extension of the wave, the absorption depth increases, and the slope of the absorption curve decreases. This leads to the decrease in the number of the absorbed quanta in the region of the rear junction and, therefore, decrease in the photosensitivity. In the region of the bandgap absorption, the injection is low and the current photosensitivity (0.9A/W) is slightly higher than the photosensitivity of the best photodiodes on the market (0.7A/W). A similar regularity is observed with the reverse-biased rear p–n junction. At that, the current photosensitivity is higher (Fig. 14, curves 1–4) than that at the reversed polarity (Fig. 14, curves 6–8). Taking into account that the near-surface n^+ layer has more free electrons ($n^+ \sim 10^{18}$ cm^3) than the rear n^+ layer ($n^+ \sim 5 * 10^{17}$ cm^3), and the number of the absorbed quanta in the near-surface junction is noticeably higher than that in the rear junction, the decrease in the height of the forward-biased near-surface barrier is also more than that of the forward-biased rear barrier. This means that the injection of the photoelectrons from the surface to the rear n^+ layer is greater, as well as the injection amplification of the photocurrent (Fig. 14, curves 1–4). For both polarities, the spectral maximum of the current spectral photosensitivity is in the wavelength region of 590 nm. At this wavelength, the absorption depth is most effectively combined with the structural parameters of the sample under consideration. In the absence of the bias voltage (Fig. 14, curve 5), the injection is absent, and the short-circuit photocurrent or the open-circuit photovoltage have small values and correspond to the difference in the heights of the oppositely directed potential barriers and make ~2 µA and 0.022 V.

8 On the Mechanism of the Selective Spectral Sensitivity

To obtain the selective sensitivity, it is necessary to have the real dependence of the change of x_m on the bias voltage regardless of its polarity, i.e. the uniform change of $x_m - x_0$ and $d - x_m$ one at the expense of the other one. This can be proved by the distribution of the number of the absorbed quanta of a given wave between the depleted regions of two barriers and the corresponding relation of the opposite photocurrents. In this case, depending on the depth of x_m or the bias voltage, the photocurrent sign may change. The larger the voltage range at which the change in the sign occurs, the larger the change range of x_m and the spectral range of the photocurrent sign change. Such a behavior is observed in all samples, but to different degrees. In the samples with the injection amplification of the photocurrent (37_n-p-n and 45_n-p-n), the voltage change range is insignificantly small. This means that the barrier widths do not change. Most likely, the injection of the photocarriers through the forward-biased barrier reduces it already at low voltages. Thus, the current is determined only by the reverse-biased barrier and does not change its sign.

A different picture is in the sample 8_Silicide-n-p. Here, the injection amplification of the photocurrent is absent, and the sign change is observed at a rather large range of the wavelengths (630–790 nm). The dependence of the inversion point on the bias voltage has a linear character (Fig. 15). When the voltage is positive, the surface barrier is reverse-biased, and with the increase in the voltage, the inversion point moves towards the long waves, and when the voltage polarity changes, it moves towards the short waves (Figs. 16 and 17). At that, the long-wave photocurrent is created mainly by the rear barrier and forms the spectral maximum in the region of the bandgap absorption of silicon, while the short-wave photocurrent created by the surface barrier, forms the short-wave maximum. The spectral minimum is created between these maxima as a result of the compensation of the oppositely directed photocurrents. The compensation factor determines the position of the short-wave spectral maximum (Fig. 16).

Fig. 15 The dependence of the inversion point on the bias voltage for the sample 8_Silicide-n-p

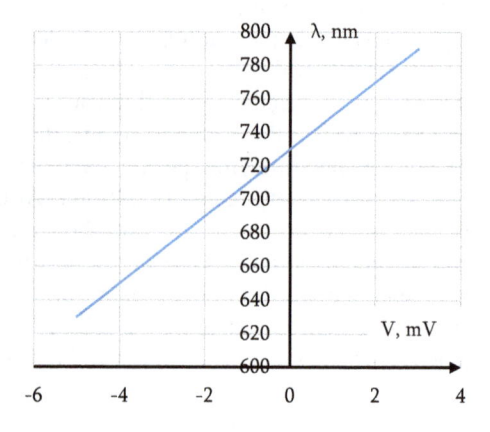

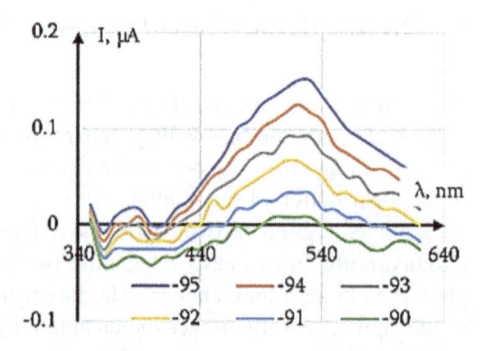

Fig. 16 The dependence of the spectral distribution of the shortwave photocurrent for the sample 8_Silicide-n-p

Fig. 17 The dependence of the spectral distribution of the longwave photocurrent for the sample 8_Silicide-n-p

Fig. 18 The dependence of the spectral distribution of the shortwave photocurrent for the sample 8_Silicide-n-p

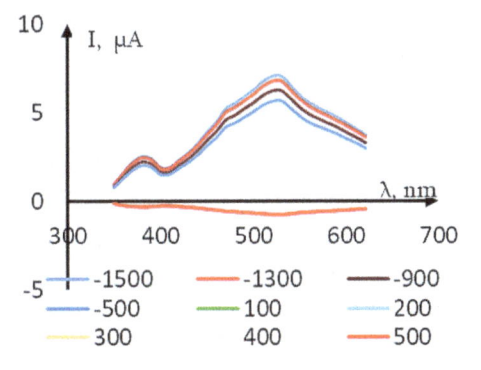

According to the experimental data, the shortwave change of the sign does not obey the linear law of the change from the bias voltage. At high voltages on the photodetector, the change in the sign of the spectral photocurrent is absent (Fig. 18). At the fixed voltage, due to the presence of the surface recombination centers, the effective absorption is low. With the increase in the wavelength, the absorption depth increases. The spectral photocurrent passing through the maximum decreases with the occurrence of the influence of the compensating rear opposite photocurrent (Fig. 18).

Thus, the linear change of the inversion point of the spectral photocurrent occurs at bias voltages of −5 –3 mV (Figs. 15 and 16). The injection amplification of the photocurrent is also absent in the sample 253_NPN. The two rather deeply placed opposite n-p and p-n junctions (Fig. 1) ensure the linear dependence of the change of the spectral photocurrent sign both in the wavelength range of 560–620 nm (with the voltage change from −33 to −55 mV) and 600–830 nm (with the voltage change from 15 to −50 mV) (Figs. 19a and 20 and Figs. 19b, 21).

Thus, if the injection of photocarriers from the out-of-the-base low-resistance regions is absent, then with the change in the voltage, the width of the depleted regions changes, and so does the inversion of the spectral photocurrent sign.

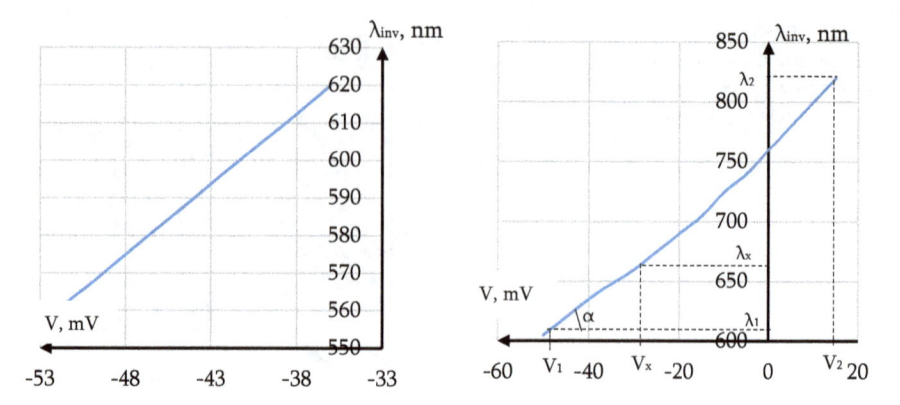

Fig. 19 The dependence of the inversion point λ_{imv} of the spectral photocurrent on the bias voltage for the sample 253_n-p-n for **a** the shortwave and **b** the longwave regions

Fig. 20 The dependence of the spectral distribution of the longwave photocurrent for sample 253_n-p-n

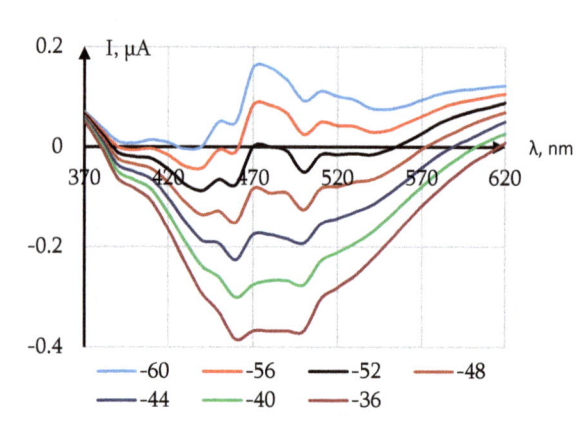

Fig. 21 The dependence of the spectral distribution of the shortwave photocurrent for sample 253_n-p-n

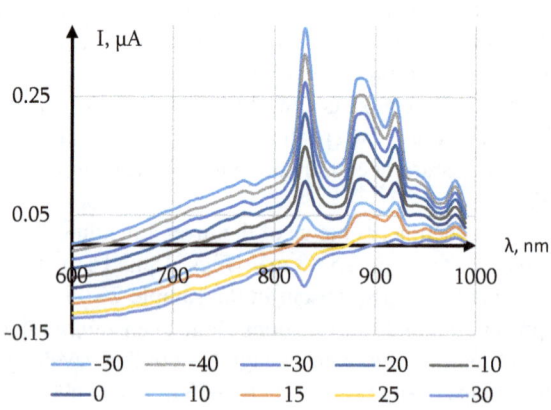

Fig. 22 The dependence of the position of x_m on the bias voltage and the range within which the inversion point changes

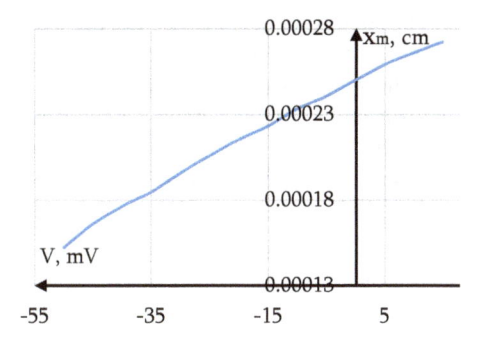

It is obvious that the unknown wavelength of the absorbed radiation can be determined with the help of the value of $tg\alpha$ in the linear region of the dependence of the inversion point on the bias voltage (Fig. 19b) by the following expression,

$$\lambda_x = \frac{V_x(\lambda_2 - \lambda_1) + \lambda_1 V_2 - \lambda_2 V_1}{V_2 - V_1}$$

Figure 22 shows the dependence of the position of x_m on the bias voltage and the range within which the inversion point changes. In the expression (12), x_m is determined by equating the current to zero and using the values of L_p, d and α. Hence, it is necessary to use those data of the absorption coefficient that correspond to the wavelengths of the connection point.

As seen from Fig. 22, the curve built in terms of the above, may be considered as linear within the limits of the measurement error, which confirms the linear dependence of x_m (V) in Eq. (3). It is the voltage range of -0.050–015 V that actively modulates the position of x_m in the sample 253_n-p-n. When developing the corresponding algorithm, the section of the current–voltage characteristic within the voltage range at which the change in the spectral photocurrent sign occurs, can be used to determine the radiation wavelength.

This investigation presented above can be used to determine the origin of any unknown wavelength in the spectrum, which may arise from the luminescence of substances, from transparent aqueous medium. In view of emergent contaminants, it is more relevant than ever before to have a characterization tool that corresponds to unknowns than knowns.

9 Conclusion

Thus, we can conclude from above the following:

1. A sufficiently long and intense wave reaches the rear junction and generates the photocarriers that decrease the height of the rear potential barrier. As a result,

2. The maximum of the spectral current photosensitivity is on the wavelength (Fig. 6 11, 590 nm) at which the highest absorption of the quanta in the region of the rear junction, as well as the decrease in the potential barrier are provided.
3. The compensating junctions in silicon structures provide the shift of the main maximum of the spectral photosensitivity from the bandgap (~ 850 nm) to the short-wave region (~ 590 nm and 530 nm, respectively, for samples 45_n-p-n and 253_n-p-n).
4. At the base narrowing (below 5.8 μm) and the impurity concentration changing in the base (to ensure the coupling of the barriers), it is possible to create **shorter**-wave silicon photodetectors (the maximum of the spectral sensitivity <530 nm) with the high sensitivity for registering the target weak optical signals.
5. In the samples without the injection amplification of the photocurrent, the inversion of the spectral photocurrent sign takes place. The inversion point linearly depends on the bias voltage, which can be used to find the unknown wavelength.
6. In this way Experimental samples have been studied. It has been shown that it is possible to obtain the spectral selective sensitivity by changing the potential barriers in the base one at the expense of another by external voltage.

Acknowledgements Research was supported by Project No. 1.3. of ERDF Project No. 1.2.1.1/18/A/006.

References

1. Wachowiak A, Slesazeck S, Jordan P, Holz J, Mikolajick T (2013) New color sensor concept based on single spectral tunable photodiode NaMLab gGmbH Dresden Nöthnitzer Straße 64, 01187 Dresden, Germany. 978-1-4799-0649-9/13/$31.00 ©2013 IEEE
2. Kalkhoran NM, Namavar F (1997) Multi-band spectroscopic photodetector array. U. S. Patent: US005671914A. http://www.google.com/patents/US5671914
3. Kautzsch Th (2013) Photocell devices and methods for spectrometric applications. Patent US 20130285187 A1
4. Kautzsch Th (2014) Photodetector with controllable Spectral response. Patent US 8916873 B2
5. Kautzsch Th, Photocell devices and methods for spectroscopic applications. Patent
6. DE 102013207801 A1 (2012)
7. Vanyushin IV, Gergel VA, Zimoglyad VA, Tishin YuI (2005) Adjusting the spectral response of silicon photodiodes by additional dopant implantation. Russ Microlectron 34(3):155–159
8. Gergel VA, Lependin AV, Tishin YI et al (2006) Boron distribution profiling in asymmetrical n+ -p silicon photodiodes and new creation concept of selectively sensitive photoelements for megapixel color photoreceivers. In: Proceedings of SPIE, vol 6260
9. Elif ÇS, David SF, Mutlu G, Ekmel Öz, Mesut S, Selim MÜn ((2014)) Improved selectivity from a wavelength addressable device for wireless stimulation of neural tissue. Front Neuroeng 1–12. https://doi.org/10.3389/fneng.2014.00005
10. Technavio releases new report on global spectroscopy market, 2016. http://www.businesswire.com/news/home/20160127005709/en/Technavio-Releases-Report-GlobalSpectroscopy-Market.

11. Global spectroscopy market 2017–2021, April 2017, https://www.technavio.com/report/global-embedded-systems-global-spectroscopy-market-2017-2021. Accessed 3 Oct 2018
12. Peng J, Hongbo X, Zhiye H, Zheming W (2009) Design of a water environment monitoring system based on wireless sensor networks. Sensors 9:6411–6434
13. Normatov PI, Armstrong R, Normatov ISh, Narzulloev N (2015) Monitoring extreme water factors and studying the anthropogenic load of industrial objects on water quality in the Zeravshan River basin. Russ Meteorol Hydrol 40(5):347–354
14. Khudaverdyan S, Dokholyan J, Kocharyan A, Kechiyantz A, Khudaverdyan D (2005) On functional potentiality of photodiode structures with a high-resistance layer. Elsevier, J Solid State Electron 49(4):634–639
15. Khudaverdyan S, Kocharyan A, Dokholyan J (2005) Photoreceiver structures with the extended functional potentiality on the CdTe base. J Phys D Appl Phys 38(2):272–275
16. Khudaverdyan S, Dokholyan J, Arustamyan V, Khudaverdyan A, Clinciu DL (2009) On the mechanism of spectral selective sensitivity of photonic biosensors. Elsevier, Nuclear Ins Methods Phys Res A. 610:314–316
17. Khudaverdyan S, Meliqyan V, Hovhannisyan T, Khudaverdyan D, Vaseashta A (2017) Identification and analysis of hazardous materials using optical spectroscopy. Opt Photon J 7:6–17. ISSN Print: 2160-8881
18. Khudaverdyan S, Hovhannisyan T, Meliqyan N, Mehrabyan N, Tsaturyan S, Khachatryan M, Vaseashta A (2016) On the model of spectral analysis of optical radiation. J Electromag Anal Appl 8:23–32 (2016.8.2003). https://doi.org/10.4236/jemaa
19. Khudaverdyan SKh, Khachatryan MG, Khudaverdyan DS, Tsaturyan SH, Vaseashta AK (2013) New model of spectral analysis of integral flux of radiation. In: NATO science for peace and security series B: physics and biophysics. Springer, pp 261–269. https://doi.org/10.1007/978-94-007-7003-4
20. Khudaverdyan S, Vaseashta A, Khachatryan M, Lapkis M, Rudenko S (2020) New method of optical spectroscopy for environmental protection and safety. In: Functional nanostructures and sensors for CBRN defence and environmental safety and security. Springer, pp 271–281. https://doi.org/10.1007/978-94-024-1909-2_19
21. Shalimova KV (1976) Semiconductor physics. Moscow, p 415
22. Vaseashta A (2005) Nanostructured materials based next generation devices and sensors. In: Vaseashta A, Dimova-Malinovska D, Marshall JM (eds) Nanostructured and advanced materials for applications in sensor, optoelectronic and photovoltaic technology. NATO science series II: mathematics, physics and chemistry, vol 204. Springer, Dordrecht. https://doi.org/10.1007/1-4020-3562-4_1
23. https://www.pveducation.org/pvcdrom/materials/optical-properties-of-silicon

Surik Khudaverdyan was born on May 06, 1952, Akhalkalaki, Republic of Georgia. In 1973, he received his degree engineering in electronic from the Yerevan Political Institute, Republic of Armenia. In 1981, his degree Ph.D. at the National Research University of Electronic Technology—MIET Zelenograd, Moscow, Russia, in the field of semiconductor physics. In 2005 he received his majoring Sc.D. degree in electronics and micro-nano electronics, in the State Engineering University of Armenia. Since 2007 she is a Professor at the State Engineering University of Armenia. From 1973 to 1987 years worked as an engineer, junior researcher, head of a sector, head of a department at the Institute of Radio physics and electronics of the National Academy of Sciences of Armenia, Ashtarak, Armenia. He was engaged in semiconductor photodetectors and digital angle converter. Currently engaged in spectral analysis of optical transparent substances and environmental safety among. He is the author of three books, over 100 articles and over 10 inventions. His research interests include semiconductor devices, multifunctional optoelectronic sensors, monitoring systems, micro- and nanoelectronics. Dr. Author was Awarded: In 2012 he Gold Medal of the State Engineering University of Armenia. Awarded the Presidential in Computer Science and Information Technology in 2013. He is a member of the Special Council for the award of a doctoral degree, a member of the editorial board of the of the Bulletin of ANPU and information technologies, electronics and radio engineering.

Ashok Vaseashta (M'79-SM'90) received Ph.D. in Materials Science and Engineering (minor in Electrical Engineering) from Virginia Polytechnic Institute and State University, Blacksburg, Virginia, USA. He is Executive Director of Research for International Clean Water Institute in VA, USA, Chaired Professor of Nanotechnology at the Academy of Sciences of Moldova and Professor, Nanotechnology and Biomedical Engineering at the Faculty of Mechanical Engineering, Transport and Aeronautics at the Riga Technical University. Prior to his current position, he served as Vice Provost for Research at the Molecular Science Research Center in Orangeburg, South Carolina. He served as visiting professor at the 3 Nano-SAE Research Centre, University of Bucharest, Romania and visiting scientist at the Helen and Martin Kimmel Center of Nanoscale Science at the Weizmann Institute of Science, Israel. He served the U.S. Department of State in two rotations, as strategic S&T advisor and U.S. diplomat. His research interests span nanotechnology, environmental/ecological science, and safety and security. His research on nanotechnology has been on improving the understanding, design, and performance of nanofibers and sensors/detectors, mainly for applications such as wearable electronics, target drug delivery, detection of biomarkers and toxicity of nano and xenobiotic materials. In the security arena, he has worked on counterterrorism, countering unconventional warfare and hybrid threats, critical-Infrastructure protection, biosecurity, dual-use research concerns, and mitigating hybrid threats. He has authored over 250 research publications, edited/authored eight books on nanotechnology, and presented many keynotes and invited lectures worldwide. He serves on the editorial board of several highly reputed international journals. He is an active member of several national and international professional organizations. He is a fellow of the American Physical Society, Institute of Nanotechnology, and the New York Academy of Sciences. He has earned several other fellowships and awards for his meritorious service including 2004/2005 Distinguished Artist and Scholar award.

Gagik Ayvazyan has earned his Ph.D. in physics and technology of semiconductor materials and devices at St. Petersburg State University of Information Technologies, Mechanics and Optics in 1989. He is Corresponding Member of International Academy of Engineering. Gagik Ayvazyan has been employed as Senior Researcher at the "Transistor" Company (Yerevan, Armenia), heading the advanced technology group (1989–2001). At present, G. Ayvazyan is an Associate Professor at National Polytechnic University of Armenia. He is also Scientific Manager of Barva Innovation Center (Talin, Armenia) and Scientific Expert of the International Commercialization Reactor (Riga, Latvia). His primary research interests are related to manufacturing and mechanics issues of silicon wafers and structures. His research spans multiple topics including wafer inspection, silicon-based solar cells and sensors, nanomanufacturing, RE technologies, acoustic and ionization methods for remote sensing of the atmosphere and weather-climate modification. Currently his research interests are focused on the structural and optical properties and applications of black silicon layers. Gagik Ayvazyan together with co-authors has published more than 100 papers in peer-reviewed journals and conference proceedings. He is inventor for 10 national and 3 international patents, as well as author of 5 collective monographs.

Mane Khachatryan was born in Yerevan, Armenia in 1987. She received the bachelor's degree (in 2008) and master's degree (in 2010) in the field of Biomedical Engineering, then the Ph.D. degree in the field of Automation Systems (in 2014) from State Engineering University of Armenia. Since 2011, she has been a Scientific Worker at "The Photoelectronic Equipment in Optical Systems of Communication" Research Laboratory, National Polytechnic University of Armenia (NPUA). Since 2016, she has been a Scientific Worker at "System Analysis" Research Laboratory, NPUA. Since 2012, she has been an Associate Professor in Chair "Communication Systems", NPUA. She is an author of 20 articles, and she is a coauthor of 1 monography. Her research interests include mathematical modelling, monitoring systems, the development of the optical signals. Dr. Khachatryan has been a responsible secretary of the Journal "Proceedings of NPUA. Information Technologies, Electronics, Radio Engineering" since 2015. She was awarded with President Prize, Republic of Armenia in the field of "Technical Sciences and Information Technologies" in 2013.

Aigars Atvars received his B.S., M.S., and Ph.D. (2008) from the University of Latvia, Riga, Latvia, working in the field of atomic and laser spectroscopy and performing mathematical modeling. From 2009 to 2014 he was a co-founder and the director of the Institute of Physical Research and Biomechanics, Latvia, where several industry-driven research projects were implemented. Since 2015 A. Atvars is a senior researcher at the University of Latvia. Current research interests of A. Atvars cover theoretical simulation of optical processes, including mathematical modeling of optical microresonators. He is active in research project preparation and implementation and collaboration with the industry.

Mihail Lapkis was born in Ukraine, Kherson, in 1948. He received his B.S. and M.S. degrees in Physics with the specialisation 'Diffusion Processes in Real Crystals' from Kharkov State University Physics Department in 1971. From 1972–1976, he worked as the Leading Engineer at Riga Scientific Research Institute—Laboratory of Thin Dielectric Films. In 1976, he took the position of Leading Technologist, Deputy Chief Technologist and Chief Technologist at the Scientific and Production Complex 'Planar', where he worked until 2000, when he co-founded the JSC "RD ALFA Microelectronics Department", which is the legal successor of Riga SRI of micro-devices and plant Invertor, and he holds the positions of Director General and Chairman of the Board to this day. Mihail is the author of 6 inventions in the field of technology and construction of linear integrated micro-circuits, as well as of 5 scientific papers in the field of mathematical physics, solid state physics, and technology of integrated microcircuits. So far, he has been a Design Manager of several IMCs (radiation-resistant) and Operations Manager responsible for improvement of construction and production technology and implementation of mass production.

Sergey Rudenko was born in Russia, Grozny, in 1953. He received his B.S. and M.S. degrees from Leningrad Polytechnic Institute named after M.I. Kalinin, Faculty of Radio Electronics, Semiconductor and Dielectric Department with the specialization 'Semiconductors and Dielectrics' in 1977. Starting from 1977, he was an engineer in Riga Scientific Research Institute of Micro-devices PA Alfa, later worked as engineer technologist of the plant Invertor of PA Alfa and, from 1987, deputy chief technologist of the plant Invertor of PA Alfa. From 1994 to 2000, he was a technical director of Bipolārs Ltd, but in 2000 became a co-founder and Technical Director of the JSC "RD ALFA Microelectronics Department". Sergey is the Chief Constructor of several IMC, author of 5 inventions in the field of technology and construction of linear integrated micro-circuits. Recognized Invention Certificates: Method of production of linear IC with complimentary bipolar and MOS-transistors. Method of production of linear IC (LIC) with complimentary bipolar and MOS-transistors. Method of production of IC with complimentary bipolar and MOS-transistors. Method of fitting of sputtering of resistors made of silicon composites. Method of production of IC with thin film resistors.

Modeling the Evolution of Surface and Groundwater Quality

Erika Beilicci, Robert Beilicci, and Mircea Visescu

Abstract Water is a key environmental factor for the life of humans, animals, and plants representing an indispensable resource for the economy. The state of water resources is important both in terms of quantity and quality. To use and effectively protect water resources and ecosystems of Europe against pollution, climate change and marine waste, it requires coordinated action at European Union (EU) level. In this respect, the EU has established a community framework for the protection of water resources management by developing water directives. An important step to develop sustainable management plans of surface and groundwater resources is forecast of their quality evolution. To achieve this, it requires modeling the phenomena that occur in surface and groundwater (viz. hydrodynamic modeling and propagation of various substances which pollute water with diffuse and/or point pollution sources and due to various accidents), which is possible using advanced hydroinformatic modeling tools with satisfactory precision. The detailed results obtained from modeling increase general understanding of the evolution of water quality and support authorities to act (in time and space), in case of pollution, according to the management plans, based on plans risk management of pollution of water resources. This chapter shows the necessity and usefulness of advanced hydroinformatics tools in modeling of the evolution of water quality. We present several modelling tools, such as MODFLOW—with ASMWIN and PMWIN variants; MIKE BY DHI and FEFLOW. In the last section, we present some examples of application of these softwares.

Keywords Water quality · Hydroinformatics · Modeling · Pollution

---

E. Beilicci (✉) · R. Beilicci · M. Visescu
Faculty of Civil Engineering, Department of Hydrotechnical Engineering, Polytechnic University Timisoara, Splaiul Spiru Haret 1A, 300022 Timisoara, Romania
e-mail: erika.beilicci@upt.ro

R. Beilicci
e-mail: robert.beilicci@upt.ro

M. Visescu
e-mail: mircea.visescu@upt.ro

1 Introduction

Water is a key environmental factor for the life of humans, plants and animals, representing an indispensable resource for the economy. The state of water resources is important both in terms of quantity and quality, both in the case of surface and groundwater. Water quantity refers to the ensuring the volume of water discharge necessary for the water supply of the population to carry out economic activities with optimal parameters and for ensuring the needs of ecosystems, to remain in a good condition. Water quality refers to the chemical, biological, physical, and radiological characteristics of water. Water quality and water quantity are both closely linked. If we have a sufficient amount of water that does not meet the requirements in terms of quality, is a challenge just as much if the water is limited in quantity, even good quality, which will not be able to meet all requirements. In both cases, the results of water use will not be optimal. In all parts of the world, the quality and quantity of water resources must be analyzed together, in order to ensuring the necessary quantities of water, with an adequate quality, for drinking, irrigation, zootechnics, ecosystems health, industry, power generation, and recreation.

Protecting and managing water resources to meet water needs requires comprehensive information and understanding of the impacts of natural conditions and anthropogenic activities on water quantity and quality [18]. To use and effective protection of water resources and water ecosystems of Europe against pollution, climate change and marine waste, requires coordinated action at EU level. In this respect, the EU has established a community framework for the protection and water resources management by developing water directives. An important step to develop sustainable management plans of surface and groundwater resources is forecast of their quality evolution. To achieve this, requires modeling the phenomena that occur in surface and groundwater (hydrodynamic modeling and propagation of various substances which pollute water in case of diffuse and/or point pollution sources present, including various accidents), which is possible using advanced hydroinformatic modeling tools for satisfactory precision. The detailed results obtained from modeling increase general understanding of the evolution of water quality and support authorities to act (in time and space), in case of pollution, according to the management plans, based on plans risk management of pollution of water resources [21].

Member States are obliged to adopt management plans and programs of measures appropriate for each body of water considering the result of analyzes, studies and results of water resources quality evolution modeling. Management plans had to be implemented in the period 2012–2015, which must then be reviewed every 6 years. These plans, drawn up at the pool/river basin aim to; (a) prevent deterioration, improvement and restoration of surface water bodies; (b) achieve good chemical and ecological properties no later than the end of 2015; (c) reduce pollution discharges and emissions of hazardous substances; (d) protection, improvement and restoration of groundwater, prevent pollution or deterioration ensuring a balance between abstraction and recharge; and (e) conservation of protected areas [21].

2 Advanced Hydroinformatic Tools for Modeling the Evolution of Surface and Groundwater Quality

The most modern methods of modeling the evolution of surface and groundwater quality are based on the use of hydroinformatics (or water informatics), which is a combination of modelling tools and Information and Communication Technologies (ICT), resulting in a unique methodological approach for physical, social and economic aspects analyze of sustainable water resources management. Hydroinformatics is a complex interdisciplinary field, which exceed traditional definition of water/environmental science and engineering, informatics/computer science (including artificial intelligence, data mining and optimization techniques) and environmental engineering, and has applications in various areas of water resources management, including:

- development and application of decision-support systems, used by involved authorities and stakeholders in water and environment domain;
- simulation and optimization models able to provide solutions to water engineering problems;
- computational tools and techniques and their effective application to managing risk and uncertainties associated with water systems, especially in context of climate changes;
- cross-disciplinary complex system approaches to water resource management (integrated water resources management);
- to improve understanding of water systems functioning, including technical, socio-economic and environmental issues [20].

Advanced hydroinformatic tools are those tools whose modelling technologies achieve their full potential in terms of practical application. The advanced tools integrate the hydraulic and hydrological models with new hydroinformatics technologies including, in principal, the following elements: standards for "open" modeling systems, logical modelling techniques, knowledge base system, systematic calibration, advanced optimization methods and decision methods [1]. The recent development of research activities in hydroinformatics make possible a combined access to powerful features, like: standard interfaces to GIS, giving the user access to necessary specific data; a choice of alternative and compatible hydrological and hydraulic modelling system; graphical interface; utilities for model calibration, validation, verification, optimization and decision making; enable specialists from other involved fields (agronomists, ecologists, meteorologists, climatologists etc.) to access integrated hydrological and hydraulic data and knowledge resources in efficient and responsible way [1].

In applying and calibrating models, the use of "good practice" codes is essential. Any modeling methodology must ensure that the data is used properly, that the models are calibrated and validated correctly and that each model is used for the purpose for which it was designed. The increasing complexity of the models has made the selection of models a key element for the success of any study, together with the establishment of a robust and rigorous quality assurance procedure [19]. The advanced hydroinformatic tools that will be selected to be applied must meet a number of basic requirements, including simulation of a wide range of chemical parameters, preferably physical-based and not conceptual, compatible with the MS operating system and GIS (ArcView), to be flexible in terms of input data required to use the model, to provide results compatible with the application of the Water Framework Directive 2000/60/EC. The Water Framework Directive is the basis for integrated management and coordination for all water resources within the European Union. The approach to water management is defined in terms of environmental effects and aims at the chemical and biological quality of water and the viability of water resources [19]. Choosing the adequate hydroinformatic tool is a key success factor for any modeling study. The increasing complexity of the models (on rivers and on hydrographic basin) has made model selection, a process of great importance for the success of any study.

Water quality models can be classified as follows:

- models on the hydrographic basin for the evaluation of the impacts, at the scale of the river basin;
- simple river and intermediate models for assessing compliance with chemical quality standards for rivers;
- complex river and intermediate models to assess the impact of pollutant discharges into rivers [19].

When modeling the evolution of water quality, the magnitude of significant anthropogenic pressures must be taken into account, such as:

- Point sources of pollution;
- Persistent sources of pollution;
- Diffuse sources of pollution;
- The effects of the change of the flow regime due to the water withdrawals for users, the regularization of the river, the hydrotechnical arrangements;
- Morphological changes (river regularization, constructions in the major riverbed).

Other pressures on the environment and water resources should be considered, where appropriate, especially land use (urban, industrial, agricultural, forestry) which may help to identify and locate the impact on them. An integrated approach is needed to allow for point and diffuse sources, runoff changes, morphological changes to be considered, such that river hydrographical basins can reach the required quality standards. The greatest difficulties are posed by diffuse sources of surface and groundwater pollution, such as agriculture, due to the difficulty in defining the problems and the number of generally large scope of elements involved. It is very difficult to directly measure the contribution of diffuse sources to the total pollutant load, and this must be evaluated by means of scientific estimation techniques or modeling using advanced hydroinformatics tools.

Advanced hydroinformatic tools require the following data for modeling, which accurately define the following:

- for surface waters: cross sections on the river; longitudinal sections along rivers and main tributaries; information defining the works on the river such as dams, gates and bridges; downstream boundary conditions; upstream discharges; water quality data and upstream sediment loading; information on point and diffuse sources; physical-chemical information of the river; the composition of the riverbed.
- for groundwater: topographic data; terrain characteristics (terrain stratification, porosity, transmittance coefficient, hydraulic conductivity coefficient); surface water levels (in rivers, lakes); aquifer thickness; data on drillings, drains.
- for river hydrographical basins: topographic data (slopes); soil data (physical-chemical characteristics, structure, chemical composition, texture, infiltration capacity); meteorological data (precipitation, temperature); land use (drainage coefficient); economic and social factors.

To effectively use these models, they must be calibrated in a number of sections, where there are field-measured data, including: water level, evacuations, chemical composition and sediment loading [19]. Water quality models based on advanced hydroinformatic tools are efficient way to simulate and predict water quality evolution and pollutant transport in water courses, aquifers, lakes and reservoirs, which can contribute to saving the cost of labors and materials for a large number of sampling and chemical experiments to determine the degree of pollution of water bodies. Water quality models are useful tools to simulate and predict the levels, distributions, and risks of chemical pollutants in each water body, in different scenarios of pollution (point source, distributed source and accidental pollution).

2.1 MODFLOW (ASMWIN)

Aquifer Simulation Model for Windows (ASMWIN) is an integrated program for groundwater flow and transport modeling developed by Wen-Hsing Chiang, Wolfgang Kinzelbach and Randolf Rausch. It is an integrated program with implementation of 2-D groundwater model for use on PC's under MS-Windows, which was originally developed for education of students in hydrogeology, civil and environmental engineering fields. ASM was published in 1989 and initially ran under MS-DOS. Since that time, ASM has been continually enhanced and now, the latest and most powerful version 6.0 of ASM runs under the Microsoft Windows operating system. The most notable changes are the use of a new, friendly graphical user interface which replaces the character- oriented interfaces of previous versions and at the same time, allows the manipulation of larger model grids which makes it suitable for professional use.

In order to perform a steady-state flow simulation we need to create a new model, assign model data (generate a model grid, specify boundary conditions and model parameters), perform flow simulation, check results and produce output. This very user-friendly implementation of a groundwater model on Windows PCs hopes to lower the limit, which otherwise prevents the widespread use of computer-based groundwater models. The version (6.0) is strongly enhanced with a finite-difference flow model, a finite-difference transport model, a professional graphical user-interface (GUI), a tool for the automatic calibration of a flow model, a particle tracking model, a random walk transport model and several other useful modeling tools [16, 17].

2.2 MODFLOW (PMWIN)

Processing Modflow was developed by Wen-Hsing Chiang and Wolfgang Kinzelbach as a part of remediation project in Northern Germany. Processing Modflow for Windows (PMWIN) is an application for modeling groundwater flow and transport processes with a modular three-dimensional finite-difference groundwater model MODFLOW of the U. S. Geological Survey. PMWIN supports the calculation of elastic and inelastic compaction of an aquifer due to changes of hydraulic heads. PMPATH uses a semi-analytical particle tracking scheme to calculate the groundwater paths and travel times and allows a user to perform particle tracking. PMPATH calculates and shows path-lines or flowlines and travel time marks simultaneously. MODPATH permits forward and backward tracking in transient flow fields as well as steady-state flow fields. The MT3D transport model uses a mixed Eulerian-Lagrangian approach to the solution of the three-dimensional advective-dispersive-reactive transport equation. The MT3D transport model can be used to simulate changes in concentration of single species miscible contaminants in groundwater considering advection, dispersion and some simple chemical reactions. The chemical reactions included in the model are currently limited to equilibrium-controlled linear or non-linear sorption and first-order irreversible decay or biodegradation.

Packages of MODFLOW that are supported by PMWIN are: Horizontal Flow Barrier Package (HFB1) for easily simulating slurry walls, the Time Variant Specified Head Package (CHD1), and the Interbed-Storage Package (IBS1) for simulating transient storage and calculating compaction and subsidence of an aquifer due to changes of hydraulic heads, Field Generator which can generate heterogeneously-distributed transmissivity or hydraulic conductivity, Water Budget Calculator for calculating water budgets and also the exchange of flows between zones [7–9].

2.3 MIKE by DHI

MIKE11 advanced hydroinformatic tool, part of the DHI software products, is a professional engineering software package for the simulation of flows, water quality and sediment transport in estuaries, rivers, irrigation systems, channels and other water bodies. MIKE11 is a user-friendly, fully dynamic, one-dimensional modelling tool. The modules used for modeling water quality evolution are Hydrodynamic

Module (HD) and ECOLab module [13]. The MIKE11 hydrodynamic module uses an implicit, finite difference scheme for the computation of unsteady flows in rivers and estuaries. The MIKE11 HD module solves the vertically integrated equations for the conservation of continuity and momentum, i.e. the Saint-Venant equations. The basic forms of the equations used in MIKE 11 are shown in Eqs. 1 and 2.

$$\frac{\partial Q}{\partial x} + \frac{\partial A}{\partial t} = q \qquad (1)$$

$$\frac{\partial Q}{\partial t} + gA\frac{\partial h}{\partial x} + \frac{\partial \left(\alpha \frac{Q^2}{A}\right)}{\partial x} + \frac{g|Q|Q}{C^2 A R} = 0 \qquad (2)$$

where: Q is discharge, x is longitudinal channel distance, A is cross-sectional area, q is lateral inflow, t is time, h is flow depth, C is the Chezy coefficient and R is the hydraulic radius [13].

ECOLab is a numerical lab for Ecological Modelling. It is an open and generic tool for customizing aquatic ecosystem models to describe water quality, eutrophication, heavy metals and ecology. The module is mostly used for modelling water quality as part of an Environmental Impact Assessment (EIA) of different human activities. ECOLab is a very powerful module for advective and dispersive modeling for a very large number of parameters. The model solves the one-dimensional mass balance equation for dissolved oxygen, suspended matter, totally dissolved gas, non-cohesive sediments and temperature exchange. ECOLab is designed for modeling chemical, biological and ecological processes and the interactions between them. Simulated parameters for water quality through this system include: Biochemical Oxygen Demand (BOD), Dissolved Oxygen (DO), bacteria, ammonia nitrites, nitrates, eutrophication, phosphates etc. [13]. MIKE11 also has strong capabilities for modeling sediment transport and riverbed morphology.

The advection-dispersion model (AD) is based on the one-dimensional (vertically and laterally integrated) equation for the conservation of mass of a substance in solution, i.e. the one-dimensional advection-dispersion equation:

$$\frac{\partial (AC)}{\partial t} + \frac{\partial (QC)}{\partial x} - \frac{\partial}{\partial x}\left(AD\frac{\partial C}{\partial x}\right) = -A \cdot K \cdot C + C_2 q \qquad (3)$$

where: C is the concentration, D the dispersion coefficient, A the cross-sectional area, K the linear decay coefficient, C_2 the source/sink concentration, q the lateral inflow, x the space coordinate and the time coordinate. The advection-dispersion equation is solved numerically using an implicit finite difference scheme which, in principle, is unconditionally stable and has negligible numerical dispersion. The equation reflects two transport mechanisms: advective (or convective) transport with the mean flow; dispersive transport due to concentrations gradients.

The main assumptions underlying the advection-dispersion equation are: the substance under consideration is completely mixed over the cross-sections, implying that a source/sink term is considered to mix instantaneously over the cross-section; the substance is conservative or subject to a first order reaction (linear decay); Fick's diffusion law applies, i.e. the dispersive transport is proportional to the concentration gradient. The module requires output from the hydrodynamic module, in time and space, in terms of discharge and water level, cross-sectional area and hydraulic radius [13].

The user can use predefined ECOLab (WQ) templates containing the mathematical descriptions of ecosystems or can choose to develop own model templates. The module is developed to describe chemical, biological, ecological processes and interactions between state variables and also the physical process of sedimentation of components can be described. The ECOLab is integrated with the advection-dispersion module [13]. As a basis for the description of the water quality conditions, the AD calculates the conservative transport of the modelled components. The WQ (water quality) processes in combination with the AD transport give the final result [11, 12].

MIKE HYDRO Basin and MIKE HYDRO River are a versatile and highly flexible advanced hydroinformatic tools for a large variety of applications.

- MIKE HYDRO Basin for: Integrated Water Resources Management (IWRM) studies; provision of multi-sector solution alternatives to water allocation and water shortage problems; reservoir and hydropower operation optimization; exploration of conjunctive use of groundwater and surface water; irrigation scheme performance improvements.
- MIKE HYDRO River for: flood analysis and flood alleviation design studies; real time flood or drought forecasting; dam break analysis; optimization of reservoir and gate operations; ecology and water quality assessments in rivers and wetlands; water quality forecasting; sediment transport and long term assessment of river morphology changes; wetland restoration studies. Features include: Water quality options using ECOLab [14].
- MIKE HYDRO has a simple physical model for the aquifer that interacts with the surface water by infiltrations from the river (river to the aquifer); groundwater supply (basin to aquifer); groundwater discharge (aquifer to river).

2.4 FEFLOW

FEFLOW (Finite Element subsurface FLOW and transport system) is an interactive groundwater modeling system for: three-dimensional and two-dimensional, areal and cross-sectional (horizontal, vertical or axisymmetric), fluid density-coupled, also thermohaline, or uncoupled, variably saturated, transient or steady state, flow, mass and heat transport, reactive multi-species transport in subsurface water resources with or without one or multiple free surfaces.

FEFLOW is the tool to simulate all the flow, mass and heat transport processes in the subsurface for analyzing interactions below the land surface by including any manner of underground structures, tunnels and other types of excavation. Model can extend by coupling FEFLOW to other MIKE surface water models to investigate groundwater-surface water interaction. FEFLOW can estimate ground subsidence and more accurately predict pumping rates with a hydromechanical coupling plugin. With FEFLOW, users can create layer-based, partially or fully unstructured meshes in 3D. Produce precise spatial representation of complex geology and geometry for rivers, fractures, pipes, tunnels and well locations. It can create powerful visuals, cross-sectional and 3D visualization—even with Oculus Virtual Reality technology. Results can be presented with high-quality snapshots or video sequences. Since its first appearance in 1979, FEFLOW has been continuously extended and improved. Currently, it is maintained and developed by DHI-WASY experts [10].

3 Case Studies for Modeling the Evolution of Surface and Groundwater Quality

3.1 Groundwater and Solute Transport Modeling in the Proximity of Arad Thermal Power Plant, Romania—ASMWIN Application

There are a significant numbers of thermal power plants around the world and new plants are put into operation almost weekly. This rapid industrialization has resulted in an increased in the use of natural resources, such as coal in case of fossil fuel burning power plants. All these power plants brought along serious environmental imbalance due to the dumping of industrial wastes. The Arad Thermal Power Plant is located in the northern part of Arad locality and was designed to run on solid fossil fuel (lignite) with the natural gas as the support for fire. The plant provides the heat carrier and over 50% of the electricity for Arad city. The plant uses around 750,000 tons of lignite per year. The control drillings situated to the north, east and south of the storage pit were put into execution to verify if there are possible polluted water leakages towards east where the tapping of groundwater for Arad city drinking water is performed. For modeling, input data consisted of terrain topography, geological and hydraulic characteristics of the soil, boundary conditions, characteristics of observation wells, chemical characteristics of the pollutant [2].

Following the hydraulic modeling, the results are presented in Figs. 1 and 2.

The results of simulation quantify evolution of concentrations with respect to time, for all points of polluted zones. This concept is important to find a technical method for limitation, reduce or eliminate pollution over time. The water samples drawn from the control drillings indicate pH, chlorides and sulphates values that exceed the accepted limits. The transport water and phreatic water both show high values for chlorides, sulphates and hardness.

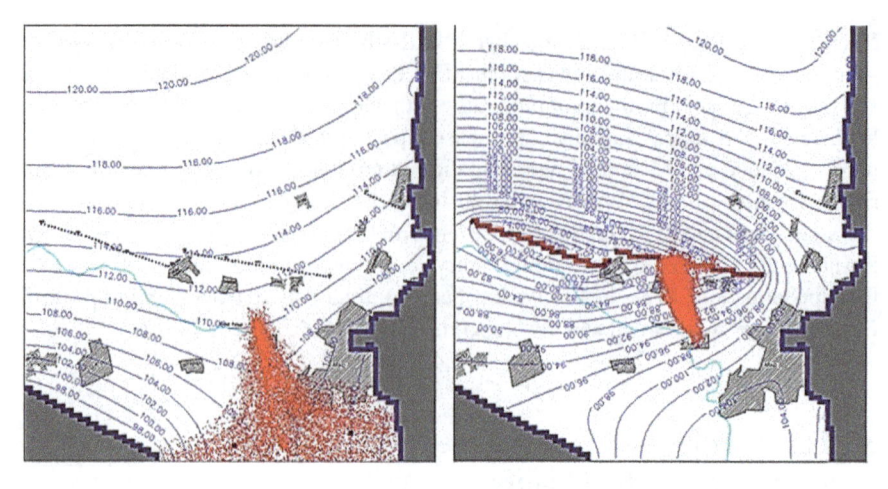

Fig. 1 Head contour and distribution of the pollutant (in deep aquifer, in shallow groundwater)

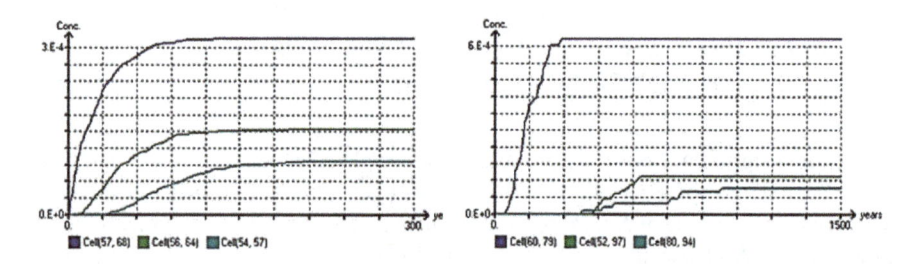

Fig. 2 Concentration in observation well (in deep aquifer, in shallow groundwater)

3.2 Pollutants Transport Simulation in Aquifers Layer, Provided from Infiltration of Wastewater Coming from Pig's Farms—ASMWIN Application

The level of pig breeding in Romania reached a maximum in 1989 to almost 14,351,000 pigs. During that period in Romania, almost 300 pig's farms in industrial system, with a total effective population of 10,351,000 pigs in individual farms bred ~4,000,000 pigs. In Timis County, the largest pig farm is owned by COMTIM (SMITHFIELD) which, between 1986 and 1989, bred over 1,463,000 heads. The pollution produced from the residues from pig farms was widespread and was felt in atmosphere, underground water, surface water and soil. The smell produced by anaerobe decompose of residues has been enormous and pathogenic agents can be transmitted from animals to people from atmosphere. Due to high level of phosphor, potassium and azoth from waste water which was evacuated in surface water will appear the eutrophication phenomena. Through infiltration of wastewater in underground, they pollute the aquifer stratum and also the catchments of water basins.

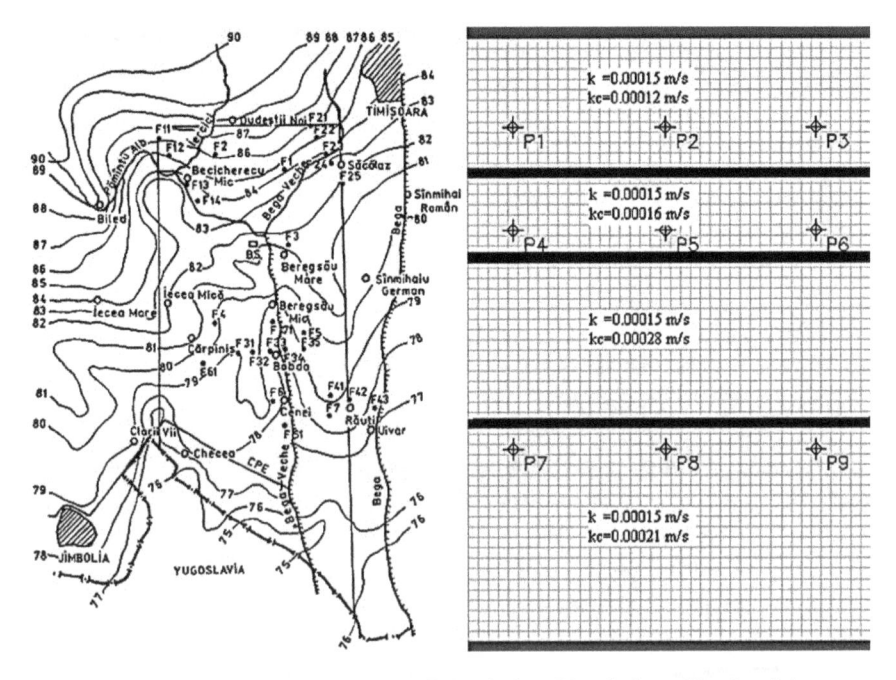

Fig. 3 Studied zone and the permeability coefficient before (k) and after calibration (kc)

The wastewater contains pathogen agents which can cause human diseases. Infiltrations effect of the wastewater in underground water and the way it affected the underground water catchments, a study on 50,000 ha area situate in south-west of Timisoara city was conducted, as shown in Fig. 3 [3].

For modeling, input data used consisted of: terrain topography, geological and hydraulic characteristics of the soil (Fig. 3), boundary conditions, characteristics of observation wells, and chemical characteristics of the pollutant (Fig. 4).

Following the hydraulic modeling, the results are presented in Fig. 5.

The pollution phenomenon of the underground water will remain for a long time, even if the source of pollution (Berecsau Mare) is terminated. The environmental protection around pig's farm can be realized through wastewater treatment plants equipped with biological step. The actual technology for treatment of wastewater, equipped only with grates and settling tanks can be improved with biological ponds after secondary settling tanks. The biological ponds can be considered an advanced step of treatment and can be equipped with aeration systems. For depollution of underground water, reactors with ions exchange are used which use synthetic resin.

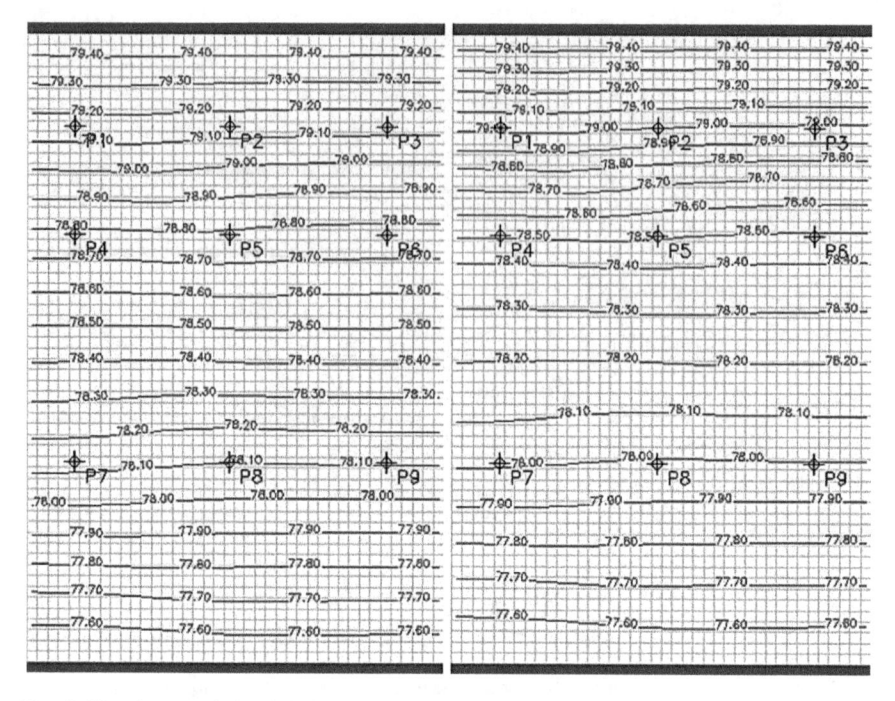

Fig. 4 The piezometric level (before calibration, after calibration)

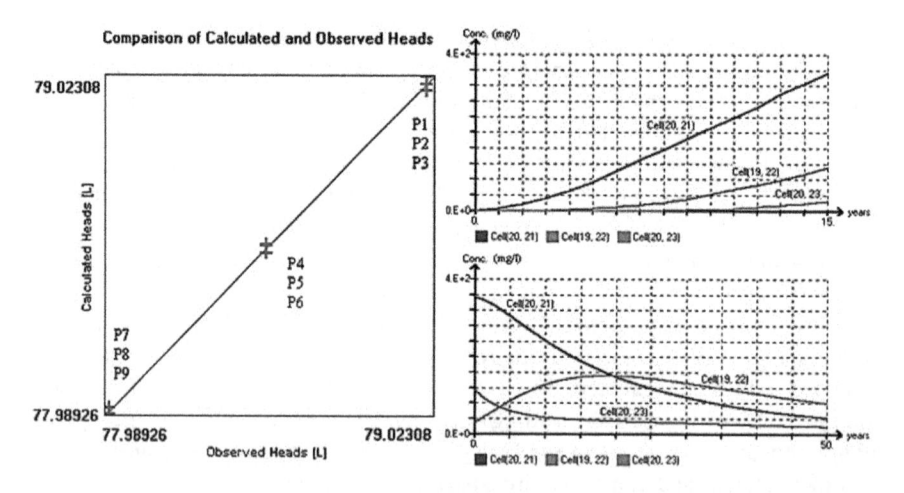

Fig. 5 Comparison of calculated and observed heads, the maximum concentration of sulphates from phreatic stratum, measured in observer wells (permanent injection 15 years and instantaneous injection 50 years)

3.3 Water Catchment Modeling System in the City of Oradea, Romania—MODFLOW Application

Oradea Water Company, is an economic unit which has the object of capturing the raw water, treat it and pumping water in the supply of Oradea city. These goals are achieved through the five water stations located on both sides of Crisul Repede River, in the north-east of the city, which has a total capacity of installed pumping 2,100 l/s, as show in Fig. 6 [4].

The entire water supply system consists of Oradea catchments, water connection pools enrichment field's infiltration, and water plant, which is within a protection zone with a total area of approx. 280 ha. The plant's physical perimeter area is hydraulically significant for analysis and modeling of pollutant transport, taking into account the characteristics of the operating conditions. This area represents the active part of the numerical model.

For modeling, input data consisted of: terrain topography, geological and hydraulic characteristics of the soil, boundary conditions, characteristics of observation wells (Fig. 7). For this particular site plan, the typical objects with role in groundwater regime are typical elements of water supply system, such as capture on Crisul Repede river, infiltration basins, primary and secondary drains. Crisul Repede serves as an active element to penetrate the aquifer.

Following the hydraulic modeling, the results are presented in Fig. 8.

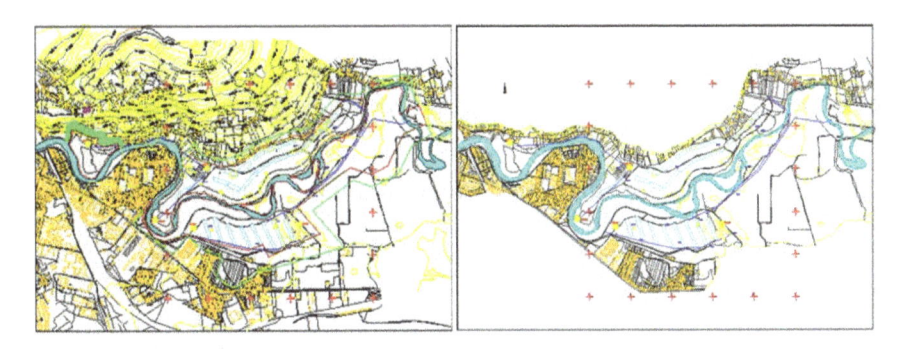

Fig. 6 Plan view capture area and modeling area

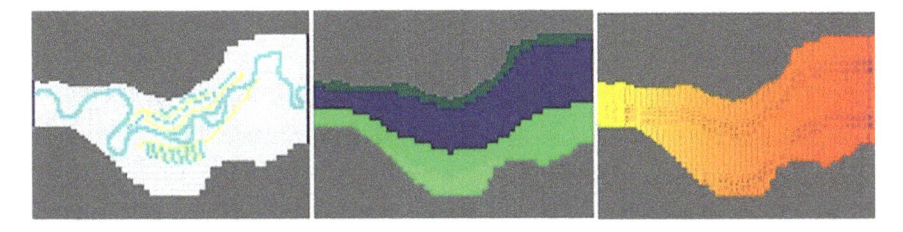

Fig. 7 Meshing and boundary conditions, permeability zones, odds

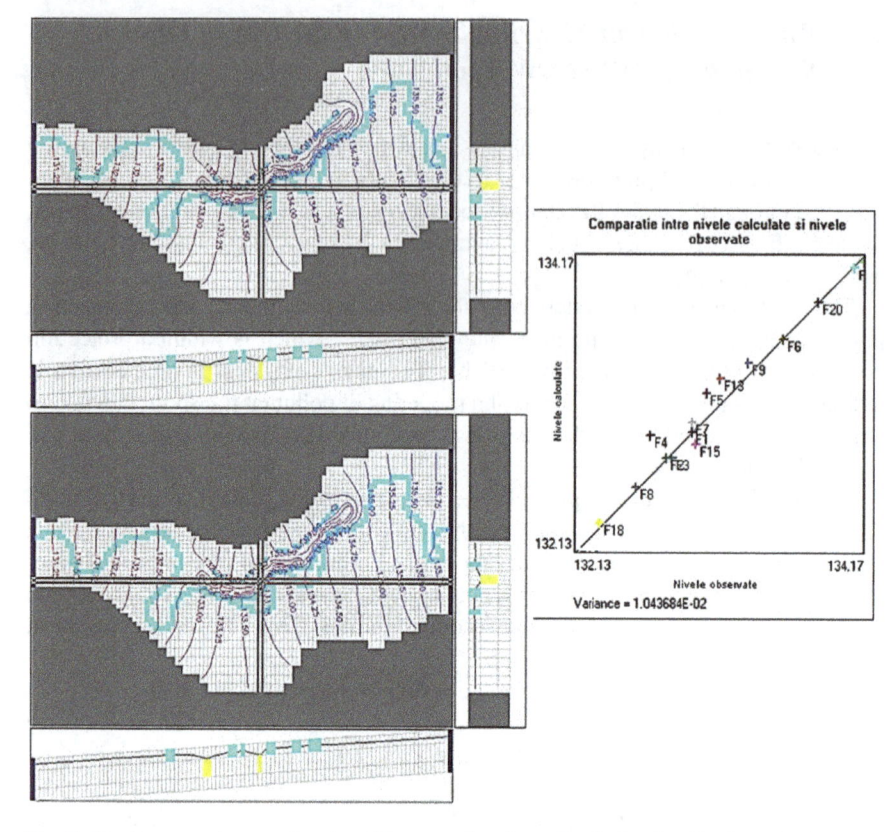

Fig. 8 Head contour Assumption 1 and 2, comparison between the levels of hydrostatic forming

The numerical model assumes that groundwater flow regime, covers the period for which there were available hydrogeological studies. The second assumption is that the modeling baseline was set with objects that replicate situation of the existing complex system (tanks, drains). To confront the results obtained in the second assumption and in accordance with the network, numerical modeling was conducted on hydrogeological wells (16 pcs), as representative from objects placed in the system. We obtained a good agreement between hydrostatic levels obtained from modeling and registered to the network of wells.

3.4 3D Modelling for Phreatic Aquifers Inland Fills. Case Study Parta Landfill—MODFLOW Application [5]

The aquifer extends several square kilometers (14 × 11.6 km), and is situated in space of river Timis, in south east respectively channel Bega in north vest. In map

Modeling the Evolution of Surface and Groundwater Quality

presented below in Fig. 9, marked with model limits are two rivers. Also, it shows landfill Parta, which is also used to model polluted transport and the pollution source is the existing landfill Parta, which is considered as a permanent source of pollution.

For modeling, we used input data, such as, terrain topography, geological and hydraulic characteristics of the soil, boundary conditions, characteristics of observation wells, pollutants characteristics. Following the hydraulic modeling, the results are presented in Figs. 10 and 11.

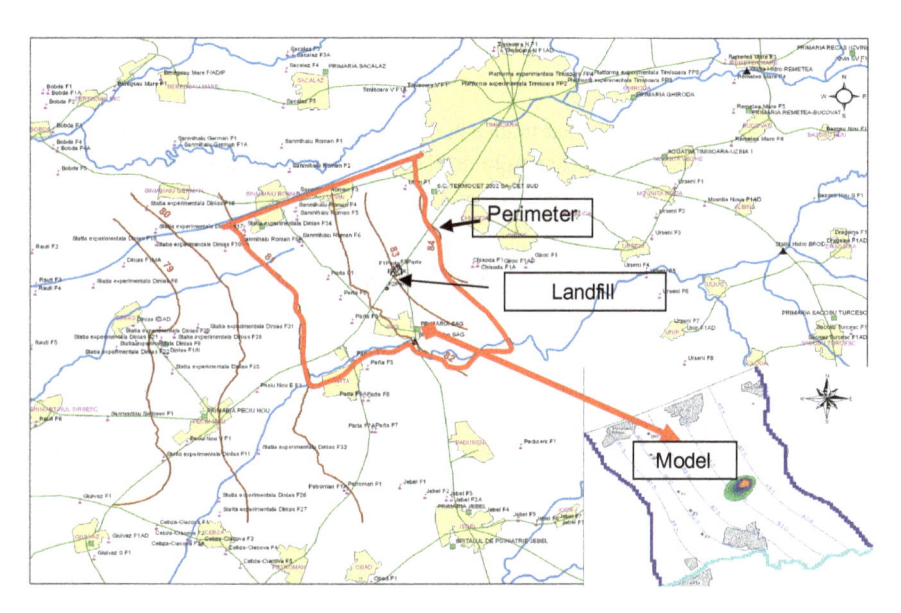

Fig. 9 Plan view of the modeling site

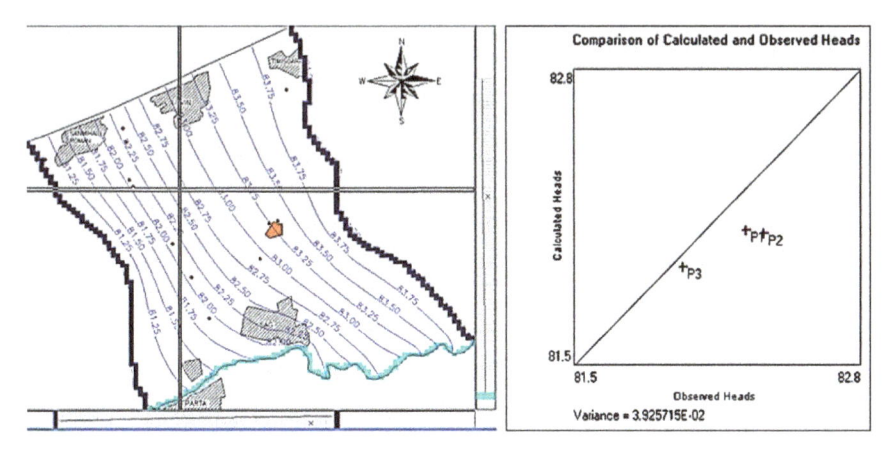

Fig. 10 Head contour and comparison between the levels of calibrations

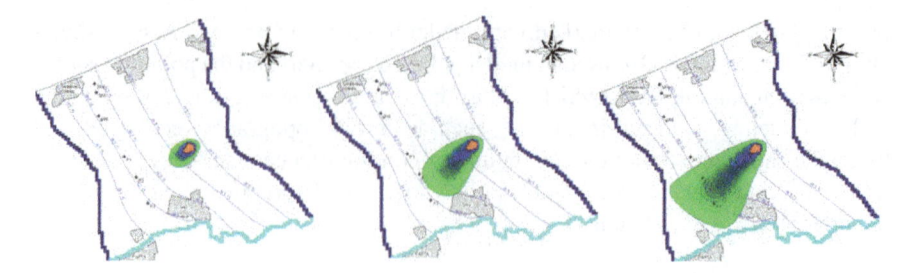

Fig. 11 Modeled concentration at 10, 50 and 100 years

Concentration is a general parameter and may serve as a base for calculation for an absolute concentration (for example mg/l) for all dissolved pollutants in water.

3.5 Sediment Transport—MIKE11 Application

To exemplify of study of water quality evolution through sediment transport in rivers using MIKE11, was considered a sector of Crasna River, located in northwestern Romania. The sector considered here has a length of 64 km, and representative cross sections are considered in the right of localities Supuru de Jos, Craidorolt, Domanesti and Berveni, the border with Hungary. Cross sections have been raised by the Romanian Waters, Somes-Tisa Water Basin Administration (Fig. 12) [6].

The change of riverbed morphology has been modelled in case of appearance of flood hydrograph during 27 July–8 August 2008, showing a maximum discharge of 88.4 m^3/s in Supuru de Jos section. The input data for the area under plan with location of cross sections consists of; cross sections topographical data and roughness

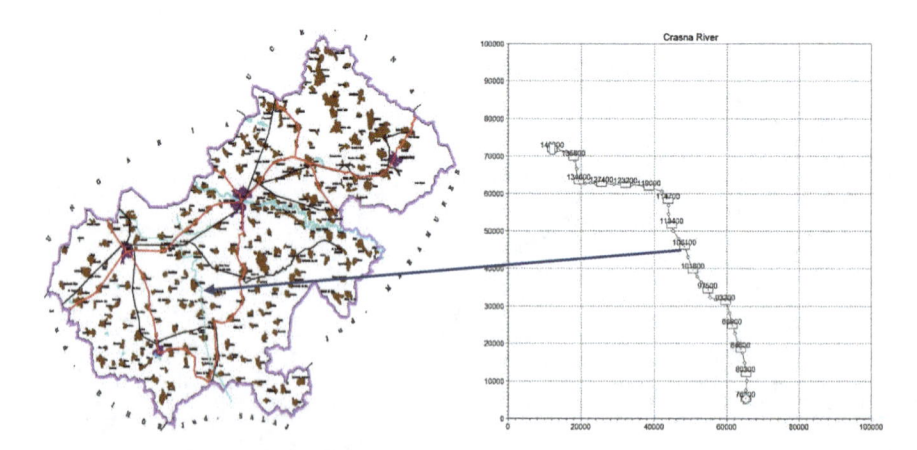

Fig. 12 Plan area and cross section locations

of riverbed; flood discharge hydrograph in section Supuru de Jos (Fig. 13). In Fig. 13 one can see, the water level variation in longitudinal profile, after simulation using hydrodynamic module.

The input data and model selection for sediment transport module can be seen in Fig. 14. The obtained results regarding the sediment transport from modelling are variations of bed load and suspended load (Fig. 15); variations of total sediment transport and bed level evolution in the longitudinal profile (Fig. 16). From Fig. 16 it is observed that the high rate variations of bottom level are between Supuru de Jos section and Craidorolt section, where the slope change is high.

The sediment transport modelling with advanced hydroinformatic tools is important for water courses which may appear problematic due to erosion or deposition in riverbed due to deterioration of the navigation conditions, silting of water intakes, worsening of water quality—like eutrophication, increase of turbidity etc. A special

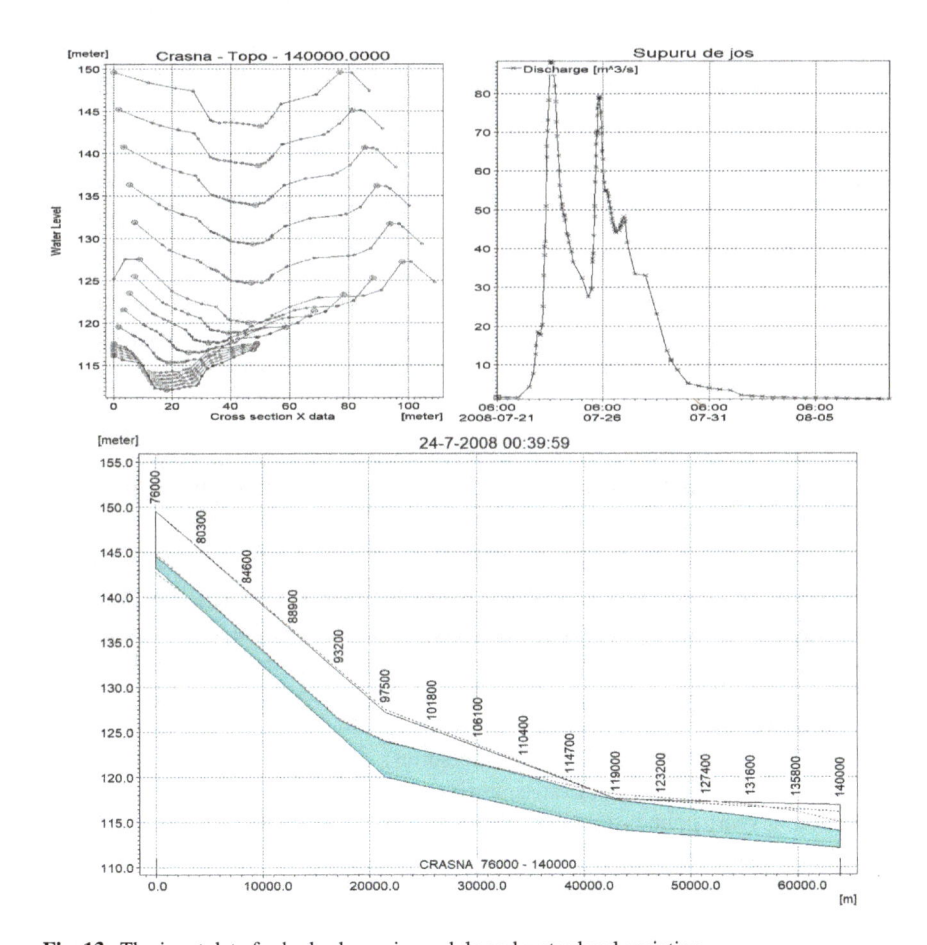

Fig. 13 The input data for hydrodynamic module and water level variation

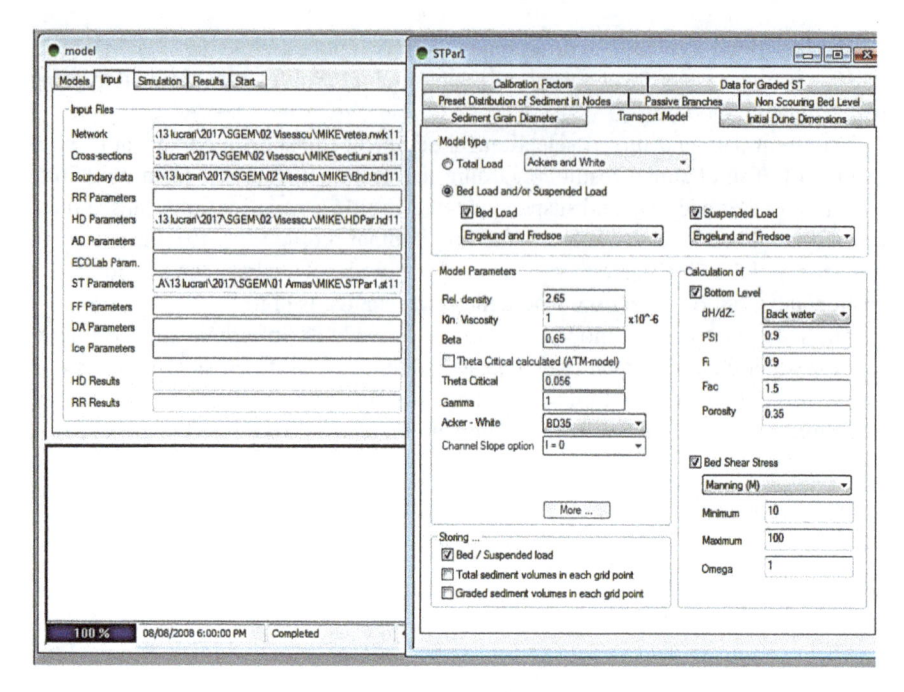

Fig. 14 Sediment transport parameter data

importance has the quality of input data needed for modelling: topographical data of cross sections, slopes, bed roughness and sediments characteristics.

3.6 Water Quality Evolution—MIKE11 ECOLab Application [15]

The water quality evolution was modelled on Bega Channel sector (City of Timisoara to Romanian—Serbian border, Bega Channel is transboundary water course), in order to Bega Channel sustainable development. The data required for modelling are: longitudinal profile of studied river sector (Fig. 17); 13 cross-sections (where was performed over time for bathymetric measurements by Banat Water Basin Administration), as shown in (Fig. 18) are; time series: discharge hydrograph—average monthly discharge for 2005 in cross-section upstream of Timisoara—duration of simulation 1 year; boundary conditions: and Q-H curve in cross-section situated downstream, on the state border (Fig. 19).

The obtained results from modeling with MIKE11–ECOLab are shows in Figs. 20, 21, 22 and 23.

Graphs show the minimum and maximum values achieved by BOD and DO. For BOD, the maximum is reached in the border section, as is for DO. From the graphs

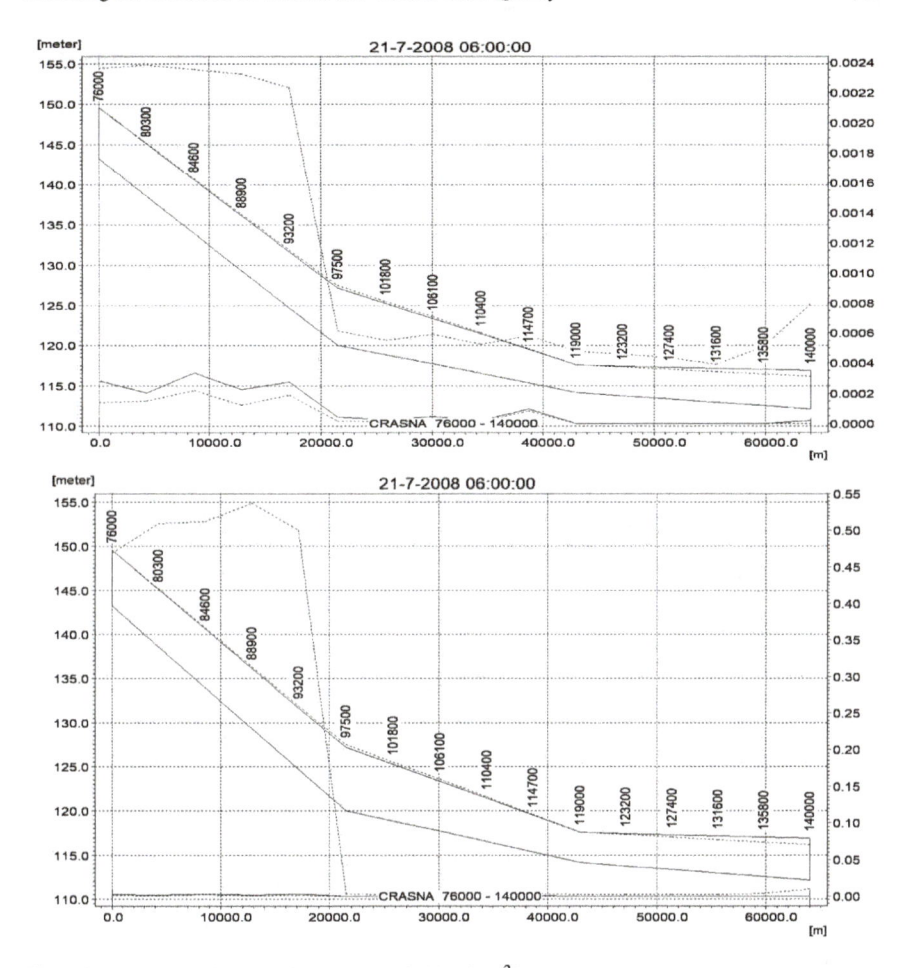

Modeling the Evolution of Surface and Groundwater Quality 709

Fig. 15 Variations of bed load and suspended load (m³/s)

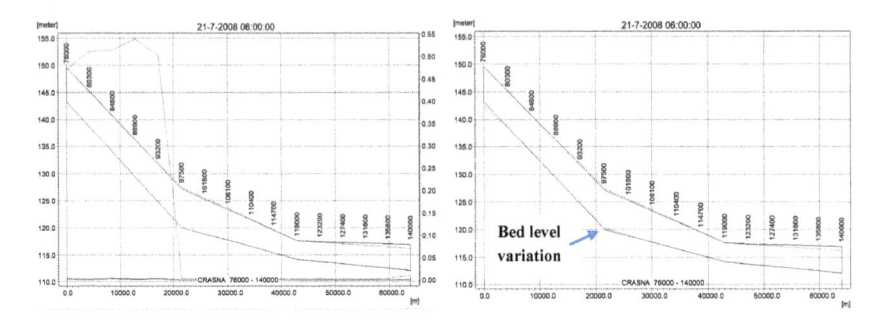

Fig. 16 Variations of total sediment transport (m³/s) and bed level evolution in the longitudinal profile

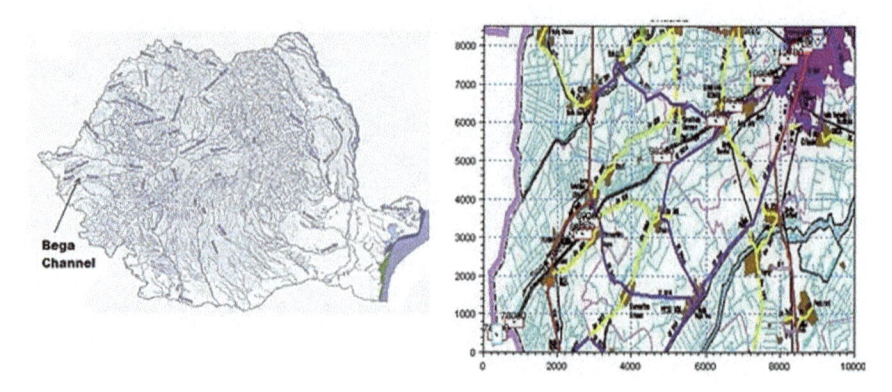

Fig. 17 Area plan of Bega Channel and studied sector

P1- km 76+000 - RO-SRB border;

P2-km 78+000; P3- km 88+000;

P4-km 89+200; P5-km 98+200;

P6-km 104+200; P7-km 106+200;

P8-km 109+800; P9-km 112+700;

P10-km 114+100; P11-km 114+400;

P12-km 115+300;

P13-km 115+900–upstream Timisoara

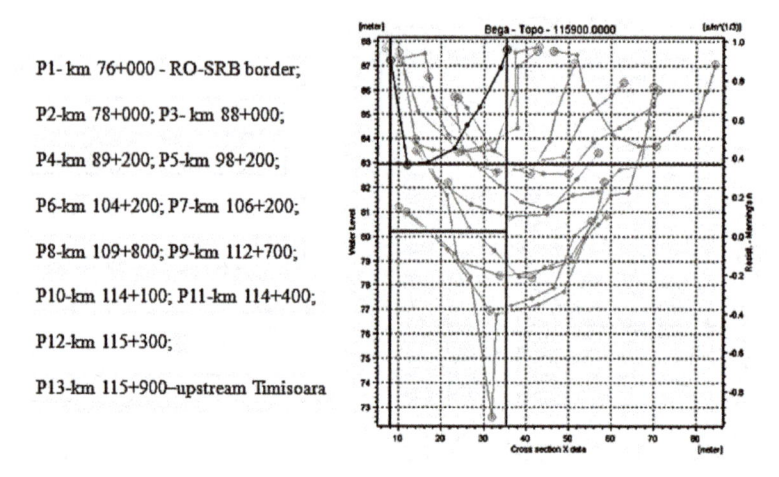

Fig. 18 Cross sections between upstream Timisoara and RO-SRB border

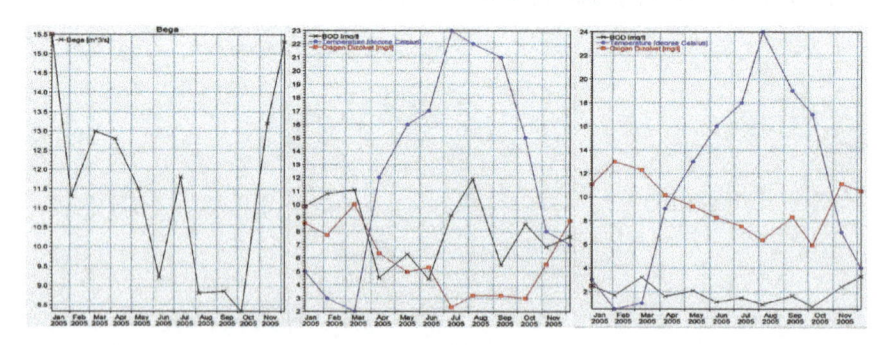

Fig. 19 Average monthly discharges hydrograph and variation of OD, BOD and temperature in upstream and downstream cross sections

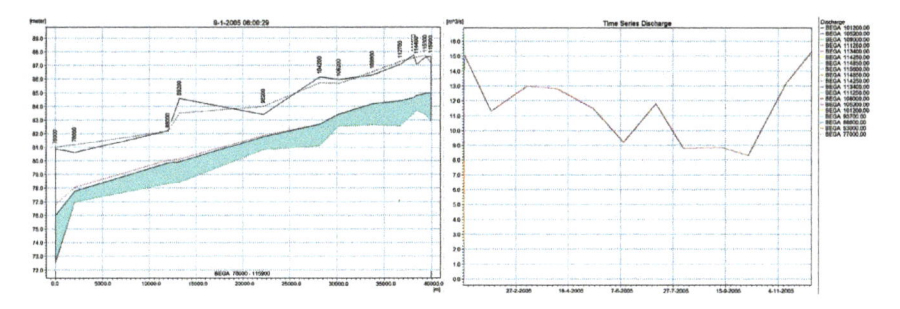

Fig. 20 Water level in longitudinal profile and discharge in cross-sections

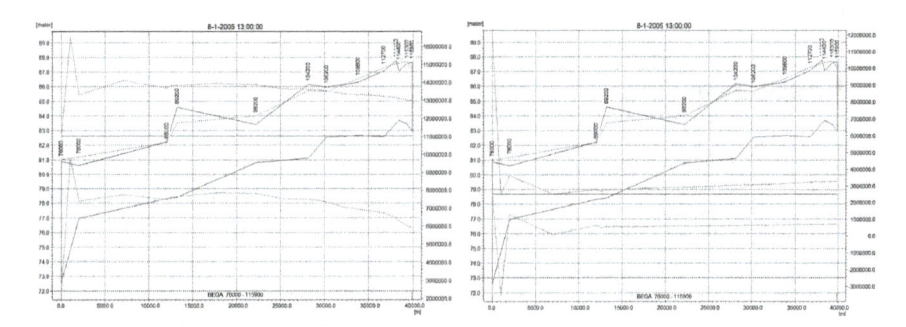

Fig. 21 Variation of dissolved oxygen and BOD

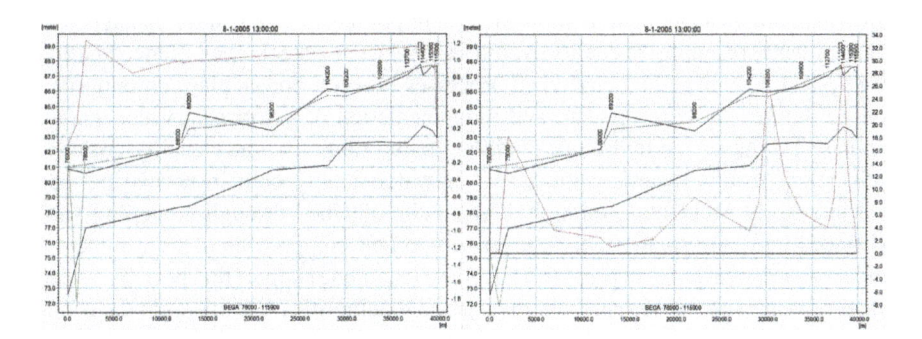

Fig. 22 BOD degradation and re-aeration

we can see the variation of all the analyzed characteristics, at each step of time, in each cross section along the studied sector. At each analyzed feature, important variations occur in the border section, probably due to the large slope variation of the bottom of the bed. The tracking of the water quality evolution on the Bega channel is of particular importance because it is a cross-border watercourse and must fall within the limits set by the international treaties between Romania and Serbia.

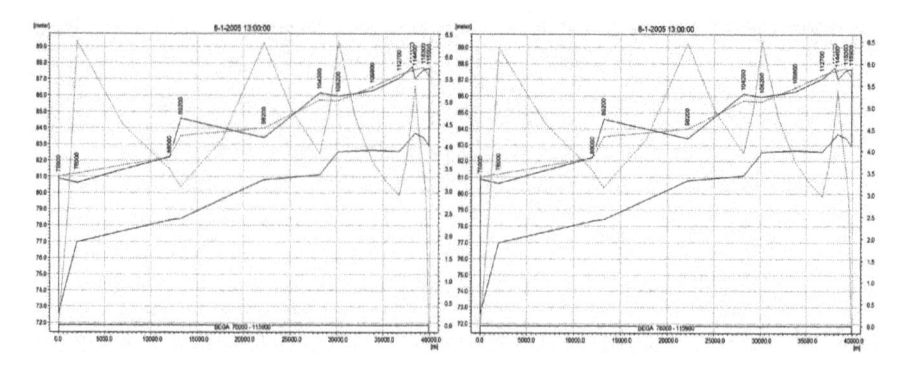

Fig. 23 Photosynthesis and respiration

4 Conclusion

To use and effective protection of water resources and ecosystems of world against pollution, climate change requires coordinated action worldwide. In particularly, EU Member States are obliged to adopt management plans and programs of measures appropriate for each water body with considering the result of different analyzes and studies. Special attention should be paid to cross-border water bodies, which, in addition to EU directives and regulations, are also subject to international treaties concluded between riparian states. Water quality evolution modelling is very important to predict the changes in surface and groundwater quality for water resources and environmental management in the world. An important step to achieve management plans of water resources is quality evolution forecast in water bodies. Understanding of physical, chemical, and biological processes in water bodies plays an important role in the design, development, and implementation of performant water quality models. The advanced hydroinformatic modeling tools for water quality provides satisfactory results regarding the status of water quality both in normal periods and in case of accidental pollution. The results obtained through modelling with advanced hydroinformatic tools are very important components of environmental impact assessment and can provide a basis and technique support for specialists from water resources and environmental management authorities to make right decisions in order to support a sustainable development of water resources.

Acknowledgements For the examples from points 3.5 and 3.6, the MIKE 11 program was acquired through projects LLP-LdV-ToI-2011-RO-002/2011-1-RO1-LEO05-5329 and POSCCE—ACTEX ID1827/SMIS48741 was applied.

References

1. Abbott MB, Refsgaard JC (1996) Distributed hydrological modelling. Kluwer Academic, Dordrecht/Boston/London, pp 1–17
2. Beilicci E, Beilicci R (2014) Groundwater and solute transport modeling in the proximity of Arad Thermal Power Plant, Romania. In: Recent advances in energy, environment, biology and ecology. Tenerife, pp 128–133. ISBN: 978-960-474-358-2
3. Beilicci R, Girbaciu A, Achim C, Podoleanu C, Carabet A (2004) Pollutants transport simulation in aquifers layer, provided from infiltration of waste water coming from pig's farms. In: The 6th international conference on hydraulic machinery and hydrodynamics, Timisoara, Romania, 21–22 October 2004, pp 695–700
4. Beilicci R, David I et al (2006) Project nr. BC 542/23.10.2006, "Politehnica" University of Timisoara, Romania, pp 1–66
5. Beilicci RF, David I, Beilicci E, Şumălan I (2008) 3D modelling for phreatic aquifers inland fills. Case study Parta landfill. Trans Hydrotech UPT 2/2008
6. Beilicci E, Visescu M, Beilicci R (2017) River-bed processes modelling. Study case—modelling on Crasna river sector. In: International multidisciplinary 17th scientific GeoConference SGEM, 29 June–5 July, Albena, Bulgaria, vol 17, no 31, pp 553–560. Hydrology and Water Resources. ISBN 978-619-7408-04-1. ISSN 1314-2704
7. Chiang WH, Kinzelbach W (1993) Processing Modflow (PM), pre- and postprocessors for the simulation of flow and contaminant transport in groundwater system with MODFLOW, MODPATH and MT3D
8. Chiang WH, Kinzelbach W (1998) 3D-groundwater modeling with PMWIN. Spinger
9. Chiang W-H, Kinzelbach W (1998) PMPATH 98. An advective transport model for Processing Modflow and Modflow. In: Harbaugh AW, McDonald MG (eds)
10. DHI, DHI-WASY Software (2014) Installation guide & demonstration exercise FEFLOW. Berlin, p 11
11. DHI, MIKE by DHI (2012) WQ templates. Horsholm, Denmark, pp 38–43
12. DHI, MIKE by DHI, ECO LAB (2014) 1D, 2D and 3D water quality and ecological modelling. In: *User guide*. Horsholm, Denmark, pp 8–15
13. DHI, MIKE 11 (2014) A modelling system for rivers and channels. In: Short introduction and tutorial. Horsholm, Denmark, pp 5–6
14. DHI, MIKE HYDRO (2014) User guide. Horsholm, Denmark, pp 15–17
15. Hausler-Cozma DP, Beilicci E, Beilicci R (2019) Modeling of water quality evolution with advanced hydroinformatic tool: case study from Bega channel sector. IOP Conf Ser: Mater Sci Eng 603:042027 603(3). 10.1088. ISSN: 1757-899X. Online ISSN: 1757-899X. Print ISSN: 1757-8981
16. Kinzelbach W, Rausch R (1995) ASM user manual
17. Kinzelbach W, Rausch R (1995) Grundwassermodellierung. Gebrüder Borntraeger, Berlin
18. Satinder A (2013) Comprehensive water quality and purification. Elsevier, pp 1–15. eBook ISBN: 9780123821836
19. Saunders A (2004) Implementarea noii directive cadru a apei in bazine pilot (WAFDIP) EuropeAid/114902/D/SV/RO RO.0107.15.02.01, pp 2–59
20. Savic D (2016) Hydroinformatic tool. http://emps.exeter.ac.uk/modules/ECMM124. Accessed Sept 2020
21. Water Framework Directive 2000/60/EC of the European Parliament and of the Council (2000)

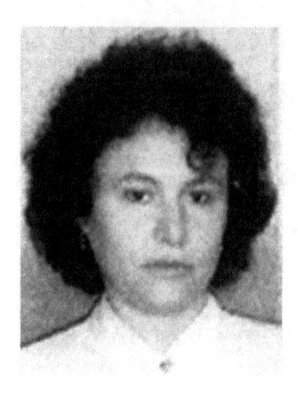

Erika Beilicci was born in Carei, Satu Mare, SM, Romania in 1972. She received the B.S. and M.S. degrees in civil engineering from the Polyethnic University of Timisoara, in 1996; B.S. degree in mathematics from the West University of Timisoara, in 1999 and the Ph.D. degree in civil engineering from Polytechnical University of Timisoara, Romania, in 2006. Since 1996 to present, she has been an Assistant Professor/Lecturer with the Hydrotechnical Engineering Department, Faculty of Civil Engineering, Polyethnic University of Timisoara. She is the author of three books and one book chapter, more than 150 research articles. Her research interests include hydrology and hydrogeology, water resources management, environmental engineering, management—climate changes—integrated resources water management, soil erosion, sediment transport. Dr. Beilicci is member of International Association of Hydrological Sciences (IAHS) from 2008; World Association of Soil and Water Conservation from 2002; European Water Resources Association from 2007; The Transylvanian Museum Society—Community of Hungarians scientists in Transylvania from 2009; Association for Multidisciplinary Research of the West Zone of Romania (ACM-V) from 2010.

Robert Beilicci was born in Carei, Satu Mare, SM, Romania in 1972. He received the B.S. degree in civil engineering from Polyethnic University Timisoara in 1995, manager certificate in Complementary study in management from Polytechnic University Timisoara in 1995, M.S. degrees in civil engineering from Polyethnic University Timisoara in 1997, B.S. degree in mathematics from West University of Timisoara in 1997, and the Ph.D. degree in civil engineering from Polyethnic University Timisoara in 2005. From 1995, he is a Lecturer with the Hydrotechnical Department, Civil Engineering Faculty, Polyethnic University Timisoara. He is the author of three books, more than 150 articles. His research interests include hydraulic processes and applications, mathematics modeling, hydraulic and pollution modeling. Dr. Beilicci is member of International Association of Hydrological Sciences (IAHS) from 2008; World Association of Soil and Water Conservation from 2002; European Water Resources Association (EWRA) from 2007; The Transylvanian Museum Society—Community of Hungarians scientists in Transylvania from 2009; Association for Multidisciplinary Research of the West Zone of Romania (ACM-V) from 2010.

Mircea Visescu was born in Timisoara, Romania, in 1984. He received the B.S degree in environmental engineering from Polyethnic University Timisoara in 2008, Ph.D. degree in civil engineering from Polyethnic University Timisoara in 2011, and in 2017 M.S. degree in civil engineering from Polyethnic University Timisoara. From 2012 to 2020 he was an Assistant Professor with the Hydrotechnical Engineering Department, Polyethnic University Timisoara. He is the author of one book, more than 30 articles. His research interests include environment impact assessment, heat exchangers, heat transfer, groundwater, geothermal energy.

Author Index

A
Adedoja, Oluwaseye Samson, 47
Alazzawi, Marwa, 291
Atvars, Aigars, 663
Ayvazyan, Gagik, 663

B
Barbeş, Lucica, 599
Bărbulescu, Alina, 599
Beilicci, Erika, 691
Beilicci, Robert, 691
Belhassan, Kaltoum, 443
Bölgen, Nimet, 197, 263, 275
Brčeski, Ilija, 333
Buzgaru, Nicusor, 583

C
Cerneaga, Constantin, 583
Chinnasamy, Pennan, 615, 637
Cococeanu, Adrian Lucian, 115, 305, 395
Constantin, Cerneaga, 583

D
Demir, Didem, 263, 275
Dobrica, Gabriel, 583
Dumitriu, Cristian-Ştefan, 599

G
Gevorgyan, Gor, 67
Ghosh, Asim K., 547
Glevitzky, Ioana, 371, 421
Grebel'na, Yuliia, 523

H
Hamam, Yskandar, 47

I
Ionete, Eusebiu Ilarian, 495
Ionete, Roxana Elena, 135, 495
Ivanenko, Kateryna, 523
Ivanov, Ognyan, 67

J
Janaćković, Goran, 243
Jawaid, Mohammad, 67

K
Karnena, Manoj Kumar, 215, 463
Kavaz, Doga, 67
Khachatryan, Mane, 663
Khadke, Leena, 637
Khalaf, Baset, 47
Khudaverdyan, Surik, 663
Kirchu, Fedor, 523
Kodrik Ivanovich, Anatoly, 523
Kodryk, Anatoly, 523
Konni, Madhavi, 463

L
Lapkis, Mihail, 663

M
Maftei, Carmen, 583
Man, Teodor Eugen, 115, 305, 395

Miricioiu, Marius Gheorghe, 135
Mishra, Jnyana Ranjan, 161
Mohanty, Smita, 161
Mukkamala, Saratchandra Babu, 463

N
Nayak, Sanjay K., 161
Niculescu, Violeta-Carolina, 135
Nikulin, Alexander, 523
Nikulin Fedorovich, Alexander, 523

P
Papatheodorou, Konstantinos, 23
Pokropivny, Alex, 523
Popa, Maria, 371, 421
Pradhan, Sukanya, 161

R
Roy, Anirban, 547
Rudenko, Sergey, 663

S
Sadiku, Rotimi, 47
Saritha, Vara, 215
Sasmazel, Hilal Turkoglu, 291
Satsangi, Aparna, 565
Saxena, Sachin, 565
Sementsov, Yurii, 523
Soamidas, Vuppulury, 565
Spiridon, Stefan-Ionut, 495
Srikanth Vemuri, R. S. S., 463
Srivastava, Aman, 615, 637

T
Titenko, Alexander, 523
Turkoglu Sasmazel, Hilal, 291

V
Vaseashta, Ashok, 3, 67, 197, 243, 263, 275, 333, 523, 663
Vasović, Dejan, 67, 243
Visescu, Mircea, 691

Subject Index

A
Absorbed waves, 663, 673
Adsorbate, 141, 146, 466–469, 474, 475, 482
Adsorbent, 135, 140, 141, 143, 146, 205, 234, 235, 263, 265, 266, 269–271, 275, 276, 279, 281–283, 285, 286, 323, 465–474, 477–482, 484, 528–530, 539
Adsorption, 73, 102, 135–137, 139–146, 151, 165, 168, 170, 171, 173, 178, 198, 202, 203, 205, 207, 209, 219, 224, 230, 233, 234, 263–266, 268–271, 276, 281–286, 299, 306–308, 323, 324, 339, 349, 362, 465–469, 471–477, 479–483, 506, 508, 524, 525, 527, 531, 533, 539, 566, 568
Agriculture, 6, 12, 18, 31, 48, 67, 69, 77, 78, 91, 124, 129, 136, 246, 337, 373, 379, 443, 444, 447–449, 451, 452, 454, 457, 459, 497, 501, 548, 549, 566, 587, 589, 615–617, 619, 623, 625, 626, 630, 631, 638, 645, 694
Anthracenes, 354, 356
Anthropogenic, 30, 35, 37, 39, 77, 78, 90, 91, 101, 116, 197, 244, 246, 247, 333, 344, 348, 349, 360, 361, 423, 425, 432, 437, 464, 500, 524, 533, 539, 599, 600, 692, 694
Aquatic Toxicity Index (ATI), 72
Artificial intelligence, 10, 73, 76, 342, 353, 693
Asset management, 408, 410, 411

B
Bacteriological, 70, 310, 312, 422, 428

Bioavailability, 344
Bioinformatics, 75–77, 96, 104
Biological, 4, 7, 13, 17, 25, 51, 69, 71, 78, 85, 89, 98, 100, 102, 116, 118, 127, 136, 138, 151, 161, 198, 215, 253, 264, 281, 284, 292, 299–301, 306, 308, 310, 316, 323, 324, 326, 341, 345, 346, 348, 351, 354, 361–364, 379, 381, 383, 390, 417, 422–425, 428, 436, 464, 465, 495, 498, 499, 505, 511, 512, 568, 601, 604, 692, 694, 697, 698, 701, 712
Biomonitoring, 75–77, 93, 94
Biosorbent, 264, 281
Biotope, 422
Brine handling, 549, 553, 559
British Columbia Water quality index (BCWQI), 71, 72, 602, 603, 606–609

C
Canadian Council of Ministers of the Environment (CCME), 72
Catalysis, 13, 67, 93, 151, 208, 472, 479, 483
Chemostratigraphy, 347, 350
Chitosan, 143, 150, 202, 203, 205, 230, 263–265, 267–270, 276, 279, 283
Climate Change, 7, 11, 23, 136, 138, 161, 216, 229, 245, 253, 337, 371, 373, 396, 443–447, 450, 452, 455, 457–459, 498, 584, 592, 638, 664, 691–693, 712
Coagulation, 102, 116, 163, 215–224, 229, 233, 235, 264, 276, 296, 298, 299, 301, 306–308, 310, 312, 319, 464, 465, 508, 566

Commercial-Off The Shelf (COTS), 17, 96
Community Water Supply Study (CWSS), 70
Contaminant detection, 50, 54, 57
Contaminants, 4, 7, 9, 13, 14, 16, 17, 19, 31, 47–51, 53–57, 60, 68, 73, 75–93, 95, 102, 103, 115, 135–137, 140–148, 150, 151, 161, 163, 176, 197, 198, 202, 209, 226, 280, 281, 283, 284, 286, 291, 305, 334–336, 353, 359, 364, 371, 372, 374, 381, 383, 384, 390, 421, 436, 447, 455, 458, 463, 464, 469, 473, 475, 482, 483, 495, 498, 505, 506, 511–514, 520, 523, 524, 531, 567, 569, 600, 663, 664, 683, 696
Contamination, 4, 7, 12, 13, 17, 19, 23, 33, 39, 48–52, 54, 55, 57–59, 68–70, 73, 74, 76, 78, 91, 92, 94–96, 102–104, 162, 198, 219, 285, 291–293, 296, 297, 299–302, 308, 310, 333–335, 337, 346, 355, 359, 361, 364, 374, 375, 377, 379–385, 390, 391, 413, 422–424, 428, 431, 432, 435–437, 445, 446, 449, 452, 496, 498, 499, 502–505, 513, 519, 520, 524, 525, 528, 529, 532, 565, 567, 573, 600, 664
Contamination Identification and Level Monitoring Electronic Display Systems (CILM-EDS), 17, 95
COVID-19, 9, 79, 91, 92, 95, 104, 282
Critical infrastructure, 47, 93, 397, 398, 400, 417, 538
Cryogel, 263, 265–271, 275–286

D

Decision-support systems, 631, 693
Decontamination, 216, 291, 301, 424, 428, 505
Delphi, 3, 10, 11, 71, 72
Desalination, 6, 19, 148, 150, 164, 336, 396, 443, 452, 455, 456, 459, 547–553, 555, 559, 560
Desertification, 585
Dichlorodiphenyltrichloroethane (DDT), 90, 357
Di-n-Butyl Phthalate (DBP), 92
Dinius WQI (DWQI), 72
Disinfection process, 291, 328, 498
DNA, 7, 95, 295, 297, 450
Dobrogea, 583, 585–588, 592, 594, 595
Domain of Interest (DOI), 95

Drinking water, 3, 10–13, 17–19, 50, 60, 67–71, 73, 77–89, 91, 92, 96, 102–104, 115, 117, 122, 125, 135, 136, 161, 215, 218, 226, 229, 243, 247, 252, 253, 255–258, 292, 294, 295, 300–302, 305, 308–310, 312, 326, 328, 338, 345, 356, 358–360, 371–373, 381, 390, 395, 396, 417, 421–425, 428–431, 435, 436, 449, 450, 453–455, 458, 459, 495–505, 512, 513, 524, 548, 565–567, 569, 570, 576, 584, 600, 617, 621, 645, 646, 699
Drought, 6, 9, 69, 91, 126, 135, 138, 396, 423, 445–447, 449–451, 454, 457, 459, 498, 519, 583–585, 594, 595, 615, 616, 618, 621, 624, 627, 630, 632, 638, 698
Dye removal, 206, 271, 467, 471, 481

E

ECOlab, 697, 698, 708
Ecology, 13, 523, 697, 698
Electron Capture Detector (ECD), 362, 363
Electronic Health Record (EHR), 97
Electrospinning, 163, 164, 198–203, 205, 207, 208
Emerging materials, 291, 299, 301, 302
EN 15975-1, 252, 253
EN 15975-2, 252, 253
Endemic, 423
Enterococci, 383, 390, 426, 428, 431, 432, 434, 437
Enterprise Resource Planning (ERP), 410, 418
Environment, 4, 6, 7, 13, 17, 18, 36, 37, 71, 72, 78, 79, 89–91, 93, 95–97, 99–101, 103, 104, 117–119, 128, 135–137, 151, 164, 215, 217, 228, 229, 243–248, 253, 257, 263, 264, 269, 275, 276, 283, 292, 295, 333–335, 338, 340, 345–347, 349–351, 355, 357, 359, 361, 363, 364, 372, 374, 378, 405, 410, 413, 422, 428, 429, 448–450, 453, 454, 457, 459, 463–465, 477, 496, 498, 501, 505, 506, 509, 518, 524–527, 572, 584, 595, 601, 602, 664, 693, 694
Environmental Impact Assessment (EIA), 697
Environmental Science, 693
Epidemic, 70, 423, 424, 428
Escherichia coli, 77, 203, 207, 294, 310, 428, 435

Subject Index

Estradiol Equivalency Factor (EEF), 90
Expanded graphite, 523, 525, 527, 528, 536, 538
Exposome, 68, 75–77, 94, 97, 104
Extreme climatic events, 615, 624

F

Feasibility, 52–54, 94, 476, 548
Fingerprinting, 333, 335, 347
Finite Element subsurface FLOW (FEFLOW), 414, 691, 698, 699
Flame Ionization Detector (FID), 362
Flocculation, 215, 217–219, 221–223, 226, 264, 296, 298, 299, 301, 306–308, 312, 319, 324, 464, 465, 508, 511, 515, 517
Floods, 17, 51, 91, 120, 396, 416, 417, 423, 445, 454, 455, 458, 573, 584, 594, 615, 616, 618, 621, 624, 626, 629, 630, 632, 638, 698, 706, 707
Focal Plane Array (FPA), 352
Forensics, 333–335, 337, 342–347, 350, 351, 353–364
Fourier Transform Infrared Spectroscopy (FTIR), 168, 183, 263, 265, 267, 268, 341, 343, 352
Free open-source, 637
Fulvic acid, 340, 344, 359

G

Gas Chromatography (GC), 362, 363
Gelatin, 263–265, 267–270, 276, 279, 283
Genetically Modified Organisms (GMOs), 91
Geographical Information System (GIS), 17, 18, 26, 94, 95, 343, 408–410, 416–418, 570, 617, 693, 694
Global Positioning System (GPS), 17, 18, 94, 95, 343, 621
Granular Activated Carbon (GAC), 73, 218
Granulometry, 320, 323, 325, 341, 344
Green Chemistry, 229
Groundwater, 23–27, 29–39, 40, 69, 77, 78, 99, 101, 102, 116–118, 121, 122, 125, 126, 136, 247, 292, 306, 309, 311, 312, 318, 322, 335, 342, 364, 373, 390, 406, 411–415, 421–425, 428, 430, 432, 434–437, 443, 444, 446–449, 452–459, 498, 500–502, 504, 568, 572, 573, 575, 584, 586, 591, 600, 615–618, 621, 622, 626–628, 630, 631, 637–640, 642, 643, 645–648, 650–655, 691–696, 698–700, 703, 704, 712
Groundwater conceptual model, 25
Groundwater exploration, 23, 26, 29, 31, 39, 631
Groundwater management, 36, 617, 638, 653

H

High-Performance Liquid Chromatography (HPLC), 362, 363
Household Water Treatment and Safe Storage (HWTS), 495–498, 505, 514, 515, 518–520
Human Biomonitoring (HBM), 94
Human Early-Life Exposome (HELIX), 75
Humatolenic, 340
Humins, 340
Hydrodynamic, 100, 416, 691, 692, 697, 698, 707
Hydroinformatics, 413, 691–696, 698, 707, 712
Hydro informatics tools, 411, 417
Hydrological, 116, 215, 256, 316, 343, 346, 351, 411, 417, 428, 455, 570, 586, 594, 600, 604–606, 615, 617, 618, 621, 622, 626–628, 630–632, 693
Hydrology, 51, 116, 417, 452, 639, 641, 654
Hyperspectral imaging, 96

I

Inductively Coupled plasma with Mass detection (ICP-MS), 352
Industrial Control Systems (ICS), 397–399, 417
Injection amplification, 664, 679–681, 684
Integrated Water Resources Management (IWRM), 698
Internet of Everything (IoET), 94
Internet of Things (IoT), 16, 68, 75, 95, 96
Ion Exchange Chromatography, 362
Isotopes, 4, 338, 342, 346–352, 363, 568

L

Laser Induced Breakdown Spectroscopy (LIBS), 96, 347
Light Imaging, Detection and Ranging (LIDAR), 96, 343
Lineament mapping, 31–35

M

Macrophytic, 341

Maintenance, 39, 40, 48, 60, 98, 102, 397–399, 407–410, 418, 422, 453, 455, 465, 496, 505, 514, 515, 518, 519, 534, 566, 568, 630

Management, 13, 16, 18, 19, 23, 24, 31, 35, 36, 38–40, 51, 69, 73, 76, 95–99, 101, 103, 104, 116–118, 216, 228, 244–247, 249–255, 257, 258, 373–375, 377, 379, 380, 391, 395–397, 403, 405, 408, 410, 411, 417, 418, 429, 447, 452, 453, 458, 459, 483, 498, 505, 528, 547, 566, 567, 583–585, 589, 594, 595, 600, 601, 615–620, 628–632, 637, 638, 652–655, 691–694, 698, 712

Megacities, 75, 443–447, 450, 452, 455–459, 524

Membranes, 7, 19, 90, 102, 135, 137, 139, 140, 148–152, 161–186, 197–199, 202–209, 264, 276, 280, 299–301, 312, 313, 320–322, 325, 339, 455, 456, 465, 467, 468, 477, 506, 511, 547, 548, 554–559, 566, 568

Metal-organic frameworks, 139, 141–143, 148, 463, 465, 471–484

Methemoglobinemia, 309, 502

Microbial Fingerprinting, 335

Microorganisms, 79, 102, 179, 203, 217, 218, 230, 233, 276, 292, 293, 296–302, 306, 308, 309, 319, 322, 326, 341, 344, 349, 357, 390, 432, 435, 437, 499, 506, 507, 600, 601

MIKE HYDRO, 698

Millennium Development Goals (MDGs), 9, 73

Modeling, 3, 93, 95, 333, 335, 343, 355, 408, 411, 413–418, 639, 691–699, 701, 703–705, 708, 712

Monitoring, 12, 17, 19, 23–25, 30, 36, 38–40, 47, 49–52, 57, 60, 68, 75–77, 93–97, 99, 101, 103, 104, 116, 117, 246, 252, 257, 310, 340, 341, 346, 350, 360, 373, 374, 386, 387, 390, 391, 398, 400, 411, 421, 453, 505, 515, 518, 585, 594, 595, 599–601, 628, 632, 664

Multi-Criteria Decision Analysis (MCDA), 98, 99

N

Nanocomposite, 149–151, 161, 163, 165, 168, 169, 176, 177, 179–181, 183, 186

Nanofibers, 138, 139, 150, 177, 198–203, 205–209

Nanofiller, 161, 168, 176

Nanomaterial, 7, 13, 17, 94, 135, 137–142, 144, 145, 147, 150, 151, 161–165, 197, 299, 300, 352, 463, 471, 538

NATO, 17

Natural coagulants, 215, 217, 222, 224, 229–231, 235

Natural water circuit, 118

NCOV-SARS2, 91, 104, 282, 360

Neutron activation analysis, 341, 352

Nitrogen- Phosphorus Detector (NPD), 362

O

Oil contamination, 364

Oregon Water Quality Index (OWQI), 71, 72, 602

Organoleptic, 306, 326

P

Parasitological, 341

Pedology, 342

Persistent Bio-accumulative and Toxic (PBT), 90

Persistent Organic Pollutants (POPs), 90

Pharmaceuticals, 7, 13, 19, 67, 73, 76, 87, 91, 93, 102, 141, 147, 151, 167, 197, 271, 281, 360, 397, 402, 464, 497, 500

Pharmaco-Kinetic (PBPK), 93

Phenanthrenes, 354

Photocurrent, 663–665, 668–684

Physico-chemical, 3, 102, 116, 219, 306, 310, 319, 323, 422, 430, 433–435, 464, 465, 567, 604

Pilot plant, 313–316, 555

PLFA, 335

Policy, 10, 12, 13, 17, 19, 40, 68, 69, 103, 104, 116, 245, 249, 250, 335, 372, 387, 396, 397, 410, 411, 443, 452, 458, 459, 584, 594, 595, 617, 631, 632

Pollution, 7, 9, 23, 26, 30, 31, 33, 35–40, 71, 75, 77, 90, 94, 98, 99, 101, 117, 126, 135–137, 146, 161, 197, 198, 229, 234, 245, 247, 256, 264, 276, 280, 281, 291–293, 295, 296, 299,

Subject Index 723

301, 309, 310, 322, 324, 326, 333–337, 340, 343, 345–350, 354–356, 358, 360, 361, 364, 372, 395, 396, 413, 418, 421–425, 427, 428, 430–432, 434, 436, 437, 443–445, 447–453, 455, 459, 463, 464, 504, 506, 524, 525, 584, 589, 599–601, 604, 691, 692, 694, 695, 699–701, 705, 712
Polychlorinated Biphenyls (PCBs), 85, 90, 146, 341, 359, 363
Polymer, 89, 148, 150, 152, 161–164, 168, 178, 179, 199–205, 209, 217, 220–222, 226–228, 230, 263, 265, 267, 270, 271, 275–280, 283, 284, 298, 300, 301, 321, 325, 508, 528, 529, 538
Polyvinylidene Fluoride (PVDF), 161–186, 202, 203, 208
Purification, 13, 68, 100–102, 104, 136, 148, 151, 161, 162, 165, 166, 171, 175, 178, 180, 197, 198, 206, 215, 224, 230, 255, 271, 275–277, 280, 281, 285, 300, 312, 313, 315, 316, 322, 351, 372, 428, 506, 527, 531, 532, 539, 575

Q

Quality of life, 68, 103, 335, 353, 372, 411, 422, 664

R

Radiochemistry, 350
Raman, 352
Redox, 100, 101, 339, 342, 344, 348, 349, 354, 357, 361, 408
Remediation, 7, 12, 13, 19, 24, 73, 74, 102–104, 140, 197, 207, 244, 285, 333, 696
Remote sensing, 26, 29, 31, 33, 35, 38, 40, 94, 590, 592, 593, 618, 621
Reverse osmosis, 102, 137, 148, 163, 164, 198, 299, 307, 309, 310, 321, 455, 516, 547, 550, 553, 558, 566
River bank filtration, 566
RNA, 95, 295, 335

S

Safe Drinking Water Act (SDWA), 71, 423
Safe Drinking Water Act (SWDA), 70, 73

Safe Drinking Water Information System (SDWIS), 73
Safeguarding, 47, 49, 51, 458
Safety, 3, 6, 7, 9, 10, 12, 13, 17, 19, 24, 49, 51, 52, 69, 92, 95, 243, 244, 246, 250–253, 257, 258, 263, 264, 271, 301, 311, 313, 316, 321, 322, 334, 335, 363, 371–375, 377–382, 387, 389, 390, 395, 398, 400, 407, 421–423, 428, 458, 478, 479, 495, 496, 498–501, 503–505, 509, 512, 518, 520, 664
Scarcity, 5, 9, 10, 17, 26, 50, 69, 76, 77, 93, 104, 161, 198, 215, 216, 246, 247, 373, 390, 423, 424, 443–445, 447, 449–452, 454, 457–459, 584, 615–617, 621, 629–632
Security, 3, 5, 9, 10, 12, 13, 16, 17, 19, 39, 49, 51, 69, 95, 99, 104, 244–246, 250, 363, 371, 373, 374, 377–380, 382, 387, 389–391, 395–399, 405, 408, 417, 418, 425, 429, 456–458, 498, 548, 549, 584, 616, 617, 629, 632, 638, 639, 653, 655
Sensor placement, 47, 49–54, 56–60
Sensor placement problem, 50
Smart water, 19, 395, 397, 418, 459
Sorbent, 282, 283, 345, 471, 523–526, 528–533, 538, 539
Spectral photosensitivity, 663, 664, 674–676, 679, 684
Spectroscopy, 96, 175, 263, 341, 347, 352, 361, 664, 665
Spring water, 123, 352, 375, 381, 385, 387, 388, 390, 429, 430, 432–436, 496
Standards, 13, 19, 39, 40, 52, 68–71, 73, 92, 97, 115, 124, 136, 137, 218, 223, 244, 246, 248, 252, 253, 255, 257, 258, 306, 310, 312, 326, 328, 335, 341, 351, 354, 358, 361, 371–374, 377, 378, 381, 382, 390, 402, 403, 410, 417, 421–423, 430, 434–436, 453, 464, 483, 497, 505, 509, 513, 514, 548, 554, 575, 584, 602, 604, 605, 616, 693, 694
Statistical Assessment, 599
Storage, 29, 82, 94, 119–121, 228, 255, 297, 301, 316, 356, 377, 380, 382, 388, 389, 413, 444, 456, 495–499, 513, 516, 518, 538, 547, 615, 621, 622, 624, 626–631, 645, 653, 696, 699
Supervisory Control and Data Acquisition (SCADA), 16, 19, 397–404, 417

Surface water, 5, 23–25, 33, 69, 77, 78, 85, 116, 121, 122, 125–127, 215, 216, 218, 219, 306, 309, 311–316, 318, 319, 322, 337, 343, 350, 360, 373, 390, 411, 422, 425, 428, 434, 443, 446, 447, 452, 453, 456–458, 498, 501, 502, 504, 565–570, 573, 575, 600, 615–617, 622, 631, 632, 645, 692, 695, 698–700

Sustainability, 3, 6, 9, 10, 12, 17, 25, 50, 55, 59, 60, 97, 104, 162, 215, 216, 235, 243, 244, 246, 247, 250, 256, 301, 335, 363, 396, 414, 464, 465, 477, 479, 496, 513, 515, 519, 547–550, 560, 568, 570, 617, 621, 628, 630

Sustainable development, 3, 9–11, 23, 50, 55, 59, 68, 73, 118, 230, 257, 335, 499, 584, 595, 601, 708, 712

Sustainable Development Goals (SDG), 3, 9–12, 50, 55, 73

Sustainable water use, 127

T

Thermohaline, 698

Total coliform, 78, 79, 431, 568, 599, 601, 604–606, 608

Total Dissolved Solids (TDS), 500, 502, 555, 572, 601

Toxic Industrial Materials (TIMs), 13, 73, 91

Tracer, 30, 334, 345–350, 355

Trademarks, 334, 335, 527

Transdisciplinarity, 8

Treatment, 13, 48, 60, 69, 73, 77, 78, 82, 84, 89, 98, 99, 102, 103, 125, 127, 135–142, 147–151, 161–164, 167, 176, 183, 198, 202, 203, 208, 209, 215–218, 222–224, 226–230, 232, 234, 256, 257, 264, 275, 276, 279–281, 291, 293, 295–302, 305, 306, 308, 310–319, 322, 323, 325–328, 336, 337, 371, 374, 375, 377, 387–390, 396, 398, 402, 408, 417, 425, 452, 453, 456, 457, 463–466, 468–473, 477–479, 482, 483, 495–498, 500, 501, 503–509, 512–515, 517–520, 524, 528, 532, 548, 559, 560, 566–569, 572, 575, 701

Turbidity, 72, 78, 79, 216, 222–224, 226, 230–235, 306, 307, 312, 316, 319, 320, 322, 323, 343, 384, 389, 408, 499, 502, 503, 509, 511, 517, 567, 568, 572, 575, 600, 707

U

Ultrafiltration, 148, 167, 174, 183, 198, 310, 312, 313, 321, 322, 516

United Nation Children Fund (UNICEF), 518

Universal Water Quality, 19, 67

Universal Water Quality Index (UWQI), 68, 69, 71, 73, 74, 92, 97, 103, 104

Unmanned Aerial Vehicle (UAV), 95, 96

U.S. Environmental Protection Agency (US EPA), 7, 447

U.S. Public Health Service (USPHS), 70

UV-VIS, 266, 351

W

Wastewater, 6, 13, 18, 19, 69, 73, 78, 82, 89, 93, 98, 99, 102, 115, 121, 124, 125, 135–140, 142, 143, 146–148, 150, 151, 161, 198, 203, 206, 209, 216, 224, 227, 228, 230, 233, 235, 244, 263, 264, 271, 275, 276, 281–283, 286, 308, 309, 336, 344, 372, 397, 398, 400, 402, 411, 428, 447, 453, 456–459, 463–466, 468–471, 477, 479, 483, 498, 502, 504–509, 511, 513–515, 524, 528, 532, 539, 558, 584, 700, 701

Waterborne, 5, 9, 70, 91, 92, 161, 198, 280, 291–295, 310, 422, 437, 448, 449, 451, 459, 497, 499, 518

Water characteristics, 127, 306, 328, 504

Water contaminantion, 17, 39, 198, 285, 296, 337, 374, 383, 390, 423, 431, 437, 449, 452, 498, 499, 513, 519, 520

Water distribution system, 16, 17, 47–51, 385, 416

Water-energy nexus, 547

Water Framework Directive, 116, 117, 411, 584, 692, 694

Water management, 18, 19, 97–99, 245, 247, 257, 258, 373, 374, 377, 395–397, 418, 447, 452, 458, 459, 589, 615–618, 628, 630, 632, 694

Water pollution, 9, 136, 146, 161, 197, 198, 276, 280, 281, 292, 293, 295, 296, 299, 336, 337, 343, 347, 356, 424, 430–432, 437, 443, 447, 449, 452, 453, 459, 464, 599

Water Process Function (WPF), 73, 103

Water purification, 68, 102, 104, 161, 162, 165, 166, 171, 175, 178, 180, 197, 198, 206, 224, 230, 271, 275, 277, 285, 300, 527, 531, 532, 539

Subject Index

Water quality, 7, 13, 17–19, 30, 47–52, 57–60, 67–74, 76, 78, 79, 90, 91, 93, 95, 96, 98–104, 116, 121, 138, 140, 218, 246, 252, 257, 295, 302, 305, 307, 310, 316, 317, 324, 335, 337, 341, 372, 373, 417, 422, 423, 428–430, 432, 435, 437, 451, 453, 496, 498–500, 503–505, 515–519, 570, 572, 584, 599–604, 606–608, 691–698, 706–708, 711, 712
Water Quality Index (WQI), 67–69, 71–74, 103, 104, 599, 601–603
Water quality monitoring, 17, 47, 49–51, 57, 60, 93, 96, 373, 600
Water resources, 7, 13, 17, 23, 24, 49, 50, 69, 76, 91, 93, 98, 104, 116, 117, 125, 127, 197, 198, 243, 244, 246, 247, 264, 281, 291, 292, 335, 337, 340, 341, 347, 348, 372, 373, 391, 396, 423, 424, 443, 444, 446, 447, 449, 450, 452, 456–459, 464, 497, 498, 519, 547, 549, 565, 566, 572, 584, 600, 601, 615–621, 627, 629–632, 638, 655, 691–694, 698, 712
Water resources management, 104, 116, 373, 391, 396, 629, 691–693, 698
Water safety, 3, 12, 24, 69, 243, 244, 246, 250–253, 257, 258, 301, 372, 423, 458, 505, 509, 520
Water scarcity, 5, 9, 10, 17, 26, 50, 69, 76, 77, 93, 104, 161, 215, 246, 247, 373, 390, 423, 424, 443–445, 447, 449–452, 454, 457–459, 584, 615–617, 621, 629–632

Water security, 51, 246, 371, 373, 374, 377, 395–397, 408, 417, 418, 457, 549, 584, 616, 629, 632, 638
Watersheds, 9, 121, 335, 505, 584, 617, 618, 622, 625, 626, 628–632, 639, 653, 655
Water treatment, 13, 69, 73, 78, 84, 98, 102, 103, 137, 138, 149–151, 162, 163, 167, 183, 198, 202, 208, 209, 215–218, 222–224, 230, 234, 264, 276, 279–281, 291, 295–302, 305, 308, 311, 313–315, 319, 323, 326, 328, 374, 377, 388, 389, 396, 398, 417, 452, 457, 463, 464, 469, 471, 472, 478, 482, 483, 495–498, 500, 501, 503–509, 513, 515, 518, 520, 548, 559, 566, 568, 572
Wetlands, 7, 117, 335, 457, 497, 698
World Health Organization (WHO), 4, 5, 71, 78, 91, 215, 229, 250, 251, 291–295, 300, 308, 373, 422, 424, 451, 452, 496, 498–502, 518, 519
World Resources Institute (WRI), 77, 373, 374

X

Xenobiotic, 92, 334
X-ray fluorescence (XRF), 341, 352

Z

Zootechnics, 692

Lightning Source UK Ltd.
Milton Keynes UK
UKHW020606180722
406005UK00002B/8

9 783030 760106